The Arcata Conference on Representations of Finite Groups

PROCEEDINGS OF SYMPOSIA IN PURE MATHEMATICS
Volume 47, Part 1

The Arcata Conference on Representations of Finite Groups

Paul Fong, Editor

AMERICAN MATHEMATICAL SOCIETY
PROVIDENCE, RHODE ISLAND

PROCEEDINGS OF THE SUMMER RESEARCH INSTITUTE
ON REPRESENTATIONS OF FINITE GROUPS
HELD AT HUMBOLDT STATE UNIVERSITY
ARCATA, CALIFORNIA
JULY 7-25, 1986

with support from the National Science Foundation, Grant DMS-8512866

1980 *Mathematics Subject Classification* (1985 *Revision*). Primary 20C15, 20C20, 20G05, 20G40; Secondary 20C05, 20C30, 20G10.

Library of Congress Cataloging-in-Publication Data

Arcata Conference on Representations of Finite Groups
(1986: Humboldt State University)
The Arcata Conference on Representations of Finite Groups.
(Proceedings of symposia in pure mathematics; v. 47)
1. Representations of groups–Congresses. 2. Finite groups–Congresses.
I. Fong, Paul, 1933- . II. Title. III. Series.
QA171.A73 1986 512′.2 87-19348
ISBN 0-8218-1479-6 (set) (alk. paper)
ISBN 0-8218-1477-x (Part 1)
ISBN 0-8218-1478-8 (Part 2)

Table of Contents

Preface

The papers in these proceedings of the 1986 Arcata Summer Institute bear witness to the extraordinarily vital and intense research in the representation theory of finite groups in the years since the 1979 Santa Cruz Summer Institute on finite groups. The confluence of diverse mathematical disciplines has brought forth work of great scope and depth. Particularly striking is the influence of algebraic geometry and cohomology theory in the modular representation theory and the character theory of reductive groups over finite fields, and in the general modular representation theory of finite groups. Noteworthy too are the continuing developments in block theory and the general character theory of finite groups. The expository and research aspects of the Summer Institute are well represented by these papers.

On behalf of the organizers of the meeting—Jonathan Alperin, Charles Curtis, Walter Feit, and myself—I should like to thank all the participants for contributing so much to the success of the meeting. I should also like to thank the authors of the many papers for their cooperation in making possible the timely publication of the proceedings.

Paul Fong

I. EXPOSITORY PAPERS

Proceedings of Symposia in Pure Mathematics
Volume 47 (1987)

Cohomology Is Representation Theory

J. L. ALPERIN

1. Introduction. Group theorists find cohomology of groups a confusing and difficult subject: a great deal of formalism is necessary to formulate the subject while the applications to group theory involve only a fraction of the machinery; connections with topology are exciting but mysterious at the same time. The history of the subject is not much help either. Early work by Schur, at the turn of the century, studying projective complex representations, that is, homomorphisms of a finite group G into $\mathrm{PGL}(n, \mathbf{C})$, introduced cohomology groups we now know as $H^1(G, \mathbf{C})$ and $H^2(G, \mathbf{C})$. Later work by Schreier on group extensions led to other second cohomology groups. The third cohomology groups were interpreted by Eilenberg-Mac Lane. The full machinery appears as a very formal calculus of many-valued functions on groups and rather arbitrary boundary operators with only low-dimensional cases actually used.

However, cohomology of groups is part of a much broader subject, homological algebra, which has a much different history. Hilbert's work on syzygies is usually taken as the starting point of this subject and this involves rings and modules. In the case of finite groups, modules are a way of looking at representations of groups. Hence, cohomology can be viewed as part of the representation theory of groups. The interpretations of the low-dimensional cohomology groups in group-theoretical terms are now nothing more than another use of representation theory in group-theoretical problems. This paper will show how accepting this view leads to a whole new area of representation theory while explaining, at the same time, some seemingly isolated parts of cohomology of groups.

Let us begin by making this more explicit. The central idea of homological algebra is that of resolutions. To see what this means in our context, we fix a finite group G and a field k of prime characteristic p. All our kG-modules will be finite-dimensional and our notation is standard [**1**]. A *projective resolution* of the kG module U is an exact sequence

$$\cdots \to P_n \to \cdots \to P_1 \to P_0 \to U \to 0$$

1980 *Mathematics Subject Classification* (1985 *Revision*). Primary 20C20, 20J06.
Supported in part by National Science Foundation Grant DMS 8421367.

of kG-modules, where each of the P_i is projective, that is, a summand of a free module. For example, if G is cyclic of order p, then there is a projective resolution of the trivial module of the following form:

$$\cdots \to kG \to \cdots \to kG \to kG \to k \to 0.$$

If G is the direct product of two cyclic groups of order p, then we have another resolution as follows:

$$\cdots kG \oplus kG \oplus kG \to kG \oplus kG \to kG \to 0.$$

These easy examples do illustrate the important point that resolutions are far from arbitrary and have a lot of structure. Indeed, since kG is an algebra, for any kG-module U there is a canonical projective resolution of U called the *minimal projective resolution* of U: it is unique up to ismorphism, in the obvious sense, and is characterized by the fact that any other projective resolution of U is isomorphic with the direct sum of the minimal resolution of U and a resolution of 0. The two examples we have just given are also minimal. We shall be demonstrating here how well these resolutions are part of modular representation theory. Since resolutions like these can be used to define cohomology groups, we have a setting in which cohomology is naturally part of representation theory.

Before proceeding further, we want to point out applications of the ideas that will be developed, applications to topics other than the homological properties of representations, where there are, of course, more applications and surprises. This gives convincing evidence of how this subject is now at the point of influencing other areas of representation theory and mathematics and has gone far beyond its beginnings. In fact, there are now applications to topology. Benson and Habeggar have contributed new results on the Moore space problem of Steenrod by using these homological methods in representation theory, and Benson and Carlson **[7]** have given a new and very thematic proof of Gunnar Carlsson's theorem on free actions of finite groups on products of spheres.

However, closer to home is a result of Puig. A kG-module U is called an endo-trivial module if there is a projective module P such that $U \otimes U^* \cong k \oplus P$, that is, U has an "inverse," its dual, modulo projectives. These modules and generalizations of them are important and arise in studying simple modules. The set of isomorphism classes of endo-trivial modules can be made into an abelian group: if U and V are endo-trivial modules, then $U \otimes V$ is the direct sum of an endo-trivial module and a projective module so the product of the isomorphism classes of U and V is the isomorphism class of the third endo-trivial module. While we would very much like to know what all these modules are, the theorem of Puig is one of the best results: *the group of isomorphism classes of endo-trivial modules is finitely generated.* The proof relies not only on the techniques we shall discuss but also needs an argument using algebraic geometry as well as Dade's classification of these modules in the special case of elementary abelian p-groups **[12]**.

The story does not end here. Benson and Carlson **[6]** have applied these homological methods to the study of the Green ring and they have given really new insights into the existence of nilpotent elements in these rings, constructing such elements in a natural way, instead of the ad hoc and limited previous ways. These ideas have spread further: Webb was the first to carry these ideas over to integral representations; more recently, Friedlander and Parshall [**14**] have developed an independent but analogous theory, with its own applications, for algebraic groups.

2. Complexity. With the importance of resolutions in our mind, we shall now develop the idea of the complexity of a module; this amounts to concentrating on the easiest and simplest invariant of the minimal projective resolution of a module and disregarding all the additional structure. In order to motivate this idea and the theorems it leads to, let us first begin with some background, that is, previous results that point the right way.

Let us begin with the phenomenon of periodicity, which first arose in the study of finite groups acting on spheres. Our group is said to be periodic (or p-periodic to be specific) if the minimal projective resolution of the trivial kG-module k

$$\to P_n \to P_{n-1} \to \cdots \to P_1 \to P_0 \to k$$

is periodic, in that the sequence of projective modules and homomorphisms is periodic. This is equivalent to the p-rank of G being one, that is, every abelian p-subgroup of G is cyclic. However, this result is related in two ways to modular representation theory. First, if p is odd then the p-rank being one forces the Sylow p-subgroups of G to be cyclic and so the whole theory of blocks with cyclic defect groups comes into play; this is some of the deepest material in all of representation theory. In fact, one can deduce from this theory, as the author and G. Janusz [**4**] did years ago, the structure of the periodic minimal resolution from the most important piece of data in the cyclic theory, namely the Brauer tree, in a very combinatorial and natural manner. However, this is not the only connection of periodicity and representation theory, as it is possible to generalize this notion of periodicity to any kG-module U in place of k in the obvious way and this too leads to a body of results. Hence, a homological result for the trivial module, with a topological connection, is just a special case of an interesting module-theoretic phenomenon.

The cohomology algebra $H^*(G,k)$ of G is a $\mathbf{Z}$-graded algebra

$$H^*(G,k) = H^0(G,k) \oplus H^1(G,k) \oplus \cdots \oplus H^n(G,k) \oplus \cdots$$

defined by the minimal projective resolution of the kG-module (in fact, $H^n(G,k) \cong \mathrm{Hom}_{kG}(P_n,k)$) so that $H^0(G,k) = k.1.$ consists of scalar multiples of the identity, and each $H^n(G,k)$ is a finite-dimensional vector space over k with the properties of being finitely generated as an algebra and of being commutative in the graded sense, so that homogeneous elements commute unless they are both of odd degree, in which case they anticommute. Since this algebra is very similar to strictly commutative algebras, it is easily possible to define its Krull dimension in a number of ways, and the striking theorem of Quillen [**16**] gives a description of this invariant: *the dimension of* $H^*(G,k)$ *equals the* p*-rank of* G. That is, the dimension is equal to e, where every abelian p-subgroup of G can be generated by e elements but at least one of these subgroups can be generated by no fewer than e elements. Again, abelian subgroups have appeared in a homological problem.

So here we have a very interesting, though apparently totally isolated, result on cohomology. Is there a general result on modules which generalizes this result and also ends the isolation of Quillen's theorem by making it a part of representation theory? Evidence for this can be found in a result of Chouinard [**11**], which relates modules and elementary abelian subgroups of G: *the kG-module U is projective if, and only if, each restriction U_E of U to an elementary abelian p-subgroup E of G is projective.* The connection with p-rank is easy: if G has p-rank e then p^e is the largest order of any elementary abelian p-subgroup of G.

Before answering the question in the positive, let us look at one more result, in a very special case, which shows the structure of a projective resolution. (It is interesting to record that this example arose in the study of simple groups, in a determination of Sylow 2-subgroups that later played a role in the discovery of the O'Nan sporadic group.) Let $p = 2$ and $G = \mathrm{GL}(3, 2)$ so kG has four isomorphism classes of simple modules, three of them represented by the modules k, V, and V^*, where V is the usual three-dimensional module and V^* is its dual module. Let P_k, P_V, and P_{V^*} be their projective covers (the zero term of their minimal projective resolutions). We form a double complex as follows (that is, for each pair of nonnegative integers we choose a kG-module, pictured at the lattice point in the first quadrant of the plane) using only a few well-chosen maps in a very periodic way:

$$\begin{array}{ccccccccc}
\downarrow & & \downarrow & & \downarrow & & \downarrow & & \\
P_k & \longleftarrow & P_V & \longleftarrow & P_{V^*} & \longleftarrow & P_k & \longleftarrow & \\
\downarrow & & \downarrow & & \downarrow & & \downarrow & & \\
P_V & \longleftarrow & P_{V^*} & \longleftarrow & P_k & \longleftarrow & P_V & \longleftarrow & \\
\downarrow & & \downarrow & & \downarrow & & \downarrow & & \\
P_{V^*} & \longleftarrow & P_k & \longleftarrow & P_V & \longleftarrow & P_{V^*} & \longleftarrow & \\
\downarrow & & \downarrow & & \downarrow & & \downarrow & & \\
P_k & \longleftarrow & P_V & \longleftarrow & P_{V^*} & \longleftarrow & P_k & \longleftarrow &
\end{array}$$

Here, as usual with double complexes, the small squares have maps which anti-commute while the composition of any two horizontal or vertical maps is zero. Therefore, when we form the direct sums of terms on the diagonals, from upper left to lower right, we get a sequence of projective modules which is a complex, that is, the composition of any two consecutive maps is zero:

$$\cdots \to P_V \oplus P_k \oplus P_{V^*} \to P_{V^*} \oplus P_V \to P_k.$$

Since the definition of P_k provides a map of P_k to k we can form a longer sequence: *This is a minimal resolution.* Hence, the $k\mathrm{GL}(3,2)$-module k has a minimal resolution which has a two-dimensional structure, so there should be some way of saying that k is two-dimensional. This is far from an isolated example. In fact, the resolution for the direct product of two groups of order p that we described in the introduction arises from a double complex; this is directly related to the Kunneth theorem. Similarly, a periodic module should be of dimension one and a projective module P, which has as its minimal resolution the sequence

$$\cdots \to 0 \to P \to P \to 0,$$

should be of dimension zero since one needs only a zero-dimensional complex, that is, a module, to construct the resolution. Hence, we should have a concept which disregards all the intriguing structure of a resolution and just measures its size.

However, this is easy to do: let G and k be general as before. The finite generation of the cohomology algebra [**13**] implies strong results about minimal resolutions. In fact, if U is any kG-module, then there is a nonnegative integer c, depending on U, such that if P_n is the nth term of the minimal projective resolution of U,

$$\dim_k(P_n) = a_0 + a_1 \cdot n + a_2 \cdot n^2 + \cdots + a_{c-1} \cdot n^{c-1}$$

for almost all n, where the a_i are periodic functions of n, rather than constants, and a_{c-1} is not zero. Here, if $c = 1$, we mean the zero polynomial, that is, P_n is almost always zero, which means, as is easily seen since projective modules for kG are also injective, that U is projective. This invariant $c = c_G(U)$ is called the *complexity* of U. It is easy to see that in the examples above this number, the complexity, does coincide with our intuitive guess at a dimension. The name is chosen because it measures the difficulty of U from a homological point of view and it also measures the sizes of complexes (double, triple, ...) used to construct projective resolutions. It is easy to see that complexity equal to zero is equivalent to projectivity; it is true that complexity one is equivalent to periodicity. Moreover, Benson and Carlson [**7**] have remarkably shown, using the full machinery of the subject which we are just beginning to describe, that a kG-module U has complexity c if, and only if, U has a projective resolution (not necessarily the minimal one) which comes from a multiply periodic c-dimensional complex and has no resolution coming from a smaller-dimensional complex.

However, the result [**2**] which starts the subject and ties together all of the material we introduced as motivation, namely, resolutions, elementary subgroups, periodicity, Quillen's theorem, and so on, is as follows:

THEOREM (ALPERIN-EVENS). *If U is a kG-module, then $c_G(U)$ equals the maximum of the complexities $c_G(U_E)$ of the restrictions of U to the elementary abelian p-subgroups E of G.*

The proof of this is along the lines one alway works in representation theory: reduce to a local situation and then apply Clifford theory, the representation theory of groups with normal subgroups. Specifically, it is easy to see that

$$\cdots \to P_V \oplus P_k \oplus P_{V^*} \to P_{v^*} \oplus P_V \to P_k \to 0.$$

$c_G(U) = c_S(U_S)$, where S is a Sylow p-subgroup of G, so one only has to prove the result for p-subgroups, by Sylow's theorem. Indeed, if

$$\cdots \to P_n \to \cdots \to P_1 \to P_0 \to U$$

is the minimal projective resolution of U then the restrictions

$$\cdots \to (P_n)_S \to \cdots \to (P_1)_S \to (P_0)_S \to U_S$$

form a projective resolution of U_S. Indeed, the restriction of a projective module to a subgroup is projective since the restriction of a free module is free (which is essentially Lagrange's theorem). Hence the dimensions of the terms of the minimal projective resolution of kS-module U_S are majorized by the dimensions of the P_n and so $c_G(U) \geq c_S(U_S)$. In a similar manner, using induction instead of restriction, starting with a minimal projective resolution of U_S one gets a projective resolution of the kG-module $(U_S)^G$, the induced module, which has the minimal projective resolution of U as a summand since U is a summand of $(U_S)^G$.

Now, having reduced to p-groups, which have plenty of normal subgroups, there is standard machinery for dealing with homological questions involving normal subgroups: spectral sequences. The methods of Quillen-Venkov [**17**] for proving Quillen's theorem can be generalized to handle the present situation. A more elementary approach, avoiding spectral sequences, is that of Kroll [**15**].

This theorem is the unifying result, does encompass a great number of previous theorems, and imples a host of new applications. For example, Chouinard's theorem is a special case. Indeed, if U is a projective module so is every restriction U_E, while if every restriction is projective then every $c_E(U_E)$ is zero, so the maximum of these values is zero and $c_G(G) = 0$ by the theorem and U is projective. This generalizes to periodic modules, so a module is periodic if, and only if, its restrictions to elementary abelian p-subgroups are periodic.

Quillen's theorem itself is a consequence, though to see this one needs some knowledge of standard homological ideas. Indeed, there is a formula for $\dim_k H^n(G,k)$ in terms of a polynomial with periodic coefficients, holding almost always, and if we call the degree of this polynomial plus one the growth rate γ of $H^*(G,k)$, then the Krull dimension d of $H^*(G,k)$ equals this growth rate γ. If U is any kG-module then $\mathrm{Ext}^*_{kG}(k,U)$ also has a growth rate γ_U and since $\mathrm{Ext}^*_{kG}(k,S)$ is a module for $\mathrm{Ext}^*_{kG}(k,k) \cong H^*(G,k)$ we have $\gamma_U \leq \gamma$. However, we have a formula (in general but given here in case k is algebraically closed) for the nth term P_n of the minimal projective resolution of k as a direct sum of copies of the projective covers of the simple kG-modules, namely,

$$P_n \cong \bigoplus (\dim_k \mathrm{Ext}^n_{kG}(k,S))P_S$$

where P_S is the projective cover of the simple kG-module S and the sum is over all such modules S, up to isomorphism. Hence, $d = c_G(k)$ and the theorem now easily implies Quillen's theorem as $c_E(k)$ is the rank of E for any elementary abelian p-subgroup E of G.

A number of different approaches now exist to this theorem and to its generalizations. One stronger theorem of Carlson [**8**] is particularly striking: *There is a positive constant C, depending only on G, such that if U is any kG-module having no projective summand then there is an elementary abelian p-subgroup E of G and a summand V of U_E such that V has no projective summand and $\dim_k V \geq C \cdot \dim_k U$.* It is easy to see how this result generalizes Chouinard's but more argument is needed to see how to recover the Alperin-Evens theorem.

3. Varieties. It should not be surprising that when one has a concept of some sort of dimension, then there will be some algebraic varieties present; in fact, in our case there are many varieties present. We shall briefly discuss these varieties, tying up with the idea of complexity and setting the stage for the survey paper by Carlson in these Proceedings [**10**]. Therefore, we henceforth assume that k is algebraically closed. The first varieties go back to Quillen's work and we shall explain them in a manner that is a little more transparent than the original way. We begin with a remark.

LEMMA. *The set N of nilpotent elements in $H^*(G, k)$ is a homogeneous ideal and $H^*(G,k)/N$ is commutative.*

To see this let M be the set of all sums of nilpotent homogeneous elements of $H^*(G, k)$, so M is an ideal since $H^*(G, k)$ is graded commutative. Every element of M is also nilpotent since a power of a sum of homogeneous elements of $H^*(G, k)$ is up to sign, by the commutativity in the graded sense, a sum of products of powers of the elements making up the sum. Finally, suppose that

$$a = a_0 + a_1 + \cdots + a_r$$

is a nilpotent element of $H^*(G, k)$ with $a_i \in H^i(G, k)$ and $a_r \neq 0$. If $a^t = 0$ then we must have $(a_r)^t = 0$ so $a_r \in M$. Hence, some power of $a_0 + \cdots + a_{r-1}$ lies in M, so continuing in this way we have $a \in M$. In particular, every nilpotent element is in M. In the case that $p \neq 2$, since every homogeneous element of odd degree has square zero, we have that $H^*(G, k)/N$ is commutative and the lemma is proved.

Therefore, the quotient $H^*(G,k)/N$ is an affine ring; that is, it is finitely generated commutative and has no nonzero nilpotent elements, and is the coordinate ring of an affine variety whose points can be identified with the elements of the set of algebra homomorphisms of $H^*(G,k)/N$ to k. However, any homomorphism of $H^*(G, k)$ to k must have N in the kernel so we can define the points of the variety $V(k)$ of the kG-module to be $\operatorname{Hom}(H^*(G, k), k)$. Of course, we are not interested in just the isolated case of the trivial module, so we need to define a variety for any kG-module M. We do this by slightly modifying the Scott-Serre approach. If U is any kG-module then $H^*(G, M \otimes U)$ is a module for

$H^*(G,k)$. Let B_M be the common annihilator of all $H^*(G, M\otimes S)$ as S runs over all simple kG-modules so B_M is an ideal. If t is the Loewy length of kG, so every kG-module has a series of length at most t with semisimple successive quotients, then $(B_M)^t$ annihilates $H^*(G, M \otimes U)$ for any kG-module U. Hence, if we let A_M be the common annihilator of all $H^*(G, M \otimes U)$ then A_M contains a power of B_M and so A_M and B_M define the same variety contained in $V(k)$. We let this be the variety $V(M)$ corresponding to M so $V(M) = \mathrm{Hom}(H^*(G,k)/A_M)$.

The idea now is to generalize results about complexity to theorems on these varieties and to use these new insights to solve further problems. Indeed, $V(M)$ is a homogeneous affine variety and the dimension of the associated projective variety equals one less than the complexity of M. Moreover, it is possible to relate the structure of $V(M)$ and the varieties $V(M_E)$ for elementary abelian p-subgroups E in such a way as to generalize the main theorem on complexity [**3**, **5**]. These results focus our attention on the question of the complexity of modules for elementary abelian p-groups: we have reduced the computation of complexity to these groups, and other question as well about the nature of the variety of a module can similarly be reduced. How do we then proceed in the case of these very special sorts of groups?

The answer lies with another variety, the rank variety of Carlson [**9**], whose history lies in work of Dade [**12**] and of Kroll [**15**], and which is defined in a completely different way from the varieties we have just defined. It is a remarkable and critical result that these new varieties are closely tied to the previous ones, in fact being isogeneous with them. We shall close our discussion of varieties with the definition of the Carlson variety. Let E be an elementary abelian p-group of order p^n with generators $x_1, x_2, \ldots, x_n$. If we set $X_i = x_i - 1$ then $X_1, \ldots, X_n$ also generate kE and satisfy the relations $X_i^p = 0$ and there is an isomorphism of a truncated polynomial algebra onto kE carrying the generators to the X_i. Let A be the affine space over k spanned by the $X_i : A = kX_1 + \cdots + kX_n$. If M is a kE-module then the Carlson rank variety $W(M)$ equals the subset of A consisting of 0 and the points $a \in A$ such that M is not projective as a module for the algebra over k generated by $1+a$; it is also a homogeneous affine variety. Since $a \in A$ we have $a^P = 0$ and so $1 + a$ is a unit of order p in kE and this algebra is a group algebra for a group of order p. The element $1+a$ is a called a shifted element of E: it is usually not an element of E. The rank variety, as opposed to the earlier varieties, is, in theory and sometimes in practice, calculable and it is a vital tool in further research.

4. Concluding remarks. The remarkable structures of projective resolutions and the roles they play in module theory have been the main motivation behind the work described. It is quite surprising that concentrating on the simplest possible invariant, the size, of resolutions should lead to so much mathematics and applications to other sorts of questions in representation theory and mathematics, as well as to a host of problems of a homological nature in modular representation theory, the latter largely not discussed here as we have been emphasizing basic ideas and connections. We have presented here just one chapter on the growing use of homological methods in representation theory, methods with quite old antecedents actually, but now playing a dominating role in some areas such as the Deligne-Lusztig construction of characters and the complexes introduced by Brown and studied by others. The future should bring even more ideas of this nature.

REFERENCES

1. J. L. Alperin, *Local representation theory*, Cambridge University Press, 1986.
2. J. L. Alperin and L. Evens, *Representations, resolutions and Quillen's dimension theorem*, J. Pure Appl. Algebra **22** (9181), 1–9.
3. ——, *Varieties and elementary abelian subgroups*, J. Pure Appl. Algebra **26** (1981), 221–227.
4. J. L. Alperin and G. Janusz, *Resolutions and periodicity*, Proc. Amer. Math. Soc. **37** (1973), 403–406.
5. G. S. Avrunin, *Annihilators of cohomology modules*, J. Algebra **69** (1981), 150–154.
6. D. J. Benson and J. F. Carlson, *Nilpotent elements in the Green ring*, J. Algebra **104** (1986), 329–350.
7. ——, *Complexity and multiple complexes*, Math. Z. (to appear).
8. J. F. Carlson, *Dimensions of modules and their restrictions over modular group algebras*, J. Algebra **69** (1981), 95–104.
9. ——, *The varieties and the cohomology ring of a module*, J. Algebra **85** (1983), 104–143.
10. ——, *Varieties for modules*, this volume.
11. L. Chouinard, *Projectivity and relative projectivity over group rings*, J. Pure Appl. Algebra **7** (1976), 278–302.
12. E. C. Dade, *Endo-permutation modules over p-groups*. II, Ann. of Math. (2) **108** (1980), 317–346.
13. L. Evens, *The cohomology ring of a finite group*, Trans. Amer. Math. Soc. **101** (1961), 224–239.
14. E. Friedlander and B. Parshall, *Support varieties for restricted Lie algebras* (to appear).
15. O. Kroll, *Complexity and elementary abelian p-groups*, J. Algebra **88** (1984), 155–172.
16. D. Quillen, *The spectrum of an equivariant cohomology ring*. I, II, Ann. of Math. (2) **94** (1971), 549–602.
17. D. Quillen and B. B. Venkov, *Cohomology of finite groups and elementary abelian subgroups*, Topology **11** (1972), 317–318.

UNIVERSITY OF CHICAGO

Proceedings of Symposia in Pure Mathematics
Volume **47** (1987)

Families of Irreducible Characters of Weyl Groups

DEAN ALVIS

1. Introduction. The purpose of this paper is to provide a survey of results concerning the families of irreducible characters of Weyl groups, and their connections with Lusztig's classification theorem for irreducible characters of reductive groups over finite fields. We begin with a brief review of notation and some results of the Deligne-Lusztig theory.

Let G be a connected, reductive algebraic group with connected center, defined over the field $\mathbf{F}_q$ of q elements, and let $F: G \to G$ be the corresponding Frobenius endomorphism. For each F-stable maximal torus T of G and each linear character $\theta: T^F \to \mathbf{C}^*$, let $R_T^G\theta$ be the generalized character of G^F defined by Deligne and Lusztig in [**4**].

Let $\mathrm{Irr}(G^F)$ be the set of irreducible characters of G^F (over a splitting field of characteristic 0). Deligne and Lusztig defined a partition of $\mathrm{Irr}(G^F)$ into certain subsets, called *geometric conjugacy classes*, one for each conjugacy class of semisimple elements in G^{*F^*} (where the pair G^*, F^* is dual to G, F). These subsets may be described as follows. For χ, $\psi \in \mathrm{Irr}(G^F)$, define $\chi \sim \psi$ if χ and ψ both appear with nonzero multiplicity in some $R_T^G\theta$. Then the geometric conjugacy classes are the equivalence classes of the equivalence relation on $\mathrm{Irr}(G^F)$ generated by $\sim$. We have a disjoint union

$$\mathrm{Irr}(G^F) = \bigcup_{(s)} \mathrm{Irr}_{(s)}(G^F),$$

where $\mathrm{Irr}_{(s)}(G^F)$ is the geometric conjugacy class corresponding to the semisimple conjugacy class (s) of G^{*F^*}.

The elements of $\mathrm{Irr}_{(1)}(G^F)$ are the *unipotent characters* of G^F; they are characterized by appearing with nonzero multiplicity in some $R_T^G 1$ (with $\theta = 1$ the trivial character of T^F). The significance of the unipotent characters is apparent from the following result.

1.1. THEOREM (LUSZTIG [**11**]). *There is a bijection*

$$\mathrm{Irr}_{(s)}(G^F) \to \mathrm{Irr}_{(1)}(Z(s)^{F^*}), \qquad \chi \mapsto \rho_\chi,$$

1980 *Mathematics Subject Classification* (1985 *Revision*). Primary 20C15.

where $Z(s)$ is the centralizer of s in G^*. Moreover, if χ appears in $R_T^G\theta$ with nonzero multiplicity m then ρ_χ appears in $R_{T^*}^{Z(s)}1$ with multiplicity m or $-m$ (where $T^* \leq Z(s)$ is dual to T), and

$$\deg(\chi) = (|G^F|/|Z(s)^{F^*}|)_{p'} \deg(\rho_\chi),$$

where $p = \mathrm{char}(\mathbf{F}_q)$.

Thus, classifying the irreducible characters of reductive groups over finite fields reduces to classifying the unipotent characters. In completing the classification of unipotent characters, Lusztig showed that $\mathrm{Irr}_{(1)}(G^F)$ may be partitioned into disjoint subsets, called *families.* In case G is split over $\mathbf{F}_q$, these families of unipotent characters are in bijective correspondence with certain subsets, also called families, of the irreducible characters of the Weyl group W of G. Moreover, if $\mathcal{F}$ is a family of irreducible characters of W and $\mathcal{U}_{\mathcal{F}}$ is the corresponding family of unipotent characters of G, then the multiplicities with which the members of $\mathcal{U}_{\mathcal{F}}$ appear in the R_T^G1 are given by a combinatorial object, called a nonabelian Fourier transform matrix, attached to $\mathcal{F}$.

The remainder of this paper is organized as follows. The second section contains a discussion of the left cells and two-sided cells of a Weyl group W. In the third section, the constructible characters of W are considered. These characters are used to define the families of irreducible characters of W. In the last section, Lusztig's classification theorem is stated, as it applies in the untwisted case to the unipotent characters of G^F.

2. Cell representations of Hecke algebras. Let W be a finite Weyl group, with set of simple reflections S, and let $l: W \to \mathbf{N}$ be the associated length function. Put $R = \mathbf{Q}[\sqrt{u}, \sqrt{u^{-1}}]$, the ring of Laurent polynomials in the indeterminate $\sqrt{u}$ over $\mathbf{Q}$. The *generic Hecke algebra* $H = H(W, S)$ of (W, S) over R is the R-algebra with free R-basis $\{T_w : w \in W\}$ and multiplication determined by

$$T_sT_w = \begin{cases} T_{sw} & \text{if } l(sw) = l(w) + 1, \\ uT_{sw} + (u-1)T_w & \text{if } l(sw) = l(w) - 1, \end{cases}$$

for $s \in S$, $w \in W$. Note $\{T_s : s \in S\}$ generates H as an R-algebra. Indeed, if $s_1 s_2 \cdots s_l$ is a reduced expression for w ($s_i \in S$, $l = l(w)$), then $T_w = T_{s_1}T_{s_2}\cdots T_{s_l}$. Put $K = \mathbf{Q}(\sqrt{u})$; it is known that $H^K = K \otimes_R H$ is split semisimple over K (Lusztig [**8**]).

Let $\leq$ be the Bruhat order on W; $y \leq w$ if some expression for y may be obtained by cancelling factors from a reduced expression for w. For $w \in W$, define

$$\varepsilon_w = (-1)^{l(w)}, \qquad u_w = u^{l(w)}, \qquad L(w) = \{s \in S \mid sw < w\}.$$

Let $\alpha \mapsto \overline{\alpha}$ be the automorphism of R given by $\sqrt{u} \mapsto \sqrt{u^{-1}}$. The rule

$$\overline{\sum_w \alpha_w T_w} = \sum_w \overline{\alpha_w} T_{w^{-1}}^{-1}$$

defines an R-semilinear automorphism of H of order two. (Note each T_w is a unit in H; this follows from the fact that $T_s^2 = uT_1 + (u-1)T_s$ for $s \in S$.)

2.1. THEOREM (KAZHDAN AND LUSZTIG [6]). *For each $w \in W$, there is a unique element of C_w of H satisfying the following two conditions:*

(i) $\overline{C_w} = C_w$,

(ii) $C_w = \sum_{y \le w} \varepsilon_y \varepsilon_w u_y^{-1} u_w^{1/2} \overline{P_{y,w}} T_y$, *where* $P_{y,w} \in \mathbf{Z}[u]$ *for* $y \le w$, $P_{w,w} = 1$, *and* $\deg P_{y,w} \le \frac{1}{2}(l(w) - l(y) - 1)$ *when* $y < w$.

The elements C_w ($w \in W$) form an R-basis for H. This basis has pleasing connections with the ideal structure of H, as we shall indicate below.

If $y < w$ and $\deg P_{y,w} = \frac{1}{2}(l(w) - l(y) - 1)$, we write $y \prec w$ and define $\mu(y, w) = \mu(w, y)$ to be the coefficient of the highest power of u in $P_{y,w}$. We write y—w if $y \prec w$ or $w \prec y$. It is known that $\mu(y, w) > 0$ when y—w (Kazhdan and Lusztig [7]). Kazhdan and Lusztig have proved the following [6, 2.3].

$$T_s C_w = \begin{cases} -C_w & \text{if } s \in L(w), \\ uC_w + \sqrt{u} \sum\limits_{\substack{y - w \\ s \in L(y)}} \mu(y, w) C_y & \text{if } s \notin L(w), \end{cases} \tag{2.2}$$

for $s \in S$, $w \in W$. This motivates the following definition.

(2.3) For $y, w \in W$, $y \le_L w$ if there is some sequence $y = x_0, x_1, x_2, \ldots, x_n = w$ such that x_i—x_{i+1} and $L(x_{i+1}) \not\subset L(x_i)$ for $0 \le i < n$.

Thus $y \le_L w$ if C_y appears with nonzero coefficient in the expression for some product $T_{s_1} T_{s_2} \cdots T_{s_l} C_w$ ($s_i \in S$) as a linear combination of the basis elements C_x ($x \in W$). The relation $\le_L$ is a preorder on W; the corresponding equivalence relation on W is denoted $\sim_L$. The equivalence classes of $\sim_L$ are the *left cells* of W.

From the discussion above, it is clear that

$$I_L(w) = \sum_{y \le_L w} RC_y \quad \text{and} \quad I'_L(w) = \sum_{\substack{y \le_L w \\ y \not\sim_L w}} RC_y$$

are left ideals of H for $w \in W$. Let X be the left cell of W containing w. Define

$$M_L(X) = I_L(w)/I'_L(w),$$

a left H-module. Clearly $M_L(X)$ is independent of the choice of $w \in X$. Also, $M_L(X)$ has free R-basis consisting of the images of the C_x for $x \in X$, and thus the H^K module $M_L(X)^K = K \otimes_R M_L(X)$ has dimension $|X|$.

Now, define $y \le_R w$ if $y^{-1} \le_L w^{-1}$, and let $\le_{LR}$ be the preorder on W generated by $\le_L$ and $\le_R$. The equivalence classes of the equivalence relation $\sim_{LR}$ corresponding to $\le_{LR}$ are the *two-sided cells* of W. If $w \in W$, then

$$I_{LR}(w) = \sum_{y \le_{LR} w} RC_y \quad \text{and} \quad I'_{LR}(w) = \sum_{\substack{y \le_{LR} w \\ y \not\sim_{LR} w}} RC_y$$

are ideals of H. Thus if Y is a two-sided cell of W and $w \in Y$, then

$$M_{LR}(Y) = I_{LR}(w)/I'_{LR}(w)$$

is an (H, H)-bimodule. Using deep connections between the left and two-sided cells of W and the theory of primitive ideals in enveloping algebras, Lusztig proved the following result [**8**, 4]; see also [**12**, 6.3].

(2.4) If x and y are in the same two-sided cell of W and $x \leq_L y$, then x and y are in the same left cell of W.

Let Y be a two-sided cell of W. From (2.4) it follows immediately that $M_{LR}(Y) = \bigoplus M_L(X)$ as left H-modules, where X ranges over the left cells contained in Y.

We now recall the notion of W*-graph* [**6**]. First, suppose Γ is a graph with set of vertices V. Assume that for each vertex $v \in V$ a subset $I(v)$ of S is given. Also suppose that integers $m(v, v')$, $m(v', v)$ are given for each edge $\{v, v'\}$ of Γ. Let $E(\Gamma)$ be the free R-module with basis $\{e_v \mid v \in V\}$. For each $s \in S$, define an R-endomorphism τ_s of $E(\Gamma)$ by the rule

$$\tau_s(e_v) = \begin{cases} -e_v & \text{if } s \in I(v), \\ ue_v + \sqrt{u} \sum\limits_{\substack{\{v,v'\} \in \Gamma \\ s \in I(v')}} m(v, v')e_{v'} & \text{if } s \notin I(v). \end{cases}$$

Then Γ (with the indicated data) is a W-graph if $T_s \mapsto \tau_s$ extends to a representation $H \to \operatorname{End}_R(E(\Gamma))$.

Assume X is a left cell of W. We construct a W-graph $\Gamma(X)$, as follows. The vertex set of $\Gamma(X)$ is the set X itself, and $\{x, y\}$ is an edge of $\Gamma(X)$ if $x, y \in X$ and x—y. Define $I(x) = L(x)$ for $x \in X$, and let $m(x, y) = \mu(x, y)$ whenever $\{x, y\}$ is an edge of $\Gamma(X)$. Then $\Gamma(X)$ is a W-graph, and $E(\Gamma(X)) \cong M_L(X)$ as H-modules.

EXAMPLE. Let

$$W = W(G_2) = \langle s, t \mid s^2 = t^2 = (st)^6 = 1 \rangle.$$

The two-sided cells of W are $\{1\}$, $\{s, t, st, ts, sts, tst, stst, tsts, ststs, tstst\}$, and $\{w_0\}$, where $w_0 = ststst$. The left cells of W are $\{1\}$, $\{s, ts, sts, tsts, ststs\}$, $\{t, st, tst, stst, tstst\}$, and $\{w_0\}$. The W-graphs corresponding to the left cells are as follows:

∅ ⓢ-ⓣ-ⓢ-ⓣ-ⓢ ⓣ-ⓢ-ⓣ-ⓢ-ⓣ s, t

(We draw a W-graph by labeling each vertex v by the elements of $I(v)$, and by labeling each edge $\{v, v'\}$ by the product $m(v, v')m(v', v)$ when this product differs from 1.)

Gyoja has shown [**5**] that every simple H^K-module is isomorphic to $E(\Gamma)^K = K \otimes_R E(\Gamma)$ for some W-graph Γ. In the example above, with $W = W(G_2)$, the left cell modules of dimension 5 decompose into simple modules as follows:

ⓢ-ⓣ-ⓢ-ⓣ-ⓢ = ⓢ + ⓢ-ⓣ + ⓢ $\overset{3}{-}$ ⓣ,

ⓣ-ⓢ-ⓣ-ⓢ-ⓣ = ⓣ + ⓢ-ⓣ + ⓢ $\overset{3}{-}$ ⓣ.

Now, let Γ be a W-graph, with corresponding H-module $E(\Gamma)$. Let $f\colon R \to \mathbf{Q}$ be the homomorphism of $\mathbf{Q}$-algebras determined by $f(\sqrt{u}) = 1$. Then $\mathbf{Q} \otimes_f H$ is isomorphic to the group algebra $\mathbf{Q}W$ of W over $\mathbf{Q}$. Thus

$$E(\Gamma)_1 = \mathbf{Q} \otimes_f E(\Gamma)$$

is a $\mathbf{Q}W$-module, called a specialization of $E(\Gamma)$ (see [**3**] for more details concerning specialization). If X is a left cell of W, we denote by $M_L(X)_1$ the $\mathbf{Q}W$-module $E(\Gamma(X))_1$.

Let $\lambda\colon H \to R$ be the R-linear map defined by

$$\lambda\left(\sum_w \alpha_w T_w\right) = \alpha_1,$$

and define $\beta\colon H \times H \to R$ by $\beta(a,b) = \lambda(ab)$. It is known that β is symmetric and associative, and that

$$\beta(T_x, T_{y^{-1}}) = \begin{cases} u_x & \text{if } x = y, \\ 0 & \text{otherwise.} \end{cases}$$

Therefore there are unique elements D_w of H, $w \in W$, satisfying

$$\beta(C_x, D_{y^{-1}}) = \begin{cases} 1 & \text{if } x = y, \\ 0 & \text{otherwise.} \end{cases}$$

We have the following (Lusztig [**11**, 5.1.14]):

(2.5) For $x, y \in W$, $C_x D_{y^{-1}} \neq 0$ if and only if $y \leq_L x$.

Let $a \to a^\frown$ be the R-linear antiautomorphism of H determined by $(T_w)^\frown = T_{w^{-1}}$. It is easy to check that $(C_w)^\frown = C_{w^{-1}}$. If I is a left ideal of H, then

$$(I^\frown)^\perp = \{a \in H \mid \beta(a, b^\frown) = 0 \text{ for } b \in I\}$$

is also a left ideal of H. In particular, if $w \in W$, then

$$J_L(W) = (I'_L(w)^\frown)^\perp = \sum_{y \sim_L w} RD_y + \sum_{y \not\leq_L w} RD_y$$

and

$$J'_L(w) = (I_L(w)^\frown)^\perp = \sum_{y \not\leq_L w} RD_y$$

are left ideals of H. If X is the left cell of W containing w, put

$$N_L(X) = J_L(w)/J'_L(w),$$

a left H-module. From the definition of the elements D_w, $w \in W$, it is easy to see that $a \mapsto \beta(a^\frown, -)$ induces an epimorphism of H-modules

$$\pi\colon J_L(w) \to \operatorname{Hom}_R(M_L(X), R),$$

where H acts on $\operatorname{Hom}_R(M_L(X), R)$ according to the formula

$$(af)(m) = f(a^\frown m)$$

for $a \in H$, $f \in \mathrm{Hom}_R(M_L(X), R)$, $m \in M_L(X)$. Since the kernel of π is $J'_L(w)$, we have a natural isomorphism of H-modules

$$N_L(X) \cong \mathrm{Hom}_R(M_L(X), R).$$

2.6. THEOREM (LUSZTIG [**11**, 12.15]). *Let X, X' be left cells of W. Then*

$$\dim_K \mathrm{Hom}_{H^K}(M_L(X)^K, M_L(X')^K) = |X \cap X'^{-1}|.$$

The idea of the proof of 2.6 is as follows. For $y \in W$, let $\Phi_y: H^K \to H^K$ be the map $a \mapsto aD_y^{-1}$. If $y \in X \cap X'^{-1}$, then Φ_y induces a homomorphism of H^K-modules $\phi_y: M_L(X)^K \to N_L(X')^K$ by 2.5. Lusztig showed that $\{\phi_y \mid y \in X \cap X'^{-1}\}$ is in fact a K-basis for $\mathrm{Hom}_{H^K}(M_L(X)^K, N_L(X')^K)$. The theorem is then proved, since we have isomorphisms of H^K-modules

$$N_L(X')^K \cong \mathrm{Hom}_K(M_L(X')^K, K) \cong M_L(X')^K,$$

where the existence of the second isomorphism follows from Lusztig [**9**, 1.7].

2.7. COROLLARY. *Let X be a left cell of W. Then*

$$\dim_K \mathrm{End}_{H^K}(M_L(X)^K) = \sum_E \langle E, M_L(X)^K \rangle = |X \cap X^{-1}|,$$

where E ranges over the simple H^K-modules (up to isomorphism), and $\langle E, M_L(X)^K \rangle$ is the multiplicity of E in $M_L(X)^K$.

It is known that

$$\sum_E \dim(E) = \sum_{\chi \in \mathrm{Irr}(W)} \deg(\chi) = n(W),$$

where $n(W)$ is the number of involutions of W. (For convenience, we consider $w = 1$ to be an involution in this discussion.) The next result provides a remarkable refinement of the formula above.

2.9. THEOREM (LUSZTIG [**11**, 12.13]). *Assume W is of classical type and X is a left cell of W. Then $M_L(X)^K$ is multiplicity free, and the number of simple components of $M_L(X)^K$ is equal to the number of involutions in X. Further, $X \cap X^{-1}$ consists entirely of involutions, and $|X \cap X^{-1}|$ is a power of 2.*

REMARK. If W is of type A, then the left cell modules $M_L(X)^K$ are simple H^K-modules, and all simple H^K-modules arise in this fashion (Kazhdan and Lusztig [**6**]). If W is of exceptional type, then there are left cells X of W for which $M_L(X)^K$ is not multiplicity free.

We have seen that the left cell modules have remarkable structural properties linking them directly with the elements of W. The left cells (and their modules) may be determined by direct calculation of the $P_{y,w}$ if $|W|$ is sufficiently small; see Kazhdan and Lusztig [**6**, 2.2c] for a recursion formula for the $P_{y,w}$. The complexity of this calculation tends to grow rapidly as $|W|$ increases, however. Techniques are known for constructing new left cells from a given left cell; see, for example, [**6**, 3.3 and 4.3]. The characters afforded by the left cell modules have been determined by Lusztig; this result will be stated in the next section.

3. Constructible characters and families. We continue with the assumption that $W = \langle S\rangle$ is a (finite) Weyl group. Let V be the Euclidean space on which W acts in its natural reflection representation, and let $S(V)$ be the $\mathbf{R}$-algebra of all polynomial functions $V \to \mathbf{R}$. Let $I(V)$ be the ideal of $S(V)$ generated by the W-invariants in $S(V)$ of positive degree. The coinvariant algebra $S(W) = S(V)/I(V)$ is a graded $\mathbf{R}W$-module; let $S_i(W)$ be the ith homogeneous component of $S(W)$. It is known that $S(W) \cong \mathbf{R}W$ as $\mathbf{R}W$-modules, and the dimension of $S_i(W)$ is equal to the number of elements w of W such that $l(w) = i$.

For $\chi \in \operatorname{Irr}(W)$, the *fake degree* $f_\chi(u)$ is defined by

$$f_\chi(u) = \sum_i m(i)u^i,$$

where $m(i)$ is the multiplicity of χ in the character afforded by $S_i(W)$. It is clear that $f_\chi(u) \in \mathbf{Z}[u]$, and $f_\chi(1) = \deg(\chi)$. The *generic degree* $d_\chi(u)$ is defined by

$$d_\chi(u) = \frac{\deg(\chi)P_W(u)}{\sum_w u_w^{-1}\psi(T_w)\psi(T_{w^{-1}})},$$

where $P_W(u) = \sum u_w$ is the Poincaré polynomial of W, and ψ is the irreducible character of H^K corresponding to χ under the specialization $\sqrt{u} \to 1$. It is known that $d_\chi(u) \in \mathbf{Q}[u]$. We write

$$\begin{aligned} d_\chi(u) &= \gamma_\chi u^{a(\chi)} + \text{terms of higher degree},\\ f_\chi(u) &= n_\chi u^{b(\chi)} + \text{terms of higher degree}, \end{aligned}$$

where $\gamma_\chi \in \mathbf{Q}^+$, $n_\chi \in \mathbf{Z}^+$. Lusztig has observed that $a(\chi) \le b(\chi)$ in all cases. If $a(\chi) = b(\chi)$, then χ is called a *special character* of W.

Assume now that $W' = \langle J\rangle$ is a standard parabolic subgroup of W $(J \subseteq S)$, and suppose $\theta \in \operatorname{Irr}(W')$. Define a character $J_{W'}^W(\theta)$ of W as follows:

$$J_{W'}^W(\theta) = \sum \langle \chi, \operatorname{Ind}_{W'}^W(\theta)\rangle \chi,$$

where the sum is over all $\chi \in \operatorname{Irr}(W)$ with $a(\chi) = a(\theta)$. The domain of $J_{W'}^W$ is extended to include arbitrary characters of W' by linearity.

The *constructible characters* of W are defined recursively, as follows. If $W = \langle 1\rangle$, then only the trivial character of W is constructible. If $W \ne \langle 1\rangle$, then the characters of W of the form

$$J_{W'}^W(\theta) \quad \text{or} \quad \operatorname{sign} \bigotimes J_{W'}^W(\theta)$$

are constructible, where W' is a proper standard parabolic subgroup of W and θ is a constructible character of W'.

For $\chi, \psi \in \operatorname{Irr}(W)$, define $\chi \approx \psi$ if χ and ψ both appear with positive multiplicity in some constructible character of W. The equivalence classes of the equivalence relation generated by $\approx$ are the *families* of $\operatorname{Irr}(W)$.

3.1. THEOREM (BARBASCH AND VOGAN [**1**, **2**]). *For $\chi, \psi \in \mathrm{Irr}(W)$, the following are equivalent.*

(i) *χ and ψ belong to the same family of* $\mathrm{Irr}(W)$.

(ii) *χ and ψ both appear with positive multiplicity in the character afforded by $M_{LR}(Y)_1$ for some two-sided cell Y of W.*

Thus there is a bijective correspondence between the families of $\mathrm{Irr}(W)$ and the two-sided cells of W. We also have the next result, which determines the characters afforded by the left cell modules of W.

3.2. THEOREM (LUSZTIG [**13**]). *Assume θ is a character of W. Then θ is constructible if and only if θ is afforded by $M_L(X)_1$ for some left cell X of W.*

The decomposition of the constructible characters into irreducible characters has been explicitly described by Lusztig (see [**10**]). Therefore the decompositions of the characters of the left cell modules and also the families of $\mathrm{Irr}(W)$ are explicitly known. We summarize below some additional properties of constructible characters and left cells.

3.3. Let X be a left cell of W, and let θ be the character of W afforded by $M_L(X)_1$. Then there is a unique special character χ_0 of W such that $\langle \theta, \chi_0 \rangle \neq 0$; for this χ_0, we have $\langle \theta, \chi_0 \rangle = 1$.

3.4. Let X, θ, χ_0 be as in 3.3, and let $\chi \in \mathrm{Irr}(W)$ be such that $\langle \theta, \chi \rangle \neq 0$. Then $a(\chi) = a(\chi_0)$.

3.5. Assume X, X' are left cells of W and χ_0, χ_0' are the corresponding special characters of W (as in 3.3). Then X and X' are contained in the same two-sided cell of W if and only if $\chi_0 = \chi_0'$.

Thus there are natural bijections

$$\begin{aligned}\{\text{families of } \mathrm{Irr}(W)\} &\leftrightarrow \{\text{two-sided cells of } W\}\\ &\leftrightarrow \{\text{special characters of } W\}.\end{aligned}$$

We conclude this section by describing a combinatorial object attached to each family of $\mathrm{Irr}(W)$. First, to each family $\mathcal{F}$ of $\mathrm{Irr}(W)$, Lusztig associates a finite group $\mathcal{G}_\mathcal{F}$. If W is of classical type, $\mathcal{G}_\mathcal{F}$ is an elementary abelian 2-group, and if W is of exceptional type, $\mathcal{G}_\mathcal{F}$ is one of the symmetric groups S_n, $n \leq 5$. (See [**11**, 13] for a geometric description of the groups $\mathcal{G}_\mathcal{F}$.) Given $\mathcal{G}_\mathcal{F}$, let $X(\mathcal{G}_\mathcal{F})$ be the set of all pairs (x, σ), $x \in \mathcal{G}_\mathcal{F}$, $\sigma \in \mathrm{Irr}(Z(x))$, where $Z(x)$ is the centralizer of x in $\mathcal{G}_\mathcal{F}$. Then $\mathcal{G}_\mathcal{F}$ acts on $X(\mathcal{G}_\mathcal{F})$ by conjugation, and we define $\mathcal{M}(\mathcal{G}_\mathcal{F}) = X(\mathcal{G}_\mathcal{F})/\mathcal{G}_\mathcal{F}$. Finally, define a mapping $\{\ ,\ \}\colon M(\mathcal{G}_\mathcal{F}) \times M(\mathcal{G}_\mathcal{F}) \to \mathbf{C}$ as follows:

$$\{(x,\sigma),(y,\tau)\} = |Z(x)|^{-1}|Z(y)|^{-1} \sum_{\substack{g \in \mathcal{G}_\mathcal{F} \\ gyg^{-1} \in Z(x)}} \sigma(gyg^{-1})\tau(g^{-1}x^{-1}g).$$

The matrix $M = [\{(x,\sigma),(y,\tau)\}]$ is called a *nonabelian Fourier transform matrix.*

4. Families of unipotent characters. We return now to the situation of the first section. Assume for convenience that G is split over $\mathbf{F}_q$. Let T_0 be an

$\mathbf{F}_q$-split maximal torus of G; F acts trivially on the Weyl group $W = N_G(T_0)/T_0$ of G.

For θ a generalized character of W, we define (following Lusztig) the "fake character" R_θ of G^F as follows:

$$R_\theta = |W|^{-1} \sum_{w \in W} \theta(w) R_w.$$

Here $R_w = R^G_{T(w)} 1$, where $T(w)$ is an F-stable maximal torus corresponding to $w \in W$ in the usual way. The R_θ are not necessarily generalized characters of G^F; they are, apparently, $\mathbf{Q}$-linear combinations of characters of G^F. The degrees of the R_χ, $\chi \in \mathrm{Irr}(W)$, are given by the fake degree polynomials according to the formula $\deg(R_\chi) = f_\chi(q)$. Using the orthogonality relations for $\mathrm{Irr}(W)$, the generalized characters $R_w (w \in W)$ may be recovered from the R_χ $(\chi \in \mathrm{Irr}(W))$ as follows:

$$R_w = \sum_\chi \chi(w) R_\chi.$$

Thus, to understand the decomposition of the R_w into unipotent characters, it is enough to understand the multiplicities $\langle \rho, R_\chi \rangle$, $\rho \in \mathrm{Irr}_{(1)}(G^F)$, $\chi \in \mathrm{Irr}(W)$. It is known that for $\chi, \psi \in \mathrm{Irr}(W)$,

$$\langle R_\chi, R_\psi \rangle = \begin{cases} 1 & \text{if } \chi = \psi, \\ 0 & \text{if } \chi \neq \psi. \end{cases}$$

Let $w \in W$. For each $\chi \in \mathrm{Irr}(W)$, let $\Gamma(\chi)$ be a W-graph such that χ is afforded by $E(\Gamma(\chi))_1$. It is known that

$$\mathrm{trace}(T_w, E(\Gamma(\chi))) = \varepsilon_w c(w,\chi) u^{(l(w)-a(\chi))/2} + \text{terms of higher degree}$$

for some $c(w,\chi) \in \mathbf{Z}$. Define

$$\alpha(w) = \sum_\chi c(w,\chi)\chi,$$

a generalized character of W. The Integrality Theorem (6.14 of [**11**]) implies the following.

4.1. THEOREM (LUSZTIG). *$R_{\alpha(w)}$ is a generalized character of G^F (not just a $\mathbf{Q}$-linear combination of characters of G^F).*

From the Disjointness Theorem (6.17 of [**11**]) we have the following.

4.2. THEOREM (LUSZTIG). (i) *If $\chi, \psi \in \mathrm{Irr}(W)$ belong to different families of $\mathrm{Irr}(W)$, then R_χ and R_ψ are disjoint in the sense that for all $\rho \in \mathrm{Irr}_{(1)}(G^F)$, either $\langle \rho, R_\chi \rangle = 0$ or $\langle \rho, R_\psi \rangle = 0$.*

(ii) *For $w \in W$, $R_{\alpha(w)}$ is either 0 or an actual character of G^F.*

Let $\mathcal{U} = \mathrm{Irr}_{(1)}(G^F)$. By (4.2) we see that $\mathcal{U}$ may be partitioned into disjoint subsets $\mathcal{U}_\mathcal{F}$, one for each family $\mathcal{F}$ of $\mathrm{Irr}(W)$; the set $\mathcal{U}_\mathcal{F}$ consists of all $\rho \in \mathcal{U}$ appearing with nonzero multiplicity in some R_χ, $\chi \in \mathcal{F}$. We remark that Lusztig has described, for each family $\mathcal{F}$, an embedding $i\colon \mathcal{F} \to \mathcal{M}(\mathcal{G}_\mathcal{F})$. (See [**11**, 4] for an explicit description of this embedding.) We can now state Lusztig's classification theorem, as it applies to $\mathcal{U} = \mathrm{Irr}_{(1)}(G^F)$.

4.3. THEOREM (LUSZTIG [**11**]). *Let $\mathcal{F}$ be a family of* $\operatorname{Irr}(W)$. *There is a bijective correspondence* $\mathcal{M}(\mathcal{G}_\mathcal{F}) \to \mathcal{U}_\mathcal{F}$, $(x,\sigma) \mapsto \rho_{x,\sigma}$, *such that*

$$\langle \rho_{x,\sigma}, R_\chi \rangle = \delta_{x,\sigma}\{(x,\sigma), i(\chi)\}, \tag{4.4}$$

where $\delta_{x,\sigma}$ *is equal to* 1 *or* -1, *and* $\delta_{x,\sigma} = 1$ *if* G *has no composition factors of types* E_7 *or* E_8.

Formula 4.4 and results of Deligne and Lusztig [**4**] provide a method for computing the value of a unipotent character of G^F at any semisimple element of G^F. In particular, we have the following (Lusztig [**11**, 4.26]):

$$\deg(\rho_{x,\sigma}) = \sum_{\chi \in \mathcal{F}} \{(x,\sigma), i(\chi)\} f_\chi(q) = \{(x,\sigma), i(\chi_0)\} q^{a(\chi_0)} g(q),$$

where $g(u) \in \mathbf{Z}[u]$ has constant term 1 and χ_0 is the special character in $\mathcal{F}$. Since $i(\chi_0) = (1,1) \in \mathcal{M}(\mathcal{G}_\mathcal{F})$, we have

$$\deg(\rho_{x,\sigma}) = (\deg(\sigma)/|Z(x)|) q^{a(\chi_0)} g(q).$$

The reader is encouraged to read Lusztig's book [**11**], where more general statements and proofs of the results of this section appear.

REFERENCES

1. D. Barbasch and D. Vogan, *Primitive ideals and orbital integrals in complex classical groups*, Math. Ann. **259** (1982), 153–199.

2. ——, *Primitive ideals and orbital integrals in complex exceptional groups*, J. Algebra **80** (1983), 350–382.

3. C. Curtis, N. Iwahori, and R. Kilmoyer, *Hecke algebras and characters of parabolic type of finite groups with* (B, N)*-pairs*, Inst. Hautes Études Sci. Publ. Math. **40** (1971), 81–116.

4. P. Deligne and G. Lusztig, *Representations of reductive groups over finite fields*, Ann. of Math. (2) **103** (1976), 103–161.

5. A. Gyoja, *On the existence of a W-graph for an irreducible representation of a Coxeter group*, J. Algebra **86** (1984), 422–438.

6. D. Kazhdan and G. Lusztig, *Representations of Coxeter groups and Hecke algebras*, Invent. Math. **53** (1979), 165–184.

7. ——, *Schubert varieties and Poincaré duality*, Proc. Sympos. Pure Math., vol. 36, Amer. Math. Soc., Providence, R. I., 1980, pp. 185–203.

8. G. Lusztig, *On a theorem of Benson and Curtis*, J. Algebra **71** (1981), 490–498.

9. ——, *Unipotent characters of the symplectic and odd orthogonal groups over a finite field*, Invent. Math. **64** (1981), 263–296.

10. ——, *A class of irreducible representations of a Weyl group*. II, Proc. Kon. Nederl. Akad. A **85** (1982), 219–226.

11. ——, *Characters of reductive groups over a finite field*, Ann. of Math. Studies, no. 107, Princeton University Press, Princeton, N.J., 1984.

12. ——, *Cells in affine Weyl groups*, Algebraic Groups and Related Topics, Adv. Stud. Pure Math., vol. 6, North-Holland, 1985, pp. 255–287.

13. ——, *Sur les cellules gauches des groupes de Weyl*, C. R. Acad. Sci. Paris Sér. I Math. **302** (1986), 5–8.

UNIVERSITY OF NOTRE DAME

Proceedings of Symposia in Pure Mathematics
Volume **47** (1987)

Modular Representations of Algebraic Groups

HENNING HAAHR ANDERSEN

This is a survey of some recent developments in the representation theory for semisimple algebraic groups in positive characteristic. By recent we mean "after 1980" although we shall also—mainly in the first section, where we introduce the notation and general setup—briefly mention a few "classical" results. We have chosen to deal with those aspects of the theory which in one way or another are closely related to the still open problem of determining the irreducible characters of the groups.

The first topic we discuss is the construction of filtrations of certain representations. Some 10 years ago, J. C. Jantzen [**24**] used the so-called contravariant form to construct filtrations of Weyl modules and he obtained several nice results by studying these. By exploring the natural homomorphisms between certain cohomology modules, the author [**4**] recently generalized Jantzen's construction, and in §2, we describe how this is done.

By analogy with the characteristic zero theory for Verma modules, the above-mentioned filtrations ought to be preserved by the homomorphisms between adjacent Weyl modules. This is the so-called Jantzen conjecture (known to hold for the types A_1, A_2, and B_2), and §3 is devoted to a close look at some of its many interesting consequences, the most important being a positive answer to Lusztig's conjecture [**34**] on the irreducible characters.

Recently it has proved very fruitful to study the representations of the subgroup schemes arising as kernels of the Frobenius endomorphisms. We give some of the highlights of this theory in §4, emphasizing the close relations between the representation theory of these subgroups and of the groups themselves. The final section illustrates how Ye [**40**] was able to take advantage of this relation in obtaining a precise description of the set of composition factors of Weyl modules.

Finally we point out that the bibliography contains, in addition to the items referred to in the text, a list of other references concerning recent developments in related areas. Although this list is of course by no means complete, we hope that it will at least help the reader to place this short survey in a somewhat broader perspective.

1980 *Mathematics Subject Classification* (1985 *Revision*). Primary 20G05, 20G10.

1. Notation and "classical" results. We let G denote a connected and simply connected semisimple algebraic group over an algebraically closed field k of characteristic $p > 0$. By $B \leq G$ we denote a Borel subgroup and $T \leq B$ is a maximal torus. Then $X(T) = X(B)$ is the corresponding character group and it contains the root system R associated to (G, T). Choose a set of positive roots R_+ in such a way that $-R_+$ corresponds to B. Then we call $X(T)_+ = \{\lambda \in X(T) \mid \langle \lambda, \alpha^\vee \rangle \geq 0,\ \alpha \in R_+\}$ the set of dominant characters. Here $\alpha^\vee$ denotes the coroot of α.

The torus T has the property that any finite-dimensional representation V of T is completely reducible. This allows us to write

$$V = \bigoplus_{\lambda \in X(T)} V_\lambda,$$

where $V_\lambda = \{v \in V \mid tv = \lambda(t)v,\ t \in T\}$, and to define the character of V, $\operatorname{ch} V$, by

$$\operatorname{ch} V = \sum_{\lambda \in X(T)} (\dim V_\lambda) e^\lambda \in \mathbf{Z}[X(T)].$$

Now let E be a finite-dimensional representation of B. Then we set

$$H^0(E) = H^0(G/B, E) = \{\varphi \colon G \to E \mid \varphi \text{ is polynomial and } \varphi(gb) = b^{-1}\varphi(g),\ g \in G,\ b \in B\}.$$

Note that G acts on $H^0(E)$ as follows: If $g \in G$, $\varphi \in H^0(E)$, then $g\varphi \colon G \to E$ takes $x \in G$ into $\varphi(g^{-1}x)$. In other words, we have that $H^0(E)$ is the G-module induced by E.

We also have the right derived functors of induction, $H^i(-) = H^i(G/B, -)$. Alternatively, $H^i(E)$ may be thought of as the ith cohomology group of the locally free sheaf on G/B induced by E. Of particular interest in what follows will be the modules $H^i(\lambda)$ for $\lambda \in X(T)$.

The following result is at least 30 years old.

THEOREM 1 [**12**]. *Let $\lambda \in X(T)_+$. Then $H^0(\lambda)$ contains a unique irreducible submodule, $L(\lambda)$, and in this way $X(T)_+$ parametrizes the finite-dimensional irreducible G-modules.*

We can now state the problem which, as mentioned in the introduction, is the central point in this survey.

PROBLEM 1. Find $\operatorname{ch} L(\lambda)$.

This problem is open. In 1979, G. Lusztig formulated a conjecture [**34**] which at least for big primes gives the answer; see Conjecture 1 below.

On the other hand, the character of $H^0(\lambda)$ is known. This is a consequence of Kempf's vanishing theorem:

THEOREM 2 [**31**]. *If $\lambda \in X(T)$, then $H^i(\lambda) = 0$ for $i > 0$.*

An alternative and much easier proof of this result was found independently by the author [**2**] and W. Haboush [**19**].

COROLLARY 1. *If* $\lambda \in X(T)_+$, *then* $\operatorname{ch} H^0(\lambda)$ *is given by Weyl's formula.*

The point here is of course that the alternating sum $\sum(-1)^i \operatorname{ch} H^i(\lambda)$ is easy to compute.

Via this corollary, Problem 1 may be transformed into the following equivalent problem.

PROBLEM 2. Let $\lambda, \mu \in X(T)_+$. Find the composition factor multiplicity $[H^0(\lambda) : L(\mu)]$ of $L(\mu)$ in $H^0(\lambda)$.

Let $W = N_G(T)/T$ be the Weyl group and define the "dot-action" of W on $X(T)$ by

$$w \cdot \lambda = w(\lambda + \rho) - \rho, \qquad \lambda \in X(T),\ w \in W.$$

Here ρ is half the sum of the positive roots. We shall also consider the affine Weyl group W_p, which is given by

$$W_p = \langle \{s_{\alpha,n} | \alpha \in R_+, n \in \mathbf{Z}\} \rangle,$$

where $s_{\alpha,n}$ is the affine reflection which takes $\lambda \in X(T)$ into $s_\alpha \cdot \lambda + np\alpha$.

With this notation we may formulate the linkage principle (for the strong, resp. very strong, resp. regular linkage principle; see [**1**, **39**, and **33**]).

THEOREM 3 [**1**]. *Let* $\lambda \in X(T)$, $\mu \in X(T)_+$, *and* $i \in \mathbf{N}$. *If* $[H^i(\lambda) : L(\mu)] \neq 0$, *then* $\mu \in W_p \cdot \lambda$.

We can now also state Lusztig's conjecture. Let us agree to set

$$h = \max_{\alpha \in R_+} \langle \rho, \alpha^\vee \rangle + 1.$$

CONJECTURE 1 [**34**, **5**]. Assume $p \geq h$ and fix $\lambda \in X(T)_+$ such that $\langle \lambda + \rho, \alpha^\vee \rangle < p$ for all $\alpha \in R_+$. Then for all $y, w \in W_p$ for which $yw_0 \cdot \lambda$, $ww_0 \cdot \lambda \in X(T)_+$, and $\langle ww_0(\lambda + \rho), \alpha^\vee \rangle < p(p - h + 2)$ for all $\alpha \in R_+$ we have

$$[H^0(ww_0 \cdot \lambda) : L(yw_0 \cdot \lambda)] = Q_{y,w}(1).$$

Here $Q_{y,w}$ denotes the "inverse" Kazhdan-Lusztig polynomial associated to $y, w \in W_p$. (More precisely the $Q_{y,w}$'s are determined by the inversion formula $\sum_z (-1)^{l(z)-l(y)} P_{y,z} Q_{z,w} = \delta_{y,w}$, where $P_{y,w}$ denotes the usual Kazhdan-Lusztig polynomial associated to $y, w \in W_p$ and l is the length function on W_p.)

As stated, the conjecture only solves Problem 2 for λ p-regular. However, there is a translation principle (due to Jantzen [**25**]; see also [**3**] for a different proof) which tells us how to deduce the general case from the p-regular case.

So far this conjecture has only been verified for groups of rank ≤ 2 (see [**35**]). Strong evidence is, however, also provided by the fact that it has been proved

[30] to be consistent with Steinberg's tensor theorem:

THEOREM 4 **[37]**. *Let $\lambda \in X(T)_+$ and write $\lambda = \lambda^0 + p\lambda^1$ for appropriate $\lambda^0, \lambda^1 \in X(T)_+$. Then*

$$L(\lambda) \cong L(\lambda^0) \otimes L(\lambda^1)^{(1)}.$$

Here (1) denotes "twist" by Frobenius endomorphism of G; see §4 below. See **[13]** for a recent proof of this theorem.

We end this section by an example showing how the situation is when G is so small that everything can be explicitly computed.

EXAMPLE 1. Let $G = \mathrm{SL}_2(k)$,

$$T = \left\{ \begin{pmatrix} t^{-1} & 0 \\ 0 & t \end{pmatrix} \mid t \in k \setminus \{0\} \right\} \quad \text{and} \quad B = \left\{ \begin{pmatrix} t & u \\ 0 & t^{-1} \end{pmatrix} \mid t \in k \setminus \{0\}, u \in k \right\}.$$

Then $X(T) \cong \mathbf{Z}$ and $X(T)_+ \cong \mathbf{N}$. If $n \in \mathbf{Z}$, we let $\lambda_n \in X(T)$ denote the character which takes $\left(\begin{smallmatrix} t^{-1} & 0 \\ 0 & t \end{smallmatrix}\right)$ into t^n. Then

$$H^0(\lambda_n) \cong S^n V \cong k[X, Y]_n,$$

where V is the natural 2-dimensional representation of G. So $H^0(\lambda_n)$ has basis $v_i = X^i Y^{n-1}$, $i = 0, 1, \ldots, n$, and it is easy to check that

$$L(\lambda_n) = \operatorname{span}\{v_i \mid \tbinom{n}{i} = 0 \ (\operatorname{mod} p)\}.$$

For $0 \leq n < p$ we thus have $H^0(\lambda_n) = L(\lambda_n)$, whereas for $n = n^0 + pn^1$, $0 \leq n^0 < p - 1$, $0 < n^1 < p$,

$$[H^0(\lambda_n) : L(\lambda_m)] = \begin{cases} 1 & \text{if } m = n \text{ or } m = pn^1 - n_0 - 2, \\ 0 & \text{otherwise.} \end{cases}$$

Since G/B in this case is the projective line, it is clear that $H^i(\lambda_n) = 0$ for $i > 1$. We can find $H^1(\lambda_n)$ via Serre duality,

$$H^1(\lambda_n) \cong H^0(\lambda_{-n-2})^*.$$

Hence, if $n \geq 0$, $\{v_i\}$ is the above basis for $H^0(\lambda_n)$, and we let $\{f_i\}$ denote the dual basis for $H^0(\lambda_n)^* \cong H^1(\lambda_{-n-2})$, then we have the two G-homomorphisms

$$H^1(\lambda_{-n-2}) \xrightarrow{\varphi} H^0(\lambda_n) \xrightarrow{\psi} H^1(\lambda_{-n-2})$$

defined by

$$\varphi(f_i) = \tbinom{n}{i} v_{n-i}, \qquad \psi(v_j) = (n-j)!j!f_{n-j}, \qquad 0 \leq i, j \leq n.$$

Note that the composite (in either order) of φ and ψ is multiplication by $n!$ Also notice that $L(\lambda_n)$ is realized as the image of φ.

2. Filtrations. Let A be a principal ideal domain and let $\mathfrak{p} \subset A$ be a prime ideal. Set Q equal to the fraction field of A and let k be a field containing $A/\mathfrak{p}$. Denote by $\nu_{\mathfrak{p}}$ the valuation corresponding to $\mathfrak{p}$.

Suppose now that $\varphi \colon M \to M'$ is a homomorphism between two A-modules and define

$$M^i = \{m \in M \mid \varphi(m) \in \mathfrak{p}^i M'\}.$$

Then

$$M = M^0 \supset M^1 \supset M^2 \supset \cdots$$

and we set

$$\overline{M}^i = (\text{image of } M^i \text{ in } M/\mathfrak{p}M) \otimes_{A/\mathfrak{p}} k \subset M/\mathfrak{p}M \otimes_{A/\mathfrak{p}} k = M \otimes_A k = \overline{M}.$$

In particular, we note that

$$\overline{M}^1 = \ker(\varphi \otimes 1).$$

LEMMA 1. *If M and M' are free of finite rank and if $\varphi \otimes Q$ is an isomorphism, then*

$$\sum_{i>0} \dim_k \overline{M}^i = \nu_\mathfrak{p}(\det \varphi).$$

It is clear that if M and M' are modules for an algebraic group H defined over A, and φ is an H-homomorphism, then in the above construction M^i is an H-submodule of M and $\overline{H}^i$ is an $H(k)$-submodule of $\overline{M}$. (Here $H(k)$ denotes the points of H over k.)

Suppose now that the semisimple group G we considered in §1 is in fact defined over A. In order to preserve the notation from §1, we denote by G_A the group over A whereas G denotes the k-points (k being algebraically closed as in §1 and containing $A/\mathfrak{p}$ as above). We will also assume that T and B are defined over A, and we use a similar notation for these.

Hence, if $\varphi\colon M \to M'$ is a G_A-homomorphism, then the above construction gives us a G_A-filtration M^i of M and a G-filtration $\overline{M}^i$ of $\overline{M}$. Also, if the assumptions of Lemma 1 hold, then by applying the lemma to each weight space we get the sum formula

$$(*) \qquad \sum_{i>0} \operatorname{ch} \overline{M}^i = \sum_\lambda \nu_\mathfrak{p}(\det \varphi|_\lambda) e^\lambda,$$

where $\varphi|_\lambda$ denotes the restriction of φ to the λ weight space.

In particular, we shall apply this to the cohomology modules $H^i(\lambda)$ considered in §1. Taking $A = \mathbf{Z}$, there does actually exist an algebraic group $G_\mathbf{Z}$ over $\mathbf{Z}$ such that $G = G_\mathbf{Z}(k)$. We assume that T and B come from the subgroups $T_\mathbf{Z} \leq B_\mathbf{Z}$ of $G_\mathbf{Z}$.

Note that the maps of φ and ψ considered in Example 1.1 are actually defined over $\mathbf{Z}$. If α is a simple root and $P_{\alpha,\mathbf{Z}} \leq G_\mathbf{Z}$ denotes the associated minimal parabolic subgroup containing $B_\mathbf{Z}$, then we have the following generalization of this example.

LEMMA 2. *Let $\lambda \in X(T)$ and suppose $n = \langle \lambda, \alpha^\vee \rangle \geq 0$. Then there exist natural $P_{\alpha,\mathbf{Z}}$-homomorphisms*

$$\varphi_\alpha\colon H^1_{\alpha,\mathbf{Z}}(s_\alpha \cdot \lambda) \to H^0_{\alpha,\mathbf{Z}}(\lambda)$$

and

$$\psi_\alpha\colon H^0_{\alpha,\mathbf{Z}}(\lambda) \to H^1_{\alpha,\mathbf{Z}}(s_\alpha \cdot \lambda)$$

such that $\psi_\alpha \circ \varphi_\alpha$ and $\varphi_\alpha \circ \psi_\alpha$ are both multiplications by $n!$ (here we have used the abbreviation $H^i_{\alpha,\mathbf{Z}}(-) = H^i(P_{\alpha,\mathbf{Z}}/B_{\alpha,\mathbf{Z}},-)$.

By applying the functor $H^i(G_\mathbf{Z}/P_{\alpha,\mathbf{Z}},-)$ to the homomorphisms in Lemma 2, we obtain $G_\mathbf{Z}$-homomorphisms

$$H^{i+1}_\mathbf{Z}(s_\alpha\cdot\lambda) \xrightarrow{T_{s_\alpha}} H^i_\mathbf{Z}(\lambda) \xrightarrow{\tilde{T}_{s_\alpha}} H^{i+1}_\mathbf{Z}(s_\alpha\cdot\lambda), \qquad i \geq 0$$

(where $H^{\cdot}_\mathbf{Z}(-) = H^{\cdot}(G_\mathbf{Z}/B_\mathbf{Z}, -)$).

Now let $w_0 = s_{\beta_1}\cdots s_{\beta_N}$ be a reduced expression for the longest word w_0 in W. Then the composite

$$\begin{aligned} T_{w_0}\colon H^N_\mathbf{Z}(w_0\cdot\lambda) &\xrightarrow{T_{s_{\alpha_N}}} H^{N-1}_\mathbf{Z}\left(s_{\alpha_{N-1}}\cdots s_{\alpha_1}\cdot\lambda\right) \to \cdots \\ &\to H^1_\mathbf{Z}\left(s_{\alpha_1}\cdot\lambda\right) \xrightarrow{T_{s_{\alpha_1}}} H^0_\mathbf{Z}(\lambda) \end{aligned}$$

is a $G_\mathbf{Z}$-homomorphism. Moreover, it follows from Theorem 1.2 (combined with Serre duality) that $H^N_\mathbf{Z}(w_0\cdot\lambda)$ and $H^0_\mathbf{Z}(\lambda)$ are free modules, and since both when tensored by $\mathbf{C}$ become irreducible $G_\mathbf{Z}(\mathbf{C})$-modules, we conclude that $T_{w_0}\otimes_\mathbf{Z}\mathbf{Q}$ is an isomorphism. In other words, we have a situation where Lemma 1 applies. Of course $\nu_p(\det T_{w_0}) = \nu_p(\prod_i \det T_{s_{\alpha_i}}) = \sum_i \nu_p(\det T_{s_{\alpha_i}})$ so that to compute this expression we have reduced the problem to studying the homomorphisms $T_{s_{\alpha_i}}$. A closer look at the "SL_2-situation" (compare Example 1 in §1) and an analysis of the functors $H^j(G_\mathbf{Z}/P_{\alpha,\mathbf{Z}},-)$ give the following result (here $\chi(\mu)$, $\mu\in X(T)$, denotes the Weyl character

$$\chi(\mu) = \sum_{w\in W}(-1)^{l(w)}e^{w(\mu+\rho)} \Big/ \sum_{w\in W}(-1)^{l(w)}e^{w(\rho)} \in \mathbf{Z}[X(T)].)$$

THEOREM 1 [**4**]. *Let $\lambda\in X(T)_+$. Then $H^N(w_0\cdot\lambda)$ has a G-filtration*

$$H^N(w_0\cdot\lambda) = H^N(w_0\cdot\lambda)^0 \supset H^N B(w_0\cdot\lambda)^1 \supset \cdots$$

with the properties

(i) $$H^N(w_0\cdot\lambda)/H^N(w_0\cdot\lambda)^1 \cong L(\lambda),$$

(ii) $$\sum_{i>0}\operatorname{ch} H^N(w_0\cdot\lambda)^i = \sum_{\alpha\in R_+}\ \sum_{\substack{m\\ 0<mp<\langle\lambda+\rho,\alpha^\vee\rangle}} \nu_p(mp)\chi(s_{\alpha,m}\cdot\lambda).$$

(See §3 for the background for this theorem.)

Note that the right-hand side of (ii) consists of terms for which we have explicit formulas. In particular, it is possible to check whether it is zero or not. This is important because we have the following immediate consequence of Theorem 1.

COROLLARY 1. *Let $\lambda\in X(T)_+$. Then $H^N(w_0\cdot\lambda)$ is irreducible if and only if the character on the right-hand side of* (ii) *in Theorem 1 is zero.*

This corollary gives an irreducibility criterion which for type A_n was made quite explicit by J. C. Jantzen already in his dissertation [**23**].

Another observation is that the right-hand side of (ii) in Theorem 1 involves only Weyl characters corresponding to dominant weights which are strictly smaller than λ. Hence in an inductive determination of the composition factors of Weyl modules the following corollary is often useful.

COROLLARY 2. *If $\lambda, \mu \in X(T)_+$, then $L(\mu)$ is a composition factor of $H^N(w_0 \cdot \lambda)$ if and only if $\mu = \lambda$ or $L(\mu)$ occurs with positive coefficient when the right-hand side of* (ii) *in Theorem* 1 *is expressed as a sum of irreducible characters. Moreover, if the coefficient of $L(\mu)$ is* 1 *(resp. n) in this expression, then $L(\mu)$ occurs with multiplicity* 1 *(resp. $\leq n$) as a composition factor of $H^N(w_0 \cdot \lambda)$.*

Via this corollary J. C. Jantzen solved Problem 2 of §1 for groups of rank 2 and for type A_3; see [**24**, **25**]. Before that B. Braden had dealt with type A_2 and he also had partial results for B_2 [**9**]. For G_2, T. A. Springer had settled the problem in characteristic 3 [**36**], whereas the characteristic 5 case which was left open by Jantzen was recently solved by J. Hagelskjær [**20**].

Recently, J. C. Jantzen has used the method described in this section to construct filtrations of the principal series modules for the finite Chevalley groups corresponding to G [**28**]. In particular, he obtains a sum formula which expresses the Brauer characters of the filtration in terms of Deligne-Lusztig characters. It is not known whether this result can be generalized to all the Deligne-Lusztig modules.

3. Jantzen's conjecture and its consequences. Let $\mathfrak{g}$ denote a complex semisimple Lie algebra of the same type as G. Choose an irreducible (finite-dimensional) $\mathfrak{g}$-module $V_{\mathbf{C}}(\lambda)$ of highest weight $\lambda \in X(T)_+$ (then λ may also be thought of as an integral dominant weight for the Cartan subalgebra of $\mathfrak{g}$ corresponding to T). We let $v \in V_{\mathbf{C}}(\lambda)$ denote a maximal weight vector and we set $V_{\mathbf{Z}}(\lambda) = U_{\mathbf{Z}}(\mathfrak{g})v$, where $U_{\mathbf{Z}}(\mathfrak{g})$ denotes the Kostant $\mathbf{Z}$-form in the universal enveloping algebra of $\mathfrak{g}$. The G-module $V(\lambda) = V_{\mathbf{Z}}(\lambda) \otimes k$ is then called the Weyl module for G with highest weight λ.

There is a so-called contravariant form on $V_{\mathbf{Z}}(\lambda)$ which Jantzen used to construct a filtration of $V(\lambda)$ [**24**] (as an analog to the Jantzen filtration of Verma modules). He proved the results in Theorem 1 of §2 for this filtration when $p \geq h$ (and for type A_n as well as some low rank types for all p) and he conjectured it to hold in general [**25**].

It follows from Theorem 2 in §1 that $V(\lambda) \cong H^N(w_0 \cdot \lambda)$, and it is not hard to prove (see [**4**]) that the filtration constructed in §2 in fact coincides with Jantzen's filtration.

Based on the rank 2 cases, J. E. Humphreys conjectured in 1982 [**22**] that filtrations similar to the Jantzen filtrations for Weyl modules should exist for the other cohomology modules $H^i(\mu)$, $\mu \in X(T)$, $i \geq 0$. The construction in §2 does actually produce such a filtration for a large subquotient of $H^{l(w)}(w \cdot \lambda)$, $w \in W$, $\lambda \in X(T)_+$ [**4**], and an attempt to incorporate the so-called nondegeneric cohomology was made in [**6**].

Suppose now that λ and λ' are adjacent p-regular dominant weights. This means that λ is obtained from λ' by reflecting in the common wall of the two alcoves in whose interiors λ and λ' are located. In this situation we have

THEOREM 1 [**3**]. *If λ and λ' are as above, then there is a unique (up to scalar) nonzero G-homomorphism $j: V(\lambda') \to V(\lambda)$.*

For type A_n this follows already from the work of R. Carter and G. Lusztig [**10**]. Recently, R. Carter and M. T. J. Payne [**11**] have obtained the result under more general conditions on λ and λ'. M. Koppinen [**33**] has a very nice and short argument which also gives results including Theorem 1.

The following conjecture is the analog of the so-called Jantzen conjecture for Verma modules which has now been proved by J. Bernstein (unpublished).

CONJECTURE 1 [**4**]. If λ and λ' are as above and $\langle \lambda + \rho, \alpha^\vee \rangle < p(p-h+2)$ for all $\alpha \in R_+$, then j is compatible with the filtrations of $V(\lambda')$ and $V(\lambda)$ in the sense that

$$j(V(\lambda')^i) \subset V(\lambda)^{i+1} \quad \text{for all } i.$$

This conjecture has some far-reaching consequences.

THEOREM 2. *Suppose Conjecture 1 holds and let λ be a dominant weight satisfying the condition therein. Then*

(i) [**4**] *Lusztig's conjecture (Conjecture 1.1) holds.*

(ii) [**4**] $V(\lambda)_j = V(\lambda)^j / V(\lambda)^{j+1}$ *is completely reducible for all j.*

(iii) [**7**] *For $\mu \in X(T)_+$, $j \in \mathbf{N}$, $[V(\lambda)_j : L(\mu)]$ is given as a coefficient of the appropriate "inverse" Kazhdan-Lusztig polynomial.*

(iv) [**5**] *For $\mu \in X(T)_+$, $i \in \mathbf{N}$, $\dim \operatorname{Ext}^i_G(L(\lambda), H^0(\mu))$ is given as the coefficient of the appropriate Kazhdan-Lusztig polynomial. In particular,*

$$\operatorname{Ext}^i_G(L(\lambda), H^0(\mu)) = 0$$

for all even or all odd i.

(v) [**7**] *If $\langle \lambda^1, \alpha^\vee \rangle \geq 1$ for all $\alpha \in R_+$, then $V(\lambda)^j = \operatorname{Soc}_G^{N-j+1} V(\lambda)$ for all j. (Here $\operatorname{Soc}^\bullet$ dontes socle series.)*

(vi) [**7**] *For λ "generic" and p large we also have $V(\lambda)^j = \operatorname{Rad}^j_G V(\lambda)$ for all j. (Here $\operatorname{Rad}^\bullet_G$ denotes the radical series.)*

It should be pointed out that as far as Theorem 2(iv) is concerned, this may be derived directly from Lusztig's conjecture. The proof of the theorem follows in most cases the same line as the proof of the corresponding statements for Verma modules; see in particular [**18**]. However, the situation is complicated by the fact that the map j from Theorem 1 is usually not injective, whereas any nonzero $\mathfrak{g}$-homomorphism between Verma modules is injective. The analog of (vi) for Verma modules is still open.

4. The Frobenius kernel. The pth power map on k induces an endomorphism on the group $\mathrm{GL}_n(k)$, namely the map which raises the entries to their pth power. Likewise, we get an induced endomorphism F on G, called the Frobenius homomorphism. In this section we shall consider the scheme-theoretic kernel $G_1 \leq G$ of F. This is a normal subgroup scheme of G, called the Frobenius kernel, with just 1 point over k. (There are analogous higher Frobenius kernels G_n for $n > 1$ corresponding to the powers of F, but we shall limit ourselves here to consider only G_1.)

Let us from now on assume that both T and B are stable under F so that we also have Frobenius kernels T_1 and B_1. We shall also consider the subgroup schemes G_1T and G_1B of G.

Let E be a B-module and define $\hat{Z}_1(E)$ to be the induced G_1B-module,

$$\hat{Z}_1(E) = H^0(G_1B/B, E).$$

Since G_1B/B is affine (namely isomorphic to U_1^{opp}, where U^{opp} is the unipotent radical of the Borel subgroup opposite to B), we see that $\hat{Z}_1$ is an exact functor. Moreover, we get

$$(*) \qquad \operatorname{ch} \hat{Z}_1(E) = \operatorname{ch} k[U_1^{\mathrm{opp}}] \operatorname{ch} E.$$

In analogy with Theorem 1.1 we have

THEOREM 1 [**21**, **27**, **14**]. *Let $\lambda \in X(T)$. Then $\hat{Z}_1(\lambda)$ contains a unique irreducible submodule $\hat{L}_1(\lambda)$ and in this way $X(T)$ parametrizes the irreducible G_1B-modules.*

This immediately gives us the following problem:

PROBLEM 1. Find $\operatorname{ch} \hat{L}_1(\lambda)$ for $\lambda \in X(T)$.

From $(*)$ we see that $\operatorname{ch} \hat{Z}_1(\lambda)$ is known. Hence Problem 1 may be reformulated as follows:

PROBLEM 2. Let $\lambda, \mu \in X(T)$. Find the composition factor multiplicity $[\hat{Z}_1(\lambda) : \hat{L}_1(\mu)]$ of $\hat{L}_1(\mu)$ in $\hat{Z}_1(\lambda)$.

It is obvious from the construction that $\hat{Z}_1(\lambda + p\lambda') \cong \hat{Z}_1(\lambda) \otimes p\lambda'$ and $\hat{L}_1(\lambda + p\lambda') \cong \hat{L}_1(\lambda) \otimes p\lambda'$ for all $\lambda, \lambda' \in X(T)$. Now set

$$X_1(T) = \{\lambda \in X(T)_+ \mid \langle \lambda, \alpha^\vee \rangle < p \text{ for all simple roots } \alpha\}.$$

This is called the set of restricted weights. Note that each $\lambda \in X(T)$ decomposes uniquely as $\lambda = \lambda^0 + p\lambda^1$ with $\lambda^0 \in X_1(T)$, $\lambda^1 \in X(T)$. In the following, upper index 0 and 1 on a weight will always refer to this decomposition.

Via the following "classical" result, we see that Problems 1 and 2 above are equivalent to Problems 1.1 and 1.2 (see also Theorem 1.4).

THEOREM 2 [**15**]. *Let $\lambda \in X_1(T)$. Then $\hat{L}_1(\lambda) \cong L(\lambda)|_{G_1B}$.*

See also [**32**] for a new proof of this result.

Via transitivity of induction we have $H^0(E) \cong H^0(G/G_1B, \hat{Z}_1(E))$ for any B-module E, and since $\hat{Z}_1$ is exact, there is a similar identity for the higher cohomology. It is not hard to see that we also have isomorphisms $H^i(G/G_1B, E^{(1)}) \cong H^i(E)^{(1)}$, $i \geq 0$, where upper index (1) denotes Frobenius twist. (If $H \to \mathrm{GL}(V)$

is a representation of an algebraic group H, then $V^{(1)}$ is the representation of H given by the composite $H \xrightarrow{F} H \to \mathrm{GL}(V)$.) Combining this last property with the tensor identity and Theorem 1, we find

$$H^i(G/G_1B, \hat{L}_1(\lambda)) \cong L(\lambda^0) \otimes H^i(\lambda^1)^{(1)}, \qquad i \geq 0.$$

In particular, it follows from this that if $\lambda^1 \in \overline{C}_0$ so that $H^0(\lambda^1) = L(\lambda^1)$ (by the strong linkage principle [**1**]), then $H^0(G/G_1B, \hat{L}_1(\lambda)) \cong L(\lambda)$ ($= 0$ if $\lambda \notin X(T)_+$).

We can now combine the above with Theorem 1.2 and the additivity of the Euler character to do the following computations for $\lambda \in X(T)_+$:

$$\begin{aligned}\operatorname{ch} H^0(\lambda) &= \sum_i (-1)^i \operatorname{ch} H^i(\lambda) = \sum_i (-1)^i H^i(G/G_1B, \hat{Z}_1(\lambda)) \\ &= \sum_i (-1)^i \sum_\mu [\hat{Z}_1(\lambda) : L_1(\mu)] \operatorname{ch} L(\mu^0) \operatorname{ch} H^i(\mu^1)^{(1)} \\ &= \sum_\mu [\hat{Z}_1(\lambda) : L(\mu)] \operatorname{ch} L(\mu^0) \chi(\mu^1)^{(1)},\end{aligned}$$

i.e., we have

THEOREM 3 [**27**]. *Let $\lambda, \mu \in X(T)_+$. Then*
(i) $[H^0(\lambda) : L(\mu)] = \sum_\nu [\hat{Z}_1(\lambda) : \hat{L}_1(\nu)][L(\lambda^0) \otimes \chi(\nu^1)^{(1)} : L(\mu)]$.
(ii) *If* $[\hat{Z}_1(\lambda) : \hat{L}_1(\nu)] = 0$ *for* $\nu^1 \notin \overline{C}_0$, *then*

$$[H^0(\lambda) : L(\mu)] = [\hat{Z}_1(\lambda) : \hat{L}_1(\mu)].$$

An immediate consequence of Theorem 3(ii) is the following result (which J. C. Jantzen originally obtained by character calculations [**26**]).

COROLLARY 1. *Suppose both λ and $\lambda + p\nu$ satisfy the assumption in Theorem* 3(ii). *Then*

$$[H^0(\lambda + p\nu) : L(\mu + p\nu)] = [H^0(\lambda) : L(\mu)] \quad \textit{for all } \mu \in X(T)_+.$$

Another property of the $\hat{Z}_1(\lambda)$'s is the following symmetry, which follows from the expression (*) above with $E = \lambda + p\nu$ [**27**],

Let $\mu \in X_1(T)$, $\lambda, \nu \in X(T)$, and $w \in W$. Then

$$(**) \qquad [\hat{Z}_1(\lambda + p\nu) : \hat{L}_1(\mu + p(\nu - \rho))] = [\hat{Z}_1(w \cdot \lambda + p\nu) : \hat{L}_1(\mu + p(\nu - \rho))].$$

In the "generric" case where the assumption in Theorem 3(ii) is satisfied, this gives a corresponding symmetry among the composition factors of $H^0(\lambda)$'s.

Let us finish this section by mentioning that there is a very close resemblance between the G-structure of $H^0(\lambda)$ and the G_1-structure of $\hat{Z}_1(\lambda)$:

THEOREM 4 [**7**]. *Let λ be as in Theorem* 3(ii). *Then*

$$\operatorname{Soc}^j_G H(\lambda) = H^0(G/G_1B, \operatorname{Soc}^j_{G_1} \hat{Z}_1(\lambda)) \quad \textit{for all } j.$$

It should be pointed out that usually the G_1-socle series for $\hat{Z}_1(\lambda)$ differs from the G_1B-socle series.

This theorem combined with Theorem 3.2(v) explains the fact observed in the rank 2 cases that the Jantzen filtration is "invariant" under translation by $p\nu$; compare Corollary 1.

5. Ye's result. For $\lambda \in X(T)$ we set $\hat{Q}_1(\lambda)$ equal to the injective G_1T-module whose socle is $\hat{L}_1(\lambda)$. Clearly, $\hat{Q}_1(\lambda) \cong \hat{Q}_1(\lambda^0) \otimes p\lambda^1$. The following "classical" result plays an important role in this section.

THEOREM 1 [**8**]. *Suppose $p \geq 3(h-1)$. If $\lambda \in X_1(T)$, then there exists a G-module $Q_1(\lambda)$ such that $\hat{Q}_1(\lambda) \cong Q_1(\lambda)|_{G_1T}$.*

J. C. Jantzen [**27**] has proved (thereby lowering the bound on p to $2(h-1)$) that for $p \geq 2(h-1)$ one may characterize $Q_1(\lambda)$ as the injective hull of $L(\lambda)$ in the category of so-called p-bounded G-modules.

It is expected that Theorem 1 holds without any restrictions on p, but since all efforts to prove this have failed so far, we assume in what follows that $p \geq 2(h-1)$.

From Jantzen's characterization of $\hat{Q}_1(\lambda)$, $\lambda \in X(T)_+$, it follows that $\hat{Q}_1(\lambda)$ is a direct G_1B-summand of $\mathrm{St}_1 \otimes H^0((p-1)\rho + w_0(\lambda^0)) \otimes p\lambda^1$. Here St_1 denotes the Steinberg module, i.e., $\mathrm{St}_1 = L((p-1)\rho) = H^0((p-1)\rho) = Q_1((p-1)\rho) = \hat{Z}_1((p-1)\rho)$. The exactness of $\hat{Z}_1$ combined with the tensor identity gives $\mathrm{St}_1 \otimes H^0((p-1)\rho + w_0(\lambda^0)) \otimes p\lambda^1 \cong \hat{Z}_1(H^0((p-1)\rho \otimes w_0(\lambda^0)) \otimes (p\lambda^1 + (p-1)\rho))$ has a G_1B-filtration in which the successive quotients have the form $\hat{Z}_1(\mu)$ for some $\mu \in X(T)$. Hence $\hat{Q}_1(\lambda)$ has such a filtration. Moreover, if $[\hat{Q}_1(\lambda) : \hat{Z}_1(\mu)]$ denotes the number of times that $\hat{Z}_1(\mu)$ occurs in this filtration, then we have the reciprocity law

$$[\hat{Q}_1(\lambda) : \hat{Z}_1(\mu)] = [\hat{Z}_1(\mu) : \hat{L}_1(\lambda)].$$

Using this, we obtain

$$\begin{aligned}\operatorname{ch} Q_1(\lambda^0)\chi(\lambda^1)^{(1)} &= \sum_i (-1)^i \operatorname{ch} H^i(G/G_1B, \hat{Q}_1(\lambda)) \\ &= \sum_i (-1)^i \sum_\mu [\hat{Q}_1(\lambda) : \hat{Z}_1(\mu)] \operatorname{ch} H^i(G/G_1B, \hat{Z}_1(\mu)) \\ &= \sum_\mu [\hat{Z}_1(\mu) : \hat{L}_1(\lambda)]\chi(\mu) \\ &= \sum_{\mu \in X(T)_+} \sum_{w \in W} (-1)^{l(w)} [\hat{Z}_1(w \cdot \mu) : \hat{L}_1(\lambda)]\chi(\mu).\end{aligned}$$

Another consequence of the fact that $Q_1(\lambda^0)$ is a G-summand of $\mathrm{St}_1 \otimes H^0((p-1)\rho + w_0(\lambda^0))$ is that for $\lambda \in X(T)_+$, $Q_1(\lambda^0) \otimes H^0(\lambda^1)^{(1)}$ has a filtration whose quotients have the form $H^0(\mu)$, $\mu \in X(T)_+$. (Note that

$$\mathrm{St}_1 \otimes H^0(\lambda^1)^{(1)} \cong H^0((p-1)\rho + p\lambda^1)$$

and use [**38**] or [**16**].) Hence it follows that when we express $\operatorname{ch} Q_1(\lambda^0)\chi(\lambda^1)^{(1)}$ as a linear combination of $\chi(\mu)$, $\mu \in X(T)_+$, then all coefficients are nonnegative, i.e.,

$$\sum_{w \in W} (-1)^{l(w)} [\hat{Z}_1(w \cdot \mu) : \hat{L}_1(\lambda)] \geq 0.$$

Now fix a p-regular weight $\lambda \in X(T)_+$ and let $\nu \in p(\lambda^1 + \rho) + X(T)_+$ with $[\hat{Z}_1(\nu) : \hat{L}_1(\lambda)] \neq 0$. Suppose ν' is maximal with respect to being strongly linked to ν. By using the symmetry $(**)$ above and some clever weight arguments, Ye shows that there exists $\mu \in X(T)_+$ for which the above equality reads

$$[\hat{Z}_1(\nu') : \hat{L}_1(\lambda)] - [\hat{Z}_1(\nu) : \hat{L}_1(\lambda)] \geq 0.$$

An easy induction together with the observation that

$$[\hat{Z}_1(w_0 \cdot \lambda^0 + p\rho + p(\lambda^1 + \rho)) : \hat{L}_1(\lambda)] = [\hat{Z}_1(\lambda) : \hat{L}_1(\lambda)] = 1$$

now proves

THEOREM 2 [**40**]. *Let* $\lambda \in X(T)$. *Then*

$$[\hat{Z}_1(\mu) : \hat{L}_1(\lambda)] \neq 0 \quad \textit{iff } \mu \in p(\lambda^1 + \rho) + W \cdot S_{\lambda^0},$$

where $S_{\lambda^0} = \{\nu \in X(T)_+ \mid \nu \textit{ is strongly linked to } p\rho + w_0 \cdot \lambda^0\}$.

This result may be viewed as a converse to the linkage principle. In fact, S. Doty and J. Sullivan [**17**] had formulated what they called an inverse linkage conjecture, and it is easy to derive this conjecture from Ye's result.

If we compare Theorem 2 with Conjecture 1.1, we see that it says (intuitively) that when we set $q = 1$ in the patterns formed by the Kazhdan-Lusztig Q-polynomials (see [**35**]), there are no holes (i.e., all nonzero Q-polynomials have positive value at 1). This is immediately observable in the rank 2 patterns pictured in [**35**] and it has recently been verified to hold in general [**29**]. Hence Lusztig's conjecture is consistent with Ye's theorem.

In conclusion, we can thus say that we now know the composition factors of all Weyl modules, but we still do not know the multiplicities with which they occur.

REFERENCES

1. H. H. Andersen, *The strong linkage principle*, J. Reine Angew. Math. **315** (1980), 53–59.

2. ——, *The Frobenius morphism on the cohomology of homogeneous vector bundles on G/B*, Ann. of Math. (2) **112** (1980), 113–121.

3. ——, *On the structure of the cohomology of line bundles on G/B*, J. Algebra **71** (1981), 245–258.

4. ——, *Filtrations of cohomology of line bundles on homogeneous spaces for Chevalley groups*, Ann. Sci. École Norm. Sup. **16** (1983), 495–528.

5. ——, *An inversion formula for theKazhdan-Lusztig polynomials for affine Weyl groups*, Adv. in Math. **60** (1986), 125–153.

6. ——, *Torsion in the cohomology of line bundles on homogeneous spaces for Chevalley groups*, Proc. Amer. Math. Soc. **96** (1986), 537–544.

7. ——, *Jantzen's filtrations of Weyl modules*, Math. Z. **194** (1987), 127–142.

8. J. W. Ballard, *Injective modules for restricted enveloping algebras*, Math. Z. **163** (1978), 57–63.

9. B. Braden, *Restricted representations of classical Lie algebras of type A_2 and B_2*, Bull. Amer. Math. Soc. **73** (1967), 482–486.

10. R. Carter and G. Lusztig, *On the modular representations of the general linear and symmetric groups*, Math. Z. **136** (1974), 193–242.

11. R. Carter and M. T. J. Payne, *On the homomorphisms between Weyl modules and Specht modules*, Math. Proc. Cambridge Philos. Soc. **87** (1980), 419–425.

12. C. Chevalley, *Les systèmes linéaires sur G/B*, Sem. Chevalley (Classification des groupes de Lie algébriques), Exp. 15, 1957.

13. E. Cline, B. Parshall, and L. Scott, *On the tensor product theorem for algebraic groups*. J. Algebra **63** (1980), 264–267.

14. ——, *On injective modules for infinitesimal algebraic groups*. I, J. London Math. Soc. **31** (1985), 277–291.

15. C. W. Curtis, *Representations of Lie algebras of classical type with applications to linear groups*, J. Math. Mech. **9** (1960), 307–326.

16. S. Donkin, *Rational representations of algebraic groups*, Lecture Notes in Math., vol. 1140, Springer-Verlag, 1985.

17. S. R. Doty and J. B. Sullivan, *On the structure of the higher cohomology modules of line bundles on G/B*, (to appear).

18. O. Gabber and T. Joseph, *Towards the Kazhdan-Lusztig conjecture*, Ann. Sci. École Norm. Sup. **14** (1981), 261–302.

19. W. J. Haboush, *A short proof of the Kempf vanishing theorem*, Invent. Math. **56** (1980), 109–112.

20. H. Hagelskjær, *Induced modules for algebraic groups*, Speciale, Aarhus University, 1983.

21. J. E. Humphreys, *Modular representations of classical Lie algebras and semi-simple groups*, J. Algebra **19** (1971), 57–79.

22. ——, *Cohomology of G/B in characteristic p*, Adv. in Math. **59** (1986), 170–183.

23. J. C. Jantzen, *Darstellungen halbeinfacher algebraischer Gruppen und zugeordnete kontravariante Formen*, Bonn. Math. Schr. No. 67 (1973).

24. ——, *Darstellungen halbeinfacher Gruppen und kontravariante Formen*, J. Reine Angew. Math. **290** (1977), 117–141.

25. ——, *Weyl modules for groups of Lie type*, Finite Simple Groups. II, Academic Press, New York, 1980, pp. 291–300.

26. ——, *Über das Dekompositionsverhalten gewisser modularer Darstellungen halbeinfacher Gruppen und ihrer Lie-Algebren*, J. Algebra **49** (1977), 441–669.

27. ——, *Darstellungen halbeinfacher Gruppen und ihrer Frobenius-Kerne*, J. Reine Angew. Math. **317** (1980), 157–191.

28. ——, *Filtrierungen der Darstellungen in der Hauptserie endlicher Chevalley-Gruppen*, Proc. London Math. Soc. **49** (1984), 445–482.

29. M. Kaneda, *On the symmetry of inverse Kazhdan-Lusztig polynomials for affine Weyl groups*, (to appear).

30. S.-I. Kato, *On the Kazhdan-Lusztig polynomials for affine Weyl groups*, Adv. in Math. **55** (1985), 103–130.

31. G. Kempf, *Linear systems on homogeneous spaces*, Ann. of Math. (2) **103** (1976), 557–591.

32. ——, *Representations of algebraic groups in prime characteristics*, Ann. Sci. École Norm. Sup. **14** (1981), 61–76.

33. M. Koppinen, *Homomorphisms between neighbouring Weyl modules*, J. Algebra **103** (1986), 302–319.

34. G. Lusztig, *Some problems in the representation theory of finite Chevalley groups*, Proc. Sympos. Pure Math., vol. 37, Amer. Math. Soc., Providence, R. I., 1980, pp. 313–318.

35. ——, *Hecke algebra and Jantzen's generic decomposition patterns*, Adv. in Math. **37** (1980), 121–164.

36. T. A. Springer, *Weyl's character formula for algebraic groups*, Invent. Math. **5** (1968), 86–105.

37. R. Steinberg, *Representations of algebraic groups*, Nagoya Math. J. **22** (1963), 33–56.

38. Wang J.-p., *Sheaf cohomology on G/B and tensor products of Weyl modules*, J. Algebra **77** (1982), 162–185.

39. W. J. Wong, *Very strong linkage for cohomology groups of line bundles on G/B*, (to appear).

40. J. -C. Ye, *Filtrations of principal indecomposable modules of Frobenius kernels of reductive groups*, Math. Z. **189** (1985), 515–527.

Additional References (*on development in related areas*)

[a] H. H. Andersen, *Extensions of modules for algebraic groups*, Amer. J. Math. **106** (1984), 489–504.

[b] ——, *Schubert varieties and Demazure's character formula*, Invent. Math. **79** (1985), 611–618.

[c] ——, *On the generic structure of cohomology modules for semi-simple algebraic groups*, Trans. Amer. Math. Soc. **295** (1986), 397–415.

[d] ——, *Extensions of simple modules for finite Chevalley groups*, J. Algebra (to appear).

[e] H. H. Andersen and J. C. Jantzen, *Cohomology of induced representations for algebraic groups*, Math. Ann. **269** (1984), 487–525.

[f] S. Donkin, *Finite resolutions of modules for algebraic groups*, J. Algebra **101** (1986), 473–488.

[g] S. Donkin and M. J. F. Mowbray, *A note on Koppinen's decomposition numbers*, (to appear).

[h] S. R. Doty, *The submodule structure of certain Weyl modules for groups of type A_n*, J. Algebra **95** (1985), 373–383.

[i] S. R. Doty and J. B. Sullivan, *On the geometry of extensions of irreducible modules for simple algebraic groups*, (to appear).

[j] E. Friedlander and B. Parshall, *On the cohomology of algebraic and related finite groups*, Invent. Math. **74** (1983), 85–117.

[k] ——, *Cohomology of Lie algebras and algebraic groups*, Amer. J. Math. **108** (1986), 235–253.

[l] ——, *Cohomology of infinitesimal and discrete groups*, Math. Ann. **273** (1986), 353–374.

[m] ——, *Geometry of p-unipotent Lie algebras*, (to appear).

[n] ——, *Support varieties for restricted Lie algebras*, (to appear).

[o] J. E. Humphreys, *On the structure of Weyl modules*, Comm. Algebra **12** (1984), 2665–2677.

[p] ——, *Cohomology of line bundles on G/B for the exceptional group G_2*, J. Pure Appl. Algebra (to appear).

[q] ——, *Non-zero* Ext^1 *for Chevalley groups* (*via algebraic groups*), J. London Math. Soc. **31** (1985), 463–467.

[r] ——, *Projective modules for* $\mathrm{Sp}(4,p)$ *in characteristic* p, J. Algebra (to appear).

[s] ——, *The Steinberg module*, (to appear).

[t] R. S. Irving, *The structure of certain highest weight modules for* SL_3, J. Algebra **99** (1986), 438–457.

[u] ——, *Projective modules in the category* $\mathcal{O}_S$: *Loewy series*, Trans. Amer. Math. Soc. **291** (1985), 733–754.

[v] J. C. Jantzen, *Representations of algebraic groups*, (to appear).

[w] M. Kaneda, *On the alcove identification operator of semi-simple algebraic groups*, (to appear).

[x] M. Koppinen, *Good bimodule filtrations for coordinate rings*, J. London Math. Soc. **30** (1984), 244–250.

[y] ——, *On stable decomposition patterns of Weyl modules*, (to appear).

[z] K. Kühne-Hausmann, *Zur Untermodulstruktur der Weyl moduln für* SL_3, Bonn. Math. Schr. No. 162, Bonn (1985).

[æ] W. J. Wong, *A filtration of Weyl modules for large weights*, (to appear).

[ø] S. El B. Yehia, *Extensions of simple modules for the universal Chevalley group and its parabolic subgroups*, Ph.D. Thesis, Warwick, 1982.

AARHUS UNIVERSITET, DENMARK

Proceedings of Symposia in Pure Mathematics
Volume 47 (1987)

Varieties for Modules

JON F. CARLSON

Suppose that G is a finite group and K is a field of characteristic $p > 0$. The question is: What do KG-modules look like? The purpose of this lecture is to give a survey of some of the results in the ongoing effort to provide answers to this question and to develop a more comprehensive module theory using the special properties of group algebras. It is something of a continuation of Alperin's lecture on application of homological algebra to group representation theory [**1**]. Indeed, one of the major tools in module theory is cohomology. The reason for this can be illustrated by the following.

First we need some notation. For H any subgroup of G and M a KG-module we can write the restriction of M to H as $M_H = \rho_H(M) \oplus f_H(M)$, where $f_H(M)$ is a projective KH-module and $\rho_H(M)$ has no projective submodules.

THEOREM 1. [**9**]. *There exists a number $B = B(G,p)$ depending only on G and p such that if M is any KG-module then*

$$\text{Dim } \rho_G(M) \leq B \cdot \max_{E \in \mathbf{EA}(G)} \{\text{Dim } \rho_E(M)\},$$

where the maximum is taken over the set $\mathbf{EA}(G)$ of elementary abelian p-subgroups of G.

The theorem implies the theorem of Chouinard as well as the Alperin-Evens theorem on complexity. For example, Chouinard's theorem says that a KG-module is projective if its restriction to every elementary abelian p-subgroup is projective. If a module M satisfies these hypotheses then the right side of the inequality in Theorem 1 is zero and M is projective.

While the statement of the theorem involves nothing more complicated than the definition of a projective module, its proof is pure homological algebra. The reason is that the first step of the proof is to reduce to the case in which G is a p-group. After all, if P is a Sylow p-subgroup of G then M is a direct summand of $(M_P)^{\uparrow G}$. But for p-groups the classical Brauer theory of blocks and characters is ineffective and other methods of analysis must be used. The original proof

1980 *Mathematics Subject Classification* (1985 *Revision*). Primary 20C20; Secondary 20J05.
This work was partly supported by the National Science Foundation.

employed the Lyndon-Hochschild-Serre spectral sequence and Serre's theorem on the vanishing of certain products of Bocksteins. Picaronny [**25**] has shown that the spectral sequence is not necessary. Also new, more representation-theoretic proofs of Serre's theorem have been circulated at this conference by Kroll [**23**] and Okuyama and Sasaki [**24**]. Nevertheless, the cohomology theory remains the essential ingredient in the proof.

In general it is impossible to classify KG-modules. However, homological methods have been used to categorize modules by associating to each an affine variety. The definitions are as follows. Let $V_G(K)$ denote the maximal ideal spectrum of the cohomology ring $H^*(G,K)$. It is a homogeneous affine variety because $H^*(G,K)/\operatorname{Rad} H^*(G,K)$ is a finitely generated commutative graded ring. For a KG-module M let $J(M)$ be the annihilator in $H^*(G,K)$ of the cohomology ring $\operatorname{Ext}^*_{KG}(M,M)$, and let $V_G(M)$ be the subvariety of $V_G(K)$ corresponding to the ideal $J(M)$. That is, $V_G(M)$ consists of all maximal ideals of $H^*(G,K)$ that contain $J(M)$.

In spite of the level of abstraction in the definition, the variety of a module is mostly computable directly from its structure, without resorting to any homological construction. The first step in such a computation is the following theorem first proved by Jon Alperin and Leonard Evens [**3**], and independently by George Avrunin [**4**]. The case $M = K$ had been proved by Quillen [**26**], with a more algebraic proof provided by Quillen and Venkov [**28**]. It was, in some sense, the logical generalization of the Alperin-Evens result on complexity [**2**] (see (9) below).

THEOREM 2.
$$V(M) = \bigcup_{E\in \mathbf{EA}(G)} \operatorname{res}^*_{G,E}(V_E(M_E)).$$

Here $\mathbf{EA}(G)$ is the set of elementary abelian p-subgroups of G and

$$\operatorname{res}^*_{G,E}\colon V_E(K) \to V_G(K)$$

is the map on varieties induced by the restriction homomorphism

$$\operatorname{res}_{G,E}\colon H^*(G,K) \to H^*(E,K).$$

The maps $\operatorname{res}^*_{G,E}$ are finite-to-one and even though the "folding" of $V_E(K)$ is completely described in [**27**], it is by no means trivial to piece together the images of these maps in $V_G(K)$. Nonetheless, for many practical purposes, the problem of computing $V_G(M)$ is reduced to that of computing $V_E(M_E)$ for E an elementary abelian p-group. In this case the variety can be related directly to the module structure by the rank variety.

Assume now that K is algebraically closed. Suppose that $E = \langle x_1,\dots,x_n\rangle$ is an elementary abelian group of order p^n. For any $\alpha = (\alpha_1,\dots,\alpha_n) \in K^n$, let $u_\alpha = 1+\sum \alpha_i(x_i-1)$. Note that u_α is a unit of order p in KE. If $\alpha(1),\dots,\alpha(r)$ are linearly independent elements in K^n, $u_i = u_{\alpha(i)}$, and $H = \langle u_1,\dots,u_r\rangle$, then $|H| = p^r$ and KG is free as a KH-module. Such an H is called a shifted subgroup of KG. Moreover if $r = n$, then the embedding $H \to KG^{\times}$ induces an isomorphism of KH onto KG. Indeed from a ring or module-theoretic standpoint either G or H could serve as the basis from which the group algebra is constructed [**10**]. For a KE-module M the rank variety of M is defined to be

$$V_E^r(M) = \{0\} \cup \{\alpha \in K^n | M_{\langle u_\alpha \rangle} \text{ is not a free } K\langle u_\alpha\rangle\text{-module}\}.$$

The rank variety is linked to the cohomological variety by the following.

THEOREM 3. [**5, 10**]. $V_E(M) = V_E^r(M)$.

My contribution was to prove the easy assertion that $V_E^r(M) \subseteq V_E(M)$. The aim was to find a direct method for computing the complexity of M. This inclusion was sufficient for that purpose since it can easily be shown that $\dim V_E^r(M) = c_G(M)$ (see [**9**]). Avrunin and Scott needed the more difficult reverse inclusion to prove a module version of Quillen's stratification theorem [**27**].

To understand Theorem 2, we should first note that $H^*(E,K)/\operatorname{Rad} H^*(E,K)$ is a polynomial ring in n variables. The generators occur in degree 1 if $p = 2$ and in degree 2 if $p > 2$. Hence $V_E(K) = K^n$. It can be shown that for every maximal ideal $\mathfrak{m}$ in $H^*(E,K)$ there exists $\alpha \in K^n$, $\alpha \neq 0$, such that $\mathfrak{m}$ contains the kernel of the restriction homomorphism to $K\langle u_\alpha\rangle$. Moreover if $\mathfrak{m}$ is not the zero element of $V_G(K)$, i.e., not the ideal of all elements of positive degree, then $\alpha \neq 0$ and α is unique up to scalar multiple. From another viewpoint we may regard $\mathfrak{m}$ as a homomorphism $\mathfrak{m}\colon H^*(E,K) \to K$ and then $\mathfrak{m}$ factors through $\operatorname{res}_{G,\langle u_\alpha\rangle}$. This is the correspondence that identifies the varieties in the theorem. See [**10**] for details. Consequently the main task in the proof of the theorem is to show that $J(M) \subseteq \mathfrak{m}$ if and only if M is not free as a $K\langle u_\alpha\rangle$-module whenever $\alpha \neq 0$ corresponds to $\mathfrak{m}$ as above.

The importance of the rank variety lies in the fact that it can often be computed without resorting to any homological algebra. The following is my favorite example simply because it was the example that I wrote down when it first occurred to me that $V_E^r(M)$ might be a variety. Suppose that $E = \langle x_1, x_2, x_3\rangle$ is elementary abelian of order 8 (hence $p = 2$). Choose arbitrary elements A, $B, C \in K$. Let $M = M(A,B,C)$ be the underlying KE-module of the representation $\rho\colon E \to \mathrm{GL}(4,K)$ given by

$$\rho(x_1) = \left[\begin{array}{cc|cc} 1 & & & \\ & 1 & & \\ \hline 1 & & 1 & \\ & 1 & & 1 \end{array}\right], \quad \rho(x_2) = \left[\begin{array}{cc|cc} 1 & & & \\ & 1 & & \\ \hline A & & 1 & \\ & B & & 1 \end{array}\right], \quad \rho(x_3) = \left[\begin{array}{cc|cc} 1 & & & \\ & 1 & & \\ \hline & C & 1 & \\ 1 & & & 1 \end{array}\right].$$

It can be checked that M is an indecomposable KE-module, regardless of the choice of A, B, and C. Also $M(A,B,C) \cong M(A',B',C')$ if and only if $(A,B,C) = (A',B',C')$. This last statement will be obvious once we have computed the varieties. For $\alpha = (\alpha_1,\alpha_2,\alpha_3) \in K^3$, $u_\alpha = 1 + \sum \alpha_1(x_1 - 1)$, we have

$$\rho(u_\alpha) = \left[\begin{array}{cc|cc} 1 & & & \\ & 1 & & \\ \hline \alpha_1 + A\alpha_2 & C\alpha_3 & 1 & \\ \alpha_3 & \alpha_1 + B\alpha_2 & & 1 \end{array}\right].$$

The key is that u_α acts freely on M if and only if $\rho(u_\alpha)$ is similar to the matrix

$$\left[\begin{array}{cc|cc} 1 & & & \\ 1 & 1 & & \\ \hline & & 1 & \\ & & 1 & 1 \end{array}\right]$$

Hence, because $u_\alpha^2 = 1$, u_α acts freely on M if and only if the rank of the matrix $\rho(u_\alpha - 1) = \rho(u_\alpha) - I$ is exactly 2. Therefore $\alpha \in V_E^r(M)$ if and only if the determinant of the 2×2 submatrix in the lower left corner of $\rho(u_\alpha)$ is zero. That is,

$$V_E^r(M) = \{\alpha \in K^3 \mid (\alpha_1 + A\alpha_2)(\alpha_1 + B\alpha_2) + C\alpha_3^2 = 0\},$$

which is a homogeneous affine variety of dimension 2.

The example illustrates two facts. The assertion that an element u_α acts freely or not freely on a module is simply equivalent to a rank condition. Hence the name of the rank variety. Second, the rank variety can be calculated directly from a matrix presentation for the module. Observe that if $C = 0$ and $A \neq B$ in the example then the variety has two components and is not irreducible. If $A \neq 0 \neq C$ and $B = 0$, then $V_E^r(M)$ is irreducible but not linear. It can be proved that any homogeneous, closed subvariety of $V_G(K)$ can occur as $V_G(M)$ for some KG-module M.

We return now to the case of an arbitrary finite group G. In what follows we list some of the properties of varieties of modules.

(4) $\quad V_G(M) = \{0\}$ if and only if M is projective.

This is an easy consequence of the facts that $\operatorname{Ext}^*_{KG}(M, M)$ is a finitely generated module over $H^*(G, K)$ [**17**] and that projective KG-modules are also injective. For rank varieties (4) was first proved by Everett Dade using remarkably elementary methods [**15**].

(5) If $0 \to M_1 \to M_2 \to M_3 \to 0$ is an exact sequence of KG-modules then

$$V_G(M_i) \subseteq V_G(M_j) \cup V_G(M_k) \quad \text{for } \{i, j, k\} = \{1, 2, 3\}.$$

(6) $$V_G(M \oplus N) = V_G(M) \cup V_G(N).$$

These two properties are obvious for the rank variety. Hence, in general, they are consequences of Theorems 2 and 3.

(7) $$V_G(M \otimes N) = V_G(M) \cap V_G(N).$$

This has an elementary proof using rank varieties [**10**], but it also follows in general from the work on the stratification [**5**]. It is interesting to notice that

(4) and (7) together imply that a tensor product $M \otimes N$ is projective if and only if $V_G(M) \cap V_G(N) = \{0\}$.

(8) $$V_G(\Omega^n(M)) = V_G(M) = V_G(M^*) \quad \text{for all } n.$$

Here $\Omega^n(M)$ is defined to be the kernel of ∂_{n-1} where

$$\cdots \longrightarrow P_2 \xrightarrow{\partial_2} P_1 \xrightarrow{\partial_1} P_0 \xrightarrow{\epsilon} M \longrightarrow 0$$

is a minimal projective resolution of M. The dual $M^* = \mathrm{Hom}_K(M, K)$ is a KG-module with the G-action given by $(gf)(m) = f(g^{-1}m)$ for all $g \in G$, $m \in M$, and $f \in M^*$. Again the first equality follows from the rank variety. The second comes from the fact that $\mathrm{Ext}^*_{KG}(M^*, M^*)$ is isomorphic to $\mathrm{Ext}^*_{KG}(M, M)$ as $H^*(G, K)$-modules.

(9) $$c_G(M) = \dim V_G(M) \qquad [\mathbf{11}].$$

Here $c_G(M)$ is the complexity of M and is defined to be the least integer $s \geq 0$ such that $\lim_{n \to \infty}((\mathrm{Dim}\ P_n)/n^s) = 0$ where (P, ϵ) is a minimal projective resolution of M. It has many other equivalent definitions such as the Krull dimension of the center of $\mathrm{Ext}^*_{KG}(M, M)$. The result stems from the fact that $\mathrm{Ext}^*_{KG}(M, M)$ is a finite-dimensional module over $H^*(G, K)$ [**17**] and the dimension of $V_G(M)$ is the Krull dimension of $H^*(G, K)/J(M)$. The rank variety was originally devised to compute complexity, and it provides an easy proof of Kroll's theorem [**22**] that $c_G(M) = r$ if and only if $n - r$ is the largest integer such that there exists a shifted subgroup having order p^{n-r} that acts freely on M.

(10) M is periodic if and only if $c_G(M) = 1$ [**16**]. Moreover if M is periodic and indecomposable, then $V_G(M)$ is a line [**8**].

A periodic module is one such that $\Omega^n(M) \cong \Omega^m(M)$ for some $n \neq m$. The last statement provided some motivation for the definition of the rank variety and led to the conjecturing of the following.

(11) Let $\bar{V}_G(M)$ be the projective variety corresponding to $V_G(M)$. If M is indecomposable then $\bar{V}_G(M)$ is connected, topologically. Equivalently, suppose that $V_G(M) = V_1 \cup V_2$ where V_1 and V_2 are closed homogeneous nonzero subvarieties of $V_G(K)$ and $V_1 \cap V_2 = \{0\}$; then $M \cong M_1 \oplus M_2$ where $V_i = V_G(M_i)$, $i = 1, 2$ [**13**].

For a recent application of these results let us consider the problem of constructing a projective resolution for a KG-module M. First we need some notation. Let ς be a nonzero element in $H^n(G, K)$ and let

$$\cdots \longrightarrow P_2 \xrightarrow{\partial_2} P_1 \xrightarrow{\partial_1} P_0 \xrightarrow{\epsilon} K \longrightarrow 0$$

be a minimal projective resolution. Then ς is represented by a unique cocycle $\hat{\varsigma}\colon P_n \to K$ such that $\hat{\varsigma}(\partial(P_{n+1})) = 0$. Hence ς is represented by a unique homomorphism $\varsigma'\colon \Omega^n(K) \to K$. Let L_ς be the kernel of ς'. The following is proved using the rank variety [**13**].

LEMMA 12. *$V_G(L_\varsigma) = V_G((\varsigma))$ where (ς) is the ideal in $H^*(G,K)$ generated by ς.*

Now consider the diagram

$$\begin{array}{ccccccccccccccc}
 & & 0 & & 0 & & & & & & & & & & \\
 & & \downarrow & & \downarrow & & & & & & & & & & \\
 & & L_\varsigma & = & L_\varsigma & & & & & & & & & & \\
 & & \downarrow & & \downarrow & & & & & & & & & & \\
0 & \longrightarrow & \Omega^n(K) & \longrightarrow & P_{n-1} & \longrightarrow & P_{n-2} & \longrightarrow & \cdots & \longrightarrow & P_0 & \longrightarrow & K & \longrightarrow & 0 \\
 & & {\scriptstyle \varsigma'}\downarrow & & \downarrow & & \| & & & & \| & & \| & & \\
E_\varsigma : 0 & \longrightarrow & K & \xrightarrow{\rho} & \hat{L}_\varsigma & \longrightarrow & P_{n-2} & \longrightarrow & \cdots & \longrightarrow & P_0 & \xrightarrow{\epsilon} & K & \longrightarrow & 0 \\
 & & \downarrow & & \downarrow & & & & & & & & & & \\
 & & 0 & & 0 & & & & & & & & & &
\end{array}$$

where the bottom sequence E_ς is the pushout along ς' of the sequence above it. Let $X^{(\varsigma)} = \cdots \circ E_\varsigma \circ E_\varsigma \circ E_\varsigma$ be the splice of E_ς with itself an infinite number of times. It is the infinite periodic complex

$$X^{(\varsigma)} : \; \cdots \longrightarrow X_1^{(\varsigma)} \longrightarrow X_0^{(\varsigma)} \xrightarrow{\epsilon} K \longrightarrow 0$$

in which $X_{kn+j}^{(\varsigma)} = P_j$ if $0 \le j \le n-2$ and $X_{kn-1}^{(\varsigma)} \cong \hat{L}_\varsigma$. The boundary map $X_{kn}^{(\varsigma)} \to X_{kn-1}^{(\varsigma)}$ is $\rho\epsilon$.

Suppose that M is a KG-module with $c_G(M) = r$. From property (9) and some commutative ring theory it can be seen that there exist homogeneous elements $\varsigma_1, \ldots, \varsigma_r \in H^*(G,K)$ such that

$$V_G(M) \cap V_G((\varsigma_1)) \cap \cdots \cap V_G((\varsigma_r)) = \{0\}.$$

For each $i = 1, \ldots, r$ let $X^{(i)} = X^{(\varsigma_i)}$ as above. Now form the tensor product complex $Y = M \otimes X^{(1)} \otimes \cdots \otimes X^{(r)}$. Here

$$Y_n = \sum_{i_1 + \cdots + i_r = n} M \otimes X_{i_1}^{(1)} \otimes \cdots \otimes X_{i_r}^{(r)}.$$

THEOREM 13 [7]. *The (total) complex Y is a projective resolution of M.*

In other words, every KG-module M has a projective resolution that is the tensor product of $r = c_G(M)$ periodic complexes. From the definition of complexity, we see that $c_G(M)$ is optimal. The proof of the theorem is quite simple. The Künneth tensor formula implies that

$$H_*(Y) = M \otimes H_*(X^{(1)}) \otimes \cdots \otimes H_*(X^{(r)}) = H_0(Y) = M.$$

Hence Y is a resolution of M, and we need only show that every Y_n is projective. But recall that if any factor in a tensor product is projective then the product is projective. Therefore the only way that Y_n could not be projective would be if some summand had the form

$$W = M \otimes X_{i_1}^{(1)} \otimes \cdots \otimes X_{i_r}^{(r)} = M \otimes \hat{L}_1 \otimes \cdots \otimes \hat{L}_r$$

for $\hat{L}_i = \hat{L}_{\varsigma_i} = \Omega^{-1}(L_{\varsigma_i})$. By (8) and Lemma 12, $V_G(\hat{L}_i) = V_G((\varsigma_i))$. So by (7) and the peculiar way we chose $\varsigma_1, \ldots, \varsigma_r$, we have that $V_G(W) = \{0\}$. Finally (4) implies that W and hence also every Y_n are projective.

The existence of projective resolutions having the form in Theorem 13 implies all sorts of filtrations on cohomology leading to spectral sequences. Only a few of the possible relations with group cohomology have been explored. Under proper circumstances such spectral sequences are multiplicative [**14**]. Unexpectedly, Roland Schmid has shown that if G has p-rank two then the spectral sequence for the cohomology of the trivial module, when constructed as above, stops at the E_2 term.

Other work related to the varieties of modules includes some results on the structure of the cohomology ring $\mathrm{Ext}^*_{KG}(M, M)$ for a module M. Such rings can be genuinely noncommutative to the point of having factor rings which are full matrix algebras of large (but always finite) dimension. An element in $\mathrm{Ext}^n_{KG}(M, M)$ is nilpotent if and only if its restriction to every cyclic shifted subgroup corresponding to a line in $V_G(M)$ is nilpotent [**12**]. Hence the radical of $\mathrm{Ext}^*_{KG}(M, M)$ can be characterized in terms of such restriction.

The modules L_ς have also been used to construct nilpotent elements in the Green ring or representation ring of KG-modules [**6**]. The trick is to begin with a group of p-rank two and choose $\varsigma \in H^*(G, K)$ of even degree such that L_ς is periodic with period exactly two. If in addition ς annihilates the cohomology ring of L_ς then the element $L_\varsigma - \Omega(L_\varsigma)$ has square zero. It can be shown that there are many large classes of such elements. The proof is fairly simple compared to the laborious tensor product calculation in [**29**].

Interestingly, the ideal of support varieties and rank varieties has also come up in the investigation of modules over restricted p-Lie algebras. Analogs to Theorem 3 and several of the listed properties have been proved in this context. See [**18**], [**19**], and [**20**] for details or [**21**] for a survey.

REFERENCES

1. J. L. Alperin, *Cohomology is representation theory*, this volume.
2. J. L. Alperin and L. Evens, *Representations, resolutions and Quillen's dimension theorem*, J. Pure Appl. Algebra **22** (1981), 1–9.
3. ——, *Varieties and elementary abelian subgroups*, J. Pure Appl. Algebra **26** (1982), 221–227.
4. G. S. Avrunin, *Annihilators of cohomology modules*, J. Algebra **69** (1981), 150–154.
5. G. S. Avrunin and L. Scott, *Quillen stratification for modules*, Invent. Math. **66** (1982), 277–286.
6. D. J. Benson and J. F. Carlson, *Nilpotent elements in the Green ring*, J. Algebra **104** (1986), 329–350.
7. ——, *Complexity and multiple complexes*, Math. Z. (to appear)
8. J. F. Carlson, *The dimensions of periodic modules over modular group algebras*, Illinois J. Math. **23** (1979), 295–306.
9. ——, *Dimensions of modules and their restrictions over modular group algebras*, J. Algebra **69** (1981), 95–104.
10. ——, *The varieties and the cohomology ring of a module*, J.Algebra **85** (1983), 104–143.
11. ——, *Complexity and Krull dimension*, Proc. Internat. Conf. on Representations of Algebras, Lecture Notes in Math., vol. 903, Springer, Berlin, 1981, pp. 62–67.

12. ——, *The cohomology ring of a module*, J. Pure Appl. Algebra **36** (1985), 105–121.

13. ——, *The variety of an indecomposable module is connected*, Invent. Math. **77** (1984), 291–299.

14. ——, *Products and projective resolutions*, this volume.

15. E. C. Dade, *Endo-permutation modules over p-groups*. II, Ann. of Math. (2) **108**(1978), 317–346.

16. D. Eisenbud, *Homological algebra on a complete intersection, with an application to group representations*, Trans. Amer. Math. Soc. **260** (1980), 35–64.

17. L. Evens, *The cohomology ring of a finite group*, Trans. Amer. Math. Soc. **101** (1961), 224–239.

18. E. Friedlander and B. Parshall, *Geometry of p-unipotent Lie algebras*, J. Algebra (to appear).

19. ——, *Support varieties for restricted Lie algebras*, (to appear).

20. J. C. Jantzen, *Kohomologie von p-Lie-Algebren und nilpotente Elemente*, Abh. Math. Sem. Univ. Hamburg **76** (1986).

21. ——, *Restricted Lie algebra cohomology*, (to appear).

22. O. Kroll, *Complexity and elementary abelian p-groups*, J. Algebra **88** (1984), 155–172.

23. ——, *A representation-theoretic proof of a theorem of Serre*, (to appear).

24. T. Okuyama and H. Sasaki, *Norm maps and Bocksteins of cohomology groups of finite groups*, (to appear).

25. C. Picaronny, *Sur un théorème de Carlson*, C. R. Acad. Sci. Paris Ser. I Math. **299** (1984), 899–902.

26. D. Quillen, *The spectrum of an equivariant cohomology ring*. I, II, Ann. of Math. (2) **94** (1971),549–602.

27. ——, *A cohomological criterion for p-nipotency*, J. Pure Appl. Algebra **1** (1971), 361–372.

28. D.Quillen and B. B. Venkov,*Cohomology of finite groups and elementary abelian subgroups*, Topology **11** (1972), 317–318.

29. J. R. Zemanek, *Nilpotent elements in representation rings*, J. Algebra **19** (1971), 453–469.

UNIVERSITY OF GEORGIA

Proceedings of Symposia in Pure Mathematics
Volume **47** (1987)

Irreducible Characters of Finite Groups of Lie Type

R. W. CARTER

1. The finite groups of Lie type are best understood as subgroups of connected reductive algebraic groups which are fixed by a Frobenius map.

So let G be a connected reductive algebraic group over an algebraically closed field K of characteristic p. Any such group G is isomorphic to a closed subgroup of the general linear group $\mathrm{GL}_n(K)$ for some n, with respect to the Zariski topology. Let

$$F_{p^e} : \mathrm{GL}_n(K) \to \mathrm{GL}_n(K)$$

be defined by $F_{p^e}(a_{ij}) = (a_{ij}^{p^e})$. Let $F_{p^e}|_G : G \to G$ be its restriction to G. This is an example of a Frobenius map on G. However, in order to obtain all the finite groups of Lie type by this method we need a more general definition of a Frobenius map. A homomorphism $F : G \to G$ is called a Frobenius map if there exists an integer $d \geq 1$ such that $F^d = F_{p^e}|_G$ for some integer $e \geq 1$ and some embedding of G in a general linear group.

Let $G^F = \{g \in G; F(g) = g\}$. Then G^F is a finite subgroup of G. Thus a Frobenius map is a surjective homomorphism of G into itself with finite fixed point set. If G is semisimple any such map of G into itself is, conversely, a Frobenius map. However, this is not always so for reductive groups.

The groups G^F which arise in this way are called the finite groups of Lie type. To each such group there is associated a real number q defined by $q = p^{e/d}$. q is uniquely determined by G and F.

Examples of such groups G^F include $\mathrm{GL}_n(q)$, $\mathrm{SL}_n(q)$, $\mathrm{PGL}_n(q)$, $\mathrm{U}_n(q^2)$, $\mathrm{Sp}_{2n}(q)$, $\mathrm{SO}_n(q)$, $G_2(q)$, etc.

2. The structure and representation theory of the groups G^F are based on the properties of two types of subgroup, the Borel subgroups and the maximal tori.

A Borel subgroup of G is a maximal connected soluble subgroup of G. Any two Borel subgroups of G are conjugate in G. If B is a Borel subgroup of G so is $F(B)$. There exist Borel subgroups of G which are F-stable, i.e., which satisfy

---

1980 *Mathematics Subject Classification* (1985 *Revision*). Primary 20Cxx, 20G40.

$F(B) = B$. For each F-stable Borel subgroup B we may form the subgroup B^F of G^F. B^F is called a Borel subgroup of G^F. Any two Borel subgroups of G^F are conjugate in G^F.

Let $K^* = K - \{0\}$ be the multiplicative group of K. An algebraic group isomorphic to $K^* \times \cdots \times K^*$ is called a torus. Any two maximal tori of G are conjugate in G. If T is a maximal torus of G so is $F(T)$. Just as in the case of Borel subgroups there exist F-stable maximal tori in G. Each F-stable maximal torus T of G gives rise to a subgroup T^F of G^F. In contrast to the situation for Borel subgroups, not every two maximal tori of G^F are conjugate in G^F.

There is one particularly important conjugacy class of maximal tori in G^F. An F-stable maximal torus of G is called maximally split if it lies in some F-stable Borel subgroup of G. If T is a maximally split torus in G, T^F is called a maximally split torus in G^F. Any two maximally split tori in G^F are conjugate in G^F.

The G^F-classes of maximal tori may be described as follows. Let T_0 be a maximally split torus. Then any F-stable maximal torus T has form ${}^gT_0 = gT_0g^{-1}$ for some $g \in G$. Since T is F-stable we have $g^{-1}F(g) \in N_0 = \mathcal{N}_G(T_0)$. Let $W = \mathcal{N}_G(T_0)/T_0$ be the Weyl group and $\pi : N_0 \to W$ be the natural map. Then $w = \pi(g^{-1}F(g)) \in W$. The element $w \in W$ obtained in this way from T may depend on the choice of g. However, any other such element is F-conjugate to w, i.e., has the form $w'^{-1}wF(w')$ for some $w' \in W$. So every F-stable maximal torus determines an F-conjugacy class in W. In fact there is a 1–1 correspondence between G^F-classes of maximal tori and F-conjugacy classes in W.

3. The major breakthrough in the representation theory of the groups G^F over an algebraically closed field of characteristic 0 was made by Deligne and Lusztig in [**3**]. Deligne and Lusztig showed how to construct a generalized character $R_{T,\theta}$ of G^F for each pair (T, θ), where T is an F-stable maximal torus of G and θ is an irreducible character of T^F. The construction of the $R_{T,\theta}$ uses the theory of l-adic cohomology with compact support. l-adic cohomology theory was introduced by M. Artin and Grothendieck with the aim of applying it to the Weil conjectures on the number of points on an algebraic variety over a finite field. It was used by Deligne and Lusztig to construct the $R_{T,\theta}$ as follows.

Let $L : G \to G$ be defined by $L(g) = g^{-1}F(g)$. A theorem of Lang-Steinberg asserts that L is surjective. Given an F-stable maximal torus T we choose a Borel subgroup B containing T. (B will not in general be F-stable.) B decomposes into a semidirect product $B = UT$ where U is normal in B and $U \cap T = 1$. Let $\tilde{X} = L^{-1}(U) = \{g \in G; L(g) \in U\}$. $\tilde{X}$ is an algebraic subset of G. Moreover, G^F acts on $\tilde{X}$ by left multiplication, T^F acts on $\tilde{X}$ by right multiplication, and these two actions commute.

Let l be a prime with $l \neq p$. Then we can form the l-adic cohomology groups with compact support $H_c^i(\tilde{X}, \overline{\mathbf{Q}}_l)$ where $\mathbf{Q}_l$ is the field of l-adic numbers and $\overline{\mathbf{Q}}_l$ is its algebraic closure. Since G^F and T^F act on the algebraic variety $\tilde{X}$ as finite groups of automorphisms, they also act on $H_c^i(\tilde{X}, \overline{\mathbf{Q}}_l)$ as groups of nonsingular linear maps. Let $H_c^i(\tilde{X}, \overline{\mathbf{Q}}_l)_\theta$ be the subspace on which T^F acts by the character θ. This is a G^F-module since the actions of G^F, T^F commute. We define

$$R_{T,\theta}(g) = \sum_{i \geq 0} (-1)^i \operatorname{trace}(g, H_c^i(\tilde{X}, \overline{\mathbf{Q}}_l)_\theta)$$

for $g \in G^F$. $R_{T,\theta}$ is thus a generalized character of G^F. It can be shown to be independent of l and independent of the choice of B containing T.

4. $R_{T,\theta}$ can also be given using other algebraic varieties besides $\tilde{X}$. Let $T_0 \subset B_0 = U_0T_0$ be a fixed pair consisting of a maximal torus and Borel subgroup containing it. Let $\mathcal{B}$ be the variety of all Borel subgroups of G. G acts on $\mathcal{B} \times \mathcal{B}$ by conjugation and the orbits are in 1–1 correspondence with elements of $W = \mathcal{N}(T_0)/T_0$. For $w \in W$ there is a unique orbit containing the pair $(B_0, {}^{\dot{w}}B_0)$ where $\dot{w} \in \mathcal{N}(T_0)$ maps to $w \in W$. A pair of Borel subgroups in this orbit are said to be in relative position w.

Let $\mathcal{B}_{w,F} = \{B \in \mathcal{B};\ B, F(B)$ are in the relative position $w\}$. Then $\mathcal{B}_{w,F}$ is an algebraic variety closely related to the variety $L^{-1}(B_0\dot{w}B_0)$. B_0 acts by right multiplication on $L^{-1}(B_0\dot{w}B_0)$ and there is a surjective morphism of varieties

$$L^{-1}(B_0\dot{w}B_0) \to \mathcal{B}_{w,F}, \qquad g \to {}^gB_0$$

whose fibres are the B_0-orbits. Thus there is a bijective morphism from $X_w = L^{-1}(B_0\dot{w}B_0)/B_0$ to $\mathcal{B}_{w,F}$.

Now each B_0-orbit on $L^{-1}(B_0\dot{w}B_0)$ contains an element of $L^{-1}(\dot{w}U_0)$. $L^{-1}(\dot{w}U_0)$ is a subvariety of $L^{-1}(B_0\dot{w}B_0)$ which is acted on by the group of automorphisms $(U_0 \cap {}^{\dot{w}} U_0)T_0^{\dot{w}^{-1}F}$ by right multiplication, where

$$T_0^{w^{-1}F} = \{t \in T_0; F(t) = w(t)\}.$$

Moreover, any two elements of $L^{-1}(\dot{w}U_0)$ which are in the same B_0-orbit on $L^{-1}(B_0\dot{w}B_0)$ are in the same orbit of $(U_0 \cap {}^{\dot{w}}U_0)T_0^{w^{-1}F}$ on $L^{-1}(\dot{w}U_0)$. Thus there is an isomorphism of varieties

$$X_w = L^{-1}(B_0\dot{w}B_0)/B_0 \cong L^{-1}(\dot{w}U_0)/(U_0 \cap {}^{\dot{w}}U_0)T_0^{w^{-1}F}.$$

We define $\tilde{X}_w$ to be the variety

$$\tilde{X}_w = L^{-1}(\dot{w}U_0)/(U_0 \cap {}^{\dot{w}}U_0).$$

Then $X_w \cong \tilde{X}_w/T_0^{w^{-1}F}$.

We now explain the connection between these varieties and the generalized characters $R_{T,\theta}$. Let T be an F-stable maximal torus and $B = UT$ a Borel subgroup containing it. $F(B) = F(U)T$ will also be a Borel subgroup containing T, and it will be convenient to take this one instead. Let $T = {}^xT_0$ where $x^{-1}F(x) = \dot{w}$ and $U = {}^xU_0$. Then there is an isomorphism between the varieties $L^{-1}(F(U))$ and $L^{-1}(\dot{w}U_0)$ given by right multiplication by x. Moreover, the group $(U \cap F(U))T^F$ acts on $L^{-1}(F(U))$ as a group of automorphisms by right multiplication and, under the above isomorphism, the orbits of $L^{-1}(F(U))$ under $(U \cap F(U))T^F$ map to the orbits of $L^{-1}(\dot{w}U_0)$ under $(U_0 \cap {}^{\dot{w}}U_0)T_0^{w^{-1}F}$. Thus we have isomorphisms of varieties

$$X_w \cong L^{-1}(\dot{w}U_0)/(U_0 \cap^{\dot{w}} U_0)T_0^{w^{-1}F} \cong L^{-1}(F(U))/(U \cap F(U))T^F,$$
$$\tilde{X}_w = L^{-1}(\dot{w}U_0)/(U_0 \cap^{\dot{w}} U_0) \cong L^{-1}(F(U))/(U \cap F(U)).$$

Now $U \cap F(U)$ is isomorphic to an affine space. It follows that

$$\begin{aligned} R_{T,\theta}(g) &= \sum_i (-1)^i \operatorname{tr}(g, H_c^i(L^{-1}(F(U)), \overline{\mathbf{Q}}_l)_\theta) \\ &= \sum_i (-1) \operatorname{tr}(g, H_c^i(\tilde{X}_w, \overline{\mathbf{Q}}_l)_\theta). \end{aligned}$$

Also $\tilde{X}_w/T^F \cong X_w$ and so the character θ of T^F determines a locally constant sheaf $\mathcal{S}_\theta$ on X_w. Thus we have

$$R_{T,\theta}(g) = \sum_i (-1)^i \operatorname{tr}(g, H_c^i(\tilde{X}_w, \overline{\mathbf{Q}}_l)_\theta) = \sum_i (-1)^i \operatorname{tr}(g, H_c^i(X_w, \mathcal{S}_\theta)).$$

5. The generalized character $R_{T,\theta}$ of G^F defined by the equivalent formulas above has some very useful properties. For example, we have $(R_{T,\theta}, R_{T',\theta'}) = 0$ if T, T' are not G^F-conjugate. We also have $(R_{T,\theta}, R_{T,\theta'}) =$ number of $w \in W(T)^F$ with $w\theta = \theta'$ where $W(T) = \mathcal{N}_G(T)/T$. θ is said to be in general position if $W(T)_\theta^F = 1$, where

$$W(T)_\theta^F = \{w \in W(T)^F; w\theta = \theta\}.$$

If θ is in general position then $(R_{T,\theta}, R_{T,\theta}) = 1$ and so $\pm R_{T,\theta}$ is an irreducible character of G^F. Thus each G^F-class of F-stable maximal tori gives rise to a family of irreducible characters of G^F. The degrees of these characters are given by

$$\pm R_{T,\theta}(1) = |G^F : T^F|_{p'}.$$

It was shown by Deligne and Lusztig that, if $u \in G^F$ is unipotent, then the value $R_{T,\theta}(u)$ is independent of u. The common values $Q_{T,u} = R_{T,\theta}(u)$ are called Green functions.

6. The characters $\pm R_{T,\theta}$ with θ in general position do not exhaust the irreducible characters of G^F. However, given any irreducible character χ of G^F we have $(R_{T,\theta}, \chi) \neq 0$ for some pair T, θ. So if we can decompose the generalized characters $R_{T,\theta}$ into a $\mathbf{Z}$-combination of irreducible characters we shall obtain in this way all the irreducible characters of G^F.

We may define an equivalence relation on the set of irreducible characters of G^F as follows. We say $\chi \sim \chi'$ if there exists a sequence

$$\chi = \chi^1, \qquad \chi^2 \cdots \chi^k = \chi'$$

of irreducible characters such that χ^i, χ^{i+1} lie in a common $R_{T,\theta}$.

We may also define an equivalence relation on the set of pairs (T, θ). We say $(T, \theta) \sim (T', \theta')$ if there exists a sequence

$$(T, \theta) = (T_1, \theta_1), \qquad (T_2, \theta_2) \cdots (T_k, \theta_k) = (T', \theta')$$

such that R_{T_i, θ_i} and $R_{T_{i+1}, \theta_{i+1}}$ contain a common irreducible component.

These two types of equivalence class are in 1–1 correspondence. They are called geometric conjugacy classes. The number of geometric conjugacy classes is $|(Z^0)^F|q^l$ where Z is the center of G, Z^0 is the connected component of Z containing 1, and l is the semisimple rank of G.

We shall now and subsequently assume that the center Z is connected and that p is a good prime for G. In this case each geometric conjugacy class of irreducible characters of G^F contains just one character of degree prime to p, and this is the character of smallest degree in the class. An irreducible character χ of G^F will be called semisimple if $\chi(1)$ is prime to p. (The definition of a semisimple character when p is a bad prime for G is different.) An irreducible character χ is called unipotent if χ is geometrically conjugate to the principal character 1. The only character which is both semisimple and unipotent is the principal character.

7. Lusztig has proved that there is a Jordan decomposition of irreducible characters which we shall now describe. We first recall the Jordan decomposition of elements. Each $g \in G$ is uniquely expressible in the form

$$g = g_s g_u = g_u g_s,$$

where g_s is semisimple, g_u is unipotent. In the finite group G^F the semisimple elements are those whose order is prime to p and the unipotent elements are those whose order is a power of p. Thus each conjugacy class of G determines a semisimple conjugacy class, containing s say, and a unipotent conjugacy class in the centralizer $C_G(s)$.

The Jordan decomposition of characters is somewhat similar. There is a duality $G \to G^*$ on the set of connected reductive groups, and each Frobenius map $F : G \to G$ has a dual Frobenius map $F^* : G^* \to G^*$. This duality makes the role of the geometric conjugacy classes particularly clear. There is a 1–1 correspondence between geometric conjugacy classes in G and F^*-stable semisimple classes in G^*.

Let χ be an irreducible character of G^F and κ be the geometric conjugacy class containing χ. Let χ_s be the semisimple character in κ. We can find an F^*-stable semisimple element s^* in the semisimple class of G^* corresponding to κ. Let $C^* = C_{G^*}(s^*)$. Then C^* is a connected reductive group and $F^* : C^* \to C^*$ is a Frobenius map on C^*. It was proved by Lusztig that there is a 1–1 correspondence $\chi \leftrightarrow \chi_u$ between irreducible characters χ of G^F in the geometric conjugacy class κ and unipotent characters χ_u of C^{*F^*}. This correspondence (which may not be uniquely determined) satisfies

$$\chi(1) = \chi_s(1)\chi_u(1).$$

Thus each irreducible character χ of G^F determines a semisimple character of G^F and a unipotent character χ_u of $(C_{G^*}(s^*))^{F^*}$ such that the degree of χ is the product of the degrees of χ_s, χ_u. In particular, if the degrees of all semisimple and unipotent characters are known, the degrees of all irreducible characters will be known.

There is a beautiful formula for the degrees of the semisimple characters, due to Deligne and Lusztig. It is

$$\chi_s(1) = |G^{*F^*} : C^{*F^*}|_{p'}.$$

Thus the degree of χ_s is determined in terms of the centralizer of a semisimple element in the dual group.

Lusztig has also obtained a formula for the degrees of the unipotent characters, but we shall not give it here. This formula, together with further details about the topics discussed here, can be found in the following books.

REFERENCES

1. G. Lusztig, *Characters of reductive groups over a finite field*, Ann. of Math. Stud., no. 107, Princeton Univ. Press, Princeton, N. J., 1984.

2. R. W. Carter, *Finite groups of Lie type: Conjugacy classes and complex characters*, Wiley, New York, 1985.

3. P. Deligne and G. Lusztig, *Representations of reductive groups over finite fields*, Ann. of Math. (2) **103** (1976), 103–161.

UNIVERSITY OF WARWICK, COVENTRY, ENGLAND

Proceedings of Symposia in Pure Mathematics
Volume 47 (1987)

The Hecke Algebra of a Finite Coxeter Group

CHARLES W. CURTIS

1. Introduction. Let (W, S) be a finite Coxeter system. We shall consider a family $\{G(q)\}$ of finite groups of Lie type, parametrized by an infinite set of prime powers $\{q\}$, each having a Tits system with the associated Coxeter system (W, S).

EXAMPLES. (i) For each prime power q, we let $G(q)$ denote a finite Chevalley group over the finite field $\mathbf{F}_q$, with Weyl group W. In this case, the groups $\{G(q)\}$ are parametrized by all prime powers q.

(ii) Let $\underset{\sim}{G}$ denote a connected, reductive algebraic group defined over a finite field $\mathbf{F}_{q_0}$. The family $\{G(q)\}$ consists of the groups of $\mathbf{F}_q$-rational points on $\underset{\sim}{G}$, for the set of prime powers q such that $\mathbf{F}_{q_0} \subset \mathbf{F}_q$, and such that each group $G(q) = \underset{\sim}{G}(\mathbf{F}_q)$ has the same (reduced) Weyl group W.

For each group in the family, we have the Bruhat decomposition

$$G(q) = \bigcup_{w \in W} B(q) w B(q),$$

where $B(q)$ is the Borel subgroup in the given Tits system of $G(q)$. The *Hecke algebra* $H(G(q), B(q))$ is the subalgebra of the complex group algebra of $G(q)$ consisting of all functions constant on the $(B(q), B(q))$-double cosets, multiplied by convolution. This algebra is anti-isomorphic to the algebra of $G(q)$-endomorphisms of the induced permutation representation $(1_{B(q)})^{G(q)}$, and its irreducible characters are the restrictions of the irreducible characters of $\mathbf{C}G(q)$ which appear with positive multiplicity in the permutation representation $(1_{B(q)})^{G(q)}$. Its *standard basis* $\{b_w\}_{w \in W}$ is the set of characteristic functions on the double cosets $B(q)\backslash G(q)/B(q)$, normalized by multiplying by $|B(q)|^{-1}$. These are multiplied according to the rule:

$$b_{s_i} b_w = \begin{cases} b_{s_i w} & \text{if } l(s_i w) > l(w), \\ q^{c_i} b_{s_i w} + (q^{c_i} - 1) b_w & \text{if } l(s_i w) < l(w), \end{cases}$$

1980 *Mathematics Subject Classification* (1985 *Revision*). Primary 20G05, 20G40.

The author's research was supported in part by National Science Foundation grant DMS-8318037-01.

where $l(w)$ denotes the length function on the Coxeter group W with respect to the set $S = \{s_1, \ldots, s_n\}$, and $\{c_1, \ldots, c_n\}$ are the *index parameters*, defined by the condition $|B(q)s_iB(q)/B(q)| = q^{c_i}$, $1 \leq i \leq n$. (In case $G(q)$ is a family of untwisted Chevalley groups, for example, all the index parameters are equal to 1.)

If we think of q as a parameter, then upon setting $q = 1$ the multiplication becomes $b_{s_i}b_w = b_{s_iw}$, for all $s_i \in S$ and $w \in W$, and the corresponding algebra is the group algebra $\mathbf{C}W$. This suggests:

DEFINITION. The *generic Hecke algebra H* of the Coxeter system (W, S) is the $\mathbf{Q}[u]$-algebra (for an indeterminate u) with the basis $\{a_w\}_{w \in W}$ and multiplication defined by:

$$a_{s_i}a_w = \begin{cases} a_{s_iw} & \text{if } l(s_iw) > l(w), \\ u^{c_i}a_{s_iw} + (u^{c_i} - 1)a_w & \text{if } l(s_iw) < l(w), \end{cases}$$

for $s_i \in S$ and $w \in W$, and for some fixed set of positive integers $c_1, \ldots, c_n$ such that $c_i = c_j$ if $s_i = {}^w s_j$.

Corresponding to the presentation of W:

$$\langle s_1, \ldots, s_n : s_i^2 = (s_is_j)^{n_{ij}} = 1 \rangle,$$

the generic algebra H has a presentation as an associative $\mathbf{Q}[u]$-algebra with generators $\{a_{s_i} : 1 \leq i \leq n\}$ and relations:

$$a_{s_i}^2 = u^{c_i}a_1 + (u^{c_i} - 1)a_{s_i} \quad \text{for } s_i \in S;$$

and

$$(a_{s_i}a_{s_j})^{k_{ij}}a_{s_i} = (a_{s_j}a_{s_i})^{k_{ij}}a_{s_j} \quad \text{if } i \neq j \text{ and } n_{ij} = 2k_{ij} + 1,$$

or

$$(a_{s_i}a_{s_j})^{k_{ij}} = (a_{s_j}a_{s_i})^{k_{ij}} \quad \text{if } n_{ij} = 2k.$$

Let F be a field, and let $f : \mathbf{Q}[u] \to F$ be a homomorphism. Then we obtain a *specialized algebra* $F \otimes_{\mathbf{Q}[u]} H$, where the action of $\mathbf{Q}[u]$ on F is given by the homomorphism f. For the homomorphism $\mathbf{Q}[u] \to \mathbf{C}$ such that $u \to 1$, the specialized algebra $H_{(1)} = \mathbf{C} \otimes_{\mathbf{Q}[u]} H$ is isomorphic to the complex group algebra $\mathbf{C}W$, while for the homomorphism $\mathbf{Q}[u] \to \mathbf{C}$ with $u \to q$, the specialized algebra $H_{(q)}$ is isomorphic to the Hecke algebra $H(G(q), B(q))$. This observation has proved to be the source of deep connections between the representation theory of the Coxeter group W and the representations of all the groups in the family $\{G(q)\}$.

References to the literature will be given in §5.

2. Parametrization of the characters in $(1_B)^G$; generic degrees. We shall give some illustrations of the ideas discussed in §1. We first introduce some notation. We let $R = \mathbf{Q}[u]$, $K = \mathbf{Q}(u)$, and let K^* be an algebraic closure of K, and R^* the integral closure of R in K^*. We shall identify H with $1 \otimes_R H$ in $K^* \otimes_R H$.

(2.1) PROPOSITION. (i) *$K^* \otimes_R H$ is a semisimple K^*-algebra.*

(ii) *For each irreducible character μ of $K^* \otimes_R H$, we have $\mu(a_w) \in R^*$ for all $w \in W$.*

(iii) *Let $f : R \to F$ be a homomorphism such that the specialized algebra $F \otimes_R H$ is semisimple. Extend $f : R \to F$ to a homomorphism $f^* : R^* \to F^*$, where F^* is an algebraic closure of F. For each irreducible character μ of $K^* \otimes_R H$, define a linear map μ_{f^*} on the specialized algebra $F^* \otimes_R H$ by setting $\mu_{f^*}(1 \otimes a_w) = f^*(\mu(a_w))$, for $w \in W$. Then μ_{f^*} is an irreducible character of the specialized algebra, and the map $\mu \to \mu_{f^*}$ is a bijection of irreducible characters,*

$$\operatorname{Irr} K^* \otimes_R H \leftrightarrow \operatorname{Irr} F^* \otimes_R H.$$

A proof can be given by defining an appropriate decomposition map and a Cartan map, and using the relation $C = {}^tDD$ as in Brauer's theory.

A corollary of the preceding result is that there exists an isomorphism of $\mathbf{C}$-algebras: $\mathbf{C}W \cong H(G(q), B(q))$ for each q.

(2.2) THEOREM. *There exists a bijection $\varphi \to \varsigma_{\varphi,q}$ from $\operatorname{Irr} W$ to the set of irreducible characters of $G(q)$ which appear with positive multiplicity in $(1_{B(q)})^{G(q)}$, for each group $G(q)$, with the property that for all standard parabolic subgroups $P_I(q) = B(q)W_I B(q)$, $I \subset S$, we have*

$$(\varsigma_{\varphi,q}, (1_{P_I(q)})^{G(q)}) = (\varphi, (1_{W_I})^W).$$

Using this parametrization, we can attempt to give information about the values of the irreducible characters $\varsigma_{\varphi,q}$ for all the groups $\{G(q)\}$, in a uniform way. For example, here is a method for calculating the degrees of the characters $\varsigma_{\varphi,q}$. For simplicity, we assume the index parameters are all equal to 1.

DEFINITION. Let $\varphi \in \operatorname{Irr} W$. The *generic degree* associated with φ is the element of K^* defined by

$$d_\varphi = \frac{(\deg \varphi) P(u)}{\sum u^{-l(w)} \mu_\varphi(a_w) \mu_\varphi(a_{w^{-1}})},$$

where $P(u)$ is the *Poincaré polynomial* of W, given by $P(u) = \sum_{w \in W} u^{l(w)}$, and μ_φ is the irreducible character of $K^* \otimes_R H$ corresponding to φ.

(2.3) THEOREM. (i) *Let $f^*_{(1)}$ and $f^*_{(q)}$ be homomorphisms from $R^* \to \mathbf{C}$ extending the maps defined by $u \to 1$ and $u \to q$, respectively. Then we have*

$$f^*_{(1)}(d_\varphi) = \deg \varphi \quad \textit{and} \quad f^*_{(q)}(d_\varphi) = \deg \varsigma_{\varphi,q} \quad \textit{for each } q.$$

(ii) *For each φ, we have $d_\varphi \in \mathbf{Q}[u]$.*

(iii) *d_φ divides $u^{l(w_0)}P(u)$ in the polynomial ring $\mathbf{Q}[u]$, (where w_0 is the element of W of maximal length.)*

Using the preceding theorem and some other information, it has been possible to compute the polynomials d_φ explicitly in all cases. We shall give three illustrations of how the generic degrees are used.

(i) We first define, for each $\varphi \in \operatorname{Irr} W$, a second polynomial p_φ, called the *fake degree*, and defined by $p_\varphi = \sum m_i u^i$, where m_i is the multiplicity with which φ occurs in the ith graded component of the coinvariant algebra S_W of W. (The coinvariant algebra S_W is the quotient $S(V)/I^+(V)$, where $S(V)$ is the symmetric algebra on the module V affording the reflection representation of W, and $I^+(V)$ is the ideal generated by the W-invariants of positive degree. It is known that the graded algebra S_W affords the regular representation of W, and hence defines a graded structure on this representation.) Let

$$d_\varphi = \alpha u^{a(\varphi)} + \text{ higher terms}, \qquad p_\varphi = \beta u^{b(\varphi)} + \text{ higher terms },$$

$\alpha, \beta \in \mathbf{Q}$. It is known that $a_\varphi > 0$ if $\varphi \neq 1$, and from the lists of generic degrees and fake degrees, it has been observed that $a_\varphi \leq b_\varphi$ for all φ. The irreducible characters φ of W for which $a_\varphi = b_\varphi$ are called *special* and are important in Lusztig's work on the decomposition of the characters $R_T(\theta)$.

(ii) Problems which do not seem to involve characters in $(1_B)^G$ can sometimes be reduced to this situation using the structure of the groups $\{G(q)\}$. One case where this idea was successfully applied was the determination by Curtis-Kantor-Seitz of the 2-transitive permutation representations of the finite Chevalley groups. Let θ be a permutation character of a finite Chevalley group $G(q)$. If the permutation representation is 2-transitive, then $\theta = (1_L)^G$, where L is the stabilizer of a point, and $\theta = 1 + \varsigma$, for some irreducible nontrivial character ς. If $\varsigma \notin (1_B)^G$, then $((1_L)^G, (1_B)^G) = 1$, and we have $G = LB$. Such a group L is called a flag-transitive subgroup, and all the possibilities for such a subgroup were classified by Seitz. So we may assume that $\varsigma = \varsigma_{\varphi,q}$ for some φ. Now let p be the characteristic. If $p \nmid \deg \theta$, then another group-theoretical argument gives the classification. On the other hand, if p divides $\theta(1)$, then $p \nmid \deg \varsigma_{\varphi,q}$. Since the generic degree d_φ is divisible by u in $\mathbf{Q}[u]$, there remain a finite number of possibilities, which can be analyzed, to complete the proof.

(iii) Another case where we can in a sense reduce to the case of $(1_B)^G$ is the solution of the decomposition problem in Harish-Chandra's theory of cuspidal characters. Let $G(q)$ be the group of rational points $\underset{\sim}{G}(\mathbf{F}_q)$ for a reductive group $\underset{\sim}{G}$ with a connected center. Let λ be an irreducible cuspidal character of a Levi subgroup $L_I(q)$ of a standard parabolic subgroup $P_I(q)$, for some $I \subset S$. The problem is to decompose the induced character $R_{L_I(q)}^{G(q)} \lambda$.

(2.4) THEOREM (HOWLETT-LEHRER, LUSZTIG). *There exists an isomorphism of* $\mathbf{C}$*-algebras*

$$\operatorname{End}_{G(q)} R_{L_I(q)}^{G(q)} \lambda = \mathbf{C} \otimes_R H_{0,q},$$

where H_0 is the generic Hecke algebra of a suitably defined Coxeter group W_0 and set of index parameters $\{c_i\}$, for the specialization $f_0 : u \to q$.

As a result, the components of the induced character can be parametrized by the irreducible characters of W_0, and the degree of the character $\varsigma_\varphi \in \operatorname{Irr} G(q)$ corresponding to $\varphi \in \operatorname{Irr} W_0$ can be shown to be

$$\deg \varsigma_\varphi = \deg R_{L_I(q)}^{G(q)} \lambda \cdot f_0(d_\varphi / P_0(u)),$$

where d_φ is the generic degree, and $P_0(u)$ the Poincaré polynomial, for the Coxeter group W_0.

3. Rationality questions; Lusztig's isomorphism theorem. It is known that the rational field $\mathbf{Q}$ is a splitting field for each indecomposable Weyl group W; in the other words, $\mathbf{Q}W$ is a split semisimple $\mathbf{Q}$-algebra. It is natural to ask whether something similar occurs for the generic Hecke algebra H.

(3.1) THEOREM. *Let W be an indecomposable Weyl group, H the generic Hecke algebra associated with W, and μ_φ an irreducible character of $K^* \otimes_R H$, corresponding to $\varphi \in \operatorname{Irr} W$. Then μ_φ is rational-valued, in the sense that $\mu_\varphi(a_w) \in R$ for all $w \in W$, with the following exceptions:*

(i) *φ is one of the characters of degree 512, in case W is of type E_7, and*

(ii) *φ is one of the 4 characters of degree 4096, in case W is of type E_8.*

In the exceptional cases, the character μ_φ is definitely not rational, and in fact $\mu_\varphi(a_{w_0})$ involves $u^{1/2}$.

The theorem is proved by a case-by-case discussion, using the result that, with the exceptions listed and the characters of degree 2 if $|W| = 12$, we have $(\varphi, (1_{W_I})^W) = (\varphi', (1_{W_I})^W)$ for $\varphi, \varphi' \in \operatorname{Irr} W$, and all $I \subseteq S$, only if $\varphi = \varphi'$.

Using the preceding theorem it was possible to prove that there exists an isomorphism of algebras over $\mathbf{Q} : \mathbf{Q}W \cong H_{\mathbf{Q}}(G(q), B(q))$, for each indecomposable Weyl group W not of type E_7 or E_8.

The search for a more conceptual relationship between the structure of a Weyl group W and its generic Hecke algebra H was taken up by Kazhdan and Lusztig. They considered the algebra $\tilde{H} = \tilde{R} \otimes_R H$, where $\tilde{R} = Q[u^{1/2}, u^{-1/2}]$, with all the index parameters equal to one, and introduced a new basis of $\tilde{H}$, which begins to exhibit the ideal structure of $\tilde{H}$ better than the standard basis. Let i be the involution of $\tilde{H}$ which acts on $\tilde{R}$ by sending $u^{1/2} \to u^{-1/2}$ and sends a standard basis element $a_w \to a_{w^{-1}}^{-1}$.

(3.2) THEOREM. *There exists a basis of R consisting of elements $\{c'_w\}_{w \in W}$ which are characterized by the conditions*

$$i(c'_w) = c'_w \quad \textit{and} \quad c'_w = u^{-l(w)/2} \sum_{y < w} P_{y,w}(u) a_y,$$

where $\leq$ is the Bruhat order on W and the $P_{y,w}$ are polynomials in $\mathbf{Z}[u]$ such that $P_{w,w} = 1$ and, if $y < w$, $\deg P_{y,w} \leq \frac{1}{2}(l(w) - l(y) - 1)$. The multiplication of basis elements $\{c'_w\}$ is given by

$$c'_s c'_w = \begin{cases} (u^{1/2} + u^{-1/2}) c'_w & \textit{if } sw < w \\ c'_{sw} + \sum\limits_{\substack{y<w \\ sy<y}} \mu(y,w) c'_y & \textit{if } sw > w \end{cases} \quad \textit{for } s \in S,\ w \in W,$$

where $\mu(y, w)$ is the coefficient of $u^{1/2(l(w)-l(y)-1)}$ in $P_{y,w}$.

One then defines preorders $\leq_L$, $\leq_R$, $\leq_{LR}$ as follows. The preorder $\leq_L$ is generated by the elementary relations: $x \leq_L y$ if for some $s \in S$, c'_x appears with a nonzero coefficient in $c'_s c'_y$. Define $x \leq_R y$ if $x^{-1} \leq_L y^{-1}$, and define the preorder $\leq_{LR}$ by the combination of the elementary relations defining $\leq_L$ and $\leq_R$. Then, for example, the elements c'_x, with $x \leq_L w$ for some fixed element $w \in W$, generate a left ideal in $\tilde{H}$.

The equivalence classes defined by $\leq_L$, $\leq_R$, and $\leq_{LR}$ are called *left cells*, *right cells*, and *two-sided cells*, respectively. We require:

$$(*) \qquad y \leq_L x \quad \text{and} \quad y \sim_{LR} x \Rightarrow y \sim_L x.$$

As far as I know, no elementary proof of this fact is known; Lusztig first proved it using the theory of primitive ideals in enveloping algebras.

Now let E be the free R-module with basis $\{e_w : w \in W\}$, affording the left regular representation of $\tilde{H}$ (with e_w corresponding to the basis element c'_w), and with the right action by $\tilde{R}W$ obtained by specializing the right action by $\{a_s : s \in S\}$ by setting $u^{1/2} \to 1$.

These actions do not commute. Define, however, for each two-sided cell X, the submodules of E:

$$E_X = \langle e_x : x \leq_{LR} w,\ w \in X\rangle, \qquad E'_X = \langle e_x : x \leq_{LR} w,\ w \in X,\ x \notin X\rangle,$$

and set $\operatorname{gr} E = \bigoplus_X (E_X/E'_X)$, which inherits a left action by $\tilde{H}$ and a right action by $\tilde{R}W$. This time it can be proved, using $(*)$, that the left action of $\tilde{H}$ on gr E and the right action by $\tilde{R}W$ commute. We then have:

(3.3) THEOREM (LUSZTIG). (i) *There is a unique homomorphism of R-algebras $\eta\colon \tilde{H} \to \tilde{R}W$ such that for all $h \in \tilde{H}$, $w \in W$,*

$$he_w - \eta(h)e_w = \sum_{y \not\sim_{LR} w} \alpha_y e_y, \qquad \alpha_y \in \tilde{R}.$$

(ii) *For each q, η induces an isomorphism of specialized algebras,*

$$H_{\mathbf{Q}(q^{1/2})}(G(q), B(q)) \cong Q(q^{1/2})W$$

The preceding theorem gives explicit connections between characters of W and characters of Hecke algebras $H(G(q), B(q))$. In the same context, Kazhdan and Lusztig defined a combinatorial object, called a W-graph, which can be used to describe certain representations of $\tilde{H}$. Gyoja has proved that every simple $\mathbf{C}W$-module arises from a W-graph. These theorems on lifting W-modules to $\tilde{H}$-modules and the problem of calculating the characters of the resulting $\tilde{H}$-modules define a major area of current research on Hecke algebras. An application of character formulas for Hecke algebras is discussed briefly in §4 (see Theorem (4.1)). See also Lusztig's article in these Proceedings.

4. Connections with algebraic geometry. In this section, G denotes a connected, reductive algebraic group, with connected center, defined and split over $\mathbf{F}_q$, with Frobenius endomorphism F. Let W be the Weyl group of G.

(i) *Zeta functions of the varieties* X_w. For each $w \in W$, let

$$X_w = \{gB \in G/B\colon g^{-1}F(g) \in Bw\,.\,B\},$$

where B is an F-stable Borel subgroup of G. Then X_w is a locally closed subvariety of the flag variety of G/B, and the finite group G^F acts on it by translation. These varieties were introduced by Deligne-Lusztig. The irreducible characters appearing with positive multiplicity in the virtual characters $R_w = \sum(-1)^i H_c^i(X_w)$, for different choices of w, are precisely the unipotent characters of G^F. For example, $R_1 = (1_{B^F})^{G^F}$, and its irreducible components are the characters $\varsigma_{\varphi,q}$ considered in §2.

Let $r > 0$, and let $H_{(q^r)}$ be the Hecke algebra of the finite group G^{F^r}, viewed as the endomorphism algebra of $(1_{B^{F^r}})^{G^{F^r}}$. Let $x \to \hat{x}$ denote the bijection given by Shintani descent, from conjugacy classes in G^F, to twisted classes in G^{F^r}. The following result was proved by Digne-Michel, Asai, and Lusztig, using the Lefschetz fixed point theorem.

(4.1) THEOREM. $\mathrm{Tr}(xF^r, \sum(-1)^i H_c^i(X_w)) = \mathrm{Tr}(\hat{x}Fa_w, (1_{B^{F^r}})^{G^{F^r}})$, *for* $x \in G^F$.

Using this theorem, precise information about the zeta functions of the varieties X_w can be obtained. For example, in the case of GL_n, we have

$$|X_w^{F^r}| = \sum_{\varphi \in \mathrm{Irr}\, W} \mu_{\varphi,q^r}(a_w)\, d_\varphi(q^r),$$

where μ_{φ,q^r} is the irreducible character of the Hecke algebra $H_{(q^r)}$ corresponding to φ, and d_φ is a generic degree.

(ii) *Refinements of the Bruhat decomposition.* Let $\{b_w\}$ denote the standard basis of the Hecke algebra $H_{(q)}$. Kawanaka proved that the multiplication is given by

$$[b_w b_{w'} : b_{w''}] = \sum_\tau q^{a(\tau)}(q-1)^{b(\tau)},$$

where τ ranges over a set of subexpressions of a fixed reduced expression of w, and $a(\tau)$ and $b(\tau)$ are integers defined by the subexpression τ. We have

$$[b_w b_{w'} : b_{w''}] = |B^F w B^F \cap w'' U^F_{w'^{-1}} (w')^{-1}|,$$
$$\text{where } U_x = U \cap {}^{xw_0}U,\ x \in W.$$

The following result can be viewed as a kind of geometric interpretation of Kawanaka's theorem. A related result, in case $w' = w_0$, was obtained by Deodhar.

(4.2) THEOREM. *Let* w, w', w'' *be elements of* W. *Define*

$$U(w, w', w'') = \{u \in U_w : uwB \cap w'' U_{w'^{-1}}(w')^{-1} \neq \varnothing\}.$$

Then $U(w, w', w'')$ *can be expressed as a disjoint union of subsets* U_τ, *parametrized by the subexpressions of* w *defined by Kawanaka. Each subset* U_τ *is a locally closed subvariety of* U_w, *and there exists an isomorphism of varieties*

$U_\tau \cong G_a^{a(\tau)} \times G_m^{b(\tau)}$, where G_a and G_m denote the additive and multiplicative groups of an algebraically closed base field.

The preceding theorem can be proved for reductive groups over arbitrary fields. A rational version of it can be used to describe the multiplication in the Hecke algebra of the Gelfand-Graev representation of the finite group G^F.

5. Notes on the literature. Other surveys of related material are contained in [**1**, **10**, **13**, **21**, **36**, **45**].

§1. The basic ideas were first presented in [**32**, **42**].

§2. Theorems (2.1)–(2.3) appear in [**6**, **17**, **18**]. Special representations are defined and investigated in [**37**, **40**]. The classification of 2-transitive permutation representations of Chevalley groups was obtained in [**18**]. The structure of the Hecke algebras of induced cuspidal representations was described in [**29**, **40**]. Earlier results for principal series representations are contained in [**4**, **28**, **34**, **43**].

Applications of the preceding results to character values and the duality operation were given in [**12**, **14**, **30**, **43**].

For calculations of generic degrees, see [**2**, **5**, **6**, **7**, **8**, **9**, **17**, **46**, **47**].

Some representations of Hecke algebras were calculated in [**17**, **18**, **24**, **25**, **27**, **35**, **44**].

Endomorphism algebras of homology representations related to $H(G, B)$ are discussed in [**19**, **20**, **48**–**51**].

§3. Theorem (3.1) appears in [**6**]. Theorem (3.2) and related results are obtained in [**33**, **40**]. The isomorphism theorem (3.3) appears in [**38**] and is applied in various ways in [**39**, **40**]. W-graphs were defined in [**33**]. In [**26**], it was proved that they yield all irreducible represention of finite Coxeter groups. An extension of the isomorphism theorem to Hecke algebras of parabolic subgroups is obtained in [**15**].

§4. The results on zeta functions of the varieties X_w appear in [**3**], [**23**] (which contains references to earlier papers of Digne-Michel), and [**40**].

The formula for the structure constants of $H(G, B)$ appears in [**31**], its peometric interpretation in [**16**]. Results on the Hecke algebra of the Gelfand-Graev representation were obtained in [**11**, **25**, **45**, **53**].

References

1. D. L. Alvis, *Families of characters of Weyl groups*, this volume.

2. D. L. Alvis and G. Lusztig, *The representations and generic degrees of the Hecke algebra of type H_4*, J. Reine Angew. Math. **336** (1982), 201–212.

3. T. Asai, *On the zeta functions of the varieties $X(w)$ of the split classical groups and the unitary groups*, Osaka J. Math. **20** (1983), 21–32.

4. C. T. Benson, *On the generic algebra of a unipotent subgroup of a finite Chevalley group*, Comm. Algebra **5** (1977), 841–862.

5. ———, *The generic degrees of the irreducible characters of E_8*, Comm. Algebra **7** (1979), 1199–1209.

6. C. T. Benson and C. W. Curtis, *On the degrees and rationality of certain characters of finite Chevalley groups*, Trans. Amer. Math. Soc. **165** (1972), 251–173; Corrections and addition, ibid. **202** (1975), 405–406.

7. C. T. Benson and D. A. Gay, *On dimension functions of the generic algebra of type D_n*, J. Algebra **45** (1977), 435–438.

8. C. T. Benson and L. C. Grove, *The symmetric square and generic algebras of Chevalley groups*, Comm. Algebra **1** (1974), 503–516.

9. C. T. Benson, L. C. Grove, and D. B. Surowski, *Semilinear automorphisms and dimension functions for certain characters of finite Chevalley groups*, Math. Z. **144** (1975), 149–159.

10. R. W. Carter, *Finite groups of Lie type (conjugacy classes and irreducible characters)*, Wiley, New York, 1985.

11. B. Chang, *Decomposition of the Gelfand-Graev characters of* $\mathrm{GL}_3(q)$, Comm. Algebra **4** (1976), 375–401.

12. C. W. Curtis, *Reduction theorems for characters of finite groups of Lie type*, J. Math. Soc. Japan **27** (1975), 666–688.

13. ——, *Representations of finite groups of Lie type*, Bull. Amer. Math. Soc. (N.S) **1** (1979), 721–757.

14. ——, *Truncation and duality in the character ring of a finite group of Lie type*, J. Algebra **62** (1980), 320–332.

15. ——, *On Lusztig's isomorphism theorem for Hecke algebras*, J. Algebra **92** (1985), 348–365.

16. ——, *A further refinement of the Bruhat decomposition*, Proc. Amer. Math. Soc. (to appear).

17. C. W. Curtis, N. Iwahori, and R. Kilmoyer, *Hecke algebras and characters of parabolic type of finite groups with BN-pairs*, Inst. Hautes Études Sci. Publ. Math. **40** (1971), 81–116.

18. C. W. Curtis, W. M. Kantor, and G. M. Seitz, *The 2-transitive permutation representations of the finite Chevalley groups*, Trans. Amer. Math. Soc. **218** (1976), 1–59.

19. C. W. Curtis and G. I. Lehrer, *A new proof of the theorem of Solomon-Tits*, Proc. Amer. Math. Soc. **85** (1982), 154–156.

20. ——, *Generic chain complexes and finite Coxeter groups*, J. Reine Angew. Math. **363** (1985), 146–173.

21. C. W. Curtis and I. Reiner, *Methods of representation theory*, vol. II, Wiley, New York, 1987.

22. V. Deodhar, *On some geometric aspects of Bruhat orderings*. I: *A finer decomposition of Bruhat cells*, Invent. Math. **19** (1985), 499–511.

23. F. Digne and J. Michel, *Fonctions L des variéties de Deligne-Lusztig et descente de Shintani*, Bull. Math. Soc. France Mém. **20** (1985), 1–144.

24. R. Dipper, *On decomposition numbers of the finite general linear groups*, Trans. Amer. Math. Soc. **290** (1985), 315–344.

25. I. M. Gelfand and M. I. Graev, *Construction of irreducible representations of simple algebraic groups over finite fields*, Dokl. Akad. Nauk SSSR **147** (1962), 529–532.

26. A. Gyoja, *On the existence of a W-graph for an irreducible representation of a finite Coxeter group*, J. Algebra **86** (1984) 422–438.

27. P. N. Hoefsmit, *Representations of Hecke algebras of finite groups with BN-pairs of classical type*, Ph.D. Thesis, University of British Columbia, 1974.

28. R. B. Howlett and R. W. Kilmoyer, *Principal series representations of finite groups with split BN-pairs*, Comm. Algebra **8** (1980), 543–583.

29. R. B. Howlett and G. I. Lehrer, *Induced cuspidal representations and generalized Hecke rings*, Invent. Math. **58** (1980), 37–64.

30. ——, *Representations of generic algebras and finite groups of Lie type*, Trans. Amer. Math. Soc. **280** (1983), 752–777.

31. N. Kawanaka, *Unipotent elements and characters of finite Chevalley groups*, Osaka J. Math. **12** (1975), 523–554.

32. N. Iwahori, *On the structure of the Hecke ring of a Chevalley group over a finite field*, J. Fac. Sci. Univ. Tokyo **10** (1964), 215–236.

33. D. Kazhdan and G. Lusztig, *Representations of Coxeter groups and Hecke algebras*, Invent. Math. **53** (1979), 165–184.

34. R. Kilmoyer, *Principal series representations of finite Chevalley groups*, J. Algebra **51** (1978), 300–319.

35. R. Kilmoyer and L. Solomon, *On the theorem of Feit-Higman*, J. Combin. Theory Ser. (A)**15** (1973), 310–322.

36. G. Lusztig, *Representations of finite Chevalley groups*, CBMS Regional Conf. Ser. in Math., vol. 39, Amer. Math. Soc., Providence, R.I., 1977.

37. ——, *A class of irreducible representations of a Weyl group*, Nederl. Akad. Wetensch. Indag. Math. **41** (1979), 323–335; II, ibid. **44** (1982), 219–226.

38. ——, *On a theorem of Benson and Curtis*, J. Algebra **71** (1981), 490–498.

39. ——, *Unipotent characters of the symplectic and odd orthogonal groups over finite fields*, Invent. Math. **64** (1981), 263–296.

40. ——, *Characters of reductive groups over a finite field*, Ann. of Math. Stud., vol. 107, Princeton Univ. Press, Princeton, N.J., 1984.

41. ——, *Left cells in Weyl groups*, Lecture Notes in Math., vol. 1024, Springer-Verlag, 1983.

42. H. Matsumoto, *Générateurs et relations des groupes de Weyl généralisés*, C.R. Acad. Sci. Paris **258** (1964), 3419–3422.

43. K. McGovern, *Multiplicities of principal series representations of finite groups with split BN-pairs*, J. Algebra **77** (1982), 419–442.

44. D. Stanton, *Some q-Krawtchouk polynomials on Chevalley groups*, Amer. J. Math. **102** (1980), 625–662.

45. R. Steinberg, *Lectures on Chevalley groups*, Yale University, 1967.

46. D. B. Surowski, *Representations of a subalgebra of a generic algebra corresponding to a Weyl group of type F_4*, Comm. Algebra **5** (1977), 873–888.

47. ——, *Degrees of irreducible characters of BN-pairs of types E_6 and E_7*, Trans. Amer. Math. Soc. **243** (1978), 235–249.

48. ——, *The reflection character of the finite symplectic and odd dimensional orthogonal groups*, Comm. Algebra **7** (1979), 1747–1757.

49. ——, *Reflection compounds and certain r-blocks of finite Chevalley groups*, Illinois J. Math. **25** (1981), 398–401.

50. ——, *A reflection sieve for finite BN-pairs of types A_n, B_n and C_n*, J. Algebra **73** (1981), 538–551.

51. ——, *Reflection compound representations of groups of types A_n, B_n or C_n*, J. Algebra **78** (1982), 239–247.

52. T. Yokonuma, *Sur la structure des anneaux de Hecke d'un groupe de Chevalley fini*, C.R. Acad. Sci. Paris **264** (1967), 344–347.

53. ——, *Sur la commutant d'une représentation d'un groupe de Chevalley fini*, C.R. Acad. Sci. Paris **264** (1967), 433–439.

54. ——, *Sur la commutant d'une représentation d'un groupe de Chevalley fini*, J. Fac. Sci. Univ. Tokyo **15** (1968), 115–129.

UNIVERSITY OF OREGON

Proceedings of Symposia in Pure Mathematics
Volume 47 (1987)

Shintani Descent and $\mathcal{L}$ Functions on Deligne-Lusztig Varieties

FRANÇOIS DIGNE

1. Shintani descent. First we recall some notations and definitions. We consider the following situation: Let G be a finite group and $\langle F\rangle$ a cyclic group acting on G. We denote by j the injective map $g \mapsto gF$ from the *set* G into the *set* $G \rtimes \langle F\rangle$. We shall denote by $\operatorname{Irr}(G)$ the set of irreducible characters of G (on the complex number field).

DEFINITION 1.1. (a) We call F-conjugacy classes (or F-classes) of G the sets $j^{-1}(c)$ where c runs over the conjugacy classes of $G \rtimes \langle F\rangle$ whose inverse images by j are nonempty.

(b) We call F-class functions on G the functions of type $f \circ j$ where f is a class function on $G \rtimes \langle F\rangle$.

(c) If g is an element of G, we call F-centralizer of g the group $C_G(j(g))$, which will be denoted by $FC_G(g)$.

With these notations we have:

PROPOSITION 1.2. *The set $\{j \circ \tilde{\chi} | \chi \in \operatorname{Irr}(G)^F\}$, where $\tilde{\chi}$ is an arbitrary extension of χ to the semidirect product $G \rtimes \langle F\rangle$, is a basis of the space of F-class functions. This basis is orthonormal for the usual scalar product $\langle f_1, f_2\rangle = |G|^{-1}\sum_{g\in G} f_1(g)\overline{f_2(g)}$.*

We shall often identify $f \circ j$ and the restriction of f to the coset $G.F$ in $G \rtimes \langle F\rangle$, and, in the following, the notation $\tilde{\chi}$ will always have a meaning similar to the one stated in 1.2.

Let us give now the general setting for defining Shintani descent.

DEFINITION 1.3. Let $\mathbf{G}$ be a group and F be an automorphism of (the abstract group) $\mathbf{G}$. We say that these data satisfy Lang's property if $\mathbf{G}^F$ is finite and if any element $g \in \mathbf{G}$ can be written $x^{-1}.{}^F x$ for some $x \in \mathbf{G}$.

One example of this situation is the case of a connected algebraic group defined over the algebraic closure $\mathbf{F}$ of a finite field $\mathbf{F}_q$, if we take for F a Frobenius endomorphism defining on the group a rational structure over $\mathbf{F}_q$. In this case

1980 *Mathematics Subject Classification* (1985 *Revision*). Primary 20G40.

$\mathbf{G}$ is the group of points over $\mathbf{F}$ of the algebraic group, and Lang's property comes from the well-known Lang's theorem.

We then define Shintani descent:

THEOREM 1.4. *Let $\mathbf{G}$ be a group and F and F' be two commuting automorphisms of $\mathbf{G}$ such that Lang's property is satisfied for both F and F'. Then*

(i) *For any $y \in \mathbf{G}$, the element $y^{-1}\,.\,{}^F y$ is in $\mathbf{G}^{F'}$ if and only if the element $y\,.\,{}^{F'}y^{-1}$ is in $\mathbf{G}^F$.*

(ii) *The correspondence:*

$$\begin{array}{ccc} y^{-1}\,.\,{}^F y & \longleftrightarrow & y\,.\,{}^{F'}y^{-1} \\ \mathbf{G}^{F'} & \longleftrightarrow & \mathbf{G}^F \end{array},$$

defines a bijection from the set of F-classes of $\mathbf{G}^{F'}$ to the set of F'-classes of $\mathbf{G}^F$.

This theorem follows from an easy calculation.

DEFINITION 1.5. We shall denote by $N_{F'/F}$ the map defined in Theorem 1.4 and we shall call Shintani descent from $\mathbf{G}^{F'}$ to $\mathbf{G}^F$ the transpose of $N_{F/F'}$. It will be denoted by $\mathrm{Sh}_{F'/F}$.

We shall sometimes use the notation $N_{F'/F}(x)$ to denote an arbitrary element of the F'-class image of the F-class of x.

We have:

PROPOSITION 1.6. *Shintani descent is an isometry for the usual scalar product (cf. 1.2).*

SKETCH OF THE PROOF. An element x of $\mathbf{G}^{F'}$ F-centralizes the element $y^{-1}\,.\,{}^F y \in \mathbf{G}^{F'}$ if and only if yxy^{-1} is in $\mathbf{G}^F$. As yxy^{-1} clearly F'-centralizes $y\,.\,{}^{F'}y^{-1}$, we have defined a bijection between $FC_{\mathbf{G}^{F'}}(x)$ and $F'C_{\mathbf{G}^F}(N_{F'/F}(x))$, so these two groups have the same number of elements, whence the result. □

REMARK 1.7. The definition of Shintani descent that we gave above differs from the original one in **[Sh]** and in **[K1]** which was in fact $i \circ \mathrm{Sh}_{F'/F}$, where i is the transpose of the inverse map on the group. We shall see later what are the advantages of each of these definitions (moreover in **[Sh]** and **[K1]** the only case considered is the case when F' is a power of F).

We consider now the case when $\mathbf{G}$ is the group $\mathrm{GL}_n(\mathbf{F})$, where $\mathbf{F}$ is the algebraic closure of the finite field $\mathbf{F}_q$. On this group we can define two Frobenius endomorphisms such that the group of fixed points is either $\mathrm{GL}_n(\mathbf{F}_q)$ or $\mathrm{U}_n(\mathbf{F}_{q^2})$. We shall denote either of these Frobenius endomorphisms by F. With these notations we can now state Shintani's **[Sh]** and Kawanaka's **[K1]** results.

THEOREM 1.8 (SHINTANI). *Assume that F is such that $\mathbf{G}^F$ is equal to $\mathrm{GL}_n(\mathbf{F}_q)$. Then for any irreducible character χ of $\mathbf{G}^{F^m}$ invariant by F, there exists an extension $\tilde{\chi}$ of χ to the semidirect product $\mathbf{G}^{F^m} \rtimes \langle F\rangle$ such that $\tilde{\chi} \circ N^{-1}_{F^m/F}$ is an irreducible character of $\mathbf{G}^F$.*

THEOREM 1.9 (KAWANAKA). *Assume that m is prime to the characteristic and that F is such that $\mathbf{G}^F$ is equal to $\mathrm{U}_n(\mathbf{F}_{q^2})$. Then for any irreducible character χ of $\mathbf{G}^{F^m}$ invariant by F, there exists an extension $\tilde{\chi}$ of χ to the semidirect product $\mathbf{G}^{F^m} \rtimes \langle F\rangle$ such that $\tilde{\chi}\circ N^{-1}_{F^m/F}$ is an irreducible character of $\mathbf{G}^F$ up to sign.*

REMARK 1.10. Let us return to the general case and assume that $F = F'$. Then for $x \in \mathbf{G}$, the elements $x^{-1}\,.\,{}^F x$ and ${}^F x\,.\,x^{-1}$ are conjugate in $\mathbf{G}$, but not necessarily in $\mathbf{G}^F$. So the map $i\circ \mathrm{Sh}_{F/F}$ is not in general the identity. It is the identity in the GL_n and U_n cases because in these groups geometric conjugacy implies rational conjugacy. This will not be the case in other groups, and the study of $\mathrm{Sh}_{F/F}$ gives a key for studying $\mathrm{Sh}_{F^m/F}$ for any m.

For instance, we have the following result:

THEOREM 1.11 [**DM2**, **DM6** AND **A1**, **A2**]. *If $\rho_0 \in \mathrm{Irr}(\mathbf{G}^{F^m})$ is a unipotent (cf. below, §2), F-invariant character in the principal series, and if $\tilde{\rho}_0$ is an extension of ρ_0 to $\mathbf{G}^{F^m} \rtimes \langle F\rangle$, then there exist roots of unity λ_ρ such that*

$$\mathrm{Sh}_{F^m/F}\tilde{\rho}_0 = \sum_{\rho\in\mathrm{Irr}(\mathbf{G}^F)} \lambda_\rho^m c_{\rho,\tilde{\rho}_0}\rho,$$

for any m such that F^m is split, where the coefficients $c_{\rho,\tilde{\rho}_0}$ are independent of m.

Moreover, with the notations of the next section, these coefficients are given by

$$c_{\rho,\tilde{\rho}_0} = |W|^{-1}\sum_{w\in W}\langle R^{\mathbf{G}^F}_{\mathbf{T}^{wF}},\rho\rangle\tilde{\chi}((wF)^{-1}),$$

where the F-invariant character χ of W corresponds to the character ρ_0 of the principal series, and its extension $\tilde{\chi}$ to $W\rtimes\langle F\rangle$ corresponds to $\tilde{\rho}_0$.

This result shows in particular that if F is split, the knowledge of $\mathrm{Sh}_{F/F}$ gives results about $\mathrm{Sh}_{F^m/F}$ for any m. There is a precise study of $N_{F/F}$ in Chapter 4 of [**DM6**], where $N_{F/F}$ is computed for any semisimple group of type A, B, C, D, and G_2, split or nonsplit.

Asai has studied $\mathrm{Sh}_{F/F}$ in many cases (cf. [**A3**]–[**A7**]), and we now state his main results:

THEOREM 1.12. *For all reductive algebraic groups with connected center, of type B_n, C_n, D_n, or ${}^2D_n$, the map $\mathrm{Sh}_{F/F}$ is explicitly known. More precisely,*

(i) *$i\circ\mathrm{Sh}_{F/F}$ commutes with Lusztig's parametrization of characters.*

(ii) *If the group is split, for any family of unipotent characters of $\mathbf{G}^F$ parametrized by a set $\mathcal{M}(\Gamma)$, associated to the finite group Γ (with Lusztig's notations, cf.* [**L**]*), we have:*

$$\mathrm{Sh}_{F/F}(\lambda_x\rho_x) = \sum_{y\in\mathcal{M}(\Gamma)}\{x,y\}\lambda_y\rho_y,$$

where ρ_y is the character of the family associated to the element $y\in\mathcal{M}(\Gamma)$, where $\{x,y\}$ is the Fourier coefficient defined in [**L**]*, and where λ_y is a root of unity associated to y ($\lambda_y = \chi(\gamma^{-1})/\chi(1)$ if $y = (\gamma,\chi)$).*

For type ${}^2D_n$, Asai got an analogous formula (cf. [**A6**]). For split exceptional types, Asai (cf. [**A5**]) got part (ii) of this result (with the exception of the so-called exceptional families of unipotent characters for type E_7 and E_8). In [**A3–A5**] and [**A7**], Asai also got similar formulas for $\mathrm{Sh}_{F_0/F}$, where F_0 is a split Frobenius endomorphism on $\mathbf{G}$ and where F and F_0 define rational structures over the same finite field.

As for methods, Asai uses Lusztig's results, $\mathcal{L}$-function methods (see §2), and Kawanaka's results, which we now state:

THEOREM 1.13 [**K2**]. *Let* $\mathbf{G}$ *be a symplectic or special orthogonal group; there exists an integer* M *depending only on the type of* $\mathbf{G}$ *such that if* m *is prime to* M *and to the characteristic and if* r *is an integer inverse to* m *modulo* M, *then for any* F*-invariant irreducible character of* $\mathbf{G}^{F^m}$, *the class function* $(i \circ \mathrm{Sh}_{F/F})^{-r}(i \circ \mathrm{Sh}_{F^m/F})(\tilde{\chi})$ *is an irreducible character of* $\mathbf{G}^{F^m}$ *up to multiplication by* $\pm$ *an* m*th root of unity.*

The map $(i \circ \mathrm{Sh}_{F/F})^{-r}(i \circ \mathrm{Sh}_{F^m/F})$ was later generalized by Kawanaka [**K3**] to nonconnected groups, and an analogous result to 1.13 is proved in that paper (see also [**D**] for a simpler definition of Kawanaka's map). McGovern [**MG**] proved the same result for groups of type G_2.

Shoji has proved the following result about Shintani descent:

THEOREM 1.14 [**Sho**]. *For some classical groups with connected center, there exists an integer* M *such that for any* m *multiple of* M, *and any* $\chi \in \mathrm{Irr}(G^{F^m})^F$ *there exists a root of unity* $\lambda_{\tilde{\chi}}$ *such that*

$$i \circ \mathrm{Sh}_{F^m/F}(\lambda_{\tilde{\chi}}\tilde{\chi}) = R_x,$$

where χ *is parametrized by* $x \in \mathcal{M}(\Gamma)$, *with Lusztig's notations, and where* R_x *is the "Fourier transform of* χ*" as defined in* [**L**, 4.24.1].

(The phrase "some classical groups" in the preceding statement depends on the existence of an element s as in the hypothesis of Lemma 2.17 of [**Sho**] being in the fixed maximal torus and being fixed by F.)

Let us now make some remarks about the various definitions of descent (or lifting). In view of Kawanaka's and Shoji's results, it seems that the good definition of Shintani descent should be $i \circ \mathrm{Sh}_{F'/F}$. But there are some advantages in considering $\mathrm{Sh}_{F'/F}$. The first one is the complete symmetry between the two Frobenius endomorphisms. The second one is that $\mathrm{Sh}_{F'/F}$ and $\mathrm{Sh}_{F/F'}$ are inverse maps. Besides, we have the following easy-to-prove result:

LEMMA 1.15. *Let* F_1, F_2, *and* F_3 *be three pairwise commuting automorphisms of a group* $\mathbf{G}$, *each one satisfying Lang's property, and such that* $F_3 = F_1 \circ F_2$. *Then we have:*

$$\mathrm{Sh}_{F_3/F_1} = \mathrm{Sh}_{F_2/F_1} \circ \mathrm{Sh}_{F_3/F_2} \circ i.$$

This shows that "i always appears somewhere." Let us remark that the lemma makes sense (i.e., under the assumptions of the lemma, F_3-classes in $\mathbf{G}^{F_2}$ are also F_1-classes). In the same spirit, one can try to compose the map Sh_{F^m/F^n} with Sh_{F^n/F^r}, where r is the remainder of m modulo n, and then to do it again, following the Euclid algorithm. Let us call $G_{m,n}$ the resulting map, from the set of F^n-class functions on $\mathbf{G}^{F^m}$ to the set of class functions on $\mathbf{G}^{F^{\gcd(m,n)}}$, composed with $(i)^k$, where k is the number of steps in the Euclid algorithm. In this situation, Gyoja has proved:

THEOREM 1.16 [G]. *Let* $\mathbf{G}$ *be a connected reductive algebraic group, and* F *a Frobenius endomorphism defining on* $\mathbf{G}$ *a rational structure over* $\mathbf{F}_q$*; assume that* p *and* q *are large enough, and that* m *is prime to* p*; then, for any Deligne-Lusztig character of the form* $\chi_m = R_{\mathbf{T}^{F^m}}^{\mathbf{G}^{F^m}}(\theta \circ \mathrm{Norm}_{\mathbf{F}_{q^m}/\mathbf{F}_q})$*, where* θ *is a character of* $\mathbf{T}^F$ *(here* $\mathbf{T}$ *is an arbitrary rational maximal torus), there exists an extension* $\tilde{\chi}_m$ *of* χ_m *such that* $\mathbf{G}_{m,n}(\tilde{\chi}_m)$ *is equal to* $\chi_{(\gcd(m,n))}$ *for any* n.

In particular this theorem implies:

COROLLARY 1.17. *With the preceding notations, we have* $i \circ \mathrm{Sh}_{F^m/F}(\tilde{\chi}_m) = R_{\mathbf{T}^F}^{\mathbf{G}^F}(\theta)$.

2. Deligne-Lusztig varieties. In this section we shall assume to have chosen an embedding of the field $\overline{\mathbf{Q}}_l$ into the complex numbers.

Let $\mathbf{G}$ be a connected reductive algebraic group over $\mathbf{F}$, with a rational structure over $\mathbf{F}_q$ defined by a Frobenius endomorphism F as above. Let $(\mathbf{T}, \mathbf{B})$ be a pair consisting of a rational maximal torus embedded in a rational Borel subgroup of $\mathbf{G}$. We denote by $\mathbf{U}$ the unipotent radical of $\mathbf{B}$. Then, following Deligne and Lusztig [DL], we define, for any $\dot{w} \in N_{\mathbf{G}}(\mathbf{T})$ representative of an element $w \in W$ the variety

$$Y_{\dot{w}} = \{x \in \mathbf{G} | x^{-1}.{}^F x \in \mathbf{U}\dot{w}\mathbf{U}\}.$$

On this variety the group $\mathbf{G}^F \times \mathbf{T}^{wF}$ acts, if we define the action of (g,t) by $x \mapsto gxt^{-1}$. So the alternating sum of the spaces of l-adic cohomology with compact support defines a virtual $\mathbf{G}^F \times \mathbf{T}^{wF}$-module. We shall denote this alternating sum by $H_c^*(Y_{\dot{w}}, \overline{\mathbf{Q}}_l)$. This module can be used to define Deligne-Lusztig induction from $\mathbf{T}^{wF}$-modules to $\mathbf{G}^F$-modules, namely, we can associate to any irreducible character θ of $\mathbf{T}^{wF}$ the $\mathbf{G}^F$-virtual module $R_{\mathbf{T}^{wF}}^{\mathbf{G}^F}(\theta) = H_c^*(Y_{\dot{w}}, \overline{\mathbf{Q}}_l)_\theta$, where this notation means the $\mathbf{T}^{wF}$ isotypic part of type θ of the cohomology. The (virtual) character of this module is called a Deligne-Lusztig character. Deligne and Lusztig have shown that any irreducible character of $\mathbf{G}^F$ is a component of some Deligne-Lusztig character. In particular:

DEFINITION 2.1. The irreducible characters of $\mathbf{G}^F$ appearing in Deligne-Lusztig characters with $\theta = 1$ are called unipotent characters.

Let us now fix some $w \in W$ and some irreducible character θ of $\mathbf{T}^{wF}$. Let k be an integer such that F^k fixes w (and then acts on $\mathbf{T}^{wF}$) and fixes θ. For any $g \in \mathbf{G}^F$, we put:

$$N_{Y_{\dot{w}}}^m(g, \theta) = \mathrm{Trace}(gF^{km} | H_c^*(Y_{\dot{w}})_\theta).$$

Then we have:

PROPOSITION 2.2. *The sum of the formal series* $\sum_{m\geq 1} N^m_{Y_{\dot w}}(g,\theta)X^m$ *is a rational fraction whose value at infinity is the opposite of* $\mathrm{Trace}(g|R^{\mathbf{G}^F}_{\mathbf{T}^{wF}}(\theta))$.

Note that the above proposition remains true if we replace F^k with any Frobenius endomorphism commuting with F, normalizing $\mathbf{T}$, $\mathbf{B}$, and fixing g and θ.

This proposition, which appears in the Deligne-Lusztig paper [**DL**], shows the interest of studying such formal series, which are the so-called $\mathcal{L}$-functions of the varieties $Y_{\dot w}$. These $\mathcal{L}$-functions are related to Shintani descent, as can be seen in the following result:

THEOREM 2.3. *Let F and F' be two commuting Frobenius endomorphisms on $\mathbf{G}$. Assume that both F and F' fix a pair $(\mathbf{T},\mathbf{B})$ as above. Consider an element $w\in W^{F'}$ and let $\dot w$ be a representative of w in $N_{\mathbf{G}}(\mathbf{T})^{F'}$. If θ is an F'-fixed linear character of $\mathbf{T}^{wF}$, then:*

$$\mathrm{Trace}(gF'|H^*_c(Y_{\dot w})_\theta)=\mathrm{Sh}_{F'/F}(\mathrm{Trace}(g\tau_{\dot w}F|\mathrm{Ind}^{\mathbf{G}^{F'}}_{\mathbf{B}^{F'}}(\Theta))),$$

where Θ is the trivial extension to $\mathbf{B}^{F'}$ of $\mathrm{Sh}_{wF/F'}(\theta)$, and where $\tau_{\dot w}$ is an operator commuting with $\mathbf{G}^{F'}$ on the induced representation (in fact the endomorphism algebra $H=\mathrm{End}_{\mathbf{G}^{F'}}(\mathrm{Ind}^{\mathbf{G}^{F'}}_{\mathbf{B}^{F'}}(\Theta))$ is a "generalized Hecke algebra" constructed on the group $W(\Theta)=\{y\in W|{}^y\Theta=\Theta\}$, and w can be written $w=yw_1$, where w_1F defines "diagram" automorphisms of $W(\Theta)$ and of H, and where y is in $W(\Theta)$. The operator $\tau_{\dot w}$ decomposes as $T_y\Psi$, where T_y is the generator of H corresponding to y, and where $\Psi\circ F$ induces on H the automorphism defined by w_1F).

This formula is proved when θ is the trivial character in [**A1**], [**A2**], and [**DM1**], [**DM6**]. It has been generalized to a general θ by Lusztig in [**L**, 2.10], when F' is split and the center of $\mathbf{G}$ is connected, and can be generalized to any F' and to the nonconnected-center case [**DM7**]. We now give, in the easier case of a trivial θ, the following consequence of 2.3:

COROLLARY 2.4. *With the notations of* 2.3, *if θ is the trivial character, we have:*

$$\mathrm{Trace}(gF'|H^*_c(Y_{\dot w})_1)=\sum_{\chi\in\mathrm{Irr}(H)^F}\tilde\chi(T_wF)(\mathrm{Sh}_{F'/F}\rho_{\tilde\chi})(gF),$$

where ρ_χ is the character of $\mathbf{G}^{F'}$ corresponding to the character χ of the endomorphism algebra, and where $\rho_{\tilde\chi}$ is an extension of ρ_χ defined by the arbitrarily chosen extension $\tilde\chi$ of χ to the semidirect product $H\rtimes\langle F\rangle$.

REMARK. In the above corollary, if the extension $\tilde\chi$ is chosen such that F acts with finite order in the corresponding representation, then the extension $\rho_{\tilde\chi}$ has the same property.

Formula 2.4 is one of the main tools in Asai's and Digne-Michel's papers, and its generalization to the case of any F'-invariant character θ of $\mathbf{T}^{wF}$ is proved and used in Chapter 3 of Lusztig's book [**L**, 3.6.3], in the connected-center case, assuming F' to be split. When $\mathbf{T}$ is replaced with a standard Levi subgroup (Lusztig variety), 2.4 has been generalized in [**Sho**] (connected-center case) and in [**DM7**] (if the center of $\mathbf{G}$ is not necessarily connected).

3. Concluding remarks. In order to use these methods and results, one needs:

(a) A good knowledge of the endomorphism algebras that appear (and which, in the nonconnected-center case, are more general than Hecke algebras).

(b) Results on eigenvalues of Frobenius endomorphisms on the cohomology (many such results appear in Lusztig's book; see, e.g., [**L**, Chapter 2]).

Using (a), (b) and the general properties of Shintani descent one can

(1) Prove general results on Shintani descent,

(2) Find the decomposition of Deligne-Lusztig characters (cf. [**L**]) and of Lusztig's generalized induction (cf. [**Sho**]). Many problems are still open in these directions, such as:

• Compute $\mathrm{Sh}_{F'/F}$ in general (F' not necessarily split).

• Get general results on the eigenvalues of a nonsplit Frobenius endomorphism on the cohomology of Deligne-Lusztig's or Lusztig's varieties.

REFERENCES

[**A1**] T. Asai, *On the zeta function of the varieties $X(W)$ of the split classical groups and the unitary groups*, Osaka J. Math. **20** (1983), 21–32.

[**A2**] ——, *Lifting and the zeta functions of the Deligne-Lusztig varieties*, Comm. Algebra **11** (1983), 2187–2201.

[**A3**] ——, *Unipotent class functions of split special orthogonal groups* SO_{2n}^+ *over finite fields*, Comm. Algebra **12** (1984), 517–615.

[**A4**] ——, *The unipotent class functions of the symplectic and odd orthogonal groups over finite fields*, Comm. Algebra **12** (1984), 617–645.

[**A5**] ——, *Unipotent class functions of exceptional groups over finite fields*, Comm. Algebra **12** (1984), 2729–2857.

[**A6**] ——, *The unipotent class functions of non-split special orthogonal groups*, Comm. Algebra **13** (1985), 845–924.

[**A7**] ——, *On the twisting operators on the finite classical groups*, Algebraic and topological Theories—To the memory of Takehiko Itiyaka, Kinokuniya, Tokyo, 1985, pp. 239–282.

[**D**] F. Digne, *Descente de Shintani et blocs selon N. Kawanaka. Séminaire sur les groupes finis.* III, Publ. Math. Paris VII (to appear).

[**DL**] P. Deligne and G. Lusztig, *Characters of reductive groups over finite fields*, Ann. of Math. (2) **103** (1976), 103–161.

[**DM1**] F. Digne and J. Michel, *Nombre de points rationnels des variétés de Deligne-Lusztig et caractères de l'algebre de Hecke*, C. R. Acad. Sci. Paris Sér. I Math. **287** (1978), 811–814.

[**DM2**] ——, *Descente de Shintani des caractères d'un groupe de Chevalley fini*, C. R. Acad. Sci. Paris Sér. I Math. **291** (1980), 571–574.

[**DM3**] ——, *Descente de Shintani des caractères de Deligne-Lusztig*, C. R. Acad. Sci. Paris Sér. I Math. **291** (1980), 651–653.

[**DM4**] ——, *Décomposition des descentes de Shintani. Séminaire sur les groupes finis.* II, Publ. Math. Paris VII **17** (1985), 124–159.

[**DM5**] ——, *Foncteur de Lusztig et fonctions de Green généralisées*, C. R. Acad. Sci. Paris Sér. I Math. **297** (1983), 89–92.

[**DM6**] ——, *Fonctions $\mathcal{L}$ des variétés de Deligne-Lusztig; descente de Shintani*, Mém. Soc. Math. France **20** (1985).

[**DM7**] ——, *Decomposition of Lusztig's induction and Shintani descent for groups with non connected center*, unpublished.

[**G**] A. Gyoja, *Liftings of irreducible characters of finite reductive groups*, Osaka J. Math. **16** (1979), 1–30.

[**HL**] R. Howlett and G. Lehrer, *Induced cuspidal representations and generalized Hecke rings*, Invent. Math. **58** (1980), 37–64.

[**K1**] N. Kawanaka, *On the irreducible characters of the finite unitary groups*, J. Math. Soc. Japan **29** (1977), 425–450.

[**K2**] ——, *Lifting of irreducible characters of finite classical groups*. I, J. Fac. Sci. Univ. Tokyo **28** (1982), 851–861.

[**K3**] ——, *Lifting of irreducible characters of finite classical groups*. II, J. Fac. Sci. Univ. Tokyo **30** (1984), 499–516.

[**L**] G. Lusztig, *Charaacters of reductive groups over a finite field*, Ann. of Math. Stud., vol. 107, Princeton Univ. Press, Princeton, N.J., 1984.

[**MG**] K. McGovern, *On the lifting theory of finite groups of Lie type*, Pacific J. Math. **113** (1984), 383–394.

[**Sh**] T. Shintani, *Two remarks on irreducible characters of finite general linear groups*, J. Math. Soc. Japan **28** (1976), 396–414.

[**Sho**] T. Shoji, *Some generalisation of Asai's result for classical groups*, Adv. Stud. Pure Math., vol. 6, Algebraic Groups and Related Topics, Kinokuniya and North-Holland, 1985, pp. 207–209.

UNIVERSITÉ PARIS VI, FRANCE

Proceedings of Symposia in Pure Mathematics
Volume **47** (1987)

Good Filtrations of Rational Modules for Reductive Groups

S. DONKIN

This article is a survey of what is known of a certain kind of filtration for rational modules for connected algebraic groups and uses to which this filtration may put. The main interest is in modules for semisimple, simply connected groups but it is often convenient, particularly in inductive situations, to consider arbitrary connected affine algebraic groups. Our groups will be over an algebraically closed ground field, for the sake of simplicity, but many of the results work over a principal ideal domain, e.g., for Chevalley group schemes over $\mathbf{Z}$ and the general linear group scheme over $\mathbf{Z}$.

To take our discussion further we require some notation. Throughout, k will denote an algebraically closed field of arbitrary characteristic. Vector spaces are always over k and the dimension $\dim V$ of a vector space V is always the dimension over k. Let G be an (affine) algebraic group over k. A finite-dimensional (left) kG-module is rational if the representation $G \to \mathrm{GL}(V)$ afforded by V is a morphism of algebraic groups. A kG-module of arbitrary dimension is rational if it is the union of finite-dimensional rational kG-modules. By a G-module we mean a rational kG-module; we denote by $\mathrm{Mod}(G)$ the category of G-modules and module homomorphisms. The most obvious examples of G-modules are the trivial one-dimensional module, denoted k, and the coordinate ring $k[G]$, which we view as a G-module via $(xf)(g) = f(gx)$ and even as a $G \times G$-module via $((x,y)f) = f(y^{-1}gx)$, $x, y, g \in G$, $f \in k[G]$. For a vector space V we write V^* for the dual space. If V is a finite-dimensional G-module then V^* has a natural G-module structure. For any G-module V we define the coefficient space $cf(V) \leq k[G]$ as follows. Let $\{v_i : i \in I\}$ be a basis of V. The coefficient functions $f_{ij}: G \to k$ are determined by the equations

$$xv_i = \sum_{j \in I} f_{ji}(x) v_j \qquad (i \in I).$$

Each coefficient function belongs to $k[G]$ and we define $\mathrm{cf}(V)$ to be the k-span of the f_{ij} for $i, j \in I$ ($\mathrm{cf}(V)$ is independent of the choice of basis). The coefficient

1980 *Mathematics Subject Classification* (1985 *Revision*). Primary 20G05.

space is a $G \times G$-submodule of $k[G]$. We have a natural surjective linear map

$$c: V \otimes V^* \to \mathrm{cf}(V)$$

taking $v \otimes \alpha$ ($v \in V$, $\alpha \in V^*$) to $c_{v,\alpha}$, where $c_{v,\alpha}(g) = \alpha(gv)$ for $g \in G$. (See [**14**] for properties of cf in the context of coalgebras and comodules.) Regarding $V \otimes V^*$ as a G-module, with trivial action on the right factor, c becomes a G-module map. If V is finite-dimensional then $V \otimes V^*$ is naturally a $G \times G$-module and c is a $G \times G$-module map.

The category $\mathrm{Mod}(G)$ is abelian and has enough injectives ($k[G]$ is an injective G-module) and we therefore have the Hochschild cohomology groups $\mathrm{Ext}^n_G(U, V)$ and $H^n(G, V) = \mathrm{Ext}^n_G(k, V)$ for $U, V \in \mathrm{Mod}(G)$, $n \geq 0$.

Now suppose that G is connected and let B be a Borel subgroup. Then B factorizes: $B = T.U$, where T is a maximal torus and $U = R_u(B)$, the unipotent radical of B. Let $X = X(T)$ be the group of homomorphisms of T into the multiplicative group of the field. Then $X(T)$ is an abelian group isomorphic to $\mathbf{Z}^r$, where r is the dimension of T. For $\lambda \in X$ let k_λ be the one-dimensional B-module on which T acts with weight λ (and on which U acts trivially). Then the B-modules k_λ ($\lambda \in X$) are pairwise nonisomorphic and exhaust all simple B-modules (up to isomorphism). Let $Y(\lambda)$ be the induced module $\mathrm{Ind}_B^G k_\lambda$ ($\lambda \in X$). The modules $Y(\lambda)$ are finite-dimensional but are very often zero; $Y(\lambda)$ is nonzero for $\lambda \in X^+$, the set of dominant weights. The formal character of $Y(\lambda)$ ($\lambda \in X^+$) is given by Weyl's Character Formula (no matter what the characteristic of k is). The socle $L(\lambda)$ of $Y(\lambda)$ is simple ($\lambda \in X^+$) and the modules $L(\lambda)$ ($\lambda \in X^+$) are pairwise nonisomorphic and exhaust all possible simple G-modules (up to isomorphism). The main problem in the rational representation theory of algebraic groups, for k of prime characteristic, is the determination of the formal character of the $L(\lambda)$ ($\lambda \in X^+$). We have nothing to say about this but rather concentrate on aspects of the theory which are characteristic-free.

DEFINITION. A good filtration of a rational G-module V is an ascending chain of submodules $0 = V_0, V_1, \ldots$ of V such that $V = \bigcup_{i=0}^\infty V_i$ and, for each $i > 0$, V_i/V_{i-1} is either 0 or is isomorphic to $Y(\lambda_i)$ for some $\lambda_i \in X^+$.

The existence of a good filtration for a G-module V depends only on the isomorphism type of V: it is independent of the choice of B (see [**7**, p. 31]). Notice that if $V \in \mathrm{Mod}(G)$ admits a good filtration then V has countable dimension. (We do not really regard this as a disadvantage as it is difficult to imagine problems of genuine interest which hinge on properties of G-modules of uncountable dimension.) If k has characteristic 0 then $L(\lambda) = Y(\lambda)$ ($\lambda \in X^+$) and therefore every $V \in \mathrm{Mod}(G)$ of countable dimension has a good filtration (a composition series!). In characteristic $p > 0$, a given G-module V may not admit a good filtration. We give two examples, before passing on to the more positive aspects of the theory.

EXAMPLE. Let $G = \mathrm{SL}_2(k)$, T be the group of diagonal matrices $t(\varsigma) = \left(\begin{smallmatrix} \varsigma & 0 \\ 0 & \varsigma^{-1} \end{smallmatrix}\right)$, $0 \neq \varsigma \in k$, and the B be the lower triangular matrices. Then $X = \mathbf{Z}\rho$

and $X^+ = \{n\rho : n \in \mathbf{Z}_{\geq 0}\}$, where $\rho(t(\varsigma)) = \varsigma$. For $r \geq 0$, we have $Y(r\rho) \cong S^r(V)$, the rth symmetric power of the natural module V of column vectors. Let e_1, e_2 be the standard basis of V. The action of G on the symmetric algebra $S(V) = k[e_1, e_2]$ is by algebra automorphisms. Let $M = ke_1^p + ke_2^p$. Then M is a G-submodule of $S(V)$. However, M does not have a good filtration (since $\dim M = 2$ and $\dim S^r V = r + 1$ we would otherwise have to have $M \cong V$ or M filtered by trivial modules).

The reason for studying good filtrations is to try to capture properties of G-modules which are "characteristic-free." However care must be taken in the interpretation of "characteristic-free," as the next example shows.

EXAMPLE. Let G, etc., be as in the example above and let $p = 2$. We shall show that $S^2(S^3V)$ does not have a good filtration. Suppose that it does. It is easy to check that the formal character $\operatorname{ch} S^2(S^3V) = \operatorname{ch} S^6V + \operatorname{ch} S^2V$ so that the sections in a good filtration of $S^2(S^3V)$ must be S^2V and S^4V. Since $(S^2V)^G = (S^6V)^G = 0$, $S^2(S^3V)$ can have no nonzero invariant: we now write one down. Let $a = e_1^3$, $b = e_1^2e_2$, $c = e_1e_2^2$, $d = e_2^3$. It is easy to check that $ad + bc$ is G-invariant.

For $\lambda \in X^+$ we set $V(\lambda) = Y(\lambda^*)^*$, where $-\lambda^*$ is the lowest weight of $Y(\lambda)$. When G is semisimple, simply connected the $V(\lambda)$'s are known as Weyl modules and may be obtained from the irreducible module of highest weight λ, for the corresponding Lie algebra over $\mathbf{C}$, by reduction of an admissible $\mathbf{Z}$-form (see [**18**, Satz 1]). One is very often interested in good filtrations of finite-dimensional G-modules with G semisimple, simply connected and here the notion is dual to that of a Weyl filtration. The first time that Weyl filtrations appeared in print was in [**25**], where Upadhyaya showed their existence in certain projective modules for $\mathrm{SL}_2(q)$. This was soon generalized by Jantzen [**19**], who also established the basic properties of Weyl filtrations.

From now on let G be reductive. One has the following elegant acyclicity result of Cline, Parshall, Scott, and van der Kallen (see [**5**, (3.3), or **7**, (2.1.4)]).

THEOREM 1. *$H^n(G, Y(\lambda) \otimes Y(\mu)) = 0$ for all $\lambda, \mu \in X^+$ and $n > 0$.*

This is the essential ingredient of the following useful criterion for a G-module V to have a good filtration. For V finite-dimensional this is due to the author [**5**] and the general result was proved by Friedlander [**13**] using his idea of a canonical filtration.

THEOREM 2. *A G-module V of countable dimension has a good filtration if and only if $H^1(G, V \otimes Y(\lambda)) = 0$ for all $\lambda \in X^+$.*

We want to consider good filtrations not only for the reductive group G but also for a parabolic subgroup P of G, which can be assumed to contain B. Let $Y_P(\lambda) = \operatorname{Ind}_B^P(k_\lambda)$ for $\lambda \in X$ and let X_P be the set of $\lambda \in X$ such that $Y_P(\lambda) \neq 0$. A P-module V may have more than one good filtration (or more usually, none at all); however, the multiplicity with which a given $Y_P(\lambda)$ occurs as a section does not vary with the choice of filtration. More precisely, we have the following, [**7**, (12.1.1)].

THEOREM 3. *Let V be a P-module which admits a good filtration. Then the multiplicity $(V : Y_P(\lambda))$ with which $Y_P(\lambda)$ $(\lambda \in X_P)$ occurs as a section in any good filtration in any good filtration in equal to* $\dim(V \otimes Y_P(\lambda_P^*))^{L_P}$.

Here $-\lambda_P^*$ is the lowest weight of $Y_P(\lambda)$ and L_P denotes a Levi factor of P. The theorem is another application of Theorem 1 (first restricting to L_P).

We now give some applications of Theorem 2. By a principal indecomposable module we mean a rationally injective, indecomposable module.

THEOREM 4. (i) *Any rationally injective P-module of countable dimension has a good filtration.*

(ii) *Every principal indecomposable G-module I has a good filtration. For the rationally injective envelope $I(\lambda)$ of $L(\lambda)$ $(\lambda \in X^+)$ we have $(I(\lambda): Y(\mu)) = [Y(\mu): L(\lambda)]$ for every $\mu \in X^+$, where $[Y(\lambda): L(\mu)]$ is the composition multiplicity of $L(\mu)$ in $Y(\lambda)$.*

Part (ii) follows from Theorems 2 and 3. It was discovered independently by L. L. Scott (announced in [**24**]) and the author [**5**]. It is a strong analogue of the Brauer Reciprocity Theorem for finite groups. For (i) see [**7**, (12.1.3)].

For algebraic groups H_1, H_2, and $V_1 \in \mathrm{Mod}(H_1)$, $V_2 \in \mathrm{Mod}(H_2)$ we denote by $V_1 \boxtimes V_2 \in \mathrm{Mod}(H_1 \times H_2)$ the vector space $V_1 \otimes V_2$ with product action.

THEOREM 5. *Let P, Q be parabolic subgroups of G containing B with unipotent radicals U_P, U_Q. The ring of invariants $J = k[G]^{U_P \times U_Q}$, as an $L_P \times L_Q$-module, has a good filtration with sections $Y_P((\lambda^*)_P^*) \boxtimes Y_Q(\lambda_Q^*)$ for $\lambda \in X^+$, each occurring exactly once. In particular $k[G]$ has a good $G \times G$-module with sections $Y(\lambda) \otimes Y(\lambda^*)$, each occurring once.*

The existence of a good filtration for $k[G]$, as a $G \times G$-module comes from Theorem 2 via the following calculation. We have for $\lambda, \mu \in X^+$

$$\begin{aligned} H^1(G \times G, k[G] \otimes (Y(\lambda) \otimes Y(\mu)) &= H^1(G \times G, \mathrm{Ind}_{\Delta(G)}^{G\times G}(k) \otimes (Y(\lambda) \boxtimes Y(\mu)) \\ &\cong H^1(G \times G, \mathrm{Ind}_{\Delta(G)}^{G\times G}(Y(\lambda) \boxtimes Y(\mu)) \\ &\cong H^1(G, Y(\lambda) \boxtimes Y(\mu)) = 0 \end{aligned}$$

using the tensor identity and Shapiro's lemma (valid since $G\times G$ is affine, see [**4**, (3.1) Theorem; **2**, (2.1) Proposition (4.5) Corollary]). Here $\Delta(G)$ is the diagonal subgroup $\{(x, x): x \in G\}$ of $G\times G$. The multiplicity statement follows in the same way using Theorem 3. This part was discovered independently by the author [**6**, **10**] and Koppinen [**21**]. For the first part of the theorem one reduces to the semisimple, simply connected case, uses the last part to get an upper bound on the invariants, and uses the fact that G may be obtained from a Chevalley group scheme over $\mathbf{Z}$ by base change to get "enough" invariants (see [**14**] for details).

Theorem 5 has an interesting connection with invariant theory, which we now explain. For a P-module V with a good filtration we define $X_P[V]$ to be the set of $\lambda \in X_P$ such that $(V: Y_P(\lambda)) \neq 0$.

THEOREM 6. *Let A be a k-algebra (not necessarily commutative) on which P acts rationally as k-algebra automorphisms. Suppose A has a good filtration.*

(i) *$X_P[A]$ is a submonoid of X_P.*

(ii) *If $(A\colon Y_P(\lambda)) \leq 1$ for all $\lambda \in X_P$ and $X_P[A]$ is a finitely generated submonoid of X_P then A is a finitely generated algebra.*

For the proof see [**11**]. Combining Theorem 5 and Theorem 6 we get that $k[G]^{U_P}$ is finitely generated, giving a new proof of a theorem of Grosshans [**16**].

We regard the following as the deepest result on good filtrations.

THEOREM 7. *Let H be a connected, affine algebraic group over k and suppose either that the characteristic of k is not 2 or that H does not involve E_7 or E_8 (i.e., the root system of $H/R_u(H)$ has no component of type E_7 or E_8).*

(i) *If $V, V' \in \mathrm{Mod}(H)$ admit good filtrations then $V \otimes V'$ admits a good filtration;*

(ii) *If $V \in \mathrm{Mod}(H)$ admits a good filtration and Q is any parabolic subgroup of H then $V|_Q$ admits a good filtration.*

For H semisimple, simply connected and the characteristic of k large compared to the rank of H, (i) was proved first by Wang Jian-pan [**26**]. The proof of Theorem 7 quickly degenerates into a messy case-by-case analysis which occupies most of [**7**]. The characteristic/root-system restrictions are definitely unnecessary (but as yet no proof exists). A proof of Theorem 7 in the remaining cases, using only the techniques of [**7**], is probably possible but would, I estimate, be at least twice as long as the whole of [**7**].

We now move on to what may be described as local properties of good filtrations. Let π be a subset of X_P. Say a P-module V belongs to π if every weight of V in X_P belongs to π. Among all submodules belonging to π of an arbitrary P-module V, there is a unique maximal one, which we denote by $O_\pi(V)$. We say that π is saturated if every element of X_P which is less than some element of π (in the natural partial order on X) belongs to π. For a suitable choice of saturated π, O_π is the functor which takes $V \in \mathrm{Mod}(G)$ to the largest submodule which is "p^n-bounded" in the sense of [**19**, 4.2]. So, though not expressed in our formulation, Jantzen's theory of p^n-bounded modules predates the present theory. (For $\pi = \{0\}$, O_π is the fixed point functor, which has an even longer history!) We have the following (see [**7**, (12.1.3), (12.1.6)]).

THEOREM 8. *Let π be a saturated subset of X_P and V a P-module which admits a good filtration. Then $O_\pi(V)$ and $V/O_\pi(V)$ admit good filtrations.*

$$(O_\pi(V)\colon Y_P(\lambda)) = \begin{cases} (V\colon Y_P(\lambda)) & \text{for } \lambda \in \pi, \\ 0 & \text{for } \lambda \in X_P \backslash \pi, \end{cases}$$

and

$$(V/O_\pi(V)\colon Y_P(\lambda)) = \begin{cases} 0 & \text{for } \lambda \in \pi, \\ (V\colon Y_P(\lambda)) & \text{for } \lambda \in X_P \backslash \pi. \end{cases}$$

Now suppose $\pi \subset X^+$ is finite and saturated. We set $A(\pi) = O_\pi(k[G])$. Then $A(\pi)$ is a finite-dimensional subcoalgebra of $k[G]$. We call the dual algebra $S(\pi) = A(\pi)^*$ a generalized Schur algebra (for reasons explained below). We refer the reader to **[8]** (see also **[3]** for algebras Morita equivalent to the $S(\pi)$) for properties of $S(\pi)$ and record here only the following interesting features.

THEOREM 9. (i) *For $S(\pi)$-modules L, M we have*

$$\mathrm{Ext}^*_{S(\pi)}(L,M) \cong \mathrm{Ext}^*_G(L,M).$$

(ii) *$S(\pi)$ has finite global dimension.*

Now consider $G = \mathrm{GL}(E)$, where E is a k-vector space with basis $e_1, e_2, \dots, e_n$. For $t_1, t_2, \dots, t_n \in k^*$ (the multiplicative group of the field) $\mathrm{diag}(t_1, t_2, \dots, t_n) \in G$ is the element given by $\mathrm{diag}(t_1, t_2, \dots, t_n)e_i = t_i e_i$, $1 \le i \le n$. Then $T = \{\mathrm{diag}(t_1, t_2, \dots, t_n) : t_i \in k^*\}$ is a maximal torus and we identify $X = X(T)$ with the set of n-tuples of integers $\lambda = (\lambda_1, \lambda_2, \dots, \lambda_n)$ by $\lambda(\mathrm{diag}(t_1, t_2, \dots, t_n)) = t_1^{\lambda_1} t_2^{\lambda_2} \cdots t_n^{\lambda_n}$. We take B to be the set of $g \in G$ which are represented by a lower triangular matrix with respect to $e_1, e_2, \dots, e_n$. Then X^+ is the set of $\lambda = (\lambda_1, \lambda_2, \dots, \lambda_n)$ such that $\lambda_1 \ge \lambda_2 \ge \cdots \ge \lambda_n$. Let c_{ij} $(1 \le i, j \le n)$ be the coordinate function of G, i.e., the coefficient functions with respect to $e_1, e_2, \dots, e_n$. Let $A(n) \le k[G]$ be the (free polynomial) algebra generated by the c_{ij} $(1 \le i, j \le n)$. Then $A(n)$ has a grading $A(n) = \bigoplus_{r=0}^\infty A(n,r)$, where $A(n,r)$ is the space of homogeneous polynomials of total degree r. The space $A(n,r)$ is a finite-dimensional subcoalgebra of $k[G]$. The representation theory of the dual algebra $S(n,r) = A(n,r)^*$ (the Schur algebra) is the subject of Green's monograph **[15]**, where it is used to compare the representation theory of general linear groups and symmetric groups. In **[9]** we prove that $S(n,r) = S(\pi)$, where $\pi = \pi(n,r)$ is the set of $(\lambda_1, \lambda_2, \dots, \lambda_n) \in X^+$ with $\lambda_n \ge 0$ (the partitions of r into at most n parts). This is used in **[9]** to obtain new results in the representation theory of $S(n,r)$ and symmetric groups; for example, we show that for $r \le n$, the blocks of $S(n,r)$ are described by the Nakayama Conjecture (for symmetric groups). In the case $S(\pi) = S(n,r)$, (ii) is also proved by Akin and Buchsbaum **[27]**.

We should now like to illustrate the algebras $S(\pi)$ by working out another special case, in which our reductive group is the group of symplectic similitudes. So we suppose that $n = 2m$ is even and put a skew linear form $(\ ,\ )$ on E, $(e_p, e_q) = \varepsilon(p,q)$, $(1 \le p, q \le n)$, where $\varepsilon(p,q)$ is $+1$ if $p + q = n+1$ with $p < q$, is -1 if $p + q = n + 1$ with $p > q$, and is 0 otherwise. Let $G_0 = \mathrm{GSp}(E)$, the group of symplectic similitudes, i.e., the $g \in G$ such that for some $c_g \in k^*$ we have $(ge, ge') = c_g(e, e')$ for all $e, e' \in E$. Let χ be the character of G_0 defined by $\chi(g) = c_g$, the coefficient of dilation, for $g \in G_0$. The kernel of χ is then the symplectic group $\mathrm{Sp}(E)$. The group of diagonal matrices $t = \mathrm{diag}(t_1, t_2, \dots, t_n)$ such that $t_i t_{n+1-i} = t_j t_{n+1-j}$ for all $1 \le i, j \le m$ is a maximal torus T_0 of G_0. Let $\varepsilon_i \in X(T_0)$ be given by $\varepsilon_i(t) = t_i$, $t = \mathrm{diag}(t_1, t_2, \dots, t_n)$. The root system Φ_0 of G_0 with respect to T_0 is then $\{\varepsilon_i - \varepsilon_j : 1 \le i, j \le n,\ i \ne j\}$ and

$\Phi_0^+ = \{\varepsilon_i - \varepsilon_j : 1 \le i, j \le n,\ i < j\}$ is a system of positive roots. Take B_0 to be the negative Borel subgroup. The root lattice X_0 is the free abelian group on $\varepsilon_1, \varepsilon_2, \ldots, \varepsilon_{m+1}$. For $\tau \in X_0$ set $Y_0(\tau) = \operatorname{Ind}_{B_0}^{G_0}(k_\tau)$. We claim that the set of dominant weights X_0^+ consists of those $\mu = a_1\varepsilon_1 + \cdots + a_m\varepsilon_m + a\varepsilon \in X_0$ with $a_1 \ge a_2 \ge \cdots a_m \ge 0$, where ε is the restriction of χ to T_0. The weights of E are $\varepsilon_1 > \varepsilon_2 \cdots > \varepsilon_n$ so that $\Lambda^i E$ has highest weight $\omega_i = \varepsilon_1 + \cdots + \varepsilon_i$ $(1 \le i \le n)$. Hence μ of the above form is the highest weight of

$$E^{\otimes(a_1-a_2)} \otimes (\Lambda^2 E)^{\otimes(a_2-a_3)} \cdots \otimes (\Lambda^m E)^{\otimes a_m} \otimes F^{\otimes a},$$

where F is the (one-dimensional) G_0-module affording χ (and a may be negative). Thus every weight of the above form is dominant. Now suppose that $\mu \in X_0^+$. Then the restriction of μ to the symplectic torus $T_0 \cap \mathrm{Sp}(E)$ is also dominant (with respect to the negative Borel $B_0 \cap \mathrm{Sp}(E)$ in $\mathrm{Sp}(E)$), e.g., by [**7**, (3.2.7) (i)]. However, the fundamental dominant weights of $T_0 \cap \mathrm{Sp}(E)$ are the highest weights of $\Lambda^i E$, $1 \le i \le m$ (see, e.g., [**7**, (4.1.1)]). Thus $\mu - b_1\omega_1 - \cdots - b_m\omega_m$ is in the kernel of the restriction map $X(T_0) \to X(T_0 \cap \mathrm{Sp}(E))$ for some nonnegative integers $b_1, b_2, \ldots, b_m$. This kernel is $\mathbf{Z}\varepsilon$ so that $\mu = b_1\omega_1 + \cdots + b_m\omega_m + a\varepsilon$ for some a and therefore has the form $a_1\varepsilon_1 + \cdots + a_m\varepsilon_m + a\varepsilon$ for some $a_1 \ge \cdots \ge a_m \ge 0$. It follows that X_0^+ is also the set of $\sum_{i=1}^n a_i\varepsilon_i$ satisfying $a_1 - a_n \ge a_2 - a_{n-1} \ge \cdots \ge a_m - a_{m+1} \ge 0$.

Let $\pi_0(n)$ be the set of $\mu \in X_0^+$ which have a representation $\mu = \sum_{i=1}^n a_i\varepsilon_i$ with $a_i \ge 0$ $(1 \le i \le n)$. For $\mu \in \pi_0(n)$ with a representation as above let $(b_1, b_2, \ldots, b_n)$ be the sequence $(a_1, a_2, \ldots, a_n)$ arranged in descending order. Then μ is a weight of $M = E^{\otimes(b_1-b_2)} \otimes (\Lambda^2 E)^{\otimes(b_2-b_3)} \otimes \cdots \otimes (\Lambda^n E)^{b \otimes b_n}$. The formal character $\operatorname{ch} M$ of M, being independent of characteristic, is a sum of various $\operatorname{ch} Y_0(\tau)$'s $(\tau \in X_0^+)$. The character of $Y_0(\tau)$ is given by Weyl's Character Formula (see, e.g., [**7**, (2.2.6)]) and it follows that if $\nu \in X_0^+$ with $\nu < \mu$ then ν is a weight of M (see [**17**, 21.3 Proposition]). But every weight of M has the form $c_1\varepsilon_1 + \cdots + c_n\varepsilon_n$ for some $c_i \ge 0$ $(1 \le i \le n)$. Hence $\pi_0(n)$ is a saturated subset of X_0^+.

Bearing in mind that $\varepsilon_i + \varepsilon_{n+1-i} = \varepsilon$ for $1 \le i \le m$, it is a routine matter to check that $\pi_0(n)$ is the set of dominant weights of the form $a_1\omega_1 + \cdots + a_m\omega_m + b\varepsilon$, $a_i \ge 0$ $(1 \le i \le m)$ and $b \ge 0$. Let $A_0(n)$ be the subalgebra of $k[G_0]$ generated by the restrictions of the c_{ij}, i.e., $A_0(n)$ is the image of $A(n)$ under the restriction map $k[G] \to k[G_0]$. Our claim is:

$$A_0(n) = O_{\pi_0}(k[G_0])$$

where $\pi_0 = \pi_0(n)$. The weights of $A_0(n)$ have the form $\sum_{i=1}^n a_i\varepsilon_i$, $a_i \ge 0$ (the restriction of c_{ij} has weight ε_j). Hence every dominant weight of $A_0(n)$ lies in π_0 and $A_0(n) \le O_{\pi_0}(k[G_0])$.

The proof of the reverse inclusion is somewhat more difficult. First note that the k-span of the restrictions to G_0 of the c_{ij} is $\mathrm{cf}(E)$. Hence $A_0(n)$ is the k-subalgebra of $k[G_0]$ generated by $\mathrm{cf}(E)$. Let $\tau = \{\varepsilon\}$ and $\sigma_i = \{0, \omega_1, \ldots, \omega_i\}$ for $1 \le i \le m$. The weights of a G_0-module are permuted by the Weyl group of G_0. Hence ε, the unique weight of F, is fixed by the Weyl group. It follows from Weyl's Character Formula that $F \cong Y_0(\varepsilon)$ and, by [**17**, 21.3 Proposition], that τ is saturated. Hence by [**10**, (1.4), (16)], $O_\tau(k[G_0])$ is one-dimensional and therefore equal to $\mathrm{cf}(F)$ (in view of the epimorphism $F \otimes F^* \to \mathrm{cf}(F)$ given at the beginning of this paper). But F embeds in $E \otimes E$, as $\sum_{p+q=n+1} \varepsilon(p,q) e_p \otimes e_q$ generates a submodule isomorphic to F. Hence $O_\tau(k[G_0]) = \mathrm{cf}(F) \leq \mathrm{cf}(E \otimes E) = \mathrm{cf}(E).\mathrm{cf}(E) \leq A_0(n)$. By considering the weights of $\bigoplus_{j=0}^{i} \Lambda^j E$ one obtains that σ_i is saturated for $1 \leq i \leq m$ and we now prove that $O_{\sigma_i}(k[G_0]) \leq A_0(n)$. The proof is by induction on i. Suppose that $i = 1$. We have $\mathrm{cf}(E) \cong E \otimes E^*$ as a $G \times G$-module and so $k + \mathrm{cf}(E) \leq O_{\sigma_i}(k[G_0])$ and therefore by dimensions we must have $O_{\sigma_1}(k[G_0]) = k + \mathrm{cf}(E) \leq A_0(n)$. Now suppose that $i > 1$ and the result holds for $i-1$. Consider the surjective map $c: \Lambda^i E \otimes (\Lambda^i E)^* \to \mathrm{cf}(\Lambda^i E)$. If $0 \neq v \in (\Lambda^i E)^{\omega_i}$ and $\alpha \in (\Lambda^i E)^*$ with $\alpha(v) \neq 0$ then $c_{v,\alpha}$ is a nonzero element of $\mathrm{cf}(\Lambda^i E)$ of weight ω_i. Hence $\mathrm{cf}(\Lambda^i E) \not\leq O_{\sigma_{i-1}}(k[G_0])$. Also $\mathrm{cf}(\Lambda^i E) \leq \mathrm{cf}(E^{\otimes i}) = \mathrm{cf}(E).\mathrm{cf}(E) \cdots \mathrm{cf}(E)$ (i times) so $\mathrm{cf}(\Lambda^i E) \leq A_0(n)$. We therefore have a $G \times G$-morphism $\varphi: \Lambda^i E \otimes (\Lambda^i E)^* \to k[G_0]$ with image in $A_0(n)$ but not in $O_{\sigma_{i-1}}(k[G_0])$. By [**1**, p. 508], there is an $\mathrm{Sp}(E)$-module epimorphism from $\Lambda^i E$ to $Y_0(w_i)$. It follows from character considerations and [**7**, Proposition 3.2.6] that there is a short exact sequence $O \to M \to \Lambda^i E \to Y_0(\omega_i) \to 0$, where M is a G_0-module belonging to σ_{i-1}. Hence φ induces a nonzero $G \times G$-map

$$\overline{\varphi}: Y_0(\omega_i) \otimes (\Lambda^i E)^* \to J = O_{\sigma_i}(k[G_0])/O_{\sigma_{i-1}}(k[G_0]).$$

We similarly obtain a short exact sequence $O \to N \to (\Lambda^i E)^* \to Y_0(\omega_i^*) \to O$, where every weight of N is less than ω_i^*. By [**10**, (1.4), (16)], J is isomorphic to $Y_0(\omega_i) \otimes Y_0(\omega_i^*)$. It follows from [**7**, (1.5.2)] that $\overline{\varphi}(Y_0(\omega_i) \otimes N) = 0$ and so $\overline{\varphi}$ induces a nonzero map $\overline{\varphi}: Y_0(\omega_i) \otimes Y_0(\omega_i^*) \to J$ which, by [**7**, (1.5.3)], is surjective. The upshot of all this is that $O_{\sigma_i}(k[G_0]) = \mathrm{cf}(\Lambda^i E) + O_{\sigma_{i-1}}(k[G_0])$ and since by the inductive hypothesis $O_{\sigma_{i-1}}(k[G_0]) \leq A_0(n)$ we obtain at last that $O_{\sigma_i}(k[G_0])$ is contained in $A_0(n)$.

Now π_0 is a submonoid of X_0^+ generated by $\varepsilon, \omega_1, \ldots, \omega_m$. One may now prove that $O_\sigma(k[G_0]) \leq A_0(n)$ for every finite saturated subset $\sigma \subset \pi_0$ using the above and the argument of [**14**, 1.3 Proposition (ii)] (bearing in mind that $k[G_0]$ has a good filtration as a $G \times G$-module with multiplicities 0 or 1, by [**20**], or [**10**, 1.4, (16)]). We omit the details. It follows that $O_{\pi_0}(k[G_0]) \leq A_0(n)$ by a local finiteness argument (see the proof of [**14**, 1.3 Proposition (ii)]). Hence $A_0(n) = O_{\pi_0}(k[G_0])$, as required.

For $r \geq 0$, let $A_0(n,r) = \mathrm{cf}(E).\mathrm{cf}(E) \cdots \mathrm{cf}(E)$ (r times), i.e., $A_0(n,r)$ is the image of $A(n,r)$ under restriction $k[G] \to k[G_0]$. Define $z(\varsigma) \in G_0$, for $\varsigma \in k^*$, by $z(\varsigma)e = \varsigma e$ for $e \in E$. Then $Z = \{z(\varsigma): \varsigma \in k^*\}$ is the center of G_0 (and G) and $A_0(n,r)$ is an eigenspace of Z; in fact $A_0(n,r)$ is the space of $f \in A_0(n)$ such that $z(\varsigma)f = \varsigma^r f$ for all $\varsigma \in k^*$. It follows that $A_0(n)$ decomposes as a direct sum: $A_0(n) = \bigoplus_{r=0}^{\infty} A_0(n,r)$ and that $A_0(n,r) = O_{\pi_0(n,r)}(k[G_0])$, where $\pi_0(n,r) \subset \pi_0(n)$ is the set of $\mu \in X_0^+$ which have an expression $\mu = a_1\varepsilon_1 + \cdots + a_n\varepsilon_n$, with $a_i \geq 0$ $(1 \leq i \leq n)$ and $a_1 + \cdots + a_n = r$.

Hence $S_0(n,r) = A_0(n,r)^*$, the symplectic version of the Schur algebra $S(n,r)$, is a generalized Schur algebra. One knows, by Theorem 9, that its representation theory is just a portion of the representation theory of G_0 and that it has finite global dimension. One could go on, using [**8**, (2.2c)], to calculate its dimension (which must anyway be independent of the characteristic of k). It would also be interesting to have a combinatorial account of $S_0(n,r)$ modeled on Green's treatment of $S(n,r)$ with perhaps an explicit basis of $S_0(n,r)$ analogous to the one for $S(n,r)$ given in [**14**, p. 21].

I am grateful to C. Musili for stimulating discussions on the above treatment of $S_0(n,r)$.

Before moving on let us formally record the result of the above discussion and [**9**, (1.3)].

THEOREM 10. *For positive integers* n *and* r, *the algebras* $S(n,r)$ *and* $S_0(n,r)$ *are generalized Schur algebras.*

Now we return to our general setup, with G an arbitrary reductive group, etc. We shall describe a recent result [**12**], which was inspired by R. W. Richardson's paper [**23**]. Let C be a commutative k-algebra. By a (C,G)-module we mean a C-module V on which G acts rationally in such a way that $c(xv) = x(cv)$, for all $c \in C$, $x \in G$, $v \in V$. We regard C as a (C,G)-module on which G acts trivially. For a (C,G)-module V and a G-module M we denote by $V \otimes |M|$ the space $V \otimes M$ regarded as a (C,G)-module with C action $c(v\otimes m) = cv \otimes m$ and G-action $x(v \otimes m) = xv \otimes xm$, for $c \in C$, $x \in G$, $v \in V$, $m \in M$. In [**23**], Richardson considered the case in which C is the algebra of invariants of $k[G]$ for the conjugating action and analyzed $k[G]$ as a (C,G)-module. For k of characteristic 0 and G semisimple, simply connected he shows that there is a (C,G)-module decomposition $k[G] = \bigoplus_{\lambda\in X^+} C \otimes |Z_\lambda|$, where Z_λ is a direct sum of $\dim Y(\lambda)^0$ copies of $Y(\lambda)$. In our attempts to find a characteristic free result of this sort we proved the following.

THEOREM 11. *Suppose that* A *is a finitely generated commutative* k*-algebra on which* G *acts rationally as algebra automorphisms and* A *has a good* G*-module filtration. Suppose that* $C = A^G$ *is a free polynomial algebra and* A *is a flat* C*-module. Then* A *has an ascending* (C,G)*-module filtration* $0 = A_0, A_1, \dots$ *where* $A_i/A_{i-1} = C \otimes |Z_i|$, *for each* $i > 0$, Z_i *is a finite direct sum of copies of* $Y(\lambda_i)$, *and* $\lambda_1, \lambda_2, \dots$ *a labeling of the elements of* X^+ *such that* $i < j$ *whenever* $\lambda_i < \lambda_j$.

The theorem may be applied in two interesting special cases. First, $A = k[G]$ with the conjugating action, G semisimple, simply connected, with the same (unnecessary) root system/characteristic restrictions as in Theorem 7. In the second case we take $A = k[\mathfrak{g}]$, the coordinate ring of the Lie algebra $\mathfrak{g}$ of G, with G almost simple and action coming from the adjoint action of G on $\mathfrak{g}$. In this case there are characteristic restrictions, the characteristic p of k must be good: $p \neq 2$ in type B, C, D; $p \neq 2, 3$ in type G_2, F_4, E_6, E_7; $p \neq 2$, 3, 5 in type E_8.

The restrictions come from the following result of Andersen, Jantzen, and the author [**1**, 4.4 Proposition].

THEOREM 12. *Suppose that G is semisimple and p is good. Then $k[\mathfrak{g}]$ has a good filtration.*

Unfortunately, some restriction on p really is necessary. Let G have type G_2 and $p = 2$. We shall show that $k[\mathfrak{g}]$ does not have a good filtration. Suppose that it does. We adopt the notation of [**7**, Chapter 6]. Then $\mathfrak{g}^*$ is isomorphic to $Y(0,1)$, so $S(Y(0,1))$ has a good filtration. It follows from Theorem 7 above [**7**, (6.3), Proposition 3.3.2, Proposition 3.3.1, and Corollary 3.2.5] that $Y(0,1)$ has a G_Ω-summand isomorphic to $Y_\Omega(3,-1)$. Now $S(Y_\Omega(3,-1))$ is a G_Ω-summand of $S(Y(0,1))$ and hence has a good filtration by Theorem 7 above and [**7**, Proposition 3.3.2 and Corollary 3.2.5]. Now the derived subgroup of G_Ω is isomorphic to $\mathrm{SL}_2(k)$ and $Y_\Omega(3,-1)$, as an $\mathrm{SL}_2(k)$-module is isomorphic to S^3V, where V is the natural module. Hence, by [**7**, Proposition 3.2.7 (iii)], we have that $S(S^3V)$ has a good filtration. In particular $S^2(S^3V)$ has a good filtration, contrary to the second example (given before Theorem 1). Hence $k[\mathfrak{g}]$ does not have a good filtration. In this way, using the second example above and more complicated, similar examples due to C. Musili, one can show that $k[\mathfrak{g}]$ does not have a good filtration in most of the bad-characteristic cases. In fact, we believe that the restriction in Theorem 12 is precise, except for a few symplectic groups in low rank (where, for $p = 2$, $k[\mathfrak{g}]$ does have a good filtration).

In the Lie algebra application of Theorem 11, we are giving a characteristic free (and nonconstructive) version of the result of Kostant [**22**], of which Richardson's theorem is a precise analogue. It is easy to recover Richardson's theorem, in characteristic 0, from our application to $k[G]$ of Theorem 11 and complete reducibility of G-modules.

Theorem 12 has an application too to the cohomology of the infinitesimal subgroups G_n of G (in which context it arose) when k has prime characteristic. In this subject there has been enormous progress in recent years. We do not attempt to describe this here but refer the reader to the papers by Friedlander-Parshall, Andersen-Jantzen, and Jantzen listed in the bibliography of [**20**]. We conclude by describing a possible general connection between this theory and good filtrations.

We assume that G is semisimple, simply connected, defined, and split over $\mathbf{F}_p$ with Frobenius morphism $F: G \to G$. If V is a G-module affording the representation $\psi: G \to \mathrm{GL}(V)$, $V^{(n)}$ denotes the k-space V viewed as a G-module with G acting via $\psi \cdot F^n$ $(n > 0)$. Then $V^{(n)}$ is a G-module on which G_n acts trivially. Conversely, if V is a G-module on which G_n acts trivially then V is isomorphic to $M^{(n)}$ for a G-module M, unique up to isomorphism: we write $M = V^{(-n)}$. Starting with a G-module V and $r \geq 0$ we obtain a G-module $H^r(G_n, V)$ on which G_n acts trivially. The following was originally raised in lectures given by the author at the University of Turku, Finland, in 1980.

CONJECTURE. *Let V be a G-module which has a good filtration. Then $H^r(G_n, V)^{(-n)}$ has a good filtration, for every $r \geq 0$, $n \geq 1$.*

As observed by H. H. Andersen and the author, it is enough to prove the conjecture for $r = 0$, $n = 1$ and all finite-dimensional V; i.e., one only has to show that if V is a finite-dimensional G-module with a good filtration and $M^{(1)} = V^{G_1}$, then M has a good filtration. The argument is as follows. We suppose the conjecture holds in the special case just described. By a direct limit argument we get that $V^{(-1)}$ has a good filtration for all V with a good filtration. Now let $n > 1$ and let V be a G-module with a good filtration. Then $V^{G_n} = (V^{G_1})^{G_n/G_1} = (M^{(1)})^{G_n/G_1}$, where $M = (V^{G_1})^{(-1)}$. However, $F: G_n \to G_n$ induces an isomorphism $G_n/G_1 \to G_{n-1}$ giving $V^{G_n} \cong (M^{G_{n-1}})^{(1)}$. But M has a good filtration by the $n = 1$ case, so the result for arbitrary n and $r = 0$ follows by induction. Suppose now $r \geq 1$ and V is a G-module with a good filtration. Embed V in an injective module E of countable dimension and put $Q = E/V$. Note that E has a good filtration by Theorem 4 and it follows from Theorems 2 and 1 that Q has a good filtration. Also, E is injective as a G_n-module, so we get an exact sequence

$$\begin{aligned} 0 \to H^0(G_n, V)^{(-n)} &\to H^0(G_n, E)^{(-n)} \to H^0(G_n, Q)^{(-n)} \\ &\to H^1(G_n, V)^{(-n)} \to 0 \end{aligned}$$

and, for $r > 1$, an isomorphism $H^r(G_n, V)^{(-n)} \cong H^{r-1}(G_n, Q)^{(-n)}$. Thus, by induction, we may assume $r = 1$. But now $H^0(G_n, V)^{(-n)}$, $H^0(G_n, E)^{(-n)}$, and $H^0(G_n, Q)^{(-n)}$ all admit good filtrations and it follows from Theorems 2 and 1 and the above exact sequence that $H^1(G_n, V)^{(-n)}$ admits a good filtration (cf. [**1**, 4.1(2)]).

There is some justification for making such a conjecture. In the trivial case $V = Y(\lambda)$ $(\lambda \in X^+)$ we have $V^{(-1)} \cong Y(\tau)$, if p divides λ and $\lambda = p\tau$, and $V^{(-1)} = 0$ otherwise. Even in the case $G = \mathrm{SL}_2(k)$ we have been unable to construct a proof (or a counterexample) though we have checked the $r = 0$, $n = 1$ case, for $p = 2$ and all V involving only weights less than 10. But the real evidence comes from the consistency with the conjecture of the wealth of results on the cohomology of G_n-modules in the work of Friedlander-Parshall, Andersen-Jantzen, and Jantzen mentioned above.

REFERENCES

1. H. H. Andersen and J. C. Jantzen, *Cohomology of induced representations for algebraic groups*, Math. Ann. **269** (1984), 487–525.

2. E. Cline, B. Parshall, and L. Scott, *Induced modules and affine quotients*, Math. Ann. **230** (1977), 1–14.

3. ——, *Algebraic stratification in representation categories*, Preprint.

4. E. Cline, B. Parshall, L. Scott, and W. van der Kallen, *Rational and generic cohomology*, Invent. Math. **39** (1977), 143–163.

5. S. Donkin, *A filtration for rational modules*, Math. Z. **177** (1981), 1–8.

6. ——, Letter to C. Procesi, Summer 1982.

7. ——, *Rational representations of algebraic groups: Tensor products and filtrations*, Lecture Notes in Math., vol. 1140, Springer-Verlag, 1985.

8. ——, *On Schur algebras and related algebras.* I, J. Algebra **104** (1986), 310–328.

9. ——, *On Schur algebras and related algebras.* II, J. Algebra, (in press).

10. ——, *Skew modules for reductive groups*, J. Algebra, (in press).

11. ——, *Invariants of unipotent radicals*, Math. Z., (in press).

12. ——, *On conjugating representations and adjoint representations of semisimple groups*, Invent. Math., (in press).

13. E. Friedlander, *A canonical filtration for certain rational modules*, Math. Z. **188** (1984/85), 433–438.

14. J. A. Green, *Locally finite representations*, J. Algebra **41** (1976), 131–171.

15. ——, *Polynomial representations of* GL_n, Lecture Notes in Math., vol. 830, Springer-Verlag, 1980.

16. F. Grosshans, *The invariants of unipotent radicals of parabolic subgroups*, Invent. Math. **73** (1983), 1–9.

17. J. E. Humphreys, *Introduction to Lie algebras and representation theory*, Graduate Texts in Math., vol. 9, Springer-Verlag, 1972.

18. J. C. Jantzen, *Darstellungen halbeinfacher Gruppen und kontravariante Formen*, J. Reine Angew. Math. **290** (1977), 157–199.

19. ——, *Darstellungen halbeinfacher Gruppen und ihrer Frobenius-Kerne*, J. Reine Angew. Math. **317** (1980), 157–199.

20. ——, *Representations of algebraic groups* (in press).

21. M. Koppinen, *Good bimodule filtrations for coordinate rings*, J. London Math. Soc. (2) **30** (1984), 244–250.

22. B. Kostant, *Lie group representations on polynomial rings*, Amer. J. Math. **85** (1963), 327–404.

23. R. W. Richardson, *The conjugating representation of a semisimple group*, Invent. Math. **54** (1979), 229–245.

24. L. L. Scott, *Representations in characteristic p*, The Santa Cruz Conference on Finite Groups, Proc. Sympos. Pure Math., vol. 37, Amer. Math. Soc., Providence, R.I., 1980, pp. 319–331.

25. B. S. Upadhyaya, *Filtrations with Weyl module quotients of the principal indecomposable modules of the group* $\mathrm{SL}(2,q)$, Comm. Algebra **7** (1979), 1469–1488.

26. Wang Jian-Pan, *Sheaf cohomology of* G/B *and the tensor products of Weyl modules*, J. Algebra **77** (1982), 162–185.

27. K. Akin and D. A. Buchsbaum, *Characteristic-free representation theory of the general linear group* II: *Homological considerations*, preprint.

QUEEN MARY COLLEGE, LONDON, ENGLAND

Proceedings of Symposia in Pure Mathematics
Volume **47** (1987)

Problems in Representation Theory Arising from Galois Modules

A. FRÖHLICH

Developments in algebraic number theory have often led to new problems, new ideas, and new results in representation theory of groups. The classical example is Brauer induction and its application to Artin L-functions. More recently one has Serre's induction theorems for orthogonal and symplectic representations with applications to root numbers and to Galois module structure of integers. The latter problem has also led to the description of class groups of group rings by character functions.

In these notes another such development will be introduced. The main emphasis is on the number-theoretic background, with the aim of exhibiting the significance of the new notions which arise. No attempt at complete generality will be made.

1. Module theory. We consider a finite-dimensional semisimple, commutative $\mathbf{Q}$-algebra A, i.e.,

$$A = \prod_i A_i, \tag{1.1}$$

the A_i being algebraic number fields. In A we are given a $\mathbf{Z}$-order $\mathscr{A}$, i.e., a $\mathbf{Z}$-lattice spanning A which is a subring of A with identity 1_A. The maximal order $\mathscr{M}$ of A has a decomposition

$$\mathscr{M} = \prod_i \mathscr{M}_i, \tag{1.2}$$

corresponding to (1.1), the $\mathscr{M}_i$ being the rings of integers in the number fields A_i. The structure theory of $\mathscr{M}$-modules is trivial; hence, given an $\mathscr{A}$-lattice M, we are led to consider the maximal $\mathscr{M}$-sublattice $M^{\mathscr{M}}$ of M. Its structure is determined by the A-module $M\mathbf{Q}$ $(= M \otimes_{\mathbf{Z}} \mathbf{Q})$ and the element

$$(M^{\mathscr{M}}) \in \mathrm{Cl}(\mathscr{M}) \tag{1.3}$$

1980 *Mathematics Subject Classification* (1985 *Revision*). Primary 11R54, 11S45.

of the class group of $\mathscr{M}$. The deviation of M from $M^{\mathscr{M}}$ is in a first approximation measured by the index $[M : M^{\mathscr{M}}]$, which is finite. We introduce a variant of this, assuming that for some n

$$MQ \simeq A^n \quad \text{(product of } n \text{ copies)}. \tag{1.4}$$

Let $\hat{\mathscr{A}} = \mathrm{Hom}_{\mathbf{Z}}(\mathscr{A}, \mathbf{Z})$, viewed as an $\mathscr{A}$-module. Write

$$j(M) = [\hat{\mathscr{A}} : \hat{\mathscr{A}}^{\mathscr{M}}]^n / [M : M^{\mathscr{M}}] \tag{1.5}$$

(a positive rational, or a rational fractional ideal). Trivially,

$$\begin{aligned} &\text{(i)} \quad j(M) \mid [\hat{\mathscr{A}} : \hat{\mathscr{A}}^{\mathscr{M}}]^n, \\ &\text{(ii)} \quad j(M) = [\hat{\mathscr{A}} : \hat{\mathscr{A}}^{\mathscr{M}}]^n \Leftrightarrow M \sim (\hat{\mathscr{A}}^{\mathscr{M}})^n \end{aligned} \tag{1.6}$$

($\sim$ means locally isomorphic everywhere). Note that $[\hat{\mathscr{A}} : \hat{\mathscr{A}}^{\mathscr{M}}]$ can be written down explicitly. But also (cf. Fröhlich, Quart. J. Math. Oxford Ser. **16** (1965), 193–232)

$$\begin{aligned} &\text{(iii)} \quad j(M) \textit{ is integral}, \\ &\text{(iv)} \quad j(M) = 1 \Leftrightarrow M \sim \hat{\mathscr{A}}^n. \end{aligned} \tag{1.6}$$

REMARK. All this goes over to any local or global ring of integers in place of $\mathbf{Z}$.

Let Γ be a finite Abelian group, $\hat{\Gamma}$ its dual group. Let I_Γ be the group of functions f on $\hat{\Gamma}$ with $f(\varphi)$ a nonzero fractional ideal in $\mathbf{Q}(\varphi)$ (the field of values of φ) so that for all $w \in \mathrm{Gal}(\mathbf{Q}(\varphi)/\mathbf{Q})$,

$$f(\varphi^w) = f(\varphi)^w.$$

Let P_Γ be the subgroup of I_Γ for which $f(\varphi)$ is always principal in $\mathbf{Q}(\varphi)$. The condition $f(\varepsilon) = (1)$ (ε = identity character) defines subgroups I^0_Γ, P^0_Γ.

Two examples (terminology will soon be justified).

Additive

$\mathscr{A} = \mathbf{Z}\Gamma$, so $\hat{\mathscr{A}} \simeq \mathbf{Z}\Gamma$

(1.7) $\mathrm{Cl}(\mathscr{M}) \simeq I_\Gamma / P_\Gamma$

(1.8) $\Gamma \simeq \mathrm{Gal}(N/K)$

N, K number fields

$n = [K : \mathbf{Q}]$

$M = \mathfrak{o}_N$ = integers in N

$\mathfrak{o}_N \mathbf{Q} \simeq \mathbf{Q}\Gamma^n$

Multiplicative

$\mathscr{A} = \mathbf{Z}\Gamma / \left(\sum_{\gamma \in \Gamma} \gamma\right)$, so $\hat{\mathscr{A}}$ = aug. ideal

(1.7) $\mathrm{Cl}(\mathscr{M}) \simeq I^0_\Gamma / P^0_\Gamma$

(1.8) $\Gamma \simeq \mathrm{Gal}(N/\mathbf{Q})$

N real

$n = 1$

$M = U_N$ = units in N mod ± 1

$U_N \mathbf{Q} \simeq \mathbf{Q}\Gamma / \left(\sum \gamma\right)$

REMARK. These are examples of a more general approach, which work particularly well! We shall stick to these special cases.

The above notations will be used throughout.

2. Generalized resolvents and regulators. F is a number field, viewed as a subfield of $\mathbf{C}$, big enough; i.e., given integral representations of Γ are defined over the ring $\mathcal{O}_F$ of integers in F, and in the additive case there are $[N:\mathbf{Q}]$ embeddings $N \to F$. Also F contains the values of the Gauss sums which arise. Specifically $\mathbf{Q}(\varphi) \subset F$, $\forall \varphi \in \hat{\Gamma}$.

Additive

Via (1.8) Γ acts on $\mathcal{O}_N$, hence on $\mathcal{O}_N\mathcal{O}_F = \mathcal{O}_N \otimes_{\mathbf{Z}} \mathcal{O}_F$, with trivial action on $\mathcal{O}_F$.

$(\mathcal{O}_N\mathcal{O}_F)_\varphi$ is the submodule on which Γ acts via $\varphi : x\gamma = x\varphi(\gamma)$.

Extend an embedding $e : N \to F$ to a map $N \otimes F \to F : (x \otimes y)^e = x^e y$. The image under e of $(\mathcal{O}_N\mathcal{O}_F)_\varphi$ is an ideal of $\mathcal{O}_F$; denote it by $\mathscr{R}(\mathcal{O}_N, \varphi)$. This is independent of the choice of e.

Multiplicative

Similarly via (1.8) Γ acts on $U_N\mathcal{O}_F$.

$(U_N\mathcal{O}_F)_\varphi$ is the submodule on which Γ acts via φ $(\varphi \neq \varepsilon)$.

Let s be an infinite place of N. Extend the map $l_s : U_N \to \mathbf{C}$ given by $l_s(u) = \log \|u\|_s$ to $U_N\mathcal{O}_F$ (via $\mathcal{O}_F \subset \mathbf{C}$). The the image of $(U_N\mathcal{O}_F)_\varphi$ under l_s is a rank one $\mathcal{O}_F$-module in a certain one-dimensional F-subspace of $\mathbf{C}$. Denote it by $\mathscr{R}(U_N, \varphi)$. This is independent of the choice of s.

REMARK. These are very special cases of generalized resolvents (additive) and generalized regulators (multiplicative).

3. Number theory. Convention: We view the group of fractional ideals of $\mathbf{Q}(\varphi)$ as embedded in that of F.

Additive

Via (1.8), $\varphi \in \hat{\Gamma}$ may be viewed as an idele class character of K, hence has a Gauss sum $\tau(N/K, \varphi)$. Lift this to a Galois Gauss sum $\Upsilon(N/K, \varphi)$ over $\mathbf{Q}$. In other words, $\Upsilon(N/K, \varphi) = \tau(N/K, \varphi)d^{1/2}$, $d = \text{discr.}(K/\mathbf{Q})$, and $d^{1/2}$ a "canonical square root."

THEOREM 1. *$t(\mathcal{O}_N, \varphi) = \mathscr{R}(\mathcal{O}_N, \varphi) \times \Upsilon(N/K, \varphi)^{-1}$ is a fractional idea in $\mathbf{Q}(\varphi)$ and $t(\mathcal{O}_N, \varphi^w) = t(\mathcal{O}_n, \varphi)^w$ for $w \in \text{Gal}(\mathbf{Q}(\varphi)/\mathbf{Q})$.*

THEOREM 2. *$t(\mathcal{O}_N, \varphi)^{-1}$ is integral.*

Multiplicative

Via (1.8), $\varphi \in \hat{\Gamma}$ may be viewed as a primitive residue class character of $\mathbf{Q}$, hence has a Dirichlet L-function $L(s, \varphi) = L(s, N, \varphi)$. Abbreviate $L'(0, N, \varphi^{-1}) = L^x(N, \varphi)$ $(\varphi \neq \varepsilon)$.

THEOREM 1. *$\mathscr{R}(U_N, \varphi)L^x(N, \varphi)^{-1} = t(U_N, \varphi)$ $(\varphi \neq \varepsilon)$ is a fractional ideal in $\mathbf{Q}(\varphi)$ and $t(U_N, \varphi^w) = t(U_N, \varphi)^w$ for $\omega \in \text{Gal}(\mathbf{Q}(\varphi)/\mathbf{Q})$.*

THEOREM 2. *$t(U_N, \varphi)^{-1}$ is integral.*

THEOREM 3. *Under the isomorphism* (1.7) $(\mathfrak{o}_N^{\mathscr{M}}) \in \mathrm{Cl}(\mathscr{M})$ *is given by* $\varphi \mapsto t(\mathfrak{o}_N, \varphi)$.

THEOREM 3. *Under the isomorphism* (1.7) $(U_N^{\mathscr{M}}) \in \mathrm{Cl}(\mathscr{M})$ *is given by* $\varphi \mapsto t(U_N, \varphi)$.

THEOREM 4. $\prod_\varphi t(\mathfrak{o}_N, \varphi)^{-1} = j(\mathfrak{o}_N)$.

THEOREM 4. $\prod_{\varphi \neq \varepsilon} t(U_N, \varphi)^{-1} = j(U_N) h_N$, *$h_N$ the class number.*

For the following example of an application of these theorems one should note that in general the invariants $j(M)$ and $(M^{\mathscr{M}})$ of a lattice M are "independent;" e.g., for every $c \in \mathrm{Cl}(\mathscr{M})$, $\exists M \sim \hat{\mathscr{A}}$ with $(M^{\mathscr{M}}) = c$.

Additive

If Γ is a p-group then $(\mathfrak{o}_N^{\mathscr{M}}) = 1$.

Multiplicative

If Γ is a p-group and h_N a power of p then $(U_N^{\mathscr{M}}) = 1$.

PROOF. Exercise.

4. Factorizability. Let $\Delta < \Gamma$ and view $\widehat{\Gamma/\Delta}$ as a subgroup of $\hat{\Gamma}$. Denote by N^Δ the fixed subfield of Δ in N under the action given by (1.8).

Additive

THEOREM 5. *Let* $\varphi \in \widehat{\Gamma/\Delta}$. $\mathscr{R}(\mathfrak{o}_N, \varphi) = \mathscr{R}(\mathfrak{o}_{N^\Delta}, \varphi)$; *hence* $t(\mathfrak{o}_N, \varphi) = t(\mathfrak{o}_{N^\Delta}, \varphi)$ (*for* Υ *this is known*).

Multiplicative

THEOREM 5. *Let* $\varphi \in \widehat{\Gamma/\Delta}$, $\neq \varepsilon$. *At all primes* $\neq 2$, $\mathscr{R}(U_N, \varphi)_p = \mathscr{R}(U_{N^\Delta}, \varphi)_p$; *hence* $t(U_{N,\varphi})_p = t(U_{N^\Delta}, \varphi)_p$. *One knows that* $L^x(N, \varphi) = L^x(N^\Delta, \varphi)$.

Remark. One can get rid of the problem at 2 *by replacing* U_N *by a slightly bigger group.*

Now let F be a local or global field. We consider the group $G_{\Gamma,F}$ of functions from the set of subgroups of $\hat{\Gamma}$ to the fractional ideals in F.

EXAMPLE 1. M a $\mathbf{Z}\Gamma$-lattice, $M\mathbf{Q} \simeq \mathbf{Q}\Gamma^n$. Then for $\Delta < \Gamma$, M^Δ satisfies analogous conditions with respect to Γ/Δ. Recall (1.5) for $\mathscr{A} = \mathbf{Z}\Gamma$; write

$$j_M(\widehat{\Gamma/\Delta}) = j(M^\Delta), \quad \text{with } \mathscr{A} = \mathbf{Z}(\Gamma/\Delta). \tag{4.1}$$

(Analogously for $\mathbf{Z}\Gamma/(\sum \gamma)$ instead).

EXAMPLE 2. Let X be a finite $\mathbf{Z}\Gamma$-module. Write

$$\mathrm{ord}_X(\widehat{\Gamma/\Delta}) = \text{order of } X^\Delta. \tag{4.2}$$

Next let $\mathscr{M}_{\Gamma,F}$ be the group of functions g on $\hat{\Gamma}$, so that

$$g(\varphi) \text{ is a fractional ideal in } F(\varphi), \tag{4.3}$$

$$g(\varphi^w) = g(\varphi)^w \text{ for } w \in \mathrm{Gal}(F(\varphi)/F). \tag{4.4}$$

(Here view $\hat{\Gamma}$ as homomorphisms into roots of unity over F.)

If g is as above, define for $H < \hat{\Gamma}$

$$Ng(H) = \prod_{\varphi \in H} g(\varphi). \tag{4.5}$$

Then N is a homomorphism

$$\mathscr{M}_{\Gamma,F} \to G_{\Gamma,F}. \tag{4.6}$$

$f \in G_{\Gamma,F}$ is *factorizable* if $f = Ng$ for some g.

From Theorems 4 and 5:

$$j_{\mathcal{O}_N} \text{ and } j_{U_N,p} h_{N,p} \ (p \neq 2) \text{ are factorizable.}$$

EXAMPLES. (i) *If M is locally free then j_M is factorizable.*

(ii) *If $M \sim \mathscr{M}$ then j_M is not factorizable for noncyclic Γ.*

(iii) If X is a finite, cohomologically trivial module, or lifted from such a module over a quotient of Γ, then ord_X is factorizable.

5. The non-Abelian context (Nelson's London Ph.D. thesis). This is the proper framework for factorizability.

Let Ω be a finite group and Mon_Ω the Grothendieck group of monomial representations of Ω (e.g., in the sense of Dress). This is free Abelian on pairs (Γ, φ), $\Gamma < \Omega$, $\varphi \in \hat{\Gamma} \bmod \Omega$-conjugacy. Define the group of "Abelian subquotients" Ab_Ω as free Abelian on pairs $(\Gamma, \Delta) \bmod \Omega$-conjugacy with $\Delta \triangleleft \Gamma < \Omega$, Γ/Δ Abelian. There is a homomorphism

$$N : \mathrm{Ab}_\Omega \to \mathrm{Mon}_\Omega,$$
$$(\Gamma, \Delta) \mapsto \sum \varphi \qquad (\varphi \in \hat{\Gamma}, \varphi | \Delta = 1).$$

In this wider sense a function on Ab_Ω is factorizable if it factorizes via N with certain Galois conditions imposed. The functions j and ord have natural extensions to Ab_Ω: e.g., if M is a $\mathbf{Z}\Omega$-lattice with $M \simeq \mathbf{Z}\Omega^n$ we view M^Δ as a $\mathbf{Z}\Gamma/\Delta$-module and have $(\Gamma, \Delta) \mapsto j(M^\Delta, \Gamma/\Delta)$. If $\Omega \simeq \mathrm{Gal}(N/K)$, N, K number fields, then this function for $M = \mathcal{O}_N$ is still factorizable.

IMPERIAL COLLEGE, LONDON AND ROBINSON COLLEGE, CAMBRIDGE, ENGLAND

Proceedings of Symposia in Pure Mathematics
Volume **47** (1987)

Holonomic $\mathcal{D}$-Modules in Representation Theory

RYOSHI HOTTA

The history appears to have begun with a brilliant application of holonomic $\mathcal{D}$-module theory to the proof of the Kazhdan-Lusztig conjecture by Brylinski-Kashiwara [**BK**] and Beilinson-Bernstein [**BeB**], which is now widely known. Though it was never the first application of so-called algebraic analysis to representations of Lie groups (remember the six-author paper [**KKMOOT**]!), it was striking enough to attract attention in its algebro-geometric transparancy related to certain problems in analysis.

Subsequently, several works in this direction have started in some other aspects of representation theory [**BoB 1, 2**; **BBM**; **HK 1, 2**; **KT**; **T 1, 2**; **G 1**]. In this lecture, I shall introduce two topics among those, in which, however, we cannot yet find the expected close relationship. One is an extension of the results [**HK 1, 2**], while the other [**T 2**] by T. Tanisaki is an extension of [**KT**, **T 2**].

1. An affine Weyl group as monodromy.

1.1. *The Grothendieck-Springer-Steinberg diagram.* Let G be a connected reductive algebraic group over the complex number field $\mathbf{C}$. We fix a Borel subgroup B and a maximal torus T such that $T \subset B$. Thus $W = N_G(T)/T$ is the Weyl group and the flag manifold $X = G/B$ is identified with the set of all Borel subgroups.

We consider the Grothendieck-Springer-Steinberg simultaneous resolution diagram:

$$\begin{array}{ccc} \tilde{G} & \xrightarrow{\ p\ } & G \\ \downarrow{\scriptstyle \theta} & & \downarrow{\scriptstyle \chi} \\ T & \xrightarrow{\ \pi\ } & T/W \end{array}$$

Here χ is the map sending $x \in G$ to the intersection of the G-orbit of its semisimple part with T and π is the natural quotient map. $\tilde{G}$ is the closed submanifold of $G \times X$ defined by

$$\tilde{G} := \{(x, B') \in G \times X; x \in B'\},$$

1980 *Mathematics Subject Classification* (1985 *Revision*). Primary 22E47, 35A30.

where X is considered as the set of all Borels and $p\colon \tilde{G} \to G$ is the projection to G. Let $G \times^B B$ denote the fiber bundle over $X = G/B$ associated to the principal B-bundle $G \to X$ and the B-action on B by the inner automorphisms. Then

$$\begin{gathered} G \times^B B \to \tilde{G}, \\ (g, x) \mapsto (gxg^{-1}, gBg^{-1}) \end{gathered}$$

is an isomorphism of varieties. Let $B \to T$ be the projection divided by the unipotent radical of B. Then the projection $G \times B \to T$ is factorized as $G \times B \to G \times^B B \simeq \tilde{G} \xrightarrow{\theta} T$ and this θ gives a smooth map. It is easily seen that $\chi \circ p = \pi \circ \theta$.

Let G_{rs} be the open set of G consisting of regular semisimple elements in G. If we put $T_{\mathrm{rs}} := G_{\mathrm{rs}} \cap T$, then $G_{\mathrm{rs}} = \bigcup_{g \in G} gT_{\mathrm{rs}}g^{-1}$. The above diagram then turns out to be very simple on G_{rs}; we have an isomorphism

$$\begin{gathered} G/T \times T_{\mathrm{rs}} \to \tilde{G}_{\mathrm{rs}}, \\ (gT, t) \mapsto (gtg^{-1}, gBg^{-1}), \end{gathered}$$

and in this identification, on $\tilde{G}_{\mathrm{rs}}$, $\theta' := \theta \mid \tilde{G}_{\mathrm{rs}}$ is simply the projection to T_{rs} and p is given by $p(gT, t) = gTg^{-1}$. Thus on $\tilde{G}_{\mathrm{rs}}$, $p' := p|\tilde{G}_{\mathrm{rs}}$ is a Galois covering with group W, i.e., p' is the quotient by the W-action

$$(gT, t)w = (gwT, w^{-1}tw) \qquad (w \in W).$$

1.2 *An affine Weyl group and local systems.* We can naturally identify the fundamental group $\pi_1(T) = \pi_1(T, e)$ at the identity $e \in T$ with the coweight lattice $\mathrm{Hom}_{\mathrm{alg\,gp}}(\mathbf{C}^\times, T)$, which is the dual lattice of the character group $T^\wedge := \mathrm{Hom}_{\mathrm{alg\,gp}}(T, \mathbf{C}^\times)$. ($\mathbf{C}^\times$ is the multiplicative group.) Let $\mathfrak{t}$ be the Lie algebra $\mathrm{Lie}\, T$ of T. Then by taking the differential we have the embedding

$$\pi_1(T) \hookrightarrow \mathrm{Hom}_{\mathbf{C}}(\mathbf{C}, \mathfrak{t}) = \mathfrak{t}.$$

Let $\mathfrak{t}^*$ be the dual space of $\mathfrak{t}$ and for $\lambda \in \mathfrak{t}^*$, let $1^\lambda \in \pi_1(T)^\wedge$ be the one-dimensional representation of $\pi_1(T)$ into $\mathbf{C}^\times$ defined by

$$1^\lambda(\gamma) := e^{2\pi\langle\lambda, \gamma\rangle} \qquad (\gamma \in \pi_1(T) \subset \mathfrak{t}),$$

where $\langle\ ,\ \rangle$ is the pairing on $\mathfrak{t}^* \times \mathfrak{t}$. Then

$$1^\lambda = 1^{\lambda'} \quad \text{if and only if } \langle\lambda - \lambda', \gamma\rangle \in \mathbf{Z}\ (\forall\gamma \in \pi_1(T)).$$

We denote by $\mathbf{C}_\lambda$ the local system of rank one on T corresponding to 1^λ. By the above, $\mathbf{C}_\lambda$ is parametrized by $\mathfrak{t}^*$ modulo integral weights. We want to consider the local system $p'_*\theta'^{-1}\mathbf{C}_\lambda$ on G_{rs} where

$$G_{\mathrm{rs}} \xleftarrow{p'} \tilde{G}_{\mathrm{rs}} \xrightarrow{\theta'} T_{\mathrm{rs}}.$$

We put $p' := p|\tilde{G}_{\mathrm{rs}}$, $\theta' := \theta|\tilde{G}_{\mathrm{rs}}$.

The fundamental group $\pi_1(G_{\mathrm{rs}})$ has the following structure. Since p' is a W-covering, we have an exact sequence

$$1 \to \pi_1(\tilde{G}_{\mathrm{rs}}) \to \pi_1(G_{\mathrm{rs}}) \to W \to 1.$$

But then $\tilde{G}_{\mathrm{rs}} \simeq G/T \times T_{\mathrm{rs}}$ is simply connected and hence

$$\pi_1(\tilde{G}_{\mathrm{rs}}) \simeq \pi_1(T_{\mathrm{rs}}).$$

On the other hand, there is the natural surjection

$$\pi_1(T_{\mathrm{rs}}) \twoheadrightarrow \pi_1(T)$$

whose kernel is denoted by K. I do not know much about K, which seems to be related to pure braid groups. Anyway, we can consider $\pi_1(T_{\mathrm{rs}})$ as a subgroup of $\pi_1(G_{\mathrm{rs}})$ and the quotient group $\pi_1(G_{\mathrm{rs}})/K$ is an extension of W by $\pi_1(T)$, which turns out to be isomorphic to the semidirect product

$$\tilde{W} := \pi_1(T) \rtimes W$$

with the natural action of W on $\pi_1(T)$. In fact, $\exp\colon \mathfrak{t} \to T$ is the universal covering space and the W-action on T lifts to that on $\mathfrak{t}$. Then W acts freely on the subset $Z := \exp^{-1} T_{\mathrm{rs}} \subset \mathfrak{t}$ and clearly $\pi_1(Z) = K$. The covering $Z \to Z/W$ then gives rise to the splitting $W \hookrightarrow \pi_1(G_{\mathrm{rs}})/K$. Thus we have the (modified) affine Weyl group $\tilde{W}$ as a quotient of $\pi_1(G_{\mathrm{rs}})$.

In conclusion, we have:

LEMMA 1.2.1. *The local system $p'_*\theta'^{-1}\mathbf{C}_\lambda$ on G_{rs} corresponds to the induced representation $\mathrm{Ind}_{\pi_1(T)}^{\tilde{W}} 1^\lambda$ of the affine Weyl group $\tilde{W} = \pi_1(T) \rtimes W$, which is regarded as the natural quotient of $\pi_1(G_{\mathrm{rs}})$.*

To decompose the above representation, we introduce the integral Weyl group W_λ for $\lambda \in \mathfrak{t}^*$:

$$W_\lambda := \{w \in W; \langle \lambda - w\lambda, \gamma\rangle \in \mathbf{Z}\ (\forall \gamma \in \pi_1(T))\}.$$

Then under the natural action of W on $\pi_1(T)^\wedge$,

$$W_\lambda = \{w \in W; w1^\lambda = 1^\lambda\}.$$

Hence

$$\mathrm{Ind}_{\pi_1(T)}^{\tilde{W}} 1^\lambda \simeq \mathrm{Ind}_{\tilde{W}_\lambda}^{\tilde{W}} \mathrm{Ind}_{\pi_1(T)}^{\tilde{W}_\lambda} 1^\lambda,$$

where $\tilde{W}_\lambda := \pi_1(T) \rtimes W_\lambda \subset \tilde{W}$. But then, if $\mathbf{C}[W_\lambda]$ denotes the regular representation of W_λ,

$$\mathrm{Ind}_{\pi_1(T)}^{\tilde{W}_\lambda} 1^\lambda \simeq 1^\lambda \otimes \mathbf{C}[W_\lambda]$$

since W_λ stabilizes 1^λ. Thus we have a decomposition

$$\mathrm{Ind}_{\pi_1(T)}^{\tilde{W}} 1^\lambda \simeq \bigoplus_{\chi \in W_\lambda^\wedge} V_\chi^* \otimes \mathrm{Ind}_{\tilde{W}_\lambda}^{\tilde{W}} (1^\lambda \otimes V_\chi),$$

where $\mathbf{C}[W_\lambda] \simeq \bigoplus_{\chi \in W_\lambda^\wedge} V_\chi^* \otimes V_\chi$ is the usual irreducible decomposition. Here we note that $\mathrm{Ind}_{\tilde{W}_\lambda}^{\tilde{W}} (1^\lambda \otimes V_\chi)$ is irreducible and the above gives the irreducible decomposition of $\tilde{W}$-representation. In other words, we have:

LEMMA 1.2.2. *We have a decomposition as local systems*

$$p'_*\theta'^{-1}\mathbf{C}_\lambda \simeq \bigoplus_{\chi\in W_\lambda^\wedge} V_\chi^* \otimes F_{(\lambda,\chi)},$$

where $F_{(\lambda,\chi)}$ *is the simple local system on* G_{rs} *corresponding to* $\mathrm{Ind}_{\tilde{W}_\lambda}^{\tilde{W}}(1^\lambda \otimes V_\chi)$. *The LHS has the* W_λ*-action and the above decomposition is also that as* W_λ*-sheaves, where* W_λ *acts on* V_χ^* *in the RHS.*

1.3. *Intersection cohomology.* In the Grothendieck-Springer-Steinberg diagram 1.1, we also have the object $Rp_*\theta^{-1}\mathbf{C}_\lambda$, which is the direct image complex of $\theta^{-1}\mathbf{C}_\lambda$, a complex of sheaves of $\mathbf{C}$-vector spaces on G considered as an object in $D^{\mathrm{b}}_{\mathrm{c}}(\mathbf{C}_G)$. Here $D^{\mathrm{b}}_{\mathrm{c}}(\mathbf{C}_G)$ is the derived category of bounded complexes of $\mathbf{C}_G$-modules (sheaves of $\mathbf{C}$-vector spaces) on G whose cohomology sheaves are constructible.

On the other hand, according to Goresky-MacPherson [**GM**], we have the intersection cohomology complex $\mathrm{IC}(F) \in D^{\mathrm{b}}_{\mathrm{c}}(\mathbf{C}_G)$ for any local system F on G_{rs}, which is a very special object in $D^{\mathrm{b}}_{\mathrm{c}}(\mathbf{C}_G)$ called a *perverse* sheaf. So we have $\mathrm{IC}(p'_*\theta'^{-1}\mathbf{C}_\lambda)$ on G, which is an extension of $p'_*\theta'^{-1}\mathbf{C}_\lambda$ in the following sense:

$$\mathcal{H}^{-n}(\mathrm{IC}(p'_*\theta'^{-1}\mathbf{C}_\lambda)|G_{\mathrm{rs}}) \simeq p'_*\theta'^{-1}\mathbf{C}_\lambda \qquad (n = \dim G).$$

PROPOSITION 1.3.1. $Rp_*\theta^{-1}\mathbf{C}_\lambda[n] \simeq \mathrm{IC}(p'_*\theta'^{-1}\mathbf{C}_\lambda)$ *in* $D^{\mathrm{b}}_{\mathrm{c}}(\mathbf{C}_G)$, *where* $[n]$ *is the shift of degree* n.

The proof is the same as Lusztig's observation [**L 1**]. As a corollary:

COROLLARY 1.3.2. *The RHS in* 1.3.1 *has the* W_λ*-action induced by that of* 1.2.2. *Then* $Rp_*\theta^{-1}\mathbf{C}_\lambda$ *acquires the* W_λ*-action and a decomposition as* W_λ*-complexes*

$$Rp_*\theta^{-1}\mathbf{C}_\lambda \simeq \bigoplus_{x\in W_\lambda^\wedge} V_\chi^* \otimes \mathrm{IC}(F_{(\lambda,\chi)}),$$

where $\mathrm{IC}(F_{(\lambda,\chi)})$ *is a simple perverse sheaf.*

These observations in affine Weyl groups are clearly connected with S. Kato's work [**Ka**].

2. Review on $\mathcal{D}$-modules.

2.1. *Holonomic modules.* We review some generalities on $\mathcal{D}$-modules. Let X be a complex manifold and $\mathcal{D}_X$ a sheaf of rings of linear differential operators with coefficients in $\mathcal{O}_X :=$ the structure sheaf of X. Let M be a coherent (left) $\mathcal{D}_X$-module; hence M corresponds to a system of linear differential equations

$$\sum_{i=1}^{m} P_{ji}(x,\partial)u_i = 0 \qquad (1 \le j \le l)$$

in the classical sense if we choose local generators $u_1, \dots, u_l$ (locally $M \simeq \sum_{i=1}^{l} \mathcal{D}_X u_i$).

A basic invariant of a coherent $\mathcal{D}_X$-module is a characteristic variety. $\mathcal{D}_X$ has filtration by order; i.e., $F_i\mathcal{D}_X$ is a subsheaf of $\mathcal{D}_X$ consisting of differential operators of order $\leq i$. Then the associated graded algebra turns out to have an isomorphism

$$\mathrm{gr}^F\mathcal{D}_X := \bigoplus_{i=0}^{\infty} F_i\mathcal{D}_X/F_{i-1}\mathcal{D}_X \simeq \pi_*\mathcal{O}_{T^*X}$$

where $\pi\colon T^*X \to X$ is the cotangent bundle and $\pi_*\mathcal{O}_{T^*X}$ is the sheaf of algebras whose stalk at $x \in X$ is $\mathcal{O}_{X,x}[\xi_1, \dots, \xi_n]$, the polynomial algebra over the local ring $\mathcal{O}_{X,x}$ generated by a coordinate system $\xi_1, \dots, \xi_n$ of the cotangent space T_x^*X at x. Actually the above isomorphism realizes the usual principal symbols.

A *good* filtration of a coherent $\mathcal{D}_X$-module M is an increasing filtration F_jM ($j \in \mathbf{Z}$), $F_jM = 0$ ($j \ll 0$), satisfying the condition: F_jM is an $\mathcal{O}_X$-coherent submodule, $F_i\mathcal{D}_XF_jM \subset F_{i+j}M$ ($\forall i, j \in \mathbf{Z}$) and $F_i\mathcal{D}_XF_jM = F_{i+j}M$ ($\forall i \geq 0, j \gg 0$). Locally M always has a good filtration. A filtration F is good if and only if the associated graded module

$$\mathrm{gr}^F M := \bigoplus_{j\in\mathbf{Z}} F_jM/F_{j-1}M$$

is coherent over the algebra $\mathrm{gr}^F\mathcal{D}_X \simeq \pi_*\mathcal{O}_{T^*X}$.

For M with good filtration F, we thus have an $\mathcal{O}_{T^*X}$-coherent graded module

$$\pi^{\cdot}(\mathrm{gr}^F M) := \mathcal{O}_{T^*X} \underset{\pi^{-1}\pi_*\mathcal{O}_{T^*X}}{\otimes} \pi^{-1}(\mathrm{gr}^F M).$$

We denote by $\mathrm{Ch}\, M$ the support of this coherent module $\pi^{\cdot}(\mathrm{gr}^F M)$, which is clearly a closed conic subvariety in T^*X. It is seen that $\mathrm{Ch}\, M$ is independent of choice of good filtrations on M. $\mathrm{Ch}\, M$ is called the *characteristic variety* (or *singular support*) of M. The following is a fundamental theorem in the whole theory (Sato-Kawai-Kashiwara).

THEOREM 2.1 [**SKK**]. *With respect to the canonical symplectic structure of T^*X, $\mathrm{Ch}\, M$ is involutive. In particular, every irreducible component of $\mathrm{Ch}\, M$ has dimension $\geq n = \dim X$ if $M \neq 0$.*

We do not recall the definition of "involutive" and only advise the reader to remember the dimension criterion. For the proof of the theorem, Gabber's purely algebraic one [**Ga**] is often recommended.

DEFINITION. A coherent $\mathcal{D}_X$-module M is called *holonomic* if $M = 0$ or $\dim \mathrm{Ch}\, M = n$.

2.2. *The Riemann-Hilbert correspondence.* In the center of the theory of holonomic $\mathcal{D}$-modules, there is the Riemann-Hilbert correspondence established by Kashiwara [**K 2**] and Mebkhout [**M**].

The classical notion of solutions to a differential equation has a simple interpretation in our algebraic language. For a coherent $\mathcal{D}_X$-module M, the sheaf of $\mathcal{D}_X$-homomorphisms

$$\mathrm{Hom}_{\mathcal{D}_X}(M, \mathcal{O}_X)$$

turns out to be the sheaf of solutions in $\mathcal{O}_X$ to the system M. Thus one can consider the "higher extensions" of solutions, using homological algebra,

$$\mathrm{Ext}^*_{\mathcal{D}_X}(M, \mathcal{O}_X),$$

or more sophisticatedly, the right derived functor

$$\mathrm{Sol}\, M := \mathrm{RHom}_{\mathcal{D}_X}(M, \mathcal{O}_X)$$

in the derived category $D^{\mathrm{b}}(\mathbf{C}_X)$.

Kashiwara [**K 1**] actually proved, if M is *holonomic*, then the complex $\mathrm{Sol}\, M$ is a perverse sheaf (up to shift, in the recent terminology). More generally, if $D^{\mathrm{b}}_{\mathrm{h}}(\mathcal{D}_X)$ denotes the derived category of bounded complexes of $\mathcal{D}_X$-modules whose cohomology $\mathcal{D}_X$-modules are holonomic, then

$$\mathrm{Sol}\, M^{\cdot} := \mathrm{RHom}_{\mathcal{D}_X}(M^{\cdot}, \mathcal{O}_X)$$

is a bounded complex of $\mathbf{C}_X$-modules whose cohomology sheaves are constructible, that is, an object in $D^{\mathrm{b}}_{\mathrm{c}}(\mathbf{C}_X)$. Thus Sol is a functor

$$\mathrm{Sol}\colon D^{\mathrm{b}}_{\mathrm{h}}(\mathcal{D}_X) \to D^{\mathrm{b}}_{\mathrm{c}}(\mathbf{C}_X).$$

The Riemann-Hilbert correspondence means finding a nice subcategory of $D^{\mathrm{b}}_{\mathrm{h}}(\mathcal{D}_X)$ by extending the classical notion of *regular singularities* such that the (contravariant) functor Sol gives an (anti-)equivalence of categories on such a subcategory. Actually Kashiwara [**K 2**] and Mebkhout [**M**] attained the right goal, each in a different way.

There are a number of equivalent forms of the definition of regularity. We introduce only one of them, which should be called rather a theorem (see [**KK**]). A holonomic $\mathcal{D}_X$-module M is *regular* if and only if

$$\mathrm{Ext}^i_{\mathcal{D}_X}(M, \mathcal{O}_X)_x \simeq \mathrm{Ext}^i_{\mathcal{D}_{X,x}}(M_x, \hat{\mathcal{O}}_{X,x}) \quad (\forall i, \forall x \in X),$$

where $\hat{\mathcal{O}}_{X,x}$ is the formalization of the local ring $\mathcal{O}_{X,x}$ (isomorphic to the formal power series algebra over $\mathbf{C}$). Then we denote by $D^{\mathrm{b}}_{\mathrm{rh}}(\mathcal{D}_X)$ the full subcategory of $D^{\mathrm{b}}_{\mathrm{h}}(\mathcal{D}_X)$ consisting of complexes with regular holonomic cohomology modules.

THEOREM 2.2 (KASHIWARA-MEBKHOUT). *Sol:* $D^{\mathrm{b}}_{\mathrm{rh}}(\mathcal{D}_X) \to D^{\mathrm{b}}_{\mathrm{c}}(\mathbf{C}_X)$ *is an (anti-)equivalence of categories.*

In the above theorem, the abelian subcategory $\mathrm{Mod}_{\mathrm{rh}}(\mathcal{D}_X)$ consisting of regular holonomic $\mathcal{D}_X$-modules corresponds via Sol to the subcategory $\mathrm{Perv}(\mathbf{C}_X)$ of $D^{\mathrm{b}}_{\mathrm{c}}(\mathbf{C}_X)$ consisting of perverse sheaves on X (up to a certain shift).

Introductory articles to this section: [**LM, O, P, Be, BS, H 3, Sc**], etc.

3. The Harish-Chandra systems of differential equations.

3.1. *Some results of Harish-Chandra.* We are going back to the situation of §1, so we have a connected reductive group G. Let $\mathfrak{g} = \mathrm{Lie}\, G$ be the Lie algebra of G, $U(\mathfrak{g})$ the universal enveloping algebra of $\mathfrak{g}$, and Z the center of $U(\mathfrak{g})$. We have fixed a pair $B \supset T$ in §1, and hence the positive root system in $\mathfrak{t}^*$ etc. There then exists the Harish-Chandra isomorphism

$$\gamma\colon Z \simeq U(\mathfrak{t})^W \simeq \mathbf{C}[\mathfrak{t}^*]^W$$

where $U(\mathfrak{t})^W$ (resp. $\mathbf{C}[\mathfrak{t}^*]^W$) is the subring of W-invariants of the enveloping algebra $U(\mathfrak{t})$ (resp. the coordinate algebra $\mathbf{C}[\mathfrak{t}]$ of $\mathfrak{t}^*$). Thus for the set of $\mathbf{C}$-algebra homomorphisms $Z^\wedge := \mathrm{Hom}_{\mathbf{C}\text{-alg}}(Z, \mathbf{C})$, we have the isomorphism of varieties

$$Z^\wedge \simeq \mathfrak{t}^*/W.$$

Denote by $\chi_\lambda \in Z^\wedge$ the element corresponding to a W-orbit of $\lambda \in \mathfrak{t}^*$. Then

$$\chi_\lambda = \chi_{\lambda'} \text{ if and only if } \lambda + \rho = w(\lambda' + \rho) \text{ for some } w \in W,$$

where ρ is the half sum of the positive roots.

The Harish-Chandra system for χ_λ is the following system of differential equations on G:

$$M_\lambda : \begin{cases} (L_z - \chi_\lambda(z))u = 0, & (z \in Z), \\ (L_a + R_a)u = 0, & (a \in \mathfrak{g}). \end{cases}$$

Here L_a (resp. R_a) is the left (resp. right) invariant vector field corresponding to $a \in \mathfrak{g}$ and L_z is the left invariant differential operator corresponding to $z \in U(\mathfrak{g})$.

M_λ is the defining equation of so-called invariant eigendistributions on a real group $G_{\mathbf{R}} \subset G$ and, in particular, the global characters of irreducible representations of $G_{\mathbf{R}}$ with infinitesimal character χ_λ are solutions to M_λ. For the importance of M_λ, see many works in [**HC**]. The concept of character sheaves by Lusztig [**L 3**] is considered a counterpart in the finite group case to these systems of differential equations (see [**L 2**, Chapter 13, (13.8)]).

Noting that L_z, L_a, $R_a \in \Gamma(G, \mathcal{D}_G)$, we have the $\mathcal{D}_X$-module

$$\mathcal{M}_\lambda := \mathcal{D}_X u \simeq \mathcal{D}_X \Big/ \left\{ \sum_{z \in Z} \mathcal{D}_X(L_z - \chi_\lambda(z)) + \sum_{a \in \mathfrak{g}} \mathcal{D}_X(L_a + R_a) \right\}.$$

Since Z is a finitely generated algebra over $\mathbf{C}$, $\mathcal{M}_\lambda$ is $\mathcal{D}_X$-coherent. Let G_{rs} be as in §1. Then the following is a reinterpretation of Harish-Chandra's results:

PROPOSITION 3.1 (HARISH-CHANDRA). *$\mathcal{M}_\lambda | G_{\mathrm{rs}}$ is an integrable connection and if $\lambda + \rho$ is regular, then*

$$\mathrm{Hom}_{\mathcal{D}_{G_{\mathrm{rs}}}}(\mathcal{M}_\lambda | G_{\mathrm{rs}}, \mathcal{O}_{G_{\mathrm{rs}}}) \simeq p'_* \theta'^{-1} \mathbf{C}_\lambda,$$

where the RHS is the local system in §1.

3.2. *$\mathcal{M}_\lambda$ as a holonomic module.* We have the extension $Rp_*\theta^{-1}\mathbf{C}_\lambda$ of the RHS in Proposition 3.1 in $D^{\mathrm{b}}_{\mathrm{c}}(\mathbf{C}_X)$. In view of the Riemann-Hilbert correspondence 2.2, we can naturally ask:

CONJECTURE 3.2.1. $\mathrm{Sol}\, \mathcal{M}_\lambda \simeq \mathrm{Rp}_* \theta^{-1} \mathbf{C}_\lambda$ *if $\lambda + \rho$ is regular.*

Actually the following is rather easily proved:

LEMMA 3.2.2. *$\mathcal{M}_\lambda$ is a holonomic $\mathcal{D}_X$-module. In fact,* $\mathrm{Ch}\, \mathcal{M}_\lambda$ *can be determined.*

The following is more difficult to prove:

THEOREM 3.2.3 (KASHIWARA). *M_λ is regular.*

The proof for the Lie algebra version is given in [**HK 1**], and for λ integral, in [**HK 2**].

Corresponding to the functor Rp_* (resp. θ^{-1}) in the sheaf theory, we have the functor $\int_p$ (resp. θ^*) in the $\mathcal{D}$-module category (see general references given in §2). Using these functors, we can make Conjecture 3.2.1 more substantial.

For $\lambda \in \mathfrak{t}^*$, let t^λ be the (multivalued local) function on T defined by the (complex) power λ. Let $\mathcal{O}_\lambda := \mathcal{D}_T t^\lambda$ be the $\mathcal{D}_T$-module generated by t^λ. This is globally defined on T;

$$\mathcal{O}_\lambda \simeq \mathcal{D}_T / \left\{ \sum_{i=1}^{r} \mathcal{D}_T(\partial_i - \lambda_i) \right\}$$

if we choose a coordinate $(t_1, \ldots, t_r)$ on $T \simeq (\mathbf{C}^\times)^r$, $\partial_i = \partial/\partial t_i$, and if $t^\lambda = t_1^{\lambda_1} \cdots t_r^{\lambda_r}$. Then $\mathbf{C}_\lambda = \operatorname{Sol} \mathcal{O}_\lambda$ on T. The $\mathcal{D}_X$-module corresponding to $\mathrm{Rp}_* \theta^{-1} \mathbf{C}_\lambda$ is then $\int_p \theta^* \mathcal{O}_\lambda$, which is known to be regular holonomic by general theory.

CONJECTURE 3.2.4. *$\int_p \theta^* \mathcal{O}_\lambda \simeq M_\lambda$ if $\lambda + \rho$ is regular. In particular, M_λ gains the W_λ-action.*

By the Riemann-Hilbert corespondence 2.2, 3.2.4 implies 3.2.1 (we want to conjecture that 3.2.4 is true as the algebraic $\mathcal{D}$-modules).

Conjecture 3.2.4 is true for λ integral. In this case, the proof is the same as the case for $\lambda = 0$, which is given in [**HK 2**]. Actually, by definition, $\mathcal{O}_\lambda \simeq \mathcal{O}_0 = \mathcal{O}_T$ if λ is integral ("translation principle"!).

4. Primitive ideals.

4.1. *Hypothesis.* Our motivation for Conjecture 3.2.4 is its following consequence, which corresponds to Corollary 1.3.2 by the Riemann-Hilbert correspondence.

HYPOTHESIS 4.1. $M_\lambda \simeq \bigoplus_{\chi \in W_\lambda^\wedge} V_\chi \otimes M(\lambda, \chi)$ for some simple $\mathcal{D}_G$-module $M(\lambda, \chi)$ corresponding to $\chi \in W_\lambda^\wedge$, if $\lambda + \rho$ is regular. Here $V_\chi \simeq V_\chi^*$ is an irreducible W_λ-module corresponding to an irreducible class $\chi \in W_\lambda^\wedge$ of the integral Weyl group W_λ for $\lambda \in \mathfrak{t}^*$.

This hypothesis is pleasantly compatible with the "translation principle" in the theory of primitive ideals, which we shall now explain.

4.2. *Classification of primitive ideals.* A two-sided ideal of $U(\mathfrak{g})$ is called *primitive* if it is the annihilator of a simple $U(\mathfrak{g})$-module. The classification theory of primitive ideals has now been completed by Borho, Brylinski, Jantzen, Duflo, Vogan and especially Joseph (see Jantzen's book [**Ja**] and [**Bo**; **J 1, 2**; **V 1, 2**]).

Let $\mathfrak{X}$ be the set of all primitive ideals of $U(\mathfrak{g})$. If $I \in \mathfrak{X}$, then there exists a unique $\chi_\lambda \in Z^\wedge$ such that $z - \chi_\lambda(z) \in I$ for any $z \in Z$. Hence we have the map

$$\Phi \colon \mathfrak{X} \to Z^\wedge \simeq \mathfrak{t}^*/W$$

which is a surjection. Duflo [**D**], using Verma modules, shows that there is a surjection

$$W_\lambda \twoheadrightarrow \Phi^{-1}(\chi_\lambda).$$

Actually it is the quotient map by "the left cell" equivalence relation of W_λ (Joseph-Vogan).

We hereafter assume $\lambda + \rho$ is regular as in 4.1. We shall explain Joseph's classification theory of $\mathfrak{X}$. For $I \in \Phi^{-1}(\chi_\lambda)$, he attached a certain W_λ-harmonic polynomial p_I called the *Goldie-rank polynomial* and showed that its W_λ-span V_I in the W_λ-harmonic polynomials is an irreducible W_λ-module, which is called the *Goldie-rank representation*.

On the other hand, under the canonical filtration of $U(\mathfrak{g})$, the associated graded ideal $\operatorname{gr} I$ is an ideal of the commutative algebra $\operatorname{gr} U(\mathfrak{g}) \simeq \mathbf{C}[\mathfrak{g}^*]$ and the zeros of $\operatorname{gr} I =: \operatorname{Ass} I$ (the *associated variety* of I) is a G-stable subvariety of the nilpotent variety in $\mathfrak{g}^*$. It was proved rather recently that $\operatorname{Ass} I$ is irreducible and hence the closure of a single nilpotent orbit $O(I)$ [**BoB 1, 2**; **J 3**; **H 1, 2**]. Further, $O(I)$ and V_I correspond to each other by the Springer correspondence (via the induction $W_\lambda \to W$). The central theorem of Joseph's works [**J 1, 2**] states:

For a nilpotent orbit O, put

$$\operatorname{Clan} O := \{I \in \Phi^{-1}(\chi_\lambda); O(I) = O\}.$$

Then

$$O(I) = O(J) \quad \text{if and only if } V_I = V_J$$

and the set of Joseph's Goldie-rank polynomials

$$\{p_I; I \in \operatorname{Clan} O\}$$

gives a basis of the Goldie-rank module V_I.

An approach to geometrically understanding this result has been taken by [**BoB 2**, **BBM**].

4.3. *Theorem.* For $I \in \Phi^{-1}(\chi_\lambda)$, by definition, I contains $z - \chi_\lambda(z)$ $(z \in Z)$. Let M_λ be the Harish-Chandra system in 3.1. Then the $\mathcal{D}_G$-module

$$M_I := \mathcal{D}_G \Big/ \left\{ \sum_{y \in I} \mathcal{D}_G L_y + \sum_{a \in \mathfrak{g}} \mathcal{D}_G (L_a + R_a) \right\}$$

is a quotient $\mathcal{D}_G$-module of M_λ. (For a minimal $I \in \Phi^{-1}(\chi_\lambda)$, $M_I = M_\lambda$). In other words $M_I = M_\lambda / I M_\lambda$.

THEOREM 4.3. *Let $I \in \Phi^{-1}(\chi_\lambda)$. Under Hypothesis* 4.1,

(i) *The W_λ-module* $\operatorname{Hom}_{\mathcal{D}_G}(M_\lambda, M_I)$ *contains the Goldie-rank representation V_I with multiplicity one and hence there exists the surjection unique up to scalar multiplications*

$$\operatorname{Hom}_{\mathcal{D}_G}(M_\lambda, M_I) \twoheadrightarrow V_I.$$

(ii) *Let $\pi_I\colon M_\lambda \twoheadrightarrow M_I$ be the natural projection. Then $[\pi_I] \in V_I$, the image of π_I by the map in* (i), *spans $\mathbf{C}p_I$, the same line spanned by Joseph's Goldie-rank polynomial p_I.*

Roughly speaking, the theorem means that

$$M_I \simeq \mathbf{C}p_I \otimes M(\lambda, \chi_I) \oplus \sum_{\chi<\chi_I} M(\lambda,\chi)^{m_\chi}$$

in the expression of M_λ in Hypothesis 4.1 (χ_I is the class of the Goldie-rank representation V_I and $<$ is a suitable order in $W_\lambda^\wedge$).

For λ integral, Hypothesis 4.1 is true and [**HK 2**] is devoted to the proof of the theorem. There we also discuss the associated varieties from a microlocal point of view ($\mathcal{E}$-module theory).

These arguments are closely related to works by Barbasch and Vogan [**BV 1, 2, 3**] that have been exposed in more classical contexts, asymptotic behavior, etc.

5. Hodge modules after Saito.

5.1. *Origin.* A *Hodge structure of weight $k \in \mathbf{Z}$* on a finite-dimensional $\mathbf{Q}$-vector space H is a finite decreasing filtration $F^\cdot$ on $H_\mathbf{C} := H \otimes_\mathbf{Q} \mathbf{C}$ such that

$$F^i \oplus \overline{F}^{k-i+1} = H_\mathbf{C} \quad (\forall i \in \mathbf{Z}),$$

where $\overline{F}^j$ is the complex conjugate of F^j. If we put $H^{p,q} := F^p \cap \overline{F}^q$ $(p+q=k)$, then $H_\mathbf{C} = \bigoplus_{p+q=k} H^{p,q}$, $H^{p,q} = \overline{H}^{p,q}$ and $F^i = \bigoplus_{i\le p} H^{p,q}$.

A *variation of Hodge structures* on a smooth variety X is a local system H over $\mathbf{Q}$ on X with a finite decreasing filtration $F^\cdot$ on the vector bundle $\mathcal{O}_X \otimes_\mathbf{Q} H$ such that each stalk $H_x := \kappa(x) \otimes_\mathbf{Q} H$ with filtration $F^\cdot := \kappa(x) \otimes_{\mathcal{O}_{X,x}} F^\cdot$ is a Hodge structure ($\kappa(x)$ is the residue field of $\mathcal{O}_{X,x}$ at x) and (Griffiths' condition) $\theta F^i \subset F^{i-1}$ for any local vector fields θ, where the action of θ is under the canonical integrable connection on $\mathcal{O}_X \otimes H$.

Thus, if we rewrite the Hodge filtration $F_\cdot$ as $F_i := F^{-i}$ (increasing filtration), then the $\mathcal{D}_X$-module $\mathcal{O}_X \otimes H$ has the good filtration $F_\cdot$ given rise to by the Hodge filtration.

From this point of view, several authors have proposed studying Hodge structures in the category of filtered $\mathcal{D}$-modules (e.g., see [**Br**]).

Recently, Morihiko Saito succeeded in constructing this kind of theory, which seems to be very significant and satisfactory [**S 1, 2, 3**]. Since it is still difficult to introduce his theory even briefly, we shall content ourselves with some thoughts on this deep theory. (We cannot even reproduce several crucial definitions.)

5.2. *Some categories of filtered $\mathcal{D}$-modules.* Keeping the concept of variations of Hodge structures in the above version, we are going into the category of regular holonomic $\mathcal{D}$-modules (an extended concept of integrable connections) with good filtrations. We also recall the Riemann-Hilbert correspondence (in the algebraic category) in the form dual to that introduced in §2.

Let X be an algebraic variety and consider $\mathcal{D}_X$-modules in the algebraic category (i.e., $\mathcal{D}_X$ is the ring of algebraic differential operators). We say that a $\mathcal{D}_X$-module M is regular holonomic (algebraically) if M extends to a regular holonomic $\mathcal{D}_{\overline{X}}$-module for some smooth compactification $\overline{X}$ of X. Define the de Rham functor DR_X by

$$\mathrm{DR}_X M := \mathrm{RHom}_{\mathcal{D}_X}(\mathcal{O}_{X_{\mathrm{an}}}, M_{\mathrm{an}})[\dim X],$$

where $\mathcal{O}_{X_{\mathrm{an}}}$ is the analytic structure sheaf of the underlying complex manifold X_{an} of X and $M_{\mathrm{an}} := \mathcal{O}_{X_{\mathrm{an}}} \otimes_{\mathcal{O}_X} M$. Then DR_X gives an equivalence of categories:

$$\mathrm{DR}_X\colon \mathrm{Mod}_{\mathrm{rh}}\, \mathcal{D}_X \simeq \mathrm{Perv}\, \mathbf{C}_X,$$

where $\mathrm{Mod}_{\mathrm{rh}}\, \mathcal{D}_X$ is the abelian category of regular holonomic $\mathcal{D}_X$-modules and $\mathrm{Perv}\, \mathbf{C}_X \subset D^{\mathrm{b}}_{\mathrm{c}}(\mathbf{C}_X)$ is the category of (algebraic) perverse sheaves over $\mathbf{C}_X$.

The abelian category $\mathrm{Perv}\, \mathbf{Q}_X$ of perverse sheaves over $\mathbf{Q}_X$ is similarly defined, and we define the category

$$\mathrm{MF}_{\mathrm{rh}}(\mathcal{D}_X, \mathbf{Q}) := \{(M, F, K, \alpha)\},$$

where $M \in \mathrm{Mod}_{\mathrm{rh}}\, \mathcal{D}_X$, F is a good filtration on M, $K \in \mathrm{Perv}\, \mathbf{Q}_X$ and

$$\alpha\colon \mathrm{DR}_X M \to \mathbf{C} \otimes_{\mathbf{Q}} K$$

is an isomorphism. The category $\mathrm{Mod}_{\mathrm{rh}}(\mathcal{D}_X, \mathbf{Q})$ is no longer abelian.

5.3. *Categories of Hodge modules.* Saito [**S 1, 2**] defined a certain full subcategory $\mathrm{MH}(X, k)^{\mathrm{p}}$ ($k \in \mathbf{Z}$) of $\mathrm{MF}_{\mathrm{rh}}(\mathcal{D}_X, \mathbf{Q})$ which turns out to be an abelian category and corresponds by the Riemann-Hilbert correspondence, to certain extensions of variations of "pure polarized" Hodge structures of weight k. This is called the category of *polarized pure Hodge modules of weight* k and behaves nicely under several basic functors $\int_f$, Lf^*, the duality functor D_X, etc.

We pick up only few among them.

(1) (Localizability). *Let* $X = \bigcup_i U)_i$ *be an open covering. Then* $E \in \mathrm{MH}(X, k)^{\mathrm{p}}$ *if and only if* $E|U_i \in \mathrm{MH}(U_i, k)^{\mathrm{p}}$ *for any* i.

(2) (Brylinski's conjecture). *If* $(M, F, K) \in \mathrm{MH}(X, k)^{\mathrm{p}}$ *is a simple object, then* $Z = \mathrm{Supp}\, M$ *is an irreducible subvariety and for some open* U *of* Z, *there exists a variation of polarized Hodge structures* H *of weight* $k + \dim Z$ *such that*

$$K = \mathrm{IC}(H) \qquad (\textit{intersection cohomology}),$$

$$(M, F, K)|X\backslash Y = \left(\int_i (\mathcal{O}_U \otimes H, F_H), i_* H[\dim Z]\right),$$

where $Y = Z\backslash U$, $i\colon U \hookrightarrow X\backslash Y$ *is the closed immersion.*

Conversely, objects of this kind coming from the intersection cohomologies belong to $\mathrm{MH}(X, *)^{\mathrm{p}}$.

(3) *For a projective morphism* $f\colon X \to Y$, *the functor*

$$\int_f^i = \mathcal{H}^i \int_f \colon \mathrm{MH}(X, k)^{\mathrm{p}} \to \mathrm{MH}(y, k + i)^{\mathrm{p}}$$

is naturally defined.

Furthermore, in [**S 3**], Saito gives an extended category corrresponding to the mixed Hodge structures. Naturally, the following category can be defined:

$$\mathrm{MHW}(X)^{\mathrm{p}} = \{(M, F, K, W) \text{ as below}\},$$

(i) $(M, F, K) \in \mathrm{MF}_{\mathrm{rh}}(\mathcal{D}_X, \mathbf{Q})$,

(ii) W is a finite increasing filtration of (M, F, K) (weight filtration) such that $\mathrm{gr}_k^W(M, F, K) \in \mathrm{MH}(X, k)^{\mathrm{p}}$.

This category does not behave well, and he defined the full subcategories $\mathrm{MHM}(X)^{\mathrm{g}}$ (*geometric mixed Hodge modules*) $\subset$ $\mathrm{MHM}(X)^{\mathrm{p}}$ (*polarized mixed Hodge modules*) in $\mathrm{MHW}(X)^{\mathrm{p}}$, both of which are abelian. For these abelian categories, $\mathrm{MHM}(X)^{\mathrm{g}}$ in particular has a nice property. Philosophically, $\mathrm{MHM}(X)^{\mathrm{g}}$ corresponds to the category of mixed perverse sheaves of geometric origin in [**BBD**].

6. Realization of the Hecke algebra of a Weyl group by Hodge module category after Tanisaki.

6.1. *Realization by constructible sheaf category on positive characteristics.* Tanisaki's construction [**T 2**] is modeled after the following well-known one (see [**Sp**]).

Let $X = G/B$ be the flag manifold over a finite field of characteristic p. Let $\mathcal{C}$ be the category of constructible $\mathbf{Q}_l$-sheaves ($l \neq p$) on $X \times X$ whose restriction to any G-orbit Y has composition factors of type $\mathbf{Q}_{l,Y}(n)$ ($n \in \mathbf{Z}$), where (n) is the Tate twist. Then the Grothendieck group $K(\mathcal{C})$ is a $\mathbf{Z}[q, q^{-1}]$-module by the action of the indeterminant q as $q^n[K] = [K(n)]$ for $K \in \mathcal{C}$.

The multiplication of $K(\mathcal{C})$ is defined as follows. Let $p_{13}\colon X \times X \times X \to X \times X$ be the projection to the $(1,3)$-factor and $r\colon X \times X \times X \hookrightarrow X \times X \times X \times X$ be defined by $r(x, y, z) = (x, y, y, z)$. Then

$$[K_1] \circ [K_2] = \sum_{\lambda_i} (-)^i [R^i p_{13*}(r^{-1}(K_1 \boxtimes K_2))]$$

defines a $\mathbf{Z}[q, q^{-1}]$-algebra structure on $K(\mathcal{C})$, which is isomorphic to the Hecke (-Iwahori) algebra $H(W)$ for the Weyl group W of G.

6.2. *Realization by Hodge modules.* Now let $X = G/B$ be the flag manifold over $\mathbf{C}$. Modeling the category $\mathcal{C}$ in 6.1, we define the full subcategory $\mathcal{A}$ of $\mathrm{MHM}(X \times X)^{\mathrm{g}}$ consisting of such objects $E = (M, F, K, W)$ that E has a G-action and $\mathrm{gr}_i^W E$ is a direct sum of objects $L \in \mathrm{MH}(X \times X)^{\mathrm{p}}$ corresponding to shifts of the intersection cohomology $\mathrm{IC}(\mathbf{Q}_Y)$ on G-orbits Y of $X \times X$ (see 5.3, (2)).

Then $\mathcal{A}$ is an abelian category and its Grothendieck group $K(\mathcal{A})$ is a $\mathbf{Z}[q, q^{-1}]$-module by the action of q defined by

$$q^n[(M, F, K, W)] = (M, F[-n], K, W[2n]),$$

where $F[-n]_i = F_{i+n}$, $W[2n]_i = W_{i-2n}$ are the shifts of the filtrations.

For E_1, $E_2 \in \mathcal{A}$, the following:

$$[E_1] \circ [E_2] = \sum_i (-)^i \left[\int_{p_{13}}^i L_{\dim X} r^*(E_1 \boxtimes E_2) \right]$$

can be defined as a formal sum of objects in $\mathrm{MHM}(X \times X)^{\mathrm{g}}$, where P_{13}, r are as in 6.1.

THEOREM 6.2 (TANISAKI [**T 2**]). (i) $[E_1] \circ [E_2] \in K(\mathcal{A})$.

(ii) *By the multiplication in* (i), $K(\mathcal{A})$ *is isomorphic to the Hecke algebra* $H(W)$.

The proof, of course, uses several deep results of Saito [**S 2**, **3**].

Forgetting everything as $\mathcal{A} \ni (M, F, \dots) \mapsto M \in \mathrm{Mod}_{\mathrm{rh}}(X \times X)$, we have a homomorphism

$$K(\mathcal{A}) \to \mathbf{C}[W] = \text{ group algebra of } W,$$

which is the specialization homomorphism at $q = 1$, and this recovers the construction in [**KT**].

6.3. *Relationship with the Hecke algebra of an affine Weyl group.* Tanisaki's construction throws light on recent works of Kazhdan-Lusztig [**KL**] and Ginsburg [**G 2**] on the Hecke algebra of an affine Weyl group in the framework of equivariant K-theory.

Let Z be the subvariety of $X \times X \times \mathfrak{g}$ defined by

$$Z := \{(xB, yB, a) \in X \times X \times \mathfrak{g}; a\colon \text{nilpotent}, a \in x\mathfrak{b} \cap y\mathfrak{b}\},$$

where $\mathfrak{b} := \mathrm{Lie}\, B$. Then G acts on Z and the multiplicative group $\mathbf{C}^\times$ acts on the nilpotent variety of $\mathfrak{g}$. Since the cotangent bundle T^*X is identified with

$$\{(xB, a) \in X \times \mathfrak{g}; a\colon \text{nilpotent}, a \in x\mathfrak{b}\},$$

we have the embedding

$$Z \hookrightarrow T^*X \times T^*X \quad \text{by } (xB, yB, a) \mapsto (xB, yB, a, a)$$

Consider the $G \times \mathbf{C}^\times$-equivariant K-group $K^{G \times \mathbf{C}^\times}(Z)$, the Grothendieck group of coherent $\mathcal{O}_{T^*X \times T^*X}$-modules with $G \times \mathbf{C}^\times$ actions and with supports in Z.

THOEREM 6.3.1 [**KL**, **G 2**]. $K^{G \times \mathbf{C}^\times}(Z)$ *is naturally isomorphic to the Hecke algebra* $H(W_a)$ *of the affine Weyl group* W_a. *Here* W_a *is the semidirect product of* W *and the weight group* $\mathrm{Hom}_{\mathrm{alg\,gp}}(T, \mathbf{C}^\times)$.

Now going back to 6.2, we have the realization of $H(W) \simeq K(\mathcal{A})$ where

$$\mathcal{A} \subset \mathrm{MHM}(X \times X)^{\mathrm{g}} \subset \mathrm{MF}_{\mathrm{rh}}(\mathcal{D}_{X \times X}, \mathbf{Q}_{X \times X}).$$

For $(M, F, \dots) \in \mathcal{A}$, take the gradation

$$\pi^{\cdot}(\mathrm{gr}^F M) = \mathcal{O}_{T^*X^2} \underset{\pi^{-1}\pi_*\mathcal{O}_{T^*X^2}}{\otimes} \pi^{-1}(\mathrm{gr}^F M)$$

(π is the projection $T^*X^2 \to X^2$), which is an $\mathcal{O}_{T^*X^2}$-coherent module with

$G \times \mathbf{C}^\times$ action. Here the $\mathbf{C}^\times$ action is defined by

$$z(\sigma(x,\xi) \otimes u_i) = \sigma(x, z^{-1}\xi) \otimes z^i u_i \qquad (z \in \mathbf{C}^\times)$$

for $\sigma(x,\xi) \in \mathcal{O}_{T^*X^2}$, $u_i \in \mathrm{gr}_i^F M$.

If we define the automorphism $\alpha\colon T^*X^2 \to T^*X^2$ by $\alpha(xB, yB, a, a') = (xB, yB, a, -a')$, then

$$\alpha(Z) = \coprod_{w \in W} T^*_{Y_w}(X^2),$$

where Y_w is a G-orbit in $X \times X$ and $T^*_{Y_w}(X^2)$ is the conormal bundle of Y_w. Since for $E = (M, F, \dots) \in \mathcal{A}$, $\pi^{\cdot}(\mathrm{gr}^F M)$ is supported in $\alpha(Z)$, we can define the map

$$\alpha^* \circ \mathrm{gr}\colon K(\mathcal{A}) \ni [E] \mapsto [\alpha^* \circ \pi^{\cdot}(\mathrm{gr}^F M)] \in K^{G \times \mathbf{C}^\times}(Z).$$

THEOREM 6.3.2 [**T 2**]. *The above map $\alpha^* \circ \mathrm{gr}$ is an injective homomorphism of $\mathbf{Z}[q, q^{-1}]$-algebras and coincides with the natural injection $H(W) \hookrightarrow H(W_a)$ given rise to by $W \hookrightarrow W_a$.*

His proof of Theorem 6.3.2 also gives an alternative proof of Theorem 6.3.1. In fact, in its original proofs [**KL, G 2**], the relations in $H(W)$ are delicate to check (for [**G2**], see the Introduction of [**KL**]).

REFERENCES

[**BV 1**] D. Barbasch and D. Vogan, *The local structure of characters*, J. Funct. Anal. **37** (1980), 27–55.

[**BV 2**] ——, *Primitive ideals and orbital integrals in complex classical groups*, Math. Ann. **259** (1982), 153–199.

[**BV 3**] ——, *Primitive ideals and orbital integrals in complex exceptional groups*, J. Algebra **80** (1983), 350–382.

[**BeB**] A. Beilinson and J. Bernstein, *Localisation de $\mathfrak{g}$-modules*, C. R. Acad. Sci. Paris **292** (1981), 15–18.

[**BBD**] A. Beilinson, J. Bernstein, and P. Deligne, *Faisceaux pervers*, Astérisque, no. 100, Soc. Math. France, Paris, 1983.

[**Be**] J. Bernstein, *Algebraic theory of D-modules*, Preprint.

[**BS**] A. Borel et al., *Algebraic D-modules*, Perspect. Math., vol. 2, Academic Press, 1987.

[**Bo**] W. Borho, *A survey on enveloping algebras of semisimple Lie algebras*, Lie Algebras and Related Topics, CMS Conf. Proc., vol. 5, Amer. Math. Soc., Providence, R.I., 1986, pp. 19–50.

[**BoB 1**] W. Borho and J.-L. Brylinski, *Differential operators on homogeneous spaces*. I, Invent. Math. **69** (1982), 437–476.

[**BoB 2**] ——, *Differential operators on homogeneous spaces*. III, Invent. Math. **80** (1985), 1–68.

[**BBM**] W. Borho, J.-L. Brylinski, and R. D. MacPherson, *Equivariant K-theory approach to nilpotent orbits*, Preprint, Inst. Hautes Étutes Sci.

[**Br**] J.-L. Brylinski, *Modules holonomes à singularités regulières et filtration de Hodge*. II, Astérisque, no. 101–102, Soc. Math. France, Paris, 1983, pp. 75–117.

[**BK**] J.-L. Brylinski and M. Kashiwara, *Kazhdan-Lusztig conjecture and holonomic systems*, Invent. Math. **64** (1981), 387–410.

[**D**] M. Duflo, *Sur la classification des idéaux primitifs dans l'algèbre enveloppante d'une algèbre de Lie semi-simple*, Ann. of Math. (2) **105** (1977), 107–120.

[**Ga**] O. Gabber, *The integrability of the characteristic variety*, Amer. J. Math. **103** (1981), 445–468.

[**G 1**] V. Ginsburg, *$\mathfrak{g}$-modules, Springer's resolutions and bivariant Chern classes*, Adv. in Math. **61** (1986), 1–48.

[**G 2**] ———, *Deligne-Langlands conjecture and representations of affine Hecke algebras*, Preprint.

[**GM**] M. Goresky and R. D. MacPherson, *Intersection homology*. II, Invent. Math. **72** (1983), 77–130.

[**HC**] Harish-Chandra, *Collected papers*, Springer-Verlag, 1984.

[**H 1**] R. Hotta, *On Joseph's construction of Weyl group representations*, Tôhoku Math. J. **36** (1984), 49–74.

[**H 2**] ———, *A local formula for Springer's representations*, Adv. Stud. in Pure Math., vol. 6, Kinokuniya and North-Holland, 1985, pp. 127–138.

[**H 3**] ———, *Introduction to $\mathcal{D}$-modules*, Lecture notes of Inst. of Math. Sci., Madras, 1987.

[**HK 1**] J. Hotta and M. Kashiwara, *The invariant holonomic system on a semisimple Lie algebra*, Invent. Math. **75** (1984), 327–358.

[**HK 2**] ———, *Quotients of the Harish-Chandra system by primitive ideals*, Geometry of Today, Rome, 1984, Progr. Math., vol. 60, Birkhäuser, 1985, pp. 185–205.

[**Ja**] J. C. Janzten, *Einhüllende Algebren halbeinfacher Lie-Algebren*, Springer-Verlag, Berlin, 1983.

[**J 1**] A. Joseph, *Goldie rank in the enveloping algebra of a semisimple Lie algebra*. I, II, III, J. Algebra **65** (1980), 269–283; ibid., 284–306; ibid. **73** (1981), 295–326.

[**J 2**] ———, *On the classification of primitive ideals in the enveloping algebra of a semisimple Lie algebra*, Lecture Notes in Math., vol. 1024, Springer-Verlag, 1983, pp. 30–76.

[**J 3**] ———, *On the associated variety of a primitive ideal*, J. Algebra **93** (1985), 509–523.

[**K 1**] M. Kashiwara, *On the maximally overdetermined system of linear differential equations*. I, Publ. Res. Inst. Math. Sci. **10** (1975), 563–579.

[**K 2**] ———, *The Riemann-Hilbert problem for holonomic systems*, Publ. Res. Inst. Math. Sci. **20** (1984), 319–365.

[**KK**] M. Kashiwara and T. Kawai, *On holonomic systems of microdifferential equations*. III (*systems with regular singularities*), Publ. Res. Inst. Math. Sci. **17** (1981), 813–979.

[**KKMOOT**] M. Kashiwara, A. Kowata, K. Minemura, K. Okamoto, T. Oshima, and M. Tanaka, *Eigenfunctions of invariant differential operators on a symmetric space*, Ann. of Math. (2) **107** (1978), 1–39.

[**KT**] M. Kashiwara and T. Tanisaki, *The characteristic cycles of holonomic systems on a flag manifold*, Invent. Math. **77** (1984), 185–198.

[**Ka**] S. Kato, *A realization of irreducible representations of affine Weyl groups*, Nederl. Akad. Wetensch. Proc. Ser. A **86** (1983), 193–201.

[**KL**] D. Kazhdan and G. Lusztig, *Proof of the Deligne-Langlands conjecture for Hecke algebras*, Invent. Math. **87** (1987), 153–215.

[**LM**] D.-T. Lê and Z. Mebkhout, *Introduction to linear differential systems*, Proc. Sympos. Pure Math., vol. 40, Amer. Math. Soc., Providence, R.I., 1983, pp. 31–63.

[**L 1**] G. Lusztig, *Green polynomials and singularities of nilpotent classes*, Adv. in Math. **42** (1982), 169–178.

[**L 2**] ———, *Characters of reductive groups over a finite field*, Ann. of Math. Stud., vol. 107, Princeton Univ. Press, Princeton, N.J., 1984.

[**L 3**] ———, *Character sheaves*. I, II, III, IV, V, Adv. in Math. **56** (1985), 193–327; ibid. **57** (1985), 226–265, 266–317; ibid. **59** (1986), 1–63; ibid. **61** (1986), 103–155.

[**M**] Z. Mebkhout, *Une équivalence de catégories et une autre équivalence de catégories*, Compositio Math. **51** (1984), 51–62, 63–88.

[**O**] T. Oda, *Introduction to algebraic analysis on complex manifolds*, Adv. Stud. Pure Math., vol. 1, Kinokuniya and North-Holland, 1982, pp. 29–48.

[**P**] F. Pham, *Singularités des systèmes différentiels de Gauss-Manin*, Progr. Math., vol. 2, Birkhäuser, 1979.

[**S 1**] M. Saito, *Hodge structure via filtered $\mathcal{D}$-modules*, Astérisque, no. 130, Soc. Math. France, Paris, 1985, pp. 342–351.

[**S 2**] ———, *Modules de Hodge polarizables*, Preprint.

[**S 3**] ———, *Mixed Hodge modules*, Preprint.

[**SKK**] M. Sato, T. Kawai, and M. Kashiwara, *Microfunctions and pseudodifferential equations*, Lecture Notes in Math., vol. 287, Springer-Verlag, 1973, pp. 265–529.

[**Sc**] P. Schapira, *Microdifferential systems in the complex domain*, Springer-Verlag, New York, 1985.

[**Sp**] T. A. Springer, *Quelques applications de la cohomologie d'intersection*, Sém. Bourbaki, exp. 589, Astérisque, no. 92–93, Soc. Math. France, Paris, 1982, pp. 249–273.

[**T 1**] T. Tanisaki, *Holonomic systems on a flag variety associated to Harish-Chandra modules and representations of a Weyl group*, Adv. Stud. Pure Math., vol. 6, Kinokuniya and North-Holland, 1985, pp. 139–154.

[**T 2**] ——, *Hodge modules, equivariant K-theory and Hecke algebras*, Preprint.

[**V 1**] D. Vogan, *A generalized τ-invariant for the primitive spectrum of a semisimple Lie algebra*, Math. Ann. **242** (1979), 209–224.

[**V 2**] ——, *Ordering of the primitive spectrum of a semisimple Lie algebra*, Math. Ann. **248** (1980), 195–203.

TOHOKU UNIVERSITY, JAPAN

Proceedings of Symposia in Pure Mathematics
Volume 47 (1987)

Characters of Solvable Groups

I. M. ISAACS

My charge from the conference organizers for this lecture was to give a general and inclusive survey of recent activity in the character theory of solvable groups. I am happy to be able to report that I am unable to accomplish this mission; there has been far too much going on to be encompassed by a one-hour lecture. I have decided, therefore, to limit my remarks to three interrelated topics: M-groups, degree problems, and π-theory, with the main focus on the first of these.

Recall that an irreducible character $\chi \in \mathrm{Irr}(G)$ is *monomial* if it is the result of inducing to G a linear character λ of some subgroup of G. (We write $\chi = \lambda^G$ in this case.) An *M-group* is one in which every irreducible character is monomial. The key result about these groups, due to K. Taketa [**18**], puts them within the purview of this lecture: An M-group is necessarily solvable.

Taketa's proof of the solvability of M-groups actually yields more. Suppose $1 = f_1 < f_2 < \cdots < f_k$ are the distinct degrees of the irreducible characters of G. If G is an M-group, Taketa's argument proves that the ith term $G^{(i)}$ of the derived series of G is contained in the kernel of every irreducible character of G with degree $\le f_i$. Therefore $G^{(k)}$ is contained in $\ker\chi$ for every $\chi \in \mathrm{Irr}(G)$, and so is trivial. This proves not only that G is solvable, but also that its derived length is bounded by k, the number of distinct irreducible character degrees of G.

Let us write $\mathrm{cd}(G) = \{f_i\} = \{\chi(1) \mid \chi \in \mathrm{Irr}(G)\}$. For M-groups, then, the derived length $\mathrm{cd}(G) \le |\mathrm{cd}(G)|$ and it is reasonable to ask if this bound also holds for arbitrary solvable groups. (This is an example of what I call a *degree problem*: obtain control of the structure of G in terms of the set $\mathrm{cd}(G)$.) Combining the results of several authors, here is what is known about the question of bounding the $\mathrm{dl}(G)$ in terms of $|\mathrm{cd}(G)|$.

THEOREM (BERGER [**1**], GARRISON [**7**], GLUCK [**8**], ISAACS [**11**]). *Let G be solvable. Then*

(a) $\mathrm{dl}(G) \le 2|\mathrm{cd}(G)|$.

(b) *If $|G|$ is odd, then* $\mathrm{dl}(G) \le |\mathrm{cd}(G)|$.

1980 *Mathematics Subject Classification* (1985 *Revision*). Primary 20C15, 20D10.
Research partially supported by a grant from the National Science Foundation.

(c) *If* $|\mathrm{cd}(G)| \le 4$, *then* $\mathrm{dl}(G) \le |\mathrm{cd}(G)|$.

Although most of the above results are fairly old now, I mention this here because of the recent work of D. Gluck, who improved my original bound by proving (a). (My inequality had a coefficient of 3.) It is still open whether or not the inequality $\mathrm{dl}(G) \le |\mathrm{cd}(G)|$ is valid for all solvable groups, and essentially nothing is known which indicates whether or not the "correct" bound on $\mathrm{dl}(G)$ is anywhere nearly as large as $|\mathrm{cd}(G)|$. Perhaps even some sort of logarithmic bound would work.

Another degree problem on which there has been substantial recent progress concerns the prime divisors of the elements of $\mathrm{cd}(G)$. Let us write $s(G)$ to denote the maximum number of different primes which divide the degree of any one irreducible character of the solvable group G. (For instance $s(G) = 1$ means that every element of $\mathrm{cd}(G)$ is a prime power, not necessarily all with the same prime.) We shall say that a prime p is *character-essential* for G if G fails to have a normal abelian Sylow p-subgroup. It is a well-known result of N. Ito that the primes which appear among the divisors of the elements of $\mathrm{cd}(G)$ when G is solvable are precisely the character-essential primes. (This has recently been generalized by G. Michler. Using the classification of simple groups, he showed that Ito's result holds for all finite groups.) We shall write $e(G)$ to denote the number of character-essential primes of G. Motivated by the result of O. Manz [**16**] that for solvable groups G, if $s(G) = 1$, then $e(G) \le 2$, B. Huppert asked if $e(G)$ is always bounded by a function of $s(G)$ for solvable groups. The answer is "yes" (Isaacs [**12**]). My bound was exponential and was improved by D. Gluck [**9**] to a quadratic bound, and then improved again (Gluck-Manz [**10**]), this time to a linear bound. (It is easy to find examples to show that it is impossible to do better than a linear bound, although the actual best possible bound is still unknown. In particular, if $s(G) = 2$, it is not known if $e(G)$ can be as large as 5.)

Let us now end our digression on degree problems and return to M-groups. Where do M-groups fit in the hierarchy of types of solvable groups? We can write

$$\text{Nilpotent} \Rightarrow \text{Supersolvable} \Rightarrow M\text{-group} \Rightarrow \text{Solvable},$$

but this, though true, is somewhat misleading, since the condition that G be an M-group is of quite a different nature from the others in that subgroups of M-groups need not be M-groups. In fact, by a result of E. C. Dade, there is absolutely no restriction at all on which solvable groups can occur as subgroups of M-groups [**3**], and this creates considerable difficulty in proving theorems about the structure of these groups. (I should mention here that there is available a result of A. Parks [**17**], which gives a purely group-theoretic characterization of M-groups; it does not refer at all to characters. So far, however, this has not proved to be of value in addressing any of the open questions in the subject.)

If we cannot say anything about which groups can occur as subgroups of an M-group, what can we say about subgroups? It is not hard to show, for instance, that a normal Hall subgroup of an M-group must, itself, be an M-group. This suggests two questions:

(i) Must every *normal* subgroup of an M-group be an M-group?

(ii) Must every *Hall* subgroup of an M-group be an M-group?

The first of these was finally answered in the negative (independently and approximately simultaneously) by R. van der Waall and E. C. Dade (see [**4**]). Their counterexample, of order $2^9 \cdot 7$, depends in a fundamental way on the properties of the prime 2, and it cannot be generalized to provide an odd order counterexample. Question (i) remains open for groups of odd order. There has been considerable evidence accumulating, however, which suggests that the following conjecture might be true.

($*$) Let G be an M-group of odd order and suppose that $N \triangleleft G$. Then N is an M-group.

Actually, it might even be true that assuming that *either* $|N|$ or $|G : N|$ is odd would be sufficient to guarantee that N is an M-group. I should mention that in [**2**] it is alleged that a proof of ($*$) exists. Actually, even more is claimed; it is asserted that given suitable oddness conditions, irreducible constituents of the restriction of a monomial character to a normal subgroup of a solvable group are always monomial. This is, in fact, false, as easy examples show. One cannot hope to prove ($*$) "one character at a time." A proof would somehow have to use the fact that all (or at least several) of the irreducible characters of G are monomial.

Conjecture ($*$) seems to provide an excellent "testing ground" for the newly developing general techniques in the character theory of solvable groups, such as the π-theory, which I will explain shortly. None of these techniques has yet succeeded in proving ($*$), but definite progress has been made. To explain this, let me remind you that an irreducible character of G is said to be *primitive* if it is not of the form ψ^G for any character ψ of a proper subgroup of G. Obviously, the only way a monomial character can be primitive is for it to be linear, to have degree 1. One who believes ($*$), therefore, would expect that if $N \triangleleft G$ where G is an M-group of odd order, then a primitive irreducible character of N should be linear. This is, in fact, a theorem [**13**].

Let me sketch, briefly, how I proved this result. Suppose that $\theta \in \mathrm{Irr}(N)$ is primitive. It definitely does not follow from the fact that all of the irreducible constituents of θ^G are monomial that θ is linear. (It is easy to produce counterexamples to this, though not, of course, in an M-group.) Using π-theory, one can reduce the problem to the case where θ has p-power degree for some (odd) prime p and also, replacing G by a possibly proper subgroup, it can be arranged that some irreducible character χ lying over θ will have p-power degree. (This too involves π-theory.) Since we have dropped down to a (possibly proper) subgroup, we can no longer assume that G is an M-group, but we can, at least, be sure that χ is monomial. I will not sketch the rest of my proof because shortly after my paper appeared, it was substantially improved by E. C. Dade. I would like to describe his result.

THEOREM (DADE [**5**]). *Let G be solvable and let $\chi \in \operatorname{Irr}(G)$ be monomial and have p-power degree for some odd prime p. Let N be any normal subgroup of G. Then every irreducible constituent of the restriction χ_N is monomial.*

Note that to finish the proof of my result, all we would need is that any primitive irreducible constituent of χ_N is linear. (That one can actually conclude monomiality without assuming primitivity in this situation came as a surprise to me.) Dade's result shows that the "one character at a time" problem disappears when that one character has prime-power degree. (The assumption that the prime is odd is essential, but the solvability hypothesis can be relaxed to p-solvability, and that is the form in which the result appears in Dade's paper.)

The key step in the proof of Dade's theorem is a powerful lemma on symplectic modules.

LEMMA (DADE [**5**]). *Let G be p-solvable and let V be an FG-module, where $F = \mathbf{Z}/p\mathbf{Z}$ and p is odd. Assume that V carries a nondegenerate alternating G-invariant F-bilinear form. Let $H \subseteq G$ have p-power index and suppose H stabilizes a maximal totally isotropic subspace of V. Then G also stabilizes such a subspace.*

To prove Dade's theorem from this lemma about "symplectic" modules, one reduces, if there exists a counterexample, to the situation where there is an extra-special p-group $E \lhd G$ and a p'-group K such that $EK \lhd G$ and such that χ_E is homogeneous, as is χ_M for every $M \subseteq E$ with $M \lhd G$.

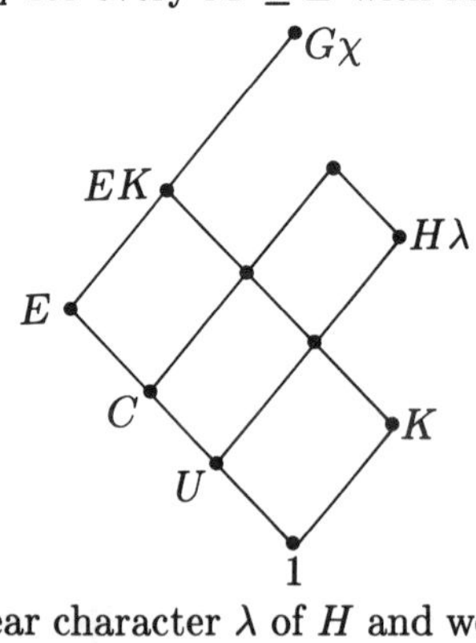

Now $\chi = \lambda^G$ for some linear character λ of H and we may assume $H \supseteq K$. It follows, since $U = E \cap H$ is normal in E, that it must be abelian. Writing $C = \mathbf{C}_E(U)$, it follows that λ_U is "fully ramified" in C. Because λ^{CH} is irreducible, $KC \lhd CH$, and p is odd, we can use the "fully ramified theory" (which I will not discuss here) to guarantee that $U = C$. Since E is extra-special, $E/\mathbf{Z}(E)$ is a symplectic G-module and $U/\mathbf{Z}(E)$ is an H-invariant maximal totally isotropic subspace. Dade's lemma now yields some $A \subseteq E$ with $A = \mathbf{C}_E(A)$ and $A \lhd G$, and this provides a contradiction.

In the proof we sketched that primitive characters of normal subgroups of odd order M-groups are linear, we twice appealed to "π-theory" to produce characters of prime-power degree. I will now explain a bit about how this magic works. Let G be solvable and let π be a set of prime numbers. We say that a character $\chi \in \mathrm{Irr}(G)$ is *π-special* if every irreducible constituent of the restriction of χ to every subnormal subgroup of G has degree and determinantal order which are π-numbers. (The *determinantal order* of a character is the order of the determinant of an associated representation when the determinant is viewed as an element of the group of linear characters.) Of course, if $\chi(1)$ is a π-number, the degrees of the irreducible constituents of the subnormal restrictions of χ will automatically be π-numbers, but the determinantal order condition must be imposed separately for all of these characters. This definition (though not the name "π-special") is due to Gajendragadkar [**6**]. One of the key results here is the following.

THEOREM (GAJENDRAGADKAR [**6**]). *Let G be solvable and suppose α and β are respectively π-special and σ-special irreducible characters of G. If $\pi \cap \sigma = \varnothing$, then $\alpha\beta$ is irreducible.*

It is easy to see in this situation that the irreducible character $\alpha\beta$ is $(\pi \cup \sigma)$-special and so we can repeat the process. This yields the following. For each prime divisor p of $|G|$, choose a $\{p\}$-special irreducible character α_p of G, allowing $\alpha_p = 1_G$. Then the product $\prod \alpha_p$ is irreducible. We say that a character of this type is *fully factored.*

It is not clear from the foregoing how many of the irreducible characters of our group are likely to be π-special for any given π, or to be fully factored. (Though, trivially, every irreducible is special for some set of primes.) The following result addresses this question.

THEOREM (ISAACS [**13**]). *Every primitive character of a solvable group is fully factored and its $\{p\}$-special factors are themselves primitive.*

We see now that to prove that a certain primitive character must be linear, for instance a character of a normal subgroup of an odd M-group, it suffices to assume that it is $\{p\}$-special and, therefore, of p-power degree. Our second application of the π-theory was to produce a monomial character of p-power degree of a subgroup of our given M-group such that this character lies over the primitive character (of the normal subgroup) which we want to prove is linear.

Recall the notation. We have $N \triangleleft G$ where G is an M-group. Also, $\theta \in \mathrm{Irr}(N)$ is primitive, and we can assume it is $\{p\}$-special for some prime p. Let χ be any irreducible character of G lying over θ and write $\chi = \lambda^G$ for some linear character of some subgroup $H \subseteq G$. (We can do this since χ is monomial because G is an M-group.) Now $\psi = \lambda^{NH}$ is irreducible and we can arrange that it lies over θ. Clearly, ψ is monomial, and we would like to show that $\psi(1) = |NH : H| = |N : N \cap H|$ is a p-power. Since every irreducible constituent of $(\lambda_{N\cap H})^N$ is G-conjugate to θ and hence is $\{p\}$-special, the following bit of π-theory yields the desired conclusion.

THEOREM (ISAACS [**13**]). *Let $H \subseteq G$ where G is solvable, and suppose $\varphi \in \mathrm{Irr}(H)$ has the property that every irreducible constituent of φ^G is π-special for some set π of primes. Then $|G : H|$ is a π-number.*

There is much more I could say about π-theory, but since time presses, I will close with a little more about my main topic, M-groups. If conjecture $(*)$ is true, then subnormal subgroups of odd order M-groups would be M-groups and so it would be reasonable to try to generalize my theorem and prove that primitive characters of such subgroups must be linear. I was unable to make this work using the method of my original proof, but using Dade's powerful techniques and some deeper applications of π-theory, I was able to prove the result for subnormal subgroups [**14**].

We asked above if Hall subgroups of M-groups must be M-groups. Less is known about this than about the corresponding question for normal subgroups, and there is no known counterexample. Using π-theory, I have proved the result analogous to those above about normal and subnormal subgroups. Primitive characters of Hall subgroups of M-groups are necessarily linear [**15**].

I will close with a question about M-groups which, through lack of courage, I will not formulate as a conjecture. If this question were to have an affirmative answer, a corollary would be that, indeed, normal subgroups of odd order M-groups are M-groups. Suppose G is an M-group and $\chi \in \mathrm{Irr}(G)$. How can we find a subgroup and a linear character of it which induces χ? We might pick any character ψ inducing χ from some proper subgroup and then repeat this and find some character inducing ψ. We continue this process until we get stuck by reaching a primitive character. Unfortunately, this primitive character need not be linear. In general, we cannot realize the monomiality of χ by proceeding at random down an inducing chain. My question is whether or not this, in fact, does work if G has odd order. In other words, if G is an odd order M-group and $\theta \in \mathrm{Irr}(H)$ is a primitive character of an *arbitrary* subgroup $H \subseteq G$ and θ induces irreducibly to G, must θ be linear?

REFERENCES

1. T. R. Berger, *Characters and derived length in groups of odd order*, J. Algebra **39** (1976), 199–207.

2. I. A. Cubarov, *Monomial characters and subgroups*, Russian Math. Surveys **33** (1978), 143–144.

3. E. C. Dade, oral communication to B. Huppert. See p. 585 of B. Huppert, *Endliche Gruppen*. I, Berlin, 1967.

4. ——, *Normal subgroups of M-groups need not be M-groups*, Math. Z. **133** (1973), 313–317.

5. ——, *Monomial characters and normal subgroups*, Math. Z. **178** (1981), 401–420.

6. D. Gajendragadkar, *A characteristic class of characters of finite π-separable groups*, J. Algebra **59** (1979), 237–259.

7. S. C. Garrison, *On groups with a small number of character degrees*, Ph.D. Thesis, University of Wisconsin, Madison, 1973.

8. D. Gluck, *Bounding the number of character degrees of a solvable group*, (to appear).

9. ——, *Primes dividing character degrees and character orbit sizes*, (to appear).

10. D. Gluck and O. Manz, *Prime factors of character degrees of solvable groups*, (to appear).

11. I. M. Isaacs, *Character degrees and derived length of a solvable group*, Canad. J. Math. **27** (1975), 146–151.

12. ——, *Solvable groups character degrees and sets of primes*, J. Algebra **104** (1986), 209–219.

13. ——, *Primitive characters, normal subgroups and M-groups*, Math. Z. **177** (1981), 267–284.

14. ——, *Characters of subnormal subgroups of M-groups*, Arch. Math. **42** (1984), 509–515.

15. ——, *Fong characters in π-separable groups*, J. Algebra **99** (1986), 89–107.

16. O. Manz, *Endliche auflösbare Gruppen deren sämtliche Charactergrade Primzahlpotenzen sind*, J.Algebra **94** (1985), 211–255.

17. A. Parks, *A group theoretic characterization of M-groups*, Proc. Amer. Math. Soc. **94** (1985), 209–212.

18. K. Taketa, *Uber die Gruppen deren Darstellungen sich sämtlich auf monomiale Gestalt transformieren*, Proc. Japan Imp. Acad. **6** (1930), 31–33.

UNIVERSITY OF WISCONSIN, MADISON

Proceedings of Symposia in Pure Mathematics
Volume 47 (1987)

The Representation Theory of the Symmetric Groups

GORDON JAMES

Any presentation of a topic as far-reaching as the representation theory of the symmetric group $\mathfrak{S}_n$ within the confines of one article will necessarily be very selective. We shall attempt to describe the fundamental theorems while also presenting an account of the more recent advances in the subject. Throughout, we concentrate upon representations, and we shall say nothing direct about the characters of $\mathfrak{S}_n$.

1. Partitions of n. A nonincreasing sequence $\lambda = (\lambda_1, \lambda_2, \dots)$ of nonnegative integers whose sum is n is called a *partition* of n. For example,

$$\begin{aligned}(4,0,0,0,\dots) &= (4),\\ (3,1,0,0,\dots) &= (3,1),\\ (2,2,0,0,\dots) &= (2,2),\\ (2,1,1,0,\dots) &= (2,1^2),\\ (1,1,1,1,\dots) &= (1^4)\end{aligned}$$

are the partitions of 4, and we normally adopt abbreviations such as those used here.

We partially order the set of partitions of n as follows:

$$\lambda \trianglerighteq \mu \text{ if and only if } \sum_{i=1}^{j} \lambda_i \geq \sum_{i=1}^{j} \mu_i \text{ for all } j.$$

This is called the *dominance order* on the set of partitions of n. The dominance order has the property that it contains every sensible total order on the set of partitions of n.

The conjugacy classes of the symmetric group $\mathfrak{S}_n$ are indexed by partitions of n, so general representation theory suggests that we should encounter $\mathfrak{S}_n$-modules indexed in a similar way. We shall discuss several (right) $\mathfrak{S}_n$-modules

---

1980 *Mathematics Subject Classification* (1985 *Revision*). Primary 20C30.

0082-0717/87 \$1.00 + \$.25 per page

labelled by the partition λ:

$$\begin{array}{ll} M^\lambda & \text{a permutation module,} \\ Y^\lambda & \text{a Young module,} \\ S^\lambda & \text{a Specht module,} \end{array}$$

and we shall see that $S^\lambda \leq Y^\lambda \leq M^\lambda$ for each partition λ of n.

Certain homomorphic images of Specht modules will also appear:

$$D^\lambda \quad \text{an irreducible module (or } D^\lambda = 0).$$

Over fields of characteristic zero, we shall have $D^\lambda = S^\lambda = Y^\lambda$.

2. The permutation module M^λ. Let F be any field, and define M^λ to be the $F\mathfrak{S}_n$-module obtained by the action of $\mathfrak{S}_n$ on the right cosets of the subgroup

$$\mathfrak{S}_{\lambda_1} \times \mathfrak{S}_{\lambda_2} \times \cdots .$$

Hence

$$\dim M^\lambda = n!/(\lambda_1!\lambda_2!\cdots).$$

3. The Specht module S^λ. Let λ' be the partition of n which is conjugate to λ. To obtain λ' we simply turn the diagram $[\lambda]$ on its side, as in the following example:

$$\lambda = (5,3,1) \qquad [\lambda] = \begin{array}{lllll} \times & \times & \times & \times & \times \\ \times & \times & \times & & \\ \times & & & & \end{array} \qquad [\lambda'] = \begin{array}{lll} \times & \times & \times \\ \times & \times & \\ \times & \times & \\ \times & & \\ \times & & \end{array} \qquad \lambda' = (3,2,2,1,1).$$

Let c^λ be the element of the group algebra $F\mathfrak{S}_n$ given by:

$$c^\lambda = \sum_{\pi \in \mathfrak{S}_{\lambda'_1} \times \mathfrak{S}_{\lambda'_2} \times \cdots} (\operatorname{sgn} \pi)\pi.$$

Possibly the most important result in the representation theory of $\mathfrak{S}_n$ is:

(3.1) $M^\lambda c^\lambda$ *has dimension* 1, *and*
$M^\mu c^\lambda \neq 0$ *only if* $\lambda \trianglerighteq \mu$.

These facts allow us "to pick from M^λ a 1-dimensional space which is special to the partition λ." Let

$$S^\lambda = M^\lambda c^\lambda (F\mathfrak{S}_n).$$

Then S^λ is a *Specht module.* If $F = \mathbf{Q}$, then

$$\{S^\lambda : \lambda \text{ is a partition of } n\}$$

is a complete set of ordinary irreducible representations of $\mathfrak{S}_n$.

It is clear why the dimension of the permutation module M^λ is independent of the field F. More surprising is that, in fact, the dimension of the Specht module

S^λ is also independent of the field. (The dimension of S^λ equals the number of so-called standard λ-tableaux; see also the Hook formula [**15**, **22**, §20].)

4. The module D^λ. It is of interest that the only facts we use in the following argument are that $M^\lambda c^\lambda$ is 1-dimensional and that $S^\lambda = M^\lambda c^\lambda(F\mathfrak{S}_n)$.

Suppose that X is an $F\mathfrak{S}_n$-submodule of M^λ. Then

$$\begin{aligned} &\textit{either} && \text{(i) } Xc^\lambda = 0 \\ &\textit{or} && \text{(ii) } Xc^\lambda \neq 0; \text{ in this case, } X \supseteq S^\lambda. \end{aligned}$$

Let $R^\lambda = \sum\{X \text{ an } F\mathfrak{S}_n\text{-submodule of } S^\lambda : Xc^\lambda = 0\}$.

Then $R^\lambda c^\lambda = 0$ and every submodule X of S^λ, other than perhaps S^λ itself, is contained in R^λ. Thus,

$$\begin{aligned} &\textit{either} && \text{(i) } S^\lambda c^\lambda = 0, \text{ which implies that } R^\lambda = S^\lambda, \\ &\textit{or} && \text{(ii) } S^\lambda c^\lambda \neq 0 \text{ and } R^\lambda \neq S^\lambda, \end{aligned}$$

and in the latter case S^λ has a unique maximal submodule, namely R^λ.

Let $D^\lambda = S^\lambda / R^\lambda$. Then

$$\begin{aligned} &\textit{either} && \text{(i) } D^\lambda = 0 \\ &\textit{or} && \text{(ii) } D^\lambda \text{ is irreducible.} \end{aligned}$$

We shall consider next a criterion which characterizes the case where D^λ is nonzero.

5. Regular and singular partitions. A partition $\lambda = (\lambda_1, \lambda_2, \dots)$ is said to be *p-singular* if

$$\lambda_{i+1} = \lambda_{i+2} = \cdots = \lambda_{i+p} > 0$$

for some $i \geq 0$. Otherwise, λ is *p-regular.*

For example, of the seven partitions of 5, just (5), (4,1), (3,2) are 2-regular. On the other hand, the partitions (5), $(3, 1^2)$, (1^5) correspond to the conjugacy classes of 2-regular elements (i.e., elements of odd order) in $\mathfrak{S}_5$. For a proof that the set of p-regular partitions of n always has the same cardinality as the set of partitions corresponding to p-regular conjugacy classes, see [**19**, §19.4], and for explicit bijections between the two sets see [**38**] and [**41**].

It turns out that the following are true:

(i) *The $F\mathfrak{S}_n$-module D^λ is zero if and only if* char $F = p$ *and λ is p-singular.*

(ii) *If* char $F = p$ *then $\{D^\lambda : \lambda$ is a p-regular partition of $n\}$ is a complete set of inequivalent irreducible $F\mathfrak{S}_n$-modules. Each element of the set is a self-dual absolutely irreducible $F\mathfrak{S}_n$-module.*

We have now encountered three types of $F\mathfrak{S}_n$-module:

$$\begin{aligned} M^\lambda &= \text{ the permutation module on } \mathfrak{S}_{\lambda_1} \times \mathfrak{S}_{\lambda_2} \times \cdots, \\ S^\lambda &= (M^\lambda c^\lambda)(F\mathfrak{S}_n), \text{ a Specht module,} \\ D^\lambda &= \begin{cases} \text{the unique top composition factor of } S^\lambda \text{ (if } \lambda \text{ is } p\text{-regular)} \\ 0 \text{ (if } \lambda \text{ is } p\text{-singular).} \end{cases} \end{aligned}$$

It is natural to ask: When does S^λ have a composition factor isomorphic to D^μ? When does M^λ have a factor isomorphic to S^μ? These and related questions we shall discuss next.

6. The multiplicity of D^μ in S^λ. We know that over fields of characteristic zero, $D^\lambda = S^\lambda$ and Specht modules for distinct partitions are nonisomorphic. Moreover, every field is a splitting field for $\mathfrak{S}_n$. We therefore assume throughout §6 that F is the field of p elements.

It happens that the Specht module S^λ over F is a p-modular reduction of the corresponding Specht module defined over $\mathbf{Q}$. In accordance with the usual definition, we let the p-modular decomposition matrix of $\mathfrak{S}_n$ be the matrix whose rows are labelled by partitions of n and whose columns are labelled by p-regular partitions of n, the entry in row λ and column μ being the composition multiplicity of D^μ in S^λ (that is, the number of factors which are isomorphic to D^μ in a composition series of S^λ). The problem of determining the decomposition matrix of $\mathfrak{S}_n$ is unsolved.

The following facts are known:

(i) *The Specht module S^λ has a composition factor isomorphic to D^μ (μ p-regular) only if $\mu \trianglerighteq \lambda$.*

(ii) *The composition multiplicity of D^λ in S^λ is* 1 *(λ p-regular).*

(A slight refinement of this result can be found in [**28**, 6.3.59iii].)

The process of writing down a p-modular decomposition matrix of $\mathfrak{S}_n$ involves choosing some order for the row and column labels. Totally order the p-regular partitions which label the columns in a sensible way (so that the total order contains the dominance order), starting with the partition (n), and use these labels (in the same order) for the first bunch of rows. The last result says that the p-modular decomposition matrix then has the form:

$$\begin{bmatrix} 1 & & & 0 \\ & 1 & & \\ & & \ddots & \\ * & & & 1 \end{bmatrix}.$$

We now list references to various pieces of information known about the decomposition numbers.

(i) The theorem known as the Nakayama Conjecture (see §9 below) tells us how the decomposition matrix splits into blocks.

(ii) Schaper's theorem (see §9 below) determines which entries in row λ of the decomposition matrix are nonzero and gives an upper bound for each entry in the row; *all this provided that the entries in every row μ with $\mu \triangleright \lambda$ have already been determined.*

(iii) Some decomposition numbers are known to be nonzero, by a theorem of Carter and Payne [**3**]. In this case, the information does not rely on knowing the values of other decomposition numbers.

(iv) A theorem of Donkin [**12**] determines some decomposition numbers in terms of decomposition numbers for symmetric groups of smaller degree.

(v) Benson [**1**] gives some techniques for dealing with the 2-modular decomposition matrix. (He uses the covering group of $\mathfrak{S}_n$.)

(vi) All the entries in the rows of the decomposition matrix corresponding to partitions with at most two parts are known [**22**, 24.15]. The same is true for the partitions of the form $(n - m, 1^m)$ (see [**39**] for p odd, [**22**, p. 93] for $p = 2$; see [**21**, p. 52] to determine the relevant column labels). The rows of the decomposition matrix for the partitions $(n - m - 1, m, 1)$ are given in [**20**] and [**42**].

The decomposition matrices of $\mathfrak{S}_n$ are known for $n \leq 13$ and all primes (see [**28**, Appendix 1E] for $p = 2, 3$; see [**32**] for $p = 5$; for $p \geq 7$ and $n \leq 13$ there are no problems since all the p-blocks have defect 0 or 1). Recently, Benson [**1**] has calculated the 2-modular decomposition matrices for $n = 14$, $n = 15$ (with one ambiguity), and $n = 16$ (with one more ambiguity).

It seems likely that when an algorithm for determining the decomposition matrix is eventually found, it will involve intricate computations for any given n. There are likely to be questions about the decomposition matrix which are still very hard to answer. We give two unsolved problems, both of which look tantalizingly easy:

PROBLEM 1. For a p-regular partition λ, we may tensor the irreducible representation D^λ with the alternating representation, and obtain another irreducible representation. This must be isomorphic to D^μ for some μ. Which μ? Mullineux [**35**] has conjectured a solution to this problem.

PROBLEM 2. The 2-modular decomposition matrix of S_{12} is known [**28**, p. 418]. Hence, or otherwise, find that of A_{12}.[1]

7. The multiplicity of S^μ in M^λ. For the moment, let F be a field of characteristic zero. Then the permutation module M^λ is completely reducible and we may ask how many direct summands of M^λ are isomorphic to the irreducible module S^μ. The answer is known. In particular, we have:

(7.1) *The permutation module M^λ has a composition factor isomorphic to S^μ if and only if $\mu \trianglerighteq \lambda$.*

In order to describe the composition multiplicity of S^μ in M^λ, we first discuss a related problem. If α is a partition of $n-k$ and β is a partition of k then $S^\alpha \otimes S^\beta$ may be regarded as a module for $\mathfrak{S}_{n-k} \times \mathfrak{S}_k$. We denote by $(S^\alpha \otimes S^\beta) \uparrow \mathfrak{S}_n$ the result of inducing this module to $\mathfrak{S}_n$.

Consider now the special case where $\beta = (k)$, so S^β is the trivial $\mathfrak{S}_k$-module. Recall that $\lambda' = (\lambda'_1, \lambda'_2, \dots)$ denotes the partition conjugate to λ. One version

[1]Note added in proof: This problem has now been solved. See Benson, "Spin modules for symmetric groups," preprint.

of a theorem known as Young's rule is

YOUNG'S RULE. *Suppose that μ is a partition of n and α is a partition of $n-k$. Then the composition multiplicity of S^μ in $(S^\alpha \otimes S^{(k)}) \uparrow \mathfrak{S}_n$ is 1 if for all i*

$$\alpha_i \leq \mu_i \quad \text{and} \quad \mu_i' \leq \alpha_i' + 1$$

and is 0 otherwise.

EXAMPLE. To calculate $(S^{(3,2)} \otimes S^{(2)}) \uparrow \mathfrak{S}_7$, note that the following objects correspond to the partitions μ relevant to Young's rule with $\alpha = (3,2)$ and $k = 2$:

$$\begin{matrix} \times\times\times 1 \\ \times\times \end{matrix} \qquad \begin{matrix} \times\times\times 1 \\ \times\times 1 \end{matrix} \qquad \begin{matrix} \times\times\times 1 \\ \times\times \\ 1 \end{matrix} \qquad \begin{matrix} \times\times\times \\ \times\times 1 \\ 1 \end{matrix} \qquad \begin{matrix} \times\times\times 1. \\ \times\times \\ 1\,1 \end{matrix}$$

Thus,

$$\left(S^{(3,2)} \otimes S^{(2)}\right) \uparrow \mathfrak{S}_7 = S^{(5,2)} + S^{(4,3)} + S^{(4,2,1)} + S^{(3,3,1)} + S^{(3,2,2)}.$$

Next, notice that $M^\lambda = (S^{(\lambda_1)} \otimes S^{(\lambda_2)} \otimes \cdots) \uparrow \mathfrak{S}_n$. Since induction is transitive, we may determine the multiplicity of S^μ as a factor of M^λ as follows. Consider

$$\left(S^{(\lambda_1)} \otimes \cdots \otimes S^{(\lambda_m)}\right) \uparrow \mathfrak{S}_{\lambda_1+\cdots+\lambda_m}.$$

When $m = 1$, this equals $S^{(\lambda_1)}$. Assume inductively that we have calculated the factors when $m = m'$. We can then apply Young's rule to obtain the factors when $m = m' + 1$. Thus, we can find the composition factors of M^λ, together with their composition multiplicities.

The construction shows:

(7.2) *The modules $S^{(n)}$ and S^λ occur once as composition factors of M^λ. All the other composition factors have the form S^μ with $(n) \rhd \mu \rhd \lambda$.*

With care, one can also deduce the "if" part of result (7.1). Notice that it is not obvious that the combinatorics involved in Young's rule imply equalities such as that between the factors of $(S^{(5)} \otimes S^{(3)} \otimes S^{(2)}) \uparrow \mathfrak{S}_{10}$ and $(S^{(2)} \otimes S^{(3)} \otimes S^{(5)}) \uparrow \mathfrak{S}_{10}$.

In view of (7.2), we have

$$S^\lambda = M^\lambda + (\text{a combination of } M^\mu\text{'s with } \mu \rhd \lambda)$$

in the Grothendieck group of $\mathfrak{S}_n$-modules. The exact rule for inverting Young's rule in this way is given by

The determinantal form: $\quad [\lambda] = \det\left([\lambda_i + i - j]\right)$.

We trust the following example will suffice to explain the meaning we attach to the statement of the determinantal form.

EXAMPLE.

$$[4,2,1] = \det \begin{pmatrix} [4] & [5] & [6] \\ [1] & [2] & [3] \\ 0 & [0] & [1] \end{pmatrix}$$

means that

$$S^{(4,2,1)} = M^{(4,2,1)} - M^{(4,3)} - M^{(5,1,1)} + M^{(6,1)}.$$

We now know how to calculate the factors of modules like $(M^\alpha \otimes M^\beta) \uparrow \mathfrak{S}_n$, since $(M^\alpha \otimes M^\beta) \uparrow \mathfrak{S}_n = (S^{(\alpha_1)} \otimes S^{(\alpha_2)} \otimes \cdots \otimes S^{(\beta_1)} \otimes S^{(\beta_2)} \otimes \cdots) \uparrow \mathfrak{S}_n$. The determinantal form shows that we may therefore, in principle, work out the factors of $(S^\alpha \otimes S^\beta) \uparrow \mathfrak{S}_n$. There is a combinatorial rule, known as the *Littlewood-Richardson rule*, for performing this calculation directly (see [**22**, §§15 and 16; **28**, §2.8; or **33**, pp. 68–73]).

We make several remarks concerning the Littlewood-Richardson rule:

(i) The Littlewood-Richardson rule turns up in many different places, for example, in the representation theory of the general linear group over arbitrary fields, in the theory of polynomial identity algebras, and in the theory of immanants. If faced with any problem in which two objects, indexed by partitions α of $n-k$ and β of k, are to be combined to produce new objects indexed by partitions of n, it is an excellent principle to suspect that the Littlewood-Richardson rule is relevant.

(ii) Even after lengthy practice, it is still easy to make slips when computing with the Littlewood-Richardson rule. In important calculations it is sensible to check that the proposed answer has the correct dimension.

(iii) Unfortunately, the Littlewood-Richardson rule is much harder to prove than was at first suspected. The author was once told that the Littlewood-Richardson rule helped to get men on the moon, but it was not proved until after they got there. The first part of this story might be an exaggeration. For several excellent articles on the Littlewood-Richardson rule and the combinatorics involved in proving it, see [**14**].

Now consider the permutation module M^λ defined over the integers. Specht modules may be defined exactly as in the case where we have a field. It turns out that M^λ has a series of submodules (a *Specht series* or a *filtration by Specht modules*)

$$0 = M_0 \subset M_1 \subset \cdots \subset M_k = M^\lambda,$$

where each subquotient M_i/M_{i-1} is isomorphic to a Specht module. (The top factor M_k/M_{k-1} is isomorphic to $S^{(n)}$ and the bottom factor M_1 is S^λ.) The number of times we have a subquotient M_i/M_{i-1} isomorphic to S^μ is, of course, given by Young's rule. Tensoring with a field F we get a Specht series for M^λ defined over F.

EXAMPLE. For $\lambda = (3, 2)$ we have

$$M^{(3,2)} = \begin{array}{l} S^{(5)} \\ S^{(4,1)} \\ S^{(3,2)} \end{array} \qquad \begin{array}{c} \text{Dimension} \\ 1 \\ 4 \\ 5 \\ 10 \end{array} \quad = \dim M^{(3,2)}.$$

For further information about Specht series for M^λ and other $F\mathfrak{S}_n$-modules, see [**4**, **5**, **6**, **13**, **22**, **30**, **43**].

8. The Young module Y^λ. We again work over an arbitrary field. Note that the construction of Young modules which we give below uses only the facts that $M^\lambda c^\lambda$ is 1-dimensional and that this 1-dimensional space generates the Specht module S^λ (cf. §4). These facts show:

(8.1) If M^λ is a direct sum of two $F\mathfrak{S}_n$-modules, $M^\lambda = Y_1 \oplus Y_{12}$, say, then for precisely one $i \in \{1, 2\}$ we have $Y_i c^\lambda \neq 0$; and for that i we have $S^\lambda \subseteq Y_i$.

Now choose an $F\mathfrak{S}_n$-module Y^λ which is minimal with respect to the properties

$$S^\lambda \subseteq Y^\lambda \subseteq M^\lambda \quad \text{and} \quad Y^\lambda \oplus Q = M^\lambda, \quad \text{for some } F\mathfrak{S}_n\text{-module } Q.$$

By (8.1), Y^λ is an *indecomposable* direct summand of M^λ; this, in turn, implies that Y^λ is uniquely defined, to within isomorphism. The module Y^λ is called a *Young module*.

Here are some properties of Young modules:

(i) *The Young modules Y^λ and Y^μ are isomorphic if and only if $\lambda = \mu$.*

(ii) *The permutation module M^λ is isomorphic to a direct sum of Young modules.*

(iii) *Precisely one component of M^λ is isomorphic to Y^λ. Every other component of M^λ is isomorphic to some Y^μ with $\mu \rhd \lambda$.*

Moreover, Y^λ has a filtration by Specht modules. We record information about the Specht factors which occur, alongside similar results we have already come across, for comparison:

$$S^\lambda \leftrightarrow D^\lambda + \sum_{\mu \rhd \lambda} a_{\lambda\mu} D^\mu \qquad (a_{\lambda\mu} \in \mathbf{Z}^+),$$

$$M^\lambda \leftrightarrow S^\lambda + \sum_{\mu \rhd \lambda} b_{\lambda\mu} S^\mu \qquad (b_{\lambda\mu} \in \mathbf{Z}^+),$$

(8.2)

$$M^\lambda \cong Y^\lambda \oplus \sum_{\mu \rhd \lambda}^{\oplus} c_{\lambda\mu} Y^\mu \qquad (c_{\lambda\mu} \in \mathbf{Z}^+),$$

$$Y^\lambda \leftrightarrow S^\lambda + \sum_{\mu \rhd \lambda} d_{\lambda\mu} S^\mu \qquad (d_{\lambda\mu} \in \mathbf{Z}^+).$$

Here "$\leftrightarrow$" means "has the same composition factors as," and $\mathbf{Z}^+$ denotes the set of nonnegative integers.

For example, if $n = 5$ and char $F = 2$, consider the following matrix.

	(1^5)	$(2,1^3)$	$(2^2,1)$	$(3,1^2)$	$(3,2)$	$(4,1)$	(5)
(1^5)	**1**	**0**	**0**	0	0	0	0
$(2,1^3)$	**0**	**1**	**0**	0	0	0	0
$(2^2,1)$	**1**	**0**	**1**	0	0	0	0
$(3,1^2)$	**2**	**0**	**1**	1	0	0	0
$(3,2)$	**1**	**0**	**1**	1	1	0	0
$(4,1)$	**0**	**1**	**0**	0	0	1	0
(5)	**1**	**0**	**0**	1	1	0	1

The information recorded in this matrix is illustrated by the statement that the columns labelled $(2^2, 1)$ and $(3, 1^2)$ tell us that

$$Y^{(2^2,1)} \leftrightarrow S^{(2^2,1)} + S^{(3,1^2)} + S^{(3,2)},$$
$$Y^{(3,1^2)} \leftrightarrow S^{(3,1^2)} + S^{(3,2)} + S^{(5)}.$$

Now, $M^{(1^n)}$ is the regular representation of $\mathfrak{S}_n$. Its components are just the principal indecomposable modules. Hence, every principal indecomposable module is a Young module. General theory therefore tells us that the matrix recording Specht factors of Young modules contains the decomposition matrix of $\mathfrak{S}_n$. In the matrix above, the entries printed in boldface give the 2-modular decomposition matrix of $\mathfrak{S}_5$. (But note the compulsory reversal of the usual order on the row labels.)

In summary, then, the following facts are known about the coefficients $a_{\lambda\mu}$, $b_{\lambda\mu}$, $c_{\lambda\mu}$, and $d_{\lambda\mu}$ in (8.2):

(i) *The coefficients* $b_{\lambda\mu}$ *are given by Young's rule* (*see* §7).

(ii) *When* μ *is a* p*-regular partition of* n, *we have* $a_{\lambda\mu} = d_{\mu'\lambda'}$.

(iii) *Since the coefficients* $b_{\lambda\mu}$ *are known, the coefficients* $c_{\lambda\mu}$ *determine the coefficients* $d_{\lambda\mu}$ *and vice versa.*

For proofs of the results in this section, and further information on Young modules, see [**16**, **18**, **25**, **26**, **31**].

9. Rim hooks. Throughout this section, we assume that we are working over a field of characteristic p.

Given a partition λ of n, we let its *nodes* be the ordered pairs (i, j) of integers which satisfy $1 \leq i \leq \lambda_i$ and $1 \leq j \leq \lambda'_j$. Viewing the nodes of λ as crosses in the plane we get the diagram $[\lambda]$ (cf. the example at the beginning of §3). The *rim* of λ consists of those nodes (i, j) for which $(i + 1, j + 1)$ is not a node.

If (a, b) is a node of λ then the (a, b) *rim hook* consists of those elements (i, j) of the rim of λ which satisfy $a \leq i$ and $b \leq j$. The *length* of a rim hook is its cardinality and the *leg length* of the (a, b) rim hook is $\lambda'_b - a$.

EXAMPLE. $\lambda = (7,6,4,3,3,1)$. The circled crosses indicate the $(3,2)$ rim hook, which has length 5 and leg length 2:

$$\begin{array}{ccccccc} \times & \times & \times & \times & \times & \times & \times \\ \times & \times & \times & \times & \times & \times & \\ \times & \times & \otimes & \otimes & & & \\ \times & \times & \otimes & & & & \\ \times & \otimes & \otimes & & & & \\ \times & & & & & & \end{array} \qquad \text{Rim hook lengths:} \quad \begin{array}{ccccccc} 12 & 10 & 9 & 6 & 4 & 3 & 1 \\ 10 & 8 & 7 & 4 & 2 & 1 & \\ 7 & 5 & 4 & 1 & & & \\ 5 & 3 & 2 & & & & \\ 4 & 2 & 1 & & & & \\ 1 & & & & & & \end{array}$$

Given a partition of n and a positive integer p it may be possible to remove a rim p-hook (that is, a rim hook of length p) to leave a set of nodes which corresponds to a partition of $n-p$. For $p=5$, there are two ways of doing this in the example above; we get $(7,6,2,2,1,1)$ by removing the $(3,2)$ rim hook, and we get $(7,6,4,2)$ by taking away the $(4,1)$ rim hook. In terms of the diagram, removing a rim p-hook amounts to "unwrapping p crosses from the rim." Having unwrapped a rim p-hook, it may be possible to "wrap it on again somewhere else."

EXAMPLE. $\lambda = (7,6,4,3,3,1)$ and $p=5$.

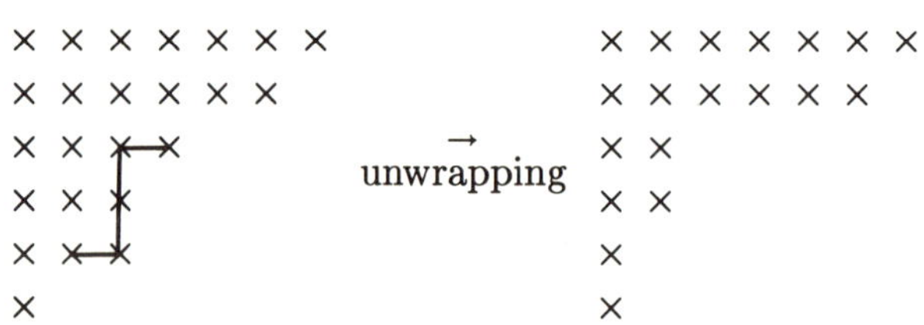

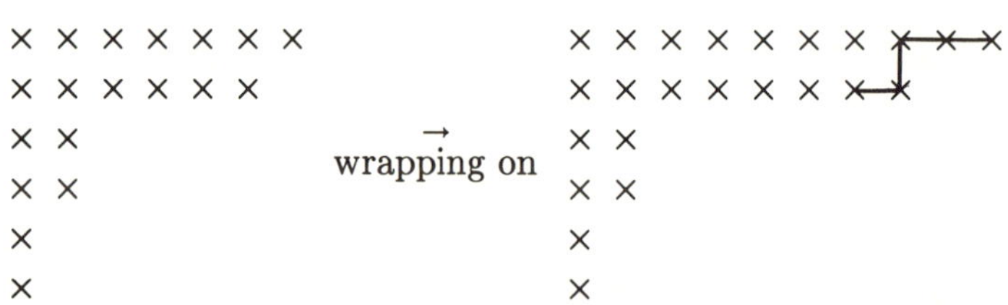

In fact, a rim 5-hook can be wrapped on to $(7,6,2,2,1,1)$ in seven ways. As well as the way shown, where the lowest added node is in row 2, it is possible to wrap on rim 5 hooks with the lowest node lying in row 1, 4, 5, 6, 8, or 11 (but not in row 3, say).

For an efficient method of dealing with rim hook calculations, using an abacus, see [**23**; **28**, §2.7].

Now, for a given n and p, the partitions of n may be divided into *p-blocks*, two partitions being in the same p-block if and only if we can maneuver from one to the other by a series of moves, each move consisting of unwrapping a skew p-hook and wrapping it back on elsewhere.

EXAMPLE. If $n = 5$ and $p = 3$, then there are three p-blocks:

$$\begin{array}{lcl} \times\times\times\times\times & \leftrightarrow & \begin{array}{l}\times\times\\ \times\times\\ \times\end{array} \quad \leftrightarrow \quad \begin{array}{l}\times\times\\ \times\\ \times\\ \times\end{array} \end{array}$$

$$\begin{array}{l}\times\times\times\times\\ \times\end{array} \quad \leftrightarrow \quad \begin{array}{l}\times\times\times\\ \times\times\end{array} \quad \leftrightarrow \quad \begin{array}{l}\times\\ \times\\ \times\\ \times\\ \times\end{array}$$

$$\begin{array}{l}\times\times\times\\ \times\\ \times\end{array}$$

We were not ill-advised in choosing the term "p-block":

The Nakayama Conjecture. *The Specht modules S^λ and S^μ are in the same p-block of $\mathfrak{S}_n$ if and only if λ and μ are in the same p-block.*

Despite its name, this result has been proved in many different ways. In [**34**], Meier and Tappe gave an economical proof using Brauer's Second Main Theorem, and they also provided a guide to the proofs which appeared before 1976. The author's favorite proof of the Nakayama Conjecture is that of Murphy [**36**, **37**].

One implication in the Nakayama Conjecture follows immediately from Schaper's Theorem, which we shall discuss next.

Let h_{ij} be the length of the (i,j) rim hook. We denote by $v_p(h_{ij})$ the greatest integer r such that p^r divides h_{ij}. A special case of Schaper's Theorem is the following result:

(9.1) *Suppose that λ is a p-regular partition. A necessary and sufficient condition for the Specht module S^λ to be irreducible is that $v_p(h_{ac}) = v_p(h_{bc})$ for all possible a, b, c.*

Thus, to see whether or not S^λ is irreducible we check whether or not "$v_p(h_{ac})$ is constant down each column of $[\lambda]$." This result was first conjectured by Carter and was proved in [**24**] and [**29**].

EXAMPLE. When $\lambda = (9,6,3)$ and $p = 2$, S^λ is irreducible. In the diagram $[\lambda]$ below, the circled nodes are those with even rim hook lengths, none of the rim hook lengths is divisible by 4.

$$\begin{array}{l}\times\otimes\times\times\otimes\times\times\times\times\\ \times\otimes\times\times\otimes\times\\ \times\otimes\times\end{array}$$

Schaper's Theorem gives information about the composition factors of every Specht module S^λ. Its statement involves rim hooks for pairs of nodes in the same column of $[\lambda]$. Once more we shall be concerned with unwrapping a rim hook and wrapping it on again elsewhere. We have seen that there are some restrictions on where a rim hook may be wrapped on to give a partition, and this accounts for the uninteresting third part of the following definition.

(9.2) Suppose that (b, c) is a node of the partition λ, and that $a < b$.

(i) Unwrap the (b, c) rim hook.

(ii) If we obtain a partition by wrapping the nodes back on with the lowest node in the added rim hook lying in row a, then let the new partition be $\lambda(a, b, c)$ and let $l(a, b, c)$ be the sum of the leg lengths of the rim hook which has been removed and the rim hook which has been added.

(iii) Otherwise, let $S^{\lambda(a,b,c)} = 0$.

EXAMPLE. Let $\lambda = (8, 8, 7, 5, 5, 3)$ and $(b, c) = (5, 2)$.

If $a = 3$ then $\lambda(a, b, c) = (10, 9, 9, 5, 2, 1)$ and $l(a, b, c) = 1 + 2 = 3$. If $a = 4$ then $S^{\lambda(a,b,c)} = 0$.

Now consider the following element of the Grothendieck group of $F\mathfrak{S}_n$:

$$(9.3)\quad \sum_{c}\sum_{a<b}(-1)^{l(a,b,c)}(v_p(h_{ac}) - v_p(h_{bc}))S^{\lambda(a,b,c)} = \sum_{\mu\ p\text{-regular}} \alpha_{\lambda\mu}D^{\mu}, \text{ say.}$$

Since $a < b$, we know that $\lambda(a, b, c)$ is obtained from λ by moving a rim hook *up*. Hence $\lambda(a, b, c) \rhd \lambda$ and this, in turn, implies that $\alpha_{\lambda\mu}$ is nonzero only if $\mu \rhd \lambda$. Now, apparently the coefficients $\alpha_{\lambda\mu}$ could be negative. However, Schaper's Theorem gives an explicit description of an $F\mathfrak{S}_n$-module whose composition factors are given by (9.3), so $\alpha_{\lambda\mu} \geq 0$.

We do not wish to introduce the further notation which would be needed to state Schaper's Theorem in its entirety. Although the following formulation is weaker than the full theorem, it still contains an impressive amount of information.

Recall from §4 that $R^{\lambda} = S^{\lambda}$ if λ is p-singular, and R^{λ} is the unique maximal submodule of S^{λ} if λ is p-regular. In the notation of (8.2) we have

$$R^{\lambda} \leftrightarrow \sum_{\mu \rhd \lambda} a_{\lambda\mu} D^{\mu} \qquad (a_{\lambda\mu} \in \mathbf{Z}^{+}),$$

and the decomposition matrix is determined by the decomposition numbers $a_{\lambda\mu}$ (λ a partition of n, μ a p-regular partition of n). We are now able to state a form of Schaper's Theorem concerning the composition factors of the Specht module S^{λ}.

SCHAPER'S THEOREM [40] (A WEAK FORMULATION). *With the notation above, we have*

$$a_{\lambda\mu} \leq \alpha_{\lambda\mu}.$$

Furthermore, there exists a positive integer $k = k(\lambda)$ such that

$$\alpha_{\lambda\mu} \le k a_{\lambda\mu}.$$

In particular, $a_{\lambda\mu}$ is nonzero if and only if $\alpha_{\lambda\mu}$ is nonzero.

To be able to calculate the coefficients $\alpha_{\lambda\mu}$ in (9.3), we only need to know the decomposition numbers for Specht modules which correspond to partitions which strictly dominate λ. If this information is available, then the inequalities in our statement of Schaper's Theorem tell us precisely which irreducible modules D^μ occur as composition factors of S^λ, and give an upper bound for their composition multiplicities.

EXAMPLE. Let $\lambda = (6,3,1)$ and $p = 2$. We indicate $v_p(h_{ij})$ for each node:

$$\begin{matrix} 3&1&0&0&0&0 \\ 2&1&0 \\ 0 \end{matrix}$$

The only contribution to the double sum in (9.3) comes from $c = 1$, since $v_p(h_{ac}) = v_p(h_{bc})$ for $c > 1$. For $c = 1$, we get

$$\begin{aligned} &\text{from } a = 1,\ b = 3 : 3S^{(7,3)}, \\ &\text{from } a = 1,\ b = 2 : \ -S^{(10)}, \\ &\text{from } a = 2,\ b = 3 : 2S^{(6,4)}. \end{aligned}$$

Now, the composition factors of S^λ when λ has just two parts are known:

$$\begin{aligned} S^{(7,3)} &\leftrightarrow D^{(10)} + D^{(8,2)} + D^{(7,3)}, \\ S^{(6,4)} &\leftrightarrow D^{(8,2)} + D^{(7,3)} + D^{(6,4)}, \end{aligned}$$

so the expression in (9.3) becomes

$$2D^{(10)} + 5D^{(8,2)} + 5D^{(7,3)} + 2D^{(6,4)}. \tag{9.4}$$

Therefore every composition factor of $S^{(6,3,1)}$, other than the top factor $D^{(6,3,1)}$, is isomorphic to one of $D^{(10)}$, $D^{(8,2)}$, $D^{(7,3)}$, or $D^{(6,4)}$; furthermore, $S^{(6,3,1)}$ does have composition factors isomorphic to each of these, and the coefficients in the expression (9.4) give upper bounds for their composition multiplicities.

The most difficult problem in evaluating the decomposition numbers of $\mathfrak{S}_n$ for $n \le 10$ is to decide whether the composition multiplicity of $D^{(10)}$ in $S^{(6,3,1)}$ is 1 or 3 over a field of characteristic 2. The bound given by the last example shows that the correct answer is 1.

The expression in (9.3) appears to be very daunting. But it can be handled very adeptly with only a little practice. The bounds which it gives on the decomposition numbers are frequently very tight. We encourage the reader to experiment with some two-part partitions.

10. Weyl modules for $GL_d(K)$. Let $\mathrm{GL}_d(K)$ be the group of nonsingular $d \times d$ matrices over the field K. To avoid special cases, we shall assume in this section that $n \le d$ and that K is infinite.

There is much interplay between the representation theory of $\mathfrak{S}_n$ over K and the representation theory of $\mathrm{GL}_d(K)$ over K. Indeed, the only known way to prove some theorems for the symmetric group involves working with $\mathrm{GL}_d(K)$. The main result of this section plainly exhibits the close connection between the two theories.

Let V be the d-dimensional vector space over K on which $\mathrm{GL}_d(K)$ acts in the natural way. Then $\mathfrak{S}_n$ acts on $V^{\otimes n} = V \otimes \cdots \otimes V$ (n factors) by place permutations, and we may use this fact to define certain $K\mathrm{GL}_d(K)$ submodules W^λ (*Weyl modules*) of $V^{\otimes n}$ for each partition λ of n. For example, the reader will be familiar with the skew-symmetric and symmetric parts of the tensor square of V:

$$W^{(1^2)} = K\{u \otimes v - v \otimes u : u, v \in V\},$$
$$W^{(2)} = K\{u \otimes v + v \otimes u, u \otimes u : u, v \in V\}.$$

A definition of W^λ for an arbitrary partition λ appears in [**2**; **17**, §5; **28**, §8.1]. The following facts are known:

(i) *For every partition λ, W^λ has a unique maximal $K\mathrm{GL}_d(K)$-submodule $W^{\lambda\max}$. Let $F^\lambda = W^\lambda / W^{\lambda\max}$.*

(ii) *The irreducible $K\mathrm{GL}_d(K)$-modules F^λ and F^μ are isomorphic if and only if $\lambda = \mu$.*

(iii) *The Weyl module W^λ has a composition factor isomorphic to F^μ only if λ and μ are partitions of the same integer and $\lambda \trianglerighteq \mu$.*

(iv) *The composition multiplicity of F^λ in W^λ is* 1.

Note that in part (iii) we have "$\lambda \trianglerighteq \mu$," not the other way round.

Recall from (8.2) that

$$Y^\lambda \leftrightarrow S^\lambda + \sum_{\mu \triangleright \lambda} d_{\lambda\mu} S^\mu.$$

A major result connecting the representation theory of $K\mathfrak{S}_n$ with that of $K\mathrm{GL}_d(K)$ is [**16**; **17**, §6.6; **18**; **25**; **26**; **31**]:

$$W^\lambda \leftrightarrow F^\lambda + \sum_{\lambda \triangleright \mu} d_{\mu\lambda} F^\mu.$$

For example, the matrix appearing in §8 contains the "decomposition numbers" for the Weyl modules corresponding to the partitions of 5. That is, the entry in row λ and column μ is the composition multiplicity of F^μ in W^λ. Thus, the rows labelled $(2^2, 1)$ and $(3, 1^2)$ tell us that

$$W^{(2^2,1)} \leftrightarrow F^{(1^5)} + F^{(2^2,1)},$$
$$W^{(3,1^2)} \leftrightarrow 2F^{(1^5)} + F^{(2^2,1)} + F^{(3,1^2)}.$$

To summarize: the matrix of decomposition numbers of those Weyl modules which are indexed by partitions of n coincides with the matrix recording the Specht factors of Young modules for $\mathfrak{S}_n$. This matrix is square and lower unitriangular, and we obtain the decomposition matrix of $\mathfrak{S}_n$ by deleting those columns whose index μ has the property that μ' is p-singular.

Finally, we should like to mention that apparently everything we have discussed in this article can be subsumed in a more general theory. There exist certain *Hecke algebras* $\mathcal{H}$ whose definition involves a parameter q. When $q = 1$, $\mathcal{H}$ is the symmetric group algebra. Now the symmetric group algebra gives results for the finite general linear group $\mathrm{GL}_n(q)$ only over fields whose characteristic divides q, by the very nature of Weyl modules, but the study of Hecke algebras eliminates this restriction. See **[9]** for a definition of $\mathcal{H}$ and for proofs of q-analogs of most of the results in §§1–7 of this article. In **[10]** there appear proofs of the q-analog of the Nakayama Conjecture and Carter's Conjecture (9.1). For applications to the representation theory of $\mathrm{GL}_n(q)$, see **[7, 8, 11, 27]**.

REFERENCES

1. D. Benson, *Some remarks on the decomposition numbers for the symmetric groups*, Preprint.

2. R. W. Carter and G. Lusztig, *On the modular representations of the general linear and symmetric groups*, Math. Z. **136** (1974), 193-242.

3. R. W. Carter and M. T. J. Payne, *On homomorphisms between Weyl modules and Specht modules*, Math. Proc. Cambridge Philos. Soc. **87** (1980), 131–145.

4. M. Clausen, *Letter-Place-Algebren und ein charakteristik-freier Zugang zur Darstellungstheorie symmetrischer und vollen linearer Gruppen*, Bayreuth. Math. Schr. **4** (1980).

5. M. Clausen and F. Stötzer, *Pictures und Standardtableaux-Grundlagen und Algorithmen*, Bayreuth. Math. Schr. **16** (1984), 1–122.

6. ———, *Pictures and skew (reverse) plane partitions*, Lecture Notes in Math., vol. 969, Springer-Verlag, 1982, pp. 100–114.

7. R. Dipper, *On the decomposition numbers of the finite general linear groups*, Trans. Amer. Math. Soc. **290** (1985), 315–344.

8. ———, *On the decomposition numbers of the finite general linear groups II*, Trans. Amer. Math. Soc. **292** (1985), 123–133.

9. R. Dipper and G. D. James, *Representations of Hecke algebras of general linear groups*, Proc. London Math. Soc. (3) **52** (1986), 20–52.

10. ———, *Blocks and idempotents of Hecke algebras of general linear groups*, Proc. London Math. Soc. (3) **54** (1987), 57–82.

11. ———, *Identification of the irreducible modular representations of* $\mathrm{GL}_n(q)$, J. Algebra **104** (1986), 266–288.

12. S. Donkin, *A note on decomposition numbers for general linear and symmetric groups*, Math. Proc. Cambridge Philos. Soc. **97** (1985), 57–62.

13. ———, *Finite resolutions of modules for reductive algebraic groups*, J. Algebra **101** (1986), 473–488.

14. D. Foata (Ed.), *Combinatoire et Représentation du Groupe Symétrique*, Lecture Notes in Math., vol. 579, Springer-Verlag, 1977.

15. J. S. Frame, G. de B. Robinson, and R. M. Thrall, *The hook graphs of the symmetric group*, Canad. J. Math. **6** (1954), 316–324.

16. J. Grabmeier, *Unzerlegbare Moduln mit trivialer Youngquelle und Darstellungstheorie der Schuralgebra*, Bayreuth. Math. Schr. **20** (1985), 9–152.

17. J. A. Green, *Polynomial Representations of* GL_n, Lecture Notes in Math., vol. 830, Springer-Verlag, 1980.

18. ———, *Functor categories and group representations*, Preprint.

19. G. H. Hardy and E. M. Wright, *An introduction to the theory of numbers*, Oxford Univ. Press, Oxford, 1960.

20. G. D. James, *Representations of the symmetric group over the field of order* 2, J. Algebra **38** (1976), 280–208.

21. ———, *On the decomposition matrices of the symmetric groups II*, J. Algebra **43** (1976), 45–54.

22. ——, *The representation theory of the symmetric groups*, Lecture Notes in Math., vol. 682, Springer-Verlag, 1978.

23. ——, *Some combinatorial results involving Young diagrams*, Math. Proc. Cambridge Philos. Soc. **83** (1978), 1–10.

24. ——, *On a conjecture of Carter concerning irreducible Specht modules*, Math. Proc. Cambridge Philos. Soc. **83** (1978), 11–17.

25. ——, *The decomposition of tensors over fields of prime characteristic*, Math. Z. **172** (1980), 161–178.

26. ——, *Trivial source modules for symmetric groups*, Arch. Math. (Basel) **41** (1983), 294–300.

27. ——, *The irreducible representations of the finite general linear groups*, Proc. London Math. Soc. **52** (1986), 236–268.

28. G. D. James and A. Kerber, *The representation theory of the symmetric group*, Encyclopedia Math. Appl. **16** (1981).

29. G. D. James and G. E. Murphy, *The determinant of the Gram matrix for a Specht module*, J. Algebra **59** (1979), 222–235.

30. G. D. James and M. H. Peel, *Specht series for skew representations of symmetric groups*, J. Algebra **56** (1979), 343–364.

31. A. A. Klyachko, *Direct summands of permutation modules*, Sel. Math. Sov. **3** (1983/84), 45–55.

32. E. Mac Aogáin, *Decomposition matrices of symmetric and alternating groups*, Trinity College Dublin Research Notes, TCD 1976–10 (1976).

33. I. G. MacDonald, *Symmetric functions and Hall polynomials*, Clarendon Press, Oxford, 1979.

34. N. Meier and J. Tappe, *Ein neuer Beweis der Nakayama-Vermutung über die Blockstrucktur Symmetrischer Gruppen*, Bull. London Math. Soc. **8** (1976), 34–37.

35. G. Mullineux, *Bijections of p-regular partitions and p-modular irreducibles of the symmetric groups*, J. London Math. Soc. **20** (1979), 60–66.

36. G. E. Murphy, *A new construction of Young's seminormal representation of the symmetric groups*, J. Algebra **69** (1981), 287–297.

37. ——, *The idempotents of the symmetric group and Nakayama's conjecture*, J. Algebra **81** (1983), 258–265.

38. J. B. Olsson, *Lower defect groups in symmetric groups*, Preprint.

39. M. H. Peel, *Hook representations of symmetric groups*, Glasgow Math. J. **12** (1971), 88–97.

40. K. D. Schaper, *Charakterformeln für Weyl-Moduln und Specht-Moduln in Primcharacteristik*, Diplomarbeit, Bonn, 1981.

41. D. Stockhofe, *Bijective Abbildungen auf der Menge der Partitionen einer natürlichen Zahl*, Bayreuth. Math. Schr. **10** (1982), 1–59.

42. K. W. To Law, *Results on the decomposition matrices for the symmetric groups*, Ph.D. Thesis, Cambridge University, 1983.

43. A. V. Zelevinsky, *Representations of finite classical groups*, Lecture Notes in Math. **869** (1981).

IMPERIAL COLLEGE, LONDON, ENGLAND

Proceedings of Symposia in Pure Mathematics
Volume 47 (1987)

Representations of Chevalley Groups in Their Own Characteristic

J. C. JANTZEN

The construction of a finite Chevalley group involves a finite field. This article deals with the representation theory of these groups over an algebraic closure of this finite field. There already exist several surveys on the subject or on parts of it [see **21**, **5**, **22**, **24**]. The present one contains some results discovered after the earlier ones had been written. Furthermore, I have decided to treat rather briefly some parts of the theory already well explained in the previous surveys and to put more emphasis on other parts. Still, there is a considerable overlap with the earlier surveys.

Usually, the representation theory of these groups is approached via that of the ambient algebraic groups. Many results are of the form: If we know some data for the algebraic group, then we know (in principle) some data for the finite group. (Here "know in principle" often means "there is an algorithm.") I shall not say much about what is known for the algebraic groups, but refer you instead to some of the surveys from above, to Henning Haahr Andersen's lectures at Arcata [**49**] or to [**33**, **34**].

1. Simple modules.

1.1. Let p be a prime number, and let k be an algebraic closure of the prime field $\mathbf{F}_p$. Let G be a connected, simply connected, and semisimple algebraic group, defined and split over $\mathbf{F}_p$. We shall look at the representations over k of the groups $G(q) = G(\mathbf{F}_q)$ of the points of G over any finite field $\mathbf{F}_q$ (with $\mathbf{F}_p \subset \mathbf{F}_q \subset k$).

Of course, we should deal with a larger class of groups. We should take any connected, reductive group H defined over $\mathbf{F}_q$ and look at any subgroup of $H(\mathbf{F}_q)$ containing the commutator subgroup. This would, however, make many statements and formulas much more complicated without giving much more insight into what is going on. Therefore I prefer to deal explicitly only with

1980 *Mathematics Subject Classification* (1985 *Revision*). Primary 20G40; Secondary 20C20, 20G05.

the special case but to give references also to generalizations to more general situations.

1.2. Let me introduce a lot of notations:

F	the Frobenius endomorphism of G,
T	a maximal torus in G, defined and split over $\mathbf{F}_p$,
$X(T)$	the group of characters on T,
R	the set of roots of G with respect to T,
S	a basis for the root system R,
R^+	the set of roots positive with respect to S,
U (resp. U^+)	the subgroup of G generated by all root subgroups corresponding to negative (resp. positive) roots,
ω_α	the fundamental weight corresponding to $\alpha \in S$,
ρ	$= \sum_{\alpha \in S} \omega_\alpha$,
$X(T)_+$	the set of dominant weights,
$X_n(T)$	$= \{\sum_{\alpha\in S} m_\alpha \omega_\alpha \mid 0 \leq m_\alpha < p^n \text{ for all } \alpha \in S\}$ (for any $n > 0$),
W	the Weyl group of G with respect to T,
w_0	the unique element in W with $w_0(S) = -S$,
V_μ	$= \{v \in V \mid tv = \mu(t)v \text{ for all } t \in T\}$, the weight space of weight μ (for any $\mu \in X(T)$ and any T-module V),
$(e(\lambda) \mid \lambda \in X(T))$	the canonical basis of the group ring $\mathbf{Z}[X(T)]$,
$\operatorname{ch}(V)$	$= \sum_{\mu \in X(T)} (\dim V_\mu) e(\mu)$, the formal character of V (for any finite-dimensional T-module V),
$s(\mu)$	$= \sum_{\nu \in W\mu} e(\nu)$ (for any $\mu \in X(T)$).

Finally we shall use two order relations, $\leq$ and $\leq_{\mathbf{Q}}$, on $X(T)$. Here $\lambda \leq \mu$ (resp. $\lambda \leq_{\mathbf{Q}} \mu$) if and only if there are $m_\alpha \in \mathbf{N}$ (resp. $m_\alpha \in \mathbf{Q}$, $m_\alpha \geq 0$) such that $\mu - \lambda = \sum_{\alpha \in S} m_\alpha \alpha$.

1.3. For each $\lambda \in X(T)_+$ there is a simple G-module $L(\lambda)$, unique up to isomorphism, with highest weight λ. So $\dim L(\lambda)_\lambda = 1$, and $L(\lambda)_\mu \neq 0$ implies $\mu \leq \lambda$. There are $m_{\lambda\mu} \in \mathbf{N}$ such that

$$\operatorname{ch} L(\lambda) = s(\lambda) + \sum_{\substack{\mu < \lambda \\ \mu \in X(T)_+}} m_{\lambda\mu} s(\mu). \tag{1}$$

THEOREM. *Let $n \in \mathbf{N}$, $n > 0$.*

(a) *Each $L(\lambda)$ with $\lambda \in X_n(T)$ is simple as a $kG(p^n)$-module.*

(b) *Any simple $kG(p^n)$-module is isomorphic to exactly one $L(\lambda)$ with $\lambda \in X_n(T)$.*

This result was first proved in the special case $n = 1$ by Curtis in [**13**] and then in general by Steinberg in [**43**]. A different approach to part (a) is mentioned very briefly by Kempf in [**37**]. It deserves to be known better, so in Appendix 1 I give some details of it. This appendix goes back to joint attempts with H. Haahr Andersen to understand Kempf's remarks.

There is a completely different method [cf. **14**] for constructing the simple $kG(p^n)$-modules, using not the ambient algebraic group G, but the Tits system of $G(p^n)$.

1.4. Let $n \in \mathbf{N}$, $n > 0$, and set $q = p^n$.

For any finite-dimensional G-module V the formal character $\mathrm{ch}(V)$ determines the Brauer character of V as a $kG(q)$-module. Indeed, any p-regular element $g \in G(q)$ is semisimple, hence conjugate in G to some $t \in T$. Taking $\dim(V_\mu)$ times $\mu(t)$ for each $\mu \in X(T)$ we get the eigenvalues of g on V with their multiplicities, and thus the Brauer character of V.

Both the $s(\lambda)$ with $\lambda \in X(T)_+$ and the $\mathrm{ch}\, L(\lambda)$ with $\lambda \in X(T)_+$ form bases of the abelian group $\mathbf{Z}[X(T)]^W$. Therefore there is a homomorphism $\chi \mapsto \tilde{\chi}$ from $\mathbf{Z}[X(T)]^W$ to the group of Brauer characters on $G(q)$ which maps any $\mathrm{ch}(V)$ as above to the Brauer character of V. The image of $s(\lambda)$, $\chi(\lambda)$, $\mathrm{ch}(V)$ will be denoted by $\tilde{s}(\lambda)$, $\tilde{\chi}(\lambda)$, $\widetilde{\mathrm{ch}}(V)$. Here $\chi(\lambda)$ is defined by

$$\chi(\lambda) = \frac{\sum_{w \in W} \det(w) e(w(\lambda+\rho)-\rho)}{\sum_{w \in W} \det(w) e(w(\rho)-\rho)} \tag{1}$$

for any $\lambda \in X(T)$. In case $\lambda \in X(T)_+$ then $\chi(\lambda)$ is the character of a "Weyl module" $V(\lambda)$ with highest weight λ. If there is $w \in W$ with $w(\lambda+\rho)-\rho \in X(T)_+$, then $\chi(\lambda) = \det(w)\chi(w(\lambda+\rho)-\rho)$; if not, then $\chi(\lambda) = 0$. So $\chi(\lambda) \in \mathbf{Z}[X(T)]^W$ always.

For $\lambda \in X(T)_+$ one knows that $\chi(\lambda)$ is the sum of $\mathrm{E}s(\lambda)$ and of an integral linear combination of all $s(\mu)$ with $\mu \in X(T)_+$, $\mu < \lambda$. Therefore there are integers $d_{\lambda\mu}, a_{\lambda\mu} \in \mathbf{Z}$ such that

$$\chi(\lambda) = \mathrm{ch}\, L(\lambda) + \sum_{\substack{\mu<\lambda \\ \mu \in X(T)_+}} d_{\lambda\mu}\, \mathrm{ch}\, L(\mu), \tag{2}$$

$$\mathrm{ch}\, L(\lambda) = \chi(\lambda) + \sum_{\substack{\mu<\lambda \\ \mu \in X(T)_+}} a_{\lambda\mu} \chi(\mu). \tag{3}$$

Note that $d_{\lambda\mu} \geq 0$ for all λ, μ as these numbers are multiplicities of composition factors. We set $d_{\lambda\lambda} = a_{\lambda\lambda} = 1$ and $d_{\lambda\mu} = a_{\lambda\mu} = 0$ whenever $\mu \not\leq \lambda$, and we thus get matrices $D = (d_{\lambda\mu})$ and $A = (a_{\lambda\mu})$. The determination of D or A, which are equivalent problems in principle, is a big open problem in the representation theory of G. Many results to come have the form: If we know D and A, then we know....

1.5. For any G-module V one constructs a new ("twisted") G-module $V^{(n)}$ by composing the given representation $G \to \mathrm{GL}(V)$ with $F^n\colon G \to G$. Obviously T acts via $p^n\lambda$ on (V_λ) when regarded as a subspace of $V^{(n)}$. Hence:

$$\mathrm{ch}(V^{(n)}) = \sum_{\mu \in X(T)} \dim(V_\mu) e(p^n\mu). \tag{1}$$

Furthermore, if V is simple, then so is $V^{(n)}$; therefore $L(\lambda)^{(n)} \simeq L(p^n\lambda)$ for all $\lambda \in X(T)_+$. More generally, one has *Steinberg's tensor product theorem*, proved in [**43**]: If $\lambda_0 \in X_n(T)$ and $\lambda_1 \in X(T)_+$, then

(2) $$L(\lambda_0 + p^n\lambda_1) \simeq L(\lambda_0) \otimes L(\lambda_1)^{(n)}.$$

As F^n is the identity on $G(q)$, any $V^{(n)}$ is isomorphic to V as a $kG(q)$-module. We get especially in the situation of (2)

(3) $$L(\lambda_0 + p^n\lambda_1) \simeq_{G(q)} L(\lambda_0) \otimes L(\lambda_1).$$

This fact enables us to compute for any G-module V the multiplicities, denoted by $[V : L(\mu)]_{G(q)}$, of its composition factors as a $kG(q)$-module, provided that we can express $\mathrm{ch}(V)$ as a linear combination of the $\chi(\lambda)$ (or $\mathrm{ch}\, L(\lambda)$) with $\lambda \in X(T)_+$ and provided that we know the matrices A and D. Indeed, we can restrict ourselves to the case $V = L(\lambda)$ with $\lambda \in X(T)_+$. There we can use induction over $\leq_{\mathbf{Q}}$. If $\lambda \in X_n(T)$, nothing is to be shown. Otherwise there are $\lambda_0 \in X_n(T)$ and $\lambda_1 \in X(T)_+$, $\lambda_1 \neq 0$ with $\lambda = \lambda_0 + p^n\lambda_1$. Any composition factor $L(\mu)$ of $L(\lambda_0) \otimes L(\lambda_1)$ as a G-module satisfies

$$\mu \leq \lambda_0 + \lambda_1 <_{\mathbf{Q}} \lambda_0 + p^n\lambda_1 = \lambda.$$

Therefore we can apply induction to $L(\mu)$. The composition factors of $L(\lambda_0) \otimes L(\lambda_1)$ as a G-module can be computed from our knowledge of A and D using Brauer's formula

(4) $$\chi(\lambda) \sum_{\mu \in X(T)} a_\mu e(\mu) = \sum_{\mu \in X(T)} a_\mu \chi(\lambda + \mu)$$

for all $\lambda \in X(T)$ and all $\sum_\mu a_\mu e(\mu) \in \mathbf{Z}[X(T)]^W$.

This proof shows for any two simple $kG(q)$-modules that we can compute the composition factors of their tensor product (knowing A and D).

Let $\lambda \in X(T)_+$ and let V be a finite-dimensional G-module such that $\nu \leq \lambda$ for each weight ν of V. The argument from above yields [cf. **47**]: If $\mu \in X_n(T)$ and $[V : L(\mu)]_{G(q)} \neq 0$, then $\mu \leq_{\mathbf{Q}} \lambda$. If $\lambda \in X_n(T)$, then $[V : L(\lambda)]_{G(q)} = \dim(V_\lambda)$. One has especially for all $\lambda, \mu, \nu \in X_n(T)$: If $[L(\lambda) \otimes L(\mu) : L(\nu)]_{G(q)} \neq 0$, then $\nu \leq_{\mathbf{Q}} \lambda + \mu$. If $\lambda + \mu \in X_n(T)$, then $[L(\lambda) \otimes L(\mu) : L(\lambda + \mu)]_{G(q)} = 1$.

Another consequence, also observed in [**47**], is that there are $m'_{\lambda\mu} \in \mathbf{N}$ with

(5) $$\widetilde{\mathrm{ch}} L(\lambda) = \tilde{s}(\lambda) + \sum m'_{\lambda\mu} \tilde{s}(\mu) \quad \text{for all } \lambda \in X_n(T),$$

where we sum over all $\mu \in X_n(T)$ with $\mu <_{\mathbf{Q}} \lambda$. The $\tilde{s}(\lambda)$ with $\lambda \in X_n(T)$ are a basis of the group of Brauer characters of $G(q)$.

By the way, the full strength of (2) is not needed for a proof of the result mentioned at the end. It is enough to know the obvious facts that $L(\lambda_0 + p^n\lambda_1)$ occurs with multiplicity one as a composition factor of $L(\lambda_0) \otimes L(\lambda_1)^{(n)}$ and that all other composition factors have a strictly lower highest weight.

1.6. The simple G-module $L((p^n - 1)\rho)$ is called the nth *Steinberg module* and denoted by St_n. One knows that $\dim(\mathrm{St}_n) = p^{n|R^+|}$ and

(1) $$\mathrm{ch}(\mathrm{St}_n) = \chi((p^n - 1)\rho).$$

Furthermore, St_n is injective (= projective) as a $kG(q)$-module and belongs to a block of its own. (For the Steinberg modules one can also consult the survey [**27**].)

Two simple $kG(q)$-modules $L(\lambda)$, $L(\mu)$ with $\lambda, \mu \in X_n(T) - \{(p^n - 1)\rho\}$ belong to the same block if and only if $\lambda|_{Z(G)(q)} = \mu|_{Z(G)(q)}$, where $Z(G) \subset T$ is the center of G. The defect groups of all these blocks (i.e., not containing St_n) are the p-Sylow subgroups of $G(q)$, e.g., $U(q)$; cf. [**15, 20**]. Each $L(\lambda)$ with $\lambda \in X_n(T)$, $\lambda \neq (p^n - 1)\rho$, has vertex $U(q)$; cf. [**16, 17**].

There is not much known about the complexity of the simple $kG(q)$-modules. However, only very few of them are periodic; cf. [**35, 28, 19**].

The Ext-groups between simple $kG(q)$-modules are known only in very few cases; cf. [**9, 10, 2, 4, 25, 1**].

2. Principal indecomposable modules. We still fix $n \in \mathbf{N}$, $n > 0$ and set $q = p^n$.

2.1. For each simple $kG(q)$-module $L(\lambda)$ with $\lambda \in X_n(T)$ let $U_n(\lambda)$ be its injective hull (and projective cover). Most of the information available on these modules is on numerical data like dimensions, characters, and Cartan invariants. Very little is known about the module structure, for example, about the Loewy series. It is only for some rather small groups that we have better information; cf. [**2, 38**].

2.2. The Steinberg module St_n is simple and injective, hence equal to its injective hull:

$$U_n((p^n - 1)\rho) = \mathrm{St}_n = L((p^n - 1)\rho). \tag{1}$$

For any finite-dimensional $kG(q)$-module V the tensor product $\mathrm{St}_n \otimes V$ is injective, so there are $m(V, \lambda) \in \mathbf{N}$ with

$$\mathrm{St}_n \otimes V \simeq \bigoplus_{\lambda \in X_n(T)} U_n(\lambda)^{m(V,\lambda)}. \tag{2}$$

Obviously,

$$\begin{aligned} m(V, \lambda) &= \dim \mathrm{Hom}_{G(q)}(L(\lambda), \mathrm{St}_n \otimes V) \\ &= \dim \mathrm{Hom}_{G(q)}(L(\lambda) \otimes V^*, \mathrm{St}_n). \end{aligned}$$

The simplicity and simultaneous injectivity of St_n implies that this number is equal to the multiplicity of St_n as a composition factor in $L(\lambda) \otimes V^*$, i.e.,

$$m(V, \lambda) = [L(\lambda) \otimes V^* : \mathrm{St}_n]_{G(q)}. \tag{3}$$

For all $\lambda \in X_n(T)$ also $\lambda^0 = (p^n - 1)\rho + w_0\lambda \in X_n(T)$ and $(\lambda^0)^0 = \lambda$. Furthermore, $L(\lambda^0)^* \simeq L((p^n - 1)\rho - \lambda)$. So the discussion in 1.5 implies that $m(L(\lambda^0), \lambda) = 1$, and if $m(L(\lambda^0), \mu) \neq 0$, then $\mu + (p^n - 1)\rho - \lambda \geq_{\mathbf{Q}} (p^n - 1)\rho$, hence $\lambda \leq_{\mathbf{Q}} \mu$. We get:

$$\mathrm{St}_n \otimes L(\lambda^0) \simeq U_n(\lambda) \oplus \bigoplus_{\substack{\mu \in X_n(T) \\ \lambda <_{\mathbf{Q}} \mu}} U_n(\mu)^{m(L(\lambda^0),\mu)}. \tag{4}$$

This implies that the $\widetilde{\mathrm{ch}}(\mathrm{St}_n)\widetilde{\mathrm{ch}}(L(\lambda))$ with $\lambda \in X_n(T)$ are a basis of the group generated by all Brauer characters of all principal indecomposable $kG(q)$-modules. We see also that this group is just the ideal generated by $\widetilde{\mathrm{ch}}(\mathrm{St}_n)$ in the ring of all Brauer characters (see **[40]** for a different approach to this result) and that the Brauer character of $U_n(\lambda)$ is the sum of $\tilde{s}(\lambda^0)\widetilde{\mathrm{ch}}(\mathrm{St}_n)$ and of a linear combination of the $\tilde{s}(\mu)\widetilde{\mathrm{ch}}(\mathrm{St}_n)$ with $\mu \in X_n(T)$, $\mu >_{\mathbf{Q}} \lambda^0$; cf. **[3]**.

2.3. The results in the last section show that we can compute (in principle) the Brauer characters of all $U_n(\lambda)$ if we know all $\mathrm{ch}\, L(\lambda)$ and all $[L(\lambda) \otimes L(\mu) : \mathrm{St}_n]_{G(q)}$ with $\lambda, \mu \in X_n(T)$. The discussion in 1.5 shows we can compute the latter data whenever we know the matrices A and D (from 1.4).

If you want to carry out the computations there are some simplifications you might benefit from. For any $\chi \in \mathbf{Z}[X(T)]^W$ the class function $\tilde{\chi}\widetilde{\mathrm{ch}}(\mathrm{St}_n)$ is a linear combination of the Brauer characters of the $U_n(\lambda)$ with coefficients say $m(\chi, \lambda)$. If $\chi = \mathrm{ch}(V)$ for some G-module V, then $m(\chi, \lambda) = m(V, \lambda)$, and in general χ is a linear combination of such $\mathrm{ch}(V)$.

For any $\chi = \sum_\mu a_\mu e(\mu) \in \mathbf{Z}[X(T)]$ set $\chi^* = \sum_\mu a_\mu e(-\mu)$. If $\chi = \mathrm{ch}(V)$ for some finite-dimensional G-module V, then $\mathrm{ch}(V)^* = \mathrm{ch}(V^*)$. Then one gets for all $\chi \in \mathbf{Z}[X(T)]^W$:

$$m(\chi, \lambda) = [L(\lambda) \otimes \chi^* : \mathrm{St}_n]_{G(q)},$$

where the right-hand side is defined via

$$\widetilde{\mathrm{ch}}(L(\lambda))\tilde{\chi}^* = \sum_{\mu \in X_n(T)} [L(\lambda) \otimes \chi^* : L(\mu)]_{G(q)} \widetilde{\mathrm{ch}} L(\mu).$$

(For $\chi = \mathrm{ch}(V)$ this is true by 2.2(3); in general one uses additivity.)

Instead of decomposing all $\mathrm{St}_n \otimes L(\lambda)$ as described in 2.2 we can therefore as well decompose all $\widetilde{\mathrm{ch}}(\mathrm{St}_n)\tilde{s}(\lambda)$ with $\lambda \in X_n(T)$. Another choice would be to take all $\widetilde{\mathrm{ch}}(\mathrm{St}_n)\tilde{\varphi}(\lambda)$ with $\lambda \in X_n(T)$ where

$$\varphi(\lambda) = \sum_\mu \dim L(\lambda)_\mu s(\mu)$$

and where we sum over all $\mu \in (W\lambda + p\mathbf{Z}R) \cap X(T)_+$. One advantage of the $\varphi(\lambda)$ over the $s(\lambda)$ is the fact that all $m(\varphi(\lambda), \mu)$ are nonnegative because $\varphi(\lambda)\, \mathrm{ch}(\mathrm{St}_n)$ is the character of a direct summand of $\mathrm{St}_n \otimes L(\lambda)$. (This is a consequence of the linkage principle for G.)

2.4. The computations described in 2.2–2.3 will give us the Brauer character of each $U_n(\lambda)$ as a linear combination of some $\tilde{s}(\mu)\widetilde{\mathrm{ch}}(\mathrm{St}_n)$ or of some $\tilde{\chi}(\mu)$. (Observe that $s(\mu)\, \mathrm{ch}(\mathrm{St}_n) = \sum_{\nu \in W\mu} \chi((p^n - 1)\rho + \nu)$.) We assume A and D to be known, so we can decompose all $\tilde{\chi}(\mu)$ with $\mu \in X(T)$ into linear combinations of the $\widetilde{\mathrm{ch}} L(\lambda)$ with $\lambda \in X_n(T)$. Therefore we can also compute the composition factors of the $U_n(\lambda)$, i.e., the Cartan matrix of $kG(q)$.

On the other hand, there is a formula (due to **[23]**) expressing each $\tilde{s}(\mu)\, \mathrm{ch}(\mathrm{St}_n)$ in terms of Deligne-Lusztig characters. (One has

$$\tilde{s}(\mu)\widetilde{\mathrm{ch}}(\mathrm{St}_n) = \frac{1}{|\,\mathrm{Stab}_W\, \mu|} \sum_{w \in W} R_w(n, \mu)$$

using some notation to be introduced only in 3.2.)

Combining the computations with the known results about the decomposition of the $R_w(n,\mu)$ into irreducible characters (mostly due to Lusztig) we get the (transpose of the) decomposition matrix of $G(q)$ with respect to p.

2.5. There is a somewhat different approach to the $U_n(\lambda)$ using modules for the nth Frobenius kernel $G_n = \ker(F^n)$ (in the scheme-theoretic sense) of G. The $L(\lambda)$ with $\lambda \in X_n(T)$ are simple for G_n, too, and have an injective hull $Q_n(\lambda)$ as a G_n-module. One can extend $Q_n(\lambda)$ to a G_nT-module such that the inclusion of $L(\lambda)$ into $Q_n(\lambda)$ is also compatible with the T-action. There is a formula for $\operatorname{ch} Q_n(\lambda)$ (cf. [**30**], 2.5(1), (2)) which implies $\operatorname{ch} Q_n(\lambda) \in \mathbf{Z}[X(T)]^W$ and which admits explicit calculation whenever we know A and D. One expects to have even a G-module structure on the $Q_n(\lambda)$. This is known to be so for p large, but it is an open problem in general (since 1973, shame on us!).

One has not only $\operatorname{ch} Q_n(\lambda) \in \mathbf{Z}[X(T)]^W$, but also

$$\operatorname{ch} Q_n(\lambda) \in \operatorname{ch}(\mathrm{St}_n)\mathbf{Z}[X(T)]^W.$$

This implies that one can express $\operatorname{ch} Q_n(\lambda)$ as a linear combination of the Brauer characters of the $U_n(\mu)$. The coefficients $[Q_n(\lambda) : U_n(\mu)]$ occurring here have been determined in [**6**] and [**30**]. The matrix of these coefficients turns out to be unipotent upper triangular, so we can invert it and get another approach to the Brauer characters of the $U_n(\mu)$ provided we know A and D.

2.6. Even in those cases where we know A and D it is rather difficult to carry out all these computations and to get closed formulas. There is a survey of the known results in [**24**]. Let me mention only two results explicitly. For $G = \mathrm{SL}_3$ and $p = 2$ one has $\dim U_n(0) = 2^{3n}(6^n - 5^n)$ (cf. [**8**]), and for $G = \mathrm{SL}_3$, $p \geq 7$ one has

$$[U_n(0) : L(0)]_{G(q)} = \left(9 + \sqrt{33}\right)^n + \left(9 - \sqrt{33}\right)^n + 6^n - 2 \cdot 8^n;$$

cf. [**48**]. Quite general results of Chastkofsky in [**7**] imply that there are always algebraic numbers $a_1, \ldots, a_r, b_1, \ldots, b_s$ (depending on G and p, but not on n) and signs $\varepsilon_i, \varepsilon'_j \in \{\pm 1\}$ such that

$$\dim U_n(0) = \sum_{i=1}^{r} \varepsilon_i a_i^n \tag{1}$$

and

$$[U_n(0) : L(0)]_{G(q)} = \sum_{j=1}^{s} \varepsilon'_j b_j^n. \tag{2}$$

In order to give you at least some idea why such a result should hold I shall describe some of Chastkofsky's arguments in Appendix 2.

3. Reduction of ordinary characters.

3.1. For any function φ on $G(q)$ let $\overline{\varphi}$ be the function that coincides with φ on the p-regular elements and that takes the value 0 on all other elements.

If φ is the character of a $\mathbf{C}G(q)$-module, then $\overline{\varphi}$ is the Brauer character of its reduction mod p.

The Deligne-Lusztig theory shows that the space of all (complex valued) class functions on $G(q)$ vanishing outside the p-regular elements is generated by all $\overline{R}_{T'}^{\theta}$, where the $R_{T'}^{\theta}$ are the Deligne-Lusztig characters. Usually, Lusztig's results on the decomposition of the $R_{T'}^{\theta}$ into irreducible characters tell you how to write any $\overline{\varphi}$ (with φ an irreducible ordinary character of $G(q)$) as a linear combination of the $\overline{R}_{T'}^{\theta}$. Therefore we shall concentrate below on the reduction mod p of the Deligne-Lusztig characters. One can find some results on the general case in [**31**, **36**].

3.2. We shall use notations and conventions for the Deligne-Lusztig characters that are not quite standard. For any $w \in W = N_G(T)/T$ choose a representative $n_w \in N_G(T)$ and (using Lang's theorem) an element $g_w \in G$ with $g_w^{-1}F^n(g_w) = n_w$. Then $T_w = g_w T g_w^{-1}$ is an F^n-stable maximal torus in G. Any $\mu \in X(T)$ yields a character $\theta(\mu) : T_w(\mathbf{F}_q) \to \mathbf{C}^\times$ by composing $t \mapsto \mu(g_w^{-1} t g_w)$ (from T_w to $k^\times$) with the embedding of $k^\times$ into $\mathbf{C}^\times$ used when defining Brauer characters. Then set $R_w(n,\mu) = \det(w) R_{T_w}^{\theta(\mu)}$. The multiplication by $\det(w)$ has the effect that $\deg(R_w(n,\mu)) > 0$.

There is a general formula (from [**30**]) for $\overline{R}_w(n,\mu)$ in terms of $\tilde{\chi}(\lambda)$ with $\lambda \in X(T)$:

$$(1) \qquad \overline{R}_w(n,\mu) = \sum_{w_1,w_2 \in W} \tilde{\gamma}_{w_1,w_2} \tilde{\chi}(w_1(\mu - w\varepsilon_{w_0 w_2}) + p^n \rho_{w_1} - \rho).$$

Several terms in this formula have to be explained. For any $w \in W$ set ρ_w equal to the sum of all ω_α with $\alpha \in S$ and $w^{-1}(\alpha) < 0$, and set $\varepsilon_w = w^{-1}\rho_w$. A theorem of Hulsurkar implies that the $W \times W$-matrix with entries

$$\det(w')\chi(-\varepsilon_{w_0 w} + \varepsilon_{w'} - \rho)$$

is unipotent upper triangular (for a suitable ordering of W). Therefore it has an inverse matrix with entries (denoted by $\gamma_{w,w'}$) in $\mathbf{Z}[X(T)]^W$. Note that $\gamma_{w,w} = 1$ for all $w \in W$.

3.3. The construction of the $\gamma_{w,w'}$ implies that there is an algorithm for computing them. If W gets bigger and bigger, then the matrix of all $\gamma_{w,w'}$ seems to get more and more complicated. For R of type A_2 one has $\gamma_{w,w'} = 0$ for all w, w' with $w \neq w'$. For R of type B_2 there are exactly four pairs $(w,w') \in W \times W$ with $w \neq w'$ and $\gamma_{w,w'} \neq 0$. These four pairs satisfy $\gamma_{w,w'} = 1$.

Consider R of type A_4, denote the simple roots by $\alpha_1, \alpha_2, \alpha_3, \alpha_4$ (with usual ordering), and set $\omega_i = \omega_{\alpha_i}$ for all i. Each $\gamma_{w,w'}$ belongs to

$$\{0, \pm 1, \pm\chi(\omega_1) \pm \chi(\omega_4)\}.$$

The number of pairs $(w,w') \in W \times W$ with $\gamma_{w,w'} = 1$ resp. with $\gamma_{w,w'} = -1$ is equal to 200 (including the 120 diagonal pairs) resp. 100. There are 10 pairs each with $\gamma_{w,w'}$ equal to $\chi(\omega_1), -\chi(\omega_1), \chi(\omega_4), -\chi(\omega_4)$. So in this case 3.2(1) yields $\overline{R}_w(n,\mu)$ as a sum of 500 terms of the form $\pm\tilde{\chi}(\lambda)$. (Note that each $\pm\chi(\omega_1)$, $\pm\chi(\omega_4)$ contributes five terms.)

This example indicates that the usefulness of 3.2(1) will be smaller for groups like $\mathrm{SL}_5(\mathbf{F}_2)$ and greater for groups like $\mathrm{SL}_5(\mathbf{F}_{9\,999\,991})$, once we know A and D in that case. For groups of type A_2 or B_2, however, formula 3.2(1) is quite useful for all p. As an example, let me describe the composition factors of all $\overline{R}_1(1,\mu)$ with $\overline{R}_1(1,\mu)$ simple for the group $\mathrm{Sp}_4(\mathbf{F}_p)$. Set $S = \{\alpha, \beta\}$ with α short and write $(r,s) = r\omega_\alpha + s\omega_\beta$. For $\mu = (r,s)$ with $0 < r,\ s, r+s < (p-1)/2$ the character $R_1(1,\mu)$ is irreducible, and we thus get each simple $R_1(1,\mu)$ exactly once. The composition factors of $R_1(1,(r,s))$ are contained in the following list, where each $L(\lambda)$, $\lambda \in X_1(T)$ occurs as often as its multiplicity in a Jordan-Hölder series. However, any term of the form $L(-1,a)$ or $L(b,-1)$ has to be deleted from the list. For $r > 1$ the list is

$$\begin{aligned}
&L(p-(r+1),p-(s+1)),\ L(r+2s,p-(s+1)),\ L(p-(r+1),r+s),\\
&L(p-(r+1),r+s-1),\ L(r+2s,p-(r+s+1)),\ L(r+2s,p-(r+s+2)),\\
&L(p-(r+2s+1),r+s),\ L(p-(r+2s+1),r+s-1),\\
&L(p-(r+2s+3),r+s),\ L(r,p-(r+s+1)),\ L(r,p-(r+s+1)),\\
&L(r-2,p-(r+s+1)),\ L(p-(r+2s+1),s),\ L(p-(r+2s+3),s),\\
&L(p-(r+2s+1),s-1),\ L(p-(r+2s+1),s-2),\ L(r,s), L(r-2,s),\\
&L(r,s-2),\ L(r,s-1).
\end{aligned}$$

For $r = 1$ we have to add one more term:

$$L(p-2(s+1),s-1).$$

In order to show the effectiveness of 3.2(1) for Sp_4 in general, let me add as an Appendix 3 for each irreducible ordinary character ξ of $\mathrm{Sp}_4(\mathbf{F}_q)$ the expression of $\overline{\xi}$ in terms of the $\tilde{\chi}(\mu)$ derived from 3.2(1).

3.4. If I had also written down in 3.3 the composition factors of the $\overline{R}_w(1,\mu)$ with $w \neq 1$ (and for $\mathrm{Sp}_4(\mathbf{F}_p)$), then there would again have been 20 factors in general. The number 20 is also the length of a Weyl module for Sp_4 "in generic position" and of certain induced modules for the Lie algebra of Sp_4.

Such equalities will occur for any G and $G(q)$; cf. [**30**, 4.3, or **21**, **5**, **22**]. The similarity between modules defined within quite different contexts seems to be even stronger and to extend also to the module structure. For example, for $\mathrm{SL}_3(\mathbf{F}_p)$ the principal series modules (having Brauer character $\overline{R}_1(1,\mu)$) have the same submodule structure as the Weyl modules for SL_3, provided we are in a generic position on both sides; cf. [**39**]. There is some more information about the submodule structure of principal series modules in [**32**] and [**39**].

Appendix 1. As mentioned in 1.3 I want to describe Kempf's proof of the simplicity of $L(\lambda)$ with $\lambda \in X_n(T)$ as a $kG(q)$-module (where $q = p^n$). The crucial result to prove is this one:

(1) *If* $\lambda \in X_n(T)$, *then* $L(\lambda)^{U(q)} = L(\lambda)_{w_0\lambda}$.

We get the simplicity of $L(\lambda)$ from (1) as follows. Choose $v \in L(\lambda)$ with $L(\lambda)_{w_0\lambda} = kv$. The p-group $U(q)$ has a nonzero fixed point in any nonzero submodule of $L(\lambda)$, so (1) implies that v is contained in each such submodule, hence that $kG(q)v$ is a simple $kG(q)$-module. We have to show that $kG(q)v = L(\lambda)$. There is a nondegenerate "contravariant" form (,) on $L(\lambda)$ with $(v,v) \neq 0$; cf. [**29**]. So $(kG(q)v)^\perp$ is a $kG(q)$-submodule of $L(\lambda)$ with $v \notin (kG(q)v)^\perp$. This implies $(kG(q)v)^\perp = 0$, hence $L(\lambda) = kG(q)v$.

In order to get (1) we shall use the corresponding result with $U(q)$ replaced by the nth Frobenius kernel U_n of U:

(2) *If* $\lambda \in X_n(T)$, *then* $L(\lambda)^{U_n} = L(\lambda)_{w_0\lambda}$.

This formula is proved rather early in the representation theory of the Frobenius kernels of G; cf. [**34**, II.3.10/15].

Let $k[U]$ be the algebra of regular functions on U. Choose $v \in L(\lambda)_{w_0\lambda}$, $v \neq 0$ as above and define $\varphi \in L(\lambda)^*$ via $\varphi(v) = 1$ and $\varphi(L(\lambda)_\mu) = 0$ for all $\mu \neq w_0\lambda$. For any $x \in L(\lambda)$ the function $\Phi_x : u \mapsto \varphi(ux)$ is an element of $k[U]$. One has $\Phi_v = 1$ and $\varphi(u'tux) = \varphi(tux) = w_0(\lambda)(t)\varphi(ux) = w_0(\lambda)(t)\Phi_x(u)$ for all $u' \in U^+$, $t \in T$, $u \in U$, $x \in L(\lambda)$. So $\Phi_x = 0$ implies $\varphi(gx) = 0$ for all $g \in \overline{U^+TU} = G$, hence $v \notin kGx$ and $x = 0$ by the simplicity of $L(\lambda)$ for G. This shows that the map $x \mapsto \Phi_x$ is injective. It is a homomorphism of U-modules, if we take the right regular representation on $k[U]$, i.e., $(uf)(u') = f(u'u)$, and a homomorphism of T-modules, if we take the conjugation representation on $k[U]$ tensored by the character $w_0\lambda$. (If one constructs $L(\lambda)$ as a submodule of a module induced from TU, it is contained in the ring of regular functions on G and one gets the embedding $L(\lambda) \hookrightarrow k[U]$ by restricting the functions to Un_0, where $n_0 \in N_G(T)$ is a representative of w_0.)

Using this embedding of $L(\lambda)$ we see that (1) will follow from (2), if we can prove for any TU-submodule V of $k[U]$:

(3) *If* $V \cap k[U]^{U_n} = k1$, *then* $V \cap k[U]^{U(q)} = k1$.

The regular functions on U invariant under U_n, resp. under $U(q)$, are the regular functions on the quotient scheme U/U_n, resp. $U/U(q)$. The Frobenius map $F^n : U \to U$ induces an isomorphism $U/U_n \tilde{\to} U$, the Lang map $L_n : U \to U$, $u \mapsto F^n(u)u^{-1}$ induces an isomorphism $U/U(q) \tilde{\to} U$. Therefore

$$k[U]^{U_n} = \{f \circ F^n \mid f \in k[U]\} = (F^n)^* k[U]$$

and

$$k[U]^{U(q)} = \{f \circ L_n \mid f \in k[U]\} = L_n^* k[U].$$

The set of weights of $k[U]$ with respect to T is $\sum_{\alpha \in S} \mathbf{N}\alpha$ and one has $k[U]_0 = k1$. If $f \in k[U]_\mu$ for some $\mu \in X(T)$, then

$$f \circ F^n \in k[U]_{q\mu}.$$

Some elementary computations show for $f \in k[U]_\mu$ that $f \circ L_n$ is the sum of $f \circ F^n$ and of weight vectors of weight strictly less than $q\mu$.

Now consider the situation as in (3); take $f \in k[U]$, $f \neq 0$, such that $f \circ L_n \in V$. Decompose $f = \sum_\mu f_\mu$ and $f \circ L_n = \sum_\mu f'_\mu$ where $f_\mu, f'_\mu \in k[U]_\mu$ for all

μ. Let ν be maximal for $f_\nu \neq 0$. The remarks above imply $f_\nu \circ F^n = f'_{q\nu}$. Because V is a T-module and $f \circ L_n \in V$, all f'_μ belong to V, so especially $f_\nu \circ F^n \in V \cap (F^n)^* k[U] = k1$, hence $\nu = 0$. The maximality of ν implies $f = f_0 \in k1$, and (3) follows.

Appendix 2. As announced in 2.6 I shall try to explain Chastkofsky's results on the behavior of $\dim U_n(0)$ as a function of n.

For any finite-dimensional G-module V and any $\lambda \in X_n(T)$ define an $X(T)_+ \times X(T)_+$-matrix $A_{n,\lambda}(V)$ via

(1) $$A_{n,\lambda}(V)_{\nu,\nu'} = [V \otimes L(\nu) : L(p^n\nu' + \lambda)]_G$$

for all $\nu, \nu' \in X(T)_+$ using the notation $[? : L(\nu)]_G$ for the multiplicity of $L(\nu)$ as a composition factor of a G-module. There is $\mu \in X(T)_+$ such that all weights μ' of V satisfy $\mu' \leq_{\mathbf{Q}} \mu$. Then $A_{n,\lambda}(V)_{\nu,\nu'} \neq 0$ implies $p^n\nu' \leq_{\mathbf{Q}} p^n v' + \lambda \leq_{\mathbf{Q}} \mu + \nu$. Almost all $\nu' \in X(T)_+$ satisfy $\mu <_{\mathbf{Q}} \nu'$ and for such ν' the inequality $p^n\nu' \leq_{\mathbf{Q}} \mu_+\nu$ implies $(p^n-1)\nu' <_{\mathbf{Q}} \nu$. Therefore, the matrix $A_{n,\lambda}(V)$ will have only finitely many entries on or above the diagonal (for a suitable ordering of $X(T)_+$). A similar statement is true for the other matrices to be constructed below. I shall not repeat the necessary arguments. It is now possible to take the trace of any $A_{n,\lambda}(V)$. The results of **[6]** and **[30]** imply for all $\lambda, \mu \in X_n(T)$:

(2) $$\operatorname{tr} A_{n,\lambda}(L(\mu)) = [Q_n(\mu) : U_n(\lambda)].$$

One has obviously

$$A_{n,\lambda}(L(\mu))_{0,\nu} = \begin{cases} 1 & \text{if } \lambda = \mu \text{ and } \nu = 0, \\ 0 & \text{otherwise.} \end{cases}$$

Let $B_{n,\lambda}(\mu)$ be the matrix $A_{n,\lambda}(L(\mu))$ with the row and column corresponding to $\nu = 0$ removed. Then for all $\lambda, \mu \in X_n(T)$

$$\operatorname{tr} B_{n,\lambda}(\mu) = \begin{cases} [Q_n(\lambda) : U_n(\mu)] & \text{if } \lambda \neq \mu, \\ 0 & \text{otherwise.} \end{cases}$$

The matrix of all $[Q_n(\mu) : U_n(\lambda)]$ with $\lambda, \mu \in X_n(T)$ is unipotent upper triangular. The inverse matrix $H_n = (h_{n,\lambda,\mu})$ satisfies $h_{n,\lambda,\lambda} = 0$ for all λ and

$$h_{n,\lambda,\mu} = \sum_{j>0} (-1)^j \sum [Q_n(\lambda) : U_n(\mu_1)][Q_n(\mu_1) : U_1(\mu_2)] \cdots [Q_n(\mu_{j-1}) : U_n(\mu)]$$

for $\lambda \neq \mu$, where we sum over all $\mu_1, \ldots, \mu_{j-1} \in X_n(T)$ with $\mu_i \neq \mu_{i+1}$ for $0 \leq i \leq j-1$ (where $\mu_0 = \lambda$, $\mu_j = \mu$). By (2) we can replace all $[Q_n(\mu_i) : U_n(\mu_{i+1})]$ by $\operatorname{tr} B_{n,\mu_i}(\mu_{i+1})$ and then drop the conditions $\mu_i \neq \mu_{i+1}$. The product $\prod_{i=0}^{j-1} \operatorname{tr} B_{n,\mu_i}(\mu_{i+1})$ can be interpreted as the trace of a matrix $B_{n,\mu_0}(\mu_1) \otimes B_{n,\mu_1}(\mu_2) \otimes \cdots \otimes B_{n,\mu_{j-1}}(\mu_j)$. So

(4) $$h_{n,\lambda,\mu} = \sum_{j \geq 0} (-1)^j \operatorname{tr} C_{n,j,\lambda}(\mu)$$

for all $\lambda, \mu \in X_n(T)$ where

(5) $$C_{n,j,\lambda}(\mu) = \sum B_{n,\lambda}(\mu_1) \otimes B_{n,\mu_1}(\mu_2) \otimes \cdots \otimes B_{n,\mu_{j-1}}(\mu)$$

(sum over all $\mu_1, \mu_2, \dots, \mu_{j-1} \in X_n(T)$) for $j > 0$ and where $C_{n,0,\lambda}(\mu)$ is the 1×1-matrix $(\delta_{\lambda\mu})$.

One has for all $\lambda_0, \lambda_1, \dots, \lambda_{r-1} \in X_n(T)$

$$L\left(\sum_{i=0}^{r-1} p^{ni}\lambda_i\right) \simeq L(\lambda_0) \otimes L(\lambda_1)^{(n)} \otimes \cdots \otimes L(\lambda_{r-1})^{(n(r-1))} \tag{6}$$

by Steinberg's tensor product theorem and

$$Q_{nr}\left(\sum_{i=0}^{r-1} p^{ni}\lambda_i\right) \simeq Q_n(\lambda_0) \otimes Q_n(\lambda_1)^{(n)} \otimes \cdots \otimes Q_n(\lambda_{r-1})^{(n(r-1))}; \tag{7}$$

cf. [**34**, II.11.15]. We get from (6) for all $\lambda = \sum_{i=0}^{r-1} p^{ni}\lambda_i$, $\mu = \sum_{i=0}^{r-1} p^{ni}\mu_i \in X_{nr}(T)$ where $\lambda_i, \mu_i \in X_n(T)$ for all i:

$$A_{nr,\lambda}(\mu) = A_{n,\lambda_0}(L(\mu_0))A_{n,\lambda_1}(L(\mu_1)) \cdots A_{n,\lambda_{r-1}}(L(\mu_{r-1})); \tag{8}$$

hence

$$B_{nr,\lambda}(\mu) = B_{n,\lambda_0}(\mu_0)B_{n,\lambda_1}(\mu_1) \cdots B_{n,\lambda_{r-1}}(\mu_{r-1}) \tag{9}$$

and

$$C_{nr,j,\lambda}(\mu) = C_{n,j,\lambda_0}(\mu_0)C_{n,j,\lambda_1}(\mu_1) \cdots C_{n,j,\lambda_{r-1}}(\mu_{r-1}). \tag{10}$$

For all $\lambda = \sum_{i=0}^{r-1} p^{ni}\lambda_i$ we get

$$\begin{aligned}
\dim U_{nr}(\lambda) &= \sum_{\mu \in X_n(T)} h_{rn,\lambda,\mu} \dim Q_{nr}(\mu) \\
&= \sum_{j \geq 0} (-1)^j \sum_{\mu \in X_{nr}(T)} \operatorname{tr} C_{rn,j,\lambda}(\mu)(\dim Q_{nr}(\mu)) \\
&= \sum_{j \geq 0} (-1)^j \sum_{\mu_0, \dots, \mu_{r-1} \in X_n(T)} \operatorname{tr}\left(\prod_{i=0}^{r-1} (\dim Q_n(\mu_i))C_{n,j,\lambda_i}(\mu_i)\right) \\
&= \sum_{j \geq 0} (-1)^j \left(\prod_{i=0}^{r-1}\left(\sum_{\mu \in X_n(T)} \dim Q_n(\mu)C_{n,j,\lambda_i}(\mu)\right)\right) \\
&= \sum_{j \geq 0} (-1)^j \operatorname{tr}(D_{n,j,\lambda_0}D_{n,j,\lambda_1} \cdots D_{n,j,\lambda_{r-1}})
\end{aligned}$$

where

$$D_{n,j,\lambda'} = \sum_{\mu \in X_n(T)} (\dim Q_n(\mu))C_{n,j,\lambda'}(\mu)$$

for all $\lambda' \in X_n(T)$.

For any $\nu \in X_n(T)$ set $\nu^{(r)} = \sum_{i=0}^{r-1} p^{ni}\nu$ for all r. Then

$$\dim U_{nr}(\nu^{(r)}) = \sum_{j \geq 0} (-1)^j \operatorname{tr}(D^r_{n,j,\nu}). \tag{11}$$

If $c_{j0}, c_{j1}, \dots$ are the nonzero eigenvalues of $D_{n,j,\nu}$, then

$$\dim U_{nr}(\nu^{(r)}) = \sum_{j \geq 0} \sum_{i \geq 0} (-1)^j c^r_{ji}. \tag{12}$$

This yields 2.6(1) as a special case. Similar arguments show for all $\mu, \nu \in X_n(T)$ that there are algebraic numbers b_{ji} with

$$[U_{nr}(\mu^{(r)}) : L(\nu^{(r)})]_{G(p^{nr})} = \sum_{j\geq 0}\sum_{i\geq 0}(-1)^j b_{ji}^r.$$

As a special case one gets 2.6(2).

Appendix 3. Let us consider the case $G = \mathrm{Sp}_4$. The ordinary irreducible characters of $G(q)$ have been determined by Srinivasan in [**42**]. We shall express these characters in terms of the $R_w(n,\mu)$ without identifying for each series of representations which of Srinivasan's characters corresponds to which Deligne-Lusztig data.

We shall use the notation $(r,s) = r\omega_\alpha + s\omega_\beta$ as in 3.3. Furthermore we shall write $R_w(\mu)$ instead of $R_w(n,\mu)$. Let s_1, resp. s_2, be the simple reflection corresponding to α, resp. β.

Assume q to be odd. Then we have the following ordinary irreducible characters:

(1) The $\chi_1(j)$ for $(q^2-1)/4$ different j in [**42**]. These are the $R_{s_1s_2}(r,s)$ with $0 < r$, $0 \leq s$, $r+2s \leq q-1$. We have

$$\begin{aligned}\overline{R}_{s_1s_2}(r,s) = {}& \tilde{\chi}(q-(r+1), q-(s+1)) + \tilde{\chi}(r+2s, q-(s+2))\\ &+ \tilde{\chi}(q-r+1, r+s-2) + \tilde{\chi}(r+2s-2, q-(r+s))\\ &+ \tilde{\chi}(q-(r+2s+3), r+s) + \tilde{\chi}(q-(r+2s+1), r+s-1)\\ &+ \tilde{\chi}(r-2, q-(r+s+1)) + \tilde{\chi}(r, q-(r+s+2))\\ &+ \tilde{\chi}(q-(r+2s+1), s) + \tilde{\chi}(q-(r+2s)+1, s-2)\\ &+ \tilde{\chi}(r-4, s+1) + \tilde{\chi}(r-2, s-1).\end{aligned}$$

(2) The $\chi_2(j)$ for $(q-1)^2/4$ different j in [**42**]. These are the $R_{s_1}(r,s)$ with $0 < r$, $0 \leq s$, $r+2s < q-1$. We have

$$\begin{aligned}\overline{R}_{s_1}(r,s) = {}& \tilde{\chi}(q-(r+1), q-(s+1)) + \tilde{\chi}(r+2s, q-(s+2))\\ &+ \tilde{\chi}(q-(r+1), r+s) + \tilde{\chi}(r+2s, q-(r+s+2))\\ &+ \tilde{\chi}(q-(r+2s+1), r+s-2) + \tilde{\chi}(q-(r+2s+1), r+s-1)\\ &+ \tilde{\chi}(r-2, q-(r+s)) + \tilde{\chi}(r-2, q-(r+s+1))\\ &+ \tilde{\chi}(q-(r+2s+1), s-2) + \tilde{\chi}(q-(r+2s+3), s)\\ &+ \tilde{\chi}(r-2, s+1) + \tilde{\chi}(r, s-1).\end{aligned}$$

(3) The $\chi_3(k,l)$ with $1 \leq k < l \leq (q-3)/2$ in [**42**]. These are the $R_1(r,s)$ with $0 < r$, $s, r+s < (q-1)/2$. We have

$$\begin{aligned}\overline{R}_1(r,s) = {}& \tilde{\chi}(q-(r+1), q-(s+1)) + \tilde{\chi}(r+2s, q-(s+1))\\ &+ \tilde{\chi}(q-(r+1), r+s) + \tilde{\chi}(r+2s, q-(r+s+1))\\ &+ \tilde{\chi}(q-(r+2s+1), r+s) + \tilde{\chi}(q-(r+2s+1), r+s-1)\\ &+ \tilde{\chi}(r, q-(r+s+1)) + \tilde{\chi}(r, q-(r+s+2)) + \tilde{\chi}(q-(r+2s+1), s)\\ &+ \tilde{\chi}(q-(r+2s+3), s) + \tilde{\chi}(r,s) + \tilde{\chi}(r-2, s).\end{aligned}$$

(4) The $\chi_4(k,l)$ with $1 \le k \le l \le (q-1)/2$ in [**42**]. These are the $R_{w_0}(r,s)$ with $0 < r,\ s,\ r+s < (q+1)/2$. We have

$$\begin{aligned}\overline{R}_{w_0}(r,s) = {} & \tilde{\chi}(q-(r+1), q-(s+1)) + \tilde{\chi}(r+2s-2, q-(s+1)) \\ & + \tilde{\chi}(q-(r+1), r+s-2) + \tilde{\chi}(r+2s-2, q-(r+s+1)) \\ & + \tilde{\chi}(q-(r+2s+1), r+s-2) + \tilde{\chi}(q-(r+2s+1), r+s-1) \\ & + \tilde{\chi}(r-2, q-(r+s+1)) + \tilde{\chi}(r-2, q-(r+s)) \\ & + \tilde{\chi}(q-(r+2s+1), s-2) + \tilde{\chi}(q-(r+2s)+1, s-2) \\ & + \tilde{\chi}(r-2, s-2) + \tilde{\chi}(r, s-2).\end{aligned}$$

(5) The $\chi_5(k,l)$ with $1 \le k \le (q-1)/2$, $1 \le l \le (q-3)/2$ in [**42**]. These are the $R_{s_2}(r,s)$ with $0 \le r$, $0 < s, r+s < (q-1)/2$ and the $R_{s_1s_2s_2}(r,s)$ with $0 < r, s, r+s < (q+1)/2$. (The last terms could also be described as $R_{s_2}(r,s)$ with $0 < s < (q+1)/2$, $r+s < q-1 < r+2s$, but it is more convenient to keep r, s small.) We have

$$\begin{aligned}\overline{R}_{s_2}(r,s) = {} & \tilde{\chi}(q-(r+1), q-(s+1)) + \tilde{\chi}(r+2s, q-(s+1)) \\ & + \tilde{\chi}(q-(r+3), r+s) + \tilde{\chi}(r+2s-2, q-(r+s+1)) \\ & + \tilde{\chi}(q-(r+2s+3), r+s) + \tilde{\chi}(q-(r+2s+1), r+s-1) \\ & + \tilde{\chi}(r-2, q-(r+s+1)) + \tilde{\chi}(r, q-(r+s+2)) \\ & + \tilde{\chi}(q-(r+2s)+1, s-2) + \chi(q-(r+2s+1), s-2) \\ & + \tilde{\chi}(r+2, s-2) + \tilde{\chi}(r, s-2)\end{aligned}$$

and

$$\begin{aligned}\overline{R}_{s_1s_2s_1}(r,s) = {} & \tilde{\chi}(q-(r+1), q-(s+1)) + \tilde{\chi}(r+2s-2, q-(s+1)) \\ & + \tilde{\chi}(q-r+1, r+s-2) + \tilde{\chi}(r+2s, q-(r+s+1)) \\ & + \tilde{\chi}(q-(r+2s)+1, r+s-2) + \tilde{\chi}(q-(r+2s+1), r+s-1) \\ & + \tilde{\chi}(r, q-(r+s+1)) + \tilde{\chi}(r-2, q-(r+s)) \\ & + \tilde{\chi}(q-(r+2s+3), s) + \tilde{\chi}(q-(r+2s+1), s) \\ & + \tilde{\chi}(r-4, s) + \tilde{\chi}(r-2, s).\end{aligned}$$

(6) The $\chi_6(k)$ resp. $\chi_7(k)$ with $1 \le k \le (q-1)/2$ in [**42**]. These are the $(R_{s_2s_1s_2}(0,s) - R_{w_0}(0,s))/2$ resp. $(R_{s_2s_1s_2}(0,s) + R_{w_0}(0,s))/2$ with $1 \le s \le (q-1)/2$. We have

$$\begin{aligned}& \tfrac{1}{2}(\overline{R}_{s_2s_1s_2}(0,s) - \overline{R}_{w_0}(0,s)) \\ & \quad = \tilde{\chi}(2s-2, q-s) - \tilde{\chi}(2s-2, q-(s+1)) \\ & \qquad + \tilde{\chi}(q-(2s+1), s) - \tilde{\chi}(q-(2s+1), s-2) + \tilde{\chi}(0, q-(s+1)) \\ & \qquad + \tilde{\chi}(0, q-(s+2)) - \tilde{\chi}(0, s-2) + \tilde{\chi}(0, s-3)\end{aligned}$$

and

$$\begin{aligned}\tfrac{1}{2}(\overline{R}_{s_2s_1s_2}(0,s)+\overline{R}_{w_0}(0,s))&\\ =\tilde{\chi}(q-1,q-(s+1))+\tilde{\chi}(q-1,s-2)&\\ +\tilde{\chi}(2s-2,q-s)+\tilde{\chi}(2s-2,q-(s+1))+\tilde{\chi}(q-(2s+1),s)&\\ +\tilde{\chi}(q-(2s+1),s-2)+\tilde{\chi}(q-2s+1,s-2)+\tilde{\chi}(q-(2s+1),s-1).&\end{aligned}$$

(7) The $\chi_8(k)$ resp. $\chi_9(k)$ with $1\le k\le (q-3)/2$ in [**42**]. These are the $(R_1(0,s)-R_{s_1}(0,s))/2$ resp. $(R_1(0,s)+R_{s_1}(0,s))/2$ with $1\le s\le (q-3)/2$. We have

$$\begin{aligned}\tfrac{1}{2}(\overline{R}_1(0,s)-\overline{R}_{s_1}(0,s))&\\ =\tilde{\chi}(2s,q-(s+1))-\tilde{\chi}(2s,q-(s+2))&\\ +\tilde{\chi}(q-(2s+1),s)-\tilde{\chi}(q-(2s+1),s-2)+\tilde{\chi}(0,q-(s+1))&\\ +\tilde{\chi}(0,q-(s+2))+\tilde{\chi}(0,s)-\tilde{\chi}(0,s-1)&\end{aligned}$$

and

$$\begin{aligned}\tfrac{1}{2}(\overline{R}_1(0,s)+\overline{R}_{s_1}(0,s))&\\ =\tilde{\chi}(q-1,q-(s+1))+\tilde{\chi}(q-1,s)&\\ +\tilde{\chi}(2s,q-(s+1))+\tilde{\chi}(2s,q-(s+2))+\tilde{\chi}(q-(2s+1),s)&\\ +\tilde{\chi}(q-(2s+1),s-2)+\tilde{\chi}(q-(2s+1),s-1)+\tilde{\chi}(q-(2s+3),s).&\end{aligned}$$

(8) The $\xi_1(k)$ resp. $\xi_1'(k)$ with $1\le k\le (q-1)/2$ in [**42**]. These are the $(R_{s_1s_2s_1}(r,0)-R_{w_0}(r,0))/2$ resp. $(R_{s_1s_2s_1}(r,0)+R_{w_0}(r,0))/2$ with $1\le r\le (q-1)/2$. We have

$$\begin{aligned}\tfrac{1}{2}(\overline{R}_{s_1s_2s_1}(r,0)-\overline{R}_{w_0}(r,0))&\\ =\tilde{\chi}(q-r+1,r-2)-\tilde{\chi}(q-(r+1),r-2)&\\ +\tilde{\chi}(r,q-(r+1))-\tilde{\chi}(r-2,q-(r+1))+\tilde{\chi}(q-(r+1),0)&\\ +\tilde{\chi}(q-(r+3),0)+\tilde{\chi}(r-2,0)+\tilde{\chi}(r-4,0)&\end{aligned}$$

and

$$\begin{aligned}\tfrac{1}{2}(\overline{R}_{s_1s_2s_1}(r,0)+\overline{R}_{w_0}(r,0))&\\ =\tilde{\chi}(q-(r+1),q-1)+\tilde{\chi}(r-2,q-1)&\\ +\tilde{\chi}(q-r+1,r-2)+\tilde{\chi}(q-(r+1),r-2)+\tilde{\chi}(r,q-(r+1))&\\ +\tilde{\chi}(r-2,q-(r+1))+\tilde{\chi}(q-(r+1),r-1)+\tilde{\chi}(r-2,q-r).&\end{aligned}$$

(9) The $\xi_3(k)$ resp. $\xi_3'(k)$ with $1\le k\le (q-3)/2$ in [**42**]. These are the $(R_1(r,0)-R_{s_2}(r,0))/2$ resp. $(R_1(r,0)+R_{s_2}(r,0))/2$ with $1\le r\le (q-3)/2$. We have

$$\begin{aligned}\tfrac{1}{2}(\overline{R}_1(r,0)-\overline{R}_{s_2}(r,0))&\\ =\tilde{\chi}(q-(r+1),r)-\tilde{\chi}(q-(r+3),r)&\\ +\tilde{\chi}(r,q-(r+1))-\tilde{\chi}(r-2,q-(r+1))+\tilde{\chi}(q-(r+1),0)&\\ +\tilde{\chi}(q-(r+3),0)+\tilde{\chi}(r,0)+\tilde{\chi}(r-2,0)&\end{aligned}$$

and

$$\tfrac{1}{2}(\overline{R}_1(r,0)+\overline{R}_{s_2}(r,0))$$
$$=\tilde{\chi}(q-(r+1),q-1)+\tilde{\chi}(q-(r+1),r)$$
$$+\tilde{\chi}(r,q-1)+\tilde{\chi}(q-(r+3),r)+\tilde{\chi}(r,q-(r+1))$$
$$+\tilde{\chi}(r-2,q-(r+1))+\tilde{\chi}(q-(r+1),r-1)+\tilde{\chi}(r,q-(r+2)).$$

(10) The $\xi_{21}(k)$ and $\xi_{22}(k)$ with $1\le k\le (q-1)/2$ in [**42**]. The $\overline{\xi}_{21}(k)=\overline{\xi}_{22}(k)$ are the

$$\frac{1}{2}\overline{R}_{s_2}\left(r,\frac{q-1}{2}-r\right)=\tilde{\chi}\left(q-(r+1),\frac{q-1}{2}+r\right)+\tilde{\chi}\left(q-(r+3),\frac{q-1}{2}\right)$$
$$+\tilde{\chi}\left(r,\frac{q-3}{2}\right)+\tilde{\chi}\left(r-2,\frac{q-1}{2}\right)$$
$$+\tilde{\chi}\left(r+2,\frac{q-5}{2}-r\right)+\tilde{\chi}\left(r,\frac{q-5}{2}-r\right)$$

with $0\le r\le (q-3)/2$.

(11) The $\xi'_{21}(k)$ and $\xi'_{22}(k)$ with $1\le k\le (q-1)/2$ in [**42**]. The $\overline{\xi}'_{21}(k)=\overline{\xi}'_{22}(k)$ are the

$$\frac{1}{2}\overline{R}_{w_0}\left(r,\frac{q+1}{2}-r\right)=\tilde{\chi}\left(q-(r+1),\frac{q-3}{2}+r\right)+\tilde{\chi}\left(q-(r+1),\frac{q-3}{2}\right)$$
$$+\tilde{\chi}\left(r-2,\frac{q-1}{2}\right)+\tilde{\chi}\left(r-2,\frac{q-3}{2}\right)$$
$$+\tilde{\chi}\left(r,\frac{q-3}{2}-r\right)+\tilde{\chi}\left(r-2,\frac{q-3}{2}-r\right)$$

with $1\le r\le (q-1)/2$.

(12) The $\xi_{41}(k)$ and $\xi_{42}(k)$ with $1\le k\le (q-3)/2$ in [**42**]. The $\overline{\xi}_{41}(k)=\overline{\xi}_{42}(k)$ are the

$$\frac{1}{2}\overline{R}_1\left(r,\frac{q-1}{2}-r\right)=\tilde{\chi}\left(q-(r+1),\frac{q-1}{2}+r\right)+\tilde{\chi}\left(q-(r+1),\frac{q-1}{2}\right)$$
$$+\tilde{\chi}\left(r,\frac{q-1}{2}\right)+\tilde{\chi}\left(r,\frac{q-3}{2}\right)+\tilde{\chi}\left(r,\frac{q-1}{2}-r\right)$$
$$+\tilde{\chi}\left(r-2,\frac{q-1}{2}-r\right)$$

with $1\le r\le (q-3)/2$.

(13) The $\xi'_{41}(k)$ and $\xi'_{42}(k)$ with $1\le k\le (q-3)/2$ in [**42**]. The $\overline{\xi}'_{41}(k)=\overline{\xi}'_{42}(k)$ are the

$$\frac{1}{2}\overline{R}_{s_1s_2s_1}\left(r,\frac{q+1}{2}-r\right)=\tilde{\chi}\left(q-(r+1),r+\frac{q-3}{2}\right)+\tilde{\chi}\left(q+1-r,\frac{q-3}{2}\right)$$
$$+\tilde{\chi}\left(r,\frac{q-3}{2}\right)+\tilde{\chi}\left(r-2,\frac{q-1}{2}\right)$$
$$+\tilde{\chi}\left(r-2,\frac{q+1}{2}-r\right)+\tilde{\chi}\left(r-4,\frac{q+1}{2}-r\right)$$

with $2\le r\le (q-1)/2$.

(14) The $\overline{\Phi}_i (1 \leq i \leq 9)$ in [**42**]. We have

$$\begin{aligned}
\overline{\Phi}_1 = \overline{\Phi}_2 &= \frac{1}{4}\left(\overline{R}_{s_1 s_2 s_1}\left(\frac{q+1}{2}, 0\right) - \overline{R}_{w_0}\left(\frac{q+1}{2}, 0\right)\right) \\
&= \tilde{\chi}\left(\frac{q+1}{2}, \frac{q-3}{2}\right) - \tilde{\chi}\left(\frac{q-3}{2}, \frac{q-3}{2}\right) + \tilde{\chi}\left(\frac{q-3}{2}, 0\right) + \tilde{\chi}\left(\frac{q-7}{2}, 0\right), \\
\overline{\Phi}_3 = \overline{\Phi}_4 &= \frac{1}{4}\left(\overline{R}_{s_1 s_2 s_1}\left(\frac{q+1}{2}, 0\right) + \overline{R}_{w_0}\left(\frac{q+1}{2}, 0\right)\right) \\
&= \tilde{\chi}\left(\frac{q-3}{2}, q-1\right) + \tilde{\chi}\left(\frac{q+1}{2}, \frac{q-3}{2}\right) + \tilde{\chi}\left(\frac{q-3}{2}, \frac{q-1}{2}\right) \\
&\quad + \tilde{\chi}\left(\frac{q-3}{2}, \frac{q-3}{2}\right), \\
\overline{\Phi}_5 = \overline{\Phi}_6 &= \frac{1}{4}\left(\overline{R}_1\left(\frac{q-1}{0}, 0\right) - \overline{R}_{s_2}\left(\frac{q-1}{2}, 0\right)\right) \\
&= \tilde{\chi}\left(\frac{q-1}{2}, \frac{q-1}{2}\right) - \tilde{\chi}\left(\frac{q-5}{2}, \frac{q-1}{2}\right) + \tilde{\chi}\left(\frac{q-1}{2}, 0\right) + \tilde{\chi}\left(\frac{q-5}{2}, 0\right), \\
\overline{\Phi}_7 = \overline{\Phi}_8 &= \frac{1}{4}\left(\overline{R}_1\left(\frac{q-1}{2}, 0\right) - \overline{R}_{s_2}\left(\frac{q-1}{2}, 0\right)\right) \\
&= \tilde{\chi}\left(\frac{q-1}{2}, q-1\right) + \tilde{\chi}\left(\frac{q-1}{2}, \frac{q-1}{2}\right) + \tilde{\chi}\left(\frac{q-1}{2}, \frac{q-3}{2}\right) \\
&\quad + \tilde{\chi}\left(\frac{q-5}{2}, \frac{q-1}{2}\right), \\
\overline{\Phi}_9 &= \frac{1}{4}\left(\overline{R}_1\left(0, \frac{q-1}{2}\right) - \overline{R}_{w_0}\left(0, \frac{q-1}{2}\right)\right) \\
&= \tilde{\chi}\left(q-1, \frac{q-1}{2}\right) - \tilde{\chi}\left(q-1, \frac{q-3}{2}\right) + \tilde{\chi}\left(0, \frac{q-1}{2}\right) + \tilde{\chi}\left(0, \frac{q-5}{2}\right).
\end{aligned}$$

(15) The θ_i $(0 \leq i \leq 13)$ in [**42**]. We have

$$\begin{aligned}
\overline{\theta}_0 &= \tilde{\chi}(0) \quad \text{(the trivial one-dimensional character)}, \\
\overline{\theta}_1 = \overline{\theta}_2 &= \frac{1}{8}\left(\overline{R}_1\left(0, \frac{q-1}{2}\right) + \overline{R}_{w_0}\left(0, \frac{q+1}{2}\right) + 2\overline{R}_{s_1}\left(0, \frac{q-1}{2}\right)\right) \\
&= \tilde{\chi}\left(q-1, \frac{q-1}{2}\right) + \tilde{\chi}\left(q-1, \frac{q-3}{2}\right), \\
\overline{\theta}_3 = \overline{\theta}_4 &= \frac{1}{8}\left(\overline{R}_1\left(0, \frac{q-1}{2}\right) + \overline{R}_{w_0}\left(0, \frac{q+1}{2}\right) - 2\overline{R}_{s_1}\left(0, \frac{q-1}{2}\right)\right) \\
&= \tilde{\chi}\left(0, \frac{q-1}{2}\right) - \tilde{\chi}\left(0, \frac{q-5}{2}\right), \\
\overline{\theta}_5 = \overline{\theta}_6 &= \frac{1}{4}\left(\overline{R}_{s_1 s_2 s_1}\left(1, \frac{q-1}{2}\right) + \overline{R}_{s_1 s_2}\left(1, \frac{q-1}{2}\right)\right) \\
&= \tilde{\chi}\left(q-2, \frac{q-1}{2}\right) + \tilde{\chi}\left(q, \frac{q-3}{2}\right),
\end{aligned}$$

$$\overline{\theta}_7 = \overline{\theta}_8 = \frac{1}{4}\left(\overline{R}_{s_1s_2s_1}\left(1, \frac{q-1}{2}\right) - \overline{R}_{s_1s_2}\left(1, \frac{q-1}{2}\right)\right)$$
$$= \tilde{\chi}\left(1, \frac{q-3}{2}\right) - \tilde{\chi}\left(1, \frac{q-5}{2}\right),$$
$$\overline{\theta}_9 = \tfrac{1}{4}(\overline{R}_1(0) - \overline{R}_{s_1s_2}(0)) = \tilde{\chi}(q-1,0) + \tilde{\chi}(0,q-1),$$
$$\overline{\theta}_{10} = \tfrac{1}{4}(\overline{R}_{s_1s_2}(0) - \overline{R}_{w_0}(0)) = \tilde{\chi}(q-3,0) + \tilde{\chi}(0,q-2),$$
$$\overline{\theta}_{11} = \tfrac{1}{8}(\overline{R}_1(0) - \overline{R}_{w_0}(0) - 2\overline{R}_{s_1}(0) + 2R_{s_2}(0)) = \tilde{\chi}(q-3,0) + \tilde{\chi}(0,q-1),$$
$$\overline{\theta}_{12} = \tfrac{1}{8}(\overline{R}_1(0) - \overline{R}_{w_0}(0) + 2\overline{R}_{s_1}(0) - 2\overline{R}_{s_2}(0)) = \tilde{\chi}(q-1,0) + \tilde{\chi}(0,q-2),$$
$$\overline{\theta}_{13} = \tilde{\chi}(q-1,q-1) \quad \text{(the Steinberg character)}.$$

Let me now mention the necessary changes for q even. Then the characters mentioned in (10) to (14) and the θ_i with $1 \le i \le 8$ do not occur; in (6) to (9) the upper bound for s or r is to be replaced by $q/2$ in (6), (8) or by $(q-2)/2$ in (7), (19). Apart from that everything else remains valid.

In order to get from these formulas the decomposition into simple modules one has to use (for $p \neq 2$)

(a) $\chi(r,s) = \operatorname{ch} L(r,s)$ for $0 \le r$, $0 \le s$, $r + 2s \le p - 3$,

(b) $\chi(r,s) = \operatorname{ch} L(r,s) + \operatorname{ch} L(r, p-(r+s+3))$ for $0 \le r$, $p-3 \le r+2s$, $r+s \le p-2$,

(c) $\chi(r,s) = \operatorname{ch} L(r,s) + \operatorname{ch} L(2p-(r+2s+4), s)$ for $p-2 < r+s$, $r+2 < 2p-3$, $r < p-1$,

(c′) $\chi(r,s) = \operatorname{ch} L(r,s)$ for $r = p-1$, $0 \le s \le (p-2)/2$,

(d) $\chi(r,s) = \operatorname{ch} L(r,s) + \operatorname{ch} L(r, 2p-(r+s+3))$ for $2p-3 < r+2s$, $s < p-1$, $r \le p-1$,

(d′) $\chi(r,s) = \operatorname{ch} L(r,s)$ for $s = p-1$, $0 \le r \le p-1$.

These formulas describe the decomposition into simple modules of all $\chi(\lambda)$ with $\lambda \in X_1(T)$. For all $\lambda \in X(T)_+$, $\lambda \notin X_1(T)$ one has to use Steinberg's tensor product theorem to get the corresponding information. For $p = 2$ one has $\chi(\lambda) = \operatorname{ch} L(\lambda)$ for $\lambda \in \{0, \omega_\alpha, \rho\}$ and $\chi(\omega_\beta) = \operatorname{ch} L(\omega_\beta) + \operatorname{ch} L(0)$.

REFERENCES

1. H. H. Andersen, *Extensions of simple modules for finite Chevalley groups*, J. Algebra (to appear).

2. H. H. Andersen, J. Jørgensen, and P. Landrock, *The projective indecomposable modules of* $\mathrm{SL}(2,p^n)$, Proc. London Math. Soc. (3) **46** (1983), 38–52.

3. J. W. Ballard, *Projective modules for finite Chevalley groups*, Trans. Amer. Math. Soc. **245** (1978), 221–249.

4. J. F. Carlson, *The cohomology of irreducible modules over* $\mathrm{SL}(2,p^n)$, Proc. London Math. Soc. (3) **47** (1983), 480–492.

5. R. W. Carter, *The relation between characteristic 0 representations and characteristic p representations of finite groups of Lie type*, Proc. Sympos. Pure Math., vol. 37, Amer. Math. Soc., Providence, R.I., 1980, pp. 301–311.

6. L. Chastkofsky, *Projective characters for finite Chevalley groups*, J. Algebra **69** (1981), 347–357.

7. ——, *Rationality of certain zeta functions associated with modular representation theory*, Finite Groups—Coming of Age (J. Mackay, ed.), Contemp. Math., vol. 45, Amer. Math. Soc., Providence, R.I., 1985, pp. 41–50.

8. L. Chastkofsky and W. Feit, *On the projective characters in characteristic* 2 *of the groups* $SL_3(2^m)$ *and* $SU_3(2^m)$, J. Algebra **63** (1980), 124–142.

9. E. Cline, B. Parshall, and L. Scott, *Cohomology of finite groups of Lie type*. I, Inst. Hautes Études Sci. Publ. Math. **45** (1975), 169–191.

10. ——, *Cohomology of finite groups of Lie type*. II, J. Algebra **45** (1977), 182–198.

11. M. Collins (ed.), *Finite simple groups*. II (Proc. Durham, 1978), Academic Press, London/New York, 1980.

12. B. Cooperstein and G. Mason (eds.), *The Santa Cruz Conference on Finite Groups* (1979), Proc. Sympos. Pure Math., vol. 37, Amer. Math. Soc., Providence, R.I., 1980.

13. C. W. Curtis, *Representations of Lie algebras of classical type with application to linear groups*, J. Math. Mech. **9** (1960), 307–326.

14. ——, *Modular representations of finite groups with split* (B, N)*-pairs*, Seminar on Algebraic Groups and Related Finite Groups (Proc. Princeton, N.J., 1968/69), Lecture Notes in Math., vol. 131, Springer-Verlag, 1970, pp. 57–95.

15. S. W. Dagger, *On the blocks of the Chevalley groups*, J. London Math. Soc. (2) **3** (1971), 21–29.

16. R. Dipper, *Vertices of irreducible representations of finite Chevalley groups in the describing characteristic*, Math. Z. **175** (1980), 143–159.

17. ——, *On irreducible modules of twisted groups of Lie type*, J. Algebra **81** (1983), 370–389.

18. S. Donkin, *A note on the characters of the projective modules for the infinitesimal subgroups of a semisimple algebraic group*, Math. Scand. **51** (1982), 142–150.

19. P. Fleischmann, *Einfache periodische Moduln der Gruppen* $SU_3(q^2)$ *in der beschreibenden Charakteristik* $p \neq 2$, Dissertation, Univ. of Essen, 1986.

20. J. E. Humphreys, *Defect groups for finite groups of Lie type*, Math. Z. **119** (1971), 149–152.

21. ——, *Modular representations of finite groups of Lie type*, Finite Simple Groups. II (Proc. Durham, 1978), Academic Press, London/New York, 1980, pp. 259–290.

22. ——, *Cartan invariants and decomposition numbers of Chevalley groups*, Proc. Sympos. Pure Math., vol. 37, Amer. Math. Soc., Providence, R.I., 1980, pp. 347–351.

23. ——, *Deligne-Lusztig characters and principal indecomposable modules*, J. Algebra **62** (1980), 299–303.

24. ——, *Cartan invariants*, Bull. London Math. Soc. **17** (1985), 1–14.

25. ——, *Non-zero* Ext^1 *for Chevalley groups* (*via algebraic groups*), J. London Math. Soc. (2) **31** (1985), 463–467.

26. ——, *Projective modules for* $\mathrm{Sp}(4,p)$ *in characteristic* p, J. Algebra **104** (1986), 80–88.

27. ——, *The Steinberg representation*, Bull. Amer. Math. Soc. (N.S.) (to appear).

28. I. Janiszczak, *Irreducible periodic modules over* $\mathrm{SL}(m,q)$ *in the describing characteristic*, Comm. Algebra (to appear).

29. J. C. Jantzen, *Weyl modules for groups of Lie type*, Finite simple groups. II (Proc. Durham, 1978), Academic Press, London/New York, 1980, pp. 291–300.

30. ——, *Zur Reduktion modulo p der Charaktere von Deligne und Lusztig*, J. Algebra **70** (1981), 452–474.

31. ——, *Zur Reduktion modulo p unipotenter Charaktere endlicher Chevalley-Gruppen*, Math. Z. **181** (1982), 97–128.

32. ——, *Filtrierungen der Darstellungen in der Hauptserie endlicher Chevalley-Gruppen*, Proc. London Math. Soc. (3) **49** (1984), 445–482.

33. ——, *Modular representations of reductive groups*, Lecture Notes in Math., vol. 1185, Springer-Verlag, 1986, pp. 118–154.

34. ——, *Representations of algebraic groups*, Academic Press, Orlando, Fla. (to appear).

35. A. V. Jeyakumar, *Periodic modules for the group* $\mathrm{SL}(2,q)$, Comm. Algebra **8** (1980), 1721–1735.

36. M. Kaneda, *A truncation formula for certain principal indecomposable modules of finite Chevalley groups*, Comm. Algebra **14** (1986), 911–922.

37. G. Kempf, *Representations of algebraic groups in prime characteristic*, Ann. Sci. École Norm. Sup. (4) **14** (1981), 61–76.

38. S. Koshitani, *The Loewy structure of the projective indecomposable modules for* $\mathrm{SL}(3,3)$ *and its automorphism group in characteristic* 3, Comm. Algebra (to appear).

39. K. Kühne-Hausmann, *Zur Untermodulstruktur der Weylmoduln für* Sl_3, Bonn. Math. Schr. **162** (1985).

40. G. Lusztig, *Divisibility of projective modules of finite Chevalley groups by the Steinberg module*, Bull. London Math. Soc. **8** (1976), 130–134.

41. ——, *Some problems in the representation theory of finite Chevalley groups*, Proc. Sympos. Pure Math., vol. 37, Amer. Math. Soc., Providence, R. I., 1980, pp. 313–317.

42. B. Srinivasan, *The characters of the finite symplectic group* $\mathrm{Sp}(4, q)$, Trans. Amer. Math. Soc. **131** (1968), 488–525.

43. R. Steinberg, *Representations of algebraic groups*, Nagoya Math. J. **22** (1963), 33–56.

44. ——, *Endomorphisms of linear algebraic groups*, Mem. Amer. Math. Soc., No. 80 (1968).

45. ——, *Lectures on Chevalley groups*, New Haven, Conn., 1968 (Yale Univ.).

46. Tuan Hsio-Fu (ed.), *Group theory, Beijing* 1984, Lecture Notes in Math., vol. 1185, Springer-Verlag, 1986.

47. W. J. Wong, *Representations of Chevalley groups in characteristic p*, Nagoya Math. J. **45** (1971), 39–78.

48. Ye Jia-chen, *On the first Cartan invariant of the groups* $\mathrm{SL}(3, p^n)$ *and* $\mathrm{SU}(3, p^n)$, Lecture Notes in Math., vol. 1185, Springer-Verlag, 1986, pp. 388–400.

49. H. H. Andersen, *Modular representations of algebraic groups*, this volume.

UNIVERSITÄT HAMBURG, WEST GERMANY

Proceedings of Symposia in Pure Mathematics
Volume 47 (1987)

Shintani Lifting and Gelfand-Graev Representations

NORIAKI KAWANAKA

The purpose of this paper is to introduce the reader to two topics—Shintani lifting and generalized Gelfand-Graev representations—in the representation theory of finite reductive groups. Although both are meaningful (and perhaps more interesting) for reductive groups over local fields, we almost always confine ourselves to the finite field case; the main exception to this is the last subsection, where we state some conjectures concerning generalized Gelfand-Graev representations of reductive groups over local fields.

ACKNOWLEDGMENT. The author would like to express his heartfelt thanks to T. Asai, A. Gyoja, K. Uno, G. Lusztig, H. Matumoto, and H. Yamashita for stimulating discussions and helpful comments.

Notations. For a set X, $|X|$ denotes its cardinality. If σ is a transformation of X, X_σ is the set of fixed points of σ. If f is, say, a **C**-valued function on X, $\text{supp}(f)$ is the support of f. For a group G and an element x of G, $Z_G(x)$ denotes the centralizer of x in G. If G is finite, $\hat{G}$ is the set of equivalence classes of irreducible representations of G and will often be identified with the set of irreducible characters. The identity component of an algebraic group $\mathbf{G}$ is denoted by $\mathbf{G}^0$.

1. Shintani lifting (or Shintani descent or base change).

1.1. *History.* In 1975 Saito [**56**], generalizing an earlier work of Doi and Naganuma, proved:

Let F be a totally real cyclic extension of $\mathbf{Q}$ of prime degree l. Then automorphic forms in l-variables (Hilbert cusp forms over F) which are "invariant with respect to the Galois group $G(F/\mathbf{Q})$" can be obtained as "liftings" of automorphic cusp forms in one variable.

The main part of the proof was establishing an identity which connects traces of "twisted" Hecke operators (acting on a space of Hilbert cusp forms) and traces

---

1980 *Mathematics Subject Classification* (1985 *Revision*). Primary 20Gxx; Secondary 20Cxx, 22Exx.

This work was partly supported by a Grant-in-Aid for Scientific Research from the Ministry of Education, Science and Culture.

0082-0717/87 $1.00 + $.25 per page

of (ordinary) Hecke operators (acting on spaces of automorphic forms in one variable). Inspired by Saito's work (in particular, by the identity just mentioned), Shintani [**58**, **59**] invented the notion of liftings of (infinite-dimensional) representations of GL_2 over local fields, and used it to give a representation-theoretical interpretation and a generalization of the result of Saito. See Langlands [**44**] and Flicker [**23**] for the further developments of lifting theory of automorphic representations. Shintani also showed [**60**] that his idea of liftings works for representations of general linear groups over finite fields. Prior to this, Glauberman [**27**] had proved a result (on characters of a finite group with a coprime action) whose formulation has a remarkable resemblance to that of Shintani.

1.2. *The norm mapping.* Let $\mathbf{G}$ be a connected algebraic group, and m a fixed positive integer. Let σ be an endomorphism of $\mathbf{G}$ such that

$$|\mathbf{G}_{\sigma^m}| < \infty. \tag{1.2.1}$$

We put $G = \mathbf{G}_{\sigma^m}$. Let $A = \langle\sigma'\rangle$ be a cyclic group of order m acting on G by

$$x^{\sigma'} = x^{\sigma}, \qquad x \in G.$$

Henceforth, we write σ for σ'. Let $x, y \in G$. Then two elements σx and σy of the coset σG in the semidirect product AG are AG-conjugate if and only if x and y are σ-conjugate in G, i.e., $x = g^{-\sigma}yg$ for some $g \in G$. We denote by $x^{G,\sigma}$ the σ-conjugate class of $x \in G$. By Lang-Steinberg's theorem [**67**], any $x \in \mathbf{G}$ can be written as $x = \alpha_x^{-\sigma}\alpha_x$ with some $\alpha_x \in \mathbf{G}$ ($\alpha_x^{-\sigma}$ stands for $(\alpha_x^{-1})^{\sigma}$). Then, as can be easily checked, $x \in G$ if and only if

$$\alpha_x\alpha_x^{-\sigma^m} \ (= \alpha_x(x^{\sigma^{m-1}}x^{\sigma^{m-2}}\cdots x^{\sigma}x)\alpha_x^{-1}) \in G_{\sigma}.$$

Thus we have a well-defined map $x^{G,\sigma} \to (\alpha_x\alpha_x^{-\sigma^m})^{G_\sigma}$ (= the G_σ-conjugacy class of $\alpha_x\alpha_x^{-\sigma^m}$) from the set $\mathbf{G}/\underset{\sigma}{\sim}$ of σ-conjugacy classes of G to the set $G_\sigma/\sim$ of conjugacy classes of G_σ. We denote this mapping by $n = n_{\sigma^m/\sigma}$. (This was introduced in [**37**].) The special case $m = 1$ is also of interest; we put

$$t = t_\sigma = n_{\sigma/\sigma} \colon G_\sigma/\sim \longrightarrow G_\sigma/\sim .$$

Let $M = \text{L.C.M.}\{\mathrm{ord}(\bar{x}); x \in G_\sigma\}$. Here $\mathrm{ord}(\bar{x})$ is the order of the image $\bar{x}$ of x in $Z_{\mathbf{G}}(x)/Z_{\mathbf{G}}(x)^0$. Then t^M =identity. Thus we have various "normlike" mappings:

$$t^l \circ n \colon G/\underset{\sigma}{\sim} \longrightarrow G_\sigma/\sim, \qquad l \in Z/MZ, \tag{1.2.2}$$

all of which are bijective and satisfy:

$$\left|t^l \circ n(x^{G,\sigma})\right| |G_\sigma|^{-1} = |x^{G,\sigma}||G|^{-1}, \qquad x \in G. \tag{1.2.3}$$

We now assume that m is admissible [**38**] for $(\mathbf{G}, \sigma)$, i.e.,

$$(m, M) = 1 \tag{1.2.4}$$

(e.g., if $G = \mathrm{Sp}_{2n}$ or SO_n, odd numbers are always admissible), and take an $r \in \mathbf{Z}$ such that

$$rm \equiv 1 \pmod{M}.$$

Then we can define the norm mapping [**38**] $N = N_{\sigma^m/\sigma}: G/\tilde{\sigma} \to G_\sigma/\sim$ by

(1.2.5) $$N = t_\sigma^{-r} \circ n_{\sigma^m/\sigma}.$$

This mapping is distinguished among the normlike mappings (1.2.2) in that it satisfies:

(1.2.6) $$N(a^{G,\sigma}) = (a^m)^{G_\sigma}$$

for any $a \in G_\sigma$ such that $(\mathrm{ord}(a), m) = 1$. (As pointed out by Digne [**17**], (1.2.6) actually holds for any $a \in G_\sigma$.) If we assume:

(1.2.4)′ $$(m, |Z_{\mathbf{G}}(x)/Z_{\mathbf{G}}(x)^0|) = 1$$

for any $x \in G_\sigma$ (this condition is slightly stronger then (1.2.4)), then we have [**38**; II, 2.4.8]:

(1.2.7) $$N(x^{G,\sigma}) = (x^{\sigma^{m-1}} x^{\sigma^{m-2}} \cdots x^\sigma x)^G \cap G_\sigma$$

for any $x \in G$. Recall that, for $\mathbf{G} = \mathrm{GL}_n$, Shintani [**60**], following the idea of Saito [**56**], defined his norm mapping by the formula (1.2.7). (In that case, any m satisfies (1.2.4)′ by [**65**; III, 3.22].)

(1.2.8) REMARK. Let $\mathbf{G}$ be a nonconnected algebraic group, and σ its endomorphism. We assume that (1.2.1) and (1.2.4) hold. Then [**38**, II] one can generalize the above construction and define the norm mapping N with similar properties as in the connected case. See also [**17**].

1.3. *Liftings.* Here $\mathbf{G}$ is a not necessarily connected (see (1.2.8)) algebraic group, σ is an endomorphism of $\mathbf{G}$, and m is a fixed positive integer. We assume (1.2.1). Let $\chi \in (\hat{G})_\sigma$. Then, by Schur's lemma, χ extends to an irreducible character $\tilde{\chi}$ of AG. Then

$$x \to \tilde{\chi}(\sigma x), \qquad x \in G$$

is a function on $G/\underset{\sigma}{\sim}$. Hence, assuming (1.2.4), we can define a class function $(N^{-1})^*(\tilde{\chi})$ on G_σ, i.e., the function on $G_\sigma/\sim$ defined by

$$a^{G_\sigma} \xrightarrow{N^{-1}} x^{G,\sigma} \to \tilde{\chi}(\sigma x), \qquad a \in G_\sigma.$$

(1.3.1) CONJECTURE. We assume (1.2.4). Let $\chi \in (\hat{G})_\sigma$. Then we have $(N^{-1})^*(\tilde{\chi}) = \varepsilon\varsigma\psi$ with $\psi = \psi(\chi) \in (G_\sigma)\hat{}$, where $\varsigma = \varsigma(\tilde{\chi})$ is an mth root of 1 and $\varepsilon = \varepsilon(\chi) = 1$ (resp. ± 1) if m is even (resp. odd).

(1.3.2) REMARK. (i) If (1.3.1) is true for a $\{\mathbf{G}, \sigma, m\}$, then $\chi \leftrightarrow \psi$ is a bijective correspondence between $(\hat{G})_\sigma$ and $(G_\sigma)\hat{}$.

(ii) If $\mathbf{G}$ is connected and (at least) reductive, then it is very likely that we always have $\varepsilon(\chi) = 1$ (see (2.4.5)).

(1.3.3) CONJECTURE. We assume that $\mathbf{G}$ is connected, but do not assume (1.2.4).

(i) If $\chi \in (\hat{G})_\sigma$ satisfies

$$t_{\sigma^m}^*(\chi) \; (= \chi \circ t_{\sigma^m}) = \chi,$$

then there exists a $\psi = \psi(\chi) \in (G_\sigma)^\frown$ such that $t_\sigma^*(\psi) = \psi$ and $(n_{\sigma^m/\sigma}^{-1})^*(\chi) = \varepsilon\varsigma\psi$, where $\varsigma = \varsigma(\tilde{\chi})$ and $\varepsilon = \varepsilon(\chi)$ are as in (1.3.1). Moreover, this gives a 1-1 correspondence between $\left\{\chi \in (\hat{G})_\sigma; t_{\sigma^m}^*(\chi) = \chi\right\}$ and $\{\psi \in (G_\sigma)^\frown; t_\sigma^*(\psi) = \psi\}$.

(ii) Let **G** be reductive and let $\psi \in (G_\sigma)^\frown$. Then $t_\sigma^*(\psi) = \psi$ if and only if ψ is a uniform function, i.e., a **Q**-linear combination of Deligne-Lusztig [**16**] virtual characters.

If $\chi \in (\hat{G})_\sigma$ and $\psi \in (G_\sigma)^\frown$ are related as in (1.3.1) or (1.3.3)(i), then we say that χ (resp. ψ) is a lifting (resp. descent) of ψ (resp. χ), and write

$$\chi = \mathrm{lift}(\psi) = \mathrm{lift}_{\sigma^m/\sigma}(\psi).$$

(1.3.4) CONJECTURE. For positive integers m, n, we have "the transitivity of lifting maps":

$$\mathrm{lift}_{\sigma^{mn}/\sigma}(\psi) = \mathrm{lift}_{\sigma^{mn}/\sigma^m} \circ \mathrm{lift}_{\sigma^m/\sigma}(\psi)$$

for any $\psi \in (G_\sigma)^\frown$ such that lift $_{\sigma^{mn}/\sigma}(\psi)$ is definable.

We now review some of the known facts supporting the conjectures (1.3.1), (1.3.3), and (1.3.4):

(1.3.5) RESULTS. (i) Shintani's original result amounts to saying that (1.3.1) is true when $\mathbf{G} = \mathrm{GL}_n(\overline{F}_q)$ and σ is an untwisted Frobenius endomorphism. On the other hand, according to Glauberman [**27**], (1.3.1) holds when **G** is finite. It is easy to see that (1.3.1) is true if **G** is commutative; in fact, $\chi = \mathrm{lift}(\psi)$ if and only if $\chi(g) = \psi \circ N(g)$ $(= \psi(g^{\sigma^{m-1}} g^{\sigma^{m-2}} \cdots g^\sigma g))$ for any $g \in G$. Other cases for which (1.3.1) is verified (mostly under an additional assumption that $(m, \mathrm{char}(F_q)) = 1$) are: $\mathbf{G} = \mathrm{GL}_n(\overline{F}_q)$ and σ is a twisted Frobenius [**37**]; $\mathbf{G} = \mathrm{SO}_n(\overline{F}_q)$, $\mathrm{Sp}_{2n}(\overline{F}_q)$, $\mathrm{O}_n(\overline{F}_q)$ [**38**]; $\mathbf{G} = \mathrm{PSO}_8(\overline{F}_q)$ and σ is a triality Frobenius [**45**]; $\mathbf{G} = G_2(\overline{F}_q)$ and σ is a Frobenius [**53**].

(i)′ By (1.2.6) and [**27**], the statement of (1.3.1) is true if $(m, |G_\sigma|) = 1$.

(ii) The if part of (1.3.3)(ii) is known to be true. This is proved by Digne-Michel [**20**] and Asai [**2**] in good characteristic and by Asai [**8**] in general. (See also 1.5.3.) If **G** is a classical group, the only-if part is also known (Asai [**8**]). Note, however, that a t_σ^*-invariant class function on G_σ need not be uniform (see [**9**]).

(ii)′ As to (1.3.3)(i), see Gyoja [**28**]; he considered liftings of Deligne-Lusztig irreducible characters of a finite reductive group. See also Feit [**22**; Theorem 9B].

(iii) If **G** is finite, then (1.3.4) is true by Glauberman [**27**].

(iv) Let **G** be connected and reductive. Then analogs of (1.3.1) and (1.3.4) for modular representations (i.e., representations of **G** over $\overline{F}_q$) are true essentially by Steinberg's "tensor product theorem." See [**37**] and [**45**].

1.4. $n_{\sigma^m/\sigma}^*$ *and Lusztig's Fourier transformation.* In this subsection, we assume that **G** is connected and reductive. Then an interesting problem is to know the mapping $\mathrm{lift}_{\sigma^m/\sigma}$ explicitly in terms of, say, Lusztig's parametrization [**47**] of the irreducible characters. Going one step further, one might also want to know explicitly the mappings $n_{\sigma^m/\sigma}^*$ for *any m*. In this respect, Asai [**3–8**] and Digne-Michel [**18**–**20**] independently made a very interesting discovery that the mappings $n^*_{\sigma^m/\sigma}$ are closely related to Lusztig's Fourier transformation [**47**]. In particular, Asai [**3**–**7**], using results of Lusztig and results in (1.3.5)(i), (i)′, determined $n^*_{\sigma/\sigma}(\chi)$ for any unipotent character χ of a finite reductive group. Moreover, as is implicitly contained in [**3**–**7**], he also determined $(n^{-1}_{\sigma^m/\sigma})^*(\tilde{\chi})$ for any m and any unipotent character χ of a classical group $G = \mathbf{G}_{\sigma^m}$. Shoji [**63**], generalizing Asai's result, determined $(n^{-1}_{\sigma^m/\sigma})^*(\tilde{\chi})$ for any $x \in \hat{G}$, when G is classical and m is "sufficiently divisible." Moreover, as a corollary of this, he obtained a decomposition formula for Lusztig's generalized induction $R^G_{L\subset P}(\cdot)$. Using this, Asai [**8**] determined $n^*_{\sigma/\sigma}(\chi)$ for any $\chi \in \hat{G}$, when G is classical. In these works of Asai, Digne and Michel, and Shoji (and also in Lusztig [**47**; Chapters 2 and 3]), certain trace identities connecting (ordinary) traces and twisted traces play an important role. Once formulated, they can be proved relatively easily (as compared to those of Saito and others mentioned in 1.1) using the Lefschetz fixed points formula.

1.5. *Other topics.*

1.5.1. *Making q (and p) smaller.* Suppose that we have a conjectural statement on representations of a finite reductive group $G = \mathbf{G}(F_q)$ with $q = p^n$ a power of a fixed prime p. Suppose that we already know that a statement is true if n is sufficiently large. Then it is natural to expect that the lifting theory might be applied to establish the truth of the conjecture for any n. For example, Asai [**10**, **11**] used this idea in showing that Lusztig's formula for the multiplicities of unipotent characters in Deligne-Lusztig virtual characters (which was proved only for large q at that time) is valid for any q. (Later, Lusztig [**47**] proved this in a more general setting without appealing to liftings.) Another example can be found in [**40**; pp. 203–205], where both liftings and generalized Gelfand-Graev characters are used to show that Hotta-Springer's formula for the Green functions of the unitary groups (which was originally proved for large p and q) is valid for any p and q. Essentially the same argument shows that Shoji-Beynon-Spaltenstein's table [**61**, **13**] for the Green functions of exceptional simple groups is valid in any good characteristic. See (2.3.3) (i).

1.5.2. *Blocks.* For a connection between liftings and Brauer's theory of blocks, see [**38**, II; **17**].

1.5.3. *Lusztig's generalized induction.* Let $\mathbf{G}$ be connected and reductive. Then it is known that $n^*_{\sigma/\sigma}$ commutes with Lusztig's generalized induction $R^G_{L\subset P}$ if $\mathbf{G}$ is classical [**8**] or if the characteristic is good [**20**].

1.5.4. *Liftings in a wide sense.* Let G be, say, a finite group and $A_1 = \langle\sigma_1\rangle$ a finite cyclic group acting on G as automorphisms. Then, in general, there might exist another cyclic subgroup $A_2 = \langle\sigma_2\rangle$ of the semidirect product A_1G such that $A_1G = A_2G$ but $G_{\sigma_1} \not\cong G_{\sigma_2}$. (This cannot happen in the situation considered in 1.2.) If this is the case, one cannot expect an intrinsic relation between $(\hat{G})_{\sigma_1}$ and $(G_{\sigma_1})^\frown$; we must, for example, replace $(G_{\sigma_1})^\frown$ with $(G_{\sigma_1})^\frown \amalg (G_{\sigma_2})^\frown \amalg \cdots$.

See Nagao [**51**], Feit [**22**], Flicker [**24**]. Perhaps we need more examples before formulating reasonable conjectures.

1.6. *From liftings to generalized Gelfand-Graev representations.* We use the notations in 1.2. Suppose we are given $\mathbf{G}$, σ, and $\chi \in (\hat{G})_\sigma$ explicitly. Then how can we calculate the values $\tilde{\chi}(\sigma g)$ $(g \in G)$ or, equivalently, the class function $(n^{-1})^*(\tilde{\chi})$ on G_σ? One of the difficulties in lifting theory is that this is by no means an easy task in general. In this respect, the following elementary lemma essentially due to Shintani [**60**] (see also [**38**]) is sometimes useful.

(1.6.1) LEMMA. *Let* $\mathbf{G}$ *be connected,* $\mathbf{H}$ *a connected* σ*-stable subgroup of* $\mathbf{G}$*, and* $H = \mathbf{H}_{\sigma^m}$*. Let* $\tilde{\mu} \in (\hat{H})_\sigma$*. Then* $\operatorname{ind}_{AH}^{AG}(\tilde{\mu})$ *is an extension of the* (*not necessarily irreducible*) *character* $\operatorname{ind}_H^G(\mu)$ *of* G*. Then we have*

$$(n_{\sigma^m/\sigma}^{-1})^*(\operatorname{ind}_{AH}^{AG}(\tilde{\mu})) = \operatorname{ind}_H^G((n_{\sigma^m/\sigma}^{-1})^*(\tilde{\mu})).$$

(*Compare with* 1.5.3.)

This lemma suggests the following problem: Does there exist a "connected" version of Brauer's induction theorem? In other words, we would like to express $\chi \in (G_\sigma)^\wedge$ as a sum

$$\chi = \sum_{(H,\mu)} a_{(H,\mu)} \operatorname{ind}_H^G(\mu), \tag{1.6.2}$$

where $a_{(H,\mu)} \in \mathbf{Z}$ and the sum is over a set of pairs $(\mathbf{H}, \mu)$ such that $H = \mathbf{H}_\sigma$ for a connected σ-stable subgroup $\mathbf{H}$ of $\mathbf{G}$ and $\mu \in \hat{H}$. It is easy to see that this cannot be done in general. For example, consider the case that $G = \mathrm{SL}_2(F_q)$ with odd q and $\chi(1) = (q \pm 1)/2$. But in [**40**, 4.2] it was shown that, when $\mathbf{G}$ is a general linear or unitary group, Deligne-Lusztig virtual characters χ can be written in the form (1.6.2) using generalized Gelfand-Graev characters. A similar thing can also be done when $\mathbf{G}$ is a simple adjoint group of exceptional type; see (2.4.4) (ii). (For a classical group, this is still a conjecture.) These results should have applications in determining the liftings of Deligne-Lusztig irreducible representations $\pm R_T^G(\theta)$ explicitly. See also the final remark in 2.4. Taking this opportunity, the author would like to correct the statement of [**40**, (4.2.2)] as follows: the integers $c_\rho^\mu = c_\rho^\mu(q)$ in [**40**, (4.2.2)] actually depend on q in general, but they still can be written as polynomials in q with integral coefficients independent of $\mathrm{char}(F_q)$. This can be proved easily by using the GL_n-case of [**41**, (3.2.7)] and an induction argument. In accordance with this, one must replace the c_ρ^μ in [**40**, (4.2.10)] with $c_\rho^\mu(-q)$. These changes do not affect other results in that paper.

2. Generalized Gelfand-Graev representations.

2.1. *History.* In 1962, Gelfand and Graev [**26**] announced the following result (and proved it for $\mathrm{SL}_n(F_q)$):

Let G be a finite Chevalley group, and U a maximal unipotent subgroup of G. Then a representation Γ (which is now called a Gelfand-Graev representation) induced from a "generic" linear character of U is multiplicity free. (See [**66**] for a precise statement and a proof of this.) In another paper [**25**], they suggested that such representations might be useful in understanding the notion of "series" of irreducible representations in an intrinsic manner, and showed how it works in the simplest case $\mathrm{SL}_2(K)$ with $K = F_q$ or $\mathbf{R}$. (We shall adopt this point of view later.)

In 1970, Jacquet and Langlands [**36**] proved the multiplicity free property (or the uniqueness of Whittaker models) for the Gelfand-Graev representation of GL_2 over local fields, and used it to define L-functions of automorphic representations. This generalizes the classical procedure (due to Hecke and Maass) of defining L-functions of automorphic forms through their Fourier expansions. See [**43**], [**57**] for the uniqueness of Whittaker models for more general reductive groups. Prior to us [**40, 41**] (see also [**52, 69**]), special cases of generalized Gelfand-Graev representations had already been considered, e.g., by Piatetski-Shapiro and Soudry [**54**] (in connection with L-functions of automorphic representations of GSp_4) and by Hashizume [**31**] (irreducible highest weight representations of real semisimple Lie groups). See also Howe [**35**] for a related idea.

2.2. *Definition of generalized Gelfand-Graev representations (GGGR).* Let $\mathbf{G}$ be a connected reductive linear algebraic group and $\mathbf{g}$ the Lie algebra of $\mathbf{G}$. We assume that $\mathbf{G}$ is defined over a finite field F_q, and denote by σ the corresponding Frobenius endomorphism. The idea is to associate with each unipotent conjugacy class o of $G = \mathbf{G}_\sigma$ an induced representation Γ_o in such a way that when o is regular unipotent and the characteristic of F_q is good [**65**], Γ_o is an (original) Gelfand-Graev representation. Since, in bad characteristic, we still have no uniform description for the unipotent classes of $\mathbf{G}$ (cf. [**14**, 5.11]), it would be reasonable, at least for the moment, to restrict ourselves to the good characteristic case. Then, according to Springer (see [**65**]), the unipotent classes of $\mathbf{G}$ correspond bijectively with the nilpotent $\mathrm{Ad}(\mathbf{G})$-orbits in $g = \mathbf{g}_\sigma$, and for the latter we can use (a positive characteristic version [**65, 41**] of) Dynkin-Kostant theory. (If, instead, we use Bala-Carter theory [**14**], we will be led to another generalization of Gelfand-Graev representations, which we do not discuss here.) Thus, with each nilpotent element A of g, one can canonically associate a σ-stable parabolic subgroup $\mathbf{P}_A$ of $\mathbf{G}$ such that the Lie algebra $\mathbf{u}_A$ of its unipotent radical $\overline{\mathbf{U}}_A$ contains A. Let $X \to X^*$ be an opposition F_q-automorphism of $\mathbf{g}$. The, via "Killing-like" form, $u_A^* = (\mathbf{u}_A^*)_\sigma$ can be identified with the dual space of $u_A = (\mathbf{u}_A)_\sigma$. Hence, by using the method of Kirillov [**42**], one can define a ($q^{(\dim \overline{\mathbf{U}}_A - \dim \mathrm{Ad}(\mathbf{P}_A)A)/2}$-dimensional) irreducible representation η_A of $U_A = (\overline{\mathbf{U}}_A)_\sigma$ corresponding to the coadjoint U_A-orbit of $A^* \in u_A^*$. The induced representation

$$\Gamma_A = \mathrm{Ind}_{U_A}^G(\eta_A)$$

is called the generalized Gelfand-Graev representation (GGGR for short) of G associated with A. (See [**41**] for the details of this construction.) We denote by γ_A the character of Γ_A. The following properties of Γ_A (or γ_A) are easy consequences of the definition:

(2.2.1) The equivalence class of Γ_A depends only on the $\mathrm{Ad}(G)$-orbit $\mathrm{Ad}(G)A$ of A.

(2.2.2) When A is regular nilpotent, Γ_A is an (original) Gelfand-Graev representation.

(2.2.3) $\dim \Gamma_A = \gamma_A(1) = |G|q^{-(\dim \mathrm{Ad}(\mathbf{G})A)/2}$.

(2.2.4) $\mathrm{supp}(\gamma_A) \subset \bar{o}_A$ (= the closure of the unipotent $\mathbf{G}$-conjugacy class o_A corresponding to $\mathrm{Ad}(\mathbf{G})A$).

The following construction was motivated by works of Piatetski-Shapiro, Novodovorsky, and others (see, e.g., [**54**]). Let A be as above, and put $L = L_A = (\mathbf{L}_A)_\sigma$ for a σ-stable Levi-subgroup $\mathbf{L} = \mathbf{L}_A$ of $\mathbf{P}_A$. Then the stabilizer of the equivalence class of η_A in L is equal to $Z_L(A)$, which is known to be reductive. Hence η_A extends to an $\tilde{\eta}_A \in (Z_L(A)U_A)^\frown$. (This is trivial if η_A is linear. If η_A is not linear, $U_A/\mathrm{Ker}\,\eta_A$ is a Heisenberg group. Hence one can define $\tilde{\eta}_A$ by a well-known argument (see, e.g., [**33**, **46**]) used in defining Weil representations, which are always actual (i.e., not just projective) representations in the finite field case.) Thus, for each $\alpha \in Z_L(A)^\frown$, one can define

$$\Gamma_{A,\alpha} = \mathrm{ind}_{Z_L(A)U_A}^G(\eta_{A,\alpha}), \tag{2.2.5}$$

where $\eta_{A,\alpha} = \alpha \otimes \tilde{\eta}_A \in (Z_L(A)U_A)^\frown$. This will be called the GGGR associated with the pair (A, α). For our later purposes, particularly interesting $\Gamma_{A,\alpha}$ are the ones satisfying

$$\alpha \,|\, (Z_{\mathbf{L}}(A)^0)_\sigma \text{ is a sum of linear representations.} \tag{2.2.6}$$

By the construction, one can easily see that

$$\mathrm{ind}_{Z_L(A)U_A}^{P_A}(\eta_{A,\alpha}) \in \hat{P}_A$$

for any $\alpha \in Z_L(A)^\frown$, and that

$$\mathrm{ind}_{U_A}^{P_A}(\eta_A) \sim \bigoplus_{\alpha \in Z_L(A)^\frown} \dim(\alpha) \cdot (\mathrm{ind}_{Z_L(A)U_A}^{P_A}(\eta_{A,\alpha})).$$

Hence we have

$$\Gamma_A \sim \bigoplus_{\alpha \in Z_L(A)^\frown} \dim(\alpha) \cdot \Gamma_{A,\alpha}. \tag{2.2.7}$$

Later we shall also need the following variation of GGGR. Let H be a subgroup of $Z_L(A)$ which consists of semisimple elements. Then, for any $\alpha \in \hat{H}$, one can define $\eta_{A,\alpha} \in (HU_A)^\frown$ in exactly the same way as above. We put

$$\Gamma_{A,\alpha} = \Gamma_{A,(H,\alpha)} = \mathrm{ind}_{HU_A}^G(\eta_{A,\alpha}). \tag{2.2.8}$$

This will also be called a GGGR.

2.3. *Characters of GGGR.* In [**40**, **41**], we have proved an explicit formula for the character of Γ_A when $\mathbf{G}$ is the adjoint group of type A_n, E_n, F_4, or G_2

in good characteristic. The same can also be done for classical groups of lower rank. The formula can be written as

$$\gamma_A = \gamma_A^0 + \gamma_A^1, \tag{2.3.1}$$

where γ_A^0 is the "uniform part" and γ_A^1 is the "Gauss sum part" of γ_A. More precisely,

$$\begin{aligned}\gamma_A^0 &= |W|^{-1} \sum_{w \in W} (-1)^{s(\mathbf{G})-s(\mathbf{T}_w)} |(\mathbf{T}_w)_\sigma| X_w^A(q) Q_w \\ &= |W|^{-1} \sum_{w \in W} \sum_{\theta \in (\hat{\mathbf{T}}_w)_\sigma} (-1)^{s(\mathbf{G})-s(\mathbf{T}_w)} X_w^A(q) r_{T_w}^G(\theta),\end{aligned} \tag{2.3.2}$$

where $s(\cdot)$ is the split rank of the indicated group, $\mathbf{T}_1$ is a σ-stable maximal torus of $\mathbf{G}$ contained in a σ-stable Borel subgroup, $W = N_{\mathbf{G}}(\mathbf{T}_1)/\mathbf{T}_1, \mathbf{T}_w$ ($w \in W$) are σ-stable maximal tori, $r_{T_w}^G(\theta)$ are the characters of the Deligne-Lusztig $R_{T_w}^G(\theta)$, Q_w ($w \in W$) are the Green functions [**16**] of G, and $X_w^A(q)$ ($w \in W$) are polynomials in q obtained from "Green polynomials" (see (2.3.3)(i)) by replacing q with q^{-1} and multiplying $q^{n(A)}$, where $n(A) = (\dim Z_{\mathbf{G}}(A) - \operatorname{rank} \mathbf{G})/2$. The Gauss sum part γ_A^1 is zero if, for any $\psi \in (Z_{\mathbf{G}}(A)/Z_{\mathbf{G}}(A)^0)^\wedge$, the Springer representation [**64**] of W associated with pair (A, ψ) exists. In particular, this is the case if $Z_{\mathbf{G}}(A)$ is connected.

(2.3.3) REMARK. (i) The formula (2.3.2) requires explanation. Let $\mathbf{G}$ be as in there. If the characteristic is large, then Deligne-Lusztig's Green functions Q_w ($w \in W$) coincide with Springer-Kazhdan's Green functions [**64**], which can be written as polynomials in q explicitly (by Springer, Hotta-Springer, Shoji, and Beynon-Spaltenstein). Thus, for any nilpotent element C of g, there exist explicitly calculable polynomials ("Green polynomials") $Q_w^C(q)$ ($w \in W$) such that

$$Q_w(u(C)) = Q_w^C(q), \qquad w \in W, \tag{2.3.4}$$

holds in large characteristic, where $u(C)$ is the unipotent element of G corresponding to C under a Springer map [**65**]. What has been proved in [**40**, **41**] is that the formula

$$\gamma_A^0(u(C)) = |W|^{-1} \sum_{w \in W} (-1)^{s(\mathbf{G})-s(\mathbf{T}_w)} |(\mathbf{T}_w)_\sigma| X_w^A(q) Q_w^C(q) \tag{2.3.2$'$}$$

holds whenever the characteristic is good (and not just in the case of large characteristic). Moreover, the formula (2.3.2)′ (together with (2.4.4) (ii)) can be used to prove that (2.3.4) is actually true in any good characteristic. Thus we eventually have (2.3.2) in good characteristic, and see that the right-hand side of (2.3.2) is explicitly calculable as we said above.

(ii) We call γ_A^1 the Gauss sum part, because (a) when $G = \mathrm{SL}_2(F_q)$ and q is odd, this is essentially a usual (quadratic) Gauss sum, and (b) in general, this can be considered as a special case of an interesting generalization of classical Gauss sums, which we call "Gauss sums of prehomogeneous vector spaces" [**29**; **4**, 3.4].

Let H, $\alpha \in \hat{H}$ and $\Gamma_{A,(H,\alpha)}$ be as in the last paragraph of 2.2. Then the calculation of the character $\gamma_{A,(H,\alpha)}$ of $\Gamma_{A,(H,\alpha)}$ can be reduced to that of γ_A by the following lemma.

(2.3.5) LEMMA. *Let $t \in H$ and $v \in U_A$ be such that $tv = vt$. We denote by $\{t_i; i \in I\}$ a set of representatives of the H-conjugacy classes of $H \cap \{gtg^{-1}; g \in G\}$. For $i \in I$, take $g_i \in G$ such that $t_i = g_i t g_i^{-1}$, and put $v_i = g_i v g_i^{-1}$. Then*

$$\gamma_{A,(H,\alpha)}(tv) = \sum_{i \in I} |Z_H(t_i)|^{-1} \varepsilon(t_i)\alpha(t_i)\gamma_{Z_G(t_i),A}(v_i),$$

where $\gamma_{Z_G(t_i),A}$ is the character of the GGGR of $Z_G(t_i)$ associated with A ($\in (\mathrm{Lie}\, Z_{\mathbf{G}}(t_i))_\sigma$) and $\varepsilon(t_i) = \pm 1$.

EXAMPLE. Let $\mathbf{G}$ be simple and of type G_2. Let $\mathrm{char}(F_q)$ be good, i.e., $\mathrm{char}(\mathbf{G}_q) \neq 2,\ 3$. Let A be a nilpotent element of $g = (\mathrm{Lie}\, \mathbf{G})_\sigma$ such that $Z_{\mathbf{L}_A}(A)$ and $Z_{L_A}(A)$ are both isomorphic to the symmetric group S_3. (Such an element A is unique up to G-conjugacy.) For $c \in S_3 = Z_{L_A}(A)$, let A_c be an element of $g \cap \mathrm{Ad}(\mathbf{G})A$ associated with c. Hence $Z_{L_A}(A_c) \cong Z_{S_3}(c)$. We put

$$H_c = Z_{L_A}(A_c) = Z_{S_3}(c), \qquad c \in S_3.$$

Then for $\alpha_c \in \hat{H}_c$ we have, by (2.3.1) and (2.3.5),

$$\gamma_{A_c,(H_c,\alpha_c)} = |H_c|^{-1} \sum_{t \in H_c} \alpha_c(t)\{X^0_{c,t} + X^1_{c,t}\},$$

where $X^i_{c,t}$ ($t \in H_c; i = 0, 1$) are class functions on G such that $X^i_{c,t}(x) = 0$ if the semisimple part of $x \in G$ is not G-conjugate to t, and that

$$X^i_{c,t}(tv) = \gamma^i_{Z_G(t),A_c}(v)$$

if v is a unipotent element of $Z_G(t)$. Using (2.3.2) one can calculate the values of $X^0_{c,t}$ explicitly. As for $X^1_{c,t}$, it can be shown (cf. [**41**; 3.3, (4.2.10)]) that

$$X^1_{c,t} = \sum_{\phi \in \hat{H}^1_c} \phi(c) Y_{t,\phi},$$

where $\hat{H}^1_c = \{\mathrm{sgn}_{S_3}\}$ (resp. $\hat{H}_c - \{\mathrm{id.}\}$) if $c = 1$ (resp. $c \neq 1$) and, in the notation of Chang and Ree [**15**], $Y_{t,\phi}$ is the function Y_1 (resp. Y_2, resp. Y_3 or Y_4) if the order of t is equal to 1 (resp. 2, resp. 3). Thus we have calculated explicitly the values of the characters of the 8 inequivalent GGGR $\Gamma_{A_c,(H_c,\alpha_c)}$ associated with the orbit $\mathrm{Ad}(\mathbf{G})A$. Similarly, when $\mathbf{G}$ is the F_4- (resp. E_8-) group in good characteristic, and $A \in g$ is a nilpotent element such that $Z_{L_A}(A) \cong S_4$ (resp. S_5), one can calculate the values of the characters of the 21 (resp. 39) inequivalent GGGR associated with the orbit $\mathrm{Ad}(\mathbf{G})A$.

2.4. *Multiplicities in GGGR.* Let G be a finite reductive group. We are interested in the following problems:

(1) Compute the multiplicities of the irreducible representations of G in various GGGR.

(2) Try to classify the irreducible representations of G using the above multiplicities.

This is our version of the (rather obscure) program of Gelfand and Graev [**25**, **26**] (see 2.1). When $\mathbf{G}$ is an adjoint group of type A_n, E_n, F_4, or G_2 in good characteristic, then, at least in principle, we can carry out the program (1) (for the Γ_A's and $\Gamma_{A,(H,\alpha)}$'s) using the results in 2.3 and Lusztig's formula [**47**] for the multiplicities of the irreducible representations in the $R_T^G(\theta)$'s. (Hence, we also rely on Lusztig's parametrization [**47**] of the irreducible characters of G, which is, of course, against the spirit of our program. Thus it is desirable to reprove our results below in a different way.) Some of such calculations were done in [**40**, **41**], an interesting by-product of which was the explicit determination of the values of the irreducible characters of G above at the unipotent elements. In general, it seems possible to calculate the whole character table of a finite reductive group by carrying out program (1) thoroughly. It would be interesting to know relations between our point of view and that of Lusztig [**48**].

In the rest of this subsection, unless otherwise stated, $\mathbf{G}$ always denotes an adjoint group of type A_n, E_n, F_4, or G_2 in good characteristic. Let u be a unipotent element of G. We put $\Gamma_u = \Gamma_A$ if A is a nilpotent element of g corresponding to u under a Springer map [**65**].

(2.4.1) THEOREM. *Let $\chi \in \hat{G}$.*

(i) *There exists a unique σ-stable unipotent class $o = o_\chi$ of* $\mathbf{G}$ *with the following properties.*

(ia) $\langle \Gamma_u, \chi\rangle \neq 0$ *for some $u \in o \cap G$;*

(ib) *If o' is a σ-stable unipotent class of* $\mathbf{G}$ *such that $\bar{o}' \not\subset \bar{o}$ ($\bar{o}$ is the closure of o in* $\mathbf{G}$*), then*

$$\langle \Gamma_v, \chi\rangle = 0 \quad \textit{for any } v \in o' \cap G.$$

(ii) *The above correspondence $\chi \to o_\chi$ coincides with the one described in Lusztig* [**47**, 13.4]. *In particular,*

(iia) *If χ is a unipotent character, then o_χ is a special unipotent class. Two unipotent characters χ and χ' lie in the same family if and only if $o_\chi = o_{\chi'}$.*

(iib) *For $\chi \in \hat{G}$, we denote by* $\operatorname{Dim}\chi$ *the degree of* $\dim\chi = \chi(1)$ *as a polynomial in q. Then*

$$\operatorname{Dim}\chi = (\dim o_\chi)/2.$$

(iii) *Let u be a unipotent element of G and χ an element of $\hat{G}$. If the* $\mathbf{G}$*-conjugacy class o_u of u is contained in $\bar{o}_\chi$, then either $\langle \Gamma_u, \chi\rangle = 0$ or the degree of $\langle \Gamma_u, \chi\rangle$ as a polynomial in q is equal to* $(\dim o_\chi - \dim o_u)/2$. *In particular, when $o_u = o_\chi$ (resp. $o_u = \{1\}$), this says that $\langle \Gamma_u, \chi\rangle$ is independent of q (resp. this reduces to* (iib)). *Moreover, if $\langle \Gamma_u, \chi\rangle$ is nonzero and is independent of q, then we must have $o_u = o_\chi$.*

(iv) *Let* $\operatorname{Inv}(g_0)$ *be the space of nilpotently supported* $\operatorname{Ad}(G)$*-invariant functions on g, $\mathcal{F}$:* $\operatorname{Inv}(g_0) \to \operatorname{Inv}(g_0)$ *the "Fourier transformation" as defined in* [**39**] *(see also* [**41**, 3.5]). *For $\chi \in G$, we define $\chi_* \in \operatorname{Inv}(g_0)$ by $\chi_*(C) = \chi(u(C))$ for* *any nilpotent element C of $\mathfrak{g}$. Then we have*

$$\mathcal{F}^{-1}(\chi_*)(A) = \pm q^{n(A)} \left\{ \langle \Gamma_{u(A)}, \chi \rangle \right\} (q^{-1})$$

for any nilpotent element A of $\mathfrak{g}$. (Here also we are considering $\langle \Gamma_{u(A)}, \chi \rangle$ as a polynomial in q). In particular, if $u(A) \in o_\chi$ (resp. $u(A) \notin \bar{o}_\chi$), the left-hand side is equal to (constant) $\times q^{n(A)}$ *(resp. zero). (Compare with* [**40**; (3.2.18) (ii)].)

(2.4.2) REMARK. (i) In a somewhat weaker form, (2.4.1) was conjectured in [**40**, (3.3.3)]. There we stated that the correspondence $\chi \to o_{D(\chi)}$ ($D =$ the duality operation) was identical with that of Lusztig. That statement was based on a preprint version of Lusztig [**47**, 13], and, hence, must be corrected as in (2.4.1)(ii).

(ii) Our proof of (2.4.1) is as outlined above; it is of case-by-case nature and relies heavily on explicit knowledge of Green polynomials [**61**, **62**, **13**] and of induce/restrict matrices [**1**] for representations of Weyl groups.

(iii) We might call the $\bar{o}_\chi$ the "wave front set" of χ, and $\mathrm{Dim}\,\chi$ the "Gelfand-Kirillov dimension" of χ by an analogy with the Lie group case (see [**12**, **32**, **34**, **68**] and their references).

By carefully examining the multiplicities $\langle \Gamma_u, \chi \rangle$ for $u \in o_\chi \cap G$, one can show (see [**40**] for the GL_n-case):

(2.4.3) THEOREM. *Let $\mathcal{U}$ be a set of representatives of the unipotent classes of G. Then $\{\gamma_u = \mathrm{trace}\,\Gamma_u; u \in \mathcal{U}\}$ is a $\mathbf{Z}$-basis of the $\mathbf{Z}$-module of unipotently supported virtual characters of G.*

(2.4.4) REMARK. (i) Compare (2.4.3) with the following elementary (and perhaps well-known) fact (the author learned a proof of this from Gyoja [**28**, Lemma 5.1] and K. Uno):

Let H be a finite group and p a prime. Let ψ be a virtual character of H. Then $\mathrm{supp}(\psi)$ is contained in the set of p-elements of H if and only if ψ can be written as a $\mathbf{Z}$-linear combination of characters induced from linear characters of p-subgroups of H.

(ii) By (2.4.3), one can write Kazhdan's virtual characters and, hence, Deligne-Lusztig's virtual characters $r_T^G(\theta)$ as $\mathbf{Z}$-linear sums of characters induced from linear characters of "almost connected" subgroups of G. See 1.6.

Theorem (2.4.1) already contains an interesting contribution to our program (2). To get a finer result, one should compute the multiplicities in $\Gamma_{A,\alpha}$'s or $\Gamma_{A,(H,\alpha)}$'s, which are finer (see (2.2.7)) than Γ_A's. This is the problem we would like to consider in the rest of this subsection.

Let u be a special unipotent element of $\mathbf{G}$, and $\overline{A}(u)$ the canonical quotient group of $Z_{\mathbf{G}}(u)/Z_{\mathbf{G}}(u)^0$ defined in Lusztig [**47**, 13.1]. Following Lusztig, we put

$$\mathcal{M}(\overline{A}(u)) = \{(x,\phi); x \in \overline{A}(u)/\sim, \phi \in Z_{\overline{A}(u)}(x)^\wedge\}.$$

Then, with each element (x,ϕ) of $\mathcal{M}(\overline{A}(u))$, we can associate a GGGR $\Gamma_{(x,\phi)}$ as follows.

From the beginning, we choose u so that it is a "good" representative (see [**41**]) of $o_u(F_q)$. Then each $x \in \overline{A}(u)/\sim$ determines an element $u_x \in o_u(F_q)$ uniquely up to G-conjugacy, and $\phi \in Z_{\overline{A}(u)}(x)^\frown$ can be naturally considered as an irreducible representation of $Z_L(u_x)$ trivial on $(Z_{\mathbf{L}}(u_x)^0)_\sigma$, where $L = L_u \subset G$ is a Levi-subgroup of the parabolic subgroup associated with u. Thus we can define a GGGR $\Gamma_{u_x,\phi}$ (see 2.2), and we put $\Gamma_{(x,\phi)} = \Gamma_{u_x,\phi}$.

Motivated by Lusztig's work [**47**], we would like to state the following

(2.4.5) CONJECTURE. Let the notations be as above. Let f_o be the family [**47**] of unipotent representations of G associated with $o = o_u$. Then there exists a unique labeling $\rho_{(x,\phi)}$ of elements of f_o such that

$$\langle \Gamma_{(x,\phi)}, \rho_{(x',\phi')}\rangle = \begin{cases} 1 & \text{if } (x,\phi) = (x',\phi'), \\ 0 & \text{otherwise,} \end{cases}$$

for any (x,ϕ) and (x',ϕ') in $\mathcal{M}(\overline{A}(u))$.

If this can be proved, we shall get an intrinsic parametrization of the elements of f_o in terms of $\mathcal{M}(\overline{A}(u))$. We must hasten to add that this parametrization will be almost the same as that of Lusztig [**47**], but is different from it when G is of type E_7 (resp. E_8) and f_o (as a family of irreducible representations of the Weyl group W) contains 512 (resp. 4096) dimensional representations. In fact, the author expects that the labeling of the elements of $\hat{W}$ contained in f_o via (2.4.5) will be the same as the one given in the following.

(2.4.6) THEOREM. *Let $o^\#$ be the "dual" [**12**] class of o, and $u^\# \in o^\#(F_q)$. For $(x,\phi) \in \mathcal{M}(\overline{A}(u^\#)) = \mathcal{M}(\overline{A}(u))$, let $\dot{x}$ be a semisimple representative of x in $Z_{\mathbf{G}}(u^\#)$. Then the Weyl group $W(\dot{x})$ of $\mathbf{M}(\dot{x}) = Z_{\mathbf{G}}(\dot{x})^0$ is a subgroup of W, and ϕ can be considered as an element of $(Z_{\mathbf{M}(\dot{x})}(u^\#)/Z_{\mathbf{M}(\dot{x})}(u^\#)^0)^\frown$. We denote by $\tau_{(u^\#,\phi)}$ the Springer representation [**64**] of $W(\dot{x})$ (normalized by $\tau_{(1,1)} = \mathrm{sgn}_{W(\dot{x})}$) if it exists. Then there exists at most one element $\alpha_{(x,\phi)}$ of $\hat{W}$ contained in f_o such that*

$$\langle \mathrm{ind}_{W(\dot{x})}^{W}\left(\mathrm{sgn}_{W(\dot{x})} \otimes \tau_{(u^\#,\phi)}\right), \alpha_{(x,\phi)}\rangle \neq 0. \tag{2.4.7}$$

If such an $\alpha_{(x,\phi)}$ exists, then it is independent of the choice of $\dot{x}$, and the multiplicity (2.4.7) *is equal to* 1. *Moreover any irreducible representation of W contained in f_o arises uniquely in this way.* (*Compare with Barbasch-Vogan* [**12**, *Lemma* 10.17].)

This can be checked case by case using [**1**]. Notice that, in the above correspondence, $(1,1) \in \mathcal{M}(\overline{A}(u))$ does not necessarily correspond to the special representation in the family, whereas in Lusztig [**47**], the pair (1,1) always goes to the special representation.

EXAMPLE. Let $\mathbf{G}$ be of type G_2 (resp. F_4, resp. E_8), and let u be a (special) unipotent element of G corresponding to A in the example in 2.3. Then the family f_{o_u} consists of 8 (resp. 21, resp. 39) elements, and is in bijective correspondence with the same number of GGGR associated with o_u in the way described in (2.4.5). (In a conversation (on the author's work [**41**]), Lusztig conjectured that a statement like this would be true. The author is very thankful to Lusztig for this comment.) As a by-product, we obtain the explicit values of the unipotent characters in f_o. At the unipotent elements, these characters were computed in [**41**, **48**]. The general formula is similar to the one given there in that Lusztig's [**47**] 8×8 (resp. 21×21, resp. 39×39) matrix $\{(x,\phi),(x',\phi')\}$ beautifully enters in; we now need all the entries of the square matrix, and not just a rectangular part of it.

Finally, we would like to point out that results like (2.4.1) or (2.4.5) are very useful in determining Shintani lifting maps explicitly.

2.5. *GGGR of reductive groups over local fields.* The purpose of this subsection is to state a conjecture on GGGR of reductive groups over local field. In doing this, we shall be optimistic as much as possible.

Let G be a connected reductive group over a local field F of good characteristic, and g the Lie algebra of G. We use the notations in 2.2 freely, when the difference of the underlying field is irrelevant. Let A be a nilpotent element of g. Then the GGGR Γ_A can be defined as the unitarily (or smoothly) induced representation $\mathrm{ind}_{U_A}^G(\eta_A)$. One can also define $\Gamma_{A,\alpha} = \mathrm{ind}_{Z_L(A)U_A}^G(\eta_{A,\alpha})$ with $\eta_{A,\alpha} = \alpha\otimes\tilde{\eta}_A$ as in 2.2. The only difference is that now $\tilde{\eta}_A$ is, in general, a projective representation of $Z_L(A)U_A$, and hence we must take α as a projective representation of $Z_L(A)$ such that

(2.5.1) $$\eta_{A,\alpha} = \alpha\otimes\tilde{\eta}_A \text{ is an actual representation of } Z_L(A)U_A.$$

(See [**21**, **46**] for the details.) For our present purpose, particularly interesting GGGR $\Gamma_{A,\alpha}$ are those satisfying

(2.5.2) $$\alpha \text{ is finite-dimensional.}$$

(Compare this with (2.2.6).) Let χ be an irreducible unitary representation of G. A continuous linear map from the representation space of χ into that of η_A (resp. $\eta_{A,\alpha}$) will be called a Whittaker map if it commutes with the U_A-action (resp. $Z_{L_A}(A)U_A$-action). The space of all Whittaker maps is denoted by $\mathrm{Wh}(\chi,\eta_A)$ (resp. $\mathrm{Wh}(\chi,\eta_{A,\alpha})$). See [**31**, **43**, **52**, **54**, **57**, **69**]. We can now state a conjecture, which is a direct analog of Theorem (2.4.1) and Conjecture (2.4.5).

(2.5.3) CONJECTURE. Let G, g, and χ be as above. Let $\mathbf{G}$ be the algebraic group over $\overline{F}$ such that $G = \mathbf{G}(F)$, and $\mathbf{g} = \mathrm{Lie}\,\mathbf{G}$.

(i) There exists a unique nilpotent $\mathbf{G}$-orbit $O = O_\chi$ in $\mathbf{g}$ with the following properties.

(ia) $\mathrm{Wh}(\chi,\eta_A)\neq\{0\}$ for some $A\in O\cap g$;

(ib) If O' is a nilpotent $\mathbf{G}$-orbit in $\mathbf{g}$ such that $\overline{O}'\not\subset\overline{O}$, then $\mathrm{Wh}(\chi,\eta_A) = \{0\}$ for any $A'\in O'\cap g$.

(ii) The nilpotent class O_χ can also be characterized by (ia) and

(2.5.4) $$\dim\mathrm{Wh}(\chi,\eta_A)<\infty \quad \text{for any } A\in O_\chi\cap g.$$

(iii) If $\mathrm{char}(F)=0$, then the closure of the set

$$\{A\in O_\chi\cap g;\, \mathrm{Wh}(\chi,\eta_A)\neq\{0\}\}$$

will coincide with the wave front set of χ. (For the notion of the wave front set of an irreducible representation, see [**34**, **32**] and the papers by Barbasch-Vogan cited there. When F is a non-Archimedean field, we mean by "wave front set" the one defined through the asymptotic expansion of characters given in Harish-Chandra [**30**].)

(iv) For an $A \in O_\chi \cap g$, such that $\mathrm{Wh}(\chi, \eta_A) \neq \{0\}$, there exists an irreducible projective representation α of $Z_L(A)$ satisfying (2.5.1), (2.5.2), and

$$\dim \mathrm{Wh}(\chi, \eta_{A,\alpha}) = 1.$$

It is probable that the multiplicity (2.5.4) can also be interpreted as a kind of "multiplicity" attached to the corresponding wave front set. See Rodier [**55**] for a result of this nature. See also [**31**, **43**, **52**, **54**, **68**, **70**] for some of the results supporting Conjecture (2.5.3).

After the conference, the author received a recent preprint [**50**] by Moeglin and Waldspurger, in which they consider (independently of the present author) the space $\mathrm{Wh}(\chi, \eta_A)$ for a reductive group G over a non-Archimedean field of characteristic zero, generalize the above-mentioned result of Rodier, and "almost" prove (2.5.3) (i)–(iii) in that case.

REFERENCES

1. D. Alvis, *Induce/restrict matrices for the exceptional Weyl groups*, Preprint.

2. T. Asai, *On the zeta functions of the varieties $X(w)$ of the split classical groups and the unitary groups*, Osaka J. Math. **20** (1983), 21–32.

3. ——, *Liftings and the zeta functions of the Deligne-Lusztig varieties*, Comm. Algebra **11** (1983), 2187–2201.

4. ——, *Unipotent class functions of split special orthogonal groups SO_{2n}^{+} over finite fields*, Comm. Algebra **12** (1984), 517–615.

5. ——, *The unipotent class functions on the symplectic and the odd orthogonal groups over finite fields*, Comm. Algebra **12** (1984), 617–645.

6. ——, *The unipotent class functions of exceptional groups over finite fields*, Comm. Algebra **12** (1984), 2729–2857.

7. ——, *The unipotent class functions of non-split finite special orthogonal groups*, Comm. Algebra **13** (1985), 845–924.

8. ——, *On the twisting operators on the finite classical groups*, Algebraic and Topological Theories, Kinokuniya, Tokyo, 1985, pp. 239–282.

9. ——, *The dimensions of spaces of all class functions invariant under the twisting operators of finite classical groups*, Algebraic Groups and Related Topics (Advanced Studies in Pure Math., vol. 6), Kinokuniya, Tokyo, and North-Holland, Amsterdam, 1985, pp. 155 174.

10. ——, *Unipotent characters of $\mathrm{SO}_{2n}^{\pm}$, Sp_{2n} and SO_{2n+1} over F_q with small q*, Osaka J. Math. **20** (1983), 631–643.

11. ——, *Unipotent characters of exceptional groups over finite fields F_q with small q*, Comm. Algebra **11** (1983), 2203–2220.

12. D. Barbasch and D. Vogan, *Unipotent representations of complex semisimple groups*, Ann. of Math. (2) **121** (1985), 41–110.

13. W. M. Beynon and N. Spaltenstein, *The computation of Green functions of finite Chevalley groups of type $E_n (n = 6, 7, 8)$*, Computer Centre Report 23, Univ. of Warwick.

14. R. W. Carter, *Finite groups of Lie type: Conjugacy classes and complex characters*, Wiley-Interscience, New York, 1985.

15. B. Chang and R. Ree, *The characters of $G_2(q)$*, Ist. Naz. Alta Mat. Sympos. Math., vol. 13, 1974, 385–413.

16. P. Deligne and G. Lusztig, *Representations of reductive groups over finite fields*, Ann. of Math. (2) **103** (1976), 103–161.

17. F. Digne, *Descente du corps de base et block, d'après N. Kawanka*, Preprint.

18. F. Digne and J. Michel, *Nombre de points rationnels d'algèbre de Hecke*, C. R. Acad. Sci. Paris Ser. A **287** (1978), 811–814.

19. ——, *Descente de Shintani des caractères d'un groupe de Chevalley fini*, C. R. Acad. Sci. Paris Ser. A **291** (1980), 571–574.

20. ——, *Functions L de variétés de Deligne-Lusztig et descente de Shintani*, Bull. Soc. Math. France Mém. No. 20 (1985).

21. M. Duflo, *Sur les extensions de représentations irréductibles des groupes de Lie nilpotents*, Ann. Sci. École Norm. Sup. **5** (1972), 71–120.

22. W. Feit, *Extensions of cuspidal characters of* $\mathrm{GL}_m(q)$, Preprint.

23. Y. Z. Flicker, *The trace formula and base change for* GL(3), Lecture Notes in Math., vol. 927, Springer-Verlag, 1982.

24. ——, *Outer automorphisms and instability: the adjoint liftings from* SL(2) *to* PGL(3), Séminaire de Theorie des Nombres, Paris 1980–1981, Prog. Math., vol. 22, Birkhäuser, 1982, pp. 57–65.

25. I. M. Gelfand and M. I. Graev, *Categories of group representations and the problem of classifying irreducible representations*, Soviet Math. Dokl. **3** (1962), 1382–1385.

26. ——, *Construction of irreducible representations of simple algebraic groups over a finite field*, Soviet Math. Dokl. **3** (1962), 1646–1649.

27. G. Glauberman, *Correspondence of characters for relatively prime operator groups*, Canad. J. Math. **20** (1968), 1465–1488.

28. A. Gyoja, *Liftings of irreducible characters of finite reductive groups*, Osaka J. Math. **16** (1979), 1–30.

29. A. Gyoja and N. Kawanaka, *Gauss sums of prehomogeneous vector spaces*, Proc. Japan Acad. Ser. A **61** (1985), 19–22.

30. Harish-Chandra, *Admissible distributions on reductive p-adic groups*, Lie Theories and Their Applications, Queen's Papers in Pure and Appl. Math. (1978), 281–347.

31. M. Hashizume, *Whittaker models for representations with highest weights*, Lectures on Harmonic Analysis on Lie Groups and Related Topics (T. Hirai et al., eds.), Lectures in Math., vol. 14, Kinokuniya, Tokyo,1982, pp. 259–293.

32. R. Hotta and M. Kashiwara, *The invariant holonomic system on a semisimple Lie algebra*, Invent. Math. **75** (1984), 327–358.

33. R. Howe, *On the character of Weil's representation*, Trans. Amer. Math. Soc. **177** (1973), 287–298.

34. ——, *Wave front sets of representations of Lie groups*, Automorphic Forms, Representation Theory, and Arithmetic, Tata Inst. Fund. Res. Studies in Math., vol. 10, Tata Inst. Fund. Res., Bombay, 1981, pp. 117–140.

35. ——, *A notion of rank for unitary representations of classical groups*, C. I. M. E. Summer School on Harmonic Analysis, Corton, 1980.

36. H. Jacquet and R. P. Langlands, *Automorphic forms on* GL(2), Lecture Notes in Math., vol. 114, Springer-Verlag, 1970.

37. N. Kawanaka, *On the irreducible characters of finite unitary groups*, J. Math. Soc. Japan **29** (1977), 425–450.

38. ——, *Liftings of irreducible characters of finite classical groups*. I, J. Fac. Sci. Univ. Tokyo **28** (1982), 851–861; II, ibid. **30** (1984), 499–516.

39. ——, *Fourier transforms of nilpotently supported invariant functions on a simple Lie algebra over a finite field*, Invent. Math. **69** (1982), 411–435.

40. ——, *Generalized Gelfand-Graev representations and Ennola duality*, Algebraic Groups and Related Topics Advanced Studies in Pure Math., vol. 6, Kinokuniya, Tokyo, and North-Holland, Amsterdam, 1985, pp. 175–206.

41. ——, *Generalized Gelfand-Graev representations of exceptional simple algebraic groups over a finite field*. I, Invent. Math. **84** (1986), 575–616.

42. A. Kirillov, *Unitary representations of nilpotent Lie groups*, Russian Math. Surveys **17** (1962), 53–104.

43. B. Kostant, *On Whittaker vectors and representation theory*, Invent. Math. **48** (1978), 101–184.

44. R. P. Langlands, *Base change for* GL(2), Annals of Math. Stud., vol. 96, Princeton Univ. Press, Princeton, N. J., 1980.

45. I. Lee, *Liftings of irreducible characters of* ${}^3D_4(F_q)$, Ph.D. Dissertation, Yale Univ., 1986.

46. G. Lion and P. Perrin, *Extension des représentations de groupes unipotents p-adiques*, Lecture Notes in Math., vol. 880, Springer-Verlag, 1981, pp. 337–356.

47. G. Lusztig, *Characters of reductive groups over a finite field*, Ann. of Math. Studies, vol. 96, Princeton Univ. Press, Princeton, N. J., 1984.

48. ———, *On the character values of finite Chevalley groups at unipotent elements*, Preprint.

49. C. Moeglin, *Modèle de Whittaker et idéaux primitifs complétement premiers dans les algèbres enveloppantes*. I, II, Preprint.

50. C. Moeglin and J. L. Waldspurger, *Modèles de Whittaker dégénérés pour des groupes p-adiques*, Preprint.

51. H. Nagao, *Some correspondences in the representation theory of finite groups*, Proc. Internat. Math. Conf. (Singapore 1981), L. H. Y. Chen et al., editors, North-Holland, Amsterdam, 1982, pp. 35–41.

52. H. Matumoto, *Gelfand-Kirillov dimension and infinitesimal Whittaker models for real semisimple Lie groups*, Preprint.

53. K. McGovern, *On the lifting theory of finite groups of Lie type*, Pacific J. Math. **113** (1984), 383–394.

54. I. I. Piatetski-Shapiro and D. Soudry, *L and ε factors for* GS_{p4}, J. Fac. Sci. Univ. Tokyo **28** (1981), 505–530.

55. F. Rodier, *Modèle de Whittaker et caractères de représentations*, Non-Commutative Harmonic Analysis, Lecture Notes in Math., vol. 466, Springer-Verlag, 1977, pp. 117–195.

56. H. Saito, *Automorphic forms and algebraic extensions of number fields*, Lectures in Math., vol. 8, Kinokuniya, Tokyo, 1975.

57. J. A. Shalika, *Multiplicity one theorem for* GL(n), Ann. of Math. (2) **100** (1974), 171–193.

58. T. Shintani, *On liftings of holomorphic cusp forms*, Proc. Sympos. Pure Math., vol. 33, part 2, Amer. Math. Soc., Providence, R. I., 1979, pp. 97–110.

59. ———, *On irreducible unitary characters of a certain group extension of* GL$(2, C)$, J. Math. Soc. Japan **29** (1977), 165–188.

60. ———, *Two remarks on irreducible characters of finite general linear groups*, J. Math. Soc. Japan **28** (1976), 396–414.

61. T. Shoji, *On the Green polynomials of a Chevalley group of type* F_4, Comm. Algebra **10** (1982), 505–543.

62. ———, *On the Green polynomials of a classical group*, Invent. Math. **74** (1983), 239–267.

63. ———, *Some generalization of Asai's result for classical groups*, Algebraic Groups and Related Topics, Adv. Stud. Pure Math., vol. 6, Kinokuniya, Tokyo, and North-Holland, Amsterdam, 1985, pp. 207–229.

64. T. A. Springer, *Trigonometric sums, Green functions of finite groups and representations of Weyl groups*, Invent. Math. **36** (1976), 173–203.

65. T. A. Springer and R. Steinberg, *Conjugacy classes*, Seminar on Algebraic Groups and Related Topics, Lecture Notes in Math., vol. 131, Part E, Springer-Verlag, 1970.

66. R. Steinberg, *Lectures on Chevalley groups*, Yale University, 1967.

67. ———, *On theorems of Lie-Kolchin, Borel, and Lang*, Contributions to Algebra, H. Bass et al., editors, Academic Press, New York, 1977, pp. 349–354.

68. D. Vogan, *Gelfand-Kirillov dimension for Harish-Chandra modules*, Invent. Math. **48** (1978), 75–98.

69. H. Yamashita, *On Whittaker vectors for generalized Gelfand-Graev representations of semisimple Lie groups*, J. Math. Kyoto Univ. **26** (1986), 263–298.

70. ———, *Finite multiplicity theorems for induced representations of semisimple Lie groups*. I,II, Preprint.

OSAKA UNIVERSITY, JAPAN

Proceedings of Symposia in Pure Mathematics
Volume **47** (1987)

Introduction to Character Sheaves

GEORGE LUSZTIG

Abstract. This is an exposition (essentially without proofs) of a geometric character theory for reductive groups, following [**3**, **4**]. We have also included disconnected groups in the treatment. (For a study of conjugacy classes in such groups, one can read [**7**].)

1. Generalities on perverse sheaves.

1.1. The basic source for the theory of perverse sheaves on algebraic varieties is Beilinson, Bernstein, Deligne, and Gabber's work [**1**]. Let k be an algebraically closed field; all algebraic varieties will be over k. We fix a prime number l, invertible in k and let $\overline{Q}_l$ be an algebraic closure of the field of l-adic numbers.

Let X be an algebraic variety. We assume known the notion of $\overline{Q}_l$-local system and $\overline{Q}_l$-constructible sheaf on X. (We shall drop the prefix $\overline{Q}_l$.) For a constructible sheaf $\mathcal{F}$ on X we denote by $\mathcal{F}_x$ the stalk of $\mathcal{F}$ at a point $x \in X$; it is a finite-dimensional $\overline{Q}_l$-vector space. We denote by $\mathcal{D}X = \mathcal{D}_c^b(X, \overline{Q}_l)$ the bounded derived category of constructible sheaves on X [**1**, 2.2.18]. For $K \in \mathcal{D}X$ we denote by $\mathcal{H}^iK$ the ith cohomology sheaf of K (a constructible sheaf on X). We denote by $K[i] \in \mathcal{D}X$ the ith shift of K, so that $\mathcal{H}^j(K[i]) = \mathcal{H}^{i+j}K$. A constructible sheaf $\mathcal{F}$ on X gives rise to an object $K \in \mathcal{D}X$ such that $\mathcal{H}^0K = \mathcal{F}$, $\mathcal{H}^iK = 0$ for $i \neq 0$; we identify this K with $\mathcal{F}$.

If $K \in \mathcal{D}X$ the support of K is the closure of the set $\{x \in X \mid \mathcal{H}_x^iK \neq 0$ for some $i\}$.

1.2. Let $D: \mathcal{D}X \to \mathcal{D}X$ be the Verdier duality. Let $\mathcal{D}X^{\leq 0}$ be the full subcategory of $\mathcal{D}X$ whose objects are those K in $\mathcal{D}X$ such that for any i, $\dim \operatorname{supp} \mathcal{H}^iK \leq -i$. Let $\mathcal{D}X^{\geq 0}$ be the full subcategory of $\mathcal{D}X$ whose objects are those K in $\mathcal{D}X$ such that $DK \in \mathcal{D}X^{\leq 0}$.

Let $\mathcal{M}X$ be the full subcategory of $\mathcal{D}X$ whose objects are those K in $\mathcal{D}X$ such that $K \in \mathcal{D}X^{\leq 0} \cap \mathcal{D}X^{\geq 0}$; the objects of $\mathcal{M}X$ are called perverse sheaves on X. We shall use the term "Sheaf" instead of "perverse sheaf."

1980 *Mathematics Subject Classification* (1985 *Revision*). Primary 20G99; Secondary 20G40.
Supported in part by the National Science Foundation.

It is known that $\mathcal{M}X$ is an abelian category [**1**, 2.14, 1.3.6] in which all objects have finite length [**1**, 4.3.1].

1.3. Let V be a locally closed smooth irreducible subvariety of dimension d of X, let $\overline{V}$ be its closure, and let $\mathcal{L}$ be a local system on V. There is a unique Sheaf K on $\overline{V}$ such that

$$\mathcal{H}^{-d}K|V \cong \mathcal{L}, \dim\operatorname{supp}\mathcal{H}^iK < -i, \quad \text{for } i \neq -d,$$

$$\mathcal{H}^{-d}DK|V = \mathcal{L}^* \ (= \text{dual of } \mathcal{L}), \dim\operatorname{supp}\mathcal{H}^iDK < -i, \quad \text{for } i \neq -d.$$

It is $K = \mathrm{IC}(\overline{V}, \mathcal{L})[d]$ where $\mathrm{IC}(\overline{V}, \mathcal{L})$ is the intersection cohomology complex of Deligne-Goresky-MacPherson on $\overline{V}$ with coefficients in $\mathcal{L}$.

If $\mathcal{L}$ is irreducible then K is an irreducible object of $\mathcal{M}\overline{V}$ and its extension to X (by 0 outside $\overline{V}$) is an irreducible object of $\mathcal{M}X$. Moreover, all irreducible Sheaves on X are obtained from a pair $(V, \mathcal{L})$ in this way [**1**, 4.3.1].

1.4. We now recall from [**1**] the definition of the functors ${}^pH^i: \mathcal{D}X \to \mathcal{M}X$ ($i \in \mathbf{Z}$). The inclusion $\mathcal{D}X^{\leq 0} \to \mathcal{D}X$ has a right adjoint ${}^p\tau_{\leq 0}: \mathcal{D}X \to \mathcal{D}X^{\leq 0}$ [**1**, 2.2.11, 1.3.3(i)]; thus we have $\operatorname{Hom}_{\mathcal{D}X}(A, {}^p\tau_{\leq 0}B) = \operatorname{Hom}_{\mathcal{D}X}(A, B)$ for all $A \in \mathcal{D}X^{\leq 0}$, $B \in \mathcal{D}X$. Let ${}^p\tau_{\geq 0}: \mathcal{D}X \to \mathcal{D}X^{\geq 0}$ be the functor $D \circ {}^p\tau_{\leq 0} \circ D$. For any $K \in \mathcal{D}X$ we have ${}^p\tau_{\leq 0}({}^p\tau_{\geq 0}K) \cong {}^p\tau_{\geq 0}({}^p\tau_{\leq 0}K)$, [**1**, 1.3.5]; this is a Sheaf denoted ${}^pH^0K$. For any integer i, we define ${}^pH^iK = {}^pH^0(K[i])$.

1.5. Let $f: X \to Y$ be a morphism between two algebraic varieties X, Y. Let $f^*: \mathcal{D}Y \to \mathcal{D}X$ (resp. $f_!: \mathcal{D}X \to \mathcal{D}Y$) be the inverse image (resp. the direct image with compact support). If $B \in \mathcal{D}Y$ and $x \in X$ we have $\mathcal{H}^i_x(f^*B) = \mathcal{H}^i_{f(x)}(B)$; if $A \in \mathcal{D}X$ and $y \in Y$, we have $\mathcal{H}^i_y(f_!A) = H^i_c(f^{-1}(y), A)$ (= hypercohomology with compact support of $f^{-1}(y)$ with coefficients in $A|f^{-1}(y)$.)

If f is smooth with connected fibers of dimension d, then for any $B \in \mathcal{M}(Y)$, $f^*B[d]$ is a Sheaf on X.

1.6. Assume that $m: H \times Y \to Y$ is an action of a connected algebraic group H on the variety Y; let $\pi: H \times Y \to Y$ be the second projection. If $K \in \mathcal{M}Y$ then $m^*K[\dim H]$, $\pi^*K[\dim H]$ are Sheaves on $H \times Y$. We say that K is H-equivariant if these two Sheaves on $H \times Y$ are isomorphic. Similarly, a local system $\mathcal{F}$ on Y is said to be H-equivariant if the local systems $m^*\mathcal{F}$, $\pi^*\mathcal{F}$ on $H \times Y$ are isomorphic.

If H acts transitively on Y and H_y is the isotropy group of $y \in Y$, then the group of components $\overline{H}_y$ of H_y acts naturally on the stalk $\mathcal{F}_y$ of an H-equivariant local system $\mathcal{F}$ on Y; this defines an equivalence between H-equivariant local systems on Y and finite-dimensional representations of $\overline{H}_y$ over $\overline{Q}_l$.

1.7. If $f: X \to Y$ is a principal H-bundle (H as in 1.6) and K is an H-equivariant Sheaf on X then there is a unique Sheaf K' on Y such that $K = f^*K'[\dim H]$.

1.8. If G is an algebraic group, we denote by G^0 its identity component. In particular, the center Z_G of G has identity component Z^0_G and the centralizer $Z_G(X)$ of a subset $X \subset G$ has identity component $Z^0_G(X)$. If G_1 is a subgroup of G, we denote by $\mathcal{N}_G(G_1)$ the normalizer of G_1 in G.

2. Quasi-admissible Sheaves

2.1. Let G be a reductive algebraic group over an algebraically closed field k, with a given connected component G^1 such that the group of components G/G^0 is cyclic with generator G^1/G^0.

2.2. We wish to define a partition of G^1 into finitely many constructible subsets.

Let $s \in G$ be a semisimple element. Let

$$T(s) = Z_G^0(Z_G^0(s)), \qquad \text{(a torus in } G^0). \tag{2.2.1}$$

Let g_s (resp. g_u) be the semisimple (resp. unipotent) part of an element $g \in G$, so that $g = g_s g_u = g_u g_s$. For $g \in G^1$, let

$$\Theta(g) = T(g_s)g = gT(g_s) \subset G^1. \tag{2.2.2}$$

For any $g' \in \Theta(g)$, we have $T(g'_s) \subset T(g_s)$; we define

$$\Theta_r(g) = \{g' \in \Theta(g) \mid T(g'_s) = T(g_s)\}. \tag{2.2.3}$$

Then $\Theta_r(g)$ is an open subset of $\Theta(g)$ containing g. For $g, \overline{g} \in G^1$, the sets $\Theta_r(g)$, $\Theta_r(\overline{g})$ are either disjoint or coincide.

The group G^0 acts on G^1 by conjugation and it permutes among them the sets $\Theta_r(g) \subset G^1$, $(g \in G^1)$, with only finitely many orbits. (Here we use the finiteness of the number of unipotent classes in a reductive group [**2**, **7**].) Hence the sets

$$Y(g) = \bigcup_{x \in G^0} x\Theta_r(g)x^{-1}, \qquad (g \in G^1), \tag{2.2.4}$$

form a partition of G^1 into finitely many (constructible) subsets.

This is the required partition.

2.3. We now give an equivalent definition of the partition in 2.2.

(2.3.1) Let (L^0, Σ) be a pair consisting of a Levi subgroup L^0 of a parabolic subgroup of G^0 and a subset $\Sigma \subset G^1$ such that:
(a) Σ normalizes L^0.
(b) Σ is a single orbit of $Z_{L^0}^0 \times L^0$, acting on G^1 by $(z, l): g \to zlgl^{-1}$.
(c) For some (or equivalently any) $g \in \Sigma$ we have $Z_{L^0}^0 Z_{L^0}^0(g_s) - Z_{L^0}^0$.

Note that, for all $g \in \Sigma$, we have $T(g_s) \subset Z_{L^0}^0$; we define

$$\Sigma_r = \{g \in \Sigma \mid T(g_s) = Z_{L^0}^0\} \tag{2.3.2}$$

(an open dense subset of Σ). Then

$$Y_{L^0,\Sigma} = \bigcup_{x \in G^0} x\Sigma_r x^{-1} \tag{2.3.3}$$

is a constructible subset of G^1, depending only on the G^0-orbit of (L^0, Σ). The subsets $Y_{L^0,\Sigma}$ for various L^0, Σ as above form a partition of G^1 which coincides with the partition defined in 2.2. Indeed, we have $Y_{L^0,\Sigma} = Y(g)$ for any $g \in \Sigma_r$; conversely, if $g \in G^1$, we define

$$L^0(g) = Z_G^0 T(g_s),$$
$$\Sigma(g) = \bigcup_{x \in L^0(g)} x\Theta(g)x^{-1}; \tag{2.3.4}$$

then $(L^0(g), \Sigma(g))$ is as in (2.3.1) and $g \in \Sigma_r$, hence $Y(g) = Y_{L^0(g),\Sigma(g)}$.

We wish to describe the closure of $Y_{L^0,\Sigma}$ in G^1. First we choose:

(2.3.5) a parabolic subgroup P of G^0 with Levi subgroup L^0 and unipotent radical U_P such that Σ normalizes P.

We then have

$$\overline{Y}_{L^0,\Sigma} = \bigcup_{x \in G^0} x\overline{\Sigma}U_P x^{-1} \tag{2.3.6}$$

where $\overline{\Sigma}$ is the closure of Σ in G^1 (or in the normalizer of L^0). (The existence of P in (2.3.5) is trivial when $G = G^0$. In the general case let $g \in \Sigma_r$, $T_1 = (T(g_s) \cap Z_G(g))^0$; one can show that $Z_{G^0}(T_1) = L^0$. We can then clearly find a 1-parameter group $\lambda: k^* \to T_1$ such that $Z_{G^0}(\lambda(k^*)) = Z_{G^0}(T_1) = L^0$. We define P to be the parabolic subgroup of G whose Lie algebra is the sum of the ≥ 0 eigenspaces of k^* acting via λ on the Lie algebra of G. It is clear that P has L^0 as a Levi subgroup and that P is normalized by g. From (2.3.1)(b) it follows that P is normalized by Σ. The proof of (2.3.6), at least for connected G, is given in [**3**, 4.3(b)].

From (2.3.6) one can deduce that:

(2.3.7) $\overline{Y}_{L^0,\Sigma}$ is a union of subsets $Y_{L',\Sigma'}$ with L', Σ' as in (2.3.1).

It follows easily that:

(2.3.8) $Y_{L^0,\Sigma}$ is locally closed in G^1.

It is easy to see that:

(2.3.9) $Y_{L^0,\Sigma}$ is a smooth, irreducible subvariety of G^1 of dimension $\dim(G/L^0) + \dim \Sigma$.

(2.3.10) If $g, g' \in G'$ are in the same piece $Y_{L^0,\Sigma}$ then $Z_G^0(g)$, $Z_G^0(g')$ are conjugate under an element of G^0.

2.4. EXAMPLES. (a) If G^0 is a torus then the partition 2.2 of G^1 consists of a single piece: G^1 itself.

(b) In general, there is a unique open piece in the partition 2.2 of G^1; when G is connected it is the set of all regular semisimple elements of G.

(c) If $G = \mathrm{PGL}_2$, the partition 2.2 of G consists of three pieces: the set of regular semisimple elements, the set of regular unipotent elements, and the set $\{1\}$.

(d) If $G = \mathrm{O}_3$, $G^1 = \mathrm{O}_3 - \mathrm{SO}_3$ (k of characteristic $\neq 2$) the partition 2.2 of G^1 consists of two pieces: one is open and the other one consists of all reflections.

(e) If $G = \mathrm{SL}_2$ (char. $k \neq 2$) the partition 2.2 of G consists of five pieces: the set of regular semisimple elements, the set of regular unipotent elements, (-1) times the previous set, the set $\{1\}$, the set $\{-1\}$.

2.5. Let (L^0, Σ), $Y = Y_{L^0,\Sigma}$ be as in 2.3. We wish to define a finite covering $\pi: \tilde{Y} \to Y$ of Y.

Let $\tilde{Y} = \{(g, xL^0) \in G^1 \times (G^0/L^0) \mid x^{-1}gx \in \Sigma_r\}$ and let $\pi: \tilde{Y} \to Y$ be the map defined by $(g, xL^0) \to g$.

Consider the finite group $\mathcal{W}_\Sigma =$ stabilizer of Σ in $\mathcal{W} = N_{G^0}(L^0)/L^0$.

We have an action of $\mathcal{W}_\Sigma$ on $\tilde{Y}$ given by $\overline{n}: (g, xL^0) \to (g, xn^{-1}L^0)$, ($\overline{n} \in \mathcal{W}_\Sigma$ is represented by $n \in N_{G^0}(L^0)$). Then $\pi: \tilde{Y} \to Y$ is a principal $\mathcal{W}_\Sigma$-bundle. Consider the diagram

$$\Sigma \xleftarrow{i} \Sigma_r \xleftarrow{\pi''} \hat{Y} \xrightarrow{\pi'} \tilde{Y} \xrightarrow{\pi} Y$$

where

$$\hat{Y} = \{(g, x) \in G^1 \times G^0 \mid x^{-1}gx \in \Sigma_r\},$$

$$\pi'(g, x) = (g, xL^0), \quad \pi''(g, x) = x^{-1}gx, \quad i(g) = g.$$

Let $\mathcal{E}$ be a G^0-equivariant local system on Y (for the conjugation action of G^0 on Y). Then $\pi'^*\pi^*\mathcal{E}$ is a G^0-equivariant local system on $\hat{Y}$ for the action of G^0, $g_0: (g, x) \to (g_0 g g_0^{-1}, g_0 x)$, which makes $\hat{Y} \xrightarrow{\pi''} \Sigma_r$ a principal G^0-bundle. Hence there is a uniquely defined local system $\mathcal{L}_r$ on Σ_r such that $(\pi'')^*\mathcal{L}_r \cong \pi'^*\pi^*\mathcal{E}$. We say that $\mathcal{E}$ is *quasi-admissible* (q-ad) if $\mathcal{L}_r$ extends to a local system $\mathcal{L}$ on Σ (which is necessarily unique) and if, moreover,

(2.5.1) $\mathcal{L}$ is equivariant for the action of $\mathcal{Z}^0_{L^0} \times L^0$ on Σ given by $(z, l): g \to z^n lgl^{-1}$, for some integer $n \geq 1$, invertible in k.

(2.5.2) Conversely, if we are given an irreducible local system $\mathcal{L}$ on Σ, satisfying (2.5.1), then $(\pi'')^*i^*\mathcal{L}$ is an L^0-equivariant local system on $\hat{Y}$ for the action of L^0, $l_0: (g, x) \to (g, g_0 l^{-1})$ which makes $\hat{Y} \xrightarrow{\pi'} \tilde{Y}$ a principal L^0-bundle. Hence there is a uniquely defined local system $\tilde{\mathcal{E}}$ on $\tilde{Y}$ such that $\pi'^*\tilde{\mathcal{E}} \cong (\pi'')^*i^*\mathcal{L}$. Then $\pi_*\tilde{\mathcal{E}}$ is a direct sum of irreducible q-ad local systems on Y and all irreducible q-ad local systems on Y are obtained in this way.

Note that any q-ad local system on Y is semisimple and any of its direct summands is again q-ad; a direct sum of two q-ad local sytems on Y is again q-ad.

We now consider some examples.

(2.5.3) If $L^0 = G^0$, a local system $\mathcal{L}$ on $Y_{L^0,\Sigma} = \Sigma$ is q-ad if and only if it satisfies (2.5.1).

(2.5.4) If G is a torus (hence $L^0 = \Sigma = Y = G$) an irreducible local system $\mathcal{L}$ on G is q-ad if and only if $rk\mathcal{L} = 1$ and $\mathcal{L}^{\otimes n} \cong \overline{Q}_l$ for some integer $n \geq 1$, invertible in k.

(2.5.5) If G^0 is a torus (hence ($L^0 = G^0$, $\Sigma = Y = G^1$) an irreducible local system $\mathcal{L}$ on G^1 is q-ad if and only if $rk\mathcal{L} = 1$, $\mathcal{L}^{\otimes n} \cong \overline{Q}_l$ for some integer $n \geq 1$ invertible in k and $L_g^*\mathcal{L} \cong R_g^*\mathcal{L}$ for some (or equivalently any) $g \in G^1$; here $L_g: G^0 \to G^1$ (resp. $R_g: G^0 \to G^1$) is left (resp. right) multiplication by g.

2.6. An irreducible Sheaf A on G^1 is said to be quasi-admissible (q-ad) if it is supported by $\overline{Y}$, for some piece $Y = Y_{L^0,\Sigma}$ in the partition 2.2 of G^1 and if $A|Y$ is isomorphic to $\mathcal{E}[d]$ for some q-ad irreducible local system $\mathcal{E}$ on Y, ($d = \dim Y$).

Then A is automatically G^0-equivariant for the conjugation action of G^0 on G^1. Now $\mathcal{E}$ is a direct summand of $\pi_*\tilde{\mathcal{E}}$, attached to a local system $\mathcal{L}$ on Σ as in (2.5.2). Hence $\mathrm{IC}(\overline{Y}, \mathcal{E})[d]$ is a direct summand of $\mathrm{IC}(\overline{Y}, \pi_*\tilde{\mathcal{E}})[d]$. We shall now give an alternative description of the last Sheaf. Choose P as in (2.3.5). Let $X = \{(g, xP) \in G^1 \times (G^0/P) \mid x^{-1}gx \in \overline{\Sigma}U_P\}$, where $\overline{\Sigma}$ is the closure of Σ in L^1. Let $\gamma: X \to \overline{Y}$ be defined by $\gamma(g, xP) = g$ (see (2.3.6)). It is a proper map. The natural map $\tilde{Y} \to X$, $(g, xL^0) \to (g, xP)$ is an isomorphism of $\tilde{Y}$ onto the open subvariety $\gamma^{-1}(Y)$ of X. Via this map, we may regard $\tilde{\mathcal{E}}$ (a local system on $\tilde{Y}$) as a local system on $\gamma^{-1}(Y)$. Hence $\mathrm{IC}(X, \tilde{\mathcal{E}})[d] \in \mathcal{M}(X)$ is well defined. (Its inverse image under $\hat{X} \to X$, $\hat{X} = \{(g, x) \in G^1 \times G^0 | x^{-1}gx \in \overline{\Sigma}U_p\}$, $(g, x) \to (g, xP)$ is equal, up to shift, to the inverse image of $\mathrm{IC}(\overline{\Sigma}, \mathcal{L})$ under $\hat{X} \to \overline{\Sigma}$, $(g, x) \to \mathrm{pr}_{\overline{\Sigma}}(x^{-1}gx)$.) We have

$$\mathrm{IC}(\overline{Y}, \pi_*\tilde{\mathcal{E}})[d] = \gamma_*(\mathrm{IC}(X, \tilde{\mathcal{E}})[d]). \tag{2.6.1}$$

The proof is similar to that of [**3**, 4.5].

It is an application of the following general result. Let $f: X' \to X''$ be any proper, dominant morphism between two irreducible varieties X', X'' of the same dimension d. Let K be an irreducible perverse sheaf on X'. Let (X'_α) be a finite stratification of X' such that $\mathcal{H}^i K|X'_\alpha \neq 0 \Rightarrow i < -\dim X'_\alpha$, $\mathcal{H}^i DK|X'_\alpha \neq 0 \Rightarrow i < -\dim X'_\alpha$ for all α except for α_0 such that X'_{α_0} is the open stratum. Assume that

(a) $\dim(X'_\alpha \times_{X''} X'_\alpha) \leq \dim X'_\alpha$ for all α.

(b) Any d-dimensional irreducible component of $X'_{\alpha_0} \times_{X''} X'_{\alpha_0}$ is mapped by f onto a dense subset of X''.

Then $f_! K \cong \mathrm{IC}(X'', \mathcal{F})[d]$ where $\mathcal{F}$ is a local system on an open subset of X''. (When $X' = X'_{\alpha_0}$ this reduces to a result of Goresky-MacPherson on "small maps.")

In our case $X' = X$, $X'' = \overline{Y}$, and we consider the stratification $\overline{\Sigma} = \bigcup_\alpha (\overline{\Sigma})_\alpha$ given by the $Z^0_{L^0} \times L^0$-orbits.

This defines a stratification

$$X = X' = \bigcup_\alpha X'_\alpha, \quad X'_\alpha = \{(g, xP) \in G^1 \times (G^0/P) \mid x^{-1}gx \in (\overline{\Sigma})_\alpha U_P\}.$$

The dimension estimates (a), (b) are verified in our case by the method [**3**, §1].

2.7. *Induction.* Let P be a parabolic subgroup of G^0 with a given Levi subgroup L^0 and with unipotent radical U_P. Assume that the set

$$L^1 = \{g \in G^1 \mid gL^0g^{-1} = L^0, \quad gPg^{-1} = P\}$$

is nonempty. Then L^1 generates a closed subgroup L of G with identity component L^0; L^1 is a component of L and we have $L/L^0 \cong G/G^0$.

We consider the diagram

$$L^1 \xleftarrow{\alpha} V_1 \xrightarrow{\beta} V_2 \xrightarrow{\gamma} G^1$$

where

$$V_1 = \{(g,x) \in G^1 \times G^0 \mid x^{-1}gx \in L^1 \cdot U_P\},$$
$$V_2 = \{(g,xP) \in G^1 \times (G^0/P) \mid x^{-1}gx \in L^1 \cdot U_P\},$$
$$\alpha(g,x) = \mathrm{pr}_{L^1}(x^{-1}gx), \qquad \beta(g,x) = (g,xP), \qquad \gamma(g,xP) = g.$$

Let A_0 be an L^0-equivariant Sheaf on L^1 (for the conjugation action of L^0 on L^1). Then A_0 is also a P-equivariant Sheaf for the action $p: l \to \mathrm{pr}_{L^1}(plp^{-1})$, $(plp^{-1} \in L^1 \cdot U_P)$. Now $\alpha^* A_0[\dim G + \dim U_P]$ is a well-defined P-equivariant Sheaf on V_1 (P acts on V_1 by $p: (g,x) \to (g, xp^{-1})$); moreover, β is a principal P-bundle. Hence $\alpha^* A_0[\dim G + \dim U_P] = \beta^* A_1[\dim P]$ for a well-defined Sheaf A_1 on V_2. (See 1.7.) We define

$$\operatorname{ind} A_0 = \gamma_! A_1 \in \mathcal{D}G^1. \tag{2.7.1}$$

Assume now that (L^0, Σ) is as in (2.3.1) and $\mathcal{L}$ is an irreducible local system on Σ as in (2.5.1); let A_0 be an irreducible Sheaf on L^1 with support $\overline{\Sigma}$, with $A_0 \mid \Sigma = \mathcal{L}[d_0]$, $d_0 = \dim \Sigma$. From (2.6.1) it follows that

(2.7.2) $\operatorname{ind} A_0 \in \mathcal{M}G^1$ and is a direct sum of q-ad Sheaves on G^1, with support $\overline{Y}_{L^0,\Sigma}$.

Using this, we see by an argument in [**3**, 4.4(b)] that

(2.7.3) If A' is a q-ad Sheaf on L^1, then $\operatorname{ind} A' \in \mathcal{M}G^1$ and is a direct sum of q-ad Sheaves on G^1.

3. Admissible sheaves.

3.1. Let C be a unipotent G^0-conjugacy class in G^1. Let Q^0 be a parabolic subgroup of G^0 such that $Q^1 = \{g \in G^1 \mid gQ^0g^{-1} = Q^0\}$ is nonempty, hence a single Q^0-orbit under left (or right) translation. Let $\overline{Q}^0 = Q^0/U_{Q^0}$, $\overline{Q}^1 = Q^1/U_{Q^0}$; then $\overline{Q}^0$ acts naturally by conjugation on $\overline{Q}^1$; let $\overline{C}$ be any $\overline{Q}^0$-orbit on $\overline{Q}^1$ and let $\overline{g} \in \overline{C}$. Let $\pi: Q^1 \to \overline{Q}^1$ be the natural map. We have

$$\dim(C \cap \pi^{-1}(\overline{g})) \le \tfrac{1}{2}(\dim C - \dim \overline{C}).$$

(For G connected, this is proved in [**3**, 1.2(i)]; the proof in the general case is similar.) Hence if $\mathcal{E}$ is a G^0-equivariant local system on C, we have

$$H_c^i(C \cap \pi^{-1}(\overline{g}), \mathcal{E}) = 0 \quad \text{for } i \ge \dim C - \dim \overline{C}.$$

We say that the pair $(C, \mathcal{E})$ is *cuspidal* if for any Q^0, $\overline{C} \subset \overline{Q}^1$, $\overline{g} \in \overline{C}$, as above with $Q^0 \ne G^0$, we have

$$H_c^{\dim C - \dim \overline{C}}(C \cap \pi^{-1}(\overline{g}), \mathcal{E}) = 0.$$

3.2. We now describe the condition that $(C, \mathcal{E})$ is cuspidal in an alternative way. Let A be a q-ad Sheaf on G^1 with support $\overline{Y}_{L^0,\Sigma}$, $L^0 \neq G^0$. Then for a suitable i, the restriction of $A[i]$ to the unipotent variety $\mathcal{U}^1$ of G^1 (assumed nonempty) is a direct sum of irreducible G^0-equivariant Sheaves on $\mathcal{U}^1$, corresponding to various pairs $(C', \mathcal{E}')$ (see 1.3) where C' is a G^0-conjugacy class in $\mathcal{U}^1$ and $\mathcal{E}'$ is an irreducible G^0-equivariant local system on C'; the pairs $(C', \mathcal{E}')$ appearing in this way from various A as above are precisely those which are not cuspidal. Hence this provides another definition of cuspidal pairs. (The proofs of these statements, for G connected, are essentially contained in [**3**, §6].)

3.3. If $(C, \mathcal{E})$ is a cuspidal pair then, for $g \in C$, $Z^0_{G^0}(g)/(Z^0_{G^0}(g) \cap \operatorname{center} G^0)$ is a unipotent group.

(For G connected, this is proved in [**3**, 2.8].

3.4. We give an (exhaustive) list of examples of unipotent classes C which are parts of cuspidal pairs $(C, \mathcal{E})$, assuming that $G = G^0$ is almost simple, simply connected of classical type, $\operatorname{char} k \neq 2$. Let $\Delta = \{1, 3, 6, \dots\}$, $\square = \{1, 4, 9, \dots\}$.

(a) $G = \mathrm{SL}_n$, C = regular unipotent class, $(p, n+1) = 1$.

(b) $G = \mathrm{Sp}_{2n}$, $n \in \Delta$, C is the class of an element with Jordan blocks of sizes $2, 4, 6, \dots$ (in the standard representation).

(c) $G = \mathrm{Spin}_n$, $n \in \square$, C is the class of an element with Jordan blocks of sizes $1, 3, 5, \dots$ or $n \in \Delta$ and C is the class of an element with Jordan blocks of sizes $1, 5, 9, \dots$ or $3, 7, 11, \dots$ (in the standard n-dimensional representation).

See [**3**, §10–14; **6**].

3.4. For general G, we consider a parabolic subgroup $Q^0 \subset G^0$ with unipotent radical U_{Q^0} such that $Q^1 = \{g \in G^1 \mid gQ^0g^{-1} = Q^0\} \neq \varnothing$. Let $Q = \mathcal{N}_G Q^0$, $\overline{Q} = Q/U_{Q^0}$, $\overline{Q}^1 = Q^1/U_{Q^0}$, let $\pi: Q^1 \to \overline{Q}^1$ be the natural map, and let $i: Q^1 \to G^1$ be the natural inclusion. Note that $\overline{Q}$ is a reductive group and $\overline{Q}_1$ is a connected component of $\overline{Q}$, generating $\overline{Q}$. For any $A \in \mathcal{D}G^1$, we define

$$\operatorname{res}_{Q^1} A = \pi_! i^* A \in \mathcal{D}\overline{Q}^1.$$

3.5. Let A_0 be a Sheaf on G^1 which is equivariant for the action of $Z^0_{G^0} \times G^0$ on G^1 given by $(z, g_0): g \to z^n g_0 g g_0^{-1}$ for some integer $n \geq 1$, invertible in k. We say that A_0 is a *cuspidal* Sheaf, if for any parabolic subgroup $Q^0 \subsetneq G^0$ such that $Q^1 \neq \varnothing$ (see 3.4) we have $\operatorname{res}_{Q^1} A_0[-1] \in \mathcal{D}^{\leq 0}(\overline{Q}^1)$ or, equivalently: $\dim \operatorname{supp} \mathcal{H}^i(\operatorname{res}_{Q^1} A) < -i$ for all i. We say that A_0 is a *strongly cuspidal* Sheaf if for any Q^0 as above we have $\operatorname{res}_{Q^1} A_0 = 0$ in $\mathcal{D}(\overline{Q}^1)$; such A_0 is, in particular, cuspidal.

3.6. Consider a subset Σ of G^1 satisfying conditions (a), (b), (c) of (2.3.1) with $L^0 = G^0$ and let $\mathcal{L}$ be a local system on Σ as in (2.5.1). Let $g \in \Sigma$ and let $g_s, g_u \in G$ be the semisimple and unipotent parts of g. Let H be the subgroup of G generated by g_u and by $Z^0_G(g_s)$; then H is a reductive group with identity component $H^0 = Z^0_G(g_s)$ and H is generated by $H^1 = g_u H^0$. Let $C \subset H^1$ be the orbit of g_u under conjugation by H^0, let $j: C \to \Sigma$ be defined by $j(y) = g_s y = y g_s$, and let $\mathcal{E}$ be the local system $j^* \mathcal{L}$ on C. Let A be the Sheaf on G^1 with support $\overline{\Sigma}$ such that $A|\overline{\Sigma} = \mathrm{IC}(\overline{\Sigma}, \mathcal{L})[d_0]$, $d_0 = \dim \Sigma$. We say that $\mathcal{L}$ is *cuspidal* if the equivalent conditions (3.6.1), (3.6.2) are satisfied:

(3.6.1) A is cuspidal in the sense of 3.5.

(3.6.2) $(C, \mathcal{E})$ is a cuspidal pair for H^1 (see 3.1).

Conversely,

(3.6.3) If A_0 is any irreducible Sheaf (see 3.5) on G^1 then there exists Σ, $\mathcal{L}$ as above such that $\operatorname{supp} A_0 = \overline{\Sigma}$, $A_0|\Sigma = \mathcal{L}[d_0]$, $d_0 = \dim \Sigma$ and $(C, \mathcal{E})$ defined as above in terms of Σ, $\mathcal{L}$ is a cuspidal pair.

This is proved in [**3**, 3.12] in a special case; but essentially the same proof is valid in general.

3.7. Let (L^0, Σ) be as in (2.3.1), let $\mathcal{E}$ be a q-ad local system on $Y_{L^0,\Sigma}$, and let $\mathcal{L}$ be the local system on Σ attached to $\mathcal{E}$ as in 2.5. We say that $\mathcal{E}$ is *admissible* if $\mathcal{L}$ is cuspidal (in the sense of 3.6).

An irreducible Sheaf A on G^1 is said to be admissible if it is supported by $\overline{Y}$ for some piece Y in the partition 2.2 of G^1 and if $A|Y$ is isomorphic to $\mathcal{E}[d]$ for some admissible irreducible local system $\mathcal{E}$ on Y, $(d = \dim Y)$.

Clearly, if A is admissible then it is quasi-admissible. By (3.6.3), any irreducible cuspidal Sheaf on G^1 is admissible.

In the setup of 2.7, if A' is an admissible Sheaf on L^1, then $\operatorname{ind} A' \in \mathcal{M}G^1$ is a direct sum of admissible Sheaves on G^1.

3.8. EXAMPLES. (a) If G^0 is a torus, any q-ad Sheaf on G^1 is admissible.

(b) If $G = \mathrm{GL}_2$, a q-ad Sheaf on G is admissible if and only if its support is G.

(c) Assume $G = \mathrm{SL}_2$ (in characteristic $\neq 2$) and let A be a q-ad Sheaf on G with $\operatorname{supp} A = \overline{Y}_{L^0,\Sigma}$, $A|Y_{L^0,\Sigma} \cong \mathcal{E}[d]$, $d = \dim Y_{L^0,\Sigma}$. Then A is admissible precisely in the following cases: $\overline{Y}_{L^0,\Sigma} = G$, or $\overline{Y}_{L^0,\Sigma}$ = unipotent variety, $\mathcal{E} \not\cong \overline{Q}_l$, or $\overline{Y}_{L^0,\Sigma} = (-1)$ times the unipotent variety, $\mathcal{E} \not\cong \overline{Q}_l$.

4. Character sheaves.

4.1. We fix maximal torus T^0 of G^0 and a Borel subgroup B of G^0 containing T^0. Let $W = \mathcal{N}_G(T^0)/T^0$, $W^0 = \mathcal{N}_{G^0}(T^0)/T^0$, $W^1 = (\mathcal{N}_G(T^0) \cap G^1)/T^0$. For any $w \in W$, we consider a representative $\dot{w}$ in $\mathcal{N}_G(T^0)$ and we denote $w\colon T^0 \to T^0$ the map $t \to \dot{w}t\dot{w}^{-1}$. We assume given an element $w \in W^1$ and a local system $\mathcal{L}$ on T^0 such that

(4.1.1) $$w^*\mathcal{L} \cong \mathcal{L}$$

and

(4.1.2) $\operatorname{rank} \mathcal{L} = 1$, $\mathcal{L}^{\otimes n} \cong \overline{Q}_l$ for some integer $n \geq 1$, invertible in k.

Let $T^1 = \dot{w}T^0 = T^0\dot{w}$; let $\mathcal{L}_w$ be the inverse image of $\mathcal{L}$ under the map $T^1 \to T^0$, $t \to (\dot{w})^{-1}t$.

From (4.1.1) we see that $\mathcal{L}_w$ is a T^0-equivariant local system on T^1 (for the conjugation action of T^0 on T^1); its isomorphism class does not depend on the choice of $\dot{w}$.

We denote the unipotent radical of B by U. We have a natural morphism $\alpha: UT^1U \to T^1$ (where $UT^1U \subset G^1$) defined by $\alpha(ut^1u') = t^1$ for $u, u' \in U$, $t^1 \in T^1$. We consider the diagram

$$T^1 \overset{\phi}{\leftarrow} \hat{Y}_w \overset{\rho}{\to} Y_w \overset{\pi}{\to} G^1$$

where

$$\hat{Y}_w = \{(g,x) \in G^1 \times G^0 \mid x^{-1}gx \in UT^1U\},$$
$$Y_w = \{(g,x) \in G^1 \times (G^0/B) \mid x^{-1}gx \in UT^1U\},$$
$$\phi(g,x) = \alpha(x^{-1}gx), \qquad \rho(g,x) = (g, xB), \qquad \pi(g,xB) = g.$$

The group B acts on T^1 by $b = tu \in T^0U = B: t^1 \to tt^1t^{-1}$, on $\hat{Y}_w$ by $b: (g,x) \to (g, xb^{-1})$, and on Y_w, G^1 trivially; these B-actions are compatible with ϕ, ρ, π. Since $\mathcal{L}_w$ is B-equivariant, $\phi^*\mathcal{L}_w$ is a B-equivariant local system on $\hat{Y}_w$; it is of form $\rho^*\tilde{\mathcal{L}}$ for a well-defined local system $\tilde{\mathcal{L}}$ on Y_w, since ρ is a principal B-bundle. We define

$$K_w^{\mathcal{L}} = \pi_!\tilde{\mathcal{L}} \in \mathcal{D}G^1.$$

We shall also define a complex $\overline{K}_w^{\mathcal{L}}$, closely related to K_w. We have a commutative diagram

$$\begin{array}{ccccc} \hat{Y}_w & \overset{\rho}{\to} & Y_w & \overset{\pi}{\to} & G^1 \\ \downarrow & & \downarrow & \nearrow_{\overline{\pi}} & \\ \hat{\overline{Y}}_w & \overset{\overline{\rho}}{\to} & \overline{Y}_w & & \end{array}$$

where

$$\hat{\overline{Y}}_w = \{(g,x) \in G^1 \times G^0 \mid x^{-1}gx \in \overline{UT^1U}\},$$
$$\hat{Y}_w = \{(g,x) \in G^1 \times (G^0/B) \mid x^{-1}gx \in \overline{UT^1U}\},$$

$\overline{\rho}(g,x) = (g, xB)$, $\overline{\pi}(g,xB) = g$, and the vertical maps are the obvious embeddings; $\overline{UT^1U}$ is the closure of UT^1U in G^1. Then $\overline{\pi}$ is a proper map, $\hat{Y}_w$ is open dense in $\hat{\overline{Y}}_w$, hence Y_w is open dense in $\overline{Y}_w$. Note also that Y_w is smooth, irreducible. Let $J_w^{\mathcal{L}} = \mathrm{IC}(\overline{Y}_w, \tilde{\mathcal{L}})$ and

$$\overline{K}_w^{\mathcal{L}} = (\overline{\pi})_! J_w^{\mathcal{L}} \in \mathcal{D}G^1.$$

By the Decomposition Theorem [**1**, 6.2.5], $\overline{K}_w^{\mathcal{L}}$ is a direct sum of irreducible Sheaves on G^1 (with shifts).

4.2. Let $\mathcal{L}$ be a local system on T^0 as in (4.1.2). Let $W'_{\mathcal{L}}$ be the set of all $w \in W$ such that $w^*\mathcal{L} \cong \mathcal{L}$ (a subgroup of W). The following two conditions for an irreducible Sheaf A on G^1 are equivalent:

(a) There exist $w \in W'_{\mathcal{L}} \cap W^1$ and $i \in \mathbf{Z}$ such that A is a constituent of ${}^pH^i(K_w^{\mathcal{L}})$.

(b) There exist $w \in W'_{\mathcal{L}} \cap W^1$ and $i \in \mathbf{Z}$ such that $A[-i]$ is a direct summand of $\overline{K}_w^{\mathcal{L}}$.

The set of (isomorphism classes of) irreducible Sheaves A on G^1 satisfying (a) or (b) above is denoted $\hat{G}^1_{\mathcal{L}}$. The irreducible Sheaves on G^1 which belong to $\hat{G}^1_{\mathcal{L}}$ for some $\mathcal{L}$ as in (4.1.2) are called *character Sheaves*; they (or rather their isomorphism classes) form a set $\hat{G}^1$.

4.3. We say that the characteristic p of k is *bad* for G if it is a prime number such that:

(a) (when G is connected) p divides the coefficient of a root expressed as a linear combination of simple roots,

(b) (for general G) there exists a unipotent element $u \in G$ normalizing T^0 and B (as in 4.1) such that p is bad (in the sense of (a)) for the connected reductive group $Z^0_G(u)$.

For example, if $p = 0$ or $p \geq 7$, then p is not bad for G; when G^0 is of classical type and $p \neq 2$, then p is not bad for G.

From now on, we shall assume that the characteristic of k is not bad for G.

4.4. We shall state several properties of character Sheaves. (The references given cover only the case where G is connected.)

(4.4.1) An irreducible Sheaf on G^1 is admissible if and only if it is a character Sheaf. (See [**4**, I.4.4(a), V23.1(b)].)

(4.4.2) A character Sheaf of G^1 is cuspidal if and only if it is strongly cuspidal, (3.5). (See [**4**, I6.9(b)].) If $A \in \hat{G}^1$ is cuspidal, then $\operatorname{supp} A = \overline{\Sigma}$ (Σ as in 3.6) and $A|\overline{\Sigma} - \Sigma = 0$. (See [**4**, V23.1(a)].)

(4.4.3) If $\mathcal{L}$, $\mathcal{L}'$ are as in (4.1.2) then $\hat{G}^1_{\mathcal{L}}$, $\hat{G}^1_{\mathcal{L}'}$ coincide if $\mathcal{L}$, $\mathcal{L}'$ are in the same W^0-orbit and are disjoint otherwise. The Verdier duality $D\colon \mathcal{M}G^1 \to \mathcal{M}G^1$ defines a bijection $\hat{G}^1_{\mathcal{L}} \xrightarrow{\sim} \hat{G}^1_{\mathcal{L}^{-1}}$. (See [**4**, III11.2].)

(4.4.4) Let $A \in \hat{G}^1_{\mathcal{L}}$, and let $\delta = \dim \operatorname{supp} A$. Then, if $A[-i]$ is a direct summand of $\overline{K}^{\mathcal{L}}_w$, $(w \in W'_{\mathcal{L}} \cap W^1)$, then $i \equiv \delta \pmod 2$. (See [**4**, V23.1(a)].) If $\mathcal{H}^i A \neq 0$, then $i \equiv \delta \pmod 2$. (See [**4**, V24.11].)

(4.4.5) In the setup of 2.7, for any $A' \in \hat{L}^1$, $\operatorname{ind} A' \in \mathcal{M}G^1$ is a direct sum of character Sheaves. (See [**4**, I4.8(b)].) In the setup of 3.4, for any $A \in \hat{G}^1$, $\operatorname{res}_{Q^1} A \in \mathcal{D}\overline{Q}^1$ is a direct sum of character Sheaves of $\overline{Q}^1$. (See [**4**, I6.9(a)].)

4.5. In the remainder of this section *we assume that* $G = G^0 = G^1$.

Let $\mathcal{L}$ be a local system on T^0 as in (4.1.2). Let $R \subset \operatorname{Hom}(T^0, k^*)$ be the set of roots of G^0 and let $\check{R} \subset \operatorname{Hom}(k^*, T^0)$ be the set of coroots. Let R^+ be the set of positive roots determined by B. Let $\check{R}_{\mathcal{L}}$ be the set of coroots $\check{\alpha}$ such that $(\check{\alpha})^*\mathcal{L} \cong \overline{Q}_l$ as local systems on k^*; let $R_{\mathcal{L}}$ be the corresponding set of roots. Let $W_{\mathcal{L}}$ be the subgroup of W generated by the reflections of W with respect to roots in $R_{\mathcal{L}}$. Then $R_{\mathcal{L}}$ is a root system with Weyl group $W_{\mathcal{L}}$, with set of positive roots $R_{\mathcal{L}}^{+} = R_{\mathcal{L}} \cap R^{+}$; let $S_{\mathcal{L}}$ be the corresponding set of simple reflections of $W_{\mathcal{L}}$.

Let V be the $\overline{Q}_l$-vector space with scalar product (:), with orthonormal basis given by the character Sheaves on G. Then ${}^pH^i(K_y^{\mathcal{L}})$ ($y \in W'_{\mathcal{L}}$) may be regarded as an element of V in which $A \in \hat{G}$ appears with coefficient equal to the multiplicity of A in a Jordan-Hölder series of ${}^pH^i(K_y^{\mathcal{L}})$.

Let $\hat{W}'_{\mathcal{L}}$ be the set of (isomorphism classes of) irreducible representations of $W'_{\mathcal{L}}$ (over $\overline{Q}_l$). We have a natural partition $\hat{W}'_{\mathcal{L}} = \coprod(\hat{W}'_{\mathcal{L}})_{\underline{c}}$ indexed by the two-sided cells $\underline{c}$ of $W_{\mathcal{L}}$ (modulo the natural action of $W'_{\mathcal{L}}/W_{\mathcal{L}}$); if $E \in \hat{W}'_{\mathcal{L}}$, we have $E \in (\hat{W}'_{\mathcal{L}})_{\underline{c}}$ if some irreducible component of $E|W_{\mathcal{L}}$ belongs to $\underline{c}$.

For $E \in \hat{W}'_{\mathcal{L}}$, let

$$R_E^{\mathcal{L}} = |W'_{\mathcal{L}}|^{-1} \sum_{y \in W'_{\mathcal{L}}} \operatorname{Tr}(y^{-1}, E) \sum_i (-1)^{i+\dim G}\, {}^pH^i(K_y^{\mathcal{L}}) \in V. \tag{4.5.1}$$

We have the following results:

(4.5.2) $(R_E^{\mathcal{L}} : R_{E'}^{\mathcal{L}}) = \delta_{E,E'}$ for $E, E' \in \hat{W}'_{\mathcal{L}}$; in particular $R_E^{\mathcal{L}} \neq 0$. (See [**4**, III14.13].)

If $A \in \hat{G}_{\mathcal{L}}$ then $(A : R_{E'}^{\mathcal{L}}) \neq 0$ for some $E \in \hat{W}'_{\mathcal{L}}$; if we have also $(A : R_{E'}^{\mathcal{L}}) \neq 0$ for $E' \in \hat{W}'_{\mathcal{L}}$ then $E, E' \in (\hat{W}'_{\mathcal{L}})_{\underline{c}}$ for some $\underline{c}$ as above. (See [**4**, III16.6(b)].)

Hence we have a natural map

$$\Phi : \hat{G} \to \text{set of all pairs } (\mathcal{L}, \underline{c}) \text{ modulo the natural action of } W. \tag{4.5.3}$$

($\mathcal{L}$ is any local system on T^0 as in (4.1.2) and $\underline{c}$ is any two-sided cell in $W_{\mathcal{L}}$); this map is defined by $\Phi(A) = (\mathcal{L}, \underline{c})$ where $(A : R_E^{\mathcal{L}}) \neq 0$ for some $E \in (\hat{W}'_{\mathcal{L}})_{\underline{c}}$.

4.6. Let $\hat{G}_{\mathcal{L},\underline{c}}$ be the fiber of Φ over $(\mathcal{L}, \underline{c})$; this is a finite, nonempty set. We can attach to $(\mathcal{L}, \underline{c})$ a unipotent class $C = C(\mathcal{L}, \underline{c})$ in G; it is characterized by the following two properties: (a) there exists $A \in \hat{G}_{\mathcal{L},\underline{c}}$ such that $C \subset \operatorname{supp} A$, (b) if C' is a unipotent class in G with $C' \subset \operatorname{supp} A'$ for some $A' \in \hat{G}_{\mathcal{L},\underline{c}}$ then $\dim C' \leq \dim C$.

Let $u \in C$ and let $\mathcal{A}(u) = Z_G(u)/Z_G^0(u)$. If $A \in \hat{G}_{\mathcal{L},\underline{c}}$ is such that $C \subset \operatorname{supp} A$, then the restriction $A|C$ is (up to shift) an irreducible G-equivariant local system on C, hence it defines an irreducible representation of $\mathcal{A}(u)$; the resulting irreducible representations of $\mathcal{A}(u)$ (for varying A) all factor through some largest quotient $\overline{\mathcal{A}}(u)$ of $\mathcal{A}(u)$. (This quotient depends on $\mathcal{L}$, $\underline{c}$.)

We can reformulate (and slightly generalize) [**4**, V23.1(c)] as follows: there exists a 1-1 correspondence between the set $\hat{G}_{\mathcal{L},\underline{c}}$ and the set $\mathcal{M}(\overline{\mathcal{A}}(u))$ consisting of all pairs (x, σ) where x is an element of $\overline{\mathcal{A}}(u)$ (defined up to conjugacy) and σ is an irreducible representation of $Z_{\overline{\mathcal{A}}(u)}(x)$. (Such a correspondence is established in *loc. cit.* together with an explicit formula for the multiplicities $(A : R_E^{\mathcal{L}})$, $A \in \hat{G}_{\mathcal{L},\underline{c}}$, $E \in (\hat{W}'_{\mathcal{L}})_{\underline{c}}$.)

4.7. EXAMPLES. (a) $G = E_8$ (char. $k \neq 2,3,5$), $\mathcal{L} = \overline{Q}_l$, $\underline{c} =$ two-sided cell of $W_{\mathcal{L}} = W$ containing the irreducible representation of degree 4480. C is the unipotent class of G containing u such that $\mathcal{A}(u) \cong \mathfrak{S}_5$. We have $\overline{\mathcal{A}}(u) = \mathfrak{S}_5$. The set $\hat{G}_{\mathcal{L},\underline{c}}$ consists of 39 character sheaves $A_{x,\sigma}$, $(x,\sigma) \in \mathcal{M}(\mathfrak{S}_5)$. We may identify $\mathcal{A}(u)$ with a complement of $Z_G^0(u)$ in $Z_G(u)$. For $x \in \mathfrak{S}_5$ let s_x be the corresponding element of $Z_G(u)$, and let $\mathcal{C}_x$ be the G-conjugacy class of $us_x = s_xu$. We have $Z_G(us_x)/Z_G^0(us_x) = Z_{\mathfrak{S}_5}(x)$, hence any irreducible representation σ of $Z_{\mathfrak{S}_5}(x)$ determines an irreducible G-equivariant local system $\mathcal{E}_\sigma$ on $\mathcal{C}_x$. The indexing of $A_{x,\sigma}$ can be chosen in such a way that for all $(x,\sigma) \in \mathcal{M}(\mathfrak{S}_5)$:

$$\begin{aligned} A_{x,\sigma}|\mathcal{C}_x &\cong \mathcal{E}_\sigma \quad \text{(up to shift)}, \\ A_{x,\sigma}|\mathcal{C}_{x'} &= 0 \quad \text{(if } x' \text{ is not conjugate to } x \text{ in } \mathfrak{S}_5). \end{aligned}$$

If $A \in \hat{G}_{\mathcal{L},\underline{c}}$ then $\mathcal{C}$ can be characterized in terms of A as follows: $\mathcal{C}$ is the unique unipotent class in G of maximal possible dimension such that for some $u \in \mathcal{C}$ and some semisimple element $s \in Z_G(u)$, the restriction of A to the conjugacy class of su is nonzero. (This seems to be a general phenomenon.) Thirteen of the 39 $A_{x,\sigma}$ are cuspidal.

(b) $G = \mathrm{SL}_n$, (n invertible in k), $\mathcal{L}$ such that $W_{\mathcal{L}} = \{e\}$, $W'_{\mathcal{L}} \cong \mathbf{Z}/n\mathbf{Z}$, $\underline{c} = \{e\}$. Then $\mathcal{C}$ is the class of a regular unipotent element u of G. We have $\mathcal{A}(u) = \overline{\mathcal{A}}(u) \cong \mathbf{Z}/n\mathbf{Z}$. (We identify $\mathbf{Z}/n\mathbf{Z}$ with the center of G; for each $x \in \mathbf{Z}/n\mathbf{Z}$ let s_x be the corresponding element of the center of G.) Let $\mathcal{C}_x = s_x \cdot \mathcal{C} =$ conjugacy class of s_xu. If σ is a character of $\mathbf{Z}/n\mathbf{Z} = Z_G(s_xu)/Z_G^0(s_xu)$ we denote $\mathcal{E}_\sigma$ the corresponding G-equivariant local system of rank 1 on $\mathcal{C}_x$. The set $\hat{G}_{\mathcal{L},\underline{c}}$ consists of n^2 character sheaves $A_{x,\sigma}$, $(x,\sigma) \in \mathcal{M}(\mathbf{Z}/n\mathbf{Z})$. The indexing of $A_{x,\sigma}$ can be chosen in such a way that for all $(x,\sigma) \in \mathcal{M}(\mathbf{Z}/n\mathbf{Z})$:

$$\begin{aligned} A_{x,\sigma}|\mathcal{C}_x &\cong \mathcal{E}_\sigma \quad \text{(up to shift)}, \\ A_{x,\sigma}|\mathcal{C}_{x'} &= 0 \quad \text{if } x' \neq x. \end{aligned}$$

$A_{x,\sigma}$ is cuspidal if and only if σ is of order n.

(c) G is general, $\mathcal{L}$ is such that $W_{\mathcal{L}} = W'_{\mathcal{L}} = \{e\}$, $\underline{c} = \{e\}$. Then $\mathcal{C}$ is the class of a regular unipotent element u of G. We have $\mathcal{A}(u) = Z_G/Z_G^0$, $\overline{\mathcal{A}}(u) = \{e\}$. The set $\hat{G}_{\mathcal{L},\underline{c}}$ consists of a single character sheaf. Its restriction to $\mathcal{C}$ is isomorphic to $\overline{Q}_l$ (up to a shift).

(d) G is general, $\mathcal{L} = \overline{Q}_l$, $W_{\mathcal{L}} = W$, $\underline{c} = \{e\}$. Then $\mathcal{C}$ is the class of a regular unipotent element of G. The set $\hat{G}_{\mathcal{L},c}$ consists of a single character sheaf: $\overline{Q}_l$.

(e) G is general, $\mathcal{L} = \overline{Q}_l$, $W_{\mathcal{L}} = W$, $\underline{c} = \{\text{longest element of } W\}$.

Then $\mathcal{C} = \{e\}$. The set $\hat{G}_{\mathcal{L},\underline{c}}$ consists of a single character sheaf; its restriction to the unipotent variety has support $\{e\}$.

(f) $G = PS_{p_4}$ (char. $k \neq 2$), $\mathcal{L}$ is such that $W_{\mathcal{L}} = W'_{\mathcal{L}}$ is of type $A_1 \times A_1$, $\underline{c} =$ two-sided cell containing the longest element of $W_{\mathcal{L}}$. Then $\mathcal{C}$ is the class of a transvection. The set $\hat{G}_{\mathcal{L},\underline{c}}$ consists of a single character sheaf; its restriction to $\mathcal{C}$ is $\overline{Q}_l$.

5. F_q-rational structures.

5.1. We now assume that k is an algebraic closure of the finite field F_q and that G (as in 2.1) has a given F_q-rational structure compatible with the group structure and such that G^1 is defined over F_q. We denote $F: G \to G$ the corresponding Frobenius map.

5.2. Consider the vector space $\mathcal{V}$ of all functions $f: (G^1)^F \to \overline{Q}_l$ which are constant on the orbits of the conjugation action of $(G^0)^F$ on $(G^1)^F$.

Then $\mathcal{V}$ has a hermitian inner product

$$\langle f, f' \rangle = |(G^1)^F|^{-1} \sum_{g \in (G^1)^F} \overline{f(g)} f'(g)$$

where $\lambda \to \overline{\lambda}$ is an automorphism of order 2 of $\overline{Q}_l$ which maps each root of 1 to its inverse.

If E is an irreducible representation of $(G^0)^F$ (over $\overline{Q}_l$) which can be extended to a representation $\tilde{E}$ of G^F, then

$$\chi_E(g) = \operatorname{Tr}(g, \tilde{E}), \qquad (g \in (G^1)^F)$$

defines an element $\chi_E \in \mathcal{V}$. (This depends on the extension $\tilde{E}$; a different choice of $\tilde{E}$ gives the same element times a root of 1.)

The functions $\chi_E \in \mathcal{V}$ for various E as above (for each E we assume $\tilde{E}$ chosen) form an orthonormal basis of $\mathcal{V}$.

One of the main properties of character Sheaves is that they also provide an orthonormal basis of $\mathcal{V}$. Let $(\hat{G}^1)^F$ be the set of those $A \in \hat{G}^1$ for which F^*A is isomorphic to A in $\mathcal{M}G^1$. For such A, an isomorphism $\phi_A: F^*A \xrightarrow{\sim} A$ is uniquely determined up to a factor in $\overline{Q}_l$, since A is irreducible. We can choose ϕ_A so that the following property is satisfied: for any $y \in Y_{L^0,\Sigma}$ (where $\operatorname{supp} A = \overline{Y_{L^0,\Sigma}}$; see (2.3.3), (4.4.1)) such that $F^n y = y$ the eigenvalues of $\phi_A^n: \mathcal{H}_y^{-d} A \to \mathcal{H}_y^{-d} A$ ($d = \dim \operatorname{supp} A$) are of the form $q^{n(\dim G - d)/2}$ times a root of 1. Such ϕ_A is uniquely determined up to multiplication by a root of 1; we assume that ϕ_A is chosen for each $A \in (\hat{G}^1)^F$.

We define for $A \in (\hat{G}^1)^F$ a function $\chi_A: (G^1)^F \to \overline{Q}_l$ by

$$\chi_A(g) = \sum_i (-1)^i \operatorname{Tr}(\phi_A, \mathcal{H}_g^i A), \quad g \in (G^1)^F.$$

The values of χ_A are cyclotomic integers. (See [**4**, V25.6] for G connected.) We have the following result.

(5.2.1) The functions $\chi_A \in \mathcal{V}$ ($A \in (\hat{G}^1)^F$) form an orthonormal basis of $\mathcal{V}$.

(See [**4**, V25.7] for G connected.)

The interest of character Sheaves for the problem of computing character values of finite groups resides in the following two facts:

(a) the functions χ_A are "computable";

(b) the functions χ_E are (conjecturally) related to the functions χ_A by a relatively simple, explicitly known, almost diagonal matrix.

Some evidence in support of (b) is presented in [**5**]; for example for $G = \mathrm{GL}_n$ with the standard F_q-structure, the two basis (χ_E), (χ_A) of $\mathcal{V}$ coincide (up to multiplication by roots of 1).

REFERENCES

1. A. A. Beilinson, J. Bernstein, and P. Deligne, *Faisceaux pervers*, Astérisque, no. 100, Soc. Math. France, Paris, 1982.

2. G. Lusztig, *On the finiteness of the number of unipotent classes*, Invent. Math. **34** (1976), 201–213.

3. ——, *Intersection cohomology complexes on a reductive group*, Invent. Math. **75** (1984), 205–272.

4. ——, *Character Sheaves* I–V, Adv. in Math. **56** (1985), 193–237; **57** (1985), 226–265; **57** (1985), 266–315; **59** (1986), 1–63; **61** (1986), 103–155.

5. ——, *On the character values of finite Chevalley groups at unipotent elements*, J. Algebra **104** (1986), 146–194.

6. G. Lusztig and N. Spaltenstein, *On the generalized Springer correspondence for classical groups*, Adv. Studies in Pure Math., vol. 6, Algebraic Groups and Related Topics, North Holland and Kinokunia, 1985, pp. 289–316.

7. N. Spaltenstein, *Classes unipotentes et sous-groupes de Borel*, Lecture Notes in Math., vol. 946, Springer-Verlag, 1982.

MASSACHUSETTS INSTITUTE OF TECHNOLOGY

Proceedings of Symposia in Pure Mathematics
Volume 47 (1987)

Finite Groups and Modular Functions

GEOFFREY MASON
(WITH AN APPENDIX BY S. P. NORTON)

1. Introduction. Since the appearance of the remarkable "Moonshine" paper of Conway and Norton at the end of the last decade, there has been a considerable amount of activity concerned with the investigation of the connections between finite groups and modular forms. Such functions appear as q-expansions $\sum a_n q^n$, one for each conjugacy class of a finite group G, and for each n the function assigning the conjugacy class to the nth Fourier coefficient a_n is to be a (possibly virtual) character of G. Such assignments

$$\{\text{conjugacy class}\} \to \{\text{modular form}\}$$

are the topic of this paper. In case the modular forms are all modular functions whose invariance groups are modular groups of genus zero, such assignments have been called "Moonshines."

It is convenient to consider the formal q-expansion

$$\Gamma_G = \sum \gamma_n q^n \tag{1.1}$$

where γ_n is the (virtual) character assigning a class to its nth coefficient a_n; Γ_G will be called a *Thompson Series*, and of course we can recover the modular form associated to an element $g \in G$ (or rather to its conjugacy class g^G) from the Thompson Series as

$$\Gamma_g = \sum \gamma_n(g) q^n. \tag{1.2}$$

Conway-Norton's Moonshine paper is concerned with a conjectured Thompson Series for the Monster Simple group M in which each Γ_g is a so-called *hauptmodul.* Among other things, each γ_n is a genuine character, so that one can replace Γ_M by an infinite-dimensional $\mathbf{Z}$-graded $\mathbf{C}M$-module

$$V = \coprod V_n, \tag{1.3}$$

1980 *Mathematics Subject Classification* (1985 *Revision*). Primary 11F22; Secondary 20C99.

Supported by the National Science Foundation.

where V_n is a $\mathbf{C}M$-module whose character is γ_n. Much effort was spent on trying to construct the module V, and a few years ago Frenkel-Lepowsky-Meurman announced some truly remarkable results in this direction. They constructed a $\mathbf{Z}$-graded $\mathbf{C}M$-module $V^\natural$ (their notation!) such that the nth graded space $V_n^\natural$ has the same dimension as V_n and, roughly speaking, many elements of M have the same character on both V_n and $V_n^\natural$. Amazingly, their work draws much of its inspiration from the theory of Kac-Moody Lie algebras, in particular the use of vertex operators in constructing affine Lie algebras—ideas which themselves first saw the light of day in parts of theoretical physics dealing with certain quantum field theories. It seems that whenever one studies these infinite graded modules, one gets ineluctably drawn into this web of ideas. Furthermore, modular forms arise naturally (as string functions) in the representation theory of Kac-Moody Lie algebras, and to date this seems the most promising avenue down which an explanation of Monstrous Moonshine might lie. Nevertheless, it is still fair to say that the meaning of the "hauptmodul" property of the module V is comprehensively not understood.

In a slightly different direction, the present author initiated a study of Thompson Series for arbitrary groups G in which the series Γ_g are modular forms of some weight, level, and character (depending on g), as opposed to the situation for M where all weights are zero. Perhaps surprisingly, one can get quite far in imitating the classical theory of modular forms; in some sense one can think of Γ_G as an "equivariant" modular form. This theory is much closer in spirit to classical number theory, and there are indications that it may be related in some way to the Langlands program, though such ideas should be viewed as rather speculative at present.

In this paper we shall rehearse once again what we understand of our subject, concentrating on the topics mentioned above. Thus we will begin with a review of the theory of modular functions (i.e., forms of weight zero), followed by a consideration of the principal conjectures of Conway-Norton pertaining to the existence of "Moonshines" for M and various other groups. Next we take up the extension of these ideas to modular forms, in particular we consider the role that Hecke operators have in both theories. Then we summarize the ideas of F-L-M leading to the construction of the module $V^\natural$ mentioned above. We should point out that we have adhered fairly strictly to the group-representation aspects of their work, in consonance with the themes of this conference; there is no doubt, however, that this gives a rather lopsided picture of their remarkable work—we can only refer the reader to the literature for a fuller account.

Finally, Simon Norton has kindly consented to include in an appendix some of his recent ideas on the Monster-modular connection. I was unfamiliar with this work at the time of my original lecture at Arcata, but the reader will see that at the cost of yet more and bolder conjectures, many of the earlier observations can be systemized into an organic whole.

2. Modular functions. In anticipation of the work to be described later, we briefly review here the concept of a modular function as we shall be using it. The reader is referred to Ogg's article [**21**] for a fuller account.

We will be concerned with some mathematical objects which have descriptions in terms of algebra, analysis, or geometry and are related in ways which Mumford [**19**] refers to as the "amazing synthesis." Specifically, the algebraic object is a *finitely generated field extension K of $\mathbf{C}$ of transcendence degree* 1; in analysis we have the notion of a *compact Riemann surface X*; in geometry, *algebraic curves $C \subseteq \mathbf{P}^n$*. In each case there is a notion of *genus*, which in the case of the Riemann surface X can be described as the number of handles of X. We write $g = g(X)$ for the genus.

The objects of genus zero will be particularly relevant, and in fact K has genus zero if it is a *purely transcendental extension* of $\mathbf{C}$, $K = \mathbf{C}(t)$ for an indeterminate t; X has $g = 0$ if it is topologically a *sphere*; C has $g = 0$ if $C = \mathbf{P}^1$.

We shall have nothing more to say about algebraic curves and will concentrate on X and K. Given X, one recovers K as the *field of meromorphic functions on X*. The specific Riemann surfaces that arise in our theory are called *modular curves* and are quotients of the upper half-plane by suitable discrete groups. In order to state the Conway-Norton conjectures precisely, we will need to understand something of this procedure, so it will be worthwhile to repeat once again the possibly familiar construction. (Details can be found at the beginning of Shimura's text [**26**].) Thus we denote the upper half-plane as

$$\mathfrak{H} = \{z \in \mathbf{C} \mid \operatorname{im} z > 0\} \tag{2.1}$$

with $\mathrm{GL}_2^+(\mathbf{R})$ the group of 2×2 matrices over $\mathbf{R}$ of positive determinant. This group acts on $\mathfrak{H}$ in the usual way, i.e.,

$$\alpha \colon z \mapsto \frac{az+b}{cz+d}, \qquad \alpha = \begin{pmatrix} a & b \\ c & d \end{pmatrix} \in \mathrm{GL}_2^+(\mathbf{R}) \tag{2.2}$$

and $\mathrm{GL}_2^+(\mathbf{R})$ also acts on the real projective line $\mathbf{R} \cup \{\infty\}$ via

$$\alpha \colon r \mapsto \frac{ar+b}{cr+d}, \qquad \infty \mapsto \frac{a}{c}, \qquad \alpha = \begin{pmatrix} a & b \\ c & d \end{pmatrix} \tag{2.3}$$

with the usual conventions concerning the symbol ∞.

Now clearly in the action of $\mathrm{GL}_2^+(\mathbf{R})$ on $\mathfrak{H}$, the scalars act trivially. Thus one can usually focus attention on the subgroup $\mathrm{SL}_2(\mathbf{R})$. This group is naturally topologized, and we will be concerned with certain *discrete groups* $\Delta \leq \mathrm{SL}_2(\mathbf{R})$ (i.e., the topology of $\mathrm{SL}_2(\mathbf{R})$ induces a discrete topology on Δ).

The action of $\mathrm{SL}_2(\mathbf{R})$ on $\mathbf{R} \cup \{\infty\}$ is familiar: it is transitive and only $\pm I$ fix as many as three points. An element $\delta \in \Delta$ is called *parabolic* if it fixes a unique point of $\mathbf{R} \cup \{\infty\}$, in which case the fixed-point is called a *cusp* of Δ. For example, if δ is a translation $\delta = \left(\begin{smallmatrix} 1 & x \\ 0 & 1 \end{smallmatrix}\right)$, $x \neq 0$, then δ is parabolic and the cusp is ∞. We set

$$\mathfrak{H}^* = \mathfrak{H} \cup \{\text{cusps of } \Delta\}, \tag{2.4}$$

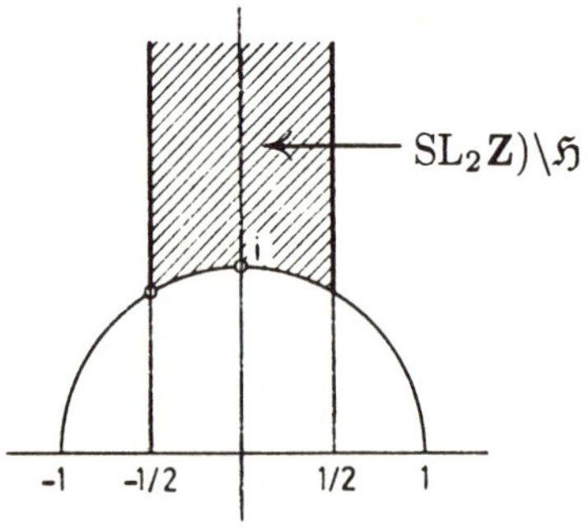

FIGURE 1

a set which depends on Δ and admits an action of Δ. It is a fact that in this situation the orbit space

$$\Delta\backslash\mathfrak{H}^* \tag{2.5}$$

can be topologized and endowed with a complex structure so that it becomes a *locally compact Riemann surface.*

The case $\Delta = \mathrm{SL}_2(\mathbf{Z})$ is a familiar and important example: the cusps of $\mathrm{SL}_2(\mathbf{Z})$ are the rationals $Q \cup \{\infty\}$, on which $\mathrm{SL}_2(\mathbf{Z})$ is transitive; thus we can take $\{\infty\}$ as the representative of $\mathrm{SL}_2(\mathbf{Z})\backslash Q \cup \{\infty\}$, in which case $\Delta\backslash\mathfrak{H}^* = (\mathrm{SL}_2\mathbf{Z})\backslash\mathfrak{H}) \cup \{\infty\}$. We can "see" that this is a sphere from Figure 1, whose appearance is obligatory in this subject.

Observe that in this case $\mathfrak{H}^*$ is compact, so that the addition of ∞ can be regarded as the one-point compactification of $\mathrm{SL}_2(\mathbf{Z})\backslash\mathfrak{H}$. Similar remarks apply to subgroups $\Delta \leq \mathrm{SL}_2(\mathbf{R})$ which are *commensurable* with $\mathrm{SL}_2(\mathbf{Z})$ in the sense that $\Delta \cap \mathrm{SL}_2(\mathbf{Z})$ has finite index in both Δ and $\mathrm{SL}_2(\mathbf{Z})$.

That is, if Δ is commensurable with $\mathrm{SL}_2(\mathbf{Z})$ then Δ is discrete, Δ has only a finite number of orbits on its cusps, and adjoining these to $\Delta\backslash\mathfrak{H}$ yields a compact Riemann surface. Unlike the case of $\mathrm{SL}_2(\mathbf{Z})$ itself, however, the genus of $\Delta\backslash\mathfrak{H}^*$ will in general not be zero.

The groups Δ which arise in the Monster-modular connection are of this type. More precisely, for an integer $N \geq 1$ let $\Gamma(N)$ be the kernel of the natural map (an epimorphism) $\mathrm{SL}_2(\mathbf{Z}) \to \mathrm{SL}_2(\mathbf{Z}/N\mathbf{Z})$ with $\Gamma(N) \leq \Gamma_0(N) \leq \mathrm{SL}_2(\mathbf{Z})$ defined as

$$\Gamma_0(N) = \left\{ \begin{pmatrix} a & b \\ c & d \end{pmatrix} \in \mathrm{SL}_2(\mathbf{Z}) \,\middle|\, c \equiv 0 \pmod{N} \right\}. \tag{2.6}$$

Now define

$$\Lambda(N) = \text{normalizer of } \Gamma_0(N) \text{ in } \mathrm{SL}_2(\mathbf{R}). \tag{2.7}$$

It is not hard to see that $|\Lambda(N) \colon \Gamma_0(N)|$ is finite, while the finiteness of $|\mathrm{SL}_2(\mathbf{Z}) \colon \Gamma(N)|$ is immediate. Thus $\Lambda(N)$ is commensurable with $\mathrm{SL}_2(\mathbf{Z})$, as indeed is any group intermediate between $\Gamma_0(N)$ and $\Lambda(N)$.

As an example, consider the *Fricke involution*

$$W_N = \begin{pmatrix} 0 & -1 \\ N & 0 \end{pmatrix} : z \mapsto -1/Nz. \tag{2.8}$$

(One must multiply by a suitable scalar to put the matrix into $\mathrm{SL}_2(\mathbf{R})$.) Evidently W_N normalizes $\Gamma_0(N)$, and by its adjunction we obtain a group

$$\Gamma_0(N)^+ = \Gamma_0(N)\langle W_N \rangle \tag{2.9}$$

(Warning: this is a slight departure from the notation of [**3**].) If N is a prime then in fact $\Gamma_0(N)^+ = \Lambda(N)$.

It is with some reluctance that I forego any further discussion of the group $\Lambda(N)/\Gamma_0(N)$, for it is here that the ubiquitious number 24 first makes its presence felt. Moreover some of the finer aspects of the Conway-Norton conjectures involve a knowledge of the structure of $\Lambda(N)/\Gamma_0(N)$, but we must refer the reader to the appropriate parts of [**3**] and [**21**] for details. Given $\Delta \leq \mathrm{SL}_2(\mathbf{R})$, commensurable with $\mathrm{SL}_2(\mathbf{Z})$, we write $X_\Delta = \Delta \backslash \mathfrak{H}^*$ and set

$$K(X_\Delta) = \{\text{meromorphic functions on } X_\Delta\}. \tag{2.10}$$

Each element $f \in K(X_\Delta)$ is a meromorphic function

$$f \colon X_\Delta \to \mathbf{C} \cup \{\infty\}$$

and $g(X_\Delta) = 0$ if, and only if, we can find an f which is 1-1 and onto, which is the same as requiring $K(X_\Delta) = \mathbf{C}(f)$. Such an f is called a generator of the function field, or a *hauptmodul.*

To explain the q-expansion of f we need to say a little about the complex structure of X_Δ. All of our discrete groups Δ will contain some $\Gamma_0(N)$; in particular they contain $\left(\begin{smallmatrix} 1 & 1 \\ 0 & 1 \end{smallmatrix}\right)$, so that we may take ∞ as a cusp. One identifies ∞ with $\lim_{y \to +\infty} iy$, so that under the map

$$q \colon z \to e^{2\pi i z}, \tag{2.11}$$

which maps $\mathfrak{H}$ onto the interior of the unit disk, ∞ is mapped to the origin. We write $q = e^{2\pi i z}$. Now in the complex structure on X_Δ we may take q as a local parameter at the point ∞. Thus any function $f \in K(X_\Delta)$, being meromorphic at each point of X_Δ, has in particular a Laurent expansion about ∞ in terms of q:

$$f(q) = \sum_{n \geq n_0} a(n) q^n, \qquad a(n_0) \neq 0. \tag{2.12}$$

If f is a hauptmodul then, being 1-1, it has a unique *pole*, and such a pole is necessarily simple. We may choose f so that the pole is at ∞, and can take the leading coefficient of the q-expansion to be 1. Thus (2.12) becomes in this case

$$f(q) = q^{-1} + \text{constant} + \sum_{n=1}^{\infty} a(n) q^n. \tag{2.13}$$

We say that $f(q)$ is *normalized* if it has the form (2.13). The constant will vary according to the location of the unique *zero* of $f(q)$, which once specified determines f uniquely. For example if $\Delta = \mathrm{SL}_2(\mathbf{Z})$ the normalized form of $f(q)$ is given by

(2.14) $$j(q) = q^{-1} + \text{constant} + 196884q + \cdots,$$

where in fact each coefficient $a(n)$ is a positive integer.

3. The Conway-Norton Moonshine conjectures. The principal conjecture of Conway-Norton concerns the Fischer-Griess Monster M. To each rational conjugacy class $\langle m^M \rangle$ there is associated a formal q-expansion

$$j_m(q) = q^{-1} + O + \sum_{n=1}^{\infty} a_m(n) q^n$$

with integer coefficients $a_m(n)$ satisfying the following:

(MI) For each $n \geq 1$ the function $H_n\colon m \mapsto a_m(n)$ is a *character* of M, called the nth *head character*.

For each m there is an integer h dividing g.c.d.$(24, o(m))$ and a discrete group Δ_m satisfying the following:

(MII) (a) $\Gamma_0(N) \leq \Delta_m \leq \Lambda(N)$ where $N = ho(m)$.

(b) $g(X_{\Delta_m}) = 0$.

(c) $K(X_{\Delta_m}) = \mathbf{C}(j_m)$.

Several remarks are in order.

(1) We could easily devise a way of associating to each m some $j_m(q)$ such that (MI) and (MII) (b), (c) held: just take $j_m(q)$ to be $j(q)$ [cf. (2.14)] for each m, in which case each H_n is a multiple of the trivial character. Thus condition (MII) (a) ensures that the association $m \to j_m(q)$ of the conjecture is nontrivial; indeed of the 172 rational conjugacy classes in M, only the two classes of order 27 have the same q-expansion.

(2) Conway-Norton specify some further properties of Δ_m, having to do with the precise structure of $\Lambda(N)/\Gamma_0(N)$. It seems that there are only 174 discrete subgroups of $\mathrm{SL}_2(\mathbf{R})$ satisfying (MII) (a)–(c) and these additional properties; 171 of them appear as above, whilst the remaining three—having $N = 25$, 49, and 50 respectively—correspond to the so-called *ghost elements*! This near bijection would have many consequences, some of which I cannot help but record for their own sakes.

(3) Corresponding to $m =$ identity we have $N = 1$, and $\Delta_m = \Gamma_0(1) = \mathrm{SL}_2(\mathbf{Z})$ and $j_m(q)$ the modular function of (2.14). Thus the first head character H_1 has degree 196, 884.

(4) With reference to (2.9), the primes p for which $\Gamma_0(p)^+$ has genus zero, i.e., $g(X_{\Gamma_0(p)^+}) = 0$, are precisely the primes p dividing $|M|$. These are the primes p with $2 \leq p \leq 31$ and $p = 41, 47, 59, 71$. Moreover for each such p there is an element $m \in M$, said to be of *type* $p+$, for which $\Delta_m = \Gamma_0(p)^+$. These

observations were made by A. Ogg, who first determined when $\Gamma_0(p)^+$ has genus 0 in [**22**].

(5) The primes p for which $\Gamma_0(p)$ itself has genus 0 are precisely those for which $p-1 \mid 24$ (this is elementary), and for each such p there is an $m \in M$, of *type* $p-$, for which $\Delta_m = \Gamma_0(p)$.

While (MI) and (MII) are perhaps the most important of the Conway-Norton conjectures, there are others of a similar type concerning other groups. Generally, in fact, there seem to be analogue for groups which are related (in various ways) to the centralizers of elements in M. In her thesis [**24**] Larissa Queen has considered some of the main examples, and we will limit ourselves to a discussion of some of the aspects of her work.

Consider first the primes p such that $p+1 \mid 24$. As we remarked in (4), for each such p the Monster has a class of elements of type $p+$, and their centralizers in M are as follows (notation for sporadic groups is standard):

p	centralizer
2	$2B$
3	$3F'_{24}$
5	$5 \times F$
7	$7 \times H$
11	$11 \times M_{12}$

Then the conjecture is that for the commutator subgroup of each of these groups the analogues of (MI) and (MII) hold with the following modification to (MII) (a):

(MII)′ (a) $\Gamma_0(N) \leq \Delta_m$ where $N = ph \cdot o(m)$ for some suitable h.

Thus Δ_m may not contain $\Gamma_0(N)$ as a normal subgroup, while the level N is now a multiple of $po(m)$ rather than $o(m)$. Moreover the function j_1 corresponding to the identity will be the normalized generator of $K(X_{\Gamma_o(p)^+})$.

In addition to 3+ and 3−, M contains a third class of elements of order 3, with centralizer $3 \times E$, and for the group E, the modified (MI) and (MII)′ conjectures should again hold. It is fair to say that these conjectures have received considerably less attention than those for M itself. Whilst this is perhaps understandable, nevertheless it is quite likely that each will shed light on the other and that a further exploration of this mini-Moonshine would prove not unprofitable; Norton's appendix perhaps gives substance to this.

Next we consider those primes p such that $p-1 \mid 24$ and the conjugacy classes of M of type $p-$ which correspond to the groups $\Gamma_0(p)$. Their centralizers in M are as follows:

p	centralizer
2	$2^{1+24}(.1)$
3	$3^{1+12}(2\,\mathrm{Sz})$
5	$5^{1+6}(2HJ)$
7	$7^{1+4}(2A_7)$
13	$13^{1+2}(\mathrm{SL}_2(3))$

For us it is not the centralizer in M which is so important; but for each such p the Conway group .0 has an element t_p of order p acting on the Leech Lattice without fixed-points. The centralizer in .0 of t_p is as follows:

p	centralizer
2	.0
3	6 Sz
5	$5 \times 2HJ$
7	$7 \times 2A_7$
13	$13 \times \mathrm{SL}_2(3)$

Denote by G_p the commutator subgroup of these centralizers. It is the groups G_p which interest us. There is a well-known connection between the groups G_p and the corresponding Monster centralizer, realized as follows: the fixed-point-free action of t_p on the Leech Lattice Λ turns Λ into a $2d$-dimensional module over $\mathbf{Z}[\exp(2\pi i/p)]$, where $24 = 2d(p-1)$. The mod p reduction yields a $2d$-dimensional $GF(p)$-space Λ_p which admits G_p. Moreover Λ_p inherits a symplectic form from the form on Λ which is also G_p-invariant, so we may lift Λ_p to an extra-special group of order p^{1+2} which admits $G_p/\mathrm{O}_p(G_p)$ as outer automorphisms.

The moonshine properties for G_p are called "Multiplicative Moonshine" by Conway-Norton, for reasons which will soon be clear. First, for $p = 3, 5, 7$ the group G_p has a pair of distinct, but algebraically conjugate, representations φ_+, φ_- of degree $2d$. For $p = 2$, $G_2 = .0$ has a unique faithful representation of degree $2d = 24$, namely that afforded by Λ, and in this case we let $\varphi_+ = \varphi_-$ be this representation. Finally, if $p = 13$ then $G_{13} = \mathrm{SL}_2(3)$ has three faithful representations of degree $2d = 2$; two are algebraically conjugate and the third is rational-valued (though not rational!). In this case we let $\varphi_+ = \varphi_-$ denote the rational-valued representation (in contradiction to that suggested by Conway-Norton in section 9 of [**3**], who prescribe that φ_+ and φ_- should be the conjugate representations). The assertions themselves are as follows:

(MM) To each rational conjugacy class $\langle g\rangle^{G_P}$ in G_p is associated the formal product

$$j_g(q) = q^{-1}\prod_{p\nmid n}\mathrm{char}_{\pm}(q^n),$$

where $\mathrm{char}_{\pm}(q) = \prod_{i=1}^{2d}(1-q\varepsilon_i)$, the ε_i's running over the eigenvalues of $\varphi_{\pm}(g)$, and where one chooses the sign in $\varphi_{\pm}(q^n)$ according as the sign of the symbol $(\frac{n}{p})$. Then one expects that each j_g is a normalized hauptmodul.

(MM) is a different kettle of fish from (MI) and (MII). First of all, having an explicit formula for $j_g(q)$ shold make it easy to verify if indeed it is a hauptmodul. Furthermore the analogue of (MI) should be obvious, namely, that the coefficients of q^n give a generalized character of G_p (generally they will not be genuine characters). Thus what is really required here (as elsewhere) is a *conceptual* explanation of the genus zero property.

Having to deal with generalized characters is disconcerting, perhaps, and it will be worth looking more closely at this.

Since $\mathrm{char}_\pm(q)$ is essentially the characteristic polynomial of $\varphi_\pm(g)$ it is easy to write down a formula for a graded, generalized module affording the character given by j_g. We will only do this explicitly for the simplest case $p = 13$. Here we find that if we let M_+ be the module affording $\varphi_+ = \varphi_-$, and set

$$V = q^{-1} \prod_{13 \nmid n} (1 - M_+ q^n + 1q^{2n}) \tag{3.1}$$

(remember that $2d = 2$) then, as a graded G_{13}-module, we have $\mathrm{char}_V(g) = j_g(q)$.

More interestingly, it is straightforward to give a genuine graded module for G_{13} which affords the character given by inverting j_g. We state this for each p:

(3.2) Let $M_\pm$ afford the representation $\varphi_\pm$, with

$$\widetilde{M_\pm} = \coprod_{(n/p)=\pm 1} M_\pm q^n$$

considered as a graded module; then the symmetric algebra (suitably graded)

$$S = qS(M_+ \oplus M_-) \tag{3.3}$$

satisfies $\mathrm{char}_S(g) = j_g(q)^{-1}$, $g \in G_p$.

We shall see this kind of module again later. But for now, we must face the fact that if we want to deal with genuine characters in this context we must accept modular functions of the form

$$j_g(q)^{-1} = q + \text{ higher powers.} \tag{3.4}$$

Thus although $j_g(q)^{-1}$ is still a generator of $K(X_{\Gamma_0(p)})$, it is not normalized in the way prescribed by (2.13). There is no apparent reason to prefer (2.13), but in fact the theory of n-tuplication—to be discussed later—suggests that (2.13) is for several reasons preferable to (3.4). Thus both normalizations have claims to ascendency and in Norton's appendix the reader will see how this antagonism can be dealt with. In any case, it would appear that the (generalized) module of (3.1), or alternatively its inverted brother (3.3), is the most elementary module which affords a "Moonshine." Why?

This completes our review of the Moonshine conjectures. We will return to some other conjectures later, but for now we turn to modular forms.

4. Modular forms. If X is a Riemann surface and $f(z)$ a holomorphic function on X then there is in general no definition of the derivative $f'(z)$, for it is not well defined with respect to a change of local variables. What *is* well defined is the *differential* $f(z)\,dz$. Now if X is also assumed compact then there are no nonconstant holomorphic functions on X; nevertheless there may well be *holomorphic differential forms* on X: to each chart with local variable z there is an assignment $f(z)\,dz$ with $f(z)$ holomorphic and such that the compatibility condition

$$f(z)\,dz = f(z_1)\,dz_1 \tag{4.1}$$

holds for overlapping charts with local variables z and z_1. Of course we can speak also of *meromorphic differentials*, where $f(z)$ is only required to be meromorphic.

It is well known that the holomorphic differentials on compact X form a vector space over $\mathbf{C}$ of dimension $g(X)$, so that genus zero corresponds to the absence of nonzero holomorphic differentials.

One can generalize all of this by considering for a nonnegative integer k those functions $f(z)$, either holomorphic or meromorphic in z, satisfying

$$f(z)(dz)^k = f(z_1)(dz_1)^k. \tag{4.2}$$

These are the differential k-forms, and for $k \geq 1$ the holomorphic k-forms form a $\mathbf{C}$-vector space of finite dimension (Riemann-Roch theorem).

Now suppose that Δ is a discrete subgroup of $\mathrm{SL}_2(\mathbf{R})$ commensurable with $\mathrm{SL}_2(\mathbf{Z})$, so that $X = \Delta\backslash\mathfrak{H}^*$ is the kind of Riemann surface we considered in §2. By pulling-back along the map $\mathfrak{H} \to \Delta\backslash\mathfrak{H}$ we find that a meromorphic 1-form ω on X corresponds to a form $f(z)\,dz$ with $f(z)$ meromorphic on $\mathfrak{H}$, with appropriate behavior at the cusps. The invariance of $f(z)\,dz$ with respect to Δ translates into

$$f(z)\,dz = f(\alpha z)(cz+d)^{-2}\,dz, \qquad \alpha = \begin{pmatrix} a & b \\ c & d \end{pmatrix} \in \Delta. \tag{4.3}$$

It turns out, because of the nature of the local parameters at the cusps, that in order for ω to be holomorphic on X, it is necessary and sufficient that the Fourier series for f in the local variable τ at a cusp κ not only have no principal part, but also have constant term equal to zero; one says that $f(\tau)$ *vanishes* at κ. Once again all of this extends to k-forms, and in order to proceed it is useful to introduce the following notation.

(4.4) For a meromorphic function $f(z)$ on $\mathfrak{H}$ and a nonnegative integer k, define

$$(f\,|_k\,\alpha)(z) = f(\alpha z)(cz+d)^{-k}(\det\alpha)^{k/2}$$

for $\alpha \in \mathrm{GL}_2^+(\mathbf{R})$.

It is readily seen that this defines a right group action of $\mathrm{GL}_2^+(\mathbf{R})$ on such functions, and preserves holomorphic functions.

We can now make the following definition:

(4.5) A *holomorphic modular form of weight k with respect to Δ* is a holomorphic function $f(z)$ on $\mathfrak{H}$ which satisfies $f|_k\alpha = f$, all $\alpha \in \Delta$ and which is "holomorphic at the cusps."

We call $f(z)$ a *cusp-form* if f *vanishes* at each cusp, that is, if in a local variable τ at a cusp κ there is a "Fourier expansion" of the form

$$f(\tau) = \sum_{n\geq 1} a_n\tau^n.$$

As a consequence of our earlier discussion, it is the cusp-forms which correspond to holomorphic differential k-forms on X.

We can replace "holomorphic" by "meromorphic" in the above and get meromorphic *modular forms*. In practice the poles of such a form will always be at the cusps, so that the function is still holomorphic on $\mathfrak{H}$.

Once again the groups Δ that we will be interested in will contain $\Gamma_0(N)$ for some N. Then as explained in §2, a form f will have an expansion at the cusp ∞ of the type $\sum_{n\geq 0} a_n q^n$, $q = e^{2\pi i z}$, $z \in \mathfrak{H}$; moreover $a_0 = 0$ if f is a cusp-form. Note also that if Δ contains the matrix $-I$ then any form f of odd weight k is necessarily zero, for application of (4.5) with $-I$ in place of α yields $f(z) = (-1)^k f(z)$.

This situation will certainly apply to the groups $\Gamma_0(N)$ (cf. (2.6)), and being unwilling to jettison forms of odd weight so lightly in this case, we generalize the situation in the following way.

Let V be the space of modular forms of weight k with respect to the group $\Gamma(N)$. Once again $\dim V$ is finite by the Riemann-Roch theorem, and as $-I \notin \Gamma(N)$ if $N \geq 2$ we may well have $V \neq 0$ for k odd. As $\Gamma(N) \trianglelefteq \mathrm{SL}_2(\mathbf{Z})$ and as (4.4) defines a group action we see that V affords a representation of $\mathrm{SL}_2(\mathbf{Z})$ with $\Gamma(N)$ contained in the kernel of the action. Then a modular form of weight k for a group Δ satisfying $\Gamma(N) \leq \Delta \leq \mathrm{SL}_2(\mathbf{Z})$ is just a vector of V which is fixed by Δ in its action on V. Thus we can generalize by taking (for example) any linear character of Δ, say ε. For the groups $\Gamma_0(N)$ the relevant linear characters are given by

$$\varepsilon\colon \begin{pmatrix} a & b \\ c & d \end{pmatrix} \mapsto d, \tag{4.6}$$

which defines a linear character $\varepsilon\colon \Gamma_0(N) \to (\mathbf{Z}/N\mathbf{Z})^*$. Thus a form $f \in V$ is a semi-invariant for $\Gamma_0(N)$ with respect to ε if it satisfies

$$f|_k \alpha = \varepsilon(d) f \quad \text{for all } \alpha \in \Gamma_0(N), \tag{4.7}$$

and we denote by $M(N, K, \varepsilon)$, resp. $S(N, K, \varepsilon)$, the space of holomorphic forms, resp. cusp-forms, f which satisfy (4.7). We say that f has *level* N, weight k, and *character* ε, and briefly refer to the triple (N, k, ε) as the *signature* of f.

We point out that the element $\gamma = \left(\begin{smallmatrix} 1 & 1 \\ 0 & 1 \end{smallmatrix}\right)$ lies not only in $\Gamma_0(N)$ but even in $\ker \varepsilon$. Thus we have $f|_k \gamma = f$ for $f \in M(N, k, \varepsilon)$, and this will ensure that the q-expansion for f will have only integral powers of q.

By looking next at some examples we will have the opportunity to introduce some of the main protagonists in our saga. In each case we will give a q-expansion and allege that it is the q-expansion of a holomorphic form; usually this will not be obvious, but we must refer the reader to the literature (e.g., [**23**, **25**, **14**]) for more details.

Suppose first that L is an *even lattice*, by which we mean that L is a free abelian group of rank d, $L \cong \mathbf{Z}^d$, equipped with a positive-definite, symmetric, integral bilinear form and an associated quadratic form f which takes only even values, $f\colon L \to 2\mathbf{Z}$. Then associated to L is its *theta-function*

$$\theta_L = \sum_{x \in L} q^{f(x)/2} \tag{4.8}$$

which can also be written as

$$\theta_L = \sum_{n=0}^{\infty} a_n q^n, \qquad a_n = |\{x \in L \mid f(x) = 2n\}|. \tag{4.9}$$

Then $\theta_L \in M(N, k, \varepsilon)$ with $k = \frac{1}{2}d$ (d is necessarily even). N and ε can be described, but we will not go into this. Note that $a_0 = 1$, so that θ_L is definitely not a cusp-form. Also, the level N will be 1 precisely when L is *unimodular*, or *self-dual*, that is $L = \mathrm{Hom}_{\mathbf{Z}}(L, \mathbf{Z})$, and this can only happen if $8 \mid d$. In fact the E_8-root lattice is self-dual. Moreover it can be shown that $M(1, 4, \mathrm{id})$ is exactly 1-dimensional. Thus

$$M(1, 4, \mathrm{id}) = \mathbf{C}\theta_{E_8}. \tag{4.10}$$

Note also, since $g(X_{\mathrm{SL}_2(\mathbf{Z})}) = 0$, that $M(1, 2, \mathrm{id}) = 0$ (cf. (4.3) and the comments preceding it); moreover as the character is necessarily trivial then $M(1, k, \mathrm{id}) = 0$ for all odd k. The first nonzero cusp-form on $X_{\mathrm{SL}_2(\mathbf{Z})}$ occurs with $k = 12$, and can be described as follows.

The *Dedekind eta-function* η is defined by the q-expansion

$$\eta(q) = q^{1/24} \prod_{n=1}^{\infty} (1 - q^n). \tag{4.11}$$

It is not a modular-form in the sense that we have defined it, but its 24th power is! One writes

$$\Delta(z) = \eta(q)^{24} = q \prod_{n=1}^{\infty} (1 - q^n)^{24} \tag{4.12}$$

and it can be shown that

$$S(1, 12, \mathrm{id}) = \mathbf{C}\Delta. \tag{4.13}$$

The occurrence of the number 24 in (4.11) and (4.12) must surely pique the interest of any simple-group theorist (I vividly recall that it did mine!), and of course no other number will work in its place. The gods beckon us on ....

A straightforward generalization of (4.12) and (4.13) is to take a positive integer multiple of 24, say m, let λ be a partition of m into parts $a_1, a_2, \ldots, a_r$, and set

$$\eta_\lambda(q) = \prod_{i=1}^{r} \eta(q_i^a). \tag{4.14}$$

As long as the number of parts r of λ is even, then η_λ will be a cusp-form of weight $r/2$.

The case (4.12) occurs when $m = 24$ and all parts of λ are equal to 1, but generally the analogue of (4.13) is false in the sense that the corresponding space of cusp-forms is of dimension greater than 1. We shall return to this issue later.

There is a further generalization of (4.14) which is of interest. We will be given a rational representation of a finite group G of determinant 1,

$$\rho\colon G \to \mathrm{SL}_n(Q) \tag{4.15}$$

with $24 \mid n$. The characteristic polynomial of $\rho(g)$ can be written in the form

$$\operatorname{char\,poly} \rho(g) = \prod_{i\geq 1}(1-t^i)^{a_i} \tag{4.16}$$

for (not necessarily positive) integers a_i, and we then set

$$\eta_g(q) = \prod_{i\geq 1}\eta(q^i)^{a_i}. \tag{4.17}$$

Then each $\eta_g(q)$ is a meromorphic modular form, the case (4.14) arising when we take $G = \mathrm{Alt}(n)$ in its usual permutation representation. Of course η_g will be holomorphic if each $a_i \geq 0$, but this is by no means necessary. The weight of η_g is equal to $\frac{1}{2}\sum a_i$, so if $\sum a_i = 0$ then certainly η_g is not holomorphic (cf. §2); note that this holds precisely when $\rho(g)$ has no eigenvalues equal to 1.

There is a rather close relationship between these theta- and eta-functions which, in our present context, can be illustrated as follows. Take L to be an even, unimodular lattice admitting G as automorphisms (that is, G preserves the associated quadratic form), and let $n = \dim L$ with ρ the representation of G afforded by $L \otimes_{\mathbf{Z}} Q$. In this situation we have not only the forms η_g but also the theta-functions θ_{L^g}, the theta-function of the fixed-sublattice L^g. We have

(4.18) Suppose that $g \in G$ acts on $L \otimes_{\mathbf{Z}} Q$ not only with determinant 1 but with spinor norm 1 when regarded as an element of $\mathrm{SO}(n, Q)$. If $24 \mid n$ then there is an integer N and a quadratic Dirichlet charcter $\varepsilon \pmod N$ such that both η_g and θ_{L^g} are forms of weight $\frac{1}{2}\sum a_i$, level N, and character ε. In particular, if η_g is holomorphic then both η_g and θ_{L^g} lie in $M(N, k, \varepsilon)$.

Finally, we may take $N = ho(g)$ with $h \mid (24, o(g))$ (cf. (MII) (a) of §3).

This result, and indeed much of the content of this section, is discussed more fully in [**15**] and [**16**]. We complete the present section by considering (4.18) in the case that L is the Leech lattice Λ and $G = .0$. In this case, G is perfect and so all of the conditions of (4.18) are satisfied. Moreover the necessary condition that $\Lambda^g \neq 0$ in order that η_g be holomorphic is also sufficient in this case (we know of no conceptual reason for this, but merely have a case-by-case analysis first done, I believe, by Atkin and reported in [**15**]). Now one of the reasons (though certainly not the only one) for studying holomorphic forms is that we obtain modular functions from the quotient of two forms of the same weight, level, and character. For example if we take the theta-function of an even unimodular lattice L of rank 24, both θ_L and Δ lie in $M(1, 12, \mathrm{id})$, and their quotient must be a modular function lying in $K(X_{\mathrm{SL}_2(\mathbf{Z})})$. It is also clear from (4.12) that Δ does not vanish on $\mathfrak{H}$, so θ_L/Δ is holomorphic on $\mathfrak{H}$; and its q-expansion being of the form $(1+\cdots)/(q+\cdots) = q^{-1}+\cdots$, θ_L/Δ must in fact be a generator of $K(X_{\mathrm{SL}_2(\mathbf{Z})})$. Choosing different L thus merely changes the constant term of the

q-expansion. For example if L is the sum of three copies of the E_8-root-lattice the constant is 744, which is the constant usually prescribed in textbooks. Taking Λ for L, the constant term is 24. This generalizes as follows:

(4.19) With respect to the Leech lattice Λ and $G = .0$, each $\theta_\Lambda g/\eta_g$ is a modular function in $K(X\Gamma_0(N))$ with N as in (4.18).

In [**3**], Conway-Norton conjectured that each of these functions appeared in the conjectures (MI), (MII) of §3, i.e., that they are normalized hauptmoduls for appropriate discrete subgroups of $\mathrm{SL}_2(\mathbf{R})$. That this is in fact false was first discovered by Koike, and the complete list of counterexamples found by K. Harada and one of his students [**9**].

5. Hecke operators. Hecke operators are linear operators which act on modular forms. They form an algebra which predates the Hecke algebras familiar to finite group-theorists, and like the latter can be described in terms of certain double cosets. We will eschew this approach, however, for we will need only a little of the extensive theory; the reader can refer to [**14**] or, for a very general treatment, [**26**], for the details.

For each integer $d \geq 1$ we define operators U_d, V_d on power series via

$$\begin{aligned} U_d&\colon \sum a_n q^n \mapsto \sum\nolimits_{d|n} a_n q^{n/d}, \\ V_d&\colon \sum a_n q^n \mapsto \sum a_n q^{nd}. \end{aligned} \tag{5.1}$$

In terms of the variable z, where as usual $q = e^{2\pi i z}$, we have

If $f(z) = \sum a_n q^n$ then

$$\begin{aligned} U_d&\colon f(z) \mapsto d^{-1} \sum\nolimits_{b=0}^{d-1} f(\tfrac{z+b}{d}), \\ V_d&\colon f(z) \mapsto f(dz). \end{aligned} \tag{5.2}$$

Now suppose that f is a modular form of level N, weight k, and character ε. For a positive integer n we define the nth Hecke operator T_n via

$$T_n f = \left(\sum_{d|n} d^{k-1} \varepsilon(d) V_d \circ U_{n/d} \right) f. \tag{5.3}$$

In particular for a prime p we have

$$T_p f = (U_p + p^{k-1}\varepsilon(p) V_p) f. \tag{5.4}$$

It is a fact that $T_n f$ is again a form of the same signature; moreover if f is holomorphic then so too is $T_n f$.

Thus the T_n span a commutative algebra of operators on both $M(N, k, \varepsilon)$ and $S(N, k, \varepsilon)$, the Hecke Algebra $\mathcal{H}$. If m and n are coprime then $T_m T_n = T_{mn}$, and in fact the relations between the T_n's is conveniently described by a formal Euler product

$$\sum_{n=1}^{\infty} \frac{T_n}{n^s} = \prod_p \left(1 - \frac{T_p}{p^s} + \frac{p^{k-1}\varepsilon(p)}{p^{2s}} \right)^{-1}. \tag{5.4}$$

We remark that if the prime p divides the level N of f then $\varepsilon(p) = 0$. Thus in this case T_p reduces to the operator U_p. In general if f is a form of level N then $U_p f$ and $V_p f$ will be forms of level pN.

To illustrate some of these ideas, consider the modular function j of (2.14). After the foregoing we know that $T_n j$ is again a modular function of j. From (5.2) we see that $T_n j$ remains holomorphic on $\mathfrak{H}$, so that it is in fact a polynomial in j, and (5.1) tells us that the q-expansion of $T_n j$ begins $q^{-n} + \cdots$, so that the polynomial is monic of degree n. An analogous discussion of the effect of the general operator T_n on other hauptmoduls has not, as far as I know, been undertaken, although it will become relevant in §7.

Let us now turn to the action of $\mathcal{H}$ on the spaces $M(N, k, \varepsilon)$ and $S(N, k, \varepsilon)$, and consider first the case that $N = 1$. There is a nondegenerate scalar product (Petersson inner product) on $S(1, k, \mathrm{id})$ with respect to which the operators T_n are hermitian. Thus $\mathcal{H}$ can be simultaneously diagonalized on $S(1, k, \mathrm{id})$, moreover the characters of $\mathcal{H}$ that arise have multiplicity one. Furthermore there is a canonical complement to $S(1, k, \mathrm{id})$ in $M(1, k, \mathrm{id})$ (the codimension is 1) which is also $\mathcal{H}$-invariant. For general N the situation is similar, though one needs the theory of Atkin-Lehner [**1**] for a complete analogy. In any case $\mathcal{H}$ can be diagonalized on $S(N, k, \varepsilon)$, and there is a canonical $\mathcal{H}$-invariant complement (the *Eisenstein series*) in $M(N, k, \varepsilon)$.

For a form $f \in M(N, k, \varepsilon)$ to be an eigenform of $\mathcal{H}$ has an important meaning most usefully expressed in terms of the Dirichlet series attached to f. Thus if f has q-expansion

$$f(z) = \sum_{n \geq 0} a_n q^n$$

the associated Dirichlet series is

$$L_f(s) = \sum_{n \geq 1} \frac{a_n}{n^s}.$$

(5.5) Suppose that f is a nonzero eigenform of $\mathcal{H}$ lying in $S(N, k, \varepsilon)$ as above. Then the leading coefficient a_1 is nonzero. Assume that the q-expansion is normalized so that $a_1 = 1$. Then f is an eigenvector for T_n with eigenvalue a_n, that is

$$T_n f = a_n f.$$

Moreover this is equivalent to $L_f(s)$ having an Euler product representation of the type

$$\sum_{n \geq 1} \frac{a_n}{n^s} = \prod_p \left(1 - \frac{a_p}{p^s} + \frac{\varepsilon(p) p^{k-1}}{p^{2s}}\right)^{-1}$$

The latter conclusion follows rather easily from (5.4). Remark also that nonzero eigenforms for non cusp-forms are even easier to describe, but we will not need this.

To illustrate, (4.13) says that $\dim S(1, 12, \mathrm{id}) = 1$, so that all nonzero elements of $S(1, 12, \mathrm{id})$ are of necessity eigenforms for $\mathcal{H}$. Evidently Δ is normalized properly, so application of (5.5) in this case leads to the following:

If $\Delta(z) = \sum_{n\geq 1} \tau(n) q^n$ then

$$\sum_{n\geq 1} \frac{\tau(n)}{n^s} = \prod_p \left(1 - \frac{\tau(p)}{p^s} + \frac{p^{\prime\prime}}{p^{2s}}\right)^{-1} \tag{5.6}$$

In proving (5.6) we have followed the path of Hecke (and even, in his own way, Mordell) in proving what was once a conjecture of Ramanujan.

There are relatively few eta-products of the type (4.14) or (4.17) which are eigenforms for $\mathcal{H}$. This situation was investigated in [**17**] and in several subsequent papers [**13**, **18**]. Note that if $\eta_g(q)$ of (4.17) is an eigenform for $\mathcal{H}$ its q-expansion must be $q + \cdots$ by (5.5), and this means that $n = 24$. This is certainly not sufficient in general, even if $\eta_g(q)$ is holomorphic. On the other hand for the Leech lattice this is essentially all that is required.

(5.7) Let the representation ρ of (4.15) be that of .0 afforded by the Leech lattice. Then one of the following holds for $g \in .0$:

(i) $\rho(g)$ has no eigenvalue 1 and η_g has weight zero.

(ii) Each $a_i \geq 0$ in (4.16) and η_g is a cusp-form and an eigenform for $\mathcal{H}$.

(iii) $\rho(g)$ has an eigenvalue equal to 1 and some $a_i < 0$. In this case η_g is an Eisenstein series (cf. comments before (5.5)) and an eigenfunction for all operators T_p, p prime, with the exception of at most one prime p. This exceptional prime exists if, and only if, $\rho(g)$ has order divisible by p but no eigenvalue which is a primitive pth root of unity.

We remark that if this exceptional prime p exists then the Dirichlet series $L_{\eta_g}(s)$ will still have an Euler product but that the Euler p-factor will not look like that of (5.6). There is no conceptual proof of (5.7), only a case-by-case analysis due essentially to Atkin and reported on at length in [**15**].

6. Holomorphic and other Thompson series. We can now refine our notion of Thompson series given in the introduction. For a finite group G with ring of virtual characters RG, we define a *holomorphic Thompson series* to be a formal q-expansion.

$$\Gamma_G = \sum_{n\geq 0} \gamma_n q^n \tag{6.1}$$

with each $\gamma_n \in RG$ and with the property that

$$\Gamma_g \in M(N(g), k(g), \varepsilon_g); \tag{6.2}$$

that is, Γ_g is a holomorphic modular form with a signature which depends on g. If each Γ_g is a cusp-form, or if Γ_g is only required to be meromorphic, we may call Γ_G a cuspidal or a meromorphic Thompson series respectively. Thus the examples in §3 were of meromorphic Thompson series in which each Γ_g has weight 0.

Now if f, g are two holomorphic forms of the same signature and $g \neq 0$ then f/g is a meromorphic modular function (weight 0, trivial character). In the same spirit, if Δ_G and Γ_G are two holomorphic Thompson series such that each pair Δ_g and Γ_g have the same signature and $\Gamma_g \neq 0$, then the Thompson series $\Pi_G = \Delta_G/\Gamma_G$ will have each of the corresponding quotients $\Pi_g = \Delta_g/\Gamma_g$ of weight zero. It was with this in mind that Thompson made a first conjecture in **[27]** concerning an explanation of the Moonshine phenomena of §3. Although Thompson's idea appears to have been discarded, its spirit underlies much of this section: how can one construct essentially distinct Thompson series Δ_G, Γ_G with the same signature (in the obvious sense that each pair Δ_g, Γ_g have the same signature)?

We start by providing two important examples of holomorphic Thompson series, which will be G-analogues of (4.8) and (4.17) respectively. First let G act on the even lattice L which we take to be unimodular for simplicity. Define

$$\Theta_G = \sum_{n\geq 0} \alpha_n q^n \tag{6.3}$$

in which α_n is the permutation representation of G realized by its action on the G-set $L_n = \{x \in L \mid f(x) = 2n\}$ (f is the quadratic form). From (4.9) it follows that $\Theta_g = \theta_{L^g}$, so Θ_G is a Thompson series. For the second example, let the notation be as in (4.15)–(4.17) with $24 \mid n$.

(6.4) There is a Thompson series Ω_G such that we have

$$\Omega_g = \eta_g(q), \qquad g \in G.$$

See **[16]**. Note that Ω_G may not be a holomorphic Thompson series. In any case we can now employ (4.18) to help answer a question posed above.

(6.5) Let the notation be as in (4.18) and suppose that $24 \mid n$ and that G acts on $L \otimes_{\mathbf{Z}} Q$ with determinant and spinor norm 1. Also for each $g \in G$ let $N(g)$ be the N of (4.18). With this choice, the Thompson series Θ_G and Ω_G have the same signature.

Because of the product representation (4.11), none of the forms η_g are zero; indeed they do not vanish on $\mathfrak{H}$. Thus we can form the quotient Θ_G/Ω_G to obtain a meromorphic Thompson series such that each of the associated forms $\theta_{L^g}|\eta_g$ is a function of weight zero and trivial character. This Thompson series has the shape $q^{-n/24}+$ higher powers, so can only be a hauptmodul if $n = 24$. Had the conjecture of Conway-Norton, that each $\theta_{\Lambda^g}|\eta_g$ is a hauptmodul in case $L = \Lambda$ is the Leech lattice, been true, this would have supplied a "Moonshine" for .0. As it is, we do not have a "Moonshine" (cf. the last paragraph of §4), only a very pleasant Thompson series.

One of the major difficulties concerning the conjectures (MI), (MII) of §3 is the construction of the graded modules involved. (Though as we saw, it is easy for conjecture (MM).) We want to say more about this aspect of the theory now, for it sheds some light on the origin of Thompson series. For more details see [**16**]. Consider first the Thompson series (6.3). The characters α_n are permutation modules with basis $L_n = \{x \in L/f(x) = 2n\}$. It is then clear that the graded module for G underlying Θ_G is the (complex) group algebra $\mathbf{C}[L]$ of L.

To understand the Thompson series Ω_G, consider first the series $(1-q)^{-1} = 1 + q + q^2 + \cdots$. This is the formal character of the symmetric algebra $S(V)$ where V is 1-dimensional. Note that $S(V)$ is naturally graded when V is given degree 1, and we prefer to write $S(Vq)$ to emphasize this fact. Because of the identity $S(A \otimes B) = S(A) \otimes S(B)$ it follows that

(6.6) $\prod_{n=1}^{\infty}(1-q^n)^{-1}$ is the formal character of the symmetric algebra on $Vq + Vq^2 + Vq^3 + \cdots$.

Because of the formal identity

$$S(Vq) \bigwedge(-Vq) = 1$$

($\bigwedge$ denotes exterior algebra) we conclude

(6.7) $\prod_{n=1}^{\infty}(1-q^n)$ is the formal character of the exterior algebra on $-(Vq + Vq^2 + \cdots)$.

Moreover,

(6.8) Let the assumptions and notation be as in (4.15)–(4.18), and let V afford the representation ρ. Then the character of g on $\bigwedge(-(Vq + Vq^2 + \cdots))$ is $q^{-n/24}\eta_g(q)$.

Thus the Thompson series Ω_G is afforded by the virtual graded module $q^{n/24} \bigwedge(-Vq + Vq^2 + \cdots))$. More to the point, application of the preceding together with (4.18) yields

(6.9) Let the notation and assumptions be as in (4.18), and set $V = L \otimes_{\mathbf{Z}} Q$. Then the Thompson series Θ_G/Ω_G is afforded by the graded module $q^{-n/24}\mathbf{C}[L] \otimes S(Vq + Vq^2 + \cdots)$.

Graded modules of this type arise as representation modules for affine Lie algebras (see [**11, 4**] for example). We have also seen that if $L = \Lambda$, the Leech lattice, then $q^{-1}\mathbf{C}[\Lambda] \otimes S(Vq + Vq^2 + \cdots)$ "nearly" affords a "Moonshine" for .0. The construction of the Monster and its graded module to be described later begins with a modification of this graded module.

Though some of these themes will be taken up in the next section, we complete the present discussion with an illustration of a different kind. We have already had cause to introduce in §5 the Dirichlet series attached to a modular form, and in the same spirit we may attach to the holomorphic Thompson series (6.1) its formal Dirichlet series

$$L_{\Gamma_G}(s) = \sum_{n \geq 1} \frac{\gamma_n}{n^s}. \tag{6.10}$$

Having done this, it is but a small step to call a *Thompson L-series* a formal Dirichlet series $\sum_{n\geq 1} \gamma_n/n^s$, $\gamma_n \in RG$ with the property that for $g \in G$, the series $\sum_{n\geq 1}\gamma_n(g)/n^s$ is an L-series. We will not dally on the question of exactly what kind of L-series we have in mind, but simply give an example which we find provocative.

Let A be a finite abelian group presented as a Galois group of a normal extension E of Q, and let χ be a (linear) character of A. Attached to this data is the Artin L-series defined as

$$L(s,\chi)=\prod_p(1-\chi(\sigma_p)/p^s)^{-1} \tag{6.11}$$

where σ_p is the Frobenius element at p and the product runs over the unramified primes. Now we can identify σ_p with an element f_p of $\hat{\hat{A}}$ (the double dual of A), in the usual way, namely, $f_p(\chi)=\chi(\sigma_p)$ for $\chi\in\hat{A}$. Writing $f_n=\prod f_{p_i}^{e_i}$ if $n=\prod p_i^{e_i}$ is the prime power decomposition of n when n is not divisible by a ramified prime and $f_n=0$ otherwise, (6.11) can be written in the form

$$\sum_{n\geq 1}\frac{f_n(\chi)}{n^s}, \tag{6.12}$$

which says that $\sum f_n/n^s$ is a Thompson L-series for the group $\hat{A}$.

7. The action of Hecke operators on Thompson series. In this section we address the question of the extent to which the theory of §5 can be reproduced for Thompson series. Once more, for the situation concerning the Monster and its associated hauptmoduls we have only a conjecture, whereas for holomorphic Thompson series the theory is in better shape. A common feature of both is the surprising role that Adams operations appear to have.

Recall that for a virtual character χ of the finite group G and for a positive integer n, the nth Adams operation ψ^n operates on χ via the formula

$$(\psi^n\chi)(g)=\chi(g^n),\qquad g\in G. \tag{7.1}$$

Then $\psi^n\chi$ is again a virtual character of G. For a Thompson series $\Gamma_G=\sum\gamma_m q^m$ we define

$$\psi^n\Gamma_G=\sum_m\psi^n\gamma_m q^m. \tag{7.2}$$

Now suppose that $\Gamma=\Gamma_M=q^{-1}+O+\sum_{n\geq 1}H_nq^n$ is the conjectured Thompson series for the Monster group M, normalized to have constant term 0. We want to be able to define a Hecke operator T_n on Γ for each $n\geq 1$. As discussed in the paragraph following (5.4), T_nj is a polynomial of degree n in j, and if Γ is really a hauptmodul "writ large," then something analogous should hold for $T_n\Gamma$ as well. The remarkable conjectured solution of this problem, due to Conway-Norton **[3]**, is as follows; for any Thompson Series Δ_G for a group G we define the operator $\tilde{T}_n$ as follows:

$$\tilde{T}_n\Delta_G=\left(\sum_{d|n}d^{-1}\psi^dV_d\circ U_{n/d}\right)\Delta_G. \tag{7.3}$$

The reader should compare this formula with (5.3). The conjecture is the following (notation as above):

(HOM) For each $n \geq 1$ there is a polynomial $P_n(\Gamma)$ in Γ of degree n, coefficients in Q, and leading term $n^{-1}\Gamma^n$, such that $\tilde{T}_n\Gamma = P_n(\Gamma)$.

It is essentially these formulae which are called *replication formulae* in [**3**], that for n being referred to as *n-tuplication.* Each such formula yields, of course, an infinity of identities between the head characters and their transforms under the Adams operations.

We note in passing that because of the complete multiplicativity of the Adams operations the relations among the operators $\tilde{T}_n$ are also given by an Euler product

$$\sum_{n\geq 1} \frac{\tilde{T}_n}{n^s} = \prod_p \left(1 - \frac{\tilde{T}_p}{p^s} + \frac{p^{-1}\psi^p}{p^{2s}}\right)^{-1} \tag{7.4}$$

cf. equation (5.4)

Now the equation in (HOM) remains true if we evaluate each side at an element $g \in M$. If $o(g)$ is coprime to n then because the head characters H_m of M are rational we have $\psi^n\Gamma_g = \Gamma_g$. Thus in this case we see that $(\tilde{T}_n\Gamma)_g$ coincides with the action of the usual Hecke operator T_n on Γ_g, so that in this case (HOM) predicts the precise polynomial in Γ_g equal to $T_n\Gamma_g$ (cf. the discussion following (5.4)). On the other hand if n is a prime p and $p \mid o(g)$ then (HOM) predicts that $U_p\Gamma_g + p^{-1}V_p\Gamma_{g^p} = P_p(\Gamma_g)$, and here Γ_g is definitely distinct from Γ_{g^p}. In this case T_p coincides with U_p, so that it is here that we depart from the usual Hecke operator and make greater use of the multiplication within M. There is also a version of (HOM) for the other conjectured "Moonshine" modules for which we refer the reader to [**3**]. In [**20**], Norton has given another setting for the polynomials $P_n(\Gamma)$. With Γ as above, we write $\Gamma(z)$ to emphasize the dependence on z (remember $q = e^{2\pi i z}$) and to distinguish it from $\Gamma(y)$ where y is a second independent variable in $\mathfrak{H}$ and $r = e^{2\pi i y}$. Norton considers the function

$$F(y, z) = \log(\Gamma(y) - \Gamma(z)) \tag{7.4}$$

which he calls the *bivarial transform* of Γ. He then asserts the following:

$$-\log r - F(y, z) = \sum_{n\geq 1} P_n(z)r^n. \tag{BM}$$

Here we have written $P_n(z)$ for $P_n(\Gamma(z))$. It is not hard to see that, as a power series in q and r, the coefficient of r^n of the l.h.s. of (BM) is a polynomial in $\Gamma(z)$, and (BM) says that it is exactly the polynomial P_n of (HOM)! Of course the polynomials P_n can be constructed in this fashion for any power series $\Gamma(z)$, independent of their being related to hauptmoduls.

We write

$$P_n(z) = n^{-1}q^{-n} + \sum_{m,n\geq 1} H_{m,n}q^m \tag{7.5}$$

so that

$$F(y,z) = \log(r^{-1} - q^{-1}) - \sum_{m,n\geq 1} H_{m,n} q^m r^n. \tag{7.6}$$

Norton proposes the existence of a bigraded algebra based on $\coprod_{m,n} V_{m,n}$ where $V_{m,n}$ affords $H_{m,n}$.

We turn now to Hecke operators in the context of holomorphic Thompson series. We will again be working in the environment described by (4.18), that is to say, L is an even, unimodular lattice admitting G as automorphisms, with ρ the representation of G afforded by $L \otimes_{\mathbf{Z}} Q$. We assume that G acts on $L \otimes_{\mathbf{Z}} Q$ with determinant and spinor norm 1.

We denote by $M(\rho)$ the space of holomorphic Thompson series which have the same signature as Θ_G (cf. (6.3)), and in case $24 \mid \dim L$ let $M^*(\rho)$ be the space of holomorphic Thompson series with the same signature as Ω_G (cf. (6.4)). By taking levels as described in (6.5), that result tells us that $M^*(\rho)$ also contains Θ_G.

Now each of the spaces $M(\rho)$, $M^*(\rho)$ are finite-dimensional, and for each $g \in G$ we have a linear transformation

$$e_g \colon M^*(\rho) \to M(N(g), k(g), \varepsilon_g) \tag{7.7}$$

given by evaluation at g of the Thompson series in $M^*(\rho)$. Rather than attempting to modify the definition of T_n as we did above, we try for a direct analogue—that is to say, a lift of T_n. Thus a Hecke operator T_n on $M^*(\rho)$ will be a linear transformation in which the following diagram commutes for each $g \in G$:

$$\begin{array}{ccc} M^*(\rho) & \xrightarrow{T_n} & M^*(\rho) \\ {\scriptstyle e_g}\downarrow & & \downarrow{\scriptstyle e_g} \\ M(N,k,\varepsilon) & \xrightarrow[T_n]{} & M(N,k,\varepsilon) \end{array} \tag{7.8}$$

There is no reason to believe a priori that such a lift exists, and indeed it doesn't. But remarkably, it exists under some rather mild additional conditions on G and the representation ρ.

We will concentrate on the existence of the lifts of T_p for p a prime. With reference to (5.4), this will be defined on Thompson series $\Gamma \in M^*(\rho)$ via

$$T_p\Gamma = (U_p + \psi_p V_p)\Gamma, \tag{7.9}$$

where ψ_p is a class-function on G given by

$$\psi_p(g) = p^{k(g)-1}\varepsilon_g(p), \tag{7.10}$$

where $(N(g), k(g), \varepsilon_g)$ is the signature of Γ_g. The lift will exist precisely when ψ_p is a virtual character of G.

Note the following properties of ψ_p: if $\dim L = 24n$ then $\psi_p(1) = p^{12n-1}$, and ψ_p vanishes on p-singular elements of G. We will need the following definition.

(7.11) Call $g \in G$ *weakly p-singular* (with respect to ρ) if $\rho(g)$ has order divisible by p but has no eigenvalue which is a primitive pth root of unity. Then we have

(7.12) Suppose that ρ can be lifted to a map of G into $\mathrm{Spin}(24n, \mathbf{R})$,

$$\begin{array}{ccc} & & \mathrm{Spin}(24n,\mathbf{R}) \\ & \nearrow & \downarrow \\ G & \xrightarrow[\rho]{} & \mathrm{SO}(24n,\mathbf{R}) \end{array}$$

and that $\rho(G)$ has no weakly 2-singular elements. Then $\psi_2 \in RG$; indeed ψ_2 is the restriction to G of the half-spin character of $\mathrm{Spin}(24n, \mathbf{R})$ afforded by the even Clifford algebra.

The situation for p odd is more complicated and proceeds as follows. The inclusion $\mathrm{Spin}(8n) \to \mathrm{Spin}(8n+1)$ induces a *monomorphism* of *real* representation rings $RO\,\mathrm{Spin}(8n+1) \to RO\,\mathrm{Spin}(8n)$. Then $RO\,\mathrm{Spin}(8n)$ is a free module over the ring $RO\,\mathrm{Spin}(8n+1)$ with two generators 1, σ where σ is the character afforded by the even Clifford algebra. We can define Adams operations ψ^p on $RO\,\mathrm{Spin}(8n)$, in which case we have a formula of the type

$$\psi^p \sigma = \theta_p \sigma + \alpha \tag{7.13}$$

for some θ_p, $\alpha \in RO\,\mathrm{Spin}(8n+1)$. The element θ_p is called the (oriented) *Bott cannibalistic class*, and we refer the reader to [**2**], [**10**], and [**28**] for more of its properties and its meaning in topology. What we need is the fact that

(7.14) θ_p is a virtual character of $\mathrm{SO}(8n)$ whose restriction to $\mathrm{SO}(8n-1)$ is of the form

$$\theta_p = p\lambda_p, \qquad \lambda_p \in R\,\mathrm{SO}(8n-1)$$

We then have

(7.15) For p odd, suppose that $\rho(G)$ has no weakly p-singular elements. If G fixes a nonzero vector of $L \otimes_{\mathbf{Z}} Q$, then $\psi_p \in RG$; indeed we can regard G as a subgroup of $\mathrm{SO}(24n-1)$ in which case ψ_p is the restriction of λ_p to G.

For a fuller discussion of (7.12)–(7.15) which include proofs, see [**15**]. We remark that we do not need the lattice L for these proofs, only the rational orthogonal G-module which it provides. These results give strong sufficient conditions for the operators T_p to exist, and the T_n may be defined in an analogous way. When they do exist, they span a commutative algebra $\mathcal{H}$ of operators on $M^*(\rho)$ in a way entirely analogous to the situation discussed in §5.

We give only a single illustration of these ideas. The reader will find more in [**15**]. Thus assume that G acts on L and satisfies the following: (a) $\mathrm{O}^2(G) = G$, (b) G has trivial Schur multiplier, (c) L affords a permutation representation of G. Then (a) assures us that G acts with determinant and spinor norm 1, while (c) ensures that no element of $\rho(G)$ is weakly p-singular for any prime. (c) also tells us that G fixes a nonzero vector of L, so that the lift of T_p exists for all odd p by (7.15). Also T_2 can be lifted because of (b) and (7.12).

An important situation where (a)–(c) hold is that of the large Mathieu group M_{24} acting on the Leech lattice Λ. In this case we have $\Omega_G \in M^*(\rho)$ since each appropriate eta-product is holomorphic. In fact each such product is a cusp-form so we may write $\Omega_G \in S(\rho)$ in an obvious extension of the usual notation. The following analogue of (4.13) holds (cf. [**17**]):

(7.16) If $G = M_{24}$ with ρ the representation afforded by Λ, then $S(\rho) = \mathbf{C}\Omega_G$.

Essentially by construction, the theory of (5.5)–(5.6) goes through for the lifted T_n's. So we get

(7.17) With $G = M_{24}$ and $\Omega_G = \sum_{n\geq 1} \alpha_n q^n$, we have

$$\sum_{n=1}^{\infty} \frac{\alpha_n}{n^s} = \prod_p \left(1 - \frac{\alpha_p}{p^s} + \frac{\psi_p}{p^{2s}}\right)^{-1}.$$

Of course evaluating at the identity yields (5.6), and more generally applying e_g yields the Euler product for η_g implicit in (5.7) (ii).

In general the T_n's, when they can be lifted, satisfy the usual Euler product relations

$$\sum_{n=1}^{\infty} \frac{T_n}{n^s} = \prod_p \left(1 - \frac{T_p}{p^s} + \frac{\psi_p}{p^{2s}}\right)^{-1}.$$

Another use of the lifted T_n's is to produce new Thompson series from old (cf. the second paragraph of §6), but we must terminate our discussion of this aspect of the theory.

8. The module $V^\natural$. We come now to the work of Frenkel-Lepowsky-Meurman. They construct a graded module $V^\natural$ for the Monster M of the form

$$V = V_{-1} \oplus 0 \oplus V_1 \oplus V_2 + \cdots \tag{8.1}$$

with each V_i a $\mathbf{C}M$-module and such that

$$\mathrm{char}_V(q) := \sum_{i\geq -1} \dim V_i q^i = J(q), \tag{8.2}$$

where $J(q)$ denotes the hauptmodul for $\mathrm{SL}_2(\mathbf{Z})$ normalized to have constant term zero. Furthermore one can compute formulae for the characters on V of elements in $C_M(t)$ where t is an involution of type 2– (cf. §3), and these appear to be hauptmoduls for appropriate discrete subgroups of $\mathrm{SL}_2(\mathbf{R})$ as predicted by the conjecture (MI, II), all normalized to have constant term 0. The reader can find thorough discussions of this work in [**4**] and [**5**]. As we intimated in the introduction, the present discussion will only deal with some of the many ideas which F-L-M have brought to bear on these issues.

Let us denote by C the centralizer $C_M(t)$ mentioned above. Already in [**3**] one can find a conjectural formula for the characters of elements of C on the conjectural Moonshine module of §3. This led Kac in [**12**] to construct a graded module for C which gave the Conway-Norton formulae, but although several ideas in this paper are relevant to the construction of $V^\natural$, nevertheless it has one serious drawback. The graded C-module W which Kac constructed had constant term W_0 equal to an irreducible C-module of degree $2^{12} \cdot 24$. Now W_0 cannot be extended to a representation of M; consequently W cannot be so extended either. One solution is to divide out by W_0 first, but then the naturality of the construction of W is lost and to date this idea has not been made to work. We therefore must go back to the drawing board and discover, as F-L-M did, a *natural* graded module for C which has character equal to $J(q)$ and *trivial* constant term. Recall that $O_2(C)$ is an extra-special 2-group of order 2^{25} with center $\langle t\rangle$. This group has a unique faithful irreducible representation of degree 2^{12}. Now let $X = X_{-1} \oplus 0 \oplus X_1 \oplus X_2 \oplus \cdots$ be the conjectured Moonshine module for M, and let $X_\pm$ be the (± 1)-eigenspaces for t on X. The space X_- must be a sum of faithful irreducible $O_2(C)$-modules; hence the degree of each space $(X_n)_- = X_n \cap X_-$ is divisible by 2^{12}. Let us provisionally write

$$\sum_{n=-1}^{\infty} \dim(X_n)_- q^n = 2^{12} a(q). \tag{8.3}$$

The first few coefficients of $a(q)$ can be computed from the tables in [**3**] giving the values of the head characters of M (a small calculator is useful!). We find that $a(q) = 24q + 2624q^2 + \cdots$ and being alerted to the prevalence of eta-type functions in these matters we observe that the coefficients of $a(q)$ are the *even* powers of q occurring in the power series

$$b(q) = q \prod_{n=1}^{\infty} (1+q^n)^{24} = q + 24q^2 + 300q^3 + 2624q^4 + \cdots . \tag{8.4}$$

In terms of the modular form $\Delta(z)$ (cf. (4.12)) we can write

$$a(q) = U_2\left(\frac{\Delta(2z)}{\Delta(z)}\right) = U_2 b(q). \tag{8.5}$$

There is also a natural graded vector space whose character is $a(q)$, as indeed there must be if X is to exist and satisfy (8.3). More precisely, if H is a 24-dimensional vector space consider the symmetric algebra

$$S(Hq^{1/2} + Hq + Hq^{3/2} + \cdots)$$

(cf. (6.6)). The involution $\theta\colon h \mapsto -h$ for $h \in H$ extends to an involution of the symmetric algebra and we have

(8.6) $a(q)$ is the character of the (-1)-eigenspace of θ on

$$S(Hq^{1/2} + Hq + Hq^{3/2} + \cdots).$$

Note that as yet we have not shown how C acts on the module of (8.6). When we consider this below the introduction of θ will appear much more natural, but it should be clear from the foregoing that somehow θ and t are to be identified as they both are to act as -1 on the module of (8.6). Having arrived at this point, we now look for a space on which θ acts trivially! The answer is to go back to our first "fake" Moonshine module in (6.9), that is to say, $q^{-1}\mathbf{C}[\Lambda]\otimes S(Hq+Hq^2+\cdots)$, where Λ is the Leech lattice and $H=\Lambda\otimes\mathbf{C}$. (We did not specify H before now beyond giving its dimension, but from now on it is $\Lambda\otimes\mathbf{C}$.)

We consider θ as being induced from the analogous involution on Λ in which case θ acts naturally on $C[\Lambda]$ and we have

(8.7) The character of the $(+1)$ eigenspace of θ on $q^{-1}\mathbf{C}[\Lambda]\otimes S(Hq+Hq^2+\cdots)$ is equal to

$$\frac{1}{2}\left\{\frac{\theta_\Lambda(z)}{\Delta(z)}+\frac{\Delta(z)}{\Delta(2z)}\right\}.$$

Our prayers are answered with

$$\frac{1}{2}\left\{\frac{\theta_\Lambda(z)}{\Delta(z)}+\frac{\Delta(z)}{\Delta(2z)}\right\}+2^{12}U_2\left(\frac{\Delta(2z)}{\Delta(z)}\right)=J(q). \tag{8.8}$$

So let T be the module affording the faithful irreducible $\mathrm{O}_2(C)$ character, and set

$$\begin{aligned} V_\Lambda &= q^{-1}\mathbf{C}[\Lambda]\otimes S(Hq+Hq^2+\cdots),\\ V'_\Lambda &= T\otimes S(Hq^{1/2}+Hq+Hq^{3/2}+\cdots), \end{aligned} \tag{8.9}$$

and use superscripts ±1 to indicate (±1)-eigenspaces of θ on these spaces (θ acts trivially on T).

Finally set

$$V^\natural = V_\Lambda^+\oplus V_\Lambda'^-. \tag{8.10}$$

Then our discussion shows that

(8.11) As a graded vector space, $V^\natural$ has character equal to $J(q)$.

To understand how we get C to act on V we need to understand the structure of C. This was originally stuided by Griess **[6]**, see also **[7]**. One introduces the proper 2-fold cover $\hat\Lambda$ of Λ defined by a bilinear 2-co-cycle

$$\varepsilon_0\colon \Lambda\times\Lambda\to GF(2), \tag{8.12}$$

which corresponds to the commutation rule

$$[e_\alpha,e_\beta]=(-1)^{\langle\alpha,\beta\rangle}\quad\text{for } \alpha,\beta\in\Lambda. \tag{8.13}$$

Here we have used $\langle\ ,\ \rangle$ for the bilinear form on Λ and used $\pm e_\alpha$ $(\alpha\in\Lambda)$ for elements of $\hat\Lambda$ in order to distinguish them from elements of Λ. (Thus $e_{-\alpha}$ is not to be confused with $-e_\alpha$.) In terms of ε_0 we have

$$e_\alpha e_\beta=(-1)^{\varepsilon_0(\alpha,\beta)}. \tag{8.14}$$

Setting $K=\{e_{2\alpha}\mid\alpha\in\Lambda\}$, we see from (8.13) that K is a central subgroup of $\hat\Lambda$ naturally isomorphic to 2Λ, and $\hat\Lambda/K$ is extra-special of order 2^{25}. It turns out that the group of automorphisms .0 of Λ can be lifted to automorphisms of $\hat\Lambda$ in the following sense:

(8.15) If $C_0=\mathrm{Aut}(\hat\Lambda)$ then C_0 is a (nonsplit) extension of $\hat\Lambda/Z(\hat\Lambda)=\hat\Lambda/\pm K$ $(\cong\Lambda/2\Lambda)$ by .0.

Now K is fixed by C_0, which therefore induces automorphisms of $\hat{\Lambda}/K$. As $\hat{\Lambda}/K \cong O_2(C)$ it follows that the representation space T of $O_2(C)$, being essentially unique, affords a *projective* representation of C_0 which lifts to a proper 2-fold cover $\hat{C}_0$ of C_0. Then $O_2(\hat{C}_0) = \langle z\rangle \times Q$ with $\langle z\rangle \cong Z_2$ and $Q \cong O_2(C)$. We may take $Q \trianglelefteq \hat{C}_0$.

With a slight abuse of notation we will set $\langle t\rangle = Z(Q)$. By construction t acts as $-I$ on T, whereas one of the other two involutions of $Z(\hat{C}_0)$, which we may take to be z, acts trivially on T. Let $y = zt$ be the third involution of $Z(\hat{C}_0)$.

Now the quotients of $\hat{C}_0$ by both $\langle z\rangle$ and $\langle y\rangle$ have the shape $2^{1+24}\cdot(.1)$, where as usual $.1 = .0/Z(.0)$. Nevertheless the two quotients are nonisomorphic; by our construction $\hat{C}_0/\langle z\rangle$ has a faithful irreducible representation of degree 2^{12} (on T), whereas the smallest such representation of $\hat{C}_0/\langle y\rangle$ is $2^{12}\cdot 24$ (arising via Clifford theory from $T\otimes\Lambda$). It is the latter group which is isomorphic to C (see [**6**]), so it is this group that we wish to represent on V. The idea is to represent the whole group $\hat{C}_0$ on $V_\Lambda \oplus V_\Lambda'$ and show that the fixed-points of y on this space coincides with V.

Dealing with V_Λ' is easy; as $\hat{C}_0/Q$ is naturally isomorphic to .0 we let it act on H in the obvious way (with Kernel Q). Together with the action on T defined above, we get an action on V_Λ'. Now note that t inverts T but acts trivially on $S(Hq^{1/2} + Hq + \cdots)$; consequently t must act on V_Λ' without nonzero fixed-points. Thus V_Λ' is the direct sum of the fixed-points of y and z. Moreover z acts on V_Λ' in the same way as θ, so that the (-1)-eigenspace of z on V_Λ'—alias $V_\Lambda'^-$—not only admits $\hat{C}_0$ since z is in its center, but admits $C \cong \hat{C}_0/\langle y\rangle$ since $V_\Lambda'^-$ is the $(+1)$-eigenspace of y.

The action on V_Λ runs as follows. We have an action of $\hat{C}_0$ on $\hat{\Lambda}$ via $\hat{C}_0/\langle t\rangle = C_0 = \mathrm{Aut}(\hat{\Lambda})$, and for any automorphism a of $\hat{\Lambda}$ there is an equation of the form

$$a(e_\alpha) = (-1)^{f_a(\alpha)}\cdot e_{a(\alpha)}, \qquad \alpha\in\Lambda. \tag{8.16}$$

Here, we have used $a(\alpha)$ to denote the natural action of a on $\Lambda \cong /Z(\hat{\Lambda})$, while of course $f_a(\alpha) = 0$ or 1 and $(-1)^{f_a(\alpha)}$ is a central group element of $\hat{\Lambda}$. In any case (8.16) induces a monomial action of C_0 on $C[\Lambda]$, namely, if we take the basis $\{e^\alpha \mid \alpha\in\Lambda\}$ of $C[\Lambda]$ we let

$$a(e^\alpha) = (-1)^{f_a(\alpha)} e^{a(\alpha)}, \qquad a\in C_0, \tag{8.17}$$

where now $(-1)^{f_a(\alpha)}$ is a *coefficient* ± 1. Thus we define $\hat{C}_0$ to act on $C[\Lambda]$ via (8.17) with t acting trivially. Together with the usual action of $\hat{C}_0$ on $S(Hq + Hq^2 + \cdots)$ we get an action of $\hat{C}_0$ on V_Λ in which t acts trivially. Again z and θ act in the same way, so that on V_Λ^+ we again get an action of $\hat{C}_0/\langle y\rangle = C$. This concludes the construction of $V^\natural$ as a C-module, which we trust will appear rather natural even without the benefit of the Kac-Moody perspective.

Given this construction it is not hard to compute the character of $g\in C$ on $V^\natural$. To do this we introduce the modified theta-function

$$\theta_a'(q) = \sum_{a(\alpha)=\alpha} (-1)^{f_a(\alpha)} q^{\langle\alpha,\alpha\rangle/2} \tag{8.18}$$

for $a \in C_0$ (or $\hat{C}_0$), where the sum runs over the fixed sublattice of a in Λ (remember that a acts on Λ via $\hat{C}_0/Q \cong .0$). The term $(-1)^{f_a(\alpha)}$ is as in (8.16). Note that if a induces an automorphism of Λ of odd order then $\theta'_a(q)$ is just the theta-function of the fixed lattice Λ^a (cf. §4).

(8.19) The character of $g \in C$ on V_Λ^+ is

$$\frac{1}{2}\left\{\frac{\theta'_g(q)}{\eta_g(q)} + \frac{\theta'_{-g}(q)}{\eta_{-g}(q)}\right\}.$$

Here, the eta-products $\eta_{\pm g}(q)$ are defined as in (4.17), with $\pm g$ identified with their images in $\hat{C}_0/Q$.

(8.20) The character of $g \in C$ on $V_\Lambda'^-$ is

$$\frac{1}{2}\left\{\frac{\mathrm{tr}_T(q)\eta_g(q)}{\eta_g(q^{1/2})} + \frac{\mathrm{tr}_T(-q)\eta_{-g}(q)}{\eta_{-g}(q^{1/2})}\right\}.$$

Beware that it is not C but $\hat{C}_0$ or $\hat{C}_0/\langle z\rangle$ that acts on the space T. But as y acts as -1 on T and the formula of (8.20) is invariant under $g \mapsto -g$, the use of the terminology $\mathrm{tr}_T(g)$ is justified.

One assumes that the sum of the terms in (8.19) and (8.20), that is, the character of g on $V^\natural$, coincides with the normalized hauptmodul j_g of conjecture (MI, II), though probably this has not been checked in every case.

Although this completes our survey of F-L-M theory, it is really only just the beginning. They go on to construct another involution σ which acts on V, and which acts on the 196,884-dimensional space V_1 in the same way as the σ of Griess's paper [**8**]. This latter result then establishes that $\langle C, \sigma\rangle \cong M$. But this completely ignores some of the main points of their work, for which we must refer the reader elsewhere. For all the brilliance of this work, it does not allow us (apparently) to compute the characters of classes of elements *outside* of C, nor does the issue of genus seem relevant. Despite this intensive tillage, there appears to be a vast acreage still waiting to be turned over and brought into production.

Proceedings of Symposia in Pure Mathematics
Volume **47** (1987)

Appendix
Generalized Moonshine

SIMON P. NORTON

In this paper we show a new way of formulating the "Moonshine" conjectures [**3**] that could lead to a clue to the explanation of the phenomenon. Nothing is proved. Notation and terminology is as in [**3**] throughout.

Let g, h be a pair of commuting elements in the Fischer-Griess Monster $M = F_1$. Our conjecture is that it is possible to associate with each such pair a modular function $F(g, h, z)$ with the following properties.

1. $F(g^a h^c, g^b h^d, z) = F(g, h, (az + b)/(cz + d))$ when $ad - bc = 1$, up to a constant factor (which will be a root of unity).

2. For any group element g and nonzero rational number f, the coefficient of $q^f = e^{2\pi i f z}$ in $F(g, h, z)$ is, as a function of h, a character of $C_M(g)$. Note that nonzero characters can occur for nonintegral f, but that generalized characters are not needed (unlike in the "Moonshine for other groups" section of [**3**]; cf. (3.3) above and the discussion which follows it).

3. Conjugation of g and h leaves the function unchanged.

4. Unless $F(g, h, z)$ is a constant function, its invariance group will be a modular group of genus zero, commensurable with the standard modular group Γ.

It is the simultaneous action of Γ on $\langle g, h\rangle$ and z that seems to suggest an explanation for the Moonshine phenomenon; but this only gives a clue to the actions of elements of Γ on z, so the reason why the full invariance group should have genus zero remains obscure. We return to this topic later.

The Moonshine functions for the Monster occur when g is 1; for other g, of course, we get the functions for $C_M(g)$. If the "up to constant factor" stipulation in condition 1 were removed, it would follow that $F(1, h, z)$ would be invariant under a congruence subgroup whose level was the order of h. This may explain why this usually happens in the Monster.

We examine the cases where g has prime order and is of type $p+$ or $p-$ (p prime); these are, respectively, typical of the cases where the invariance group of $F(1, g, z)$ contains and does not contain the Fricke element.

In the former case, it follows from condition 1 that the power series for $F(g, 1, z)$ is obtainable from that for $F(1, g, z)$ by replacing z by z/p; this means that the new interpretation of moonshine for $C_M(g)$ is virtually identical to the old one. The same happens when g is of type $p-$, except that here we must invert the function as well as replacing z by z/p. It turns out that as a result of this the coefficients become genuine characters (rather than generalized ones). Also the value at the infinite cusp becomes zero, and by applying condition 1 to negative f we find that the same is true for $F(g,h,z)$ for all h.

The behavior of $F(g,h,z)$ at the cusp m/n can be seen to be solely dependent, up to a constant factor, on the conjugacy class of $g^m h^n$; if this is not a Fricke element, the value at the cusp will be finite. There are g and h such that for all coprime m, n $g^m h^n$ is non-Fricke; it is in these cases that $F(g,h,z)$ will be a constant function.

An interesting case occurs when $\langle g,h \rangle$ is a $p+$ pure group of order p^2. Here the invariance group of $F(g,h,z)$ will have level p and be of index p in Γ; this can only happen when $p < 11$, which are precisely the cases when such groups $\langle g,h \rangle$ exist.

The replication formulae can be put into invariant form; for example the duplication formula says that $F(g^2,h,z/2) + F(g^2,gh,(z-1)/2) + F(g,h^2,2z)$ is a quadratic polynomial in $F(g,h,z)$.

If g and h do not commute, conditions 1, 3, and 4 can still be made to make sense; the first is replaced by $F(g,gh,z) = F(g,h,z+1)$ and $F(h,g^{-1},z) = F(g,h,-1/z)$ (which together are equivalent to the original condition when g and h commute), while the others are left unchanged. Under these conditions we have not been able to make sense of condition 2. In general the invariance group will not be a congruence group. If there are 2+-elements t, u, v such that $tu = g$ and $uv = h$, it is easy to see that this remains true after applying any combination of the transformations in the revised version of condition 1. Therefore, by the 6-transposition property of the Monster, after such transformations the orders of g and h will remain at most 6. It is possible to define a trivalent polyhedron such that the two transformations correspond to rotations about the centers of the faces and edges respectively, with the order of g equalling the number of sides in the corresponding face. As any trivalent polyhedron whose faces are at most hexagons will have genus 0 or 1, and the latter only when all the faces are hexagons, we can see a possible link between the Monster (as a 6-transposition group which is likely to be maximal in a strong sense) and the genus zero property. Unfortunately we have not been able to extend this link to the circumstances of the original conjecture.

References

1. A. Atkin and J. Lehner, *Hecke operators on* $\Gamma_0(m)$, Math. Ann. **185** (1970), 134–160.

2. R. Bott, *Lectures on* $K(X)$, Benjamin, New York, 1969.

3. J. Conway and S. Norton, *Monstrous moonshine*, Bull. London Math. Soc. **11** (1979), 308–339.

4. I. Frenkel, J. Lepowsky, and A. Meurman, *An* E_8*-approach to* F_1, Contemp. Math., vol. 45, Amer. Math. Soc., Providence, R. I., 1985, pp. 99–120.

5. ——, *A Moonshine module for the Monster*, Vertex Operators in Mathematics and Physics, Math. Sci. Res. Inst. Publ., vol. 3, Springer-Verlag, New York, 1985, pp. 231–273.

6. R. Griess, *The structure of the 'Monster' simple group*, Proc. Conf. on Finite Groups, Academic Press, New York, 1976, pp. 113–118.

7. ——, *The Monster and its nonassociative algebra*, Contemp. Math., vol. 45, Amer. Math. Soc., Providence, R. I., 1985, pp. 121–157.

8. ——, *The friendly giant*, Invent. Math. **69** (1982), 1–102.

9. K. Harada, private communication.

10. D.Husemoller, *Fibre bundles*, 2nd ed., Springer-Verlag, New York, 1975.

11. V. Kac, *Infinite dimensional Lie algebras*, Progr. Math., vol. 44, Birkhäuser, Boston, 1983.

12. ——, *A remark on the Conway-Norton conjecture about the "Monster" simple group*, Proc. Nat. Acad. Sci. U.S.A. **77** (1980), 5048–5049.

13. M. Koike, *On McKay's conjecture*, Nagoya Math. J. **95** (1984), 85–90.

14. S. Lang, *Introduction to modular forms*, Grundlehren Math. Wiss., vol. 222, Springer-Verlag, New York, 1976.

15. G. Mason, *Finite groups and Hecke operators*, Preprint.

16. ——, *Frame shapes and rational characters of finite groups*, J. Algebra **89** (1984), 237–246.

17. ——, *M_{24} and certain automorphic forms in*, Contemp. Math., vol. 45, Amer. Math. Soc., Providence, R. I., 1985, pp. 223–244.

18. D. Dummit, J. McKay, and H. Kisilevsky, *Multiplicative products of η-functions*, Contemp. Math., vol. 45, Amer. Math. Soc., Providence, R. I., 1985, pp. 89–98.

19. D. Mumford, *Curves and their Jacobians*, Univ. of Michigan Press, Ann Arbor, Mich., 1978.

20. S. Norton, *More on Moonshine*, Comp. Group Theory, Academic Press, London, 1984, pp. 185–193.

21. A. Ogg, *Modular functions*, Proc. Santa Cruz Conf. on Finite Groups, Proc. Sympos. Pure Math., vol. 37, Amer. Math. Soc., Providence, R. I., 1980, pp. 521–532.

22. ——, *Hyperelliptic modular curves*, Bull. Soc. Math France **102** (1974), 449–462.

23. ——, *Modular forms and Dirichlet series*, Benjamin, New York, 1969.

24. L. Queen, *Some relations between finite groups, Lie groups and modular functions*, Thesis, Cambridge Univ., 1980.

25. B. Schoeneberg, *Elliptic modular functions*, Springer-Verlag, New York, 1974.

26. G. Shimura, *Introduction to the arithmetic theory of automorphic functions*, Publ. Math. Soc. Japan, no. 11, Princeton Univ. Press, Princeton, N. J., 1971.

27. J. Thompson, *Finite groups and modular functions*, Bull. London Math. Soc. **11** (1979), 347–351.

28. T. Tom Dieck, *Transformation groups and representation theory*, Lecture Notes in Math., vol. 766, Springer-Verlag, New York, 1979.

UNIVERSITY OF CALIFORNIA AT SANTA CRUZ

UNIVERSITY OF CAMBRIDGE

Proceedings of Symposia in Pure Mathematics
Volume **47** (1987)

Representation Theory and Translation Planes

GEOFFREY MASON

1. Introduction. This paper, an expansion of the talk I delivered at Arcata on the same topic, has two separate but related themes. The first, in consonance with the general tenor of the conference, is to explain how representation theory enters into the study of translation planes. This is well known to finite geometers; nevertheless it seems fair to say that, with a few notable exceptions, the geometers have not made much use of representation theory in order to further the goals of their subject. Indeed one might ask: what are the goals of the theory of translation planes? The second of our themes is a response to this question. We introduce a class of of modules which generalizes the idea of a projective module, and which is particularly relevant to the applications of representation theory to translation planes.

Following that we suggest a program, based on the study of these modules, which might lead to some important insights into the nature of a translation plane—specifically concerning the "size" of the group of collineations of the plane. In practice, the case of a plane of characteristic 2 is easier to work with and we shall deal almost exclusively with this case in the present paper. The odd characteristic situation presents further difficulties which we shall only allude to.

I have often wished for the opportunity to present these ideas to people in representation theory, for I believe that not only can the geometers benefit from a more systematic use of representation theory in their subject, but that conversely translation planes can provide some interesting questions for those in representation theory. I am thus doubly indebted to the organizers of this conference for giving me the chancc to do so.

2. Projective planes. The next three sections include, for the sake of completeness, some standard facts from the theory. Fuller expositions can be found in Lüneburg [**4**] or Hughes and Piper [**6**].

The definition of a *projective plane* is quite familiar. We have a set of points and a set of lines (which we think of as distinguished subsets of the set of points)

1980 *Mathematics Subject Classification* (1985 *Revision*). Primary 20C99, 51E15.
Supported by a grant from the National Science Foundation.

satisfying the following axioms:

(2.1)
(a) Two distinct points lie on a unique line.
(b) Two distinct lines contain a unique common point.
(c) There is a "quadrangle," i.e., a set of four points, no three on a line.

Condition (c) is included to avoid certain "degenerate" possibilities, and indeed a set of points and lines satisfying (a), (b) but not necessarily (c) is referred to as a *possibly degenerate* plane. Such configurations arise, for example, as the fixed structure of a *collineation* (i.e., automorphism) of the plane.

We will be concerned exclusively with finite planes, in which case there is an integer n, the *order of the plane*, such that

(2.2)
(i) Each line contains $n+1$ points.
(ii) Each point is on $n+1$ lines.
(iii) There are n^2+n+1 point and also n^2+n+1 lines.

Now let α be a collineation of the projective plane Π, and denote by $F(\alpha)$ the fixed structure of α (i.e., the points and lines of Π fixed by α).

If $F(\alpha)$ is itself a nondegenerate plane (i.e., if it contains a quadrangle) we say that α is a *planar collineation*, or simply planar. Otherwise, $F(\alpha)$ is degenerate, and among such collineations the *elations* and *homologies* play a special role. These are collineations which fix a line l together with each of the points on l. Now if α fixes two distinct lines in a pointwise fashion it is easy to see that $\alpha = 1$, so if $\alpha \neq 1$ the line l is unique and is called the axis of l. One can show that in this case there is also a unique point P, the center of α, with the property that α fixes each line containing P. Then α is called an elation if $P \in l$, and a homology if $P \notin l$.

There are some critical results, due to R. Baer, concerning collineations α of Π which have order 2. Namely, in this case on knows that one of the following hold:

(2.3)
(i) n is even and α is an elation.
(ii) n is odd and α is an homology.
(iii) $n = m^2$ is a square and α is planar with $F(\alpha)$ a subplane of order m.

Now fix a line l of Π. We call l a *special line* if there are the maximal possible number of elations of Π which have axis l. Bearing in mind that the center of an elation is unique, and that in fact a nonidentity elation fixes no points not on its axis, it is easy to see that this maximal possible number (including the identity) is n^2. Evidently the set of all elations with axis l is a group T, and if l is a special line it can be shown that n is a prime power and that T is elementary abelian. Let us record these facts:

(2.4) If Π has a special line l then n is a power of a prime, $n = p^f$, and the group of elations with axis l is elementary abelian of order p^{2f}.

We complete this section by recalling the *Artin-Zorn* theorem.

(2.5) If Π has two distinct special lines then every line is special.

Moreover, we observe that the standard *Desarguesian plane*, coordinatized by the finite field $GF(p^f)$, is (up to isomorphism) the unique plane of order p^f in which every line is special.

2. Translation planes and spreads. A *translation plane* is simply a projective plane Π which has a special line l, which from now on we call *the translation axis*. After the last few paragraphs of §2 we know that as long as Π is not Desarguesian, Π has a unique translation axis. Now the group T of elations of Π with axis l (often called translations in the present context) is elementary abelian of order p^{2f} for some prime p; moreover T acts regularly on the p^{2f} points of Π not on l. These latter points are called *affine points* since they are points of the *affine plane* which results when one expunges l and the points on it and considers only the incidence structure which remains.

The notion of a *spread*, which is crucial for our purposes, arises as follows. Because of the preceding comments we can identify the affine points of Π with the elements of T, the identity of T corresponding to a distinguished affine point O. As T is elementary abelian of order p^{2f}, this identifies the affine points with elements of the $2f$-dimensional vector space V over $GF(p)$ with 0 becoming the zero vector. Furthermore, the lines of Π which contain 0 (more precisely the affine lines through 0, after removing the points on l) are seen to correspond to a certain set of f-dimensional subspaces of V. We denote this set, which evidently has $1+p^f$ elements, by $\mathcal{S}$, and call the elements of $\mathcal{S}$ *components*. Now in view of the axioms (2.1) the components of $\mathcal{S}$ *partition* V in the sense that each nonzero vector in V lies in exactly one component of $\mathcal{S}$. The $\mathcal{S}$ is then a spread (in V) in the following sense:

(3.1) Given a vector space W of even dimension $2d$ over $GF(q)$, a spread in W is a set $\mathcal{S}$ of d-dimensional subspaces of W which satisfy the following:

(i) $W = \bigcup S,\ S \in \mathcal{S}$.

(ii) If $S_1, S_2 \in \mathcal{S}$ with $S_1 \neq S_2$ then $S_1 \cap S_2 = 0$.

Our original $\mathcal{S}$ was thus a spread in the $GF(p)$-space V, but spreads in $GF(q)$-spaces, q not necessarily prime, also arise as follows: with V and $\mathcal{S}$ as above, define the *kernel* K of $(V, \mathcal{S})$ to be the set of all endomorphisms of the abelian group V which fix $\mathcal{S}$, i.e.,

(3.2) $$K = \{\kappa \in \operatorname{End}(V) \mid \kappa(S) \leq S, \text{ all } S \in \mathcal{S}\}.$$

Then K is a finite field (generally, when finiteness is ignored, K will be a division ring) also called the *kernel of* Π. Suppose that $K \cong GF(q)$.

Now the nonzero elements of K induce collineations of Π which fix each element of the translation axis l together with 0. More precisely, K^* is isomorphic to the group of homologies of Π with axis l and center 0. In any case it is clear at this point that V can be considered as a $GF(q)$-space and that the components of the spread of V are $GF(q)$-subspaces of one half the $GF(q)$-dimension of V.

We remark that all of these ideas go back to André, who also established the following result:

(3.3) Up to isomorphism, there is a bijection between
{Spreads in $2d$-dimensional $GF(q)$-space}
and
{Translation planes of order q^d with $GF(q)$ in their kernel}

4. The translation complement. We can now make precise how representation theory impinges on the theory of translation planes. Let Π be a translation plane corresponding to a spread $\mathcal{S}$ in the $GF(q)$-space V as above, with l the special line, and consider the group J of collineations of Π which fixes l (setwise, not elementwise). If Π is not Desarguesian, then l is unique, so in this case we are considering the full group of collineations of Π. Of course the group T of translations fixes l, and we have $T \trianglelefteq J$. In fact J is a split extension

$$J = T.M$$

where M is the subgroup of collineations fixing the point 0. We also identify the multiplicative group of the kernel K of Π with the group of homologies of Π with axis l and center 0, so that also $K^* \trianglelefteq M$.

We call M the *translation complement* and note from the above that

(4.1) M induces a group of K-semilinear transformations of V which preserves $\mathcal{S}$.

Now suppose that $1 \neq G = G'$ is a group of collineations of Π which fixes both l and 0. Then $G \leq M$ and G commutes with K^* since K^* is cyclic, so in this case G induces K-linear transformations of V and, naturally, still preserves $\mathcal{S}$. Of course, any subgroup of $C_M(K^*)$, the *linear translation complement*, has this property, and this leads to

(4.2) Given a (finite) group G, there is, up to isomorphism, a bijection between

$$\left\{\begin{matrix}\text{Faithful representations of } G \text{ on a } GF(q)\text{-space}\\ \text{such that } G \text{ preserves a spread}\end{matrix}\right\}$$

and

$$\left\{\begin{matrix}\text{Translation planes with } GF(q) \text{ in their kernel and}\\ \text{(a group isomorphic to) } G \text{ in their linear translation}\\ \text{complement}\end{matrix}\right\}.$$

The main problem for representation theory in this area thus becomes

(4.3) Develop a theory of $GF(q)G$-modules V such that G preserves a spread of V.

5. Examples. We start with four illustrative examples of spreads constructed from a suitable representation.

(A) Take V to be 2-dimensional over $GF(q)$ with $\mathcal{S}$ the set of 1-dimensional subspaces. In this case, of course, $\mathrm{GL}_2(q)$ acts on V and preserves $\mathcal{S}$; the plane Π is Desarguesian.

Our other examples will be in characteristic 2, in line with our later interest in this case.

(B) Let V be 4-dimensional over $GF(q)$, $q = 2^{2n+1}$, with $G \cong \mathrm{Sz}(q) = {}^2B_2(q)$. Regarding V as a $GF(q)G$-module, take

$$\mathcal{S} = \left\{ \begin{array}{c} \text{2-subspaces of } V \text{ of the form } C_V(t) \\ \text{for some involution } t \text{ of } G \end{array} \right\}.$$

A little knowledge about G and its action on V shows that $\mathcal{S}$ is a spread, yielding the *Lüneburg-Tits* planes.

(C) Let $G \cong L_2(q)$, $q = 2^{2n+1}$, with N the natural 2-dimensional $GF(q)G$-module, and take

$$V = N \otimes N^{\sigma}$$

for some nonidentity element $\sigma \in \mathrm{Gal}\,(GF(q))$. Take R to be a subgroup of G of order 3 and define a spread $\mathcal{S}$ as follows:

$$\begin{aligned} \mathcal{S} = \{&G\text{-conjugates of } C_V(R)\} \cup \{G\text{-conjugates of } [V, R]\} \\ &\cup \{G\text{-conjugates of } L\}, \end{aligned}$$

where L is a B-invariant 2-dimensional subspace of V and B is a Borel subgroup. That this is a spread requires a little work, and the resulting planes, due to *Ott* and *Schaeffer*, are related to the theory of twisted cubics.

(D) Take $G \cong L_3(2)$ and let W be either 4-dimensional indecomposable $GF(2)G$-module. Then $V = W \oplus W$ supports a G-invariant spread.

If we take W to contain a trivial G-submodule then it is clear that $C_V(G)$ is 2-dimensional and will correspond to a subplane of Π containing 7 points.

On the other hand, if there is no trivial submodule then G will fix only 0 among the affine points of Π.

We note that both possibilities for W can occur according to the following principle; if V supports a spread $\mathcal{S}$ then there is a dual spread $\mathcal{S}^*$ in V^* defined as follows: for each $S \in \mathcal{S}$, let S^* be the set of linear functionals in V^* which have S in their kernel. In the case at hand it is clear that $\mathcal{S}$ and $\mathcal{S}^*$ are nonisomorphic spreads since the corresponding G-modules are not isomorphic.

We should also mention that there are many spreads which admit $G \cong$ (P)$\mathrm{SL}_2(q)$, q even or odd, and which are contained in G-modules of the shape $N \oplus N^{\sigma 1} \oplus N^{\sigma 2} \cdots$, where N is as in (C) and the σ again lie in $\mathrm{Gal}\,(GF(q))$. The examples we have offered are a little more exotic than this, but by and large the known translation planes are quite straightforward.

The main problem here is as follows:

(5.1) Given a finite group G, is there a finite translation plane Π such that G lies in the linear translation complement of Π?

We shall restrict ourselves to quasisimple groups (perfect central extensions of simple groups) throughout the rest of the present paper, at least as far as (5.1) goes. Although *not a single* group G can presently be shown to have a negative answer to (5.1), nevertheless if we take G to be quasisimple then the only known possibilities for G are the rank 1 groups $(\mathrm{P})\mathrm{SL}_2(q)$ and $\mathrm{Sz}(q)$ (moreover example (B) and its dual are the only known examples where $\mathrm{Sz}(q)$ occurs). Thus we may ask

(5.2) Let G be a quasisimple group other than $(\mathrm{P})\mathrm{SL}_2(q)$ or $\mathrm{Sz}(q)$. Is there a translation plane Π such that G lies in the translation complement of Π?

Roughly, we are asking if a (proper) translation plane is a "rank 1" object. In the remainder of the paper we suggest a program for an initial attack on (5.2) based on representation theory.

6. Dispersive versus projective modules . Until further notice we will assume that we are working with modules over fields of characteristic 2.

Suppose then that Π is a translation plane with kernel $K = GF(q)$, and that Π corresponds to the K-space V with spread $\mathcal{S}$, $q = 2^m$. Finally, let G be a subgroup of the linear translation complement. We ask ourselves the following question: considering V as a KG-module, what additional conditions are imposed on the representation module V by dint of the fact that G preserves a spread in V? That there *are* extra conditions can be seen from (2.3). For if $\alpha \in G$ is an involution then either α is an elation of Π, in which case the axis of α is the (unique) component of $\mathcal{S}$ fixed by α, or else m is even and $F(\alpha)$ is a subplane of Π of order $\sqrt{q}$. In either case we have

(6.1) If $\alpha \in G$ is an involution then $\dim C_V(\alpha) = \frac{1}{2} \dim V$.

In terms of modules this becomes

(6.2) For each subgroup $S \leq G$ of order 2, the restriction of V to S is free, that is, $V|S$ is free.

By repeated application of (6.1) and (2.3) we can enhance (6.1) significantly, to an assertion concerning elementary abelian 2-groups in G. To formulate this we make a definition:

(6.3) Let K be a field of characteristic 2, E an elementary abelian 2-group, and V a finite-dimensional KE-module. We say that V is *E-dispersive* or a *dispersive E-module* in case the following condition holds:

Whenever we have a pair of subgroups

(∗) $1 \leq A \leq B \leq E$ with $|B : A| = 2$ then either $C_V(B) = C_V(A)$ or else $\dim C_V(B) = \frac{1}{2} \dim C_V(A)$,

Moreover if K is as above, G any group, and V a KG-module, we say that V is a *dispersive module* in case V is E-dispersive for each elementary abelian 2-group $E \leq G$.

Finally, if condition (∗) holds for any pair of subgroups $1 \leq A \leq B \leq S$ with $|B : A| = 2$ where S is a 2-Sylow of G, we say that V is a *strongly dispersive G-module.*

One can show

(6.4) Let V, K, and G be as at the beginning of this section. Then V is a dispersive G-module. Moreover if G is *planar*, i.e., the fixed points of G on Π form a subplane, then V is a strongly dispersive G-module.

I remark that it is unknown to me whether in general, in the above situation, V can be dispersive yet not be strongly dispersive. Certainly there are dispersive modules which are not strongly dispersive (in fact the smallest example which could fail, a 12-dimensional $Z_4 \times Z_4$-module, does so!), but I know of no example where a spread is preserved. The easy proof of (6.4) can be found in [**5**]. As a riposte for (4.3) we may now suggest

(6.5) Develop a theory of dispersive modules.

One direction in which one might go, and the one we shall take here, is to compare dispersive modules with projective modules. For it should be apparent that "dispersive" is a generalization of "projective." Indeed consider condition (*) once again and suppose that for each $A \leq B$ with $|B : A| = 2$ the second condition holds, that is, $\dim C_V(B) = \frac{1}{2} \dim C_V(A)$. This is simply another way of stating that V is a free E-module. But then $V|E$ is free for each elementary abelian 2-group $E \leq G$, whence V is a free S-module for $S \in \mathrm{Syl}_2(G)$ and a projectivej G-module by Chouinard's theorem [**2**]. Hence

(6.6) Every projective KG-module is dispersive.

Note that projectivity is determined by elementary 2-groups by Chouinard's result, whereas the thrust of the remark following (6.4) is that this is not so for dispersive modules. Nevertheless, there are large classes of groups for which the notions of dispersive and projective coincide.

(6.7) Suppose that $F^*(G) = O(G)$. Then a faithful KG-module V is dispersive if, and only if, it is projective.

To see why (6.7) is true, note that Chouinard's theorem allows us to assume that $G = O(G)E$ for some elementary abelian 2-group E. Moreover we may assume that $V|F$ is free for each hyperplane $F \leq E$. Now as E is faithful on $O(G)$ there is a prime r and a subgroup $R \leq O(G)$ of order r such that E normalizes, but does not centralize, R. Thus $F = C_E(R)$ is a hyperplane of E and if $x \in E - F$ then $R\langle x\rangle$ is dihedral of order $2r$. Now $R\langle x\rangle$ acts on the subspace $C_V(F)$. If the action is not faithful then R acts trivially on $C_V(F)$, so we can apply the $P \times Q$-lemma to the action of $F \times R$ on V to conclude that R centralizes not only $C_V(F)$ but also V itself. This being impossible, it follows that $R\langle x\rangle$ is faithful on $C_V(F)$. Thus E does not centralize $C_V(F)$, so by definition of dispersive we must have $\dim C_V(E) = \frac{1}{2} \dim C_V(F)$.

As $V|F$ is free this forces $V|E$ to be free, and (6.7) is proved.

We can generalize (6.7) rather significantly as follows. First consider the following hypothesis:

(6.8) (*Condition* $\mathcal{N}$). For each elementary abelian 2-group $E \leq G$ there is a group Q of odd order, depending on E, such that E normalizes, but does not centralize, Q.

The proof of (6.7) then establishes

(6.9) If G satisfies (condition И) then each faithful dispersive KG-module is projective.

We have the following result, due to G. Seitz (unpublished).

(6.10) Let G be a Chevalley group of odd characteristic, but not one defined over $GF(9)$ or $GF(p)$ for p a Fermat or Mersenne prime. Then G satisfies (condition И).

We remark that in (6.10), G is not necessarily quasisimple but can be any quotient of the corresponding universal group which has $Z(G)$ of odd order. Although (6.10) is generally false for the excluded fields, I have no doubt that (6.9) remains true for all but a handful of groups such as $L_2(5)$ and $L_2(9)$. As far as Chevalley groups are concerned, this reduces us to the case of Chevalley groups of characteristic 2, at least if we are concerned with analogs of (6.7).

To complete this section we state the question which relates (5.2) with the foregoing.

(6.11) Let G be a Chevalley group of Lie rank ≥ 2. Is it true that every dispersive KG-module is projective?

An affirmative answer would establish that if a quasisimple Chevalley group G of Lie rank ≥ 2 acts on a translation plane Π in the translation complement then the underlying module V is a projective G-module.

By the way, we've already seen in example D of §5 that $L_3(2)$ contradicts (6.11), so for this purpose we think of $L_3(2)$ as having rank 1 via the exceptional isomorphism $L_3(2) \cong \mathrm{PSL}_2(7)$.

7. Groups of characteristic 2. We investigate some special cases of (6.11) in the characteristic 2 situation. To simplify matters we will assume that $q = 2^n$, and that $G = G(q)$ is a Chevalley group over $GF(q)$ which is untwisted and whose root system Σ is simply laced of rank l. My ignorance is such that we will restrict attention to irreducible $GF(q)G$-modules V.

In this situation one knows that V is a twisted tensor product

$$V = B_1^{\sigma_1} \otimes B_2^{\sigma_2} \otimes \cdots \otimes B_n^{\sigma_n} \tag{7.1}$$

where $\{\sigma_1, \sigma_2, \ldots, \sigma_n\} = \mathrm{Gal}\,(GF(q))$ and each B_i is one of the 2^l *basic* irreducible $GF(q)G$-modules corresponding to a subset of nodes of the Dynkin diagram. Among these, the *basic Steinberg* module St of dimension 2^N (N = number of positive roots of Σ) plays a special role, and if each $B_i = \mathrm{St}$ then $V = \mathrm{St}_q$, *the* Steinberg module which is the unique projective irreducible $GF(q)G$-module.

If $l \geq 2$ and V is dispersive, and if one believes (6.11), then one expects that $V \cong \mathrm{St}_q$. Here is some evidence to support such a view.

(7.2) With the above assumptions, the following hold.

(a) Every subgroup $S \leq G$ with $|S| = 2$ satisfies $V|S$ free if, and only if, in the tensor product (7.1) some B_i is isomorphic to St.

(b) If V is strongly dispersive then each B_i is either trivial or isomorphic to St.

Several remarks are in order. First, (a) is false for multiply laced diagrams. Also, (b) shows how much stronger the assumption of dispersion is (more precisely, how much easier it is to use). With a strongly dispersive module one can apply induction far more readily, for example, to the parabolic subgroups of G in the above situation. Thus for strongly dispersive G-modules many questions reduce to questions in the case $l = 2$, in which case $G \cong \mathrm{SL}_3(q)$. This sort of reduction also occurs in the proof of the following, the main result of [**5**].

(7.3) Let V and G be as above and suppose that V supports a G-invariant spread. Then $l = 1$, that is, $G \cong \mathrm{SL}_2(q)$.

This gives an answer to (5.2) in a special case, of course, but sidesteps (6.11) to some extent.

8. The rank 1 case. We continue with the notation and assumptions of §7 but assuming now that $l = 1$, so that $G \cong \mathrm{SL}_2(q)$. (In fact most of the results below can be proved along similar lines for the group $\mathrm{Sz}(q)$ also, but we will not treat this case here.)

As G has a unique nontrivial basic module N, the natural module, we can rewrite (7.1) as

$$V = N \otimes N^{\sigma_2} \otimes \cdots \otimes N^{\sigma_p}, \tag{8.1}$$

where $p \leq n$ and we take $\sigma_1 = 1$. It is perhaps surprising that the determination of those V which are dispersive seems to be quite difficult!

We offer some partial results, mainly without proof:

(8.2) V is dispersive if $p \leq 2$ or if $n - p \leq 1$.

(8.3) Suppose that $p \geq 3$ and consider the following assertions.

(a) V is a dispersive G-module.

(b) $\langle \sigma_2, \ldots, \sigma_p \rangle = \mathrm{Gal}(GF(q))$.

(c) $V|U$ is free for each four-group $U \leq G$. Then the following implications hold: (a)⇒(b)⇒(c)⇒(b). Moreover if $p = 3$ then (b)⇒(a).

Concerning the implication (a)⇒(b), if (b) fails then $\langle \sigma_2, \ldots, \sigma_p \rangle$ is a proper subgroup of $\mathrm{Gal}(GF(q))$, hence fixes a subfield $GF(q_0)$, $q_0 \geq 4$, and the corresponding subgroup $G_0 = \mathrm{SL}_2(q_0)$. Moreover we have

$$V|G_0 \cong N_0 \otimes N_0 \otimes \cdots \otimes N_0 \quad (p \text{ times}),$$

N_0 being the natural module for G_0. One can show on the basis of the work of Alperin [**1**] that if E_0 is a 2-Sylow of G_0 then $N_0 \otimes \cdots \otimes N_0$ (p times) is not a dispersive E_0-module if $p \geq 3$, so (a) is also false.

The following definition is relevant.

(8.4) (Existence of a set of *projectors*). Let K be a field of characteristic 2, E an elementary abelian 2-group, and V a KE-module. We say that V has *projective depth d and a set of projectors $\mathcal{P}$*, if $d \geq 1$ is an integer and $\mathcal{P}$ a set of subgroups of E with the following properties:

(a) Each $P \in \mathcal{P}$ has rank d.

(b) $V|P$ is free for each $P \in \mathcal{P}$.

(c) $C_V(E) = C_V(P)$ for each $P \in \mathcal{P}$.

(d) If $D \leq E$ and $V|D$ is free then there is $P \in \mathcal{P}$ such that $D \leq P$.

(e) Each involution of E lies in at least one $P \in \mathcal{P}$.

In the tradition of Bourbaki, the next result follows from the definitions.

(8.5) Let V, K, E be as above, and suppose that V is a dispersive E-module. Then the following hold:

(i) V has a set of projectors and a projective depth.

(ii) Suppose further that $\dim C_V(E) = 1$. Then $\dim V = 2^d$ for some d and the projective depth of V is d.

A partial converse of (8.5) is

(8.6) Let V, K, E be as above and assume that $\dim V = 2^n$. If $V|F$ is free for every subgroup $F \leq E$ of rank $n-1$ then V is E-dispersive and has projective depth $n-1$ or n.

We also have

(8.7) Let $G \cong \mathrm{SL}_2(q)$ with V as in (8.1). Then V has a set of projectors and projective depth p, w.r.t a 2-Sylow of G.

Compare this with (8.5) (ii). Note that the case $p = n-1$ of (8.2) is a consequence of (8.6) and (8.7). Since not all of the modules V are dispersive (cf. (8.3)), the existence of a set of projectors $\mathcal{P}$ is not sufficient to ensure that V is dispersive—but it is close! Thus (8.6) tells us that V is dispersive if *all* 2-subgroups of G of rank p are projectors, but in general $\mathcal{P}$ will not be rich enough to carry the day.

Exactly which of the modules V in (8.1) are dispersive remains open, though I am fairly sure of an affirmative answer to the following question.

(8.8) Suppose that $q = 2^n$ with n prime, $G \cong \mathrm{SL}_2(q)$. Is it true that each of the irreducible $GF(q)G$-modules V is dispersive? In fact, is it true that $V|F$ is free for *every* elementary 2-subgroup $F \leq G$ of rank p? (If so, then V is dispersive by (8.6).)

9. Other characteristics. In this section we briefly consider the situation where Π is a translation plane with kernel $GF(q)$ and q is *odd*. In this situation (6.1) and (6.2) still hold, so because we are in odd characteristic we automatically get $V|E$ free for each elementary abelian 2-group. (Here, as usual, V is the $GF(q)$-space associated to Π.) But one is interested in the p-subgroups acting on Π, where p is a prime and $q = p^f$. The precise analog of (6.1) is false—for instance when $p = 3$ there are planes Π in which $\dim C_V(\alpha) = \frac{1}{2} \dim V$ or $\dim C_V(\alpha) = \frac{1}{3} \dim V$. It was Foulser who found the correct generalization of (6.1) in **[3]**:

(9.1) Let α be planar, have order p, and lie in the linear translation complement of Π. Then there is an integer m with $2 \leq m \leq p$, such that when considered as a linear transformation of V in Jordan canonical form, every Jordan block of α has size m.

It is possible for different α's acting on the same Π to have different values for m, though I'm not sure if there are any examples with m anything other than 2 or p. By the way, it is usually possible to show that one can restrict attention to the situation in which all elements of order p *are* planar, so this hypothesis in (9.1) is not too bothersome.

If α satisfies the conclusions of (9.1) we say that α is *uniform with index* m. Thus what is required is a theory of $GF(q)$-representations in which elements of order p are uniform. Can the theory of §§6–8 be duplicated in this case?

The situation in which G lies in the linear translation complement of Π and has order prime to p is different again. The representation-theoretic consequences of having a spread have not been systematically investigated. Recently, however, my student Mark Eastman and I have investigated the situation in which G is a symmetric group, or more generally a direct product of symmetric groups. We have established the following result:

(9.2) Let Π be a translation plane with kernel $GF(q)$ and suppose that Π admits a group $G \cong \Sigma_{n_1} \times \Sigma_{n_2} \times \cdots \times \Sigma_{n_r}$ in the linear translation complement, with each $n_i \geq 6$. Assume that $GF(q)$ contains "enough roots of unity" in the following sense: for each i and for each prime $p \leq n_i$, $p \mid q-1$.

Then the underlying $GF(q)G$-module is a free G-module and G is planar in the sense that the G-invariant points of Π form a subplane.

This exemplifies the sort of result that we would like to see established in a wider context. The condition on q in (9.2) is extremely useful, but should ultimately be expendable. In the proof of (9.2) we make use of several results concerning the values of various irreducible characters of G on certain types of elements, mainly consequences of the Murnaghan-Nakayama rule. Further results of this type would ease the difficulty of the proof.

Finally, we ask if the conclusion of (9.2) holds in the context of §6–8 so that Π is a translation plane with kernel $K = GF(2^f)$ and that G is a Chevalley group of rank ≥ 2 (not $L_3(2)$) and any characteristic, lying in the linear translation complement of Π. Let V be the associated KG-module. Can the following steps be carried out?

(I) Use the theory of dispersive modules to show that V is a projective KG-modulc.

(II) Use the theory of 2-modular representations and geometry to show that $V|L$ is free for each $L \leq G$ of odd order. Thus V is then a free KG-module.

(III) Show that G is planar.

(IV) Hope.

REFERENCES

1. J. Alperin, *Projective modules for* SL$(2, 2^n)$, J. Pure Appl. Algebra **15** (1979), 219–234.

2. L. Chouinard, *Projectivity and relative projectivity over group rings*, J. Pure Appl. Algebra **7** (1976), 278–302.

3. D. Foulser, *Planar collineations of order p in translation planes of order* p^r, Geom. Dedicata **5** (1976), 393–409.

4. H. Lüneburg, *Translation planes*, Springer-Verlag, New York, 1980.

5. G. Mason, *Irreducible translation planes and representations of Chevalley groups in characteristic* 2, Lecture Notes in Pure and Appl. Math., vol. 82, Dekker, 1983.

6. D. Hughes and F. Piper, *Projective planes*, Springer-Verlag, New York, 1973.

UNIVERSITY OF CALIFORNIA, SANTA CRUZ

Proceedings of Symposia in Pure Mathematics
Volume 47 (1987)

Modular Representation Theory and the Classification of Finite Simple Groups

GERHARD O. MICHLER

Introduction. At the Santa Cruz Conference on Finite Groups in 1979 W. Feit gave a survey lecture on "Some consequences of the classification of finite simple groups" [**14**], in which he stated several long-standing problems which might perhaps follow from the classification theorem. Most of them were asked by R. Brauer in his famous article [**7**] or are related to the Alperin-McKay conjecture. According to J. L. Alperin [**1**] all these questions are part of the main problem of block theory: If G is a finite group, then give simple rules which determine, in terms of p-local subgroups, the values of the characters on the p-singular elements.

In this survey lecture I would like to report about some progress which has been made on these problems since the Santa Cruz Conference. Since the classification theorem has been proved, much emphasis has been put on the study of the modular representations of the finite simple groups of Lie type. For these groups one can apply the powerful tools of the Deligne-Lusztig character theory presented in the books by Carter [**9**] and Lusztig [**33**]. As an example the important article [**19**] by Fong and Srinivasan has to be mentioned. After having found the block distribution of the irreducible characters of the general linear groups $\mathrm{GL}(n,p)$ and unitary groups $\mathrm{U}(n,q^2)$ they were able to show Brauer's height zero conjecture for these groups. It was this paper which has inspired a lot of further research on blocks of groups of Lie type over fields with characteristic coprime to the describing characteristics of the group. The character tables [**10**] of the sporadic simple groups and the work on the modular representations of finite groups of Lie type ensure that a reduction of a problem to the case of finite simple groups can often lead to a complete solution of the problem in general by means of the classification theorem.

In §1 we give an example of this method by means of a summary of the recent solution of Blau and the author [**6**] of Alperin's conjecture: If the Sylow p-subgroup P of the finite group G is a T. I. set, then the number of irreducible

1980 *Mathematics Subject Classification* (1985 *Revision*). Primary 20C20; Secondary 20C05.

characters of G not vanishing on the nonidentity elements of P equals the number of characters of $N_G(P)$; see [**2**, Conjecture C, p. 370]. In fact, several other problems could be solved for groups with a T. I. Sylow p-subgroup. One difficulty in the proofs of all these questions is that in general we still do not have any strong information on Brauer's [**7**] Problem 19: "Characterize the number of blocks of defect zero of a finite group G by group-theoretical properties of G." Even with the use of the classification of the finite simple groups there is no complete solution of this problem in general. Therefore we restrict ourselves in §2 to the finite simple groups. Then we present the recent results by Willems [**44**] and the author [**37**] yielding the following theorem: Every finite simple group of Lie type has p-blocks of defect zero for every prime number p. Some consequences of this result are also given.

§3 is devoted to W. Feit's solution [**16**] of the problem: What trees can be Brauer trees of a block with a cyclic defect group? He showed that only relatively few trees can occur, which he already had conjectured in [**14**]. On the other side, the Brauer trees of some finite simple groups are still not known.

§4 summarizes recent work of G. Hiss [**27**, **28**] on the finite groups G, all of whose irreducible modular characters can be lifted to irreducible ordinary characters of G. Using the classification theorem of finite simple groups and the Deligne-Lusztig theory, he showed the converse of a well-known theorem of Fong-Swan and Isaacs by proving that a finite group G is solvable, if for every prime $p > 0$ each absolutely irreducible p-modular character of G is liftable to a p-rational ordinary irreducible character of G.

In [**14**] W. Feit wrote: "The proof of a statement which consists of checking all cases is often a proof without understanding and so is not very satisfactory .... Nevertheless the method can be very powerful as a device for discovering results and has very respectable historical precedents." In the final section another such example is given. About ten years ago K. Erdmann [**12**] found the structure of the indecomposable projective modules of the principal 2-block of a finite group G with a dihedral Sylow 2-subgroup using the classification theorem of Gorenstein and Walter for these groups. In §5 her recent work [**13**] giving a completely representation-theoretic proof of these results is described. It is based on a highly original application of the Auslander-Reiten techniques in the representation theory of finite-dimensional algebras.

It is my hope that further such theoretical developments in the representation theory of finite groups will follow from a detailed study of concrete, mostly covering groups of automorphism groups of finite simple groups.

Throughout this article, R denotes a complete discrete valuation ring with residue class field $F = R/\pi R$ of characteristic 0. We assume that F and S are splitting fields for G and all its subgroups; see Landrock [**31**, p. 47]. Let $|G| = p^a q$, with $(p,q) = 1$. $\mathrm{Irr}_S(G)$ and $\mathrm{Irr}_F(G)$ denote the set of irreducible ordinary characters χ of G and the set of irreducible modular characters φ of G, respectively. Let $k(G) = |\mathrm{Irr}_S(G)|$ and $l(G) = |\mathrm{Irr}_F(G)|$. Let B be a p-block of G with defect group $\delta(B) =_G D$. If $|D| = p^{d(B)}$, then $d = d(B)$ is called the defect of B. By Theorem 4.5 of [**15**, p. 150] p^{a-d} divides the degree $\chi(1)$ of each irreducible character $\chi \in \mathrm{Irr}_S(B) = \{\chi \in \mathrm{Irr}_S(G) \mid \chi$ belongs to $B\}$. If p^{a-d+h} is the precise power of p dividing $\chi(1)$, then $h = ht\chi$ is called the *height* of the irreducible character $\chi \in \mathrm{Irr}_S(B)$.

$\mathrm{Irr}^0_S(B) = \{\chi \in \mathrm{Irr}_S(B) \mid ht\chi = 0\}$. It follows from Lemma 4.4 and Theorem 4.5 of [**15**, p. 150], that $\mathrm{Irr}^0_S(B) \neq \emptyset$ for every p-block B of a finite group G. The block invariants $k(B) = |\mathrm{Irr}_S(B)|$, $l(B) = |\mathrm{Irr}_F(B)| = |\{\varphi \in \mathrm{Irr}_F(G) \mid \varphi$ belongs to $B\}|$, and $k_0(B) = |\mathrm{Irr}^0_S(B)|$ are of particular interest in the study of modular representation theory of finite groups. For further notation and terminology we refer to the books by Carter [**9**], Feit [**15**], Gorenstein [**23**], Isaacs [**29**], Landrock [**31**], and Lusztig [**33**]. The character tables of finite simple groups used here are contained in the CAS-system [**39**] of J. Neubüser, H. Pahlings, and W. Plesken (TH Aachen, Federal Republic of Germany) or the Atlas [**10**]. The complete statement of the classification theorem of the finite simple groups and all relevant definitions and notations can be found in Gorenstein's book [**24**].

1. Modular representations of finite groups with a T. I. Sylow p-subgroup. Let p be a fixed prime, and let G be a finite group with a T. I. Sylow p-subgroup P. That is, two different conjugates of P have only the identity element in common. At the Santa Cruz Conference on Finite Groups, J. L. Alperin [**2**] stated the following conjecture: If P is a T. I. set, then the number of ordinary irreducible characters of G which do not vanish on P equals the number of ordinary irreducible characters of the normalizer $N_G(P)$ of P. Using the classification of the finite simple groups, H. Blau and the author solved this problem in their article [**6**]. In fact, they show the block version of Alperin's conjecture.

THEOREM 1.1. *Let G be a finite group with a T. I. Sylow p-subgroup P. Let B be a p-block of G with defect group $\delta(B) =_G P$ and Brauer correspondent b in $N = N_G(P)$. Then*

(a) $k(B) = k(b)$;

(b) $k_0(B) = k_0(b)$;

(c) $l(B) = l(b)$;

(d) $k(G) = k(N) + z(G)$, *where $z(G)$ denotes the number of p-blocks of defect zero in G.*

This result has several other consequences. It implies that Brauer's height zero conjecture holds for groups with T. I. Sylow p-subgroups.

COROLLARY 1.2. *Let G be a finite group with a T. I. Sylow p-subgroup P. Let B be a p-block of G with defect group $\delta(B) =_G P$.*

Then every irreducible character χ of B has height $ht\chi = 0$ if and only if P is abelian.

There are other partial results on the height zero conjecture. Gluck and Wolf proved it for p-solvable groups in [**21**] using the classification of the finite simple groups. It also holds for blocks with a normal defect group by Reynolds

[42]. It has been verified for the unitary and general linear groups by Fong and Srinivasan [19]. Deriziotis and the author [11] showed it for the Steinberg triality ${}^3D_4(q)$. Recently, T. R. Berger and R. Knörr proved in [3] the following reduction theorem for one part of the height zero conjecture.

THEOREM 1.3 (BERGER-KNÖRR). *Suppose that for all quasi-simple groups the irreducible ordinary characters in blocks with abelian defect groups are all of height zero. Then the same is true for all groups.*

Here a group $G \neq 1$ is called quasi-simple if G is its own commutator subgroup G' and every proper normal subgroup of G is contained in the center $Z(G)$ of G. The reduction of the minimal counterexample to Theorem 1.1 cannot be performed so far. Also, automorphism groups of quasi-simple groups have to be studied.

Brauer [7] conjectured that the following corollary of Theorem 1.1 holds in any group G without any condition on the Sylow p-subgroup P of G.

COROLLARY 1.4. *Let G be a finite group with a T. I. Sylow p-subgroup P such that P is not normal in G. Let B be a p-block of G with defect group $\delta(B) =_G P$. Then:*

(a) $k(B) \leq |P|$,

(b) $k_0(B) \leq |P/P'|$.

This conjecture of Brauer's appears to be very difficult even for groups with a normal Sylow p-subgroup P. For this case Gluck [22] and Knörr [30] have found solutions when P has a complement U which is of odd order or supersolvable. Olsson [40] verified Brauer's conjecture for the general linear, unitary, and symmetric groups. Deriziotis and the author [11] showed it for ${}^3D_4(q)$. So far there has not been found any reduction to the case of simple groups.

Theorem 1.1 also supplies some further support for a long-standing conjecture in the theory of finite-dimensional algebras.

Two finite-dimensional F-algebras B and b are called *stably equivalent* if their module categories Mod B and Mod b are equivalent modulo their projective objects and projective homomorphisms. It has been conjectured for a long time that two such algebras have the same number $l(B) = l(b)$ of nonisomorphic nonprojective simple modules. Generalizing work by P. Gabriel and Ch. Riedtmann [20], R. Martinez-Villa showed in [34] that two stably equivalent algebras of finite representation type have the same number of nonprojective simple modules.

Block ideals B of group algebras with a T. I. Sylow p-subgroup P of a finite group G as defect group and their Brauer correspondents b in $N = N_G(P)$ yield important examples of stably equivalent indecomposable algebras (see [15, p. 118]), which are not necessarily of finite representation type. Of course, the verification of the above conjecture in the theory of algebras would imply assertions (a), (c), and (d) of Theorem 1.1 immediately. Unfortunately, we are very far from a general solution of this conjecture.

Furthermore, I would like to mention that the classification of the finite simplegroups with a noncyclic T. I. Sylow p-subgroup given in the following result requires at present the classification of all finite simple groups; see Gorenstein-Lyons [**25**].

PROPOSITION 1.3. *Let G be a nonabelian simple group with a noncyclic T. I. Sylow p-subgroup P. Then G is isomorphic to one of the following groups:*
(a) $\mathrm{PSL}_2(q)$, *where* $q = p^n$ *and* $n \geq 2$;
(b) $\mathrm{PSU}_3(q^2)$, *where* $q = p^n$;
(c) $p = 2$ *and* $G \simeq {}^2B_2(2^{2m+1})$;
(d) $p = 3$ *and* $G \simeq {}^2G_2(3^{2m+1})$ *and* $m \geq 2$;
(e) $p = 3$ *and* $G \simeq \mathrm{PSL}_3(4)$ *or* M_{11};
(f) $p = 5$ *and* $G \simeq {}^2F_4(2)'$ *or* $\mathrm{M^cL}$;
(g) $p = 11$ *and* $G \simeq J_4$.

This classification theorem was used before by Berger, Landrock, and the author in [**4**] in order to show that a faithful complex character χ of a finite group G with a nonnormal T. I. Sylow p-subgroup P has degree $\chi(1) > \sqrt{|P|}-1$. This result verifies a conjecture of H. S. Leonard [**32**]. The precise bounds of the minimal degrees for such faithful characters have been determined by H. I. Blau [**5**].

All assertions of Theorem 1.1 and its Corollaries 1.2 and 1.4 can be considered as generalizations of the theory of blocks B with cyclic defect groups. But the deepest result on blocks with cyclic defect groups does not generalize. Whereas in such a p-block B all decomposition numbers are zero or one, the principal 5-block B_0 of the sporadic McLaughlin group $\mathrm{M^cL}$ has an irreducible character of degree 3520 in which the modular irreducible character of degree 21 occurs twice as a constituent. Of course, the Sylow 5-subgroup $\delta(B_0) = P$ of $\mathrm{M^cL}$ is a T. I. set. This example is due to Dr. Hiss (Aachen).

Alperin's conjecture is related to the Alperin-McKay conjecture. The recent work on this problem has been reported by Feit [**14**], the author [**36**], and Olsson and the author [**38**]. Most recently the Alperin-McKay conjecture has been verified for the sporadic simple groups by Th. Ostermann [**41**] and independently by R. A. Wilson of Cambridge University.

2. Existence of blocks of defect zero. Blocks of defect zero have the following beautiful ring-theoretical characterization.

THEOREM 2.1 (R. BRAUER). *The p-block of the finite group G has defect $d(B) = 0$ if and only if its block ideal B is isomorphic to a full ring $(F)_n$ of $n \times n$ matrices over the field F.*

Similar characterizations hold in characteristic zero.

On the other side, there is not much general information on Brauer's [**7**] problem: "Characterize the number $z(G)$ of blocks of defect zero of finite groups G by group-theoretical properties of G." Theorem 1.1 (d) asserting that

$z(G) = k(G) - k(N)$ for groups with a T. I. Sylow p-subgroup is an example of such a solution of the problem in a special case.

A finite simple group G does not have p-blocks of defect zero necessarily. In fact, the whole group algebra FG can be its principal p-block. By Brockhaus and the author **[8]** and Harris **[26]** this only can happen if $p = 2$ and G is one of the simple Mathieu groups M_{22} or M_{24}.

As can be seen from the CAS character table system **[39]** each of the 26 sporadic simple groups G has p-blocks of defect zero for each prime $p > 3$. The sporadic Suzuki group Suz and the Conway group $\cdot 3$ do not possess 3-blocks of defect zero. Furthermore, there are no 2-blocks of defect zero in the sporadic simple groups $M_{12}, M_{22}, M_{24}, J_2, \mathrm{HS}, \mathrm{Suz}, \mathrm{Ru}, \cdot 1, \cdot 3$, and BM.

Already R. Brauer **[7]** mentioned that the symmetric group $\mathfrak{S}_n$ of degree $n \geq 5$ has 2-blocks of defect zero if and if only if $n = \binom{k}{2}$ for a natural number k.

Using the theory of modular forms, A. O. L. Atkin and J. Olsson have recently proved the following (still unpublished) result for the alternating groups $\mathfrak{A}_n$.

THEOREM 2.2. (a) *For every prime number $p \geq 5$ there exists an integer k such that $\mathfrak{A}_n$ has p-blocks of defect zero for every $n \geq k$.*

(b) *There are infinitely many numbers n such that $\mathfrak{A}_n$ does not have any 3-blocks of defect zero.*

(c) *For each 3-group D there exists a natural number n such that all 3-blocks B of $\mathfrak{A}_n$ have a defect group $\delta(B)$ containing D.*

Furthermore, Atkin and Olsson have shown that for the primes $p = 5$, 7, and 11 the number $z(\mathfrak{A}_n)$ of p-blocks of $\mathfrak{A}_n$ with defect zero is nonzero for all natural numbers n.

The finite simple groups of Lie type behave much better. Using the Deligne-Lusztig generalized characters, the author **[37]** and W. Willems **[44]** proved the following result for odd primes p and for $p = 2$, respectively.

THEOREM 2.3. *Let G be a finite simple group of Lie type. Then G has a p-block of defect zero for every prime p.*

Although the alternating groups $\mathfrak{A}_n$, $n \geq 5$, do not necessarily contain p-blocks of defect zero, they always have p-blocks B whose defect groups $\delta(B)$ are not Sylow p-subgroups of $\mathfrak{A}_n$; see **[8]**. Therefore, as shown by the author **[36]**, the classification theorem of finite simple groups and Theorem 2.3 imply

THEOREM 2.4. *Let $p \neq 2$. Then every finite simple group G has at least one p-block B whose defect group $\delta(B)$ is not a Sylow p-subgroup of G.*

In the author's article **[36]** proofs are given for the following applications of Theorem 2.4.

THEOREM 2.5. *The finite group G has an abelian normal Sylow p-subgroup P if and only if $p \nmid \chi(1)$ for every irreducible ordinary character χ of G.*

Because of Ito's theorem (see Isaacs **[29]**) only the sufficiency of the degree condition had been an open problem.

THEOREM 2.6. *The finite group G has a normal Sylow p-subgroup P if and only if $p \nmid \varphi(1)$ for every irreducible modular character φ of G.*

In an unpublished paper T. Okuyama has given a proof of Theorem 2.6 for the case $p = 2$ which is independent of the classification theorem.

Further applications of Theorem 2.4 are given in the author's articles [**35**] and [**36**].

3. Blocks with cyclic defect groups. The theory of blocks with cyclic defect groups belongs to the best developed parts of modular representation theory; see [**15**, Chapter VII]. Nevertheless, the deep theorems of R. Brauer and E. C. Dade do not suffice to answer R. Brauer's [**7**] Problem 24: Let G be a finite group with a cyclic Sylow p-subgroup D, and let $N = N_G(D)$. Determine when $k(G) = k(N)$.

By Dade's theorem all irreducible characters of a p-block B with a cyclic defect group D of a finite group G have height zero. Therefore Brauer's first main theorem, Dade's theorem, and Theorem 2.5 yield the following answer to Brauer's Problem 24.

THEOREM 3.1. *Let G be a finite group with a cyclic Sylow p-subgroup D, and let $N = N_G(D)$. Then $k(G) = k(N)$ if and if only D is normal in G.*

Similarly, the following question of W. Feit [**14**] appears to be extremely difficult to tackle without the classification theorem of finite simple groups: What trees can be Brauer trees of a block with a cyclic defect group? In his paper [**16**], where this question is answered, Feit writes: "As far as I know not a single tree can be eliminated without using this classification." In order to state Feit's theorem the following notations and definitions are needed.

Let τ be a tree and let P_0 be a vertex of τ. Then for every natural number n the union of n copies of τ with the vertices P_0 identified is denoted by $(\tau, P_0)^n$. An isomorphism from a tree τ to a tree τ' is a bijection from the set of all objects of τ onto the set of all objects of τ which preserves edges, vertices, and incidence.

THEOREM 3.2 (W. FEIT [**16**]). *Let G be a finite group and let τ be the Brauer tree of a block B of G with a cyclic defect group D. Then τ is isomorphic to $(\gamma, P_0)^n$ for some natural number n and some graph γ, where P_0 is the exceptional vertex (if there is an exceptional vertex in τ). Furthermore, one of the following assertions holds:* (a) *γ has at most* 248 *edges or* (b) *γ is an open polygon.*

In [**18**] Fong and Srinivasan have determined the Brauer trees of the r-blocks B with cyclic defect groups of the general linear groups $\mathrm{GL}(n, q)$, where the prime $r \nmid q$. They have proved that the Brauer tree τ of such an r-block B is an open polygon with the exceptional vertex at one end. Using their work, W. Feit shows in [**16**] that neither the natural number n in Theorem 3.2 nor the number of edges of γ needs to be bounded.

4. Groups with liftable modular irreducible characters. An irreducible modular character φ of a finite group G is called liftable if there is an irreducible ordinary character χ of G such that $\chi = \varphi$ as a Brauer character. If G is a p-solvable group, then the Fong-Swan theorem asserts that each irreducible Brauer character φ of G is liftable. Therefore in a solvable group G the p-modular irreducible Brauer characters are liftable for every prime number p. It had been an open question for some time whether the converse was also true. In his thesis [**27**] G. Hiss showed that $\mathrm{SL}_2(5)$ is a counterexample. Using the classification theorem of the finite simple groups he proved

THEOREM 4.1 (HISS). *Suppose that for every prime p all the p-modular irreducible characters of the finite group G and of all its factor groups are liftable, then G is solvable.*

Although the hypothesis on the factor group in Theorem 4.1 is very unnatural, it cannot be omitted. Hiss [**27**] showed that in $\mathrm{SL}_2(5)$ all 2-modular irreducible characters are liftable, but not in $\mathrm{PSL}_2(5)$.

An ordinary irreducible character χ of the finite group G is called p-rational if its values lie in the field $Q(\omega)$, where Q is the field of rational numbers and where ω is some nth root of unity with p not dividing n.

Refining the Fong-Swan theorem, Isaacs had shown that for every irreducible Brauer character φ of a p-solvable group G there exists a p-rational irreducible ordinary character χ of G such that $\chi = \varphi$ as a Brauer character; see [**15**, p. 419]. In his recent paper [**28**] G. Hiss was able to prove that the converse of the theorem of Fong-Swan-Isaacs holds.

THEOREM 4.2 (HISS). *Let G be a finite group. Suppose that for all prime numbers p, every absolutely irreducible p-modular character of G is liftable to a p-rational ordinary irreducible character. Then G is solvable.*

The proof depends on the classification theorem of the finite simple groups. Hiss [**28**] also remarked that a local version of this theorem does not hold, because every absolutely irreducible 3-modular character of $\mathfrak{A}_5$ is liftable to a unique 3-rational ordinary irreducible character.

5. Blocks with dihedral defect groups. Using the Gorenstein-Walter theorem classifying the finite groups G with a dihedral Sylow 2-subgroup (see [**23**, p. 462]), K. Erdmann determined in [**12**] the structure of the indecomposable projective modules of the principal 2-block B_0 of such a group G.

In her recent paper [**13**] she found a self-contained, purely module-theoretic proof for a more general theorem by application of the Auslander-Reiten theory on almost split sequences.

Let B be a 2-block of a finite group with a dihedral defect group D. Then B is an F-algebra of tame representation type with $l(B) \leq 3$. The cases $l(B) = 1$ and $l(B) = 2$ had been dealt with before by Broué and Puig and by Donavan, respectively; see [**13**]. If $l(B) = 3$, then K. Erdmann shows that B has at most two simple FG-modules M with a periodic projective resolution.

THEOREM (K. ERDMANN). *Let B be a 2-block of the finite group G with dihedral defect group $\delta(B)$. Denote the Jacobson radical of the group algebra FG by J. Let M be a simple B-module with projective cover $P(M) = P$. Then the following assertions hold:*

(a) *If M is periodic, then $P(M)$ is uniserial.*

(b) *If M is not periodic, then $PJ/M \cong L_1 \oplus L_2$, where each L_i is a uniserial FG-module.*

REFERENCES

1. J. L. Alperin, *The main problem of block theory*, Proc. Conf. on Finite Groups, Academic Press, New York, 1976, pp. 341–356.

2. ____, *Local representation theory*, Proc. Sympos. Pure Math., vol. 37, 1980, pp. 364–384.

3. T. R. Berger and R. Knörr, *On Brauer's height zero conjecture*, Osaka J. Math. (to appear).

4. T. R. Berger, P. Landrock, and G. O. Michler, *Minimal degrees of faithful characters of finite groups with a T. I. Sylow p-subgroup*, Proc. Amer. Math. Soc. **99** (1987), 15–21.

5. H. I. Blau, *On linear groups with a cyclic or T. I. Sylow subgroup*, Preprint.

6. H. I. Blau and G. O. Michler, *Modular representation theory of finite groups with T. I. Sylow p-subgroups*, Preprint.

7. R. Brauer, *Representations of finite groups*, Lectures in Math., vol. 1, Wiley, New York, 1963, pp. 133–175.

8. P. Brockhaus and G. O. Michler, *Finite simple groups of Lie type have non-principal p-blocks, $p \neq 2$*, J. Algebra **94** (1985), 113–125.

9. R. W. Carter, *Finite groups of Lie type*, Wiley, New York, 1985.

10. J. H. Conway, R. T. Curtis, S. P. Norton, R. A. Parker, and R. A. Wilson, *Atlas of finite groups*, Clarendon Press, Oxford, 1985.

11. D. I. Deriziotis and G. O. Michler, *Character table and blocks of finite simple triality groups ${}^3D_4(q)$*, Trans. Amer. Math. Soc. (to appear).

12. K. Erdmann, *Principal blocks of groups with dihedral Sylow 2-subgroups*, Comm. Algebra **5** (1977), 665–694.

13. ____, *Algebras and dihedral defect groups*, Proc. London Math. Soc. (3) **54** (1987), 88–114.

14. W. Feit, *Some consequences of the classification of finite simple groups*, Proc. Sympos. Pure Math. **37** (1980), 175–182.

15. ____, *The representation theory of finite groups*, North-Holland, Amsterdam, 1982.

16. ____, *Possible Brauer trees*, Illinois J. Math. **28** (1984), 43–56.

17. P. Fong, *On the characters of p-solvable groups*, Trans. Amer. Math. Soc. **98** (1961), 263–284.

18. P. Fong and B. Srinivasan, *Blocks with cyclic defect groups in* $\mathrm{GL}(n,q)$, Bull. Amer. Math. Soc. (N.S.) **3** (1980), 1041–1044.

19. ____, *The blocks of finite general linear and unitary groups*, Invent. Math. **69** (1982), 109–153.

20. P. Gabriel and Ch. Riedtmann, *Group representations without groups*, Comment. Math. Helv. **54** (1979), 240–287.

21. D. Gluck and T. R. Wolf, *Brauer's height conjecture for p-solvable groups*, Trans. Amer. Math. Soc. **282** (1984), 137–152.

22. D. Gluck, *On the kGV-problem*, J. Algebra **89** (1984), 46–55.

23. D. Gorenstein, *Finite groups*, Harper & Row, New York, 1968.

24. ____, *Finite simple groups*, Plenum Press, New York, 1982.

25. ____, *On finite groups of characteristic 2-type*, Mem. Amer. Math. Soc. No. 276 (1982).

26. M. E. Harris, *Finite groups with exactly one p-block*, Preprint.

27. G. Hiss, *Über die Liftbarkeit modularer Darstellungen endlicher Gruppen*, Dissertation, Universität Freiburg, 1983.

28. ——, *A converse to the Fong-Swan-Isaacs theorem*, J. Algebra (to appear).

29. I. M. Isaacs, *Character theory of finite groups*, Academic Press, New York, 1976.

30. R. Knörr, *On the number of characters in a p-block of a p-solvable group*, Illinois J. Math. **28** (1984), 181–210.

31. P. Landrock, *Finite group algebras and their modules*, Cambridge University Press, Cambridge, 1983.

32. H. S. Leonard, *On finite groups whose p-Sylow subgroup is a T. I. set*, Proc. Amer. Math. Soc. **19** (1968), 667–670.

33. G. Lusztig, *Characters of reductive groups over a finite field*, Ann. of Math. Studies, vol. 107, Princeton Univ. Press., Princeton, N.J., 1984.

34. R. Martínez-Villa, *The stable equivalence for algebras of finite representation type*, Comm. Algebra **13** (1985), 991–1018.

35. G. O. Michler, *Non-solvable groups with cyclic Sylow p-subgroups have non-principal p-blocks*, J. Algebra **83**, (1983) 179–188.

36. ——, *Brauer's conjecture and the classification of finite simple groups*, Lecture Notes in Math., vol. 1178, Springer-Verlag, 1986, pp. 129–142.

37. ——, *A finite simple group of Lie type has p-blocks with different defects, $p \neq 2$*, J. Algebra **104** (1986), 220–230.

38. G. O. Michler and J. Olsson, *Character correspondences in finite general linear, unitary and symmetric groups*, Math. Z. **184** (1983), 203–233.

39. J. Neubüser, H. Pahlings, and W. Plesken, *CAS, design and use for the handling of characters of finite groups*, Proc. of the Conference on Computational Group Theory, Academic Press, London, 1984, pp. 195–247.

40. J. B. Olsson, *On the number of characters in blocks of finite general linear, unitary and symmetric groups*, Math. Z. **186** (1984), 41–47.

41. Th. Ostermann, *Irreduzible Charaktere von Sylownormalisatoren der sporadischen einfachen Gruppen*, Dissertation, Universität Essen, 1986.

42. W. F. Reynolds, *Blocks and normal subgroups of finite groups*, Nagoya Math. J. **22** (1963), 15–32.

43. G. R. Robinson, *The number of blocks with a given defect group*, J. Algebra **84** (1983), 493–502.

44. W. Willems, *Blocks of defect zero in finite simple groups of Lie type*, J. Algebra (to appear).

UNIVERSITY OF ESSEN, FEDERAL REPUBLIC OF GERMANY

Proceedings of Symposia in Pure Mathematics
Volume 47 (1987)

Cohomology of Algebraic Groups

BRIAN J. PARSHALL

Let G be a semisimple, simply connected algebraic group defined over $\mathbf{F}_p$. In this paper we present a overview of the theory of the cohomology groups $H^*(\Gamma, V)$, where Γ is taken to be either G itself, a finite Chevalley subgroup $G(q)$, or an infinitesimal subgroup G_r, and where V is a suitable representation in characteristic p.

We begin in §1 by surveying pertinent results from the finite group case. Historically, the many explicit calculations made for finite Chevalley groups provided much of the empirical information which inspired subsequent research. §2 discusses the rational cohomology of linear algebraic groups and relates this theory to that of finite groups. In §3 we consider the cohomology theory for infinitesimal groups. Here, too, the motivation and initial applications lay in the realm of finite groups. But there was an unexpected bonus: the complete calculation of the cohomology algebra of the restricted enveloping algebra of the Lie algebra of a semisimple group (subject to a restriction on the size of p). This result in turn suggested the possibility of developing a theory of support varieties for restricted Lie algebras, analogous to (but more geometric than) the theory of support varieties for finite groups. We treat this problem in §4. Finally, in §5, we indicate a few open questions worthy of further study.

Special thanks go to Eric Friedlander for helpful discussions during the writing of this paper and for his preprint [**24**]. We are also very grateful to Leonard Scott for collaborating on the proof of Theorem 2 below.

1. Finite group calculations. Fix an algebraically closed field k of positive characteristic p. Let G be a semisimple, simply connected algebraic group defined over $\mathbf{F}_p$ with Borel subgroup B having unipotent radical U and maximal torus T (all taken as defined over $\mathbf{F}_p$). For $q = p^d$, let $G(q)$ denote the finite subgroup of G of $\mathbf{F}_q$-rational points.

1980 *Mathematics Subject Classification* (1985 *Revision*). Primary 20G15, 20G10.
Research supported by the National Science Foundation

In the 1960s and early 1970s many explicit calculations of the (Eilenberg-Mac Lane) cohomology groups $H^n(G(q), V)$ for $n \leq 2$ and V a "small" $kG(q)$-module were made. The method often employed can be roughly described as follows. For simplicity we assume G is split over $\mathbf{F}_q$. Because the index $[G(q) : B(q)]$ is prime to p, the natural restriction map

$$\operatorname{res}^{G(q)}_{B(q)}(V) : H^n(G(q), V) \to H^n(B(q), V)$$

is an injection for all n [**11**, XII, Theorem 10.1]. Since the finite torus $T(q)$ is a p'-group, it has only degree zero cohomology, and an elementary Hochschild-Serre spectral sequence argument yields

$$H^n(B(q), V) \cong H^n(U(q), V)^{T(q)},$$

the fixed points of $T(q)$ acting on the cohomology of $U(q)$. Using a T-stable central series for U with sections isomorphic to root groups U_α, we reduce to the study of $H^*(U_\alpha(q), \lambda)^{T(q)}$ for positive roots α and weights λ of $T(q)$ in V. Since the latter cohomology groups can be completely determined, upper bounds for the dimension of $H^n(B(q), V)$ and hence of $H^n(G(q), V)$ can be found.

For example, suppose $n = 1$ and let Ψ be a set of positive roots such that $U(q) = \langle U_\alpha(q) : \alpha \in \Psi \rangle$. For any $G(q)$-module V satisfying $V^{B(q)} = 0$,

$$\dim H^1(G(q), V) \leq \sum_\lambda \dim V_\lambda - \dim V_0, \tag{1}$$

where the sum is over all weights λ of $T(q)$ in V satisfying

$${}^\tau\lambda = \alpha|_{T(q)}$$

for some $\alpha \in \Psi$ and some field automorphism $\tau \in \operatorname{Gal}(\overline{\mathbf{F}}_p/\mathbf{F}_p)$ [**14**, Corollary 2.9].[1] Here V_λ and V_0 denote the λ-weight and 0-weight spaces of $T(q)$ in V, respectively.

On the other hand, for small n, lower bounds can sometimes be found by *ad hoc* methods. Thus, for $n = 1$, the elements of $H^1(G(q), V) \cong \operatorname{Ext}^1_{G(q)}(k, V)$ parametrize isomorphism classes of extensions $0 \to V \to W \to k \to 0$ of the trivial module k by V, so that known indecomposable $G(q)$-modules provide nonvanishing 1-cohomology. The papers [**3**], [**14**], [**15**], and [**39**][2] provide many calculations in the above spirit (and supply references to other low-degree calculations).

The image of $\operatorname{res}^{G(q)}_{B(q)}(V)$ in $H^*(B(q), V)$ can be interpreted as the space of "stable classes" [**11**, XII, Propositions 9.3, 9.4]. The following stability result illustrates a striking feature of the cohomology of finite groups of Lie type.

[1] A similar proof to that in [**14**] gives the inequality (1) for the dimension of the rational cohomology group $H^1(G, V)$, where V is a rational G-module and the summation is over all weights of T in V which are simple roots.

[2] We correct and augment the tables of [**39**]. The cohomology dimensions for types ${}^2A_n(q)$, ${}^2D_n(q)$, ${}^2E_6(q)$ remain valid for all q and n, except $\dim H^1({}^2A_n(4), V(\lambda_i)) = 1$ for $n \geq 3$, $n \neq 5$, and $i = 3$, $n-2$; and $\dim H^1({}^2A_5(4), V(\lambda_3)) = 2$ (unpublished results of W. Jones). Also, $\dim H^1(D_{2n+1}(2^d), V(\lambda_2))$ should be corrected to 1.

THEOREM 1 [**15**]. *Let G be a semisimple, simply connected algebraic group defined over $\mathbf{F}_p$. Let $q = p^d$ and let V be a finite-dimensional rational G-module. For a fixed integer $n \geq 0$, the restriction map*

$$\mathrm{res}^{G(q)}_{B(q)}(V) : H^n(G(q), V) \to H^n(B(q), V)$$

is an isomorphism for all sufficiently large d (depending on V and n).

Although this result follows from considerations of the next section (cf. the remarks following Theorem 4), the original argument involved an interesting Hecke algebra aspect which merits mention here. Again, for simplicity, assume that G is split over $\mathbf{F}_p$. Let $\mathcal{H}$ be the free $\mathbf{Z}$-algebra with basis $\{T_w, w \in W\}$ (W = Weyl group of G) and relations (for α simple)

$$T_{s_\alpha} \cdot T_w = \begin{cases} T_{s_\alpha w} & \text{if } w^{-1}(\alpha) > 0, \\ (q-1)T_w + qT_{s_\alpha w} & \text{if } w^{-1}(\alpha) < 0 \end{cases}$$

[**6**, Chapter IV, §2, Ex. 23]. Let $\varepsilon : \mathcal{H} \to Z$ be the natural augmentation defined by $\varepsilon(T_w) = q^{l(w)}$. For $m \in H^0(B(q), V) = V^{B(q)}$, set

$$T_w \cdot m = \overline{U_w(q)} w \cdot m, \tag{2}$$

where U_w denotes the subgroup of U generated by those root subgroups U_β with $w^{-1}(\beta) < 0$ and, for a subgroup H of $G(q)$, $\overline{H}$ denotes the sum in the integral group algebra $\mathbf{Z}[G(q)]$ of the elements of H. It is readily checked by direct calculation that (2) defines an action of the Hecke algebra $\mathcal{H}$ on $H^0(B(q), V)$ and so an action on $H^*(B(q), V)$ by a standard dimension shifting argument. As observed in [**14**, §6], a cohomology class $\gamma \in H^n(B(q), V)$ is stable (i.e., ${}^w\gamma|_{{}^wB(q)\cap B(q)} = \gamma|_{{}^wB(q)\cap B(q)}$ for all $w \in W$) if and only if $h \cdot \gamma = \varepsilon(h)\gamma$ for all $h \in \mathcal{H}$. If γ satisfies

$${}^s\gamma|_{{}^sB(q)\cap B(q)} = \gamma|_{{}^sB(q)\cap B(q)} \tag{3}$$

for a simple reflection $s = s_\alpha$, then $T_s \cdot \gamma = [B : B(q) \cap {}^sB(q)]\gamma = \varepsilon(T_s)\gamma$. Since $\mathcal{H}$ is generated by $\{T_s : s = s_\alpha \text{ simple}\}$, the proof of the theorem thereby reduces to showing that (3) holds for all simple reflections $s = s_\alpha$ if $d \gg 0$. This is carried out in [**15**, §4].

In contrast to Theorem 1, we mention the following beautiful result due to Webb. It suggests a possible method (involving parabolic subgroups) for calculating the $G(q)$-cohomology for those values of q where Theorem 1 is not valid.

THEOREM 2 [**48**]. *Assume G is a semisimple, simply connected algebraic group defined over $\mathbf{F}_p$. Let V be a finite-dimensional $G(q)$-module. For any integer $n \geq 0$, we have*

$$\sum (-1)^{|J|} \dim H^n(G_J(q), V) = 0,$$

where the summation runs over all subsets J of the set of restricted simple roots and where, for such a subset J, $G_J \supset B$ denotes the associated parabolic subgroup of G.

In fact, Webb proves a more general result having to do with simplicial actions of arbitrary finite groups on simplicial complexes. He then deduces the above theorem from the case of $G(q)$ acting on the Tits building. The theorem may also be obtained as follows: We assume G is split (although essentially the same proof works in general). Let Π be the set of simple roots and for $J \subset \Pi$, let ξ_J be the dominant weight satisfying

$$(\xi_J, \alpha^\vee) = \begin{cases} q-1 & \text{for } \alpha \notin J, \\ 0 & \text{for } \alpha \in J. \end{cases}$$

Let S_J be the irreducible rational G-module of high weight ξ_J (and also denote its irreducible restriction to $G(q)$ by S_J). Then the induced module $\mathrm{ind}_{U(q)}^{G(q)}(k)$ contains a unique indecomposable summand Y_J whose head is isomorphic to S_J, and $\mathrm{ind}_{G_J(q)}^{G(q)}(k) \cong \bigoplus_{K \supset J} Y_K$ [**46**, Lemma 3.4]. By Möbius inversion and the Shapiro-Eckmann lemma, it follows for any $G(q)$-module V that

$$\begin{aligned} \dim H^n(G(q), V \otimes Y_J) &= \sum_{K \supset J} (-1)^{|K|} \dim H^n(G(q), V \otimes \mathrm{ind}_{G_K(q)}^{G(q)}(k)) \\ &= \sum_{K \supset J} (-1)^{|K|} \dim H^n(G_K(q), V), \end{aligned}$$

for all $J \subset \Pi$. Since $Y_\varnothing$ is the (projective) Steinberg module for $G(q)$, the desired formula results by setting $J = \varnothing$.

2. Algebraic group cohomology. In addition to the stability result given above in Theorem 1, early empirical results indicated that the dimension of $H^1(G(p^d), V)$ stabilizes as $d \to \infty$ for many finite-dimensional rational G-modules V. (This is vaguely suggested by the inequality (1) noted above, since the right-hand side becomes constant as $d \to \infty$.) In an attempt to explain these phenomena, attention naturally turned to the cohomology of the ambient algebraic group G. Theorem 4 below provides the connection between the two theories.

Let G be a linear algebraic group defined over the algebraically closed field k. Let $\mathbf{M}(G)$ be the category of rational G-modules. If $V \in \mathrm{Ob}(\mathbf{M}(G))$ has injective hull $I(V)'$ in the category of kG-modules, then the largest rational submodule $I(V)$ of $I(V)'$ is clearly the injective hull of V in the category $\mathbf{M}(G)$. Therefore, the category $\mathbf{M}(G)$ contains enough injectives, and the usual constructions of homological algebra are possible. In particular, the (rational) cohomology group $H^n(G, V)$ is defined as $R^nF_G(V)$, where R^nF_G is the nth right derived functor of the fixed point functor $F_G = (\text{-})^G$. Hochschild [**32**] first introduced and systematically studied these cohomology groups. (Another possible approach to rational cohomology for semisimple groups is suggested by [**20**, 4.6; **43**, 3.3].) For small values of n, these groups have the usual interpretations; e.g., for V, W rational G-modules with V finite-dimensional, $H^1(G, V^* \otimes W) \cong \mathrm{Ext}_G^1(V, W)$ classifies isomorphism classes of rational module extensions of V by W.

We now assume, as before, that G is a semisimple, simply connected algebraic group defined over $\mathbf{F}_p$. Let $X^+ = X^+(T)$ be the set of dominant weights defined by B. Let $B^\circ \supset T$ be the Borel subgroup opposite to B. For $\lambda \in X^+$, we denote by $\lambda|^G$ the rational G-module obtained by inducing the one-dimensional B^o-module defined by λ from B^o to G. Then $\lambda|^G$ is a finite-dimensional G-module whose dimension is, in fact, given by Weyl's dimension formula

$$\dim \lambda|^G = \prod_{\alpha>0} (\lambda+\rho,\alpha^\vee)/(\rho,\alpha^\vee)$$

(where ρ denotes the half-sum of the positive roots). Furthermore, the socle of $\lambda|^G$ is the irreducible G-module $S(\lambda)$ of high weight λ. For more information concerning these modules we refer to [**17**, **38**].

For a vector space V and integer r, let $V^{(r)}$ denote the k-space obtained from V by making $c \in k$ act as scalar multiplication by $c^{p^{-r}}$. If V is a rational G-module, then, for $r > 0$, $V^{(r)}$ identifies with the "twisted" rational G-module obtained by making G act on V through σ^r, where $\sigma : G \to G$ is the Frobenius endomorphism on G.

The following theorem provides a fundamental step in understanding the relationship between the cohomology of the finite Chevalley groups $G(q)$ and that of the ambient algebraic group G.

THEOREM 3 [**21**, Theorem 2.1, Corollary 3.3]. *Let G be a semisimple, simply connected algebraic group defined over $\mathbf{F}_p$ as above.*

(a) *Let V be a rational G-module. Then the restriction map*

$$\operatorname{res}^G_B(V) : H^*(G,V) \to H^*(B,V)$$

induces an isomorphism on rational cohomology.[3]

(b) *For dominant weights $\lambda, \mu \in X^+$, the module $\lambda|^G \otimes \mu|^G$ is G-acyclic, i.e., $H^n(G, \lambda|^G \otimes \mu|^G) = 0$ for all $n > 0$.*

In proving the above result, the connection between rational cohomology and the sheaf cohomology of vector bundles on the projective variety G/B plays a key role. In fact, if V is a rational B-module, let L_V be the corresponding induced homogeneous module for the structure sheaf $\mathcal{O}_{G/B}$ defined as follows: Over an open $U \subset G/B$, L_V has sections $\Gamma(U, L_V)$ consisting of all morphisms $f : \pi^{-1}(U) \to V$ satisfying $f(x \cdot b) = b^{-1} \cdot f(x)$ for $x \in \pi^{-1}(U)$, $b \in B$ (where $\pi : G \to G/B$ denotes the quotient map). Thus, in particular, for a rational B-module V, the rational G-module $H^0(G/B, L_V)$ identifies with the module obtained by inducing V from B to G. Part (a) is obtained from the Grothendieck spectral sequence

$$(4) \qquad E_2^{**} = H^*(G, H^*(G/B, L_V)) \Rightarrow H^*(B,V).$$

Observe first that if V is a rational G-module, then

$$H^n(G/B, L_V) \cong V \otimes H^n(G/B, L_k).$$

[3] Actually, this result is true when G is an arbitrary connected linear algebraic group (the same proof applies). B can also be replaced by any parabolic subgroup P [**40**] or, more generally, by an infinitesimal thickening PG_r (in the sense of [**18**, 4.2c]) of P (one reduces to the parabolic case by means of an elementary spectral sequence argument).

On the other hand,

$$H^n(G/B, L_k) \cong \begin{cases} k & \text{if } n = 0, \\ 0 & \text{if } n > 0, \end{cases}$$

by a special case of Kempf's vanishing theorem [**41**]. Thus, the spectral sequence (4) collapses, yielding the desired result. (Note that (4) implies the isomorphism $H^*(B, \mathcal{O}_G(G) \otimes V) \cong H^*(G/B, L_V)$ for all rational B-modules V.) For the proof of part (b) see [**21**, Corollary 3.3].

Theorem 3 suggests that a comparison between $H^*(G,V)$ and $H^*(G(q),V)$ can be obtained as follows. For simplicity we assume that G is split over $\mathbf{F}_p$. Embed $H^*(G(q),V)$ into $H^*(B(q),V)$ and compare $H^*(B,V)$ to $H^*(B(q),V)$. Let $\beta_1, \cdots, \beta_N$ be the roots of T in B. A comparison of corresponding spectral sequences shows that $\operatorname{res}^B_{B(q)}(V) : H^n(B,V) \to H^n(B(q),V)$ is an isomorphism for all $n \leq m$ provided that for all weights λ of T in V restriction induces an isomorphism of weight spaces

$$\left[\bigotimes_{i=1}^{N} H^{s_i}(U_{\beta_i}, k)\right]_{-\lambda} \to \left[\bigotimes_{i=1}^{N} H^{s_i}(U_{\beta_i}(q), k)\right]_{-\lambda|_{T(q)}} \tag{5}$$

for $s_1 = \cdots + s_N \leq m$ and an injection for $s_1 + \cdots + s_N = m+1$. Since

$$H^*(G_a, k) \cong \varprojlim_d H^*(G_a(p^d), k)$$

for the additive group $\mathbf{G}_a$ (cf. [**21**, Theorem 4.1]), the analysis of (5) is entirely arithmetic and ultimately leads to parts (a) and (b) of the following theorem.

THEOREM 4. *Let G be a semisimple, simply connected algebraic group defined over $\mathbf{F}_p$. Let V be a finite-dimensional rational G-module.*

(a) *For a fixed n, the groups $\{H^n(G(p^d),V)\}$ achieve a stable value $H^n_{\text{gen}}(G,V)$ as $d \to \infty$. We call $H^*_{\text{gen}}(G,V) \cong \varprojlim_d H^n(G(p^d),V)$ the generic cohomology of G at V.*

(b) *For a fixed n, $H^n_{\text{gen}}(G,V) \cong H^n(G,V^{(r)})$ if r is sufficiently large.*

(c) $H^*(G,V) \subset H^*_{\text{gen}}(G,V)$.

(d) *Generic cohomology can be interpreted as a derived functor. In fact, $H^*_{\text{gen}}(G,V) \cong H^*_{\text{dis}}(G(\overline{\mathbf{F}}_p),V)$, the discrete cohomology of the group $G(\overline{\mathbf{F}}_p)$ of $\overline{\mathbf{F}}_p$-rational points acting on the finite-dimensional $G(\overline{\mathbf{F}}_p)$-module V, where $\overline{\mathbf{F}}_p =$ algebraic closure of $\mathbf{F}_p$.*

Parts (a) and (b) are proved in [**21**, §§5, 6] for split groups. In addition, estimates are provided for integers e and f which guarantee that the restriction map $H^n(G,V^{(e)}) \to H^n(G(p^{e+f}),V^{(e)})$ is an isomorphism. (See also Theorem 6 below.) For the quasi-split groups, (a) and (b) have been established by Avrunin [**4**]. Note that Theorem 1 follows easily from these results.

Part (c) appears in [**16**] (with special cases proved earlier in [**21**]).

Cetjin [**13**] and Wang Jian-pan [**47**] showed independently that generic cohomology has an interpretation as a derived functor on a suitable category. The appropriate category consists of those kG-modules expressible as sums of submodules W such that $W^{(n)}$ is a rational G-module for some $n > 0$ (depending on W).

The striking fact that $H^*_{\text{gen}}(G, V) \cong H^*_{\text{dis}}(G(\overline{F}_p), V)$ for a finite-dimensional rational module V is due to van der Kallen (1986). His argument runs as follows. Let $C(n, V, q)$ (resp., $C(n, V, \infty)$) be the space of n-cochains $G(q)^n \to V$ (resp., $G(\overline{\mathbf{F}}_p)^n \to V$), and denote the corresponding sets of n-coboundaries (resp., n-cocycles) by $B(n, V; q)$ and $B(n, V; \infty)$ (resp., $Z(n, V; q)$ and $Z(m, V; \infty)$). Then $C(n, V; \infty) \cong \varprojlim C(n, V, q)$ and $Z(n, V; \infty) \cong \varprojlim Z(n, V; q)$. Since the spaces $C(n, V; q)$ and $Z(n, V; q)$ are finite-dimensional, the usual Mittag-Leffler condition on inverse systems is satisfied, so that $B(n, V; \infty) \cong \varprojlim B(n, V; q)$ and $H^n_{\text{dis}}(G(\overline{\mathbf{F}}_p), V) \cong \varprojlim H^n(G(q), V)$.

3. Infinitesimal theory. We begin with some preliminary remarks concerning group-schemes. If only as an aid to intuition, it is often convenient to adopt a functorial point of view. Thus, if $K = \operatorname{Spec} A$ is a affine group scheme over k, we prefer to view K as a representable functor

$$K \colon \mathbf{Com}_{k\text{-alg}} \longrightarrow \mathrm{Gr}$$

from the category $\mathbf{Com}_{k\text{-alg}}$ of commutative k-algebras to the category $\mathbf{Gr}$ of groups given by

$$K(B) = \mathrm{Hom}_{\mathbf{Com}_{k\text{-alg}}}(A, B).$$

This makes sense because the Hopf algebra structure on A endows $K(B)$ with a natural group structure. From this vantage point, a rational K-module (i.e., an A-comodule) is a k-vector space V such that for every $B \in \mathrm{Ob}(\mathbf{Com}_{k\text{-alg}})$ the B-module $B \otimes V$ has a $BK(B)$-module structure, functorial in B. As in the case of linear algebraic groups, the category $\mathbf{M}(K)$ of rational K-modules contains enough injectives so that the rational cohomology groups $H^*(K, V)$ are defined for every $V \in \mathrm{Ob}(\mathbf{M}(K))$. For a detailed discussion of these matters, see [**23**, Chapter II; **18**].

For our purposes below, the relevant group-schemes arise as certain kernels of endomorphisms of linear groups. Let G be a connected, linear algebraic group defined over $\mathbf{F}_p$ with Frobenius endomorphism $\sigma : G \to G$. For an integer $r \geq 1$, we define G_r to be the group-scheme theoretic kernel of σ^r. Thus, for any commutative k-algebra B, $G_r(B) = \operatorname{Ker}[\sigma^r(B) : G(R) \to G(R)]$. For each r, G_r is a normal, connected subgroup-scheme of G, infinitesimal in the sense that its representing coordinate algebra is finite-dimensional over k.

It is possible to give an alternative, highly useful description of the category $\mathbf{M}(G_1)$ of rational modules for the first infinitesimal subgroup G_1. Recall that in positive characteristic, the Lie algebra $\mathcal{G}$ of G is equipped with a $[p]$-operator $X \mapsto X^{[p]}$ $(X \in \mathcal{G})$ which makes it into a restricted Lie algebra in the sense of Jacobson [**34**, Chapter, V, §7]. Let $V(\mathcal{G}) = U(\mathcal{G})/\langle X^{[p]} - X : X \in \mathcal{G}\rangle$ be the corresponding restricted enveloping algebra, realized as a finite-dimensional quotient algebra of the universal enveloping algebra $U(\mathcal{G})$ of $\mathcal{G}$. Then the coordinate algebra of G_1 is a finite-dimensional Hopf algebra whose dual algebra identifies with $V(\mathcal{G})$ [**23**, II, §7.4], and hence $\mathbf{M}(G_1)$ is isomorphic to the category $V(\mathcal{G})$-mod of $V(\mathcal{G})$-modules.

EXAMPLE. Consider the special case $G = \mathrm{SL}_n$. We view G as the functor

$$G : \mathbf{Com}_{k\text{-alg}} \longrightarrow \mathbf{Gr}, \qquad R \longmapsto \mathrm{SL}_n(R).$$

Then G_r is the functor assigning to each commutative k-algebra B the group

$$G_r(B) = \{[a_{ij}] \in \mathrm{SL}_n(B) : a_{ij}^p = \delta_{ij}\}.$$

If V is a rational G_1-module, $X \in sl_n = \mathrm{Lie}(\mathrm{SL}_n)$ acts on V in the following way. Let $k[\varepsilon]$ be the k-algebra of dual numbers ($\varepsilon^2 = 0$) and consider $I_n + \varepsilon X \in (\mathrm{SL}_n)_1(k[\varepsilon])$. For $v \in V$, define $X \cdot v$ by means of the equation

$$(I_n + \varepsilon X)(1 \otimes v) = (1 \otimes v) + (\varepsilon \otimes X \cdot v).$$

Given a rational G-module V, we have for each $r \geq 1$ a Hochschild-Serre spectral sequence [**17**, 4.5]

$$E_2^{**} = H^*(G/G_r, H^*(G_r, V^{(r)}) \Rightarrow H^*(G, V^{(r)}).$$

By definition, G_r acts trivially on $V^{(r)}$. It is possible to show that the quotient group G/G_r (which requires some care to define [**23**, III, §3]) identifies with G [**74**, Lemma 3.6], and the above spectral sequence can be rewritten as

$$E_2^{**} = H^*(G, H^*(G_r, k)^{(-r)} \otimes V) \Rightarrow H^*(G, V^{(r)}). \tag{6}$$

In light of Theorem 4, this suggests the importance of the cohomology ring $H^*(G_r, k)$ in determining the finite group cohomology $H^*(G(q), V)$ for a semisimple group G and a rational G-module V.

For a restricted Lie algebra $\mathcal{G}$ defined over k, we define its nullcone to be the closed, homogeneous subvariety

$$\mathcal{N} \equiv \mathcal{N}(\mathcal{G}) = \{X \in \mathcal{G} : X \text{ is } [p]\text{-nilpotent}\}$$

of $\mathcal{G}$. Let A be the coordinate algebra of $\mathcal{N}$, and consider the natural G-equivariant surjective k-algebra map $S(\mathcal{G}^*) \to A$ from the algebra of polynomial functions on $\mathcal{G}$ to A. We grade $S(\mathcal{G}^*)$ by giving $\mathcal{G}^*$ homogeneous degree 2, and we give A the induced grading (which, of course, is concentrated in even degrees). With this notation, we can state the following result.

THEOREM 5 [**26**, **2**]. *Let G be a semisimple, simply connected algebraic group defined and split over $\mathbf{F}_p$.*

(a) *Assume that $p > h$, the Coxeter number of G. There is a G-equivariant isomorphism $H^*(G_1, k) \cong A^{(1)}$ of graded k-algebras.*

(b) *Assume that p is good for G in the sense of* [**45**, *p.* 178]. *As a rational G-module, A^{2n} has a filtration with sections isomorphic to induced modules $\lambda|^G$ ($\lambda \in X^+$). The multiplicity of $\lambda|^G$ as a section in A^{2n} equals*

$$d_n(\lambda) = \sum_{w \in W} sn(w) p_n(w(\lambda + \rho) - \rho),$$

where the partition function $p_n(\mu)$ is defined as the number of ways the weight μ can be written as a sum of exactly n positive roots (repetitions allowed).

The key step (due to Friedlander and the author) in the proof of the first part of the above theorem hinges on the application of a spectral sequence (discovered by May [**42**])

$$E_2^{2m,n}(M) = S^m(\mathcal{H}^*)^{(1)} \otimes H^n(U(\mathcal{G}), M) \Rightarrow H^{2m+n}(V(\mathcal{H}), M), \tag{7}$$

relating the cohomology of the restricted enveloping algebra $V(\mathcal{H})$ to that of the universal enveloping algebra $U(\mathcal{H})$ of a restricted Lie algebra $\mathcal{H}$ when $p \neq 2$. Using this spectral sequence for $\mathcal{H} = \mathcal{G}$ and $M = k$, the above theorem was proved for $p \geq 3h-1$ (> 23 in type G_2) in [**26**, Theorem 1.4]. An essential feature involved the use of the G-action on the spectral sequence and the determination of the cohomology algebra $H^*(U(\mathcal{G}), k)$ of the universal enveloping algebra $U(\mathcal{G})$ (cf. [**26**, Theorem 1.2]). Later, Andersen and Jantzen [**2**] gave a somewhat different argument (still using (7)) which permitted the better bound on p. In fact, for G simple they proved for $p > h$ that

$$H^i(G_1, (w\rho - \rho + p\lambda)|^G)^{(-1)} \cong (S^{(i-l(w))/2}(\mathcal{U}^{o^*}) \otimes \lambda)|^G$$

for $\lambda \in X^+$, $w \in W$, and $w\rho - \rho + p\lambda \in X^+$ (cf. [**2**, Corollary 3.7]). Here $\mathcal{U}^o = \mathrm{Lie}(U^o)$, the Lie algebra of the unipotent radical of B^o, and $|^G$ is the induction functor from rational B^o-modules to rational G-modules. (When G is of exceptional type E or F, it is required that $\lambda = 0$ or that λ be strongly regular in the sense that $(\lambda + \rho, \alpha^\vee) \geq h - 1$ for all simple roots α.) When $\lambda = 0$ and $w = 1$ this result implies the desired calculation of $H^*(G_1, k)$. Finally, the second part of the theorem is obtained for large p in [**26**, Corollary 2.5] and for p good in [**2**, 4.4, 4.6].

Making use of Theorem 5, the program outlined above for obtaining information concerning finite group cohomology can be carried out. We first introduce some notation. Assume, for simplicity, that G is of simple type. For a dominant weight λ, let $c(\lambda)$ denote the largest coefficient of λ when λ is expressed as a linear combination of simple roots. For a dominant weight λ, we will consider the p-adic expansion

$$\lambda = \lambda_0 + p\lambda_1 + \cdots + p^{d-1}\lambda_{d-1}, \tag{8}$$

where each λ_i lies in the restricted range (i.e., $0 \leq (\lambda_i, \alpha^\vee) \leq p - 1$ for all simple roots α; it is not assumed that $\lambda_{d-1} = 0$). Finally, let f be the index of the root lattice in the weight lattice of G and let ν (respectively, μ) be the maximal root (resp., maximal short root) in G.

THEOREM 6. *Let G be a simple, simply connected algebraic group defined and split over $\mathbf{F}_p$ as above. Let n be a nonnegative integer and λ a dominant weight with p-adic expansion* (8). *Assume for each i, $0 \leq i < d$, that*

$$(\lambda_i + (n/2)\nu, \mu^\vee) < p - 2h + 2 \quad \text{and} \quad c(\nu)fn + c(f\lambda_i) < p - 1.$$

Then for any integer m satisfying $2m < n$ *we have*

$$\dim H^{2m-1}(G(F_{p^d}), S(\lambda)) = 0$$

and (*in the notation of Theorem* 5)

$$\dim H^{2m}(G(F_{p^d}), S(\lambda)) = \sum_{m_0+\cdots+m_{d-1}=m} \prod_{i=0}^{d} d_{m_i}(\lambda_i).$$

Friedlander proved this in [**24**]. The theorem extends (and corrects some minor arithmetical errors of) [**28**, Theorem 3.3], which treated the case $d = 1$. The proof goes by induction on d and turns on the crucial steps of first observing that $H^n(G_r, k)^{(-r)} \cong H^n(G_1, k)^{(-1)}$ for n in a range [**25**, 4.3], then employing the spectral sequence (6) (which collapses for $d = 1$ in a range thanks to Theorem 5(b) and Theorem 3(b)) to relate the computation of $H^*(G, S(\lambda)^{(r)})$ to that of $H^*(G_r, k)^{(-r)}$, and finally using arithmetic considerations. It is interesting to note that if $\iota_d : G \to G^d = G \times \cdots \times G$ (d copies) is the diagonal embedding whose composition with projection onto the ith coordinate is σ^{i-1}, then the above theorem implies that ι_d induces an isomorphism on cohomology in a range for all irreducible G^d-modules having sufficiently small dominant weights.

4. Support varieties. In this section we discuss some of the recent work on support varieties for restricted Lie algebras. This subject has drawn considerable inspiration from the analogous situation for finite groups, as developed by various authors, e.g., [**1**, **5**, **7**–**10**]. Motivated by a remarkable projectivity criterion for representations of elementary p-groups due to Dade [**22**, Lemma 11.8], Carlson was led to formulate a conjecture relating the "rank variety" for such a representation to its associated "cohomological support variety." By recasting the problem in terms of an equivalent one involving abelian restricted Lie algebras (with trivial $[p]$-operator), Avrunin and Scott [**5**] obtained the first proof of Carlson's conjecture. Inspired by this work, but especially by Theorem 5 above, E. Friedlander and the author formulated the notion of a cohomological support variety for general restricted Lie algebras and proved Theorem 8 below, which extends the Avrunin-Scott theorem to arbitrary restricted Lie algebras. With the aid of this result, many of the theorems for finite groups generalize easily to the case of Lie algebras (e.g., Theorem 10 below).

We work with a general restricted Lie algebra $\mathcal{G}$. Given a linear functional φ on $\mathcal{G}$, we make the Lie algebra $\mathcal{G} \times k$ into a restricted Lie algebra by defining

$$(X, c)^{[p]} = (X^{[p]}, \varphi(X)^p).$$

Following Hochschild [**33**], this provides a map

$$\Phi^2 : \mathcal{G}^* \to H^2(V(\mathcal{G}), k)^{(-1)},$$

which extends to a k-algebra homomorphism

$$\Phi^* : S(\mathcal{G}^*) \to H^{\mathrm{ev}}(V(\mathcal{G}), k)^{(-1)} \subset H^*(V(\mathcal{G}), k)^{(-1)}.$$

If M is a finite-dimensional $V(\mathcal{G})$-module, then the $H^*(V(\mathcal{G}),k)^{(-1)}$-module $H^*(V(\mathcal{G}),M)^{(-1)}$ can be viewed as a module for the symmetric algebra $S(\mathcal{G}^*)$ via the map Φ^*. Using the spectral sequence (7) (and a slightly different argument when $p=2$), the following finiteness result can be established.

THEOREM 7 [**28**, Theorem 1.4]. *Let $\mathcal{G}$ be a finite-dimensional restricted Lie algebra and let M be a finite-dimensional $V(\mathcal{G})$-module. Then $H^*(V(\mathcal{G}),M)^{(-1)}$ is a finite $S(\mathcal{G}^*)$-module. In particular, the cohomology algebra $H^*(V(\mathcal{G}),k)$ is a Noetherian k-algebra.*

Let $|\mathcal{G}|$ be the affine algebraic variety associated to the algebra $H^{\mathrm{ev}}(V(\mathcal{G}),k)$ of even-dimensional cohomology. Thus, if R denotes the radical of $H^{\mathrm{ev}}(V(\mathcal{G}),k)$, then $|\mathcal{G}|$ has coordinate algebra $H^{\mathrm{ev}}(V(\mathcal{G}),k)/R$. The algebra homomorphism Φ^* induces a finite morphism

$$\Phi : |\mathcal{G}| \to \mathcal{G} = \operatorname{Spec} S(\mathcal{G}^*)$$

of affine varieties. If M is a finite-dimensional $V(\mathcal{G})$-module, we define the (cohomological) support variety $|\mathcal{G}|_M$ of M to be the support in $|\mathcal{G}|$ of the $H^{\mathrm{ev}}(V(\mathcal{G}),k)^{(-1)}$-module $H^*(V(\mathcal{G}),M^*\otimes M)$. Because $|\mathcal{G}|$ is not well understood for a general restricted Lie algebra $\mathcal{G}$, it is often easier to work with the closed subvarieties $\Phi(|\mathcal{G}|_M)$ of $\mathcal{G}$. These can be described explicitly as follows. Let I_M be the kernel of the composite map

$$S(\mathcal{G}^*) \xrightarrow{\Phi^*} H^{\mathrm{ev}}(V(\mathcal{G}),k)^{(-1)} \xrightarrow{\tau} H^*(V(\mathcal{G}),M^*\otimes M)^{(-1)},$$

where τ is induced by the natural algebra inclusion $k \to M^*\otimes M \cong \operatorname{End}(M)$ identifying $c\in k$ to the scalar transformation $c\cdot \mathrm{Id}_M$. Then $\Phi(|\mathcal{G}|_M) = V(I_M)$, the variety defined by the ideal $I_M \subset S(\mathcal{G}^*)$.

EXAMPLES. If $\mathcal{G} = \operatorname{Lie}(G)$ for G as in Theorem 5(a), then $|\mathcal{G}| \cong \mathcal{N}(\mathcal{G})$, the nullcone of $\mathcal{G}$. If $\mathcal{P} = \operatorname{Lie}(P)$ for $P\subset G$ a parabolic subgroup of G, then again $|\mathcal{P}| \cong \mathcal{N}(\mathcal{P})$ [**28**, Remark 2.7c].

Jantzen has proved the following beautiful theorem, which represents a considerable generalization of Theorem 5. The difficult part of the proof involves the case $\mathcal{G} = gl_n$. The general result then falls readily by choosing an embedding $\mathcal{G} \subset \mathrm{gl}_n$.

THEOREM 8 [**35**, SATZ 2.14]. *Let $\mathcal{G}$ be a restricted Lie algebra. Then*

$$\Phi(|\mathcal{G}|) = \{X\in\mathcal{G} : X^{[p]} = 0\}.$$

When $\mathcal{G}$ is the Lie algebra of a semisimple group G, $X^{[p]}=0$ for all $X\in\mathcal{N}$ if p is larger than the Coxeter number h of G. Thus, in the above examples, Φ is an isomorphism onto its image.

In order to state the next result we introduce the important notion of the rank variety $R(M)$ of a $V(\mathcal{G})$-module M. If $X\in\mathcal{G}$, let $\langle X\rangle$ denote the restricted subalgebra of $\mathcal{G}$ generated by X. Then we set

$$R(M) = \{X\in\mathcal{G} : X^{[p]}=0 \text{ and } M|_{V(\langle X\rangle)} \text{ is not projective}\} \cup \{0\}.$$

The following result links the geometric notion of the cohomological support variety with the representation-theoretic notion of the rank variety.

THEOREM 9 [**30**, COROLLARY 1.4]. *For a restricted Lie algebra $\mathcal{G}$ and a finite-dimensional $V(\mathcal{G})$-module M, we have*

$$\Phi(|\mathcal{G}|_M) = R(M).$$

This result is established in the special case where $\mathcal{G}$ is $[p]$-unipotent in [**29**]. The argument proceeds by induction on $\dim \mathcal{G}$ and uses an amalgamation of the Hochschild-Serre spectral sequence with the spectral sequence (7) above. The case in which $\mathcal{G}$ is the Lie algebra of a linear algebraic group G then follows from the result [**30**, Theorem 1.2]: $\Phi(|\mathcal{G}|_M)$ identifies with the orbit $G\cdot\Phi(|\mathcal{B}|_M)$ under the adjoint action of G on ($\mathcal{G}$ like), where $\mathcal{B} = \mathrm{Lie}(B)$ is a Borel subalgebra. The general theorem finally ensues as a consequence of Theorem 8 above.

REMARKS (a) It is easy to see that a finite-dimensional $V(\mathcal{G})$-module M is projective if and only if $|\mathcal{G}|_M \cong \Phi(|\mathcal{G}|_M) = \{0\}$. The above theorem thus gives an important way of testing the projectivity of modules: M is projective if and only if $M|_{V(\langle X\rangle)}$ is projective for all $X \in \mathcal{G}$ satisfying $X^{[p]} = 0$. Note that for such an $X \neq 0$, $V(\langle X\rangle) \cong k[C_p]$, where C_p is the cyclic group of order p. The reader should compare the above projectivity criterion with Chouinard's theorem [**12**] for a finite group G: a finite kG-module M is projective if and only if it is projective for every elementary abelian p-subgroup E of G.

(b) Simple considerations involving the Jordan canonical form show that if M, N are nonprojective $V(\langle X\rangle)$-modules for $X \in \mathcal{G}$ with $X^{[p]} = 0$, then $M\otimes N$ is not projective. It follows from Theorem 9 that for arbitrary restricted $\mathcal{G}$-modules M and N we have $\Phi(|\mathcal{G}|_{M\otimes N}) = \Phi(|\mathcal{G}|_M) \cap \Phi(|\mathcal{G}|_N)$ (cf. [**30**, 2.1] for a slightly different argument).

(c) Suppose G is a simple, simply connected algebraic group and $p \neq 2$ in types B_l, C_l, or F_4 and $p \neq 3$ in type G_2. Then the G-stable subvariety $\Phi(|\mathcal{G}|)$ of $\mathcal{G}$ has a unique nonzero minimal orbit $G\cdot e_\alpha$, where $e_\alpha \in \mathcal{G}$ is a root vector corresponding to a long root α [**44**]. Thus, in the case, a finite-dimensional rational G-module M is a projective $V(\mathcal{G})$-module if and only if $M|_{V(\langle e_\alpha\rangle)}$ is projective for some long root vector e_α. It is also possible to obtain the main projectivity criterion (in the special case $r = 1$) of [**19**] from Thecrem 9 (cf. [**29**, **30**] for details). Note further that if M and N are rational G-modules, then $M \otimes N$ is projective for $V(\mathcal{G})$ if and only if either M or N is projective.

EXAMPLE. Suppose $G \cong \mathrm{Sp}_{2n}$ and $p = 2$.

Let $\lambda_1, \ldots, \lambda_n$ be the fundamental dominant weights, labeled according to the above Dynkin diagram. Let $M = S(\lambda_1 + \cdots + \lambda_{n-1})$ and $N = S(\lambda_n)$ be the irreducible G-modules of high weights $\lambda_1+\cdots+\lambda_{n-1}$ and λ_n, respectively. Then $\dim M \leq \dim(\lambda_1 + \cdots + \lambda_{n-1})|^G = 2^{n^2-n}$ (using Weyl's dimension formula). Also, since $p = 2$, N has the same dimension (namely, 2^n) as $S(\lambda_n)$ for the group of type B_n. Since the irreducible Steinberg module St of high weight ρ has dimension 2^{n^2} and occurs as a section in $V\otimes W$, it follows that $M\otimes N \cong \mathrm{St}$. By [**44**], the $\mathcal{N}\backslash\{0\}$ contains the two minimal orbits $G\cdot e_{\alpha_1}$ and $G\cdot e_{\alpha_n}$, while

$\mathcal{N}\backslash(G \cdot e_{\alpha_1} \cup G \cdot e_{\alpha_n} \cup \{0\})$ contains a unique minimal orbit. Therefore, since $\Phi(|\mathcal{G}|_V) \cap \Phi(|\mathcal{G}|_W) = \Phi(|\mathcal{G}|_{\mathrm{St}}) = \{0\}$, $\Phi(|\mathcal{G}|_M)$ equals $\overline{G \cdot e_{\alpha_1}}$ or $\overline{G \cdot e_{\alpha_n}}$, and similarly for $\Phi(|\mathcal{G}|_N)$. To decide which is which, assume that $n > 3$ and suppose that $\Phi(|\mathcal{G}|_N) = \overline{G \cdot e_{\alpha_n}}$, so that N is a projective $V(\langle e_\alpha \rangle)$-module for each short root α. If G' denotes the Levi subgroup of G with simple roots $\{\alpha_1, \dots, \alpha_{n-1}\}$, it follows from Remark (c) above that N is a projective $V(\mathcal{G}')$-module, where $\mathcal{G}' = \mathrm{Lie}(G')$. Let $\mathcal{U}' = \mathrm{Lie}(U')$, where U' is the unipotent radical of a Borel subgroup of G'. Since the restricted enveloping algebra $V(\mathcal{U}')$ is a local algebra over which $V(\mathcal{G}')$ is free, we obtain that N is a free $V(\mathcal{U}')$-module. Therefore, $2^n = \dim N \geq \dim V(\mathcal{U}') = 2^{n(n-1)/2}$, which is absurd for $n > 3$. Thus, we have

$$\Phi(|\mathcal{G}|_M) = \overline{G \cdot e_{\alpha_n}} \quad \text{and} \quad \Phi(|\mathcal{G}|_N) = \overline{G \cdot e_{\alpha_1}}.$$

We leave to the interested reader the verification that this is still the case in the exceptional cases $n = 2$ or 3.

As a consequence of Theorem 9 above, we obtain the following result, whose proof is largely an adaptation to the present situation of similar results for finite groups due to Carlson [**10**].

THEOREM 10 [**30**, Theorem 2.2]. *Let $\mathcal{G}$ be a restricted Lie algebra over k. Any closed, homogeneous subvariety of $\Phi(|\mathcal{G}|)$ is of the form $\Phi(|\mathcal{G}|_M)$ for some finite-dimensional $V(\mathcal{G})$-module M. If M is a finite-dimensional $V(\mathcal{G})$-module which is indecomposable, then the projective variety* $\mathrm{Proj}(\Phi(|\mathcal{G}|_M))$ *is connected.*

For example, if $M = k$, the trivial module, then $\mathrm{Proj}(\{X \in \mathcal{G} : X^{[p]} = 0\})$ is connected for any restricted Lie algebra $\mathcal{G}$ (cf. [**30**, Corollary 2.3]). Little else appears to be known concerning the structure of the variety $\{X \in \mathcal{G} : X^{[p]} = 0\}$ in general (e.g., when it is irreducible). Also, if $\mathcal{G} = \mathrm{Lie}(G)$ for a semisimple algebraic group, it is not known if any G-stable, closed, homogeneous subvariety of $\Phi(|\mathcal{G}|)$ has the form $\Phi(|\mathcal{G}|_M)$ for some finite-dimensional rational G-module. However, a recent result of Jantzen for SL_n, which we now describe, points to an answer in the affirmative.

Let G be a semisimple, simply connected algebraic group defined over k. For a subset J of the set of simple roots Π (relative to the fixed Borel subgroup B) let R_J be the subroot system generated by J of the root system R of G. Let $G_J = L_J . U_J \supset B$ be the associated parabolic subgroup with Levi factor L_J and unipotent radical U_J. Write $\mathcal{U}_J$ for the Lie algebra of U_J. For a dominant weight λ, let R_λ be the set of all roots α satisfying $(\lambda + \rho, \alpha^\vee) \equiv 0 \bmod p$. Thus, R_λ is a closed subsystem of R. If p is good [**45**, p. 178], it can be shown that there exists an element $w \in W$ and a subset $J \subset \Pi$ such that $w(R_\lambda) = R_J$. In this case we define $\mathcal{N}(\lambda) = G \cdot \mathcal{U}_J$, the (closed) subvariety consisting of all nilpotent elements of $\mathcal{G}$ conjugate under the adjoint action of G on $\mathcal{G}$ to an element of $\mathcal{U}_J$. Then we can state the following result, which determines the image under Φ of the cohomological support varieties for the induced modules $\lambda|^G$ in case $G = \mathrm{SL}_n$.

THEOREM 11 [**36**, **37**]. *Assume* $G = \mathrm{SL}_n$. *Let* $\lambda \in X^+$ *be a dominant weight and put* $W = \lambda|^G$. *Then, with the above notation, we have*

$$\Phi(|\mathcal{G}|_W) = \mathcal{N}(\lambda).$$

The inclusion $\Phi(|\mathcal{G}|_W) \supset \mathcal{N}(\lambda)$ is proved in [**36**, 3.7(1)] (as largely a consequence of Weyl character and dimension formula considerations). On the other hand, it is shown in [**35**, 4.14(1)] that if p is at least equal to the Coxeter number of G, then $\Phi(|\mathcal{G}|_W)$ is properly contained in $\mathcal{N}$ for any dominant weight λ for which R_λ is not empty. Using this fact, the inclusion $\Phi(|\mathcal{G}|_W) \subset \mathcal{N}(\lambda)$ is obtained in [**37**, 2.6] by reducing to an appropriate G_J and using a kind of BGG (or relative Koszul) resolution. We also remark that Jantzen asks if the stated conclusion of Theorem 11 remains true for general semisimple groups G when p is good.

Finally, we mention that [**31**] contains a development of support varieties for nonrestricted $\mathcal{G}$-modules.

5. Open questions. It is appropriate to conclude with several open questions which are worthy of further study. Throughout this section, we let G be as in §1.

1. Understand $H^n(G(q), V)$ better for small values of q and large values of n. Perhaps Webb's theorem (mentioned in §1) will be useful here.

2. For $r > 1$, understand the cohomology of G_r better. For example:

(a) Is the k-algebra $H^*(G_r, k)$ Noetherian? This is known to be true for $r \leq 2$ (subject to some restriction on the size of p when $r = 2$) [**28**, Theorem 1.11]. The result is also known in general for solvable groups [**28**, Proposition 1.12].

(b) Let $H^* = \varinjlim H^*(G_r, k)^{(-r)}$. For p suitably restricted, this limit is stable [**27**] in any given cohomological degree. Determine this stable cohomology ring!

(c) Perhaps a more modest problem would be to understand the projectivity criterion of [**19**, Main Theorem] for $r > 1$ from the point of view explored in §4 above.

3. Relate the theory discussed in §4 for the restricted Lie algebra $\mathcal{G}$ directly to the support variety for the finite Chevalley groups $G(q)$. Note that $\mathcal{N}$ is G-equivariantly isomorphic to the unipotent variety $\mathcal{U} \subset G$ of all unipotent elements (although not canonically) (cf. [**45**]). A somewhat easier question is the following: If M is a rational G-module such that $M|_{V(\mathcal{G})}$ is projective, is M a projective $G(p)$-module?

REFERENCES

1. J. Alperin and L. Evens, *Representations, resolutions, and Quillen's dimension theorem*, J. Pure. Appl. Algebra **22** (1981), 1–9.

2. H. Andersen and J. Jantzen, *Cohomology of induced representations of algebraic groups*, Math. Ann. **269** (1984), 487–525.

3. G. Avrunin, *A vanishing theorem for second degree cohomology*, J. Algebra **53** (1978), 382–388.

4. ——, *Generic cohomology for twisted groups*, Trans. Amer. Math. Soc. **268** (1981), 247–253.

5. G. Avrunin and L. Scott, *Quillen stratification for modules*, Invent. Math. **66** (1982), 277–286.

6. N. Bourbaki, *Groupes et algèbres de Lie*. IV,V,VI, Hermann, 1968.

7. J. Carlson, *The complexity and varieties of modules*, Proc. Oberwolfach Conf. on Integral Representations and Applications, Lecture Notes in Math., vol. 882, Springer-Verlag, 1981.

8. ——, *Dimensions of modules and their restriction over modular group algebras*, J. Algebra **69** (1981), 95–104.

9. ——, *The varieties and the cohomology ring of a module*, J. Algebra **85** (1983), 104–143.

10. ——, *The variety of an indecomposable module is connected*, Invent. Math. **77** (1984), 291–299.

11. H. Cartan and S. Eilenberg, *Homological algebra*, Princeton University Press, Princeton, N.J., 1956.

12. L. Chouinard, *Projectivity and relative projectivity over group rings*, J. Pure Appl. Algebra **7** (1976), 287–302.

13. H. Cetjin, *Some cohomology of* $\mathrm{GL}_n(\mathbf{F}_q)$ *with coefficients and a category for generic cohomology*, Ph. D. Thesis, Northwestern University, 1985.

14. E. Cline, B. Parshall, and L. Scott, *Cohomology of finite groups of Lie type*. I, Inst. Hautes Études Sci. Publ. Math. **45** (1975), 169–191.

15. ——, *Cohomology of finite groups of Lie type*. II, J. Algebra **45** (1977), 182–198.

16. ——, *Detecting rational cohomology of algebraic groups*, J. London Math. Soc. **28** (1983), 293–300.

17. ——, *Cohomology, hyperalgebras, and representations*, J. Algebra **63** (1980), 98–123.

18. ——, *A Mackey imprimitivity theory for algebraic groups*, Math. Z. **182** (1983), 447–471.

19. ——, *On injective modules for infinitesimal algebraic groups*, J. London Math. Soc. **31** (1985), 277–291.

20. ——, *Stratifying algebraic representation categories*, J. Algebra (to appear).

21. E. Cline, B. Parshall, L. Scott, and W. van der Kallen, *Rational and generic cohomology*, Invent. Math. **39** (1977), 143–169.

22. E. Dade, *Endo-permutation modules*, Ann. of Math. (2) **108** (1978), 317–346.

23. M. Demazure and P. Gabriel, *Groupes algébriques*, Masson & Cie, Pairs, 1970.

24. E. Friedlander, *Cohomology of irreducible modules with large weights*, this volume.

25. E. Friedlander and B. Parshall, *Cohomology of algebraic and related finite groups*, Invent. Math. **74** (1983), 85–117.

26. ——, *Cohomology of Lie algebras and algebraic groups*, Amer. J. Math. **108** (1986), 235–253.

27. ——, *Limits of infinitesimal group cohomology*, Ann. of Math. Stud. (to appear).

28. ——, *Cohomology of infinitesimal and discrete groups*, Math. Ann. **273** (1986), 353–374.

29. ——, *Geometry of p-unipotent Lie algebras*, J. Algebra **109** (1987), 25–45.

30. ——, *Support varieties for restricted Lie algebras*, Invent. Math. **86** (1986), 553–562.

31. ——, *Modular representations of Lie algebras*, (to appear).

32. G. Hochschild, *Cohomology of linear algebraic groups*, Illinois J. Math. **5** (1961), 492–579.

33. ——, *Cohomology of restricted Lie algebras*, Amer. J. Math. **76** (1954), 555–580.

34. N. Jacobson, *Lie algebras*, Interscience, New York, 1962.

35. J. Jantzen, *Kohomologie von p Lie Algebren und nilpotente Elemente*, Abh. Math. Sem. Univ. Hamburg **76** (1986).

36. ——, *Restricted Lie algebra cohomology*, Proc. of Utrecht 1986 Conf. in Honor of T. A. Springer (to appear).

37. ——, *Support varieties for Weyl modules*, Bull. London Math. Soc. **19** (1987), 238–244.

38. ——, *Representations of algebraic groups*, (to appear).

39. W. Jones and B. Parshall, *On the 1-cohomology of finite groups of Lie type*, Proc. Conf. Finite Groups, W. R. Scott and F. Gross, editors, Academic Press, New York, 1976.

40. W. van der Kallen, *Infinitesimally central extensions of* Spin_7 *in characteristic* 2, J. Pure Appl. Algebra **14** (1979), 39–49.

41. G. Kempf, *Linear systems on homogeneous spaces*, Ann. of Math. (2) **103** (1976), 557–591.

42. J. May, *The cohomology of restricted Lie algebras and Hopf algebras*, J. Algebra **3** (1966), 123–146.

43. B. Parshall, *Simulating algebraic geometry with algebra.* II: *Stratifying algebraic representation categories*, this volume.

44. N. Spaltenstein, *Classes unipotentes et sous-groupes de Bores*, Lecture Notes in Math., vol. 946, Springer-Verlag 1982.

45. T. Springer and R. Steinberg, *Conjugacy classes*, Lecture Notes in Math., vol. 131, Springer-Verlag 1968.

46. N. B. Tinberg, *Some indecomposable modules of groups with split* (B, N)*-pairs*, J. Algebra **61** (1979), 508–526.

47. Wang Jian-pan, *Quasi-rational modules and generic cohomology*, Northeastern Math. J. (Dongbei Shuxue) **1** (1985), 90–100.

48. P. Webb, *A local method in group cohomology* (to appear).

UNIVERSITY OF ILLINOIS, URBANA

Proceedings of Symposia in Pure Mathematics
Volume 47 (1987)

Double Cosets and Modular Representation Theory

GEOFFREY R. ROBINSON

Introduction. Modular representation theory of finite groups is the study of the relationship between the internal structure of finite groups and the structure of their group algebras over certain rings. There are many approaches to the subject. The approach that Brauer took when he began the subject was to investigate the relationship between "ordinary" (complex) characters and representations over fields of prime characteristic. The approach initiated by Green was to study the relationship between indecomposable modules for the group algebra and indecomposable modules of certain subgroups.

A common feature of both approaches is the attempt to reduce questions about the "global" properties of the group algebra of a finite group to questions about properties of group algebras of "local" subgroups. However, there are natural questions about group algebras whose answers are not "locally" determined. One such problem is the determination of the number of full matrix algebra summands of the group algebra kG (G a finite group, k an algebraically closed field of prime characteristic), which relates to problem 19 of Brauer [**1**].

The subject matter of this article is recent work by myself and others which was initiated by the results of [**13**]. In [**13**], a very explicit connection was established between the double-coset structure of a finite group relative to its Sylow p-subgroup and certain basic questions about the structure of the group algebra kG (for k, G as above). In particular, a solution to Brauer's problem 19 was given by a special case of the main theorem of [**13**], and in conjunction with Brauer's First Main Theorem (see §2.2), enough information was provided to determine the number of indecomposable algebra summands of kG in group-theoretic terms.

In response to a question of Olsson the methods of [**13**] were extended to obtain results about the number of isomorphism types of simple kG-modules of certain kinds from group-theoretic information. This time, in [**14**], it was necessary to study the double-coset structure of a finite group relative to Sylow p-subgroups of centralizers of arbitrary p-regular elements.

1980 *Mathematics Subject Classification* (1985 *Revision*). Primary 20C20.

The methods of **[13]** and **[14]** were heavily character-theoretic and computational. In the meantime, Külshammer **[8]** showed how to derive the results of **[13]** by ring-theoretic methods, and Broué **[3]** produced an elegant approach which derived the results of **[13]** using a bilinear form (influenced to some extent by a suggestion of Green).

In **[9]**, Külshammer simplified many of the calculations of **[14]** and proved other, new results by invoking the theory of lower defect groups (which had been the subject of renewed interest with the papers of Broué **[2]**, Burry **[5]**, and Olsson **[11]**), again using a self-contained ring-theoretic approach. Recently, Broué and I **[4]** have used the theory of bilinear forms and Hecke algebras (which in this context just means endomorphism rings of permutation modules) to derive the results of **[13]** by another method and place them in a more general context.

The connection between the matrix of **[13]** and Hecke algebras was first noticed by Landrock (although J. A. Green had also suggested the possibility of an endomorphism ring approach to the results of **[13]** to Broué and myself), and Landrock **[10]** has recently found a connection between the matrix of **[13]** and projective summands of certain permutation modules.

The approach I adopt in this presentation is influenced by all of the above, to one degree or another, but especially by discussions with Broué. The unifying theme is that of Hecke algebras, which provide a bridge between group-theoretic properties and the structure of the group algebra.

The results of §1.1 are due to Green **[6]**. The first person to study links between Hecke algebras and modular representation theory was L. L. Scott **[15]**. In §1.3, I use the Reynolds ideal, which was first studied by Reynolds **[12]**. I learned the approach to the Reynolds ideal taken in 1.3 from Broué (the result 1.3.1 is well known).

Even more explicit computations than those of 2.2.1 can be found in Külshammer **[9]** or Robinson **[14]**. The results of §2.2 are some of the original results of **[13]**, while those of §2.3 encompass some of the results of Külshammer **[9]** and Robinson **[14]**.

Other papers related to the subject matter of this article are Iizuka-Watanabe **[7]**, Shi **[16]**, and Wada **[17]**.

1. Techniques. G denotes a finite group, p a prime. We let R denote a complete discrete valuation ring of characteristic zero, whose quotient field contains a primitive $|G|$th root of unity, and π denote the unique maximal ideal of R. The residue class field R/π, which we denote by k, is assumed to be of characteristic p.

The group algebra kG is a direct sum of indecomposable two-sided ideals known as blocks. The unit elements of these blocks are precisely the primitive idempotents of $Z(kG)$. The natural epimorphism from $Z(RG)$ onto $Z(kG)$ yields a bijection between primitive idempotents of $Z(RG)$ and $Z(kG)$ (given a primitive idempotent $e \in Z(kG)$ with a preimage $E \in Z(RG)$, the element $\lim_{n\to\infty} E^{p^n}$ is easily seen to be an idempotent preimage of e, where the limit is taken with respect to the metric on RG which arises naturally from the valuation ring structure of R).

1.1. *G-algebras.* Let A be an algebra (either a finite-dimensional algebra over k, or an R-algebra which is free and finitely generated as an R-module) with unit element, on which G acts as a group of algebra automorphisms. Then A is known as a G-algebra. We denote by A^G the subalgebra of G-fixed points of A, and when H is a subgroup of G, we define the map $\mathrm{Tr}_H^G\colon A^H \to A^G$ by $\mathrm{Tr}_H^G(\alpha) = \sum_{x\in H/G} \alpha^x$ for all $\alpha \in A^H$ (where H/G denotes any right transversal to H in G). Then it is clear that $\mathrm{Tr}_H^G(A^H)$ is a two-sided ideal of A^G, and we denote this ideal by A_H^G. When M is an RG (or kG)-module, we define M_H^G analogously.

When i is a primitive idempotent of A^G, there is a subgroup, H, of G which is minimal subject to $i \in A_H^G$. Furthermore, H is unique up to conjugacy and is a p-group. We say that H is a defect group for the idempotent i.

REMARKS. When $A = kG$ (G acting by conjugation), the defect group of the unit element of a block is the same p-group as the defect group of that block in Brauer's original definition. A block has defect group $\{1_G\}$ if and only if it is isomorphic to a full matrix algebra over k.

When M is a kG-module, the algebra $\mathrm{End}_k(M)$ is a G-algebra in a natural fashion. A primitive idempotent of $\mathrm{End}_k(M)^G$ ($= \mathrm{End}_{kG}(M)$) corresponds to an indecomposable summand of M, and a defect group for the primitive idempotent is just a vertex for the corresponding indecomposable summand.

1.2. *Hecke algebras.* For a subset, X of G, we let $\hat{X} = \sum_{x\in X} x$ in RG (or in kG). When H is a subgroup of G, the left ideal $RG\hat{H}$ of the group ring RG is isomorphic as a left RG-module to $\mathrm{Ind}_H^G(R)$ (where H has trivial action on R). We let $E_R(H)$ denote $\mathrm{End}_R(RG\hat{H})$ and $E_k(H)$ denote $\mathrm{End}_k(kG\hat{H})$. Then $E_R(H)^G$ has a natural R-basis $\{\omega_x\colon x \in H\backslash G/H\}$, where $\omega_x(\hat{H}) = \widehat{HxH}$ for $x \in H\backslash G/H$, and there is a natural epimorphism from $E_R(H)^G$ onto $E_k(H)^G$.

For $x \in H\backslash G/H$, we have $\omega_x = \mathrm{Tr}^G_{H\cap xHx^{-1}}(\alpha_x)$ where $\alpha_x(\hat{H}) = x\hat{H}$, $\alpha_x(g\hat{H}) = 0$ for $g \in G\backslash H$ [so that $\alpha_x \in E_R(H)^{H\cap xHx^{-1}}$]. Hence we can easily see that:

LEMMA 1.2.1. *For $x, y \in H\backslash G/H$, we have:*

$$\mathrm{trace}(\omega_x\omega_y) = \begin{cases} 0 & \text{if } y^{-1} \notin HxH, \\ [G : H\cap yHy^{-1}] & \text{if } y^{-1} \in HxH. \end{cases}$$

The left multiplicative structure of RG gives rise to a natural algebra morphism $\sigma_H : RG \to E_R(H)$. It is clear that $\sigma_H(Z(RG)) \subset Z(E_R(H)^G)$ and that for any subgroup, L, of G we have $\sigma_H(RG_L^G) \subset E_R(H)_L^G$. We let t denote the trace function on $E_R(H)$ (or $E_K(H)$) and $^-$ denote images in $E_k(H)$.

We record for future use:

1.2.2. *Let $u, v \in G$ and $z \in C_G(H)$. Then we have:*

(i) $\sigma_H(\hat{u}^G) = \sum_{x\in H\backslash G/H} |u^G \cap xH|\omega_x$ *(where u^G denotes the conjugacy class of u).*

(ii)

$$t(\sigma_H(\hat{u}^G)\sigma_H(v)\omega_z) = \sum_{x \in H\backslash G/H} \frac{|C_G(v)|}{|H \cap xHx^{-1}|} |u^G \cap xH| \; |(v^{-1})^G \cap zxH|.$$

(iii) *If v and z are p-regular, and H is a p-group, then*

$$t(\sigma_H(v)\omega_z) = 0 \textit{ if } v \textit{ is not conjugate to } z^{-1},$$
$$t(\sigma_H(v)\omega_z) = [C_G(z) : H] \textit{ if } v \textit{ is conjugate to } z^{-1}.$$

PROOF. (i) We may certainly write $\sigma_H(\hat{u}^G) = \sum_{x \in H\backslash G/H} \lambda_x \omega_x$ where each $\lambda_x \in R$. We then have $\sigma_H(\hat{u}^G)[\hat{H}] = \sum_{x \in H\backslash G/H} \lambda_x \widehat{HxH}$. On the other hand, for $x \in H\backslash G/H$, the coefficient of x in $\sigma_H(\hat{u}^G)[\hat{H}]$ $(= \hat{u}^G \cdot \hat{H})$ is easily seen to be $|u^G \cap xH|$.

(ii) We note that $t(\sigma_H(\hat{u}^G)\sigma_H(v)\omega_z) = [G : C_G(v)]^{-1} t(\sigma_H(\hat{u}^G)\sigma_H(\hat{v}^G)\omega_z)$. It is easy to check that $\omega_x \cdot \omega_z = \omega_{zx}$ for any $x \in H\backslash G/H$. By part (i) we have $\sigma_H(\hat{u}^G) = \sum_{x \in H\backslash G/H} |u^G \cap xH| \omega_x$, and from part (i) and the above remark we have $\sigma_H(\hat{v}^G)\omega_z = \sum_{x \in H\backslash G/H} |v^G \cap z^{-1}xH| \omega_x$.

Using 1.2.1, we see that

$$t(\sigma_H(\hat{u}^G)\sigma_H(\hat{v}^G)\omega_z) = \sum_{x \in H\backslash G/H} |u^G \cap xH| |v^G \cap z^{-1}x^{-1}H| [G : H \cap xHx^{-1}].$$

The result follows, after noting that

$$\begin{aligned} |v^G \cap z^{-1}x^{-1}H| &= |(v^{-1})^G \cap Hxz| = |(v^{-1})^G \cap xzH| \\ &= |(v^{-1})^G \cap xHz| = |(v^{-1})^G \cap zxH|. \end{aligned}$$

(iii) This follows immediately from (ii), taking $u = 1_G$.

1.3. *The Reynolds ideal.* Let z be a p-regular element of G. The p'-section of z is $\{x \in G : \text{the } p'\text{-part of } x \text{ is conjugate in } G \text{ to } z\}$ and is denoted by S_z. The Reynolds ideal of $Z(kG)$ is the k-span of the p'-section sums in kG, and we denote it by $\mathrm{Rey}(kG)$. There is another natural basis for $\mathrm{Rey}(kG)$, namely $\{\sum_{g \in G} \beta(g^{-1})g : \beta \text{ is the } (k\text{-valued}) \text{ character of a simple } kG\text{-module}\}$.

We remark that if λ is a symmetric k-linear form on kG with $J(kG) \subset \ker \lambda$, then $\sum_{g \in G} \lambda(g^{-1})g \in \mathrm{Rey}(kG)$ (because λ induces a symmetric, k-linear form on $kG/J(kG)$, and because the trace form is (up to scalar multiples) the only symmetric k-linear form which can be defined on a full matrix algebra).

LEMMA 1.3.1. (i) *$\mathrm{Rey}(kG)$ is an ideal of $Z(kG)$.*
(ii) $\mathrm{Ann}_{kG}(\mathrm{Rey}(kG)) = J(kG)$.

PROOF. Let τ be an irreducible k-representation of G affording character β. Since the k-span of $\{\tau(g) : g \in G\}$ is the full $\deg \tau \times \deg \tau$ matrix algebra over k, an element $a \in kG$ satisfies $\beta(ag) = 0$ for all $g \in G$ if and only if $\tau(a)$ is the zero matrix.

Now for $h \in H$, $\sum_{g \in G} \beta(g^{-1})g \,.\, h = \sum_{g \in G} \beta(hg^{-1})g$, so for any $a \in kG$, $\sum_{g \in G} \beta(g^{-1})g \,.\, a = \sum_{g \in G} \beta(ag^{-1})g$. Hence a annihilates $\sum_{g \in G} \beta(g^{-1}) \,.\, g$ if and only if $\tau(a)$ is the zero matrix. Since τ was arbitrary, (ii) is proved.

For $a \in Z(kG)$, $\sum_{g\in G}\beta(g^{-1})g\,.\,a = \sum_{g\in G}\beta(ag^{-1})g$. Now the mapping $\lambda : kG \to k$ defined by $\lambda(x) = \beta(ax)$ is a symmetric k-valued linear form on kG with $J(kG) \subset \ker\lambda$. Thus $\sum_{g\in G}\lambda(g^{-1})g \in \mathrm{Rey}(kG)$, so $\sum_{g\in G}\beta(ag^{-1})g \in \mathrm{Rey}(kG)$, and (i) is proved.

The link with Hecke algebras is provided by:

PROPOSITION 1.3.2. *Let z be a p-regular element of G, and let $H \in \mathrm{Syl}_p(C_G(z))$. Then for any $g \in G$, we have*

$$t(\overline{\sigma_H(g)})(\overline{\omega}_z) = \begin{cases} |C_G(z)|_{p'}1_k & \text{if } g \text{ is in the } p'\text{-section of } z^{-1}, \\ 0 & \text{otherwise.} \end{cases}$$

PROOF. The form $\lambda : kG \to k$ defined by $\lambda(x) = t(\overline{\sigma_H(g)\overline{\omega}_z})$ for $g \in G$ (and extending by k-linearity) is a symmetric k-linear form with $J(kG) \subset \ker\lambda$ (for note that $\overline{\omega}_z$ commutes with $\overline{\sigma_H(g)}$ for each $g \in G$). Thus $\sum_{g\in G}\lambda(g^{-1})g \in \mathrm{Rey}(kG)$. The result now follows from Lemma 1.2.2, part (iii).

COROLLARY 1.3.3. *Let v, z be p-regular elements of G, and let $H \in \mathrm{Syl}_p(C_G(z))$. Then the coefficient of $\hat{S}_v$ in $\hat{S}_z\hat{u}^G$ (in kG) is given by*

$$\sum_{x\in H\backslash G/H}\left[\frac{|C_G(v)|}{|H\cap xHx^{-1}|}|u^G \cap xH|\;|v^G\cap zxH|\right]1_k(|C_G(z)|_{p'}1_k)^{-1}.$$

PROOF. Since $\hat{S}_z\hat{u}^G \in \mathrm{Rey}(kG)$, it suffices to compute the coefficient of v in $\hat{S}_z u^G$, which is the coefficient of 1_G in $\hat{S}_z\hat{u}^G v^{-1}$, which is in turn $|u^G v^{-1} \cap S_{z^{-1}}|1_k$.

By 1.3.2, the coefficient we seek is $(|C_G(z)_{p'}1_k)^{-1}t(\overline{\sigma_H(\hat{u}^G)\sigma_H(v^{-1})\overline{\omega}_z})$, and the result now follows from 1.2.2, part (ii).

PROPOSITION 1.3.4. *In the formula of* 1.3.3, *it is only necessary to sum over those $x \in H\backslash G/H$ such that:*

(i) *There is a conjugate of u in xH, say u', which centralizes $H \cap xHx^{-1}$.*

(ii) *There is a conjugate of v in xH, say v', such that $H\cap xHx^{-1}$ contains a Sylow p-subgroup of $C_G(v')$.*

PROOF. We note that $H \cap xHx^{-1}$ $(= H\cap zxHx^{-1}z^{-1})$ permutes conjugates of v within zxH. The length of the orbit of v' is $[H\cap xHx^{-1} : C_H(v')]$. Thus

$$\frac{|C_G(v)|}{|H\cap xHx^{-1}|}|v^G\cap zxH|$$

is an integer and is divisible by p unless there is a conjugate of v, say v', in zxH, with $C_H(v') \in \mathrm{Syl}_p(C_G(v'))$ (from which it follows that $H\cap xHx^{-1}$ contains a Sylow p-subgroup of $C_G(v')$ since $C_H(v') \le H\cap v'Hv'^{-1} = H\cap xHx^{-1}$).

Similarly, $H\cap xHx^{-1}$ permutes conjugates of u within xH, so $|u^G\cap xH|$ is divisible by p unless there is a conjugate of u, say u', in xH which centralizes $H\cap xHx^{-1}$.

2. Block-theoretic invariants.

2.1. *Projection onto the semisimple part.* The algebra $Z(kG)$ is a direct sum (as k-vector space) of $J(Z(kG))$ and a semisimple commutative algebra, A say (which is unique and is the k-span of the primitive idempotents of $Z(kG)$). We let s denote projection from $Z(kG)$ onto A with kernel $J(Z(kG))$.

The next result is quoted without proof and is most easily proved using block orthogonality relations.

LEMMA 2.1.1. *Let e be an idempotent of $Z(kG)$, $\hat{u}^G$ be a class sum in kG. Then:*

(i) *If u is p-regular, $e\,.\,\hat{u}^G$ is a k-linear combination of p-regular elements.*

(ii) *If u is p-singular, $e\,.\,\hat{u}^G$ is a k-linear combination of p-singular elements.*

Now we are in a position to prove:

PROPOSITION 2.1.2. *Let v, z be p-regular elements of G, $u \in G$. Then the coefficient of v in $s(\hat{u}^G)\,.\,\hat{z}^G$ is:*

$$\sum_{x \in H\backslash G/H} (|C_G(z)|_{p'} 1_k)^{-1} \left[\frac{|C_G(v)|}{|H \cap xHx^{-1}|} |u^G \cap xH|\ |v^G \cap zxH|\right] 1_k,$$

(where $H \in \mathrm{Syl}_p(C_G(z))$).

PROOF. Since $\mathrm{Rey}(kG) \subset \mathrm{Ann}_{kG}(J(kG))$, we have $\hat{u}^G\,.\,\hat{S}_z = s(\hat{u}^G)\,.\,\hat{S}_z$. Now $\hat{S}_z = \hat{z}^G+$ some combination of p-singular class sums, and $s(\hat{u}^G)$ is a k-linear combination of idempotents, so by 2.1.1 we have $\hat{u}^G\,.\,\hat{S}_z = s(\hat{u}^G)\,.\,\hat{S}_z = s(\hat{u}^G)\,.\,\hat{z}^G+$ p-singular terms (and $s(\hat{u}^G)\hat{z}^G$ is a combination of p-regular elements). The result follows now from 1.3.3.

COROLLARY 2.1.3. *Suppose that $u, v \in G$ are p-regular, and that v appears with nonzero coefficient in $s(\hat{u}^G)$. Then there is some $x \in H\backslash G/H$ and there are conjugates u', v' of u, v, respectively, with $u', v' \in xH$, such that $C_H(v') \in \mathrm{Syl}_p(C_G(v'))$, and $C_H(v') \leq H \cap xHx^{-1} = C_H(u')$ (where $H \in \mathrm{Syl}_p(G)$).*

PROOF. From 1.3.4 and 2.1.2 (taking $z = 1_G$).

2.2. *Enumerating blocks.* Let D be a p-subgroup of G. In this section, we show how to calculate the number of blocks of kG with defect group D (i.e., the number of primitive idempotents of $Z(kG)$ lying in $kG_D^G\backslash(\sum_{Q<D} kG_Q^G)$) from group-theoretic information. We state without proof:

BRAUER'S FIRST MAIN THEOREM. *There is a bijection between blocks of kG with defect group D and blocks of $kN_G(D)$ with defect group D.*

For the remainder of this section, then, we assume that $D \lhd G$. We fix $H \in \mathrm{Syl}_p(G)$. Note that $kC_G(D)$ is a G-algebra.

PROPOSITION 2.2.1. (i) $\sum_{Q<D} kG_Q^G \subset J(Z(kG))$.

(ii) $s(kC_G(D)_D^G) \subset kC_G(D)_D^G$.

PROOF. $\sum_{Q<D} kG_Q^G$ is the k-span of $\{\hat{u}^G : u \in G$ and D strictly contains a Sylow p-subgroup of $C_G(u)\}$. Choose $u \in G$ with $s(\hat{u}^G) \neq 0$. Then by 2.1.3,

there is some $x \in H\backslash G/H$ and some $u' \in u^G \cap xH$ with $C_H(u') = H \cap xHx^{-1}$. In particular, $u' \in C_G(D)$, so that $u \in C_G(D)$. Thus (i) is proved. If, furthermore, $D \in \mathrm{Syl}_p(C_G(u))$, and v appears with nonzero coefficient in $s(\hat{u}^G)$, then v is p-regular (by 2.1.1), and by 2.1.3 again we may assume that, for x above, there is a conjugate, v' say, of v in xH with $C_H(v') \in \mathrm{Syl}_p(C_G(v'))$ and $C_H(v') \leq H \cap xHx^{-1} = C_H(u')$. Then as $D \in \mathrm{Syl}_p(C_G(u'))$, we have $x \in C_G(D)H$, so $v' \in C_G(D)H$. Since v' is p-regular and $HC_G(D)/C_G(D)$ is a p-group, we have $v' \in C_G(D)$. Now $D \leq C_H(v') \leq H \cap xHx^{-1} = C_H(u') \leq D$, since $D \in \mathrm{Syl}_p(C_G(u'))$. Thus $D \in \mathrm{Syl}_p(C_G(v'))$, so $D \in \mathrm{Syl}_p(C_G(v))$, and (ii) is proved.

The next result strengthens a theorem of Green:

COROLLARY 2.2.2. *Suppose that D is a defect group for some block of kG. Then there is a p-regular element, v', of G with $D \in \mathrm{Syl}_p(C_G(v'))$ such that $H \cap v'Hv'^{-1} = D$.*

PROOF. To say that D is a defect group for some block of kG means that some primitive idempotent of $Z(kG)$ lies in $kG^G_D \backslash \sum_{Q<D} kG^G_Q$. By 2.2.1, this is equivalent to saying that some primitive idempotent of $Z(kG)$ lies in $kC_G(D)^G_D$. This in turn is equivalent to saying that $s(\hat{u}^G) \neq 0$ for some $u \in G$ with $D \in \mathrm{Syl}_p(C_G(u))$. The result now follows from the proof of 2.2.1.

REMARK. Corollary 2.2.2 remains true when D is not normal in G.

THEOREM 2.2.3. *The number of blocks of kG with defect group D is the rank of the matrix NN^T, where N is the matrix whose rows are indexed by G-conjugacy classes of p-regular elements, u, such that $D \in \mathrm{Syl}_p(C_G(u))$, whose columns are indexed by (H,H)-double coset representatives x such that (whenever possible) x is chosen so that*

(i) $H \cap xHx^{-1} = D$,

(ii) $D \in \mathrm{Syl}_p(C_G(x))$,

(iii) *x is p-regular,*

and where the (u^G, x)-entry of N is $|u^G \cap xH| 1_k$. (If there is no such x, there are no blocks of kG with defect group D.)

PROOF. By 2.2.1, the k-span of the primitive idempotents of kG^G_D is the k-span of $\{s(\hat{u}^G) : D \in \mathrm{Syl}_p(C_G(u))\}$. Since $s^2 = s$, and since $s(\hat{u}^G)$ only involves p-regular class sums, this is also the k-span of $\{s(\hat{u}^G) : u$ is p-regular and $D \in \mathrm{Syl}_p(C_G(u))\}$.

For p-regular u, v with $D \in \mathrm{Syl}_p(C_G(u)) \cap \mathrm{Syl}_p(C_G(v))$, the coefficient of v in $s(\hat{u}^G)$ is

$$\sum_{\{x \in H\backslash HC_G(D)/H : H \cap xHx^{-1} = D\}} (|C_G(v)|_{p'} |u^G \cap xH|\ |v^G \cap xH|) 1_k (|G|_{p'} 1_k)^{-1}$$

using 1.3.3, 1.3.4, and the proof of 2.2.1. This is $([G : C_G(v)]_{p'} 1_k)^{-1}$ times the (u^G, v^G)-entry of NN^T, so the result follows by elementary linear algebra.

2.3. *Other block-theoretic invariants.* Proposition 2.1.2 can be used to express other block-theoretic invariants in terms of group-theoretic information.

When D is a (not necessarily normal) p-subgroup of G, it is easy to see that the k-span of the idempotents in kG_D^G is the k-span of $\{s(\hat{u}^G) : u \text{ is } p\text{-regular, and } C_D(u) \in \mathrm{Syl}_p(C_G(u))\}$. Brauer proved:

PROPOSITION 2.3.1. *Let B be a block of kG. Then the number of isomorphism types of simple B-modules is the dimension of the k-span of $\{1_B\hat{u}^G : u \in G$ is p-regular, and $C_P(u) \in \mathrm{Syl}_p(C_G(u))\}$ where P is a defect group for 1_B.*

From this, we obtain:

PROPOSITION 2.3.2. *Let D be a p-subgroup of G. Then the total number of (isomorphism types) of simple kG-modules in blocks whose defect groups are contained (up to conjugacy) in D is the dimension of the k-span of $\{s(\hat{u}^G)\hat{z}^G :$ u, z are p-regular, $C_D(u) \in \mathrm{Syl}_p(C_G(u))$, and $C_D(z) \in \mathrm{Syl}_p(C_G(z))\}$.*

REMARK. Proposition 2.1.2 allows us to calculate products of the form $s(\hat{u}^G)\hat{z}^G$. We may simplify the calculations by invoking 1.3.4. It is shown in **[14]** that we only need to consider p-regular elements u such that $C_D(u)$ ($= D_u$, say) is a Sylow p-subgroup of $C_G(u)$ and there is some $H \in \mathrm{Syl}_p(G)$ with $H \cap H^u = D_u$, $N_H(D_u) \in \mathrm{Syl}_p(N_G(D_u))$ (so that $H \cap H^u$ is a tame Sylow intersection).

To obtain more subtle information, we need to discuss lower defect groups. Let B be a block of kG, and Q be a p-subgroup of G. We say that Q is a lower defect group for B if $B_Q^G \neq \sum_{P<Q} B_P^G$. The multiplicity of Q as a lower defect group for B is $\dim_k(B_Q^G) - \dim_k(\sum_{P<Q} B_P^G)$.

Let $G_{p'}$ denote the set of p-regular elements of G. Then the multiplicity of Q as a lower defect group for B in the p-section of the identity is

$$\dim_k(1_B \,.\, (kG_{p'})_Q^G) - \dim_k(\sum_{P<Q} 1_B \,.\, (kG_{p'})_P^G)$$

and is denoted by $m_B^{(1)}(Q)$. Analogous definitions can be made for other p-sections.

We quote without proof (see, e.g., Olsson **[11]**):

PROPOSITION 2.3.3. *Let B be a block of kG, p^n be a power of p. Then the multiplicity of p^n as an elementary divisor of the Cartan matrix for B is $\sum_{(Q)} m_B^{(1)}(Q)$, where Q runs over a full set of representatives for the conjugacy classes of subgroups of order p^n of G. In particular, the number of isomorphism types of simple B-modules is $\sum_{(Q)} m_B^{(1)}(Q)$, where Q runs over a full set of representatives for the G-conjugacy classes of p-subgroups of G.*

We have:

PROPOSITION 2.3.4. *Let u, v, z be p-regular elements with $u \in C_G(Q)$, $Q \in \mathrm{Syl}_p(C_G(z))$, and $Q \in \mathrm{Syl}_p(C_G(v))$. Then the coefficient in $\hat{v}^G$ in $s(\hat{u}^G) \,.\, \hat{z}^G$ is*

$$\left[|[(v^{-1})^{N_G(Q)}(u^G \cap C_G(Q))] \cap Z(Q)z| \,.\, |C_G(v)_{p'}|\right] 1_k[|C_G(z)|_{p'}1_k]^{-1}.$$

PROOF. From 2.1.2 (with Q in place of H), we see that the coefficient we seek is

$$\sum_{x \in Q/N_G(Q)} [|C_G(z)|_{p'} 1_k)^{-1}(|C_G(v)|_{p'} |u^G \cap xQ|\ |v^G \cap zxQ|]1_k$$

(for when $Q \cap Q^x < Q$, the term

$$\frac{|C_G(v)|}{|Q \cap xQx^{-1}|} |u^G \cap xQ|\ |v^G \cap zxQ|$$

is divisible by p).

As in 1.3.4, we only need to consider $x \in C_G(Q)$, and we clearly only need to consider p-regular x. For such p-regular $x \in C_G(Q)$, $|u^G \cap xQ| = 0$ unless x is conjugate to u and 1 if x is conjugate to u, since x is the only p-regular element in xQ. Hence the coefficient we seek is

$$\sum_{x \in u^G \cap C_G(Q)} [|C_G(zx)|_{p'} 1_k)^{-1}(|C_G(v)|_{p'} 1_k](|v^G \cap zxQ| 1_k)$$
$$= (|C_G(z)|_{p'} 1_k)^{-1}(|C_G(v)|_{p'} 1_k)|v^G \cap [z(u^G \cap C_G(Q))Q]1_k.$$

Now $v^G \cap z(u^G \cap C_G(Q))Q \leq v^G \cap QC_G(Q) \leq v^G \cap C_G(Q)$, as v is p-regular. Also $v^G \cap C_G(Q) = v^{N_G(Q)}$, since $Q \in \mathrm{Syl}_p(C_G(v))$. The result now follows easily.

Combining 2.3.4 with Külshammer's result in [**9**] yields:

PROPOSITION 2.3.5. *Let Q, P be p-subgroups of G with $Q \leq P$. Then $\sum_B m_B^{(1)}(Q)$, where B runs over all blocks of kG with a defect group contained in P, is the rank of the matrix whose rows are indexed by ordered pairs (u^G, z^G) of conjugacy classes of p-regular elements with $Q \in \mathrm{Syl}_p(C_G(z))$, $Q \leq C_P(u)$, $Q \in \mathrm{Syl}_p(C_G(u))$, whose columns are indexed by conjugacy classes, v^G, of p-regular elements with $Q \in \mathrm{Syl}_p(C_G(v))$, and whose typical entry is*

$$|[(v^{-1})^{N_G(Q)} u^G] \cap Z(Q)z| 1_k.$$

REMARKS. Once more, we can restrict to those u's such that $C_P(u)$ is a tame Sylow intersection as in the remarks after 2.3.2. Külshammer [**9**] also discusses how to determine multiplicities of lower defect groups in other p-sections. In [**14**], it is shown that (in principle, at least) all primitive idempotents of $Z(kG)$ can be determined from knowledge of products of the form $s(\hat{u}^G)\hat{v}^G$ (u, v p-regular). Once these idempotents are determined, the methods described in this section can be used to determine the number of simple modules, and the multiplicities of all lower defect groups, of any block of kG.

REFERENCES

1. R. Brauer, *Representations of finite groups*, Lectures in Math., vol. 1, Wiley, New York, 1963, pp. 133–175.

2. M. Broué, *Brauer coefficients of p-subgroups associated with a p-block of a finite group*, J. Algebra **56** (1979), 265–383.

3. ——, *On a theorem of G. Robinson*, J. London Math. Soc. (2) **29** (1984), 425–434.

4. M. Broué and G. R. Robinson, *Bilinear forms and G-algebras*, J. Algebra **104** (1986), 377–396.

5. D. W. Burry, *Scott modules and lower defect groups*, Comm. Algebra **10** (1982), 1855–1872.

6. J. A. Green, *Some remarks on defect groups*, Math. Z. **107** (1968), 133–150.

7. K. Iizuka and A. Watanabe, *On the number of blocks with a given defect group*, Kumamoto J. Sci. **9** (1973), 55–61.

8. B. Külshammer, *Bemerkungen über die Gruppenalgebra als symmetrische Algebra*. III, J. Algebra **88** (1984), 279–291.

9. ——, *Bemerkungen über die Gruppenalgebra als symmetrische Algebra*. IV, J. Algebra **93** (1985), 310–323.

10. P. Landrock, *On the modular Hecke algebra of a Sylow p-subgroup*, J. Algebra **104** (1986).

11. J. B. Olsson, *Lower defect groups*, Comm. Algebra **8** (1980), 289–297.

12. W. F. Reynolds, *Sections and ideals of centers of group algebras*, J. Algebra **20** (1972), 176–181.

13. G. R. Robinson, *The number of blocks with a given defect group*, J. Algebra **84** (1983), 493–502.

14. ——, *Counting modular irreducible characters*, J. Algebra **90** (1984), 556–566.

15. L. L. Scott, *Modular permutation representations*, Trans. Amer. Math. Soc. **175** (1973), 101–121.

16. S. Shi, *On the number of p-blocks with given defect groups and some applications of the p-power homomorphism*, Group Theory, Beijing 1984, Tuan Hsio-Fu, ed., Lecture Notes in Math, vol. 1185, Springer-Verlag, Berlin, 1986, pp. 314–339.

17. T. Wada, *On the existence of p-blocks with given defect groups*, Hokkaido Math. J. **6** (1977), 243–248.

University of Manchester Institute of Science and Technology, England

Proceedings of Symposia in Pure Mathematics
Volume 47 (1987)

Recent Progress on the Isomorphism Problem

LEONARD L. SCOTT

A long-standing problem, posed first by Graham Higman in 1940 [**Hi**] and later by Brauer [**Br**], is the "isomorphism problem for integral group rings." Given finite groups G and H, is it true that $\mathbf{Z}G \simeq \mathbf{Z}H$ implies $G \simeq H$?

Many authors have worked on this question but progress has been difficult [**Sa**]. Perhaps the best positive result was that of Whitcomb in 1968 [**Wh**], who showed that the implication $G \simeq H$ holds for G metabelian. Dade [**D**] showed there were counterexamples, even in the metabelian case, if $\mathbf{Z}$ was replaced by the family of all fields.

We began our investigation also looking for counterexamples. Zassenhaus had conjectured, perhaps as early as 1960, a much stronger result, describing in part just *how* the isomorphism might come about: Observe first that the group of units of SG, for S any commutative ring, has the form $S^\times \times V$ where $V = V(SG)$ is the group of *normalized* units (those with augmentation 1) in SG. So not much is lost by restricting attention to normalized unit groups (subgroups of V), and indeed every group basis for SG is isomorphic by projection to a normalized subgroup. This is one of the first observations of Higman's thesis [**Hi**], where it was also shown that the elements of normalized finite unit groups in $\mathbf{Z}G$ are linearly independent and generate a pure $\mathbf{Z}$-submodule. Zassenhaus (cf. [**Za**] or [**Se**]) conjectured that, if H is any normalized finite unit group in $\mathbf{Z}G$ with $\mathbf{Z}G = \mathbf{Z}H$ (that is, H is a normalized group basis for $\mathbf{Z}G$), then H must be conjugate to G by a unit in $\mathbf{Q}G$.

Note that the conjectured conjugacy is in $\mathbf{Q}G$, and not $\mathbf{Z}G$. It turns out that even the dihedral group of order 8 has two normalized group bases for its group ring over $\mathbf{Z}$, which are not conjugate in $\mathbf{Z}G$, but indeed are conjugate in $\mathbf{Q}G$ [**Wh**]. Nevertheless, Klaus Roggenkamp and I still felt that the conjecture was much too strong and about six years ago made a first intensive search for counterexamples. The remainder of this paper, except as otherwise noted, is a report of our joint work.

1980 *Mathematics Subject Classification.* Primary 20C05; Secondary 20D15, 13F30.
This paper was written with support from the National Science Foundation.

Without going into details, let me just say that this first attempt was rebuffed fairly severely by the many examples we examined. We began to believe that the Zassenhaus conjecture should be taken much more seriously and that for p-groups, at least, a much stronger result in the same spirit might be true. In April of 1984, we completed the proof of the following result [**RS**].

THEOREM 1. *Let G be a finite p-group, and let $\mathbf{Z}_p$ denote the ring of p-adic integers. Then there is only one conjugacy class of groups of order $|G|$ in the normalized units of $\mathbf{Z}_pG$.*

Berman and Rossa [**BR**] and Whitcomb [**Wh**] independently had raised the question of whether a result like Theorem 1 might be true. Whitcomb attributes his version of the problem to his thesis advisor, John Thompson, who suggested it as part of an inductive approach to the isomorphism problem for p-groups. Whitcomb himself, though, apparently made the crucial observation that the two dihedral group bases mentioned above were conjugate over the 2-adics. Unaware of Whitcomb's work, we had made the very same observation, which really initiated our work on Theorem 1.

Actually, we proved the same result as Theorem 1 in [**RS**] for much more general coefficient domains than $\mathbf{Z}_p$, including all discrete valuation domains of characteristic 0 having residue field of characteristic p. As a consequence, we proved the following strong version of the Zassenhaus conjecture for finite nilpotent groups.

COROLLARY. *Let G be a finite nilpotent group and S an integral domain of characteristic* 0 *in which no prime divisor of $|G|$ is invertible. Let H be a finite group of normalized units in SG for which $|H| = |G|$. Then H is conjugate to G by a unit in KG, where K is the quotient field of S.*

W. Kimmerle has observed [**K**] that it is not really weaker in the Zassenhaus conjecture to consider the cases when H is known to be isomorphic to G, since in any event one knows $H \times G$ is isomorphic to $G \times H$, and one can then argue that a suitable unit in $K(G \times H) \cong KG \otimes_K KH$ carries G into a copy of H.

This same philosophy works for matrix rings and was very useful in guiding us to the formulation of the corollary below. The corollary gives, for nilpotent groups, a positive answer to the group ring isomorphism problem even when the ring isomorphism hypothesis is replaced by Morita equivalence.

COROLLARY 2. *Let G be a finite nilpotent group and S an integral domain of characteristic* 0 *in which no prime divisor of $|G|$ is invertible. Let H be a finite group for which the category of SG-modules is S-linearly equivalent to the category of SH-modules* (as occurs if $SG \simeq SH$ as S-algebras, or if a full matrix algebra over SG is S-isomorphic to a full matrix algebra over SH). *Then $G \simeq H$.*

In the proof we show $\tilde{S}G \simeq \tilde{S}H$ where $\tilde{S}$ is a semilocal Dedekind domain containing the subring of S generated by the relevant coefficients, satisfying the same hypotheses. This reduction works for any finite group G in the presence of an equivalence of categories.

Over **Z** we announced in **[RS]** a positive answer to the isomorphism problem for finite groups which are abelian by nilpotent, and gave the proof in a special case. This result, improved to more general coefficients, will appear in **[RS2]**. The Zassenhaus conjecture is, however, quite another matter, and we are currently checking a group of order 14,400 which appears to be a counterexample to the analogue of the first corollary.[1]

Many further new results have been proved since **[RS]**, but before taking them up we give a brief sketch of the proof of Theorem 1.

I would like to take this opportunity to mention that Theorem 1 and its generalization and corollaries, together with work of Fröhlich **[Fr]**, give a fairly complete picture of the automorphism and Picard groups of SG, when G is finite nilpotent, for many coefficient domains S. This is discussed in further detail in Roggenkamp's article **[R]** in these proceedings.

I would also like to repeat thanks to Bob Sandling, who read a preliminary version of the critical Chapter 3 in **[RS]**, and to Everett Dade, who went through the final manuscript.

PROOF OF THEOREM 1—SOME REMARKS. Why should the isomorphism problem have a positive answer?

One answer is that questions about isomorphisms tend to reduce to questions about automorphisms (recall Kimmerle's observations, for example) and the automorphisms of group rings may be nicer than one might expect. The latter view is supported by the following proposition, which is (1.4.1) in **[RS]**.

PROPOSITION. *Let S be a commutative ring of characteristic 0, in the sense that no nonzero element of S has finite additive order. Let H be a finite group acting by permutations on a set X, and form the permutation module SX for SH. Then $H^1(H, SX) = 0$.*

In particular, if H is contained in a group G, and we let H act on G by conjugation, then $H^1(H, SG) = 0$. For $G = H$ a finite group, this gives $H^1(SG, SG) = 0$ in the sense of augmented algebras, or, in other words, every S-linear derivation from SG to itself is inner.

PROOF. Without loss we may assume H is transitive on X. Thus SX is induced from the trivial module S for a subgroup L of H. By the Eckmann-Shapiro lemma, we have $H^1(H, SX) \cong H^1(L, S)$. The latter consists of group homomorphisms from L to the additive group of S, since L acts trivially. However, our characteristic 0 assumption implies there are none of these, except for the trivial map, so $H^1(H, SX) = 0$.

When $G = H$ the identification of $H^1(SG, M)$, for an SG-bimodule M, with $H^1(G, M)$ for the corresponding "conjugation action" $m \mapsto gmg^{-1}$ ($m \in M$, $g \in G$) on M is well known and easy to describe: A derivation $D\colon SG \to M$ gives a 1-cocycle $\gamma\colon G \to M$ defined by $\gamma(g) = D(g)g^{-1}$. Conversely, cocycles can

[1] Shortly after this paper was written we found a related group which—again, in work still being checked—appears to be a counterexample to the Zassenhaus conjecture itself. This group and the one above (which may also be a counterexample) is metabelian.

be used to define S-linear derivations, and coboundaries correspond precisely to inner derivations. This concludes the proof of the proposition.

The above property of derivations suggests a recipe for showing a given automorphism α of SG is inner: First take the logarithm of α, if possible, and show $\log\alpha$ is a derivation. Then apply the above property to obtain $\log\alpha = \operatorname{ad} a$ (the map $x \mapsto ax - xa$) for some $a \in SG$. Then form $\exp a$, if possible, and try to show that α is a conjugation by $\exp a$.

If $S = \mathbf{Z}_p$, for example, then infinite series make sense, but there is no reason for $\log\alpha$ to exist or stabilize SG. For G a finite p-group we can arrange inductively that α is trivial modulo $(c-1)SG$ for c a central element of order p in G. (We illustrate this below in the simple case of the dihedral group of order 8.) One could form $\log\alpha$ at this point and write $\log\alpha = \operatorname{ad} a$, but $\exp a$ need not exist, and efforts to get around this are hampered in particular by the fact that $c-1$ is a zero divisor. So in Chapter 2 of [**RS**] we translate the whole question into a problem concerning an automorphism α of a quotient order Λ in which the image π of $c-1$ is not a zero divisor. It is no longer true that derivations are inner for this order, but π-multiples of derivations are, and we prove in Chapter 5 of [**RS**] the following general result for orders, somewhat in the spirit of Fröhlich [**Fr**, Theorem 9].

Simplifying both hypothesis and conclusion, the theorem below asserts at least that, on an order Λ such as we have, an automorphism α which is the identity modulo $\pi \operatorname{rad}^2 \Lambda$ must be inner.

THEOREM 2. *Assume that S is a complete discrete valuation domain of characteristic 0 and residue field of characteristic $p > 0$. Suppose Λ is an S-free S-algebra of finite rank over S and that π is a nonzero-divisor in the center of Λ such that π^{p-1} belongs to $p\Lambda$, and such that πD is inner for every $S[\pi]$-linear derivation D from Λ to itself. Assume also that $\Lambda = Z(\Lambda) + \operatorname{rad} \Lambda$ (which holds, for example, if Λ is local). Put $\tilde{\mathfrak{p}} = \mathfrak{p} + \pi S[\pi]$ where $\mathfrak{p}$ is the maximal ideal of S.*

Let α be an $S[\pi]$-automorphism of Λ which is the identity modulo $\pi\Lambda$, and put $\alpha = 1 + \pi\phi$ for some $\phi \in \operatorname{End}_S \Lambda$. Assume also that

$$\phi^n(\Lambda) \subseteq \operatorname{rad}^2 \Lambda + \tilde{\mathfrak{p}}\Lambda \qquad (\textit{some } n > 0)$$

or, equivalently (as shown in the proof), *that $\phi^m \to 0$ in the p-adic topology as $m \to \infty$.*

Then the automorphism α of Λ is inner.

In the basic $S = \mathbf{Z}_p$ case of Theorem 1 we have been discussing, one obtains an automorphism that is trivial modulo $\pi \operatorname{rad} \Lambda$, not $\pi \operatorname{rad}^2 \Lambda$. The difference may not seem like much, but it is a whole world, and even the very technical formulation of Theorem 2 is only a start toward overcoming the difficulties.

Theorem 2 does at least give some shape to what we must do in order to prove Theorem 1. In the early summer of 1983, we were able to treat the case of class 2 p-groups with commutator quotient of exponent p using the theorem.

The idea was simply to exploit the many connections between Lie filtrations of the group and natural sections of the group ring. (A well-known example is the isomorphism of the commutator quotient of the group G with $I(\mathbf{Z}G)/I(\mathbf{Z}G)^2$, where $I(\mathbf{Z}G)$ is the augmentation ideal. The image of the augmentation ideal of SG in Λ above is in fact equal to r ad Λ. We used such identifications to "see" the effect of various modifications of α with inner automorphisms in moving $\phi(\Lambda)$ from rad Λ to $\text{rad}^2\,\Lambda$.)

One advantage of class 2 groups is that the center is a relatively large part of the group. In proving Theorem 1 it is a fairly easy part of the initial reductions to get the automorphism involved equal to the identity on the center. When, almost a year later, we finally completed the proof of the $S = \mathbf{Z}_p$ version of the theorem with a very difficult induction, we had found a substitute for the center in the general case: We were able to modify the automorphism so that it was the identity on

$$\Omega = C_G(Z(G, C)),$$

where C is a fixed subgroup of order p in the center of G, and $Z(G, C)$ is the subgroup of G containing C whose image in G/C is the center of that group. The group Ω is a huge subgroup of G; in particular it contains the Frattini subgroup, so that G/Ω is abelian of exponent p.

The proof for this step (which is the critical Chapter 3 in **[RS]**) is too technical to be outlined here, but the theme is the connection between additive and multiplicative 1-cohomology, on suitable ideal sections, for a group acting on a ring. This idea is very powerful, and in the Appendix we suggest a proof using it, without any reference to Theorem 2, of an even stronger result than Theorem 1 (cf. the Subgroup Rigidity theorem in the next section.) We illustrate the technique in a very simple case by giving a complete proof of Theorem 1 for the dihedral group of order 8.

This was, in fact, one of the very first cases we were able to handle, and the success of the 1-cohomology arguments here encouraged us to have hope for Theorem 1 very early **[S]**.

PROOF OF THEOREM 1 FOR THE DIHEDRAL GROUP OF ORDER 8. We take $S = \mathbf{Z}_2$ and $G = D_8$. Let $C = \langle c \rangle$ denote the central subgroup of order 2 in G, and suppose H is a normalized subgroup of order $|G|$ in SG. Then standard results (all discussed from first principles in **[RS]**, incidentally) imply that H is a group basis for SG, that C is also a (central) subgroup of H, and that $G/C = H/C$ in the factor ring SG/C. (The last assertion is a consequence here of the fact that G/C is abelian. In the reduction for Theorem 1 in the general case one can also arrange this, using induction and the conjugacy assertion of the theorem. Indeed, this is what makes the formulation of Theorem 1 such a powerful inductive tool.)

Now let $s(G, C)$ denote the ring $SG/I(SC)I(SG)$, which we called the "small group ring" in **[RS]**; its ancestry dates back to Higman's thesis **[Hi]**. Elementary arguments show there is a natural exact sequence

$$0 \to C - 1 \to s(G, C) \to SG/C \to 0,$$

where by abuse of notation we confuse the set $C - 1$ in SG with its image in $s(G, C)$, where it is an ideal of square 0 naturally isomorphic as an additive group to C. The (isomorphic) images of the groups G and H coincide in $s(G, C)$, so that we now know G and H are isomorphic. (A similar conclusion is drawn at this point in the general argument.) Now, however, our task is to prove they are conjugate by a unit in S.

Let α be any group isomorphism from G to H, which clearly induces an augmentation-preserving automorphism of the S-algebra $SG = SH$. Also, α induces an automorphism of the group G as embedded in $s(G, C)$. Composing the ring automorphism α with the inverse of this group automorphism, we may assume the ring automorphism α is the identity on $s(G, C)$.

Let $\underline{C}$ denote the sum of all the elements of C, and put $\Lambda = SG/\underline{C}G$. Note that the image in Λ of the element c is -1. (Here the case $p = 2$ is special, and much easier.) Also, $\Lambda/2\Lambda = \Lambda/(1 - c)\Lambda \cong \mathbf{F}_2G/C$, and SG is the pull-back of Λ and SG/C over $\mathbf{F}_2G/C$. The image is $I(SG)$ in Λ is rad Λ, since the quotient of Λ by that image is $\mathbf{F}_2$. Thus the automorphism α of SG induces an automorphism (which we also call α) on Λ which is the identity modulo 2 rad Λ. Using the pull-back, one obtains easily that α is inner on SG if and only if it is inner on Λ. (Note that units in $\mathbf{F}_2G/C$ are the images of central units in SG/C.)

In Λ write $\alpha(x) = \mu(x)x$ for $x \in G$. Then μ is a multiplicative 1-cocycle on G with values in the group $1 + 2$ rad Λ. That is, μ satisfies the equation

$$\mu(xy) = \mu(x)\, {}^x\mu(y) \qquad (x, y \in G).$$

The superscript means conjugation: ${}^xz = xzx^{-1}$. Write $\mu(x) = 1 + 2\gamma(x)$ for $x \in G$. Then the multiplicative cocycle equation above implies that γ is an additive cocycle modulo 2 $\mathrm{rad}^2\, \Lambda$:

$$\gamma(xy) \equiv \gamma(x) + x\gamma(y) \quad \text{modulo } 2\ \mathrm{rad}^2\, \Lambda$$

for x, y in G. This is the *connection between multiplicative and additive 1-cohomology* mentioned above. We use it here to prove the following lemma [S]:

LEMMA. *Suppose the group T of order 2 acts via automorphisms on a $\mathbf{Z}_2$-order Λ, and μ is a multiplicative 1-cocycle for this action. Suppose further that*

$$\mu(t) \equiv 1 \quad \textit{modulo } 2 \text{ rad } \Lambda$$

for all t in T. Then $\mu \sim 1$. (That is, μ is a coboundary.)

For the proof, write $\mu = 1 + 2\gamma$ as above. As noted, γ is an additive 1-cocycle modulo 2 $\mathrm{rad}^2\, \Lambda$, thus defines an element $\bar{\gamma}$ in $H^1(T, \mathrm{rad}\, \Lambda/2\, \mathrm{rad}^2\, \Lambda)$. Since the order of T is just 2, we have $2\bar{\gamma} = 0$ in this cohomology group. If the coboundary $t \mapsto b - {}^tb$ $(t \in T)$ associated to an element $b \in \mathrm{rad}\, \Lambda$ is, modulo 2 $\mathrm{rad}^2\, \Lambda$, equal to $\bar{\gamma}$, then $1 + b$ is a unit and we can modify the multiplicative cocycle μ with the multiplicative coboundary associated with this unit. After an easy calculation, we find that the new μ takes values in $1 + 2\,\mathrm{rad}^2\,\Lambda$. Repeating the argument, we can replace $\mathrm{rad}^2\,\Lambda$ here with $\mathrm{rad}^3\,\Lambda$, and so on. Passing to a limit, we find that the original μ is a coboundary, and the lemma is proved.

To complete the proof of Theorem 1 for $G = D_8$, first let T be a subgroup of order 2 in G/C. The action of G on Λ and the cocycle μ both factor through G/C, so the lemma applies to T. Thus μ is a coboundary on T. An elementary calculation now shows that the restriction of α to the inverse image $\tilde{T}$ of T in G/C is effected by conjugation by a unit in Λ. We can pull this unit back to a unit in SG whose image in SG/C is central, so induces a group automorphism by conjugation in $s(G,C)$. Modifying α on SG by this conjugation and group automorphism, we may assume α is the identity on $\tilde{T}$. (The group automorphism is not really necessary here, but that is deeper; cf. [**RS**, Chapter 4].)

We now find that the new μ takes values in the fixed point ring $\tilde{\Lambda}$ of $\tilde{T}$ in Λ. On $\tilde{\Lambda}$, the action of G and the cocycle μ factor through $G/\tilde{T}$. The lemma can now be applied to the group $G/\tilde{T}$ of order 2 and the order $\tilde{\Lambda}$. Arguing as above, we now get that the restriction of α to G is effected by conjugation by a unit in SG or, in other words, that α is inner. This completes the proof of Theorem 1 for the dihedral group of order 8.

Newer results and open questions. In my talk at the conference the following improvement of Theorem 1 was conjectured:

SUBGROUP RIGIDITY THEOREM (WEISS, ROGGENKAMP-SCOTT). *Let G be a finite p-group. Then every normalized finite subgroup of $\mathbf{Z}_pG$ is conjugate to a subgroup of G, uniquely determined up to conjugation in G.*

Our claim to this result is in a proof for $p = 2$ [**RS1, R**], and a sketch for the general case; cf. the Appendix below. Before we felt ready to claim the theorem in full generality, however, Al Weiss independently proved it with a different method. His idea is to prove by a suitable induction that the two-sided action of $H \times G$ on $\mathbf{Z}_pG$, for H a finite normalized group of units, is in fact a permutation module. This approach sidesteps many of the difficulties we have had in modifying ring automorphisms by group automorphisms.

The "rigidity" terminology comes from the analogy with Weil rigidity we drew in [**RS**] and discussed at this conference: The proposition in the first section shows $H^1(H, SG) = 0$ if H is a subgroup of the finite group G acting by conjugation on the group ring SG over an integral domain S of characteristic 0. This is analogous to a result proved by Weil with H a (mildly restricted) cocompact discrete subgroup of a semisimple Lie group, and SG the Lie algebra of the latter. Weil showed that his result does indeed severely restrict the conjugacy classes in the Lie group of subgroups isomorphic to H. (This was the inspiration for later well-known results of Mostow and Margulis.) As we remarked in [**RS**], it would be very interesting indeed to have a proof of the above theorem based on some geometric property of the quotient "variety" V/G, where V is the group of normalized units.

The theorem also implies a very intriguing identification of (the points of) the characteristic p cohomology varieties of V and G, in the sense of Quillen [**Q**]; cf. Roggenkamp's article [**R**] in these proceedings.

I would like to mention at this point that a question of Keith Dennis [**De**] is still open for $\mathbf{Z}G$ (and for $\mathbf{Z}_pG$) when G is a finite p-group. Dennis asked when a group might have a normal complement in the group of normalized units of its group ring. Though many negative answers are now known [**RS0**], the question served to stimulate activity on the isomorphism problem. (In a different direction, I would like to mention that much of my own interest in the isomorphism problem began with [**S0**], which answers in the negative a question somewhat similar to the isomorphism problem for transitive permutation representations of a fixed finite group G: If two transitive permutation modules for $\mathbf{Z}G$ are isomorphic, with G a finite group, are the underlying G-sets isomorphic? There are counterexamples of degree 203 for G =PSL(2,29).)

Also open is the isomorphism problem for finite p-groups over $\mathbf{F}_pG$, though such nice conjugacy properties as one finds in Theorem 1 just don't exist even in the simplest cases, and there are often many apparently unrelated group bases. It would be interesting to know if the Steenrod operations on the cohomology ring of $\mathbf{F}_pG$ are determined by the ring alone, together with its augmentation. Because of Theorem 1, we know that the analogous question for $\mathbf{Z}_pG$ is true, and it is well known that the cohomology ring itself is determined by the ring $\mathbf{F}_pG$ and its augmentation. (It would of course be interesting to have a more direct proof for $\mathbf{Z}_pG$!)

Weiss's method is very strong. Using it, we have been able to complete the proof (the details have been written down, but not thoroughly checked) of the following result, which we said at the conference we could prove modulo questions about conjugacy of defect groups in blocks. (We mentioned especially principal blocks and solvable groups with $O_{p'}(G) = 1$ for some prime p. The essential feature for the arguments, however, is that such groups satisfy automatically the hypothesis below. A well-known consequence of the hypothesis is that $\mathbf{Z}_pG$ is a single (principal) block.)

THEOREM. *Let G be a finite group which has a normal p-subgroup containing its own centralizer in G. Suppose H is a normalized group of units in* $\mathbf{Z}G$ *with* $\mathbf{Z}G = \mathbf{Z}H$. *Then H is conjugate to G by a unit in* $\mathbf{Z}_pG$, (*and thus also by a unit in the localization* $\mathbf{Z}_{(p)}G$; *cf.* [**RS**, (1.2.4)]). *In particular, the isomorphism problem and Zassenhaus conjecture have a positive answer for G.*

The same result holds if $\mathbf{Z}$ is replaced by its semilocalization $\tilde{\mathbf{Z}}$ at the prime divisors of $|G|$. (Explicitly, $\tilde{\mathbf{Z}}$ consists of all rational numbers with denominator not divisible by any prime dividing $|G|$.)

One hopes both results above will eventually be established for more general coefficient domains (though such generalizations are sometimes very difficult to find), and we have not tried to give the sharpest available results in this direction at this time. We did, however, want to mention $\tilde{\mathbf{Z}}$ above explicitly. The reason is that it has been conventional wisdom in the community of researchers working in this area that $\tilde{\mathbf{Z}}$ is as good as $\mathbf{Z}$. But, as mentioned earlier, Klaus and I are currently checking a group G of order 14,400 such that $\tilde{\mathbf{Z}}G$ has a second, isomorphic normalized group basis not conjugate to G in the units of $\mathbf{Q}G$.[2]

This group, of course, does not satisfy the hypothesis of the above theorem, though, like every solvable finite group, it is the subdirect product of groups satisfying the hypothesis. When Klaus and I conducted our counterexample search some six years ago, we assumed it would be a very easy matter to reduce the isomorphism problem or Zassenhaus conjecture in the solvable case to groups satisfying the hypothesis above, and the theorem as stated almost entirely explains our lack of success (in finding counterexamples). It now seems apparent, however, that our assumption about reductions was wrong, and there is something quite substantial to understand here. The following corollary may help the reader grasp the issue for the isomorphism problem.

COROLLARY. *Let G and H be finite solvable groups for which $\mathbf{Z}G \simeq \mathbf{Z}H$. Then $G/O_{p'}(G) \simeq H/O_{p'}(H)$ for each prime p.*

In other words, the isomorphism problem for solvable groups reduces to a (perhaps very subtle) question as to how the group as a whole is put together from its pieces $G/O_{p'}(G)$. This question, of putting the pieces together, may well have a negative answer, and certainly there are serious problems with the Zassenhaus conjecture. But, even so, a satisfying theory might still be achieved by understanding the underlying patchwork process in a precise way.

Note that the corollary implies G and H have isomorphic Sylow p-subgroups for each prime p. In the course of our work on the solvable case, Klaus and I encountered the following natural question about defect groups:

QUESTION. *Let B be a block over $\mathbf{Z}_p$ of a finite group G. Is the defect group of B well defined, up to conjugation and suitable normalization, independently of the group G?* (In particular, are any two defect groups for B isomorphic?)

Here "conjugation" refers to conjugation by a unit of B, rather than conjugation in any group. (This at least makes sense, if one replaces the usual defect group by its projection on B.) "Suitable normalization" is undefined, and finding a correct definition should be considered part of the problem. The principal block has a natural augmentation, and normalization of unit groups makes sense in the manner we explained at the beginning of this paper. The correct definition for general blocks is less obvious. Külshammer pointed out at the conference that, if P is a finite p-group, then $\mathbf{Z}_pP \otimes M_n(\mathbf{Z}_p)$ may easily be a block of several finite groups G, with the defect group P not always embedded in the obvious way. (The notation M_n refers here to $n \times n$ matrices.) However, if "normalization" is taken in the broad sense of applying a natural quotient homomorphism in a subgroup of units, the answer to our question appears nevertheless to be

[2]See the previous footnote, which indeed does tentatively announce a recent example using $\mathbf{Z}$ in place of $\tilde{\mathbf{Z}}$.

positive for these examples. The parenthetic simpler formulation of the defect group question is based on a remark of Jon Alperin.

A prototype positive answer to the defect group question is our theorem above, or rather its consequence for Sylow p-subgroups of the groups involved on groups with a normal p-subgroup containing its centralizer. As noted above, the p-adic group ring is just a single (principal) block in this case. It seems likely that the proof of this theorem will supply a similar positive answer to the defect group question for all blocks of solvable groups.

In the nonsolvable case there is not yet any empirical evidence one way or another. The question is, nevertheless, still reasonable, and a positive answer might be very helpful in reducing the general isomorphism problem to patchwork considerations, as in the solvable case. One can probably restrict attention here to the principal block and images of its defect group (a Sylow p-subgroup of a finite group G, as projected on the principal block of $\mathbf{Z}_pG$) under an augmentation-preserving automorphism of $\mathbf{Z}G$.

One reason that conjugacy of p-groups, which effectively means their identification, can be so powerful, is an argument of Coleman [**C**]. Although he only discusses p-group rings over modular fields, his argument shows that *automizers* (normalizers modulo centralizers) *of p-subgroups of G are the same in the units of $\mathbf{Z}_pG$ as they are in G.* The same is true for the projections of G and these p-subgroups on blocks, and the coefficients $\mathbf{Z}_p$ could well be any DVR with residue field of characteristic p, or such a field itself. Apparently Puig has recently observed a similar result.

For the general finite group G, a great deal of information about $G/O_{p'}(G)$ (which is the projection of G on the principal block) is obtainable from knowledge of a Sylow p-subgroup P of $G/O_{p'}(G)$ and automizers of its subgroups, especially in the context [**Sa**] of the isomorphism problem. For example, the quotient of $G/O_{p'}(G)$ by its generalized Fitting subgroup is determined, using a Sylow p-subgroup of the latter, modulo outer automorphisms of components [**A**]. We suggest that a positive answer to even a modest version of the defect group conjugacy question, for principal blocks, might well lead to obtaining the corollary above for the general finite group.

A number of researchers have expressed interest in the defect group question on its own merits. There is potential interaction, for example, with Puig's theory [**P**] of source algebras and with the general issue of finding some way to bound the number of (Morita equivalence classes of) blocks with a given defect group. Also, we mention that Weiss's paper [**W**], though ostensibly concerned with subgroup rigidity, contains results of fundamental interest for the representation theory of p-group permutation modules. This contact between other areas of representation theory and the theory related to the isomorphism problem is certainly welcome and very encouraging.

Another area for potential interaction is in the theories of automorphisms and of conjugacy of finite subgroups in the general arithmetic group, the group of units of $\mathbf{Z}G$ being a prime example. Indeed, there is a nice hierarchy with units in group rings the most special, then units in orders, and then arithmetic groups. Our Theorem 2 is an instance where thinking in terms of the most general of these can lead to effective techniques in the more special situations. Similarly, results at the most concrete level may suggest something more general, and we have tried to emphasize this in our discussion of subgroup rigidity.

Appendix. The following argument is taken verbatim from [**S1**]. It gives a detailed proof, modulo arguments of [**RS**], especially Chapter 3, for subgroup rigidity in $\mathbf{Z}_pG$ in case $|G| = p^3$. Then a detailed suggestion as to how the proof might be modified for $|G| = p^n$ is given. The reader is assumed to be familiar with the arguments and notation in Chapter 3 of [**RS**] and the general philosophy of reduction to the order Λ illustrated in our discussion above of Theorem 1. This philosophy is, briefly, as follows: To prove by induction, say, that a normalized torsion unit $u \in \mathbf{Z}_pG$ is conjugate to an element of G, we may assume this is true in $\mathbf{Z}_pG/C$ where C is a central subgroup of order p. This leads to a conjugacy problem, which begins the discussion below, in the order $\Lambda = \mathbf{Z}_pG/\underline{C}\mathbf{Z}_pG$ where $\underline{C}$ denotes the sum of the elements of C. The element ς below is the image of c in Λ, the notation $\underline{I}(G)$ is an abbreviation for the augmentation ideal $I(\mathbf{Z}_pG)$, and all unexplained references, especially the phrase "as usual," refer to Chapter 3 of [**RS**].

Let $|G| = p^3$, $C = \langle c\rangle$, ς, $\pi = \varsigma - 1$ be as usual, and let $t \in G$, $t \notin C$, $t^p \in C$.

Consider a torsion[3] unit $u \in (1 + \pi \operatorname{rad} \Lambda)t$ and write $u = (1 + \pi a)t$, with $a \in r$ ad Λ.

We may of course assume G is noncommutative, so choose $s \in G$ with ${}^ts = cs$. Observe that the s^it^j $(0 \le i,\ j \le p-1)$ form an $R = \mathbf{Z}_p[\pi]$ basis for Λ. (We are working here originally over $\mathbf{Z}_p$.) In fact, $R\langle t\rangle = \Lambda_0$ and $s^i\Lambda_0 = \Lambda_{\varsigma,i}$ in our previous notation. *This time, however,* we want to consider the decomposition

$$\Lambda = \Lambda_0 + (s-1)\Lambda_0 + \cdots + (s-1)^{p-1}\Lambda_0.$$

Observe that

$${}^t(s-1)^i = (\varsigma s - 1)^i = (\varsigma(s-1) + \pi)^i \equiv \varsigma^i(s-1)^i + i\varsigma^{i-1}\pi(s-1)$$

modulo $\pi^2\Lambda$. Thus

$$\begin{aligned}{}^t(s-1)^i - (s-1)^i &\equiv \frac{\varsigma^i - 1}{\varsigma - 1}\pi(s-1)^i + i\varsigma^{i-1}\pi(s-1)^{i-1}\\ &\equiv i\pi((s-1)^{i-1} + (s-1)^i).\end{aligned}$$

LEMMA 1. *We may assume a above lies in $(s-1)^{p-1}\Lambda_0 + \pi\Lambda$.*

PROOF. Suppose we have $a \equiv \sum_{i=k}^{p-1}(s-1)^ix_i$ modulo $\pi\Lambda$, with $x_i \in \Lambda_0$. Put

$$v = \frac{1}{k+1}(s-1)^{k+1}x_k.$$

[3]Remarkably, the torsion assumption is irrelevant for Lemma 1 and Lemma 2!

Thus ${}^t v - v \equiv \pi((s-1)^k + (s-1)^{k+1})x_k$ modulo $\pi^2\Lambda$. Conjugating u by $1+v$ gives

$$\begin{aligned}
{}^{(1+v)}u &= u + [v,u](1+v)^{-1} \\
&\equiv u + [v,t](1+v)^{-1} \bmod \pi^2\Lambda, \qquad \text{since } \Lambda/\pi\Lambda \text{ is commutative,} \\
&= (1+\pi a)t + [v,t]t^{-1}(1+{}^t v)^{-1}t \\
&\equiv (1+\pi a + [v,t]t^{-1}(1+v)^{-1})t, \\
&\qquad\qquad \text{since } [v,t] \in \pi\Lambda \text{ and } {}^t v \equiv v \bmod \pi\Lambda \\
&= (1+\pi a - ({}^t v - v)(1+v)^{-1})t \\
&= (1+\pi a - (\pi(s-1)^k + (s-1)^{k+1})x_k)(1+v)^{-1})t \\
&\equiv (1+\pi a - \pi(s-1)^k x_k)t \bmod \pi^2\Lambda + \pi(s-1)^{k+1}\Lambda,
\end{aligned}$$

since $v \in (s-1)^{k+1}\Lambda + \pi\Lambda$, and the latter is an ideal in Λ. In fact, $(s-1)^{k+1}\Lambda + \pi\Lambda = (s-1)^{k+1}\Lambda_0 + \cdots + (s-1)^{p-1}\Lambda_0 + \pi\Lambda$. We now have that ${}^{(1+v)}u \in (1+\pi^2\Lambda + \pi(s-1)^{k+1}\Lambda)t$ modulo $\pi^2\Lambda + \pi(s-1)^{k+1}\Lambda$, or simply that ${}^{(1+v)}u \in 1 + \pi^2\Lambda + \pi(s-1)^{k+1}\Lambda$ and the lemma follows.

LEMMA 2. *We may even assume* $a \in (s-1)^{p-1}\Lambda_0 + \pi^n\Lambda$ (*any* $n \geq 1$).

PROOF. Choose m, k with $a \in (s-1)^{p-1}\Lambda_0 + \pi^m(s-1)^k\Lambda + \pi^{m+1}\Lambda$, $m \geq 1$, $p-2 \geq k \geq 0$, and write

$$a = (s-1)^{p-1}x_0 + \pi^m \sum_{i=k}^{p-2}(s-1)^i x_i + \pi^{m+1}y$$

with all $x_i \in \Lambda_0$ and $y \in \Lambda$. Put

$$v = \frac{1}{k+1}(s-1)^{k+1}x_k.$$

Thus $-[v,t]t^{-1} = {}^t v - v \equiv \pi((s-1)^k + (s-1)^{k+1})x_k$ modulo $\pi^2\Lambda$, as before. Consider now conjugating u by $1+\pi^m v$, which gives

$$\begin{aligned}
{}^{(1+\pi^m v)}u &= u + \pi^m[v,u](1+\pi^m v)^{-1} \\
&\equiv u + \pi^m[v,u] \text{ modulo } \pi^{2m+1}\Lambda \text{ since } [\Lambda,\Lambda] \subseteq \pi\Lambda \\
&\equiv u + \pi^m[v,t] \text{ modulo } \pi^{m+1}(s-1)^{k+1}\Lambda + \pi^{m+2}\Lambda \\
&\qquad \text{since } u = t + \pi a t \in t + \pi(s-1)^{p-1}\Lambda + \pi^{m+1}\Lambda, \\
&\qquad \text{and the ideal } (s-1)^{k+1}\Lambda + \pi\Lambda, \text{ which contains } v \\
&\qquad \text{and } (s-1)^{p-1}\Lambda, \text{ is, of course, also a Lie ideal.} \\
&= (1+\pi a + \pi^m[v,t]t^{-1})t \\
&\in (1+\pi((s-1)^{p-1}\Lambda_0 + \pi^m(s-1)^{k+1}\Lambda + \pi^{m+1}\Lambda))t,
\end{aligned}$$

which proves the lemma, since the process is repeatable.

To *complete the proof*, using $u^p = t^p$ (which follows easily from the fact that $1+(c-1)I(G)$ is torsion-free in $\mathbf{Z}_pG)^\times$ and $u^p \in (1+(c-1)I(G))t^p$, with t^p central). Take n large.

We have

$$((1+\pi a)t)^p = t^p,$$

or

$$(1+\pi a)(1+\pi\, {}^{t}a)\cdots(1+\pi\, {}^{t^{p-1}}a) = 1.$$

Any product $(\pi\, {}^{t^i}a)(\pi\, {}^{t^j}a)$ lies in $\pi^2 p\Lambda$: Without loss, $i=1$, so

$$\pi a \pi\, {}^{t^j}a \equiv \pi^2 (s-1)^{p-1} x_0\, {}^{t^j}((s-1)^{p-1}x_0) \mod \pi^n\Lambda.$$

Here x_0 is in $\Lambda_0 = R\langle t\rangle$, so it suffices to check that $(s-1)^{p-1}\, {}^{t^i}(s-1)^{p-1} \in p\Lambda$. However,

$${}^{t^i}(s-1) = \varsigma^i(s-1) = \varsigma^i(s-1) + \frac{\varsigma^i - 1}{\varsigma - 1}\pi.$$

Since $(s-1)^p \in p\Lambda$, any term in ${}^{t^i}(s-1)^{p-1}$ involving $(s-1)$ gives an element of $p\Lambda$. However, the only other term in the binomial expansion has a factor $\pi^{p-1} \in p\Lambda$. This proves our claim about the product.

It now follows that

$$\pi \sum_{i=0}^{p-1} {}^{t^i}a \equiv 0 \mod \pi^2 p\Lambda.$$

In particular, the projection of a on Λ_0 is in $\pi\Lambda$. Since $(s-1)^{p-1} \equiv \sum_{i=0}^{p-1} s^i \pmod p$, this projection is x_0, modulo $\pi\Lambda$. Thus $x_0 \in \pi\Lambda$ and it of course *follows that* $a \in \pi\Lambda$, since $a \equiv (s-1)^{p-1}x_0 \mod \pi^n\Lambda$.

Application of the main arguments of Chapter 3 (but in one of the less difficult cases) *now completes the proof.* (The projection of a on each $\Lambda_{\varsigma,i}$ behaves very well, being not only in $\operatorname{rad}\Lambda$, but even in $\pi\Lambda$.)

The general p^n case. Here we have a subgroup H of G containing C and a group U of normalized units in $\mathbf{Z}_pG$ conjugate (even equal, as we may assume) to H modulo $I(C)I(G)$. We would have a common normal subgroup Ω for U and H, and an element $t \in H$, centralizing H modulo Ω, with $t \notin \Omega$ and $t^p \in \Omega$. Λ^Ω, etc., would be as usual.

If t centralizes $C_G(H/C) \cap C_G(\Omega)$, we proceed immediately as in Chapter 3. If not, *then there is a group element* $s \in C_G(H/C) \cap C_G(\Omega)$ [not] centralized by t; thus, arranging notation, *we have* ${}^ts = cs$.

Consequently, we may write $\Lambda^\Omega_{\varsigma,i} = s^i\Lambda^\Omega_0$. Also, since $sts^{-1} = s({}^ts)^{-1}t = c^{-1}t$, we have, if $X \subseteq G \subseteq A$ is a t-stable Ω-orbit (under conjugation), that ${}^sX$ is t-stable (being st-stable). Thus $s(\Lambda^\Omega_0) = \Lambda^\Omega_0$ and so $\sum_{i=0}^{p-1} s^i\Lambda^\Omega_0 = \sum_{i=0}^{p-1}\Lambda^\Omega_{\varsigma,i}$ is a ring. We didn't know *that* before!!

Call this new ring Γ. It is of course also an R-algebra, and we have an RH-module decomposition $\Lambda^\Omega = \Gamma \oplus \Lambda^\Omega_f$. Let $u = (1+\pi a)t$. Write $a = b + \pi^m e$ where $b \in \Gamma$ and $e \in \Lambda^\Omega_f$, $m \geq 0$. Then $u^p = t^p$ gives, on expansion, that $\pi^{m+1}\operatorname{Tr}(e) \in \Gamma + \pi^{m+2}\Lambda^\Omega$. Thus $\operatorname{Tr}(e) = 0 \mod \pi\Lambda^\Omega_f$, and we can do our usual modification to get a new u with m increased by 1. Taking a limit, *we can assume that* $a \in \Gamma$.

Write $a = \sum_{i=0}^{p-1} s^i x_i$ with $x_i \in \Lambda_0^\Omega$. Applying one of our standard Chapter 3 arguments, we obtain that $s^i x_i$ centralizes H modulo π, thus x_i centralizes H modulo π. Put $\Lambda_{00}^\Omega = \Lambda_0^\Omega \cap C_\Lambda(H + \pi\Lambda/\pi\Lambda)$, a subring of Λ_0^Ω stabilized by s. Thus $\Gamma' = \sum_{i=0}^{p-1} s^i \Lambda_{00}^\Omega$ is a subring containing a. Now one can just argue as in the p^3 case, all conjugations being achieved without modifying H modulo $\pi\Lambda$. To be sure, in Lemma 1, the proof must go more slowly, since $\Gamma'/\pi\Gamma'$ need not be commutative. More precisely, one should assume that a belongs to $(s-1)^k(\text{rad}\,\Gamma')^m + (s-1)^{k+1}\Gamma' + \pi\Gamma'$, $m \geq 0$, and try to modify n to increase m by one. The beginning of the calculation would be

$$\begin{aligned} {}^{(1+v)}u &= u + [v,u](1+v)^{-1} \\ &= u + ([v, 1+\pi a]t + (1+\pi a)[v,t])(1+v)^{-1} \\ &\equiv u + (\pi[v,a] + [v,t]t^{-1})(1+v)^{-1} \end{aligned}$$

modulo $\pi((s-1)^k(\text{rad}\,\Gamma')^{m+1} + (s-1)^{k+1}\Gamma' + \pi\Gamma')$, since the latter right ideal contains $\pi a(r\,\text{ad}\,\Gamma') \supseteq \pi a[\Gamma', t]$, of which $\pi a[v,t]$ is a member. One should use a slightly different version of v than in the p^3 case. Namely, if $a \equiv (s-1)^k x$ modulo $(s-1)^{k+1}\Gamma' + \pi\Gamma'$, where $x \in (r\ \text{ad}\,\Gamma')^m$, then put

$$v = \frac{1}{k+1}(s-1)^{k+1}x.$$

Thus

$$\begin{aligned} {}^t v - v &= \frac{1}{k+1}\left[({}^t(s-1)^{k+1} - (s-1)^{k+1})x + (s-1)^{k+1}({}^t x - x)\right] \\ &\equiv \frac{1}{k+1}({}^t(s-1)^{k+1} - (s-1)^{k+1})x \ \operatorname{mod} \pi(s-1)^{k+1}\Gamma', \\ &\qquad\qquad \text{since } t \text{ (even all of } H) \text{ centralizes } \Gamma' \text{ modulo } \pi\Gamma'. \\ &\qquad\qquad \text{(Basis considerations show } \Gamma' \text{ is pure in } \Lambda) \\ &\equiv (s-1)^k x \operatorname{mod} \pi(s-1)^{(k+1)}\Gamma' + \pi^2\Gamma'. \end{aligned}$$

Also note that

$$\begin{aligned} [v,a] &= (s-1)^{k+1}[x, \frac{a}{k+1}] + [(s-1)^{k+1}, \frac{a}{k+1}]x \\ &\equiv (s-1)^k[(s-1)^{k+1}, \frac{x}{k+1}]x \ \operatorname{mod}(s-1)^{k+1}\Gamma' + \pi\Gamma \\ &\equiv 0 \ \operatorname{mod}(s-1)^k(\text{rad}\,\Gamma')^{n+1} + (s-1)^{k+1}\Gamma' + \pi\Gamma'. \end{aligned}$$

The remainder of the proof for the new Lemma 1 is as usual.

REFERENCES

[**A**] M. Aschbacher, *Finite group theory*, Cambridge Univ. Press, 1986.

[**BR**] S. D. Berman and A. R. Rossa, *Integral group rings of finite and periodic groups* (Russian), Algebra and Math Logic and Studies in Algebra (Russian), Izad. Kiev. Univ., Kiev, 1966, pp. 44–53; as reported in MR 35–541.

[**Br**] R. Brauer, *Representations of finite groups*, Lectures on Modern Mathematics. I (T. L. Saaty, ed.), Wiley, New York, 1963, pp. 135–175.

[**C**] D. B. Coleman, *On the modular group ring of a p-group*, Proc. Amer. Math. Soc. **15** (1964), 511–514.

[**D**] E. C. Dade, *Deux groupes finis distincts ayant la même algèbra de groupe sur tout corps*, Math. Z. **119** (1971), 345–348.

[**De**] K. Dennis, *The structure of the unit group of group rings*, Proc. Ring Theory Conference, Oklahoma, Marcel Dekker, New York, 1976.

[**Fr**] A. Fröhlich, *The Picard group of noncommutative rings, in particular of orders*, Trans. Amer. Math. Soc. **180** (1973), 1–95.

[**Hi**] G. Higman, *Units in group rings*, D. Phil. Thesis, Oxford Univ., 1940.

[**K**] W. Kimmerle, personal communication.

[**P**] L. Puig, *Pointed groups and construction of characters*, Math. Z. **176** (1981), 265–292.

[**Q**] D. Quillen, *The spectrum of an equivariant cohomology ring*. II, Ann. of Math. (2) **94** (1971), 573–602.

[**RS0**] K. W. Roggenkamp and L. Scott, *Units in group rings: splittings and the isomorphism problem*, J. Algebra **96** (1985), 277–291.

[**RS**] ——, *Isomorphisms of p-adic group rings*, Ann. of Math. (to appear).

[**RS1**] ——, *The isomorphism problem for integral group rings of finite nilpotent groups*, Proc. of St. Andrews '85 (to appear).

[**RS2**] ——, *The isomorphism problem for finite abelian by nilpotent groups* (in preparation).

[**R**] K. W. Roggenkamp, *Picard groups of integral group rings of nilpotent groups*, this volume.

[**Sa**] R. Sandling, *The isomorphism problem for group rings, a survey*, Proc. 1984 Oberwolfach Conf. on Orders and Their Applications, Lecture Notes on Math., vol. 1148, Springer, Berlin, 1985, pp. 256–289.

[**S0**] L. Scott, *Integral equivalence of permutation representations*, unpublished manuscript (lecture presented to the Chicago summer institute on finite group theory), 1971.

[**S**] ——, *Units in p-adic group rings* (abstract of joint work with K. Roggenkamp), Amer. Math. Soc. Abstr. 798–20–55, October 1982.

[**S1**] ——, Letter of November 30, 1985, to Klaus Roggenkamp.

[**Se**] S. K. Sehgal, *Torsion units in integral group rings*, Proc. Nato Institute on Methods in Ring Theory, Antwerp, D. Riedel, Dordrecht, 1983, pp. 497–504.

[**W**] A. Weiss, *p-adic rigidity of p-torsion*, Annals of Math. (to appear).

[**Wh**] A. Whitcomb, *The group ring problem*, Ph. D. Thesis, Univ. of Chicago, 1968.

[**Za**] H. Zassenhaus, *On the torsion units of group rings*, Studies in Math. (in honor of A. Almeida Costa), Institute de Alda Cultura, Lisbon, 1974, pp. 119–126.

UNIVERSITY OF VIRGINIA

Proceedings of Symposia in Pure Mathematics
Volume **47** (1987)

Representations and Maximal Subgroups

GARY M. SEITZ

1. Introduction. Since the classification of the finite simple groups one of the main goals in finite group theory has been to obtain a good understanding of the subgroup structure of finite groups. One approach is to study the maximal subgroups and in this article we report on recent progress in this direction. This could be regarded as a follow-up to L. Scott's article [**19**] in the proceedings of the AMS Conference at Santa Cruz.

As indicated in (1.1) below, the general study of maximal subgroups reduces to the case of simple (or almost simple) groups, and here Scott argued that with sufficient information from representation theory it should be possible to describe the maximal subgroups of finite simple groups. As evidence he cited Dynkin's fundamental classification [**7**] of the maximal closed, connected subgroups of classical algebraic groups over the complex numbers.

For example, consider Dynkin's analysis of the maximal closed connected subgroups, X of $\mathrm{SL}(n, \mathbf{C})$. The maximal reducible subgroups are just the maximal parabolic subgroups, so we focus on irreducible subgroups. Then X must be semisimple, and if X is not simple, it is contained in a commuting product $\mathrm{SL}(r, \mathbf{C}) \circ \mathrm{SL}(s, \mathbf{C})$ for $n = rs$. So assume X is simple. What Dynkin did was use the representation theory of simple Lie algebras over $\mathbf{C}$ in order to determine explicitly *all* closed connected overgroups of X in $\mathrm{SL}(V)$. Of course, X may be contained in $\mathrm{Sp}(V)$ or $\mathrm{SO}(V)$, but aside from these there are remarkably few containments possible. Hence, he determined all cases where X is *not* maximal. It is in this sense that he determined the maximal subgroups of $\mathrm{SL}(n, \mathbf{C})$, and we adopt this point of view in the discussion to follow.

The connection between representation theory and maximal subgroups is much more interesting than just an application of known results from representation theory to group theory. For the representation theory is far from complete and even in the case of Dynkin's work, where the representation theory of simple complex Lie algebras was in relatively good shape, the analysis of maximal subgroups was a major project. Nonetheless, there has been significant progress since the appearance of Scott's article: an analog of Dynkin's work in

1980 *Mathematics Subject Classification* (1985 *Revision*). Primary 20D05, 20E28, 20G05.

characteristic p, nice results on the maximal subgroups of symmetric and alternating groups, and a good start to the understanding of the maximal subgroups of finite classical groups. But the proofs rely on a blend of group theory and representation theory, and there may be feedback in the other direction. Techniques used in the analysis of maximal subgroups may lead to useful methods in representation theory.

A discussion of the maximal subgroups is just a discussion of primitive permutation representations. The following result gives the general shape of a primitive permutation group G and for many purposes reduces the study of these groups to the case of G an almost simple group: $X \leq G \leq \mathrm{Aut}(X)$, with X a simple group.

(1.1). O'NAN-SCOTT THEOREM (SEE [**19**, **3**]). *Let Ω be a finite set, $\alpha \in \Omega$, and G a primitive subgroup of S_Ω. Let N be a minimal normal subgroup of G. Then one of the following holds:*

(a) $G = NG_\alpha$ *and* $N \cap G_\alpha = 1$.

(b) *There is a simple group T such that $N = T_1 \times \cdots \times T_r$, with each $T_i \cong T$ and either*

(i) (*wreathed case*) *T is the socle of a primitive group X and $G \leq X \text{ wr } S_r$.*

(ii) (*diagonal case*) *$N \cap G_\alpha = D_1 \times \cdots \times D_s$, where $r = st$, and D_1 is a diagonal subgroup of $S_1 \times \cdots \times S_t$, D_2 is diagonal in $S_{t+1} \times \cdots \times S_{2t}$, etc.*

We note that in (1.1)(a) it need not be the case that N is elementary abelian (see Aschbacher-Scott [**3**]).

In view of (1.1) we shall assume from now on that G is a simple group, although in fact one must work with almost simple groups. We concentrate on the alternating groups and groups of Lie type. For many of the sporadic groups the maximal subgroups have now been determined (see the Atlas and the work of R. Wilson). Indeed, the Atlas describes the maximal subgroups of many "small" groups of Lie type and alternating groups, some of which must be treated separately from the general theory described below. We also mention the work of Aschbacher [**2**], which essentially determines all maximal subgroups containing a Sylow p-subgroup P, assuming $|P| \geq p^2$. But the analysis of sporadic groups is by no means complete. In particular, it would be desirable to develop general techniques for dealing with questions about these groups, so as to be able to avoid excessive case analysis.

In §2, we indicate some general results, useful in the results to follow. Some of these are of independent interest in representation theory. Then in §§3–6 we describe progress for various types of groups. §7 describes some miscellaneous results.

2. General techniques. There are a number of standard results from representation theory which are important to the study of subgroups (e.g., the Steinberg tensor product theorem). In this section we describe a few results which are perhaps less well known.

(2.1) *Degrees of irreducible representations.* Let $G = G(q)$ be of Lie type in characteristic p and M a nontrivial irreducible projective representation over a field of characteristic other than p. Then

$$m(G,p) \leq \dim(M) \leq |G : B|c(q).$$

The lower bound was given in Landazuri-Seitz [**11**]. Here $m(G,p)$ is a polynomial in q (except for a few special cases where p divides the order of the Schur multiplier of G) and the degree of this polynomial is best possible.

For the upper bound B is a Borel subgroup of G and $c(q) \to 1$ as $q \to \infty$. Indeed, $c(q) = 1$ unless G is either an odd-dimensional unitary group or a Suzuki or Ree group, and in any case $c(q) \leq 2$. For sufficiently large q, there are complex irreducible characters of degree $|G : B|c(q)$, so the bound is best possible. It would be nice to have an elementary proof, as my proof uses Lusztig's Jordan decomposition of complex irreducible characters and the determination of the degrees of the unipotent cuspidal characters.

The next two results concern representations of groups of Lie type $G = G(p^a)$ in the natural characteristic. Let G act irreducibly on V, where V is an irreducible, though not necessarily absolutely irreducible, module over a finite field k of characteristic p. The first result is a slight extension of a nice result of S. Smith [**27**] and the second shows that with essentially four exceptions, V is a primitive module.

(2.2) *Restriction to parabolics* (Theorem 1 of [**21**]). Let P be a parabolic subgroup of G with Levi factor L and set $Q = O_p(P)$. Then L acts irreducibly on $V(Q)$, the fixed space of Q, and there is an L-invariant subspace V_1 of V such that $V = V(Q) \oplus V_1$.

(2.3) *Primitivity of representations* (Theorem 2 of [**21**]). V is imprimitive if and only if $G/Z(G)$ is one of the groups $L_2(5)$, $L_2(7)$, $L_3(2)$, or $PSp_4(3)$, V is the Steinberg module, and $k \geq F_{p^2}$ in all but the second case, where there is no restriction on k.

It is perhaps worth mentioning the idea of the proof of (2.3), as this already combines results in group theory and representation theory. Assume V is imprimitive, say $V = V_1 \oplus \cdots \oplus V_k$, with G permuting $\Delta = \{V_1, \ldots, V_k\}$. Let B be a Borel subgroup of G and U the unipotent radical of B. Then the sum of subspaces in each B-orbit of Δ has nontrivial fixed space for U. By (2.2) there is a unique such orbit. Hence, B is transitive and $G = BX$, where X is the stabilizer in G of V_1. But then [**24**] gives all possibilities for X, from which it is possible to complete the proof.

Using lower bounds such as those in (2.1), the classification theorem for finite simple groups, and some representation theory for groups of Lie type in the natural characteristic, Liebeck established the following result showing that the only unknown maximal subgroups of finite classical groups have relatively small order. Similar bounds exist for exceptional groups [**17**].

(2.4) *Orders of subgroups of classical groups* (Liebeck [**14**]). Let $G = G(q)$ be a simple classical group with natural (projective) module V having dimension n

over $GF(q)$. Let H be a maximal subgroup of G. Then either H is of known type with well-described (projective) action on V or $|H| < q^{2n+4}$.

3. Symmetric and alternating groups. We begin with a reduction theorem, which first appeared in Scott's article [**19**].

REDUCTION THEOREM (O'NAN-SCOTT [**19**]). *Let D be a subgroup of S_n, not containing A_n. Then one of the following holds*:

(a) $D \leq S_m \times S_k$ *(intransitive)*, $n = m + k$.

(b) $D \leq S_m \operatorname{wr} S_k$ *(imprimitive)*, $n = mk$.

(c) $D \leq S_m \operatorname{wr} S_k$ *(product action)*, $n = m^k$ $(m \geq 5)$.

(d) $D \leq V \cdot \mathrm{GL}(V)$, *where* $n = p^k = |V|$.

(e) $D \leq (G \operatorname{wr} S_k)\operatorname{Out}(G)$, *where G is simple and* $n = |G|^{k-1}$.

(f) $D = N_{S_n}(X)$, *where X is a simple group and D is primitive.*

The next step is to determine which groups in the list are actually maximal. For (a)–(e) this is carried out in Liebeck-Praeger-Saxl [**15**] and, independently, by Aschbacher. With very few exceptions the indicated groups are maximal. Moreover, it is shown that there are very few cases where a group of type (f) is contained in one of type (a)–(e).

A more difficult problem is to determine which groups of type (f) are maximal. Note that in (f), X will act transitively. Conversely, fix a simple group X, with transitive action on cosets of subgroup H. Embed X in S_n for $n = |X : H|$ and assume $N_{S_n}(X)$ acts primitively (this is automatic if H is chosen maximal in X). The issue is whether or not $\overline{X} = N_{S_n}(X)$ is maximal in S_n (similarly for A_n).

At first this appears an impossible task. But suppose $\overline{X}$ is not maximal. Excluding the handful of cases where $\overline{X}$ is contained in a group of type (a)–(e), there must be a properly larger group $\overline{Y} = N_{S_n}(Y)$ of type (f). Choose such a containment with Y minimal. It follows from the Schreier conjecture (now a theorem) that $X \leq Y \leq A_n$. Since $\overline{X}$ is transitive, $\overline{Y} = \overline{XY}_\alpha$ for $\alpha \in \Omega$, the underlying set. That is, $\overline{Y}$ can be factored into a product of maximal subgroups, with one of the factors being an almost simple group.

Using (2.1), (2.2), and similar bounds for other types of groups, one can argue that with few exceptions, one of the groups X, Y_α is of known type. Using precise information on this subgroup (order, index, etc.) it is actually possible to obtain all possibilities for the pair (X, Y). This program was carried out completely by Liebeck-Praeger-Saxl [**15**]. There is an explicit list of all cases where maximality does not occur. For example, if X is one of the groups $E_6(q)$, $E_7(q)$, or $E_8(q)$ and H is any maximal subgroup of X, then $N_{S_n}(X)$ is maximal in S_n, where $n = |X : H|$.

Thus the maximal subgroups of S_n and A_n are determined, at least in the sense of the introduction. Given a particular subgroup with known permutation action, it is possible to determine whether or not the subgroup is maximal, and when it is not maximal it is possible to determine all overgroups.

4. Classical algebraic groups. We next discuss the characteristic p analog of Dynkin's work. The results here are needed in the next section on finite classical groups. As in the last section we first state a reduction theorem.

Let V denote a finite-dimensional vector space over an algebraically closed field of prime characteristic p. Assume V to have either the trivial form or a nondegenerate bilinear or quadratic form and let G be the corresponding classical group.

(4.1) REDUCTION THEOREM [**20**]. *Let $G = \mathrm{SL}(V)$, $\mathrm{Sp}(V)$, or $\mathrm{SO}(V)$, and let X be maximal among proper closed connected subgroups of G. Then one of the following holds*:

(a) $G = \mathrm{SL}(V)$, $X = \mathrm{Sp}(V)$ *or* $\mathrm{SO}(V)$.

(b) $G = \mathrm{Sp}(V)$, $p = 2$, $X = \mathrm{SO}(V)$.

(c) $G = \mathrm{SO}(V)$, $p = 2$, *X is the stabilizer of a nonsingular* 1*-space.*

(d) *X is a maximal parabolic subgroup of G.*

(e) *V is nondegenerate, $V = W \oplus W^{\perp}$, and $X = \mathrm{stab}_G(W)^0$.*

(f) *$V = V_1 \otimes V_2$, and $X = I(V_1) \circ I(V_2)$ for suitable isometry groups $I(V_1)$ and $I(V_2)$.*

(g) *X is simple, $V|X$ is irreducible and tensor indecomposable.*

A bit more work shows that except for a few very special cases the groups in (a)–(f) are maximal among closed, connected subgroups (see [**20**]). To find the maximal closed, connected subgroups of classical algebraic groups we must decide which groups of type (g) are maximal.

Choose X as in (g) and assume X is not maximal. Then X is properly contained in a closed, simple group Y. Consequently, we have a triple (X, Y, V) such that $X < Y < \mathrm{SL}(V)$, $V|X$ is irreducible, and $V|X$ is tensor indecomposable. In recent work all such triples were found. Testerman [**28**] did this for Y an exceptional group and I settled the cases where Y is classical [**20**]. One form of the main result is the following theorem, slightly more general than is actually needed for the determination of maximal subgroups.

(4.2) THEOREM [**20**, **28**]. *Let Y be a simple algebraic group in characteristic p. Let $\phi : Y \to \mathrm{SL}(V)$ be an irreducible, rational representation, which is tensor indecomposable. If X is a closed, connected subgroup of Y with $V|X$ irreducible then one of the following holds*:

(a) $X = Y$.

(b) $\phi(Y) = \mathrm{SL}(V)$, $\mathrm{Sp}(V)$, *or* $\mathrm{SO}(V)$.

(c) *The triple $(\phi(X), \phi(Y), V)$ is explicitly known.*

The list of triples of type (c) appears as Table 1 in [**20**]. For each triple (X, Y, V) we indicate the precise embedding of X in Y together with the high weight of each of the modules $V|X$ and $V|Y$. This then is the characteristic p version of Dynkin's work [**7**].

There are lots of examples, but they are of a very specialized nature. A direct check yields the following remarkable result.

COROLLARY (Corollary 4 of [**20**]). *Let X be a simple algebraic group and $\phi\colon X \to \mathrm{SL}(V)$ an irreducible, rational representation with V tensor indecomposable. Choose $I(V) = \mathrm{SL}(V)$, $\mathrm{Sp}(V)$, or $\mathrm{SO}(V)$ minimal with respect to containing $\phi(X)$, If Y is any closed, connected subgroup of $I(V)$ with $\phi(X) < Y < I(V)$, then with just one exception Y is maximal among closed, connected subgroups of $I(V)$.*

The following is a brief outline of the general approach used to establish (4.2). The main idea is that the irreducibility assumption should be manifested within the group embedding, where things are less sensitive to changes in the characteristic. The modules $V|X$ and $V|Y$ are described by giving the high weight and this is indicated by certain nonnegative integers above the nodes of the Dynkin diagram. In the course of the proof we determine both the embedding of X in Y and the labellings

Fix maximal tori $T_X \le T_Y$ of X and Y and let $\Sigma(X)$, $\Sigma(Y)$ denote the corresponding root systems. Choose a parabolic subgroup P_X of X with $T_X \le P_X$, and write $P_X = Q_X L_X$, where Q_X is the unipotent radical of P_X and L_X is the Levi factor containing T_X. A theorem of Borel and Tits [**4**] shows there exists a parabolic subgroup $P_Y = Q_Y L_Y$ of Y such that $P_X \le P_Y$ and $Q_X \le Q_Y$.

In (2.8) of [**20**] it is shown that, under reasonably general hypotheses, there is a unique P_Y such that $L_Y = C_Y(Z(L_X)^0)$. In particular, $T_Y \le L_Y$ and $L_X \le L_Y$. Thus, we have a uniquely determined system of parabolics of Y, one for each parabolic of X containing T_X. Further, parabolics of Y obtained in this way all contain T_Y, hence are built on a common system of root groups.

Take P_X corresponding to an end node α of the Dynkin diagram of X and such that some fundamental root of $\Sigma(L_X)$ has nonzero label. Then L_X' is a simple group, although usually L_Y' will not be, and we write $L_Y' = L_1 \cdots L_k$, a commuting product of simple groups.

A trivial consequence of Smith's theorem [**27**] (i.e., (2.2) for algebraic groups) shows that with P_X, P_Y as above, $V(Q_X) = V(Q_Y)$ (the fixed points of Q_X and Q_Y) and L_X, L_Y are each irreducible on this space. Consequently, we can argue by induction. This gives information on the fixed point spaces and also on the embedding of L_X in L_Y. But the information is limited. For L_Y can have any number of simple factors, only certain of which will act nontrivially on the fixed point space. Induction gives no information regarding projections to factors of L_Y which act trivially on $V(Q_Y)$.

An argument with dual modules shows $[V, Q_X] = [V, Q_Y]$ and L_X, L_Y are irreducible on $V/[V, Q_X]$, with prescribed high weights. Indeed, by taking parabolics to be opposites of standard parabolics the labelling of $\Sigma(Y) = \bigcup \Sigma(L_i)$ is just the labelling of L_Y in its action on $V/[V, Q_Y]$. So induction gives information on the labelling of nodes in $\Sigma(L_Y)$. Let γ be a fundamental root in $\Sigma(Y) - \Sigma(L_Y)$. It is not difficult to show the label of γ must be zero unless γ adjoins $\Sigma(L_i)$ for some i.

There is a decomposition $[V, Q_Y]/[V, Q_Y, Q_Y] = \bigoplus V_\gamma(Q_Y)$, the sum ranging over the nodes γ above, and each term a weight space for the torus $Z(L_Y)^0$. One proves $\dim(V_\gamma(Q_Y)) \leq \dim(V/[V, Q_X]) \cdot \dim(Q_X)$, the right side determined completely by $V|X$. In addition there is a method for producing composition factors of these modules, with particular high weights. Combining this with the inductive information we can obtain lower bounds for the dimensions of the weight spaces. Taken together, the bounds yield further restrictions on high weights. Also, the bounds give restrictions on the rank of all factors of L adjoining the node γ.

Next we consider internal L_X and L_Y modules—namely the composition factors of Q_X and Q_Y, respectively. In each case the composition factors are isomorphic to certain sums of root groups for the maximal tori T_X, T_Y, and have natural descriptions as irreducible high weight modules. The embedding $Q_X \leq Q_Y$ forces the irreducible L_X modules obtained in this way to somehow occur within certain L_Y modules. This does not appear particularly strong since one can imagine embedding a small unipotent group in a huge one, so that almost anything could be arranged regarding L_X and L_Y composition factors. However, the equality $[V, Q_X] = [V, Q_Y]$ forces the embedding of Q_X in Q_Y to be highly nontrivial and imposes severe restrictions on the various composition factors that occur.

More precisely, choose L_i nontrivial on $V/[V, Q_Y]$ and let γ be a fundamental root adjoining $\Sigma(L_i)$. There is a normal subgroup K_Y^γ of P_Y such that $K_Y^\gamma \leq Q_Y$ and the quotient $Q_Y^\gamma = Q_Y/K_Y^\gamma$ can be regarded as an irreducible, rational L_Y-module and such that the only factors of L_Y which are nontrivial on this module are those adjoining γ. Also, Q_Y^γ is the direct sum of the images of certain T_Y-root subgroups, each image being a subspace.

One shows that the image $Q_X^\gamma = Q_X K_Y^\gamma / K_Y^\gamma$ is an L_X-submodule of Q_Y^γ. The equality $[V, Q_X] = [V, Q_Y]$ implies this submodule is nontrivial and in fact $U_{-\alpha}$ is not contained in K_Y^γ. A careful analysis of the submodule Q_X^γ, using the above restriction on other Levi factors adjoining γ, yields very precise information. In most cases we obtain the exact structure of the factors adjoining γ together with the precise projection of L_X to each factor. Further, it is usually possible to get detailed information regarding factorizations of elements of $U_{-\alpha}$ into products of elements of T_Y-root groups, all read modulo K_Y^γ. Write

$$U_{-\alpha}(x) \equiv U_{\beta_1}(f_1(x)) \cdots U_{\beta_t}(f_t(x)).$$

Conjugating by elements of T_X one can usually determine the functions $f_i(x)$ and show that $\beta_i | T_X = q_i \alpha$, where q_i is a certain power of p. In fact, q_i usually turns out to be one of the twists appearing in the tensor product decomposition of $V|X$ (which was not assumed to be tensor indecomposable). So in this way we have obtained further information on the embedding of L_X in L_Y, the embedding of Q_X in Q_Y, and the restriction to T_X of T_Y-weights of V. Changing parabolic subgroups yields further information, though it is often sufficient to work in just one parabolic.

At this point the work in [**20**] and that in [**28**] diverge. In my work, where Y is assumed to be a classical group, I next look at the classical module, M, of Y and analyze the action of X on that module. A comparison of actions eventually shows that with relatively few exceptions, $V|X$ is tensor indecomposable. It follows that at most one L_i is nontrivial on $V/[V, Q_Y]$. It is then proved that this factor usually occurs at the end of the Dynkin diagram and is the unique factor of L_Y nontrivial on $M/[M, Q_Y]$. From the known projection of L_X to L_i, we obtain information regarding $M|X$. By changing parabolics we eventually determine $M|X$. In turn this determines L_Y and we see that in fact the only factors of L_Y are those adjoining γ. By this stage we have both the labelling and the embedding and the result is proved.

Testerman does not have the advantage of the classical module, though she does know that the rank of Y is small. Consequently, she is able, though this is very difficult, to dig deeper into the unipotent radical, obtaining further restrictions.

Hopefully some of the above methods, especially the sort of analysis done by Testerman, will be useful for other questions in representation theory. Indeed, the methods apply even if $Y = \mathrm{SL}(V)$, $\mathrm{Sp}(V)$, or $\mathrm{SO}(V)$. If one considers an arbitrary rational irreducible representation of X, it is possible to get nontrivial information on embeddings of parabolic subgroups, which, in turn, gives information about the module. In particular, it may be possible to use information on parabolic embeddings to obtain results on dimensions of weight spaces and to predict those primes for which Weyl modules are reducible. Testerman is currently considering such problems.

5. Finite classical groups. Let G be a finite classical group with underlying (projective) module V defined over the finite field F. Here we proceed as in §4, although the situation is much more complicated and the results less complete.

First the reduction theorem, due to Aschbacher. In [**1**] Aschbacher defines a family, $\mathfrak{A}(G)$, of subgroups of G. The family consists of stabilizers of certain subspaces, sets of subspaces, tensor product decompositions, normalizers of subclassical groups defined over subfields and extension fields, and normalizers of certain irreducible groups of prime power order. He then proves

(5.1) REDUCTION THEOREM [**1**]. *If $D \leq G$, then either $D \leq A$ for some $A \in \mathfrak{A}(G)$ or $D = N_G(Y)$ where Y is a simple group with projective action on V being absolutely irreducible and defined over no proper subfield of F.*

The proof of the above theorem is a nice blend of group theory and representation theory as one must analyze the group-theoretical consequences of various techniques from representation theory.

As in the other cases the next step is to determine which of the groups in $\mathfrak{A}(G)$ are actually maximal in G. This is done in Kleidman-Liebeck [**10**], the analysis depending heavily on the classification of finite simple groups. A different approach to most of the cases is given by Li [**12**, **13**], where he deals with arbitrary fields.

To complete the analysis we fix a quasisimple group X acting absolutely irreducibly on the finite module V and such that $N_G(X)$ is not contained in a subgroup $A \in \mathfrak{A}(G)$. We must determine all quasisimple subgroups Y with $X < Y < G$. The problem splits into two subcases. Write $G = G(p^c) = I(n, p^c)$, where $n = \dim(V)$ and the underlying field of V has order p^c or p^{2c} if V is unitary. Set $\bar{c} = c$ or $2c$, accordingly.

The generic case. Here $X = X(p^a)$, a group of Lie type in characteristic p. First we apply the following result relating the numbers a and c. We need the following notation.

Suppose M is an absolutely irreducible module for G over a field of characteristic p. If $X \neq {}^2A_r(p^a)$, ${}^2D_r(p^a)$, ${}^2E_6(p^a)$, or ${}^3D_4(p^a)$, set $\bar{a} = \bar{a}(M) = a$. In the other cases let $a^\sim = 2a$, $2a$, $2a$, or $3a$, respectively, and set $\bar{a} = a$ or $a^\sim$, according to whether or not $M \cong M^\sigma$, σ denoting the p^a power map.

(5.2) THEOREM (Corollary 6 of [**21**]). *Let $X = X(p^a)$ be a subgroup of the classical group $G = G(p^c) = I(n, p^c)$. Then one of the following holds:*

(a) $N_G(X) \leq A$ *for some* $A \in \mathfrak{A}(G)$.

(b) *There exist integers* $1 < r$, r_1 *and an integer* $a_1 \in \{a, a^\sim\}$, *such that* $\bar{a} = r\bar{c}$, $n = m^{r_1}$, *and* $N_G(X) \leq N_G(I(m, p^{\bar{a}_1})) < G$, *for a suitable isometry group* $I(m, p^{\bar{a}_1})$.

(c) $\bar{a} = \bar{c}$.

The proof of the above result involves a careful analysis of the Steinberg tensor product theorem. In view of this result we enlarge $\mathfrak{A}(G)$ to include the groups $N_G(I(m, p^{\bar{a}_1}))$ appearing in (b). Denote this by $\underline{\mathfrak{A}}(G)$.

The next result restricts the types of overgroups of X in G.

(5.3) THEOREM [**16**]. *If $X(p^a) < Y < G(p^c)$ and if X is not contained in an element of $\underline{\mathfrak{A}}(G)$, then one of the following holds:*

(a) $Y = Y(p^b)$, *Lie type in characteristic* p.

(b) *The triple* (X, Y, V) *is explicitly known.*

(c) *Finitely many exceptions.*

The above result relies on the lower bounds in (2.1), (2.3), (2.4), and (8.1) of [**20**]. The exceptions in (c) are nearly all cases where Y is a sporadic simple group and X is a relatively small group of Lie type. These cases could be resolved given sufficient information on the modular representations of sporadic groups.

The theorem shows that with relatively few exceptions overgroups of X will again be of Lie type in characteristic p. So we now consider embeddings of the form $X(p^a) < Y(p^b) < G(p^c)$, with G a classical group and X not contained in an element of $\underline{\mathfrak{A}}(G)$. Then the first theorem applied to both X and Y implies that $\bar{a} = \bar{b} = \bar{c}$.

For convenience let us pass from the containment $X < Y < G$ to $X \to Y \to \mathrm{SL}(\overline{V})$, where now X and Y are taken to be universal groups and $\overline{V} = \overline{k} \otimes V$, where $\overline{k}$ is the closure of F_p (recall that X, Y are absolutely irreducible on V, so irreducible on $\overline{V}$). Letting $\overline{X}$, $\overline{Y}$ be the corresponding simply connected groups over $\overline{k}$, the representations $X \to \mathrm{SL}(\overline{V})$ and $Y \to \mathrm{SL}(\overline{V})$ can be lifted to $\overline{X} \to \mathrm{SL}(\overline{V})$ and $\overline{Y} \to \mathrm{SL}(\overline{V})$.

It is a consequence of Theorem 4 of [**21**] that when Y is a classical group this can all be arranged so that we obtain $\overline{X} \to \overline{Y} \to \mathrm{SL}(V)$. So in this case we can apply the main result of [**20**] to obtain all possibilities for $(\overline{X}, \overline{Y}, \overline{V})$. Since $\overline{a} = \overline{b} = \overline{c}$, we also obtain all triples (X, Y, V).

We do not know if there is an analog of Theorem 4 of [**21**] for Y an exceptional group. This is the main obstruction to the completion of the generic case. It may be possible to skirt the issue by proving a version of Testerman's work for the finite exceptional groups, but it would be much better to avoid this.

Nongeneric cases. Consider now embeddings $X < G(p^c)$, where X is of Lie type in characteristic other than p, an alternating group, or a sporadic group. This is largely unexplored territory and there are a variety of open problems. Assume X is not contained in some member of $\underline{\mathfrak{A}}(G)$.

As above we wish to find all triples (X, Y, G) with $X < Y < G$ and Y quasisimple. We may take Y minimal for this.

Recently I have begun an investigation of the case $X = X(r^a)$, Lie type in characteristic $r \neq p$. The most likely configuration is $Y = Y(r^b)$, also of Lie type in characteristic r, and this is the case I study. So the problem is to determine all triples $(X(r^a), Y(r^b), V)$ with V a module in characteristic $p \neq r$, irreducible under the action of X.

This is an interesting configuration even for complex characters, and we mention the following example. Let $X = \mathrm{Sp}(2n, r^{st})$ and $Y = \mathrm{Sp}(2nt, r^s)$, with the usual embedding of X in Y. In each case the minimal degree for a nontrivial complex irreducible character is $(r^{snt} - 1)/2$ (see [**26**]). Moreover, irreducible characters of this degree are modular irreducible for all primes $p \neq r$. Hence, we have a triple (X, Y, V) for each prime $p \neq r$.

Assume $X(r^a) < Y(r^b) < G(p^c)$ with Y a classical group. I have recently obtained the following result.

(5.4) THEOREM. *Let $X = X(r^a) < Y(r^b) = Y$, with Y classical and $r^a > 3$. Assume X, Y are absolutely irreducible on a nontrivial module of characteristic other than r. Then $X \in \underline{\mathfrak{A}}(Y)$.*

The restriction on r^a can probably be removed with more work. At this point we consider the various types of groups in $\underline{\mathfrak{A}}(Y)$ as candidates for X. The methods used to prove (5.4) suffice to give a list of all pairs (X, Y) for which a V exists in some characteristic $p \neq r$. There are only a few of these. Details will appear in a future article. However, our methods are not sufficient to determine all possible modules V for a given pair (X, Y).

The proof uses both the upper and lower bounds in (2.1) to reduce to a small range of embeddings and then a careful analysis of the restriction of representations to parabolics, together with detailed information on the embeddings of unipotent radicals similar to that used in the proof of (4.2).

I'll close this section with one problem regarding alternating groups. Determine all triples (A_n, A_m, V) with V an absolutely irreducible module for both alternating groups. The embedding $A_n < A_m$ is likely a transitive embedding of A_n in A_m and V can be a module over a field of arbitrary characteristic.

6. Exceptional groups. Currently there is no satisfactory reduction theorem for either exceptional algebraic groups or finite exceptional groups. Indeed, it is quite possible that a reduction theorem will be the last step of the analysis of maximal subgroups, rather than the first step.

A natural approach is to regard certain small modules for G as analogs of the classical module for a classical group and try to establish reduction theorems as in [**1**]. The goal here is to identify certain interesting subspaces of the appropriate module and argue that nearly all maximal subgroups appear as stabilizers of such subspaces. There has been some success for the 27-dimensional modules for E_6 (Aschbacher, Cohen), but to date the results are incomplete.

The above approach is reasonable provided there is evidence to suggest that most maximal subgroups will act reducibly on the appropriate small module. The evidence is supplied by recent work of Testerman [**28**], who showed that with very few exceptions a closed, connected subgroup of an exceptional algebraic group G acts reducibly on any nontrivial finite-dimensional rational G-module.

Another approach would be to give a purely internal analysis of maximal subgroups. The local maximal subgroups of finite exceptional groups have been determined in this way (see [**5**]). Hence, one can reduce to the normalizer of a commuting product of simple groups. This could be regarded as a very weak reduction theorem, but it is only a first step.

7. Some other results. Several of the results indicated above were stronger than actually needed for describing the maximal subgroups. For example, some of the results determine all overgroups of certain types of groups. In recent years there have been a number of such results and these are quite useful in applications. They suggest that it may be possible to work from the bottom up (study overgroups) rather than from the top down (maximal subgroups).

Let $G = G(p^c)$ be of Lie type in characteristic p and X a proper subgroup of G. Suppose $R \neq 1$ is an abelian normal r-subgroup of X, for some prime r. There is no hope of finding all overgroups of X in G. After all X could have prime order. However, a lot can be said about the overgroups of $N_G(R)$, a group containing X.

If $r \neq p$, then, typically, R is contained in a maximal torus T of G so $N_G(R) \geq T$. In [**22**] (for classical groups see also [**23**]) all overgroups of groups of Lie type containing maximal tori are determined, given certain restrictions on p^c. The subgroups can all be described in terms of the root system and Weyl group of the overlying algebraic group. On the other hand, if $r = p$, then the main result of [**25**] shows that for p odd, all overgroups of $N_G(X)$ contain nontrivial normal p-subgroups and are hence contained in proper parabolic subgroups. In particular, all maximal overgroups are parabolic.

There are many results regarding the subgroup structure of simple groups which we have failed to mention as they do not fit into the general program above but which are nonetheless useful. To mention a few, Cooperstein [**6**] and Kantor [**8**] have analyzed subgroups of groups of Lie type generated by long root elements. Also, maximal subgroups of odd index of almost simple groups have been determined, independently, by Kantor [**9**] and Liebeck-Saxl [**18**].

REFERENCES

1. M. Aschbacher, *On the maximal subgroups of the finite classical groups*, Invent. Math. **76** (1984), 469–514.

2. ——, *Overgroups of Sylow subgroups in sporadic groups*, Mem. Amer. Math. Soc. No. 343 (1986).

3. M. Aschbacher and L. Scott, *Maximal subgroups of finite groups*, J. Algebra **92** (1985), 44–80.

4. A. Borel and J. Tits, *Elements unipotent et sous-groupes paraboliques de groupes reductives*, Invent. Math. **12** (1971), 95–104.

5. A. Cohen, M. Liebeck, J. Saxl, and G. Seitz, *The local maximal subgroups of finite groups of Lie type*, (to appear).

6. B. Cooperstein, *Subgroups of exceptional groups of Lie type generated by long root elements. I: Odd characteristic*, J. Algebra **70** (1981), 270–282; II: *Characteristic* 2, J. Algebra **70** (1981), 283–298.

7. E. Dynkin, *Maximal subgroups of the classical groups*, Amer. Math. Soc. Transl. (2) **6** (1957), 245–378.

8. W. Kantor, *Subgroups of classical groups generated by long root elements*, Trans. Amer. Math. Soc. **248** (1979), 347–379.

9. ——, *Primitive permutation groups of odd degree and an application to finite projective planes*, (to appear).

10. P. Kleidman and M. Liebeck, *The subgroup structure of the finite classical groups*, (to appear).

11. V. Landazuri and G. Seitz, *On the minimal degrees of projective representations of the finite Chevalley groups*, J. Algebra **32** (1974), 418–443.

12. S. Li, *Overgroups in* $\mathrm{GL}(nr, F)$ *of certain subgroups of* $\mathrm{GL}(n, K)$, (to appear).

13. ——, *Overgroups in* $\mathrm{GL}(U \otimes W)$ *of certain subgroups of* $\mathrm{GL}(U) \otimes \mathrm{GL}(W)$, (to appear).

14. M. Liebeck, *On the orders of maximal subgroups of the finite classical groups*, Proc. London Math. Soc. (3) **50** (1985), 426–446.

15. M. Liebeck, C. Praeger, and J. Saxl, *A classification of the maximal subgroups of the finite symmetric and alternating groups*, J. Algebra (to appear).

16. M. Liebeck, J. Saxl, and G. Seitz, *Overgroups of irreducible subgroups of* $\mathrm{GL}(V)$, (to appear).

17. M. Liebeck and J. Saxl, *On the orders of maximal subgroups of the finite exceptional groups of Lie type*, (to appear).

18. ——, *The primitive permutation groups of odd degree*, J. London Math. Soc. **31** (1985), 250–264.

19. L. Scott, *Representations in characteristic p*, Proc. Sympos. Pure Math., vol. 37, Amer. Math. Soc., Providence, R.I., 1981, pp. 319–332.

20. G. Seitz, *the maximal subgroups of classical algebraic groups*, Mem. Amer. Math. Soc. (to appear).

21. ——, *Representations and maximal subgroups of finite groups of Lie type*, Geom. Dedicata (to appear).

22. ——, *The root subgroups for maximal tori in finite groups of Lie type*, Pacific J. Math. **106** (1983), 153–244.

23. ——, *On the subgroup structure of classical groups*, Comm. Algebra **10** (1982), no. 8, 875–885.

24. ——, *Flag-transitive subgroups of Chevalley groups*, Ann. of Math. (2) **97** (1973), 27–56.

25. ——, *Unipotent subgroups of groups of Lie type*, J. Algebra **84** (1983), 253–278.

26. ——, *Some representations of classical groups*, J. London Math. Soc. **10** (1975), 115–120.

27. S. Smith, *Irreducible modules and parabolic subgroups*, J. Algebra **75** (1982), 286–289.

28. D. Testerman, *Certain embeddings of simple algebraic groups*, Ph.D. Thesis, Univ. of Oregon, 1985.

UNIVERSITY OF OREGON

Proceedings of Symposia in Pure Mathematics
Volume **47** (1987)

Green Functions of Reductive Groups over a Finite Field

TOSHIAKI SHOJI

Abstract. The aim of this article is to give a survey on the theory of Green functions of reductive groups, centering the determination of Green functions and emphasizing the relationship with the theory of character sheaves developed by Lusztig [**17**].

0. Introduction. Let G be a connected reductive algebraic group defined over a finite field $\mathbf{F}_q$ of characteristic p, and let $F: G \to G$ be the corresponding Frobenius map of G. We denote by G_{uni} the variety of unipotent elements in G and by G^F_{uni} the set of its F-fixed points. Let $R^G_T(\theta)$ be the virtual representation of G^F defined by Deligne and Lusztig [**9**; see also **7**, **8**], associated with an F-stable maximal torus T and an irreducible character θ of T^F on $\overline{\mathbf{Q}}_l$. The Green function Q^G_T is defined, according to [**9**], as the restriction of the character of $R^G_T(\theta)$ to G^F_{uni}, i.e., $Q^G_T(u) = \operatorname{Tr}(u, R^G_T(\theta))$ for a unipotent element u in G^F. Then Q^G_T turns out to be a G^F-invariant function on G^F_{uni} independent of θ.

One of the important properties of Green functions is the following orthogonality relations.

The orthogonality relations of Green functions [**9**, Theorem 6.9].

$$\frac{1}{|G^F|} \sum_{u \in G^F_{\text{uni}}} Q^G_T(u) Q^G_{T'}(u) = \frac{|N_G(T,T')^F|}{|T^F|\,|T'^F|}, \tag{0.1}$$

where $N_G(T,T') = \{n \in G \mid n^{-1}Tn = T'\}$.

These orthogonality relations are closely related to the orthogonality relations of $R^G_T(\theta)$ [**9**, Theorem 6.8] in connection with the following character formula of $R^G_T(\theta)$, which describes the character values of $R^G_T(\theta)$ in terms of Green functions and is, in fact, another important property of Green functions.

The character formula [**9**, Theorem 4.2]. *Let $g = su = us$ be the Jordan decomposition of $g \in G^F$, where $s \in G^F$ is semisimple and $u \in G^F$ is unipotent.*

1980 *Mathematics Subject Classification* (1985 *Revision*). Primary 20G40, 20C15.

Then

$$\text{(0.2)} \qquad \operatorname{Tr}(g, R_T^G(\theta)) = |Z_G^0(s)^F|^{-1} \sum_{\substack{x \in G^F \\ x^{-1}sx \in T^F}} Q_{xTx^{-1}}^{Z_G^0(s)}(u)\theta(x^{-1}sx).$$

Hence the determination of the characters of $R_T^G(\theta)$ is reduced to the determination of Green functions of various connected reductive subgroups of G. Moreover, it is known that Green functions are invariant by an isogeny of G, and that their determination is reduced to the case where G is an almost simple adjoint group. In this article, we shall discuss a general algorithm for computing Green functions and some applications of this procedure.

1. Green functions and Springer representations. It is not so easy to get precise information about Green functions directly from the definition as in the Introduction. However, after Springer [**28**] and Kazhdan [**13**], a cohomological interpretation of Green functions in terms of Weyl group representations is known, which is more convenient for our purpose. (For the general reference concerning these subjects, see also [**31**].) Let B be a Borel subgroup of G and, for each $g \in G$, consider the variety $\mathcal{B}_g = \{xB \in G/B \,|\, x^{-1}gx \in B\}$. Let T be a maximal torus of G contained in B, and $W = N_G(T)/T$ the corresponding Weyl group of G. Then one can construct representations of W on the l-adic cohomology groups $H^i(\mathcal{B}_g, \overline{\mathbf{Q}}_l)$, the so-called Springer representations of W. In fact, Springer [**28**] orginally constructed them in the case where p is large enough and g is a unipotent element (or rather a nilpotent element in the Lie algebra of G). After that Lusztig [**14**] gave a general construction using the intersection cohomology theory, which works for any $g \in G$ and for arbitrary $p \geq 0$. We note that Springer representations of W are independent of $\mathbf{F}_q$-structures of G.

We now consider an $\mathbf{F}_q$-structure in G and assume that both B and T are F-stable. Let us take $g \in G^F$. Then F acts naturally on $\mathcal{B}_g$ and we have an induced map

$$F : H^i(\mathcal{B}_g, \overline{\mathbf{Q}}_l) \to H^i(\mathcal{B}_g, \overline{\mathbf{Q}}_l).$$

The following theorem gives an interpretation of Green functions in terms of Springer representations of W.

THEOREM 1.1 (SPRINGER [**28**], KAZHDAN [**13**]). *Assume that p and q are large enough. Then for each $u \in G_{\mathrm{uni}}^F$,*

$$\text{(1.1)} \qquad Q_{T_w}^G(u) = \sum_{i \geq 0} (-1)^i \operatorname{Tr}(Fw, H^i(\mathcal{B}_u, \overline{\mathbf{Q}}_l)),$$

where T_w is an F-stable maximal torus of G obtained from T by twisting by $w \in W$.

REMARK 1.2. Springer [**28**] has defined certain functions on the set of nilpotent elements $\mathfrak{g}_{\mathrm{nil}}^F$ in $\mathfrak{g}^F$, where $\mathfrak{g}$ is the Lie algebra of G, using the trigonometric sums on $\mathfrak{g}$, and has given a cohomological interpretation of these functions as in the above theorem. Kazhdan [**13**] has proved that, in fact, these functions coincide with Green functions of G under the indentification of $\mathfrak{g}_{\text{nil}}^F$ and G_{uni}^F via an exponential map. Hence, the restriction on p and q, which in part arises from the assertion of the existence of the exponential map, is essential. However, since both sides of (1.1) make sense for arbitrary $p > 0$, it is natural to expect that (1.1) holds for any $p > 0$. In fact, Kawanaka [**11, 12**] has shown, using his results on generalized Gelfand-Graev representations, that (1.1) holds in the case that either $G = \mathrm{GL}_n$ or G is an exceptional group subject to the condition that p is a good prime (with no restriction on q). Moreover, using his theory of character sheaves [**17**], Lusztig [**18**] has recently shown (1.1) for Chevalley groups in general under the same condition on p.

2. Springer representations and character sheaves. We now review briefly the construction of Springer representations due to Lusztig [**14**]. Let

$$\tilde{G} = \{(g, xB) \in G \times G/B \mid x^{-1}gx \in B\}$$

and $p : \tilde{G} \to G$ be the projection on the first factor. Let G_{reg} be the set of regular semisimple elements in G ($x \in G$ is called regular semisimple if $Z_G(x)$ is a maximal torus of G). Then G_{reg} is open dense in G, and $p^{-1}(G_{\text{reg}})$ is isomorphic to $\{(g, xT) \in G \times G/T \mid x^{-1}gx \in T_{\text{reg}}\}$, where $T_{\text{reg}} = T \cap G_{\text{reg}}$. W acts on the latter set by $w : (g, xT) \to (g, xw^{-1}T)$ ($w \in W$, $g \in G$, $xT \in G/T$), and one can regard $p_0 : p^{-1}(G_{\text{reg}}) \to G_{\text{reg}}$ as an unramified covering with group W (p_0 is the restriction of p on $p^{-1}(G_{\text{reg}})$). Then the direct image sheaf $p_{0*}\overline{\mathbf{Q}}_l$ of the constant sheaf $\overline{\mathbf{Q}}_l$ on $p^{-1}(G_{\text{reg}})$ turns out to be a W-equivariant local system (i.e., locally constant sheaf each of whose stalks is a finite-dimensional $\overline{\mathbf{Q}}_l$-space) on G_{reg} with respect to the trivial action of W on G_{reg}. We now consider the intersection cohomology complex $\mathrm{IC}(G, p_{0*}\overline{\mathbf{Q}}_l)$ on $G = \overline{G}_{\text{reg}}$. By the functoriality of intersection cohomology, W acts on $\mathrm{IC}(G, p_{0*}\overline{\mathbf{Q}}_l)$, as morphisms in the category of perverse sheaves $\mathcal{M}(G)$ of G (see Springer's survey [**30**] in these Proceedings and also [**2, 5, 6**] for the theory of perverse sheaves). Then Lusztig observed that $\mathrm{IC}(G, p_{0*}\overline{\mathbf{Q}}_l) \simeq Rp_*\overline{\mathbf{Q}}_l[n]$ in $\mathcal{M}(G)$, where $n = \dim G$ and $[\cdot]$ means a shift of complex. Hence, we have a W-action on each stalk at $g \in G$ of the cohomology sheaf of $Rp_*\overline{\mathbf{Q}}_l$, i.e., on $\mathcal{H}_g^i(Rp_*\overline{\mathbf{Q}}_l)$. On the other hand, as $p^{-1}(g) \simeq \mathcal{B}_g$, we see that $\mathcal{H}_g^i(Rp_*\overline{\mathbf{Q}}_l) \simeq H^i(\mathcal{B}_g, \overline{\mathbf{Q}}_l)$ by the proper base change theorem. Thus we have a W-action on $H^i(\mathcal{B}_g, \overline{\mathbf{Q}}_l)$, which is the Springer representation of W.

In view of the above construction, (1.1) is rewritten using the complex $Rp_*\overline{\mathbf{Q}}_l$. However, we can expect further,

CONJECTURE 2.1. For each $g \in G^F$,

$$\mathrm{Tr}(g, R_{T_w}^G(1)) = \sum_{i \geq 0} (-1)^i \, \mathrm{Tr}(Fw, \mathcal{H}_g^i(Rp_*\overline{\mathbf{Q}}_l)). \tag{2.1}$$

In fact, the right-hand side, regarded as a function on G^F, is a class function on G^F and is verified, using the result of [**17**], to satisfy the same character formula as (0.2) for $R_{T_w}^G(1)$. Hence Conjecture 2.1 is equivalent to (1.1) and so is actually true for Chevalley groups in good characteristic.

We note here, since each stalk of the local system $p_{0*}\overline{\mathbf{Q}}_l$ on G_{reg} is a regular W-module, $p_{0*}\overline{\mathbf{Q}}_l$ is decomposed as

$$p_{0*}\overline{\mathbf{Q}}_l \simeq \bigoplus_{\chi}(V_\chi \otimes \mathcal{L}_\chi), \tag{2.2}$$

where χ runs over all the irreducible characters of W, $\mathcal{L}_\chi$ is a simple local system on G_{reg} corresponding to $\chi \in W^\wedge$, and V_χ is the space of multiplicity of $\mathcal{L}_\chi$ and in fact is an irreducible W-module realizing χ. Then, by the functoriality of intersection cohomology, we have

$$\mathrm{IC}(G, p_{0*}\overline{\mathbf{Q}}_l) \simeq \bigoplus_{\chi \in W^\wedge} (V_\chi \otimes \mathrm{IC}(G, \mathcal{L}_\chi)). \tag{2.3}$$

Put $K_\chi = \mathrm{IC}(G, \mathcal{L}_\chi)[-n]$. Then, up to shift, K_χ is a G-equivariant simple perverse sheaf on G.

We now assume, for simplicity, that G is of split type and define R_χ for each $\chi \in W^\wedge$ by

$$R_\chi = \frac{1}{|W|}\sum_{w \in W} \chi(w) R^G_{T_w}(1). \tag{2.4}$$

Then Conjecture 2.1 is equivalent to

CONJECTURE 2.2. For each $g \in G^F$,

$$\mathrm{Tr}(g, R_\chi) = \sum_{i \geq 0}(-1)^i\, \mathrm{Tr}(F, \mathcal{H}^i_g(K_\chi)). \tag{2.5}$$

REMARK 2.3. Up to shift, K_χ is an example of character sheaves, which are a certain class of G-equivariant simple perverse sheaves on G defined in [**17**]. On the other hand, R_χ is an example of almost characters of G^F defined in [**15**] that form another orthonormal basis of the space of all class functions of G^F apart from a basis consisting of irreducible characters of G^F. Therefore Conjecture 2.2 is a special case of a more general conjecture due to Lusztig [**15**, **17**], which asserts that the alternating sum of traces of F on each stalk of character sheaves in an appropriate sense gives all the values of almost characters, up to scalar, as in (2.5). In [**18**], he has verified this general conjecture for unipotent elements g in G^F in the case where G is a finite Chevalley group, with a mild restriction on p.

3. The Borho-MacPherson theorem. From the viewpoint of Green functions, it is very important to study the restriction of $Rp_*\overline{\mathbf{Q}}_l$ on the unipotent variety G_{uni}. In fact, the theorem of Borho and MacPherson [**6**] gives a precise description of this complex. Before stating the theorem, we need some notations.

For each unipotent element u in G, let $C = C_u$ be the conjugacy class containing u. Let $A_G(u)$ be the centralizer $Z_G(u)$ of u in G modulo the connected centralizer $Z^0_G(u)$ of u, and take $\phi \in A_G(u)^\wedge$, an irreducible character of $A_G(u)$. To each pair (C, ϕ) as above, we can associate a G-equivariant simple local system $\mathcal{L}_{C,\phi}$ on C such that each stalk of $\mathcal{L}_{C,\phi}$ is an $A_G(u)$-module isomorphic to ϕ. Consider the intersection cohomology complex $\mathrm{IC}(\overline{C}, \mathcal{L}_{C,\phi})$ on $\overline{C}$ ($\overline{C}$ is the

Zariski closure of C in G), and take $j_{C*}(\mathrm{IC}(\overline{C}, \mathcal{L}_{C,\phi}))$, where $j_C : \overline{C} \hookrightarrow G$ is the inclusion. Hence, $j_{C*}(\mathrm{IC}(\overline{C}, \mathcal{L}_{C,\phi}))$ is a complex of sheaves on G whose support is in $\overline{C}$ and, in fact, is a simple object in $\mathcal{M}(G_{\mathrm{uni}})$.

We can now state

THEOREM 3.1 (BORHO AND MACPHERSON [**6**]).

$$Rp_*\overline{\mathbf{Q}}_l[2d]|_{G_{\mathrm{uni}}} \simeq \bigoplus_{(C,\phi)} (V_{C,\phi} \otimes j_{C*}(\mathrm{IC}(\overline{C}, \mathcal{L}_{C,\phi}))),$$

as objects in $\mathcal{M}(G_{\mathrm{uni}})$, *where* $d = \dim G/B$. *Moreover,* $V_{C,\phi}$, *the multiplicity space with respect to the corresponding simple object, turns out to be a* W*-module. Then we have the Springer correspondence:* $V_{C,\phi}$ *is an irreducible* W*-module if it is nonzero, and all the irreducible representations of* W *are obtained in this way by a unique pair* (C, ϕ).

By taking stalks on both sides, we can deduce several results from the theorem. Since $Z_G(u)$ acts on $\mathcal{B}_u$ by left translation, we have an induced action of $A_G(u)$ on $H^i(\mathcal{B}_u, \overline{\mathbf{Q}}_l)$, which commutes with the action of W. Put $d_u = \dim \mathcal{B}_u$. Then

COROLLARY 3.2. *Let* $C = C_u$. *Then*

$$H^{2d_u}(\mathcal{B}_u, \overline{\mathbf{Q}}_l) \simeq \bigoplus_{\phi \in A_G(u)^\wedge} (V_{C,\phi} \otimes \phi)$$

as $W \times A_G(u)$*-modules.*

Corollary 3.2 is nothing but the original form of the Springer correspondence proved in Springer [**28**].

We get some information on the structure of W-modules on lower cohomologies as follows.

COROLLARY 3.3. *Let* $C = C_u$. *Then*
(i) $\langle V_{C,\phi}, H^i(\mathcal{B}_u, \overline{\mathbf{Q}}_l)\rangle_W = 0$ *if* $i \neq 2d_u$,
(ii) $\langle V_{C,\phi}, H^i(\mathcal{B}_x, \overline{\mathbf{Q}}_l)\rangle_W = 0$ *for any* $i \geq 0$ *if* $x \notin \overline{C}$,
where $\langle\ ,\ \rangle_W$ *means the inner product as characters of* W.

4. The generalized Springer correspondence. Let $\mathcal{N}_G$ be the set of all the pairs (C, ϕ), where C is a unipotent conjugacy class in G and $\phi \in A_G(u)^\wedge$ for $u \in C$. By the Springer correspondence as above, we have an injection

$$\Phi : W^\wedge \hookrightarrow \mathcal{N}_G.$$

In fact, this map is not surjective in general, and there exist missing pairs (C, ϕ) such that $V_{C,\phi} = 0$. However, introducing the notion of cuspidal pairs, Lusztig [**16**] extended Φ to a bijection onto $\mathcal{N}_G$, the so-called generalized Springer correspondence, as follows. Let P be a parabolic subgroup of G with Levi decomposition $P = LU_P$, where L is a Levi subgroup of P and U_P is the unipotent radical of P. For $u \in G_{\mathrm{uni}}$ and $v \in L_{\mathrm{uni}}$, let us define

$$Y_{u,v} = \{gZ_L^0(v)U_P \mid g \in G, g^{-1}ug \in vU_P\}.$$

Put $d_{u,v} = \frac{1}{2}(\dim Z_G(u) - \dim Z_L(v))$. Then it is verified that $\dim Y_{u,v} \le d_{u,v}$. Since $Z_G(u)$ acts on $Y_{u,v}$ by left translation, we have an action of $A_G(u)$ on the set of irreducible components of $Y_{u,v}$ of dimension $d_{u,v}$. We denote by $S_{u,v}$ the permutation representation of $A_G(u)$ thus obtained. Now a pair (C, ϕ) is called a cuspidal pair if ϕ does not appear in $S_{u,v}$ for any $v \in L$ $(L \ne G)$. Then the injection $\Phi : W^\wedge \to \mathcal{N}_G$ can be extended to a natural bijection

$$\tilde{\Phi} : \coprod (N_G(L)/L)^\wedge \xrightarrow{\sim} \mathcal{N}_G,$$

where the sum is taken over all cuspidal pairs (C_1, ϕ_1) in some L (up to G-conjugate). It is shown that $N_G(L)/L$ is a Coxeter group for such an L. Note in the case $L = T$, $S_{u,1}$ is nothing but the representation of $A_G(u)$ on $H^{2d_u}(\mathcal{B}_u, \overline{\mathbf{Q}}_l)$. $\tilde{\Phi}$ is called the generalized Springer correspondence. In the following, we shall indicate briefly how to get the map $\tilde{\Phi}$, which is a natural generalization of Borho-MacPherson's theorem (Theorem 3.1). Let (C_1, ϕ_1) be a cuspidal pair for a Levi subgroup L of P, and put $S = Z(L)^0 C_1$, where $Z(L)^0$ is the connected center of L. Let us define an open dense subset S_{reg} of S by $S_{\text{reg}} = \{g \in S \mid Z_G^0(g_s) \subset L\}$, where g_s is the semisimple part of g. Now put $Y = \bigcup_{g \in G} g S_{\text{reg}} g^{-1}$. Then Y is a locally closed smooth irreducible subvariety of G, stable by conjugation. We can associate in a natural way, as in [**16**, 3.2], to the simple local system $\mathcal{L}_{C_1, \phi_1}$ on C_1 a semisimple local system $\mathcal{E}$ on Y. It can be proved that the endomorphism algebra of $\mathcal{E}$ is isomorphic to the group algebra $\overline{\mathbf{Q}}_l[N_G(L)/L]$ of $N_G(L)/L$. Then we have, as in (2.3),

$$\mathrm{IC}(\overline{Y}, \mathcal{E}) \simeq \bigoplus_{\chi \in (N_G(L)/L)^\wedge} (V_\chi \otimes \mathrm{IC}(\overline{Y}, \mathcal{E}_\chi)),$$

where $\mathcal{E}_\chi$ is the simple local system on Y corresponding to χ in the decomposition of $\mathcal{E}$ on Y as in (2.2) and V_χ is an irreducible $N_G(L)/L$-module realizing χ. Then Lusztig [**16**] has shown that the restriction of $\mathrm{IC}(\overline{Y}, \mathcal{E}_\chi)$ to $\overline{Y} \cap G_{\text{uni}}$ coincides with $\mathrm{IC}(\overline{C}, \mathcal{L}_{C,\phi})$ extended by 0 on $(\overline{Y} \cap G_{\text{uni}}) - C$, up to shift, for some $(C, \phi) \in \mathcal{N}_G$ such that $C \subset \overline{Y}$. The correspondence $\chi \to (C, \phi)$ gives the map $\tilde{\Phi} : (N_G(L)/L)^\wedge \to \mathcal{N}_G$.

The Springer correspondence has been determined in [**28**, **10**, **20**, **21**, **1**, and **24**] in the case where p is a good prime. On the other hand, all the cuspidal pairs were classified in [**16**] and, extending the previous results, the generalized Springer correspondence was determined, including the case of bad characteristic, by Lusztig and Spaltenstein [**19**] in the case of classical groups and by Spaltenstein [**25**] in the case of exceptional groups except for a few cases.

We give here some examples assuming that p is good. In the case $G = \mathrm{GL}_n$, unipotent classes $C = C_\lambda$ are parametrized, via Jordan's normal form, by partitions λ on n. In this case, $A_G(u) = \{1\}$ for any u. On the other hand, W is isomorphic to the symmetric group $\mathfrak{S}_n$ of degree n and $W^\wedge$ is parametrized by partitions of n. We denote by χ_λ the irreducible character of W corresponding to λ. Here we assume χ_λ =principal character if $\lambda = (n)$, single partition. Then the Springer correspondence is given by $\chi_\lambda \leftrightarrow (C_\lambda, 1)$.

Next, consider the case $G = \mathrm{Sp}_{2n}$ or SO_N. Then G has at most one cuspidal pair (C, ϕ). The condition that C corresponds to the cuspidal pair is as follows: $n = \frac{1}{2}d(d+1)$ for some $d \geq 1$ and Jordan's normal form of u in C has the type $(2, 4, \ldots, 2d)$ in the case $G = \mathrm{Sp}_{2n}$, and $N = d^2$ for some $d \geq 1$ and the type of u in C is $(1, 3, \ldots, 2d-1)$ in the case $G = \mathrm{SO}_N$. In these cases, $A_G(u)^\wedge$ may be naturally identified with an $\mathbf{F}_2$-vector space $\mathbf{F}_2[u]$ of dimension d with canonical basis if $G = \mathrm{Sp}_{2n}$, and with the quotient $\widetilde{\mathbf{F}_2[u]}$ of d-dimensional $\mathbf{F}_2$-space $\mathbf{F}_2[u]$ by a line defined by the sum of the canonical basis if $G = \mathrm{SO}_N$, respectively. Then $\phi \in A_G(u)^\wedge$ is given by $\phi = (1, 0, 1, \ldots, 0, 1)$ (resp. $\phi = (0, 1, 0, \ldots, 0, 1)$) if d is odd (resp. d is even) in the case $G = \mathrm{Sp}_{2n}$, and $\phi = (1, 0, 1, \ldots)$ in the case $G = \mathrm{SO}_N$. We then pass to the adjoint group $\overline{G}$. The condition that the above ϕ comes from $\bar{\phi} \in A_{\overline{G}}(\bar{u})^\wedge$ ($\bar{u} \in \overline{C}$ is the element in $\overline{G}$ corresponding to u) is that the sum of the coordinates of ϕ is equal to 0. In this case the pair $(\overline{C}, \bar{\phi})$ is cuspidal and all the cuspidal pairs in $\overline{G}$ are obtained in this way. Hence the smallest case in which a cuspidal pair $(\overline{C}, \bar{\phi})$ appears in $\overline{G}$, and so $|\mathcal{N}_G| - |W^\wedge| = 1$, is PSp_{12}, SO_9, and PSO_{16}, respectively. In general, various L appear in the generalized Springer correspondence and $|\mathcal{N}_G| - |W^\wedge|$ becomes larger as the rank of G becomes large.

In the case of exceptional groups, $|\mathcal{N}_G| = |W^\wedge|$ if G is an adjoint group of type E_6 or E_7. If G is of type G_2, F_4, or E_8, then $|\mathcal{N}_G| - |W^\wedge| = 1$ and the unique missing one is the cuspidal pair. They are given as follows. There exists a unique u in G such that $A_G(u) \simeq \mathfrak{S}_3$, $\mathfrak{S}_4$, or $\mathfrak{S}_5$ according as $G = G_2$, F_4, or E_8, respectively. Then (C_u, ε) is the cuspidal pair, where ε is the sign character of $\mathfrak{S}_i$ $(i = 3, 4, 5)$.

5. Determination of Green functions. In the remainder of this note, we assume that p is a good prime. We assume, for simplicity, that G is of split type unless otherwise stated. We shall show in this section how the Green functions (or rather the right-hand side of (1.1)) can be determined.

For each $\chi \in W^\wedge$, put

$$Q_\chi = \frac{1}{|W|} \sum_{w \in W} \chi(w) Q^G_{T_w}. \tag{5.1}$$

Then from the orthogonality relations of $\chi \in W^\wedge$, the determination of $Q^G_{T_w}$ ($w \in W$) is equivalent to the determination of Q_χ ($\chi \in W^\wedge$). Moreover, by Borho-MacPherson's theorem (Theorem 3.1), (5.1) can be expressed as

$$Q_\chi(g) = \sum_{i \geq 0} (-1)^i \operatorname{Tr}(F, \mathcal{H}^i_g(j_{C*}(\mathrm{IC}(\overline{C}, \mathcal{L}_{C,\phi})))) \tag{5.2}$$

for each $g \in G^F_{\mathrm{uni}}$, where (C, ϕ) is the pair corresponding to $\chi \in W^\wedge$ via the Springer correspondence.

The next step is to determine a good representative $u \in G^F$ for each unipotent class C_u in order to fix a Frobenius action. We shall call $u \in G^F_{\mathrm{uni}}$ *a split element* if each irreducible component of $\mathcal{B}_u$ is stabilized by F. (This condition is equivalent to saying that F acts on $H^{2d_u}(\mathcal{B}_u, \overline{\mathbf{Q}}_l)$ by a scalar multiplication by q^{d_u}.)

REMARK 5.1. Suppose p is good. Then it is known [**22**, **23**, **3**] that for each F-stable unipotent class C, there exists a unique split element u up to G^F-conjugate except the following case; i.e., G is of type E_8, $q \equiv -1 \pmod 3$ and u is of type $D_8(a_3)$ (Carter's notation [**7**]). In this exceptional case, split elements do not exist in C^F. If u is split, it is verified that F acts trivially on $A_G(u)$.

Although some minor change is needed in the arguments below for the exceptional case, essentially the similar argument works as well in this case. So, in the following, we consider only the case where a split element exists for each F-stable unipotent class.

Let $C = C_u$ be a unipotent class with u split. Then the G^F-conjugacy classes in C^F correspond bijectively to the conjugacy classes in $A_G(u)$ as follows. For each $a \in A_G(u)$, there exists $g \in G$ such that $g^{-1}F(g) = \dot{a}$ ($\dot{a}$ is a representative of a in $Z_G(u)$). Then $u_a = gug^{-1}$ ($a \in A_G(u)$, up to conjugate) gives a representative of the G^F-conjugacy classes of C^F. We note that if we identify $\mathcal{B}_u$ with $\mathcal{B}_{u_a}$ via g, the action of F on $\mathcal{B}_{u_a}$ is replaced by the action of aF on $\mathcal{B}_u$.

Now the following proposition is a direct consequence of (5.2), Corollary 3.2, Corollary 3.3, and the definition of split elements.

PROPOSITION 5.2. *Let $C = C_u$ with u split and suppose (C, ϕ) is the pair corresponding to $\chi \in W^\wedge$. Then*

(i) $Q_\chi(u_a) = q^{d_u}\phi(a)$ *for* $a \in A_G(u)$,

(ii) $Q_\chi(g) = 0$ *if* $g \notin \overline{C}^F$.

Let $I = \mathcal{N}_G$ and put $I_0 = \{(C, \phi) \in I \mid V_{C,\phi} \neq 0\}$. We denote by $\mathcal{V}$ the space of $\overline{\mathbf{Q}}_l$-valued G^F-invariant functions on G^F_{uni}. For each $i = (C, \phi) \in I$, $u \in C$ split, let us define a G^F-invariant function ψ_i on G^F_{uni} by

$$\psi_i(x) = \begin{cases} \phi(a) & \text{if } x \text{ is } G^F\text{-conjugate to } u_a, \\ 0 & \text{if } x \notin C^F. \end{cases}$$

Then $\{\psi_i \mid i \in I\}$ gives rise to a basis of $\mathcal{V}$. Let $\mathcal{V}_0$ be the subspace of $\mathcal{V}$ spanned by ψ_i ($i \in I_0$). In the following, we shall denote Q_χ by Q_i if $i = (C, \phi)$ corresponds to $\chi \in W^\wedge$ via the Springer correspondence. Then we have

THEOREM 5.3. *$Q_i \in \mathcal{V}_0$ if $i \in I_0$.*

In fact, this statement is equivalent to

$$\sum_{i \geq 0} (-1)^i \operatorname{Tr}(Fw, H^i(\mathcal{B}_u, \overline{\mathbf{Q}}_l)_\phi) = 0, \tag{5.3}$$

for each $w \in W$ if $(C_u, \phi) \notin I_0$, i.e., if $H^{2d_u}(\mathcal{B}_u, \overline{\mathbf{Q}}_l)_\phi = 0$, where $H^i(\mathcal{B}_u, \overline{\mathbf{Q}}_l)_\phi$ is the ϕ-isotypic subspace of $H^i(\mathcal{B}_u, \overline{\mathbf{Q}}_l)$.

In the case of classical groups, it can be proved [**23**], by analyzing the variety $\mathcal{B}_u$, that if $H^{2d_u}(\mathcal{B}_u, \overline{\mathbf{Q}}_l)_\phi = 0$, then $H^i(\mathcal{B}_u, \overline{\mathbf{Q}}_l)\phi = 0$ for any $i \geq 0$. In the case of exceptional groups, there is nothing to prove for the adjoint groups of types E_6 and E_7. For the groups of types G_2, F_4, and E_8, using the property $|\mathcal{N}_G| - |W^\wedge| = 1$, (5.3) can be checked in the course of the computation of Green functions by a computer [**22**, **3**].

REMARK 5.4. In [**17**, II], Lusztig defined the "generalized Green functions" as the alternating sum of a Frobenius map on the complex $\mathrm{IC}(\overline{Y}, \mathcal{E})$ in (4.1) with respect to a suitable $\mathbf{F}_q$-structure, which is a G^F-invariant function on G^F_{uni} and a natural generalization of Green functions. In fact, in [**17**, II], he proved a character formula which asserts that the determination of certain class functions of G^F associated with character sheaves (cf. Remark 2.3) is reduced to the determination of generalized Green functions. He also proved the orthogonality relations of generalized Green functions. In this scheme of generalized Green functions and the generalized Springer correspondence, Theorem 5.3 can be proved rather easily without using ad hoc arguments. See also Remark 6.5.

We now give a total order in I_0 such that

(i) $i < j$ if $C_1 \subsetneq \overline{C}_2$, where $i = (C_1, \phi_1)$, $j = (C_2, \phi_2)$, and

(ii) for a fixed C, the set $\{i \in I_0 \mid i = (C, \phi)\}$ consists of an interval.

By Theorem 5.3, Q_i $(i \in I_0)$ may be expressed as

$$Q_i = \sum_{j \in I_0} P_{ji}\psi_j, \qquad P_{ij} \in \overline{\mathbf{Q}}_l. \tag{5.4}$$

Put $P = (P_{ij})$, the matrix with respect to the above order. Since the Springer correspondence is already determined, the determination of Q_i is equivalent to the determination of the matrix P.

Let us define an inner product on $\mathcal{V}_0$ by

$$\langle f, g \rangle = \frac{1}{|G^F|} \sum_{x \in G^F_{\mathrm{uni}}} f(x)\overline{g(x)}.$$

Then by the orthogonality relations of Green functions (0.1), we have

$$\langle Q_\chi, Q_{\chi'} \rangle = \frac{1}{|W|} \sum_{w \in W} \chi(w)\chi'(w) \frac{1}{|T^F_w|}. \tag{5.5}$$

We shall define the matrices Π and Λ by

$$\Pi = (\langle Q_i, Q_j \rangle), \qquad \Lambda = (\langle \psi_i, \psi_j \rangle).$$

From (5.4), we have an equality of matrices

$${}^tP\Lambda\overline{P} = \Pi. \tag{5.6}$$

Now, from (5.5), Π is a known matrix. Moreover, from the definition of ψ_i, and Proposition 5.2, we see that Λ and P are block matrices (each block corresponds to each class C) of the following type.

Λ: a diagonal block matrix and each of the diagonal blocks is nonsingular.

P: an upper triangular block matrix and each of the diagonal blocks is a known scalar matrix.

Now we see that the equality (5.6) determines uniquely the matrices P and Λ. Thus the Green functions can be determined in this way.

REMARKS 5.5. (i) The above algorithm can be applied as well to the case where F is nonsplit. In fact, the notion of split elements can be extended to this case also and results similar to those in Proposition 5.2 [**23**, 4.7] hold by replacing Q_χ by $Q_{\tilde{\chi}}$, where $\tilde{\chi}$ is an extension of χ in $W^\wedge$ to $\tilde{W} = W\langle\sigma\rangle$ (semidirect product of W with the cyclic group $\langle\sigma\rangle$ with generator σ such that $\sigma w \sigma^{-1} = F(w)$), and

$$Q_{\tilde{\chi}} = \frac{1}{|W|} \sum_{w \in W} \tilde{\chi}(\sigma w) Q^G_{T_w}.$$

(ii) The method employed in [**22**, **23**, and **3**] is somewhat different from the one discussed here. In fact, the algorithm in this section, which is a modified one and is simpler than the original one [**22**, **23**, **3**], is due to Lusztig. In [**17**, V], extending the method of computing Green functions, he showed that there exists an algorithm for computing the generalized Green functions, which leads us to the determination of the values of all the irreducible characters of G^F (up to his conjecture; cf. Remark 2.3).

6. Some applications. Using the algorithm for the determination of Green functions, we can deduce several results as follows. All the statements in this section are actually true for sufficiently large p, and are valid for a good prime p if we admit Conjecture 2.1.

(A) *The structure of* $H^i(\mathcal{B}_u, \overline{\mathbf{Q}}_l)$. Assume that G is of split type. Then the parametrization of unipotent classes and the structure of their centralizers $Z_G(u)$ are independent of p, and all the classes are F-stable. Hence, we can parametrize the unipotent classes in G^F simultaneously for various finite fields $\mathbf{F}_q$ by choosing a split element $u = u_q$ in C^F with respect to $\mathbf{F}_q$ for each class C. Then we have the following.

PROPOSITION 6.1. *Assume G is of split type and does not contain a factor of type E_8. Then for each unipotent class C, $u \in C^F$ split and $a \in A_G(u)$, there exists a polynomial $\sum_{i\ge 0} a_i(w; C, a)T^i$ such that*

$$Q^G_{T_w}(u_a) = \sum_{i \ge 0} a_i(w; C, a) q^i,$$

where $w \mapsto a_i(w; C, a)$ is a virtual character of W independent of p and q.

REMARK 6.2. In the case where G is of type E_8, a similar result holds. However, because of the existence of nonsplit elements, there exist polynomials of two different types according as $q \equiv 1 \pmod 3$ or $q \equiv -1 \pmod 3$. In the case where G is of nonsplit type, Green functions are described in a similar way, using the characters of the extended Weyl group $\tilde{W}$ instead of W as in Remarks 5.5 (i).

By the way, Springer has proved that $\mathcal{B}_u$ satisfies the Weil conjecture.

THEOREM 6.3 (SPRINGER [**29**]). *Let $u \in G^F_{\text{uni}}$. Then all the absolute values of eigenvalues of F on $H^i(\mathcal{B}_u, \overline{\mathbf{Q}}_l)$ are $q^{i/2}$.*

Combining Proposition 6.1 and Theorem 6.3, we have

COROLLARY 6.4. (i) *$H^i(\mathcal{B}_u, \overline{\mathbf{Q}}_l) = 0$ if i is odd.*

(ii) *Let $u = u_1$ be a split element. Then $w \mapsto a_i(w; C, 1)$ is the character of $H^{2i}(\mathcal{B}_u, \overline{\mathbf{Q}}_l)$. Hence the W-module structure of $H^{2i}(\mathcal{B}_u, \overline{\mathbf{Q}}_l)$ is completely determined by coefficients of q^i in $Q^G_{T_w}(u)$.*

REMARK 6.5. Recently Lusztig has proved the vanishing of odd cohomologies of $\mathcal{B}_u$ in the case where G is defined over $\mathbf{C}$ by using the geometry of $\mathcal{B}_u$. He has also proved the fact that if $H^{2d_u}(\mathcal{B}_u, \mathbf{C})_\phi = 0$, then $H^i(\mathcal{B}_u, \mathbf{C})_\phi = 0$ for any $i \geq 0$ in the case of exceptional groups, G_2, F_4, and E_8 (cf. Theorem 5.3).

Concerning the structure of $A_G(u) \times W$-module $H^{2i}(\mathcal{B}_u, \overline{\mathbf{Q}}_l)$, Spaltenstein obtained the following result. Suppose W contains -1. Then $\chi \in W^\wedge$ is called even (resp. odd) if $\chi(-1) = \chi(1)$ (resp. $\chi(-1) = -\chi(1)$). We shall define the parity $p(\chi)$ for each $\chi \in W^\wedge$ by $p(\chi) = 0$ (resp. $p(\chi) = 1$) if χ is even (resp. odd). In the following, put $A_G(u)^\wedge_0 = \{\phi \in A_g(u)^\wedge \mid (C_u, \phi) \in I_0\}$. For each $\phi \in A_G(u)^\wedge_0$, we denote by $\chi_{u,\phi}$ the irreducible character of W corresponding to $V_{C,\phi}$.

THEOREM 6.6 (SPALTENSTEIN [**27**]). *Suppose W contains -1. Suppose $H^{2i}(\mathcal{B}_u, \overline{\mathbf{Q}}_l)_{\phi\otimes\chi} \neq 0$ for $\chi \in W^\wedge$ and $\phi \in A_G(u)^\wedge_0$.*

Then $i \equiv d_u + p(\chi) + p(\chi_{u,\phi})$ (mod 2).

(B) *Ennola-type duality for Green functions.* Assume that G is almost simple. If W does not contain -1, i.e., G is of type A_n, D_{2n+1}, or E_6, G has two different $\mathbf{F}_q$-structures, split Frobenius F_0 and nonsplit Frobenius F such that $F_0^2 = F^2$. Then G^{F_0}-conjugacy classes of $G^{F_0}_{\text{uni}}$ and G^F-conjugacy classes of G^F_{uni} correspond bijectively by choosing split elements for each class C, which is both F_0 and F-stable. We denote this correspondence by $u^+_a \leftrightarrow u^-_a$, where $u^+ \in G^{F_0}$ and $u^- \in G^F$. Moreover, we denote by $Q^{G^+}_T$ (resp. $Q^{G^-}_T$) the Green function on $G^{F_0}_{\text{uni}}$ (resp. G^F_{uni}). As in Proposition 6.1, we can regard Q^G_T as a polynomial in q, which we shall write as $Q^G_T(\cdot)(q)$ to indicate the "variable q." Let w_0 be the longest element in W. Now we have the following "$q \leftrightarrow -q$" duality.

THEOREM 6.7. (i) *Let (G, F) be one of the type B_n, C_n, D_{2n}, ${}^2D_{2n}$, ${}^3D_4$, E_7, and F_4. Then there exists an involution τ on the set of conjugacy classes in $A_G(u)$ for each u such that*

$$Q^G_{T_{ww_0}}(u_a)(q) = Q^G_{T_w}(u_{\tau(a)})(-q).$$

(ii) *Let G be one of the type A_n, D_{2n+1}, and E_6. Then there exists an involution τ as above such that*

$$Q^{G^+}_{T_{ww_0}}(u^+_a)(q) = Q^{G^-}_{T_w}(u^-_{\tau(a)})(-q).$$

Theorem 6.7 was proved by Hotta and Springer [**10**] for type A_n, by the author [**22**] for other classical groups, and by Beynon and Spaltenstein [**3**] for type E_6. Kawanaka [**12**] has observed this for other cases using the explicit table of Green functions. In particular, even in the case of type E_8, Ennola-type duality holds in a modified form [**12**], i.e., just as in (ii) of Theorem 6.7, as the duality between polynomials of two types corresponding to $q \equiv 1$ or $-1 \pmod 3$.

(C) *The second orthogonality relations.* Kawanaka [**12**] has proved the following orthogonality relations, which may be regarded as a "transposed version" of (0.1).

THEOREM 6.9 (KAWANAKA [**12**]). *Let C, C' be two unipotent classes in G, and let us choose u in C^F so that F acts trivially on $A_G(u)$, and similarly for $u' \in C'^F$. We denote by u_a $(a \in A_G(u))$ the twisted element of u by a, and the same for u'_b $(b \in A_G(u'))$. Then*

$$\frac{1}{|W|}\sum_{w\in W}|T_w^F|Q_{T_w}^G(u_a)Q_{T_w}^G(u'_b)$$
$$= \begin{cases} \left(\sum_{\phi\in A_G(u)_0^\wedge}\phi(a)\overline{\phi(b)}\right)|Z_G^0(u'_b)^F| & \text{if } C = C', \\ 0 & \text{if } C \neq C'. \end{cases}$$

Added in proof. The result mentioned in Remark 6.5 is in the following preprint: C. DeContini, G. Lusztig, and C. Procesi, *Homology of the zero-set of a nilpotent vector field on a flag manifold.*

REFERENCES

1. D. Alvis and G. Lusztig, *On Springer's correspondence for simple groups of type E_n $(n = 6, 7, 8)$*, Math. Proc. Cambridge Philos. Soc. **92** (1982), 65–72.

2. A. A. Beilinson, J. Bernstein, and P. Deligne, *Faisceaux pervers*, Astérisque, no. 100, Soc. Math. France, Paris, 1982.

3. W. M. Beynon and N. Spaltenstein, *Green functions of finite Chevalley groups of type E_n $(n = 6, 7, 8)$*, J. Algebra **88** (1984), 584–614.

4. ——, *The computation of Green functions of finite Chevalley groups of type E_n $(n = 6, 7, 8)$*, Univ. of Warwick Computer Center Report No. 23 (1982).

5. A. Borel et al., *Intersection cohomology*, Progr. Math., vol. 50, Birkhäuser, 1984.

6. W. Borho and R. MacPherson, *Représentation des groupes de Weyl et homologie d'intersection pour les variétés nilpotentes*, C. R. Acad. Sci. Paris Ser. A **292** (1981), 707–710.

7. R. W. Carter, *Finite groups of Lie type: Conjugacy classes and complex characters*, Wiley Interscience, New York, 1985.

8. ——, *Irreducible characters of finite groups of Lie type*, this volume.

9. P. Deligne and G. Lusztig, *Representations of reductive groups over finite fields*, Ann. of Math. (2) **103** (1976), 103–161.

10. R. Hotta and T. A. Springer, *A specialization theorem for certain Weyl group representations and an application to the Green functions of the unitary groups*, Invent. Math. **41** (1977), 113–127.

11. N. Kawanaka, *Generalized Gelfand-Graev representations and Ennola duality*, Algebraic Groups and Related Topics, Adv. Stud. Pure Math., vol. 6, Kinokuniya, Tokyo, and North-Holland, Amsterdam, 1985, pp. 175–206.

12. ——, *Generalized Gelfand-Graev representations of exceptional simple algebraic groups over a finite field.* I, Invent. Math. **84** (1986), 575–616.

13. D. Kazhdan, *Proof of Springer's hypothesis*, Israel J. Math. **28** (1977), 272–286.

14. G. Lusztig, *Green polynomials and singularities of unipotent classes*, Adv. in Math. **42** (1981), 169–178.

15. ——, *Characters of reductive groups over a finite field*, Ann. of Math. Stud., vol. 107, Princeton Univ. Press, Princeton, N.J., 1984.

16. ——, *Intersection cohomology complex on a reductive group*, Invent. Math. **75** (1984), 205–272.

17. ——, *Character sheaves*. I, Adv. in Math. **56** (1985), 193–237; II, Adv. in Math. **57** (1985), 226–265; III, Adv. in Math. **57** (1985), 266–315; IV, Adv. in Math. **59** (1986), 1–63; V, Adv. in Math. **61** (1986), 103–155.

18. ——, *On the character values of finite Chevalley groups at unipotent elements*, J. Algebra **104** (1986), 146–194.

19. G. Lusztig and N. Spaltenstein, *On the generalized Springer correspondence for classical groups*, Algebraic Groups and Related Topics, Adv. Stud. Pure Math., vol. 6, Kinokuniya, Tokyo, and North-Holland, Amsterdam, 1985, pp. 289–316.

20. T. Shoji, *On the Springer representations of the Weyl groups of classical algebraic groups*, Comm. Algebra **7** (1979), 1713–1745; **7** (1979), 2027–2033.

21. ——, *On the Springer representations of Chevalley groups of type F_4*, Comm. Algebra **8** (1980), 409–440.

22. ——, *On the Green polynomials of Chevalley groups of type F_4*, Comm. Algebra **10** (1982), 505–543.

23. ——, *On the Green polynomials of classical groups*, Invent. Math. **74** (1983), 239–267.

24. N. Spaltenstein, *Appendix*, Math. Proc. Cambridge Philos. Soc. **92** (1982), 73–78.

25. ——, *On the generalized Springer correspondence for exceptional groups*, Algebraic Groups and Related Topics (Advanced Studies in Pure Math., vol. 6), Kinokuniya and North-Holland, 1985, pp. 317–318.

26. ——, *Caractères unipotents de* ${}^3D_4(\mathbf{F}_q)$, Comment. Math. Helv. **57** (1982), 676–691.

27. ——, *On Springer representations of Weyl groups containing* -1, Arch. Math. (Basel) **44** (1985), no. 1, 26–28.

28. T. A. Springer, *Trigonometric sums, Green functions of finite groups and representations of Weyl groups*, Invent. Math. **36** (1976), 173–207.

29. ——, *A purity result for fixed point varieties in flag manifolds*, J. Fac. Sci. Univ. Tokyo Sect. IA **31** (1984), 271–282.

30. ——, *Perverse sheaves and representation theory*, this volume.

31. B. Srinivasan, *Representations of finite Chevalley groups*, Lecture Notes in Math., vol. 764, Springer-Verlag, 1970.

SCIENCE UNIVERSITY OF TOKYO, JAPAN

Proceedings of Symposia in Pure Mathematics
Volume 47 (1987)

Constructing Representations From Group Geometries

STEPHEN D. SMITH

Abstract. An exposition of some results from ordinary and coefficient homology on finite group geometries, yielding information on projective and irreducible modular representations of the group.

Introduction. Here "geometry" is to be understood in the sense of abstract finite projective geometry (though the techniques have natural analogues in the algebraic geometry of algebraic groups). I have in mind the study of buildings and other related geometries, which has hopes of providing a uniform approach to finite simple groups, both Chevalley and sporadic, the area usually known as "groups and geometries." In particular, these geometries are associated with a prime characteristic p, and the representations of the title will therefore always be p-modular representations.

Before going on to representation theory I should perhaps refer the reader interested in fuller details of the underlying geometric developments to one of the survey articles that have become available in recent years. It is perhaps easiest, or laziest, to mention my own from the Montreal meetings of 1982 [**Sm1**] and 1986 [**Sm2**].

I have chosen three topics to illustrate the applicability of the techniques. The first explores the connection of ordinary homology of a geometry with projective modules for its group. The second uses coefficient homology to provide a geometric construction of suitable "small" irreducible modules—a technique based on buildings which can often be extended to sporadic geometries. The final section describes a use of local techniques for representations of a group which is finitely generated, but not finite.

1. Ordinary homology and (relatively) projective modules. The original motivation comes from the "generic" case of simple groups: let G be a finite Lie-type group, of rank n and characteristic p. Then the natural geometry Δ

1980 *Mathematics Subject Classification* (1985 *Revision*). Primary 20C20; Secondary 20J05, 20F32.

for G is the corresponding Tits building [**Ti1**]. We may regard Δ as a simplicial complex, so that its homology groups $H_i(\Delta)$ with coefficients in $\mathbf{F}_p$ afford $\mathbf{F}_pG$-modules. The fundamental result:

(1.1) THEOREM (SOLOMON-TITS [**So**]). *Assume G, Δ as above with $n \geq 2$. Then*

$$H_i(\Delta) \cong \begin{cases} \mathbf{F}_p & \text{for } i = 0; \\ 0 & \text{for } i \neq 0,\ n-1; \\ \mathrm{St} & \text{for } i = n-1. \end{cases}$$

Here St denotes the Steinberg $\mathbf{F}_pG$-module, and we observe that a projective module has appeared in top homology.

The question immediately arises: is there an analogue for sporadic simple groups? For beginning with Buekenhout [**Bu**], various geometries for sporadic groups have been investigated, usually via resemblances to buildings. In a few cases, top homology does produce a projective (but not usually irreducible) module: For example, in the celebrated geometry Δ with Dynkin diagram of type C_3 for the alternating group A_7 discovered by Neumaier [**N**], Ronan and Smith [**RSm3**] verified that the 56-dimensional module $H_2(\Delta)$ is in fact the projective cover $P(20)$ of the $\mathbf{F}_2A_7$-irreducible of largest dimension 20. Similarly, for the 2-local geometry Δ of M_{24} [**RSm1**], the module $H_2(\Delta)$ of dimension $21,504 = 2^{10} \cdot 21$ was verified [**RSm3**] to be the projective cover $P(1792)$ of the $\mathbf{F}_2M_{24}$-irreducible of largest dimension 1792. At present I know of four other similar examples.

But typically one finds for sporadic geometries that the top homology module is not in fact projective. Ronan in [**R1**] made these and other computations for a wide variety of examples. What eventually emerged was the observation that the modules often have dimension divisible not by $|G|_p$, but nonetheless by a high power of p. So the earlier question was refined to the search for an explanation of this phenomenon.

For a small example, consider the 2-local geometry Δ of M_{12}. This can be exhibited in an easy combinatorial representation: as *points* take the various 4-subsets of the set of 12 letters that M_{12} acts on; as *lines*, take the M_{12}-conjugacy class of 4|4|4 partitions which are stabilized by a Sylow 2-group. One computes that $H_1(\Delta)$ has dimension $496 = 2^4 \cdot 31$ while $|M_{12}|_2 = 2^6$. The numbers are more compelling in the larger example of the 2-local geometry Δ for the Conway group $\cdot 1$ [**RSm1**]; here one finds that $|\dim H_2(\Delta)|_2 = 2^{18}$, while $|\cdot 1|_2 = 2^{21}$.

The approach to the above question now proceeds via an apparent digression: we consider combinatorics of finite groups of earlier interest to topologists. We let G denote an arbitrary finite group, and p a prime dividing $|G|$. We obtain a complex Δ_p by taking as vertices all the nontrivial p-subgroups of G, with higher simplexes defined by chains under inclusions. For the study of this complex, we introduce (a variant of) the reduced *Lefschetz* $\mathbf{F}_pG$*-module* using the chain groups

$C_i(\Delta_p)$:

$$\tilde{L}(\Delta_p) := \sum_{i=-1}^{\dim \Delta_p} (-1)^i C_i(\Delta_p).$$

By the Hopf trace formula, this yields the same character as the usual Lefschetz module defined by homology H_i. Furthermore, we want the dimension of this module, the reduced *Euler characteristic*:

$$\tilde{\chi}(\Delta_p) := \sum_{i=1}^{\dim \Delta_p} (-1)^i \dim(C_i(\Delta_p)).$$

The story begins with:

(1.2) THEOREM (BROWN [**Br**]). $\tilde{\chi}(\Delta_p) \equiv 0 \pmod{|G|_p}$.

This observation was given a representation-theoretic interpretation in the celebrated paper of Quillen:

(1.3) THEOREM (QUILLEN [**Q**]). *$\tilde{L}(\Delta_p)$ is a (virtual) projective $\mathbf{F}_pG$-module.*

Consequently we obtain a projective module for any arbitrary group G. This construction may not, however, be very simple to implement; for evidently the dimension of the complex is $\log_p |G|_p - 1$, which is uncomfortably large in many examples of interest. This difficulty is eased sometimes by other results of Quillen, showing equivalence of Δ_p to certain natural subcomplexes. In the building case of (1.1), he obtained:

(1.4) THEOREM (QUILLEN [**Q**]). *Let G, Δ be as in* (1.1). *Then Δ_p is homotopy-equivalent to Δ. In particular, $\tilde{L}(\Delta_p) \cong H_{n-1}(\Delta)$ = the Steinberg module.*

This result can refine our view of the earlier question about sporadic geometries. In one case, namely that of M_{24} mentioned above, Ronan and Webb (unpublished) checked that Δ_2 is homotopy-equivalent to the 2-local geometry Δ, so that an analogue of (1.4) holds. In the other projective example we mentioned, namely A_7, one finds that $\tilde{\chi}(\Delta_2) - 176 > \tilde{\chi}(\Delta) = 56$. In fact, Ronan and Smith with Webb (see [**RSm3**]) checked that $\tilde{L}(\Delta_2)$ is the sum of the projective covers $P(14) \oplus P(20) \oplus P(20)$. Actually, this seems fitting, since A_7 acts on two conjugacy classes of geometries Δ with $\tilde{L}(\Delta) \cong P(20)$, and also on a building Δ_b of type A_3 (from the inclusion $A_7 < A_8 \cong \mathrm{GL}_4(\mathbf{F}_2)$) with $\tilde{L}(\Delta_b) \cong P(14)$. However, for most sporadic geometries Δ we observed that $\tilde{L}(\Delta)$ is not even projective, so there can be no direct analogy of the homotopy equivalence in (1.4).

But let us return to the study of general complexes. In order to extend Quillen's result to group actions on other complexes, Peter Webb analyzed the properties of Δ_p needed in the proof. A particular intermediate result is:

(1.5) LEMMA (BROWN [**Br**]). *For $P \leq G$ with $|P| = p$, the fixed subcomplex $(\Delta_p)^P$ is contractible.*

Standard techniques of algebraic topology then lead to:

(1.6) COROLLARY. *For $H \leq G$ with $O_p(H) \neq 1$ and $H/O_p(H)$ cyclic, we have $\tilde{\chi}(\Delta_p^H) = 0$.*

These were the conditions used by Webb in making his extension:

(1.7) THEOREM (WEBB [**We**]). *Suppose G acts on a typed complex Δ satisfying* (1.5) *or* (1.6) *with Δ in place of Δ_p. Then $\tilde{L}(\Delta)$ is a (virtual) projective $\mathbf{F}_pG$-module.*

We can use the condition (1.5) to cast further light on our earlier M_{12} examples. For M_{12} contains two conjugacy classes of elements of order 2, with representatives z (2-central) and t. A simple calculation shows that Δ^z is contractible (here, a tree), while Δ^t is not even connected. Thus (1.5) fails, and we cannot apply (1.7).

The stage is now set for the result which will complete the explanation. Thévenaz had the idea to state Webb's result in a version relative to those subgroups which *do* satisfy (1.6):

(1.8) THEOREM (THÉVENAZ [**Th**]). *Let G, Δ be as in* (1.7)*, with $\mathcal{S}$ a collection of p-subgroups closed under subgroups and conjugacy. Assume* (1.6) *holds with Δ in place of Δ_p for those H with $O_p(H) \notin \mathcal{S}$. Then $\tilde{L}(\Delta)$ is projective relative to $\mathcal{S}$. In particular, if T is a member of $\mathcal{S}$ of largest order, then $\tilde{\chi}(\Delta) \equiv 0$ $(\mathrm{mod} |G|/|T|)$.*

So we return to the consideration of M_{12}. We find there are four-groups T whose involutions are all conjugate to t; but any 2-group of order ≥ 8 contains a conjugate of z (indeed $|M_{12} : M_{11}|_2 = 4$, and M_{11} has just one class of involutions, represented by z). Thus we set $\mathcal{S} = \{\langle t \rangle, T\}^G$, and apply (1.8) to conclude $\tilde{\chi}(\Delta)$ must be divisible by $2^{6-2} = 2^4$, as in fact occurs. A similar analysis can be applied to the case of $\cdot 1$ mentioned earlier. There are 3 classes of involutions z, x, t with z and x satisfying (1.5). Also these two are the only classes represented in $\cdot 2$, with $| \cdot 1 : \cdot 2|_2 = 8$, so we can apply (1.8) with $|T| \leq 8$ to conclude $\tilde{\chi}(\Delta)$ must be divisible by $2^{21-3} = 2^{18}$, which is exactly right.

Clearly Thévenaz's result gives a computable lower bound for the "degree of projectivity" of a sporadic geometry, and it provides the exact p-power in a number of other examples which have been computed. Thus it seems appropriate to say that the relative projectivity of Lefschetz modules for these geometries has been decisively settled.

2. Coefficient homology and irreducible modules. The previous section should demonstrate applications of the relatively elementary techniques of ordinary homology. Next we proceed to *coefficient* homology, where to each simplex σ of a geometry Δ we assign not just a copy of the field $\mathbf{F}_p$, but some more general $\mathbf{F}_p$-vector space $\mathcal{F}_\sigma$ which can vary with σ, with suitably generalized boundary maps. The collection of all such local coefficient spaces $\{\mathcal{F}_\sigma : \sigma \in \Delta\}$ is called a *presheaf* $\mathcal{F}$; it defines a chain complex with homology, just as in the constant-coefficient case of the previous section. In order to keep this introduction brief, I will refer the reader for definitions and details to one of the references [**L**, **CL**, **RSm2**].

The presheaf formalism had of course been developed earlier by topologists; the techniques were introduced into representation theory by Lusztig [**L**] in his work on the discrete series of GL_n. The ideas were then extended from GL_n to the other Lie-type groups by a number of authors, notably Alvis [**A**], Curtis [**C**], Curtis-Lehrer [**CL**], and Deligne-Lusztig [**DL**]; the theory of "duality" is perhaps the most well-known development. The above works were principally concerned with producing complex representations, and so considered presheaves whose terms $\mathcal{F}_\sigma$ are complex vector spaces. Ronan pursued the case where the terms are vector spaces in characteristic p, producing modular representations in the natural characteristic in which the Lie-type group is defined. The theory has been developed and extended by analogy to more general geometries in a series of papers, many joint with the present author [**RSm2**]. I mention that the natural-characteristic analogue of the above "duality" theory is developed by Ronan in [**R2**].

A strong initial motivation for this natural-characteristic theory was provided by the observation of a close connection with the irreducible modular representations. We introduce:

(2.1) DEFINITION. *Let G be a Lie-type group in characteristic p, and V an $\mathbf{F}_pG$-module. Then the fixed point sheaf $\mathcal{F}_V$ assigns to each parabolic $P = UL$ of G the subspace V^U fixed by the unipotent radical U of P.*

Then the main result of [**Sm3**] can be restated as:

(2.2) THEOREM [**RSm2**]. *If V is an irreducible module, $\mathcal{F}_V$ is an irreducible presheaf.*

Indeed, the presheaf viewpoint leads to a connection going both ways:

(2.3) THEOREM [**RSm2**]. *There is a 1 : 1 correspondence between irreducible modules and irreducible presheaves, given by $V \mapsto \mathcal{F}_V$, and $\mathcal{F} \mapsto$ (unique irreducible quotient of $H_0(\mathcal{F})$).*

It is an interesting but complicating feature of the natural-characteristic theory that $H_0(\mathcal{F}_V)$ can in general be strictly larger than V. To see this in a simple case, consider the example of $G := \mathrm{Sp}_4(2) \cong S_6$ acting on the $\mathbf{F}_2$-permutation module P of dimension 6. Let E denote the hyperplane of even subsets, and V the 4-dimensional irreducible quotient—a natural symplectic module. In the larger module E, we find that one maximal parabolic stabilizes the 1-dimensional subspace $\langle 1234\rangle$, while one of the other type stabilizes the 2-dimensional subspace $\langle 1234, 3456\rangle$. Similar remarks hold in the quotient V, and we see that the presheaf $\mathcal{F}_V$ is supported by both the irreducible V and the larger module E. (In fact, (2.4) below shows $H_0(\mathcal{F}_V) \cong E$.) A useful way of thinking about this distinction is to observe that the vectors 1234, 1235, 1236 sum to zero in V, but not in E; these vectors correspond to points of a 2-space in V which is *not* an isotropic line in the symplectic structure. Thus E is the module presented by generators and relations corresponding only to the points and lines of the symplectic geometry, while the irreducible module V satisfies additional "nongeometric" relations on the generators.

Now for general V, it will not be true that $H_0(\mathcal{F}_V)$ is equal to V, or even that it is comparable in dimension with V. But in practice, one often has to deal only with the smaller modules V for any given group G, and for these it is more realistic to hope that $H_0(\mathcal{F}_V)$ is not much larger than V itself. So the question to which this section is devoted is: what is a suitable class of "small" modules for which such a bound can be obtained?

The motivation for this question is not only aesthetic; applications arise in local analysis in group theory. One situation of frequent occurrence involves a local subgroup M where $M/O_p(M)$ is some known Lie-type group G, and the module action of G on some possibly reducible section X of $O_p(M)$ must be studied. It often happens that inductive information identifies generating subspaces of X which are fixed by unipotent radicals of parabolics of G, and that these subspaces exhibit the structure of $\mathcal{F}_V$ for some irreducible V. Then knowledge of $H_0(\mathcal{F}_V)$ provides a corresponding bound on the size of X. I first encountered such situations in Timmesfeld's large-extraspecial work [**Tm1**]; similar situations arise in more recent work of Timmesfeld [**T2**], Stroth, and others. I propose an answer to the question by suggesting that a suitable (as well as important) class is provided by the quasiminimal modules—and describe an "apartment method" which seems to deliver the desired results.

So we confine attention for the time being to the case that G is a finite Lie-type group in characteristic p, and V an irreducible $\mathbf{F}_pG$-module. Recall that V is said to be *minimal* if the set of weights that occur in V consists just of the single Weyl-group conjugacy class $\{(\lambda)W\}$ determined by the highest weight λ. (Readers unfamiliar with the weight theory may wish to consult [**H**, p. 72] or [**St**].) The slightly larger class of *quasiminimal* modules have weights lying in the single conjugacy class $\{(\lambda)W\}$ plus possibly the zero-weight (for example, adjoint modules with one root-length).

The minimal modules provided the first practical application of the natural-characteristic techniques of [**RSm2**]. Ronan's analysis focused on a fixed apartment of the building for G. He observed that the 1-dimensional weight spaces for the class $(\lambda)W$ correspond to certain vertices of the apartment; the problem is to show that these spaces span (via geometric relations) the conceivably larger space given by the 1-spaces for all conjugates $(\lambda)G$ under the whole group.

The procedure which follows is what I have dubbed the "apartment method." Let A denote a fixed apartment. In Tits's theory [**Ti**], we may fix a *root* R (or "half-apartment," half of A regarded as a sphere) and consider a neighboring apartment A' which intersects A exactly in R. The proof reduces to:

Inductive step: Show the 1-spaces defined on A' are dependent on those for A.

This is sufficient, because any vertex is contained in some apartment, and the set of apartments is connected under the common-root relation. The result obtained was:

(2.4) THEOREM [**RSm2**]. *Let G be a Lie-type group and V a minimal module, as above. Then $H_0(\mathcal{F}_V) \cong V$, with the exception G is of type BC_n in characteristic 2, and V the $2n$-dimensional symplectic module, when $H_0(\mathcal{F}_V)$ is the $(2n+1)$-dimensional orthogonal module.*

In the exceptional case, the minimal module V for G viewed as of type C_n is paired under the isomorphism $B_n(2^\alpha) \cong C_n(2^\alpha)$ with the natural orthogonal module for type B_n—which is only quasiminimal. The proof in [**RSm2**] for the quasiminimal orthogonal modules was simply a slight variant of the above method: in addition to 1-spaces assigned to vertices of the apartment, an additional 1-space was determined and assigned to the apartment "at large." This required only a mild elaboration of the above inductive step.

It soon became clear that other quasiminimal modules could be treated by this technique. A particularly interesting case is given by the 7-dimensional Cayley-algebra module C for groups of type G_2. Here (see [**RSm2**]) at least for the fields $\mathbf{F}_2$ and $\mathbf{F}_3$, $H_0(\mathcal{F}_C)$ is larger than C. At present, it is still open whether $H_0(\mathcal{F}_C) \cong C$ in larger characteristic. Using the apartment method, Segev and I were able to obtain a partial result:

(2.5) THEOREM [**SS**]. *For G of type G_2 and C the Cayley-algebra module with irreducible sheaf $\mathcal{F}$, $H_0(\mathcal{F})/(\text{one further relation}) \cong C$.*

This result can also be deduced fairly easily by reducing the extra relation to results concerning "ideal lines" in Ronan's thesis [**R3**]. However, the calculation of [**SS**] is direct, and its structure suggested the next application of the method:

(2.6) THEOREM [**Sm 1, SV**]. *For G of type A_2 over a finite field and A the adjoint module with irreducible sheaf $\mathcal{F}$, $H_0(\mathcal{F}) \cong A$.*

This particular problem arose in Timmesfeld's original approach to a geometric classification problem [**Tm2**]; at that time Ronan had a short geometric-spanning proof for the case of a prime field (cf. the $L_4(2)$ case in [**RSm3**]), but the proof for an arbitrary finite field seems to require a more computational treatment using the apartment method. The result has proved useful for other purposes. For example, Ronan (private communication) was able to apply it to the subgroups $L_3(q)$ of the group $G_2(q)$ in establishing:

(2.7) THEOREM (SEE [**V**]). *For G of type G_2, and A its adjoint module with irreducible sheaf $\mathcal{F}$, $H_0(\mathcal{F}) \cong A$.*

In my view, the situation for the group $G_2(q)$ represented by (2.5) and (2.7) is very strange—the 14-dimensional adjoint module can be handled without undue difficulty, while the same problem for the smaller 7-dimensional "natural" module can so far be treated only by adding further relations. (We remark here that the adjoint module for G_2 is not actually quasiminimal, since there are two classes of weights corresponding to the two different root lengths.)

In the most recent development, I have learned from Helmut Völklein [**V**] that he can use the result (2.6) for type A_2 to obtain an analogous result for most of the other adjoint modules (that is, those of types ADE and G_2, but not C_n, F_4). So at this time, it appears that the H_0-problem for many of the quasiminimal modules has been solved.

Finally, I will mention possible extensions of the apartment method to non-building geometries. The reader will have noticed that the method itself does not make use of any of the structural properties (such as sphericality) of the apartments in buildings; it is based simply on the connectivity of the apartments under the root-relations. Consequently, we can formulate a simple-minded analogue of the method stated for an arbitrary geometry Δ. We must simply have a set (typically a conjugacy class) $\mathcal{A}$ of substructures called "apartments," with a further set $\mathcal{R}$ of substructures of members of $\mathcal{A}$ called "roots," satisfying:

(i) Every simplex of Δ is contained in some $A \in \mathcal{A}$.

(ii) $\mathcal{A}$ is connected under the edge-relation joining A, $A' \in \mathcal{A}$ when $A \cap A' \supseteq R \in \mathcal{R}$.

Thus given a presheaf $\mathcal{F}$ on Δ, if one can establish the "inductive step" stated before (2.4) above, then the spaces attached to vertices of a fixed $A \in \mathcal{A}$ in fact span $H_0(\mathcal{F})$.

In this generality, it should be visible that we have the freedom to consider various possibilities for $\mathcal{A}$, $\mathcal{R}$ on an ad hoc basis—possibly from a study of the structure of the geometry Δ, or else as suggested by the structure of the presheaf $\mathcal{F}$.

As a test case, I considered the group A_7 with its sporadic C_3 geometry over $\mathbf{F}_2$. The diagram of type C_3 suggests we might for apartment seek a cube or octahedron; in fact there are no such substructures, but it is possible to find a deformed octahedron A—one with opposite vertices (but not edges) identified. Then roots are just determined as $A \cap A^g$ for suitable "root involutions" g. Using these apartments and verifying in each case the appropriate inductive step, I was able to produce a geometric spanning argument for each of the natural irreducible presheaves of Δ. (The H_0 calculations had already been performed "directly" in [**RSm3**].)

3. Failure of factorization and amalgams. This section is a partial inversion of the title of the talk, in that geometry will finally be recovered from module structure.

It seems appropriate to mention the module-condition which has perhaps been most important in local group theory, namely, failure of factorization. This was first developed by Thompson in his N-groups paper and has been crucial in arguments involving weak closure ever since. The condition is:

(FF) Let V be an $\mathbf{F}_pG$-module and A an elementary abelian p-group in G satisfying: $|V : C_V(A)| \leq |A|$.

This is satisfied, for example, if A is generated by a transvection on V. More generally, the condition arises when M is a p-local subgroup of G, with $M/O_p(M)$ faithful on some subspace V of $\Omega_1 ZO_p(M)$, and A is the image in $M/O_p(M)$ of a maximal-rank elementary subgroup of G which does not centralize V.

Probably the most important result for application to groups of characteristic 2-type is the theorem of Cooperstein-Mason [**Co**], which determined the pairs (G, V) with G a Lie-type group in characteristic 2 and V an (FF)-module for G over $\mathbf{F}_2$. Perhaps it's not instructive to cite the list here, but it is worthwhile to remark that the list is fairly short and consists typically of smaller-dimensional modules which arise often in practice.

With the above as preface, we proceed to an apparent digression, on amalgams. The idea here is to study groups which "locally" resemble Lie-type groups of rank 2. An *amalgam* (P_1, P_2) is defined by two finite groups P_1, P_2 satisfying:

(1) $P_1 \cap P_2 := B = N_{P_i}(U)$ for U a common Sylow p-group of P_i $(i = 1, 2)$,

(2) $\overline{P}_i := O^{p'}(P_i/O_p(P_i))$ is a rank-1 Lie-type group $(i = 1, 2)$,

(3) B contains no normal subgroup of $\langle P_1, P_2 \rangle$.

The interest of the condition arises from the generation of the rank-2 Lie-type groups, as well as certain sporadic groups like M_{12}, by such subgroups.

Now the most general group generated by such subgroups is the infinite amalgamated product $\tilde{G} = P_1 *_B P_2$, so it is not realistic to seek a complete classification of finite quotients G of such $\tilde{G}$. What is amazing is that it is possible, beginning only with the structure of the rank-1 quotients $\overline{P}_i$, to determine all possibilities for the full generating subgroups P_i—essentially, to list the possibilities for $O_p(P_i)$ and the action of $\overline{P}_i$. The first result was obtained by Goldschmidt [**G**] for the case $P_i \cong L_2(2)$; other cases were taken up by various authors, culminating in the classification of rank-2 amalgams by Delgado and Stellmacher [**DGS**], which lists all the possible amalgams (P_1, P_2). The method of proof is geometric in that it studies the graph on the cosets of P_1 and P_2, with edges determined by nonempty intersection.

The juxtaposition of the two topics just introduced leads naturally to the question: When can an amalgam-group $\tilde{G}$ have an FF-module V?

This question is unusual in that $\tilde{G}$ is infinite and might have infinite-dimensional FF-modules V. However, recent work of Chermak and Delgado shows that only the well-known examples of Cooperstein-Mason arise. The result is:

THEOREM (CHERMAK-DELGADO [**CD**]). *Let* $\tilde{G} = P_1 *_B P_2$ *for one of the rank-2 amalgams* (P_1, P_2), *and* V *an* $\mathbf{F}_p\tilde{G}$*-module with* (FF).
Then

(i) $$\dim V < \infty,$$

(ii) $$G := \tilde{G}/\ker_{\tilde{G}}(V) \textit{ is a rank-2 Chevalley group,}$$

(iii) V *is an* FF*-module for* G *in the Cooperstein-Mason list.*

I chose to mention this result, since I believe the method of proof is very instructive. To begin with, the FF-condition is so strong as to determine tight upper bounds on $\dim V$. Identification of G then amounts to recognition of the building. Here the approach is to consider the natural presheaf on the graph of the amalgam: V_1 is determined by fixed points $V^{O_p(P_1)}$, with similar determination of V_2 by P_2, and V_B by $B = P_1 \cap P_2$, with analogous spaces determined by the stabilizers of other vertices and edges. Now the finite-dimensionality of V implies that linear relations must hold among the various conjugates (ostensibly infinitely many) of the subspaces V_1 and V_2. The proof amounts to showing these relations are those of the building for a rank-2 group G.

The method seems similar to that of [**Sm5**] on M_{24}, where local methods determine the dimension of a module V as 11, which forces relations that determine the Golay code and the action of M_{24}.

REFERENCES

[**A**] D. Alvis, *The duality operation in the character ring of a finite Chevalley group*, Bull. Amer. Math. Soc. (N.S.) **1** (1979), 907–911.

[**Br**] K. S. Brown, *Euler characteristics of groups: the p-fractional part*, Invent. Math. **29** (1975), 1–5.

[**Bu**] F. Buekenhout, *Diagrams for geometries and groups*, J. Combinatorial Theory Ser. A **27** (1979), 121–151.

[**CD**] A. Chermak and A. Delgado, *Failure of factorization modules for* (B, N)*-pairs of rank two*, Preprint, Kansas State University, 1986.

[**Co**] B. Cooperstein, *An enemies list for factorization theorems*, Comm. Algebra **6** (1978), 1239–1288.

[**C**] C. Curtis, *Truncation and duality in the character ring of a finite group of Lie type*, J. Algebra **62** (1980), 320–332.

[**CL**] C. Curtis and G. Lehrer, *Homology representations of finite groups of Lie type*, Papers in Algebra, Analysis, and Statistics, R. Lidl, editor, Contemp. Math., vol. 9, Amer. Math. Soc., Providence, R.I., 1981.

[**DGS**] A. Delgado, D. Goldschmidt, and B. Stellmacher, *Groups and graphs*, Birkhäuser, New York, 1985.

[**DL**] P. Deligne and G. Lusztig, *Duality for representations of a reductive group over a finite field*, J. Algebra **74** (1982), 284–291.

[**G**] D. Goldschmidt, *Automorphisms of trivalent graphs*, Ann. of Math. (2) **111** (1980), 377–406.

[**L**] G. Lusztig, *The discrete series of* GL_n *over a finite field*, Ann. of Math. Stud., vol. 81, Princeton University Press, Princeton, N.J., 1974.

[**N**] A. Neumaier, *Some sporadic geometries related to* $PG(3,2)$, Arch. Math. **421** (1984), 89–96.

[**Q**] D. G. Quillen, *Homotopy properties of the poset of non-trivial p-subgroups of a group*, Advances in Math. **28** (1978), 101–128.

[**R1**] M. A. Ronan, *Coverings of certain finite geometries*, Finite Geometries and Designs, London Math. Soc. Lectures Notes, vol. 49, Cambridge Univ. Press, Cambridge, 1981, pp. 316–331.

[**R2**] ——, *Duality for presheaves on chamber systems, and a related chain complex*, Preprint, University of Illinois at Chicago, 1986.

[**R3**] ——, *A geometric characterization of Moufang hexagons*, Invent. Math. **57** (1980), 316–331.

[**RSm1**] M. A. Ronan and S. D. Smith, *2-local geometries for some sporadic groups*, Proc. Sympos. Pure Math., vol. 37 (1979 Santa Cruz Proc.), B. Cooperstein and G. Mason, editors, Amer. Math. Soc., Providence, R.I., 1980, pp. 283–289.

[**RSm2**] ——, *Sheaves on buildings and modular representations of Chevalley groups*, J. Algebra **96** (1985), 319–346; *Universal presheaves on finite group geometries and modular representations*, J. Algebra **101** (1986), 135–154.

[**RSm3**] ——, *Computation of 2-modular sheaves and homology for* $L_4(2)$, A_7, $3S_6$ *and* M_{24}, Preprint, University of Illinois at Chicago, 1986.

[**SS**] Y. Segev and S. D. Smith, *Apartments and the Cayley algebra module for* $G_2(k)$, Comm. Algebra **14** (1986), 493–506.

[**Sm1**] S. D. Smith, *Residual geometries for sporadic and classical groups—a survey*, Contemp. Math., vol. 45 (1982 Montreal Proc.), John McKay, editor, Amer. Math. Soc., Providence, R.I., 1985, pp. 315–334.

[**Sm2**] ——, *Geometric methods in group theory*, Proc. of 1986 Montreal Conference on Algebraic Combinatorics (to appear).

[**Sm3**] ——, *Irreducible modules and parabolic subgroups*, J. Algebra **75** (1982), 286–289.

[**Sm4**] ——, *A geometric presentation for the adjoint module of* $\mathrm{SL}_3(q)$, Preprint, University of Illinois at Chicago, 1986. (This paper is subsumed in [**SV**] below.)

[**Sm5**] ——, *Sheaf homology for truncated buildings of type* $M_{24}/\mathrm{Sp}_8(2)$, J. Algebra **95** (1985), 1–19.

[**SV**] S. Smith and H. Völklein, *A geometric presentation for the adjoint module of* $\mathrm{SL}_3(k)$, Preprint, University of Illinois at Chicago, 1987.

[**So**] L. Solomon, *The Steinberg character of a finite group with a BN-pair*, Theory of Finite Groups, R. Brauer and C.-H. Sah, editors, Benjamin, New York, 1969, pp. 213–221.

[**St**] R. Steinberg, *Lectures on Chevalley groups*, Yale University Lecture Notes (mimeograph), 1968.

[**Th**] J. Thévenaz, *Permutation representations arising from simplicial complexes*, Preprint, École Normale Supérieure, Paris, 1985.

[**Tm1**] F. G. Timmesfeld, *Finite simple groups in which the generalized fitting subgroup of the centralizer of some involution is extraspecial*, Ann. of Math. (2) **107** (1978), 297–369. See also §§2 and 7 of *Tits geometries and parabolic systems in finitely generated groups*, Math. Z. **184** (1983), 377–396, 449–487.

[**Tm2**] ——, *Classical Tits chamber systems of rank* 3 *and characteristic two*, Preprint, University of Giessen, 1984.

[**Ti**] J. Tits, *Buildings of spherical type and finite BN-pairs*, Lecture Notes in Math., vol. 386, Springer-Verlag, 1974.

[**V**] H. Völklein, *On the geometry of the adjoint representation of a Chevalley group*, Preprint, University of Florida, 1986.

[**We**] P. Webb, *A local method in group cohomology*, Preprint, University of Manchester, 1983; *Complexes, group cohomology, and an induction theorem for the Green ring*, Preprint, University of Manchester, 1985.

UNIVERSITY OF ILLINOIS, CHICAGO

Proceedings of Symposia in Pure Mathematics
Volume **47** (1987)

Perverse Sheaves and Representation Theory

T. A. SPRINGER

This note gives a very brief review of the basic facts about intersection cohomology and their application to character theory of finite groups of Lie type. These facts are best stated in the context of perverse sheaves, so these are discussed first. Next come the basic results about the intersection cohomology of Deligne-Lusztig varieties, established in the first chapters of [**4**].

Finally there is a very brief discussion of the description of characters in terms of the character sheaves introduced in [**5**]. See also [**6**].

1. Perverse sheaves. X is an algebraic variety over an algebraically closed field k (no restrictions on k, unless explicitly stated).

(a) *Sheaves on* X. Some familiarity with sheaves is assumed (see for example [**3**, Appendix]). All sheaves will be sheaves of vector spaces over a suitable field E, fixed once and for all. It depends on the topology ("topos") one works with: over $\mathbf{C}$, with the "classical" topology of X, one can take $E = \mathbf{Q}$ or $\mathbf{C}$. If $k = \overline{\mathbf{F}}_q$, the algebraic closure of the finite field $\mathbf{F}_q$, one works with the étale topology, and with sheaves of vector spaces over an extension E of a field of l-adic numbers, where $l \neq \operatorname{char} \mathbf{F}_q$. In that case the "open" sets are étale coverings of open sets in X. A sheaf attaches to any "open" set U of X a vector space $S(U)$, subject to standard conditions. For every closed point $x \in X$ one has a stalk S_x and for any "open" set $U \ni x$ there is a linear map $S(U) \to S_x$. The sheaves on X form an abelian category, which has sufficiently many injectives. If $s \in S(U)$ then the set of $x \in U$ such that the image of S in S_x is 0 is open in U. Its complement is the *support* $\operatorname{supp} s$ of s. The support $\operatorname{supp} S$ of S is the complement of the largest U such that $S_x = 0$ for all $x \in U$. One can restrict a sheaf S on X to an "open" subset U of X, in an obvious way. Also, one can restrict S to a locally closed subset U of X (the intersection of an open and a closed subset). All such restrictions are denoted $S|_U$.

(b) *Finiteness conditions.* A sheaf S is *constant* if for all $x \in X$ and any connected "open" set $U \ni x$ the map $S(U) \to S_x$ is bijective. In particular, if X is connected the stalks S_x and S_y in two closed points x, y are canonically

1980 *Mathematics Subject Classification* (1985 *Revision*). Primary 20G15; Secondary 20G40.

isomorphic. A sheaf S is *locally constant* or is a *local system* if for any $x \in X$ there is an "open" neighborhood U of x such that $S|_U$ is constant. There is an obvious notion of irreducibility for local systems.

A sheaf S is *constructible* if (i) all stalks S_x are finite-dimensional, (ii) there is a partition $X = \coprod_i X_i$ into a finite number of locally closed subsets such that each closure $\overline{X}_i$ is contained in a union of some X_j and that all restrictions $S|_{X_i}$ are locally constant.

(c) *Functorialities.* If $f\colon X \to Y$ is a morphism of varieties and S a sheaf on X, there is a direct image sheaf f_*S on Y, with $(f_*S)(V) = S(f^{-1}V)$, if V is open for Y. There is another sheaf $f_!S$, the direct image sheaf with proper supports. Over $\mathbf{C}$, with the classical topology, we have

$$(f_!S)(V) = \{s \in S(f^{-1}V) \mid \text{the induced map } \operatorname{supp}(s) \to V \text{ is proper}\}.$$

If T is a sheaf on Y there is an inverse image or pullback f^*T on X such that $(f^*T)_x = T_{fx}$ (there is also $f^!T$, which won't be needed). $f_*, f_!$ are left exact functors, but they are usually not exact. f^* is exact.

(d) *Complexes of sheaves.* Consider a complex of sheaves $A = (A^i, d^i)$, where $d^i\colon A^i \to A^{i+1}$, $d^i \circ d^{i-1} = 0$. The cohomology sheaves are $H^i(A) = \ker d^i/\mathrm{im}^{i-1}$. Such complexes form an additive category (which usually won't be abelian).

A map of complexes $f\colon A \to B$ is a *quasi-isomorphism* if the induced homomorphism $H^i(A) \to H^i(B)$ is bijective for all i. We form a new category $\mathcal{D}X$ whose objects are the complexes of sheaves, but whose morphisms are such that quasi-isomorphisms of complexes become isomorphisms in $\mathcal{D}X$.

The precise definition is thus: a morphism $A \to B$ in $\mathcal{D}X$ is a triple (C, f, g) where $f\colon C \to A$ is a quasi-isomorphism of complexes and $g\colon C \to B$ a map of complexes. One should view this triple as representing a quotient gf^{-1}. (Of course, such a definition makes sense for any abelian category.) Denote by $\mathcal{D}_c^bX$ the full subcategory of the complexes $A \in \mathcal{D}X$ such that (i) $H^i(A) = 0$ for $|i|$ large, (ii) $H^i(A)$ is constructible for all i. If $f\colon X \to Y$ is a morphism and $A \in \mathcal{D}_c^bX$ define a complex f_*A by taking an injective resolution $A \to I$ and defining $f_*A = f_*I$. This is unique up to quasi-isomorphism. One defines similarly $f_!$ and f^* for complexes. If Y is a point, then

$$H^i(f_*A) = H^i(X, A), \quad \text{the hypercohomology of } A,$$
$$H^i(f_!A) = H_c^i(X, A), \quad \text{the hypercohomology of } A \text{ with proper support.}$$

There is also a duality operation D on $\mathcal{D}_c^bX$, with nice properties. We shall not go into any details about it. For the case of the classical topology see [**2**, Chapter V, §7].

(e) *Perverse sheaves.* $A \in \mathcal{D}_c^bX$ is called a perverse sheaf if:

(i) $\dim \operatorname{supp} H^i(A) \le -i$,

(ii) $\dim \operatorname{supp} H^i(DA) \le -i$, for all i.

Assume from now on that all components of X have the same dimension.

THEOREM (*see* [**1**, nos. 1, 2]). *The perverse sheaves form an abelian category* $\mathcal{M}X$. *It is artinian and noetherian.*

The irreducible objects of $\mathcal{M}X$ can be described by the following data. Let $Y \subset X$ be a closed irreducible subvariety and $\mathcal{E}$ an irreducible local system on a smooth open piece U of Y. For each pair $(Y, \mathcal{E})$ there is a unique (up to quasi-isomorphism) simple object $\Sigma(Y, \mathcal{E})$ in $\mathcal{M}X$, whose support lies in Y and such that

$$\Sigma(Y, \mathcal{E})|_U = \mathcal{E}[\dim Y]$$

(where $\mathcal{E}$ denotes the complex with degree zero sheaf $\mathcal{E}$ and all other sheaves 0, and where we use the notation $A[n]^i = A^{n+i}$).

If $IC(Y, \mathcal{E})$ is what one usually calls the intersection cohomology complex of Y relative to $\mathcal{E}$ and if $i\colon Y \to X$ is the inclusion then

$$\Sigma(Y, \mathcal{E}) = i_*(IC(Y, \mathcal{E})[\dim Y].$$

An important and deep result is the *decomposition theorem* (*see* [**1**, nos. 5, 6]) of which we state the following version: *If* $f\colon X \to Y$ *is a proper morphism and if* A *is an irreducible perverse sheaf with finite monodromy then* f_*A *is a direct sum of complexes* $B[n]$, *where* B *is perverse irreducible on* Y. The condition of "finite monodromy" means that if $A = \Sigma(Y, \mathcal{E})$, as above, there is an étale covering $\varphi\colon V \to U$ such that $\varphi^*\mathcal{E}$ is constant. In particular this is so if $\mathcal{E}$ is constant. There is an appropriate notion of direct sum involved in the theorem. One should view this theorem as a semisimplicity result. There is an important complement over finite fields, to be stated presently.

(f) *Over finite fields.* Let $k = \overline{\mathbf{F}}_q$, and denote by F the Frobenius automorphism of $k/\mathbf{F}_q$. If X is defined over $\mathbf{F}_q$ then F operates on X. If $A \in \mathcal{M}X$ is defined over $\mathbf{F}_q$ there is an isomorphism, also denoted by $F\colon F^*A \xrightarrow{\sim} A$. We then have a notion of weight of complexes $A \in \mathcal{M}X$: A has weight $\leq w$ if for all i all absolute values of eigenvalues of F on $H^i(A)_x$ are $\leq q_x^{(i+w)/2}$, where $x \in X(k)$ and $q_x = \operatorname{card} \mathbf{F}_q(x)$. Further A has weight $\geq w$ if DA has weight $\leq -w$ and A is *pure* of weight w if it has weight $\geq w$ and $\leq w$.

Now consider an irreducible perverse sheaf $\Sigma(Y, \mathcal{E})$ as before, and suppose that $\mathcal{E}$ is defined over $\mathbf{F}_q$, in which case we have an F-action on $\mathcal{E}$, whence a notion of purity.

THEOREM (*see* [**1**, no. 5]). *If* $\mathcal{E}$ *is pure then* $\Sigma(Y, \mathcal{E})$ *is a pure complex.*

This is the complement to the decomposition theorem (which is needed to prove that theorem).

The following corollary is exploited by Lusztig in [**4**, **5**]. Let X be a complete variety which is defined over $\mathbf{F}_q$. The intersection cohomology groups $ICH^i(X, E)$ are the hypercohomology groups

$$H^i(X, \Sigma(X, E)[-\dim X]) = H^{i+\dim X}(X, \Sigma(X, E)).$$

Here $\Sigma(X, E)$ is taken to be pure of weight $-\dim X$.

COROLLARY. *$ICH^i(X, \mathcal{E})$ is pure of weight i.*

This is obtained by using the decomposition theorem for the map $X \to$ point, and the purity theorem.

2. Intersection cohomology of Deligne-Lusztig varieties. We shall review here the main results of the first chapters of [4]. We first discuss the geometric ingredients.

(a) *Local systems on tori.* Let k be an algebraically closed field of characteristic p and T an algebraic torus over k, with character group X. If n is an integer > 0 not divisible by p we have a separable nth power isogeny $n\colon T \to T$. This is a Galois covering with group ${}_nT = \{t \in T \mid t^n = 1\}$. We fix an isomorphism ψ of the group $\mu(k)$ of roots of unity in k onto a subgroup of the multiplicative group of our field E of 1(a) (which we assume to be sufficiently large). Now take $\lambda \in X$. Then $\chi(t) = \psi(\lambda(t))$ defines a character of the finite group ${}_nT$. It defines a local system on T, namely the χ-part of the direct image n_*E. We write this local system as $\mathcal{L}(\varsigma)$, where $\varsigma = n^{-1}\lambda \in \tilde{X} = \mathbf{Z}_{p'} \otimes_{\mathbf{Z}} X$ if $p > 0$, where $\mathbf{Z}_{p'}$ is the ring of rational numbers with denominator prime to p (resp. $\tilde{X} = \mathbf{Q} \otimes_{\mathbf{Z}} X$ if $p = 0$). This is well defined; i.e., the local system in question depends only on ς. This construction is nothing but an application of quite familiar facts from Galois theory of fields (Kummer extensions). We have $\mathcal{L}(\varsigma + \varsigma') \simeq \mathcal{L}(\varsigma) \otimes \mathcal{L}(\varsigma')$ and $\mathcal{L}(\varsigma) \simeq E$ (the constant sheaf) if and only if $\varsigma \in X \subset \tilde{X}$.

If $k = \mathbf{F}_q$ and if T is defined over $\mathbf{F}_q$ then F operates on T and one has $F^*\mathcal{L}(\varsigma) \simeq \mathcal{L}(\varsigma)$ if and only if $F^\vee\varsigma - \varsigma \in X$ (where $F^\vee$ is the transpose of F, operating in the character group X). So the group of these ς is isomorphic to $X/(F^\vee - 1)X$. As one knows, this group is isomorphic to the character group of the finite group of rational points T^F (see [3, p. 83]).

(b) *Deligne-Lusztig varieties.* Let $k = \mathbf{F}_q$ and assume that G is a connected reductive linear algebraic group over $\mathbf{F}_q$. Let B and T be an F-stable Borel group and maximal torus in G, with $T \subset B$. Denote by U the unipotent part of B and by $W = N_GT/T$ the Weyl group of (G, T). If $w \in W$ we denote by $\dot{w} \in N_GT$ a representative. We denote by $\mathcal{B}$ the flag manifold G/B.

The Deligne-Lusztig variety $\mathcal{B}_w$ ($w \in W$) is the locally closed subvariety of $\mathcal{B}$ defined by

$$\mathcal{B}_w = \{gB \mid g^{-1}Fg \in BwB\}.$$

Let $T_w = \{t \in T \mid wF(t) = t\}$; this is a finite group. For each E-valued character θ of T_w, Deligne and Lusztig construct a local system $\mathcal{S}_\theta$ on $\mathcal{B}_w$ of rank 1 and their basic objects are the virtual G^F-modules

$$R_{w,\theta} = \sum_i (-1)^i H_c^i(\mathcal{B}_w, \mathcal{S}_\theta).$$

See [3, Chapter 7] for further details.

The Deligne-Lusztig variety $\overline{\mathcal{B}_w}$ is the closure of $\mathcal{B}_w$ in $\mathcal{B}$,

$$\overline{\mathcal{B}_w} = \{gB \mid g^{-1}Fg \in \overline{BwB}\}.$$

We shall now give some results about the intersection cohomology complexes $IC(\overline{\mathcal{B}_w}, \mathcal{S}_\theta)$. Their hypercohomology groups give virtual modules which are better than the $R_{w,\theta}$ because of the purity of intersection cohomology.

(c) *Intersection cohomology of Deligne-Lusztig varieties, a reduction.* The first thing is to describe the stalk of the cohomology groups of $IC(\overline{\mathcal{B}_w}, \mathcal{S}_\theta)$. This requires some constructions.

We put

$$G_w = \{g \in G \mid g^{-1}Fg \in BwB\}.$$

Clearly, B operates on G_w by right multiplications. It is readily seen that $\mathcal{B}_w$ is the quotient of G_w by the B-action; let $\varphi\colon G_w \to \mathcal{B}_w$ be the quotient map. Next define $\psi\colon G_w \to BwB$ by $\psi g = g^{-1}Fg$ and define $\pi\colon BwB \to T$ by

$$\pi(u\dot{w}tu') = t,$$

where $u, u' \in U$. The map π is well defined, but depends on the choice of $\dot{w}$. We have a commutative diagram

$$\begin{array}{ccc} G_w & \xrightarrow{\ \psi\ } & BwB \\ {\scriptstyle\varphi}\big\downarrow & & \big\downarrow{\scriptstyle\pi} \\ \mathcal{B}_w & & T. \end{array}$$

Put $\overline{G_w} = \{g \in G \mid g^{-1}Fg \in \overline{BwB}\}$. We have morphisms $\overline{\varphi}\colon \overline{G_w} \to \overline{\mathcal{B}_w}$, $\overline{\psi}\colon \overline{G_w} \to \overline{BwB}$, similar to the ones above.

The Weyl group W acts on the character group X of T in the usual way: $(w\chi)(t) = \chi(w^{-1}t)$ and this action extends to $\tilde{X}$ (introduced in 2(a)). Now let θ be a character of T_w and let $\mathcal{S}_\theta$ be the corresponding Deligne-Lusztig local system on $\mathcal{B}_w$.

PROPOSITION (*compare* [**2**, *Theorem* 2.4]). (i) There exists $\varsigma \in \tilde{X}$ such that $\varphi^*\mathcal{S}_\theta \simeq \psi^*\pi^*\mathcal{L}(\varsigma)$. Its class modulo X is unique and we have $F^\vee\varsigma - w\varsigma \in X$; (ii) If $g \in G_w$ we have $IC(\overline{\mathcal{B}_w}, \mathcal{S}_\theta)_{\overline{\varphi}g} = IC(\overline{BwB}, \pi^*\mathcal{L}(\varsigma))_{\overline{\psi}g}$.

This proposition shows that we are reduced to studying the intersection cohomology of $\overline{BwB}$, for suitable local systems on BwB. This is a question which is not particular to finite fields any more and which has an interest of its own (the determination of multipilicities in Verma modules with rational highest weight can be reduced to this question, over $\mathbf{C}$).

(d) *Intersection cohomology of* $\overline{BwB}$. The algebraically closed field k is again arbitrary. We assume G, etc. to be as in (b), but over k. Let $\varsigma \in \tilde{X}$ (see 2(a)).

Denote by W'_ς the subgroup of W defined by

$$W'_\varsigma = \{w \in W \mid w\varsigma - \varsigma \in X\}.$$

Let W_ς be the subgroup of W'_ς generated by the reflections in W defined by the roots α of the root system of W such that $\alpha^\vee(\varsigma) \in \mathbf{Z}$, where $\alpha^\vee$ denotes the coroot. Then W_ς is a normal subgroup of W'_ς. The Borel group B defines Bruhat orders $\leq$ and $\leq_\varsigma$ on W and W_ς; let l and l_ς be the corresponding length functions.

In each coset modulo W there is a unique element of minimal length. We fix $w \in W$ and write $w = w_1 w'$, where w_1 is this shortest element and $w' \in W_\varsigma$. It is known that $\overline{BwB} = \coprod_{x \leq w} BxB$. Write $I(w, \varsigma) = IC(\overline{BwB}, \pi^*(\varsigma))$.

We review some results established in [**4**, Chapter 2].

PROPOSITION 1. *The restriction of $I(w, \varsigma)$ to BxB is zero unless $x = w_1 w''$ with $w'' \in W_\varsigma$ and $w'' \leq_\varsigma w'$.*

To describe the nonzero restrictions of $L(w, \varsigma)$ we first remark that $g \mapsto ut^n g u_1^{-1}$ defines a transitive action of $U \times U \times T$ (with a suitable group structure) on all BxB and that $\pi^*(\mathcal{L}(\varsigma))$ is equivariant for that action (because $n\varsigma \in X$). It follows that the restrictions of $L(w, \varsigma)$ to the BxB are also equivariant. Hence their cohomology sheaves are described by the stalks in the point $\dot{x}$ together with a representation of the isotropy group of the point x, which has ${}_nT$ (see 2(a)) as a quotient. The representation will factor through that quotient, hence comes from a character of ${}_nT$. We shall also need the Kazhdan-Lusztig polynomials $P_{x,y}$ for W_ς (see [**3**, p. 386]), which we write as

$$P_{x,y}(t) = \sum_{i \geq 0} n_{x,y,i} t^i.$$

Here $x \leq_\varsigma y$ and $P_{x,x} = 1$; moreover $\deg P_{x,y} \leq \frac{1}{2}(l_\varsigma(y) - l_\varsigma(x) - 1)$ if $x <_\varsigma y$.

PROPOSITION 2. (i) *Let $x = w_1 w''$, $w'' \leq_\varsigma w'$ as above. Then $H^i(I(w, \varsigma))_x = 0$ unless i is even. In that case we have*

$$\dim H^i(I(w, \varsigma))\dot{x} = n_{w'', w', i} - \tfrac{1}{2}(l(w) - l(w'') - l(w') - l_\varsigma(w'')),$$

(ii) ${}_nT$ *operates on $H^i(I(w, \varsigma))_{\dot{x}}$ as multiplication by the character χ of* 2(*a*).

Next let G be defined over $\mathbf{F}_q$ and assume that ς is as in the proposition of 2(c). Then the Frobenius F operates on $H^i(I(w, \varsigma))_{\dot{x}}$, if $\dot{x} \in G^F$. For a suitable choice of representatives $(\dot{x})_{x \in W}$ we then have

PROPOSITION 3. *F operates on $H^i(I(w, \varsigma))_{\dot{x}}$ as multiplication by $\varepsilon q^{i/2}$ where ε is a root of unity.*

(e) *Global intersection cohomology of Deligne-Lusztig varieties.* In [**4**, Chapter 3] Lusztig describes the global intersection cohomology groups $IC\ H^i(\overline{\mathcal{B}_w}, \mathcal{S}_\theta)$, as a consequence of the results described so far (and some more group-theoretical ones). For simplicity assume that G has connected center. Then $W'_\varsigma = W_\varsigma$. We also assume that the F-stable maximal torus T is split. We keep the other notations. In particular, $\mathcal{S}_\theta$ and ς are related according to 2(c), and w_1 is a shortest coset representative. For each class E of irreducible representations of W_ς consider the (rational) virtual representation of G^F

$$R_E = |W_\varsigma|^{-1} \sum_{\substack{x \in W_\varsigma \\ i \geq 0}} (-1)^i \operatorname{Tr}(x, E) H^i_c(\mathcal{B}_{w_1 x}, \mathcal{S}_\theta).$$

Next introduce the generic Hecke algebra $H(W_\varsigma)$, an algebra over $\mathbf{Q}[t^{1/2}, t^{-1/2}]$ (see [**3**, pp. 354, 386]). If E is as above, it defines a class of irreducible representations $E(t)$ of $H(W_\varsigma)$.

THEOREM [**4**, *Theorem* 3.8.]. *We have the following identity of virtual G^F-modules with coefficients in $Q[t^{1/2}, t^{-1/2}]$:*

$$t^{(l(w,x)-l_\varsigma(x))/2}\sum_E \operatorname{Tr}\left(\left(\sum_{y\in W_\varsigma} P_{y,x}(t)T_y\right), E(t)\right) R_E$$
$$= \sum_{i\geq 0}(-1)^i IC\ H^i(\overline{\mathcal{B}}_{w_1 x}, S_\theta)t^{i/2}.$$

Here (T_y) is the standard basis of the Hecke algebra $H(W_\varsigma)$. The important point is that the coefficient of $t^{i/2}$ on the left-hand side is an *actual* representation of G^F, up to sign. This point is further exploited in the later chapters of [**4**]. To go further one needs more information about representations of Weyl groups and Hecke algebras. We shall not go into this here.

3. Character sheaves. These are certain perverse sheaves, introduced in [**5**] (see also [**6**]). They can be used to produce (sufficiently many) characters of finite groups of Lie type. We shall discuss this only very briefly. Again k is arbitrary. The notations are as before. Put

$$Y_w = \{(g, xb) \in G \times \mathcal{B} \mid x^{-1}gx \in BwB\},$$
$$\overline{Y_w} = \{(g, xB) \in G \times \mathcal{B} \mid x^{-1}gx \in \overline{BwB}\},$$
$$Z_w = \{(g, x) \in G \times G \mid z^{-1}gx \in BwB\}, \qquad \tau\colon Z_w \to Y_w(\text{projection}).$$

Let $\tilde{\pi}_w\colon Z_w \to T$ be defined by $\tilde{\pi}_w(g, x) = \pi(x^{-1}gx)$, where π is as in 2(c). One sees without trouble that there is a (unique up to isomorphism) local system $\tilde{\mathcal{L}}(\varsigma)$ on Y_w such that $\tilde{\pi}^*\tilde{\mathcal{L}}(\varsigma) = \tau^*\tilde{\mathcal{L}}(\varsigma)$ if and only if $w\varsigma - \varsigma \in X$. Then we have an intersection cohomology complex $IC(\overline{Y}_w, \tilde{\mathcal{L}}(\varsigma))$. Let $\rho\colon \overline{Y}_w \to G$ be the projection. By the decomposition theorem 1(e) the complex $\rho_*(IC(\overline{Y}_w, \tilde{\mathcal{L}}(\varsigma)))$ is a direct sum of complexes $A[n]$, where A is irreducible perverse on G. These perverse sheaves A are called *character sheaves* (for varying w, G). They are G-equivariant. In [**5**] one will find a classification of these, which is rather similar to the classification of irreducible characters of finite groups of Lie type of [**4**].

If G is, moreover, defined over $\mathbf{F}_q$ one can consider G-equivariant perverse sheaves A such that $F^*A \simeq A$. Let $\varphi\colon F^*A \xrightarrow{\sim} A$ be an isomorphism and define a class function $\chi_{A,\varphi}$ on G^F by

$$\chi_{A,\varphi}(x) = \sum(-1)^i \operatorname{Tr}(\varphi, H^i(A)_x).$$

Lusztig proves that these $\chi_{A,\varphi}$ for A a character sheaf lead to an orthonormal basis of the space of class functions on G^F, and he conjectures that this is the same basis as the one he exhibited in [**4**]. (I understand that Lusztig knows how to prove this conjecture in good characteristics.)

(b) EXAMPLE. Take $w = e$. Then $Y_e = \overline{Y}_e = \{(g, xB) \mid x^{-1}gx \in B\}$. The morphism ρ is now what is called "Grothendieck's simultaneous resolution map." Choose T to be any F-stable maximal torus of G and $B \supset T$ a Borel group, which need not be G-stable. Take $\varsigma \in \tilde{X}$ with $F^\vee\varsigma - \varsigma \in X$ and consider $A = \rho_*(\tilde{\mathcal{L}}(\varsigma))$.

This turns out to be a perverse sheaf which does have the property that $F^*A \simeq A$. For a proper choice of the isomorphism φ the corresponding class function $X_{A,\varphi}$ should be the Deligne-Lusztig character $R_{w,\theta}$ (where w is determined by T and θ comes from ς). If q and p are large this is true, by a result of Kazhdan.

REFERENCES

1. A. A. Beilinson, J. Bernstein, and P. Deligne, *Faisceaux pervers*, Astérisque, no. 100, Soc. Math. France, Paris, 1982.

2. A. Borel et al., *Intersection cohomology*, Prog. Math., vol. 50, Birkhäuser, 1984.

3. R. W. Carter, *Finite groups of Lie type; conjugacy classes and characters*, Wiley, New York, 1985.

4. G. Lusztig, *Characters of reductive groups over a finite field*, Ann. of Math. Stud., no. 107, Princeton Univ. Press, Princeton, N.J., 1984.

5. ——, *Character sheaves* I–V, Adv. in Math. **56** (1985), 193–237; **57** (1985), 226–315; **59** (1986), 1–63; **61** (1986), 103–155.

6. ——, *An introduction to character sheaves*, this volume.

RIJKSUNIVERSITEIT UTRECHT, THE NETHERLANDS

Proceedings of Symposia in Pure Mathematics
Volume **47** (1987)

Toward Lusztig's Classification of Characters

BHAMA SRINIVASAN

0. Introduction. Let G be a connected reductive algebraic group defined over F_q, $F : G \to G$ be a Frobenius morphism of G, and G^F the finite group of F-fixed points of G. Let $T \subset B$ be an F-stable maximal torus and Borel subgroup, respectively, and let $W = N(T)/T$ be the Weyl group. We then have a bijection (see [**3**] or [**17**])

$$\{F\text{-stable maximal tori in } G\}/G^F \leftrightarrow \{F\text{-conjugacy classes in } W\}.$$

Thus, for $w \in W$ we can choose a representative T_w from the corresponding class of F-stable maximal tori in G. In their famous paper [**5**] Deligne and Lusztig defined, for each $w \in W$, a variety X_w by

$$X_w = \{gB \mid g^{-1}F(g) \in BwB\}$$

(see [**15**, p. 30]). Via the l-adic cohomology with compact support of the X_w they defined virtual representations $R^G_{T_w}(1) = R_w$ of G^F. The irreducible constituents of the R_w are the unipotent representations (or the unipotent characters, according to the context) of G^F. More generally, they defined a covering $\tilde{X}_w$ such that T_w acts on $\tilde{X}_w$ and $\tilde{X}_w/T_w \simeq X_w$. Then, if $\theta \in \mathcal{H}om(T_w^F, \overline{Q}_l^*)$, there is a corresponding l-adic sheaf $\mathcal{F}_\theta$ on X_w. By considering the l-adic cohomology groups of the X_w with coefficients in the sheaves $\mathcal{F}_\theta$ one can then define [**15**, p. 181] virtual representations $R^G_{T_w}(\theta)$ of G^F. The properties of these virtual representations are described in [**3**] and [**17**].

In 1982 Lusztig completed the decomposition of the $R^G_{T_w}(\theta)$ into irreducible constituents and gave a classification of the irreducible representations when G has a connected center. This work is described in [**15**]. Later he removed the condition that the center be connected (see [**16**]).

The aim of this paper is to give an account of some of the ideas which were developed by Lusztig during the period 1976–1982 and which led to the classification. In particular, it is hoped to show how intersection cohomology played a part in the classification.

1980 *Mathematics Subject Classification*. Primary 20G05, 20C15.

1. Hecke algebras. In the early 1970s the theory of Harish-Chandra induction for finite groups of Lie type was developed by Harish-Chandra and Springer (see [**3**]). The main result of this theory states the following:

Let ρ be an irreducible representation of G^F. Then there exists a pair (L^F, ς), unique up to G^F-conjugacy, where $P = LV$ is an F-stable parabolic subgroup of G, L is an F-stable Levi factor, and V is the unipotent radical of P, such that ρ is a constituent of $\mathrm{Ind}_{P^F}^{G^F}(\tilde{\varsigma})$ and $\tilde{\varsigma}$ is the pullback to P^F of a cuspidal representation ς of L^F. Here an irreducible representation ς of G^F is cuspidal if for any proper parabolic subgroup P as above, $\mathrm{Res}_{V^F}^{G^F}(\varsigma)$ does not contain the trivial representation.

In particular if ρ is unipotent then the corresponding ς is also unipotent. We are then reduced to considering two problems:

(i) Classify the cuspidal unipotent representations.

(ii) Describe the endomorphism algebra $\mathrm{End}_{G^F}(\mathrm{Ind}_{P^F}^{G^F}(\tilde{\varsigma}))$, where $\tilde{\varsigma}$ is cuspidal.

The second problem leads to the idea of Hecke algebras, which we will discuss at two levels. At one level we have the theory developed by Curtis-Iwahori-Kilmoyer, Howlett-Lehrer, and Lusztig in order to study the representations induced from cuspidal representations of parabolic subgroups. Let $\mathcal{W}$ be a finite Coxeter group. There is a generic Hecke algebra H over $Q[u^{1/2}, u^{-1/2}]$ (where u is an indeterminate) associated with $\mathcal{W}$, and H has a standard basis $\{S_w \mid w \in \mathcal{W}\}$ (cf. [**6**, §1, or **3**, 10.11]). Certain specialized algebras of such algebras are isomorphic to $\mathrm{End}_{G^F}(\mathrm{Ind}_{B^F}^{G^F}(1))$, or to the group algebra of W. Indeed, it turns out that the algebras $\mathrm{End}_{G^F}(\mathrm{Ind}_{P^F}^{G^F}(\tilde{\varsigma}))$ where ς is cuspidal unipotent are also specializations of certain generic Hecke algebras (see [**10**]). (There are analogous results in the nonunipotent case, but the group $\mathcal{W}$ need not be a reflection group; see [**3**, Chapter 10].) By describing these endomorphism algebras Lusztig classified the noncuspidal unipotent representations in every case [**9**, **10**, **11**]. Then he classified the cuspidal unipotent representations by a counting argument. For example, in the classical groups there is at most one cuspidal unipotent representation.

The theory of generic Hecke algebras yields that for any E in $\hat{\mathcal{W}}$, where $\hat{\mathcal{W}}$ is the set of isomorphism classes of irreducible representations of a Weyl group $\mathcal{W}$, there is a generic degree polynomial $d_E(u) \in Q[u]$. This polynomial can be used to compute the dimensions of the constituents of induced representations, and in particular the dimensions of all the unipotent representations may be obtained (cf. [**3**, Chapter 13]). We define here an integer a_E associated with $E \in \hat{\mathcal{W}}$ which plays an important role in Lusztig's work by

$$d_E(u) = \alpha u^{a_E} + \text{ terms involving higher powers of } u.$$

The classification of unipotent representations can be described combinatorially for classical groups. For $\mathrm{GL}(n, q)$ and $\mathrm{U}(n, q)$ they are parametrized by partitions of n. For groups of type B_n, C_n (resp. D_n) they are parametrized by equivalence classes under shift of symbols of rank n and odd defect (resp.

even defect) (see [**3**, p. 466]). Indeed, a symbol is a generalization of a pair of partitions.

The second level at which we discuss Hecke algebras pertains to the work of Kazhdan and Lusztig [**6**, **7**]. Although originally intended as a study of the representations of Hecke algebras, this work has surprising and far-reaching connections with algebraic geometry and representations of Lie groups. Furthermore, the idea of using intersection cohomology rather than l-adic cohomology first arose in [**7**]. Kazhdan and Lusztig introduce a new basis $\{C_z \mid z \in \mathcal{W}\}$ for the algebra H defined by

$$C_z = \sum_{y \le z} (-1)^{l(z)-l(y)} u^{-l(y)+l(z)/2} P_{y,z}(u^{-1}) S_y,$$

where the $P_{y,z}$ are the now-famous Kazhdan-Lusztig polynomials. For any E in $\hat{\mathcal{W}}$ they construct a H-module $E(u)$ such that $\mathrm{Tr}(S_w, E(u)) \in Z[u^{1/2}]$. They also define left cells and two-sided cells in $\mathcal{W}$ and representations of $\mathcal{W}$ corresponding to left cells (see the article by Alvis in this volume); these concepts have also become important in the theory of enveloping algebras.

2. Decomposition of Deligne-Lusztig representations. In this section we describe the decomposition of the R_w, $w \in W$, defined in §0. If $G = \mathrm{Gl}(n, q)$ this is easy to describe. In this case let

$$R_E = \frac{1}{|W|} \sum_{w \in W} \mathrm{Tr}(w, E) R_w, \qquad \text{where } E \in \hat{W}.$$

Then the R_E are precisely the unipotent representations. In general when F is a split Frobenius (i.e., $F(t) = t^q$ for every $t \in T$) we define the class function R_E on G^F in the same way. (For the general case, see [**15**, p. 67].) From now on we will assume F split, for convenience. We can then ask how far the R_E are from being the unipotent representations. Lusztig showed that there is a definite pattern of decomposition of the R_E into irreducible constituents, which is described by "families." He has partitioned $\hat{W}$ into subsets called families, which can be described explicitly in each case (see [**3**, p. 408]). For each family $\mathcal{F}$ there is associated a finite group $G_{\mathcal{F}} = \Gamma$. When G is a classical group, for example, Γ is trivial or an elementary abelian 2-group; when G is exceptional, Γ is S_1, S_2, S_3, S_4, or S_5. Now let

$$M(\Gamma) = \{(x, \sigma) \mid x \in \Gamma, \sigma \in \hat{C}_\Gamma(x)\}/\Gamma.$$

Then $\mathcal{F}$ can be embedded in $M(\Gamma)$. The main theorem of Lusztig [**15**, 4.23] in the case of the unipotent representations of a split group can then be stated as follows:

The irreducible constituents of the R_E, where E runs over a family $\mathcal{F}$, are parametrized by the elements of $M(\Gamma)$. The multiplicity of each irreducible constituent of R_E in R_E is described by means of a pairing on $M(\Gamma)$.

As a special case, those unipotent representations which correspond to members of $\mathcal{F}$ are in the principal series. At the other end we have the cuspidal representations.

EXAMPLE. G of type D_4. As we mentioned earlier, the unipotent representations are parametrized by symbols. The irreducible representations of W are parametrized by symbols of defect 0. We have a family $\mathcal{F}$ whose members are the symbols

$$\Lambda_1 = \begin{pmatrix} 0 & 2 \\ 1 & 3 \end{pmatrix}, \quad \Lambda_2 = \begin{pmatrix} 0 & 1 \\ 2 & 3 \end{pmatrix}, \quad \Lambda_3 = \begin{pmatrix} 0 & 3 \\ 1 & 2 \end{pmatrix},$$

and $M(\Gamma)$ consists of these symbols along with the symbol

$$\Lambda_4 = \begin{pmatrix} 0 & 1 & 2 & 3 \\ & - & & \end{pmatrix}.$$

Denote by ρ_Λ the unipotent representation corresponding to the symbol Λ. We then have

$$R_{\Lambda_3} = \tfrac{1}{2}[-\rho_{\Lambda_4} + \rho_{\Lambda_3} - \rho_{\Lambda_2} + \rho_{\Lambda_1}].$$

In the next three sections we will describe some of the ideas which were used in the classification of the irreducible representations and the decomposition of the Deligne-Lusztig representations. We have already noted that Harish-Chandra induction was one of the first tools to be used.

3. Eigenvalues of Frobenius. Let w be a Coxeter element of the Weyl group W. Then F acts on the cohomology group $H_c^i(X_w, \overline{Q}_l)$, and we might be able to use the action of F to decompose the representations of G^F on the cohomology. Since T_w is an anisotropic torus we can expect to see some cuspidal representations realized in these groups. Using this principle, Lusztig [**8**, **10**] constructed several unipotent cuspidal representations. (If F is not split one has to use a power of F acting on the cohomology.) However, in the case of the Coxeter element it turns out that the closure $\overline{X}_w$ (in the flag manifold) of X_w is smooth, and the deep results of Deligne on the eigenvalues of F can be used. In general, the variety X_w is smooth, but not proper; its closure is proper but not smooth. Thus the results of Deligne are not available. A cohomology theory which works for singular varieties was what was needed, and the intersection cohomology theory developed by Goresky and Macpherson is such a theory.

4. Intersection cohomology. As mentioned above, Goresky and Macpherson (see [**2**]) developed intersection cohomology theory for complex varieties with a view to constructing a cohomology theory for singular varieties with good properties such as Poincaré duality. Later, Deligne gave an algebraic version, which is described in [**7**]. The Purity theorems of Deligne-Gabber (see [**1**]) give us the required results on the eigenvalues of F which were lacking in l-adic cohomology. Furthermore, Kazhdan and Lusztig [**7**] found a remarkable connection between the polynomials $P_{y,z}$ (see §1) and the intersection cohomology of Schubert varieties, which are the closures of the BwB. Since the varieties X_w are closely related to the BwB, we can expect these polynomials to play a role in connecting the intersection cohomology and the l-adic cohomology. This will be explained below.

Let X be an algebraic variety over F_q and U a smooth, open, dense subvariety of X. If $\mathcal{F}$ is a locally constant l-adic sheaf of finite rank on U, there is defined a Deligne-Goresky-Macpherson complex $\mathrm{IC}(X, \mathcal{F})$ of l-adic sheaves on X. From this we get (i) the cohomology sheaves of $\mathrm{IC}(X, \mathcal{F})$, i.e., for each i an l-adic sheaf $\mathcal{H}_i(X, \mathcal{F})$, and (ii) the hypercohomology groups $\mathbf{H}^i(X, \mathcal{F})$ for each i, which are vector spaces over $\overline{Q}_l$. When $\mathcal{F} = \overline{Q}_l$, the constant sheaf, we will write $\mathcal{H}^i(X)$ or $\mathbf{H}^i(X)$. We consider the variety $\overline{X}_w$, which is a union of the X_y for $y \leq w$. The open dense subvariety U is equal to X_w in this case. Then there is a spectral sequence connecting the $\mathcal{H}^i$, $\mathbf{H}^i$, and H_c^i (see [**2**, p. 48]) which leads to the equation

$$\sum_i (-1)^i \mathbf{H}^i(\overline{X}_w) = \sum_{y \leq w} \sum_{i,j} H_c^i(X_y, \mathcal{H}^{2j}(\overline{X}_w)).$$

Arguments similar to those of Kazhdan and Lusztig for the Schubert varieties yield the equation

$$\sum_i \dim \mathcal{H}_{B'}^{2i}(\overline{X}_w) u^i = P_{y,w}(u), \qquad (B' \in X_y,\ y \leq w) \tag{4.1}$$

Here X_y is regarded as a subvariety of the flag manifold, i.e., the variety of Borel subgroups. The sheaf $\mathcal{H}^j(\overline{X}_w)$ is constant over each X_y and 0 for odd j. Then $\mathcal{H}_{B'}^{2i}(\overline{X}_w)$ denotes the stalk of the sheaf $\mathcal{H}^{2i}(\overline{X}_w)$ at B' (see [**13**, 4.3]). These results were used by Lusztig to construct various representations of G^F; they will be described in §5.

5. Shintani descent. We recall from §0 that if T_w is an F-stable maximal torus of G then there is a virtual representation $R_{T_w}^G(\theta)$ corresponding to a character of T_w^F. Since the "nonsplit torus" T_w^F becomes "split" when we extend the ground field to some F_{q^m}, we can ask if there is a connection between R_w and $\mathrm{Ind}_{B^{F^m}}^{G^{F^m}}$. Such a connection is proved by a "Shintani descent identity" (after Shintani, who first studied a "norm map" from G^{F^m} to G^F), which has been proved in various versions by Asai, Digne-Michel, and Lusztig (see Digne, this volume). We give the simplest version here.

If $x \in G^F$ we can write $x = aF^m(a^{-1})$ for some $a \in G$ by Lang's theorem. Let $\hat{x} = a^{-1}F(a)$; then $\hat{x} \in G^{F^m}$. The map $x \to \hat{x}$ sets up a bijection

$$\{\text{Conjugacy classes in } G^F\} \leftrightarrow \{F\text{-conjugacy classes in } G^{F^m}\},$$

where $y, z \in G^{F^m}$ are F-conjugate if there is an $s \in G^{F^m}$ such that $s^{-1}yF(s) = z$. The identity then says:

$$\mathrm{Tr}\left(xF^m, \sum_{i \geq 0} (-1)^i H_c^i(X_w)\right) = \mathrm{Tr}(\hat{x}FS_w, \mathrm{Ind}_{B^{F^m}}^{G^{F^m}}(1)).$$

Recall that S_w is a standard basis element of $\mathrm{End}_{G^F}(\mathrm{Ind}_{B^{F^m}}^{G^{F^m}}(1))$.

The right-hand side can be evaluated using the Hecke algebra. For the left-hand side one has to study the eigenvalues of F^m on cohomology. Lusztig analyzes a more general version of this identity [**15**, Chapters 2 and 3] using the following:

(i) the relationship between the $\mathbf{H}^i(\overline{X}_w)$ and the $H_c^i(X_w)$, via the polynomials $P_{y,z}$ (see (4.1));

(ii) the Deligne-Gabber Purity theorem [**1**], which states that the eigenvalues of F on $\mathbf{H}^i(\overline{X}_w)$ have absolute value $q^{i/2}$;

(iii) the representation theory of the Hecke algebra corresponding to the induced representation on the right-hand side.

We then get, for any $h \in \mathbf{Z}$, and for a given $w \in W$, that

$$(-1)^h \sum_{y \leq w} \sum_{i,j} (-1)^i P_{y,w,j} H_c^i(X_y)^{(h/2-j)} \tag{5.1}$$

is an actual G^F-module. Here, for any k, $H_c^i(X_y)^{(k)}$ is the part of $H_c^i(X_y)$ on which F has eigenvalues with absolute value $q^{k/2}$, and $P_{y,w,j}$ is the coefficient of u^j in $P_{y,w}(u)$. Furthermore, along the same lines we get

$$(-1)^h \sum_{E \in \hat{W}} \sum_{y \leq w} \sum_{j} P_{y,w,j} \operatorname{Tr}(S_y, E(u), h/2 - j) R_E \tag{5.2}$$

is an actual G^F-module. We recall from §1 that $E(u)$ is the representation of the Hecke algebra H corresponding to E. Then $\operatorname{Tr}(S_y, E(u), k)$ is the coefficient of u^k in $\operatorname{Tr}(S_y, E(u))$. Thus we get various integral linear combinations of the R_E to be proper representations.

6. Representations of the Weyl groups: Cells. It was seen in (5.2) that certain linear combinations of the R_E are proper representations of G^F. This leads to the idea that one might define certain virtual representations of α of W such that the R_α, which can be defined from the R_E by linearity, are proper representations of G^F. Lusztig [**13**, 1.15.1] defined one such virtual representation of W as follows.

$$\alpha_x = (-1)^{l(x)} \sum_{E \in \hat{W}} \operatorname{Tr}(S_x, E(u), (l(x) - a_E)/2) E, \qquad x \in W.$$

The following remarkable result, the proof of which involves the results stated in §5, then holds ([**15**], 6.14, 6.17);

The R_{α_x} are actual representations of G^F.

The α_x which are proper (rather than virtual) representations of W are called cells in [**3**] and constructible representations by Lusztig. Lusztig showed recently (C. R. Acad. Sci. Paris Sér. I **302** (1986), 5–8) that these are the same as the left cell representations of W which were mentioned in §1. In the same spirit one has [**15**, 6.17]:

R_E, R'_E are disjoint if E, E' are not in the same two-sided cell (i.e., E, E' are not related by the relation LR; see [**15**, 5.15]).

EXAMPLE. In the example of D_4 given in §2, the sum of the representations of the Weyl group parametrized by Λ_1 and Λ_2 is of the form α_x; similarly for Λ_1 and Λ_3.

The powerful information given by the existence of these proper representations and the disjointness theorem is a major tool in the decomposition of the R_E, which is done case by case for each type of group. In the case of the nonunipotent characters the ideas are similar but the hypercohomology groups $\mathbf{H}^i(\overline{X}_w, \mathcal{F})$ with nonconstant coefficient sheaves have to be used.

7. Historical remarks. The classification of unipotent representations was done by Lusztig for classical groups in [**9**] and for exceptional groups in [**10**] and [**11**]. The decomposition of the R_E for classical groups, using intersection cohomology, but for large q, was done in [**13**] and [**14**], and for exceptional groups, again for large q, in [**12**]. In [**15**] all the $R^G_{T_w}(\theta)$ were decomposed for groups G with a connected center, without using the prior classification of unipotent representations and for all q. Some results from the theory of primitive ideals in enveloping algebras were used here, but this has now been eliminated. The condition that G has a connected center has also been eliminated [**16**].

REFERENCES

1. A. A. Beilinson, J. Bernstein, and P. Deligne, *Faisceaux pervers*, Astèrisque, no. 100, Soc. Math. France, Paris, 1982.
2. A. Borel et al., *Intersection cohomology*, Prog. Math., no. 50, Birkhauser, 1984.
3. R. Carter, *Finite groups of Lie type*, Wiley, New York, 1985.
4. P. Cartier, *Travaux de Lusztig*, Seminaire Bourbaki **658** (1986).
5. P. Deligne and G. Lusztig, *Representations of reductive groups over finite fields*, Ann. of Math. (2) **103** (1976), 103–161.
6. D. Kazhdan and G. Lusztig, *Representations of Coxeter groups and Hecke algebras*, Invent. Math. **53** (1979), 165–184.
7. ——, *Schubert varieties and Poincaré duality*, Proc. Sympos. Pure Math., vol. 36, Amer. Math. Soc., Providence, R.I., 1980, pp. 185–203.
8. G. Lusztig, *Coxeter orbits and eigenspaces of Frobenius*, Invent. Math. **28** (1976), 101–159.
9. ——, *Irreducible representations of finite classical groups*, Invent. Math. **43** (1977), 125–175.
10. ——, *Representations of finite Chevalley groups*, CBMS Regional Conf. Ser. in Math., no 39, Amer. Math. Soc., Providence, R.I., 1978.
11. ——, *Unipotent representations of a finite Chevalley group of type E_8*, Quart. J. Math. Oxford Ser. **30** (1979), 315–338.
12. ——, *On the unipotent representations of exceptional groups over finite fields*, Invent. Math. **60** (1980), 173–192.
13. ——, *Unipotent characters of the symplectic and odd orthogonal groups over a finite field*, Invent. Math. **64** (1981), 263–296.
14. ——, *Unipotent characters of the even orthogonal groups over a finite field*, Trans. Amer. Math. Soc. **272** (1982), 733–751.
15. ——, *Characters of reductive groups over finite fields*, Ann. of Math. Stud., vol. 103, Princeton Univ. Press, Princeton, N.J., 1976, pp. 103–161.
16. ——, *Characters of reductive groups over finite fields*, Proc. Internat. Congr. Math., Warsaw, 1983.
17. B. Srinivasan, *Representations of finite Chevalley groups*, Lecture Notes in Math., vol. 764, Springer-Verlag, 1979.

UNIVERSITY OF ILLINOIS, CHICAGO

Proceedings of Symposia in Pure Mathematics
Volume 47 (1987)

Tensor Product Theorems

ROBERT STEINBERG

Introduction. The following fundamental, although elementary, theorem dates from the time of Burnside (see, e.g., [**C-R**, Theorem 10.33]).

THEOREM 0. *If $G = H \times K$, a direct product of groups, then every (absolutely) irreducible (finite-dimensional) representation of G is expressible uniquely as a tensor product of irreducible representations of H and of K.*

We have seen this theorem used many times at this conference, applied to abstract representations, polynomial representations, and so on. What we want to discuss in this paper are some cases in which groups which are not direct products nevertheless turn out to have irreducible representations which are tensor products. Other results of this nature have been obtained by Clifford [**Cl**], updated in [**C-R**, §11], Gajendragadkar [**G**], and Ferguson and Turull [**F-T**], among others, but, unlike those results, the theorems that we shall discuss in this paper have the striking feature that they apply to all irreducible representations (continuous, abstract, ..., as the case may be) of the groups involved.

Continuous representations. The oldest theorem of this kind that we are aware of is the following one.

THEOREM 1. *Let G be a simply connected complex Lie group. Then every continuous representation of G is expressible uniquely as the tensor product of one that is holomorphic and one that is antiholomorphic.*

We shall establish only the existence of the decomposition since that is our main theme. The following proof is taken from [**Se2**, Exp. 22]. Let $f\colon G \to \mathrm{GL}_m(\mathbf{C})$ be the given irreducible representation of G. With G and $\mathrm{GL}_m(\mathbf{C})$ viewed as Lie groups over $\mathbf{R}$ the continuity of f implies its differentiability, and then the differential yields a homomorphism of $\mathbf{R}$-algebras: $(*)$ $df\colon g \to gl_m(\mathbf{C})$. To keep track of the fact that g is also an algebra over $\mathbf{C}$ we write J for the $\mathbf{R}$-linear map of multiplication by i. We extend the scalars from $\mathbf{R}$ to $\mathbf{C}$ to

1980 *Mathematics Subject Classification* (1985 *Revision*). Primary 20G05, 22E46; Secondary 22E40.

get a homomorphism of C-algebras: (**) $(df)^{\mathbf{C}}\colon g^{\mathbf{C}} \to gl_m(\mathbf{C})$. Let g_1 and g_2 denote the i and $-i$ eigenspaces of J acting on $g^{\mathbf{C}}$. These are C-algebras, $g^{\mathbf{C}}$ is their direct sum, and they are, respectively, isomorphic and complex-conjugate isomorphic to g via the natural imbedding of g into $g^{\mathbf{C}}$. If G_1 and G_2 denote the simply connected complex Lie groups whose Lie algebras are g_1 and g_2, the factorization $g \to g^{\mathbf{C}} = g_1 + g_2 \to \mathrm{gl}_m(\mathbf{C})$ of df leads to a factorization $G \to G_1 \times G_2 \to \mathrm{GL}_m(\mathbf{C})$ of f. By the above remarks the components of the first map are respectively holomorphic and antiholomorphic while the second map is holomorphic and hence is a tensor product of holomorphic representations of G_1 and G_2 by Theorem 0. The required result follows.

The process of going from G to $G_1 \times G_2$ above is called restriction of the base field (from $\mathbf{C}$ to $\mathbf{R}$). If $L \supset K$ is a finite extension of fields, the process produces from a variety V defined over L a variety of $[L\colon K]$ times the dimension defined over K. The new variety $R_{L/K}V$ has the property that there is a natural bijection between $V(L)$ and $R_{L/K}V(K)$. If V is defined over K, then $R_{L/K}V$ is isomorphic over L to V^n $(n = [L\colon K])$ and the bijection becomes $x \to (\sigma_1(x), \sigma_2(x), \ldots, \sigma_n(x))$ in terms of the set $\{\sigma_i\}$ of distinct K-imbeddings of L into $\overline{K}$. In the case of Theorem 1 we have $K = \mathbf{R}$, $\overline{K} = L = \mathbf{C}$, $\{\sigma_i\} =$ {identity, complex conjugation}, and the principal result may be restated as follows.

THEOREM 1′. *Every continuous representation of* $G(\mathbf{C})$ *factors through a holomorphic representation of* $(R_{\mathbf{C}/\mathbf{R}}G)(\mathbf{C})$.

Polynomial representations in positive characteristics. The basic result here is the following one.

THEOREM 2. *Let G be a simply connected semisimple algebraic group defined over the field of p elements, and let X_p denote the set of dominant weights λ that satisfy $0 \le (\lambda, \check{\alpha}) \le p-1$ for all simple coroots $\check{\alpha}$ and M_p the corresponding set of irreducible polynomial representations of G. Then every irreducible polynomial representation of G is expressible uniquely as a tensor product $\bigotimes_{i=0}^{\infty}(g_i \circ F^i)$ (finite product, $g_i \in M_p$, $F =$Frobenius). The corresponding result for the finite group $G(\mathbf{F}(q))$ $(q = p^n)$ is also true with the product truncated to $\bigotimes_{i=0}^{n-1}$.*

This result was first proved for the type $G = \mathrm{SL}_2$, by Brauer and Nesbitt [**B-N**] in the finite case and by Chevalley [**Se1**, Exp. 20] in the algebraic case, and then for arbitrary types of groups by us in [**St1**]. Simultaneously, Wong treated the types SL_n and Sp_n and later in [**Wo**] extended his methods to apply to all types, with the emphasis throughout on the finite groups. Since then other proofs and expositions have been given by Ballard [**B**], Borel [**B-A**], Cartier [**C**], Cline, Parshall, and Scott [**C-P-S**], Kempf [**K**], and Donkin and Jantzen, the last two unpublished.

The [**C-P-S**] proof is probably the simplest one for the algebraic groups, but it does not apply to the finite groups. We sketch here our original proof since it applies to both kinds of groups and also to converses to Theorems 1, 3, and 5 of the present paper, since it depends mainly on the fact that the endomorphisms F^i of the base field are distinct.

By the classification of the irreducible polynomial representations of G according to their highest weights each of which is expressible uniquely in the form $\sum \lambda_i p^i$ ($\lambda_i \in X_p$), it is enough to show that the tensor product given in the theorem is irreducible. Now the action of G on this product leads, via a twisted version of the differential, to an action of g in which tX acts on the ith factor as $F^i(t)X$ whenever X is defined over $\mathbf{F}(p)$ and t is any scalar, and it is enough to show that this action is irreducible. Since the endomorphisms F^i are distinct, they are linearly independent. It easily follows that we need only show that g is irreducible on each factor and this is a theorem of Curtis [**Cu**]: each element of M_p is infinitesimally irreducible; i.e., the corresponding representation of g given by the differential is also irreducible.

In view of the last-mentioned result we have the following reformulation of Theorem 2 in terms of restriction of the base field.

THEOREM 2′. *Every irreducible polynomial representation of G factors through an infinitesimally irreducible representation of $R_{\mathbf{F}(p^n)/\mathbf{F}(p)}G$ for some (sufficiently large) n depending on the given representation.*

Finally, we note that under very special circumstances the elements of M_p can themselves be expressed as tensor products. This occurs, in the case of simple groups, when p is expressible as the square of the ratio of the lengths of two roots ($p = 2$ for type B_n, C_n, or F_4; $p = 3$ for type G_2). Then each f in M_p is uniquely expressible as $f' \otimes f''$ with the highest weight of f' (respectively f'') supported by the long (respectively short) simple roots [**St1**, Theorem 11.1]. According to a result of Seitz [**S**], for irreducible polynomial representations of semisimple algebraic groups, no further decompositions are possible. This includes the case of characteristic 0, where no decompositions at all are possible.

Abstract representations. The principal result here, not stated in its most general form, is due to Borel and Tits [**B-T**].

THEOREM 3. *Let G be a simple algebraic group defined over an infinite field k and of positive k-rank. Let $f\colon G(k) \to \mathrm{PGL}_m(k')$ be an irreducible projective representation with $m \geq 2$ and k' an algebraically closed field. Then there exist distinct imbeddings σ_i of k into k', finite in number, and corresponding irreducible polynomial representations g_i of $\sigma_i^\circ(G)$ such that $f = \bigotimes_i (g_i \circ \sigma_i^\circ)$ on $G(k)$.*

Here σ_i° denotes the isomorphism of groups induced by the change of base field σ_i.

If G is simply connected, then in many cases Theorem 3 is true with projective representations replaced by linear representations throughout [**Ti2**]. In any case it is true up to multiplication by a representation of $G(k)$ by scalars, as easily follows from Theorem 3. Thus we have the remarkable fact (in the present case) that if G is quasisplit, i.e., contains a Borel subgroup B defined over k, then the principal result about irreducible polynomial representations of $G(k)$ holds also for abstract ones: $B(k)$ fixes a unique line and the representation is determined by the action of $B(k)$ on that line.

The proof of Theorem 3 depends on the following result (see [**B-T**, Theorem 8.1]).

THEOREM A. *In Theorem 3 replace* PGL_m *by any simple adjoint* k'*-group* G' *and* f *by any homomorphism of* $G(k)$ *into* $G'(k')$ *whose image is dense. Then there exists an imbedding* σ *of* k *into* k' *and a* k'*-isogeny of algebraic groups* g *of* $\sigma^\circ(G)$ *into* G' *with* $dg \neq 0$*, both unique, such that* $f = g \circ \sigma^\circ$ *on* $G(k)$.

We sketch proofs for the above theorems in the quasisplit case. Let $B = TU$ be a Borel subgroup of G with T a maximal torus and U a maximal unipotent subgroup, all defined over k. These are characterized abstractly, up to conjugacy, as follows. B is maximal solvable and has only itself as a subgroup of finite index. U is the derived group of B. T is maximal nilpotent and every subgroup of finite index of T is of finite index in its normalizer. It follows that $\overline{f(B(k))}$ (Zariski closure) is a Borel subgroup of G' in Theorem A and that it equals $\overline{f(T(k))} \cdot \overline{f(U(k))}$, as B equals TU in G. The group $U(k)$ captures the additive group of k and $T(k)$ the multiplicative group, and the action of T on U binds the two together as in the distributive law. From this we get the required imbedding of k into k' and then the other parts of Theorem A (for more details see [**B-T**] or [**St2**]).

Now $\overline{f(G(k))}$ in Theorem 3 is connected by what has been said above, and it acts irreducibly on $P(k'^m)$. Hence it is reductive, hence also semisimple and adjoint (by Schur's lemma), hence it is a product of simple algebraic groups $\prod G_i$. The identity map yields an irreducible polynomial representation of this group which is therefore, by Theorem 0, a tensor product of similar representations h_i of the components G_i. Theorem A applied to the components of the factorization $G \to \prod G_i(k') \xrightarrow{\text{id}} \mathrm{PGL}_m(k')$ of f then yields the required conclusions of Theorem 3.

We note that Theorem 1 follows from Theorem 3. Further, if $k = \mathbf{R}$ and the values of f are all in $\mathrm{PGL}_m(\mathbf{R})$, or if $k = \mathbf{Q}$, then the only possibility for σ_i in Theorem 3 is the obvious one, so that we get the strong conclusion that f itself is polynomial.

An interesting problem is to extend the results of this section to groups of k-rank 0, where the known results, for general k, are fragmentary (cf. [**B-T**, §9], where such results over locally compact fields are discussed).

Representations of arithmetic subgroups. The principal result here, which we do not present in its most general form, is due to Margulis [**M**, Theorem 11]. (For an exposition of Margulis's results see [**Ti1**] or [**Z**].)

THEOREM 4. *Let* G *be a simply connected simple algebraic group defined over* $\mathbf{Q}$ *of* $\mathbf{R}$*-rank* ≥ 2 *and* Γ *an arithmetic subgroup of* $G(\mathbf{Q})$. *Let* $f\colon \Gamma \to \mathrm{GL}_m(C)$ *be an irreducible representation. Then* f *is expressible uniquely as a tensor product* $g \otimes h$ *such that* g *is polynomial and* h *has kernel of finite index in* Γ.

Here "arithmetic" means that, for some faithful realization of G by matrices (as a group defined over $\mathbf{Q}$), Γ equals $G(\mathbf{Z})$ or some subgroup of $G(\mathbf{Q})$ commensurable to this group. Actually, Margulis states his results for discrete subgroups and it is one of his major results that, under very general conditions, such subgroups are always arithmetic.

As is well known to people working in this area, Theorem 4 is essentially equivalent to:

THEOREM B. *In Theorem 4 the representation f is polynomial on a subgroup of finite index of Γ, even if f is not irreducible.*

Clearly Theorem 4 implies Theorem B in case f is irreducible. For a proof of the converse we follow Serre [**Ser**, p. 502]. Let g be the rational representation of Γ given by Theorem 4 and Γ' the subgroup of finite index, which may be taken to be normal, on which g agrees with f. Set $h(x) = g(x)^{-1}f(x)$ for all $x \in \Gamma$. We claim that $h(x)$ and $g(y)$ commute for all $x, y \in \Gamma$. Granted the claim, we see that h is also a representation Γ, and its kernel contains Γ'. It follows that the map $(x, y) \to g(x)h(y)$ defines a representation of $\Gamma \times \Gamma$ which, by Theorem 0, can be expressed as a tensor product of representations of the two factors. Since f factors through this representation via the diagonal imbedding of Γ in $\Gamma \times \Gamma$, the tensor product decomposition of f in Theorem 4 and its required properties easily follow. Returning to the claim, we have for $x \in \Gamma$ and $y \in \Gamma'$ that

$$\begin{aligned} h(x)^{-1}g(y)h(x) &= f(x)^{-1}g(x)g(y)g(x)^{-1}f(x) = f(x)^{-1}g(xyx^{-1})f(x) \\ &= f(x)^{-1}f(xyx^{-1})f(x) = f(y) = g(y), \end{aligned}$$

so that $h(x)$ and $g(y)$ commute. This continues to hold for $y \in \Gamma$ since g is a polynomial map and Γ' is dense in G, and hence also in Γ, in the Zariski topology, whence the claim and the equivalence of the two theorems.

Theorem B is a very hard theorem, but a relatively simple proof can be given when G is a split or quasisplit group. We now sketch such a proof, in several steps, taken from [**St3**, Theorem 6], in case $G = \mathrm{SL}_n (n \geq 3)$ and $\Gamma = \mathrm{SL}_n(\mathbf{Z})$.

(1) Let x, y, z in $\mathrm{GL}_m(\mathbf{C})$ be such that $z = (x, y)$ (commutator) and z commutes with x and y.

(a) $\quad (x_u, y) = z_u, (x, y_s) = z_s \quad$ and $\quad (x_u, y_s) = 1.$

(b) $\quad z_s^q = 1, \quad$ i.e., z^q is unipotent, for some q dividing $m!$.

To prove the second equation in (a), for example, write $xyx^{-1} = zy$ and take semisimple parts. Then, taking the commuting semisimple elements z_s and y_s in diagonal form as we may, we see that x acts on y_s via a permutation of the diagonal entries. If q is the order of this permutation then $y_s = x^q y_s x^{-q} = z_s^q y_s$, whence $z_s^q = 1$.

Now for $i \neq j$ let E_{ij} denote the corresponding $n \times n$ matrix unit and x_{ij} the unipotent element $1 + E_{ij}$ of $\Gamma = \mathrm{SL}_n(\mathbf{Z})$.

(2) Set $Y_{ij} = \log f(x_{ij})_u$ (using the usual series for $\log(1 + T)$, a polynomial when $1 + T$ is unipotent and hence T is nilpotent). Then the following relations

hold:

$$[Y_{ij}, Y_{kl}] = \begin{cases} Y_{il} & \text{if } i \neq l,\ j = k, \\ 0 & \text{if } i \neq l,\ j \neq k. \end{cases}$$

For the first case we use the relation $(x_{ij}^r, x_{jl}^s) = x_{il}^{rs}$ $(r, s \in \mathbf{Z})$. By (1a) this relation continues to hold if we apply f and take unipotent parts. If we then replace $f(x_{ij}^r)_u$ by $\exp rY_{ij} = 1 + rY_{ij} + \cdots$, and similarly for the other terms, and equate the coefficient of rs on the two sides of the resulting relation, we get the first case of (2). The proof of the second case is similar but easier.

It is not difficult to prove that the relations of (2) form an abstract definition of sl_n under the correspondence $E_{ij} \Leftrightarrow Y_{ij}$. Since the group SL_n is simply connected as an algebraic group there exists a polynomial representation $g\colon \mathrm{SL}_n(\mathbf{C}) \to \mathrm{GL}_m(\mathbf{C})$ such that $dg\colon E_{ij} \to Y_{ij}$ for all i and j. We have $g(x_{ij}) = g(\exp E_{ij}) = \exp Y_{ij} = f(x_{ij})_u$. The exp series here, like the log series earlier, is finite.) We also have $f(x_{ij}) = (f(x_{ik}), f(x_{kj}))$ $(k \neq i, j; k$ exists since $n \geq 3)$, so that $f(x_{ij}^q)$ is unipotent by (1b). It follows that f agrees with g on the elements x_{ij}^q and similarly on their conjugates and hence on the subgroup Γ' of Γ generated by all of these elements. By the congruence subgroup theorem [**B-L-S**] Γ' is the kernel of reduction mod q, hence is of finite index in Γ, as required.

It follows from Theorem B that every representation of Γ is completely reducible, and hence also that the irreducible representations of Γ are known once those with kernels of finite index are known, since the polynomial ones are known in terms of the highest weight classification, Weyl's character formula, and so on. However, the former are unknown even in simple cases like $\Gamma = \mathrm{SL}_n(\mathbf{Z})$, where the irreducible representations of $\mathrm{SL}_n(\mathbf{Z}/p^a\mathbf{Z})$ (p prime) have to be determined. These representations are needed for the harmonic analysis of the topological group $\mathrm{SL}_n(\mathbf{Q}_p)$ For $a = 1$ we have one of the classical essentially simple finite groups and a good deal is known about its representations (see [**L**], where earlier references may be found). For $a \geq 2$, however, only the case $n = 2$ has been solved, by Shalika [**Sh**] and Tanaka [**T**] independently. Therefore we pose here the problem of determining the representations in all cases, and similarly for the other simple algebraic groups.

We close with an extension of Theorem 4 to number fields. Again the general case is due to Margulis [**M**] but a special case occurred in the work of Serre [**Ser**, Theorem 5, Corollary 3]. For split or quasisplit groups the theorem can be proved along the lines of the proof of Theorem 4 for $G = \mathrm{SL}_n$ sketched above.

THEOREM 5. *In Theorem 4 let* $\mathbf{Q}$ *be replaced by* k*, a number field, and* Γ *by an arithmetic subgroup of* $G(k)$*. Let* $\sigma_1, \sigma_2, \ldots, \sigma_r$ *be the distinct imbeddings of* k *into* $\mathbf{C}$*. Then there exist unique representations* $g_1, g_2, \ldots, g_r$ *and* h *of* Γ *such that each* g_i *is polynomial,* h *has kernel of finite index in* Γ*, and* $f = (g_1 \circ \sigma_1) \otimes \cdots \otimes (g_r \circ \sigma_r) \otimes h$*. In other words,* f *factors through* $g \otimes h$ *with* h *as above and* g *a polynomial representation of* $R_{k/\mathbf{Q}}G$*.*

REFERENCES

[**B**] J. Ballard, *Clifford's theorem for algebraic groups and Lie algebras*, Pacific J. Math. **106** (1983), 1–15.

[**B-A**] A. Borel et al, *Seminar on algebraic groups and related finite groups*, Lecture Notes in Math., vol. 131, part A, Springer-Verlag, 1970.

[**B-L-S**] H. Bass, M. Lazard, and J.-P. Serre, *Sous-groupes d'indice fini dans $SL(n, \mathbf{Z})$*, Bull. Amer. Math. Soc. **70** (1964), 385–392.

[**B-N**] R. Brauer and C. Nesbitt, *On the modular characters of groups*, Ann. of Math. (2) **42** (1941), 556–590.

[**B-T**] A. Borel and J. Tits, *Homomorphismes "abstraits" de groupes algébriques simples*, Ann. of Math. (2) **97** (1973), 499–571.

[**C**] P. Cartier, *Représentations linéaires des groupes algébriques semi-simples en charactéristique non nulle*, Sém. Bourbaki **255**, (1962–63).

[**Cl**] A. Clifford, *Representations induced in an invariant subgroup*, Ann. of Math. (2) **38** (1937), 533–550.

[**C-P-S**] E. Cline, B. Parshall, and L. Scott, *On the tensor product theorem for algebraic groups*, J. Algebra **63** (1980), 264–267.

[**C-R**] C. Curtis and I. Reiner, *Methods of representation theory with applications to finite groups and orders*. Vol. I, Wiley-Interscience, New York, 1981.

[**Cu**] C. Curtis, *Representations of Lie algebras of classical type with applications to linear groups*, J. Math. and Mech. **9** (1960), 307–326.

[**F-T**] P. Ferguson and A. Turull, *Prime characters and factorization of quasi-primitive characters*, Math. Z. **190** (1985), 583–604.

[**G**] D. Gajendragadkar, *A characteristic class of characters of finite π-separable groups*, J. Algebra **59** (1979), 237–259.

[**K**] G. Kempf, *Representations of algebraic groups in prime characteristics*, Ann. Sci. École Norm. Sup. (4) **14** (1981), 61–76.

[**L**] G. Lusztig, *Characters of reductive groups over a finite field*, Ann. of Math. Studies, vol. 107, Princeton Univ. Press, Princeton, N. J., 1984.

[**M**] G. Margulis, *Discrete subgroups of motions of manifolds of non-positive curvature*, Proc. Internat. Congr. Math. (Vancouver, 1974), Vol. 2, Canad. Math. Congress, Montreal, 1975, pp. 21–34.

[**S**] G. Seitz, *The maximal subgroups of classical algebraic groups*, Mem. Amer. Math. Soc. (to appear).

[**Se1**] Séminare C. Chevalley, *Classification des groupes de Lie algébriques*, 2 vol., Inst. Henri Poincaré, Paris, 1956–58.

[**Se2**] Séminare "Sophus Lie," Inst. Henri Poincaré, Paris, 1954–55.

[**Ser**] J.-P. Serre, *Le problème des groupes de congruence pour* SL_2, Ann. of Math. (2) **92** (1970), 489 527.

[**Sh**] J. Shalika, *Representations of the two by two unimodular group over local fields*. I, II, Seminar at the Institute for Advanced Study, Princeton, 1966.

[**St1**] R. Steinberg, *Representations of algebraic groups*, Nagoya Math. J. **22** (1963), 33–56.

[**St2**] ——, *Abstract homomorphisms of simple algebraic groups* (after A. Borel and J. Tits), Sém. Bourbaki **435** (1972–73).

[**St3**] ——, *Some consequences of the elementary relations in* SL_n, Contemp. Math., no. 45, Amer. Math. Soc., Providence, R. I., 1985, pp. 335–350.

[**T**] S. Tanaka, *Irreducible representations of the binary modular congruence groups* mod p^λ, J. Math. Kyoto Univ. **7** (1967), 123–132.

[**Ti1**] J. Tits, *Travaux de Margulis*, Sém. Bourbaki **482** (1975–76).

[Ti2] ——, *Groupes de Whitehead de groupes algébriques simples sur un corps*, Sém. Bourbaki **505** (1976–77).

[W] A. Weil, *Adèles and Algebraic groups*, Prog. Math, vol. 23, Birkhäuser, 1982.

[Wo] W. Wong, *Irreducible modular representations of finite Chevalley groups*, J. Algebra **20** (1972), 355–367.

[Z] R. Zimmer, *Ergodic theory and semisimple groups*, Birkhäuser, Boston, 1984.

UNIVERSITY OF CALIFORNIA AT LOS ANGELES

Proceedings of Symposia in Pure Mathematics
Volume **47** (1987)

Coding Theory and Representation Theory

HAROLD N. WARD

ABSTRACT. The purpose of this paper is to present some of the relations between algebraic coding theory and group representation theory. My own interest in coding theory began with an investigation of the connection between the Weil representation of $\mathrm{SL}(2,q)$ and the quadratic residue codes, a connection known fairly early in the study of these codes [**1**]. The topics presented here follow my interests, but I have included a sketch of basic ideas both for reference and for an introduction to coding theory. The book of van Lint [**21**] presents a concise survey, while that of MacWilliams and Sloane [**24**] is comprehensive. A good reference for combinatorial connections is the notes of Cameron and van Lint [**6**]. Other surveys will be mentioned in the text.

1. Basic definitions and results. All the codes in this survey are *block codes.* Such a code is a nonempty subset C of the *ambient space* F^n of *words* of *length* n based on a finite *alphabet* F. The size $|F|$ of F will be denoted q, since F will usually be a finite field. The number n is also called the length of C itself, and the members of C are called *codewords.* The *distance* $d(a,b)$ between two words is the number of positions in which a and b differ. In practice, a word of C is transmitted and the received word is *decoded* as the nearest codeword to it, if there is one (if not, one may wish to make a choice among contenders to avoid a decoding failure). In this method of *maximum likelihood decoding,* the original transmitted codeword will be recovered if the number of errors that have occurred is at most $[\frac{1}{2}(d_{\min}-1)]$, where $d_{\min}$ is the *minimum distance* of C, the smallest distance between different codewords. C is said to be *e-error correcting* for any e with $e \leq [\frac{1}{2}(d_{\min}-1)]$.

The governing bound on code sizes is the *sphere-packing bound:* if C is an e-error correcting code of length n, and $|F| = q$, then

$$|C| \leq q^n/\{1 + n(q-1) + \cdots + \binom{n}{e}(q-1)^e\}.$$

The denominator is just the number of words within distance e of a fixed word, that is, the size of a "sphere" in F^n. The bound has asymptotic forms. For

1980 *Mathematics Subject Classification* (1985 *Revision*). Primary 20C05; Secondary 94B05.

0082-0717/87 $1.00 + $.25 per page

example, in the binary case, $q = 2$, let R be the *rate* of the code, $n^{-1}\log_2|C|$, and let ε be the error rate e/n. Then with

$$H(x) = -x\log_2 x - (1-x)\log_2(1-x),$$

the binary *entropy function*, $R \leq 1 - H(\varepsilon)$, asymptotically in n. This bound was improved by linear programming techniques based on the use of association schemes introduced into coding theory by Delsarte [**11**]; that use itself is a whole chapter of representation-theoretic methods in combinatorics. One such improvement is: $R \leq H[\frac{1}{2} - \sqrt{2\varepsilon(1-2\varepsilon)}]$, asymptotically.

A *perfect code* is one meeting the sphere-packing bound. Much is known about such codes; see the survey by van Lint [**20**].

2. Linear codes. When the alphabet F is the finite field $GF(q)$, the ambient space F^n becomes the standard vector space of n-tuples over F. A code C that is a subspace of F^n is a *linear* code over F; an (n,k) linear code has length n and dimension k. The *weight* $w(a)$ of a word a is the number of its nonzero entries, and one has $d(a,b) = w(a-b)$. For linear codes, the minimum weight (of nonzero words) and the minimum distance $d_{\min}$ coincide. The *weight distribution* of C is the listing of the numbers A_w of words in C of each weight w; the A_w may be displayed by the homogeneous *weight enumerator*:

$$W_C(X,Y) = \sum_w A_w X^{n-w}Y^w.$$

Two standard matrix displays of subspaces have coding applications:

(a) A *generator matrix* G of a linear code C is one having C as its row space. Careful construction of G leads to an upper bound on $|C|$ called the *Griesmer bound*: take as first row of G a word c in C of weight $d_{\min}$, the minimum weight. In any other word of C, some digit occurs at least $d_{\min}/q$ times in the positions where c has nonzero entries. If that word is not a scalar multiple of c, its nonzero entries must involve at least $d_{\min}/q$ further positions. An induction leads to the bound:

$$n \geq d_{\min} + \lceil d_{\min}/q\rceil + \cdots + \lceil d_{\min}/q^{k-1}\rceil$$

($\lceil x\rceil$ is the least integer not less than x). The weaker result $n \geq d_{\min} + k - 1$ is the *Singleton bound*; codes meeting it are called *optimal* or *maximum distance separable* and have many combinatorial features.

(b) C may be given as the null space of a matrix H called a *check matrix* for C; the rows of H give *check equations* for membership in C. H is a generator matrix for the *orthogonal* or *dual* code $C^\perp$ of C, relative to the standard inner product on F^n. The weight enumerators of C and $C^\perp$ are connected by the *MacWilliams identity*:

$$W_{C^\perp}(X,Y) = q^{-k}W_C(X+(q-1)Y, X-Y),$$

a consequence of the Poisson summation formula in the Fourier analysis of F^n.

Cosets of C are classified by the *syndromes* $Ha^{\text{transpose}}$ of their members, a. Maximum likelihood decoding may be systematized by selecting from each coset a word of minimum weight, the *coset leader.* A received word is decoded as $a-e$, e the coset leader whose syndrome is that of a.

Proper construction of H will provide a code with a guaranteed minimum weight; the numerical restrictions involved constitute the *Gilbert-Varshamov bound.* To guarantee minimum weight at least d, construct H a column at a time using columns of length r; no $d-1$ of them may be dependent. If c columns have been produced, a $c+1$st column can be chosen if

$$1 + c(q-1) + \cdots + \binom{c}{d-2}(q-1)^{d-2} < q^r.$$

For the left side is the number of formal linear combinations of $d-2$ or fewer of the c columns that must be avoided to ensure no $d-1$ columns of the $c+1$ are dependent. Thus, if

$$1 + (n-1)(q-1) + \cdots + \binom{n-1}{d-2}(q-1)^{d-2} < q^r,$$

there is an (n,k) linear code with $d_{\min} \geq d$ and $k \geq n-r$. An asymptotic form for $q=2$ is that $R \geq 1 - H(2\varepsilon)$, $\varepsilon \leq 1/4$.

This bound is not constructive. It was a long-standing problem to provide a construction that would even bound R and ε away from 0. A solution was given by Justesen [**16**] using *concatenated* codes, codes over large fields whose digits are in turn encoded over small fields. Recent work of Goppa and others (surveyed by Hirschfeld [**15**]) using elliptic curves has produced codes over large fields for which (R,ε) lies on a line segment above the Gilbert-Varshamov boundary. In addition, Delsarte and Piret [**13**] have shown how to use concatenation to produce sequences of codes achieving channel capacity.

3. Cyclic codes. The group of linear transformations of F^n preserving the distance function is the monomial group of transformations permuting the standard basis up to scalar multiples. The group of a linear code C is the subgroup leaving C invariant as a whole. If this group includes the cyclic shift, C is called a *cyclic code.* There are several ways to view such a code. One is to take F^n as the quotient ring $F^n[X]/(X^n-1)$, the powers $1, x, \ldots, x^{n-1}$ of the image x of X serving as the standard basis. Cyclic codes are then ideals of this ring $F[x]$, the shift being multiplication by x. These ideals are principal, and the *generator polynomial* of a cyclic code C is a monic polynomial $g(X)$ of least degree in $F[X]$ whose image generates the ideal C (X^n-1 is taken as generator polynomial of the 0 code). The *check polynomial* $h(X)$ of C is $(X^n-1)/g(X)$. Its reversal, scaled to be monic, is the generator polynomial of $C^{\perp}$, while $h(X)$ itself is the generator polynomial of the annihilator of C in $F[x]$. The condition $h(x)c = 0$ for c to be in C provides check equations for C.

Another point of view is that $F[x]$ is the group algebra of the cyclic group $\langle x \rangle$. When n is prime to q, the semisimple case, a cyclic code C can be specified by the character of the representation of $\langle x \rangle$ it affords. That amounts to giving the absolutely irreducible characters that do not vanish on C, or if one prefers, the ones that do. This *root set* R corresponds to the powers of a primitive nth root of unity, ς, over F, that are roots of the generator polynomial $g(X)$. For a polynomial $f(X)$, the corresponding word $f(x)$ is in C if and only if $f(X)$ is 0 on the members of R. If $\varsigma^m \in R$, then $\varsigma^{mq} \in R$; and $k = n - |R|$.

Cyclic codes are easily implemented electronically, they have good decoding algorithms (see McEliece's description [**25**]), and their "quality" can be controlled by the famous BCH (Bose, Ray-Chaudhuri, Hocquenghem) bound. This says that if the root set R contains $d-1$ consecutive powers of ς, then $d_{\min} \geq d$; it is a direct consequence of the nonsingularity of a Vandermonde matrix. Thus one can guarantee a certain minimum weight by specifying a "string" of roots.

For example, the binary Golay code is a $(23, 12)$ cyclic code over $GF(2)$. Its root exponents are $1, 2, 3, 4, 6, 8, 9, 12, 13, 16, 18$ (the squares and also the powers of $2 \bmod 23$). Its minimum weight is 7, and it is a perfect 3-error correcting code: $2^{11} = 1+23+\binom{23}{2}+\binom{23}{3}$. It is essentially the only perfect 3-error correcting code with a field alphabet; for related results, again see the survey of van Lint [**20**]. The BCH bound yields only $d_{\min} \geq 5$. One way to show $d_{\min} \geq 7$ is to extend the code to length 24 by augmenting each word with a digit equal to the sum of its digits. This *extended Golay code* can be shown to have word weights divisible by 4. Its group is the Mathieu group M_{24} (that of the Golay code is M_{23}); a historical exposition is given by Thompson [**29**].

Recently Roos, van Lint, and Wilson [**27**, **23**] have abstracted the combinatorics suggested by the Vandermonde matrix to improve the BCH bound. For example, let f be a nonzero polynomial whose roots form the set S. Define the family I of *S-independent sets* (in the field of the roots) recursively: $\varnothing \in I$; if $A \subseteq S$ and $A \in I$ but $b \notin S$, then $A \cup \{b\} \in I$; and if $A \in I$ and $c \neq 0$, then $cA \in I$. It is easy to show that if $A \in I$, then the number of terms in f (its weight) is at least $|A|$. The minimum weight of the Golay code emerges this way: let f be a polynomial whose image $f(x)$ is a word in the Golay code of minimum weight, and suppose f has weight at most 6. S cannot contain ς^5, by the BCH bound itself (S does contain ς, ς^2, ς^3, ς^4 ς^6); nor can S contain any of $\varsigma^{2^m \cdot 5}$. Then these sets (displayed by powers of ς) are S-independent: $\varnothing$, $\{5\}$, $\{4, 5\}$, $\{1, 2, 5\}$, $\{8, 9, 12, 14\}$, $\{12, 13, 16, 18, 5\}$, $\{2, 3, 6, 8, 18, 5\}$. Consequently f has weight 6. But then ς^0 is also a root, and $\{0, 1, 3, 4, 6, 16, 5\}$ is S-independent, a contradiction.

4. Group algebra codes. These codes generalize cyclic codes in the group algebra setting. Their ambient space is the group algebra FG of a finite group G, the members of G forming the standard basis. A *group algebra code* in FG (or G-code, for short) is a one-sided ideal in FG. This automatically puts G in the group of the code, acting on the appropriate side. Two general situations have been widely studied.

For the first, let G be Abelian and let q be relatively prime to $|G|$. Then as for the semisimple cyclic codes, a G-code C is determined by the "root" set R of absolutely irreducible characters χ of G that vanish on C. The absolutely irreducible constituents of C are the members of $\hat{G} - R$ in the character group $\hat{G}$ of G. The idempotent of C is e:

$$e = |G|^{-1} \sum_{g} \left\{ \sum_{\chi \in \hat{G} - R} \chi(g^{-1}) \right\} g,$$

and e is useful in presenting C. In the binary case e will be a sum of orbits of the map $g \to g^2$. The van Lint-Roos-Wilson approach to minimum weights carries over (see also the report by Camion [**7**]).

The generalized quadratic residue codes form one family of such codes (van Lint and MacWilliams [**22**]). Let G be an elementary Abelian group, realized as the additive group $GF(m)$ where m is a power of an odd prime l not dividing q. The members of $\hat{G}$ are the functions χ_a, $a \in G$, given by

$$\chi_a(b) = \xi^{\mathrm{trace}(ab)},$$

the trace from $GF(m)$ to $GF(l)$, and ξ a primitive lth root of unity over F (but not necessarily in F). Let δ be the quadratic character on $GF(m)^{\#}$: $\delta(a) = 1$ if a is a square, -1 if not. Then let $R = \{\chi_a \mid \delta(a) = 1\}$, $R' = \{\chi_a \mid \delta(a) = -1\}$. Both R and R' serve as root sets for what are known as quadratic residue codes. For the construction to work, F must contain the roots of $X^2 + X + (1 - \delta(-1)m)/4$; this is automatic if m is a square.

Much work has been done on determining the minimum weight of these codes. The *square-root bound* guarantees $d_{\min} \geq \sqrt{m}$ if $\delta(-1) = 1$, and $d_{\min} \geq \sqrt{m - 3/4} + 1/2$ if $\delta(-1) = -1$. When m is even, equality holds. This is related to a setting of the codes in this case involving inversive planes [**10**]. For other cases the bound seems to fade with increasing length, the actual minimum weight being much larger.

An additional digit may be adjoined to the codewords in such a way that the extended code is invariant under a monomial action by $\mathrm{SL}(2, m)$; the Golay code of length 23 is an example. One may construct the extended codes by beginning with this monomial action. It is induced from the character $\left[\begin{smallmatrix} a & b \\ 0 & a^{-1} \end{smallmatrix}\right] \to \delta(a)$ of the subgroup of $\mathrm{SL}(2, m)$ stabilizing ∞ in the action on the projective line over $GF(m)$. With the restriction above on F, the irreducible constituents over F of this induced representation are the extended quadratic residue codes when q is odd. (When q is even, the representation does not split and involves some trivial sections along with the codes.) This approach is due to Gleason [**1**] and is related to the Weil representation [**30**].

Several lines of investigation started with the quadratic residue codes:

(a) The codes can be defined formally when F is a field of characteristic 0; the results are called global quadratic residue codes [**2**]. Reading the words whose entries are in an appropriate integral domain modulo a prime ideal will produce the finite field codes. This provides a uniform approach to problems of minimum weight, followed in recent work of Newhart [**26**].

(b) The induced representation method applies to other groups. Calderbank and Wales [**5**] formalized what is involved and used it to create global codes of length 176 for the Higman-Sims group. In the reduction modulo 2, many of the combinatorial aspects of that group appear in an integrated way.

(c) More generally, given a permutation group G, one can ask for associated monomial groups whose permutation part is G, with codes whose group contains the monomial group. The code is said to admit G. This problem was studied by Knapp and Schmid [**19**]. For example, they and Klemm [**18**] showed that the extended quadratic residue codes are essentially the only codes admitting $\mathrm{PSL}(2,m)$ on the projective line (m prime to q); the anomalies involve one-dimensional sections.

(d) Some of the extended quadratic residue codes involve larger groups than $\mathrm{PSL}(2,m)$. The four known cases are:

$m = 11$, $q = 3$, admitting M_{12},

$m = 23$, $q = 2$, admitting M_{24},

$m = 7$, $q = 2$, admitting $2^3\,\mathrm{PSL}(2,7)$ (as the affine group $\mathrm{AGL}(3,2)$),

$m = 5$, $q = 4$, admitting A_6.

The classification of finite simple groups implies there are no others [**14**].

(e) The extended quadratic residue codes are equivalent to their duals—equal, in fact, when $\delta(-1) = -1$. There is a large body of work on self-dual codes in general (see the survey by Sloane [**28**], for example), classifying them and determining their groups. More will be said in §5.

The second type of group algebra code widely studied is that for which G is a p-group and q is a power of the prime p. The initial work by Berman [**4**] determined the minimum weights of the powers of the radical of FG when G is Abelian. Among recent authors are Camion, Charpin, Poli, Ventou, and Wolfmann.

For an illustration, let G be elementary Abelian again, taken as the additive group of $GF(p^m)$. Write the members of FG as sums $\sum \alpha_g(g)$, $g \in G$, with $(g)(h) = (g+h)$. (Many authors use an exponential notation: $X^gX^h = X^{g+h}$.) Let $G^\# = G - \{0\}$ and view $G^\#$ as the multiplicative group of $GF(p^m)$, cyclic of order $p^m - 1$. If $a \in GF^\#$, with a written as $\sum' \alpha_g(g)$ to indicate a sum over $G^\#$, let $\bar{a} = \sum \alpha_g(g)$, where $\alpha_0 = -\sum' \alpha_g$.Given a cyclic code C in $FG^\#$, one can ask, when will the extended code $\overline{C}$ be an ideal in FG? This happens exactly when $\overline{C}$ is invariant under the affine group of permutations $(x) \to (bx+c)$, $b \neq 0$, b and c in $GF(p^m)$. The answer was given by Kasami, Lin, and Peterson [**17**]: let $G^\# = \langle \varsigma \rangle$. The roots of cyclic codes in $FG^\#$ can be given by powers of ς, or more conveniently, by the exponents. Let R be such an exponential root set for the code C, with $0 \notin R$. In the polynomial interpretation, the value of a polynomial preimage of the word $\sum \alpha_{\varsigma^i}(\varsigma^i)$ at ς^j will be $\sum \alpha_{\varsigma^i}\varsigma^{ij}$, that is, $\sum' \alpha_g g^j$. Thus the extended code can be described by the "root set" $\overline{R}$ of exponents j for which $\sum \alpha_g g^j = 0$, $j = 0$ now included in accordance with the meaning of α_0. Thus $\overline{R} = R \cup \{0\}$. Let $i \leq_p j$ mean that each digit of i written base p is at most the corresponding digit of j. Then their result is that the extended code is an ideal if and only if for each j in $\overline{R}$, $i \leq_p j$ implies $i \in \overline{R}$ also.

For example, let $w_p(j)$ be the sum of the digits of j written base p. Then if

$$\overline{R}_r = \{j \mid w_p(j) \leq m(p-1) - r - 1\},$$

for $r < m(p-1)$, $\overline{R}_r$ is such a root set. The words of the corresponding code $\overline{C}_r$, viewed as polynomial functions on G (a vector space over $GF(p)$), are the polynomials of degree at most r, variables z being reduced by the relations $z^p = z$. These $\overline{C}_r$ are forms of Reed-Muller codes, and $\overline{C}_r = (\operatorname{rad} FG)^{m(p-1)-r}$. (For related results, see the paper of Charpin [**8**] and its references.)

These modular group algebra codes lead naturally to the last topic.

5. Divisible codes. Let $F = GF(q)$, q a power of the prime p, and let V be an m-dimensional vector space over F. As above, any function $f\colon V \to F$ is a polynomial function. Its degree (more properly, its *reduced* degree) is computed taking into account the relations $z^q = z$ on the variables used. The functions of degree at most r form the rth order *generalized Reed-Muller* codes, viewed as group algebra codes based on the additive group of V. Extending earlier work of Artin, Chevalley, and Warning, Ax [**3**] proved that if $e = [(m-1)/r]$, then $q^e | w(f)$, when f has degree r, and this is the highest power of p dividing the weights of all functions of degree at most r.

When $e \neq 0$, these codes are *divisible* codes [**31**]: all word weights are multiples of some nontrivial common divisor. Self-dual codes over small fields are further examples. Thus if $q = 3$, the weights in such a code are multiples of 3, as $w(a) \equiv a \cdot a \pmod 3$. More generally, one can ask about formally self-dual codes, those having the same weight enumerator as their duals, that are also divisible. Gleason and Pierce (see Sloane's survey [**28**] for two proofs, and the article [**30**] for another view) showed there are only certain possibilities for q and the divisor Δ (the "type" numbers are neatly mnemonic):

I: $q = 2$, $\Delta = 2$,
II: $q = 2$, $\Delta = 4$,
III: $q = 3$, $\Delta = 3$,
IV: $q = 4$, $\Delta = 2$,
V: q arbitrary, but the code equivalent to one obtained by duplicating the digits of all words of half the length; $\Delta = 2$.

If $W(X,Y)$ is the weight enumerator of a formally self-dual code, the MacWilliams relations imply

$$W(X,Y) = W\left(\frac{X + (q-1)}{\sqrt{q}}, \frac{X - Y}{\sqrt{q}}\right).$$

If the code is also divisible, then $W(X,Y) = W(X, \varepsilon Y)$ for $\varepsilon^\Delta = 1$. Thus $W(X,Y)$ will be invariant under a group of transformations containing those suggested. Writing $W(X,Y)$ in terms of basic invariants leads to relations on the coefficients A_w. Setting $A_i = 0$ to raise the possible minimum weight can only go so far, and bounds emerge. The "extremal" weight enumerator for the bound is uniquely determined. For large lengths these enumerators have some negative coefficients, and thus there are only finitely many possible codes. Much effort (with much success) has gone into finding the corresponding extremal codes. For those cases not yet settled, one attack has been to show the group of the code to be limited in size, much as for dealing with the hypothetical projective plane of order 10.

When the divisor of a divisible code is a power p^e of the characteristic, e is called the *exponent* of the code. For group algebra codes, one might hope to relate e to module properties of the code. Let $q = p$, for simplicity, and let C be a left ideal in FG, C divisible with exponent e. Then let

$$\overline{w}(c) = p^{-e}w(c) \pmod p$$

for c in C; $\overline{w}$ is the *modular* weight function. This is a nonconstant G-invariant function on C (clearly available for more general divisible codes). It can be polarized [**32**] to produce invariant multilinear forms.

To illustrate the ideas, let $F = GF(2)$. For α in F, $w(\alpha) = (1 - (-1)^{\alpha})/2$, and $(-1)^{\alpha} = 1 - 2w(\alpha)$. Then

$$\begin{aligned} w(\alpha_1 \dots \alpha_m) &= \prod w(\alpha_i) \\ &= 2^{-m} \prod (1 - (-1)^{\alpha_i}) \\ &= 2^{-m} \left\{ 1 - \sum (-1)^{\alpha_i} + \sum_{i<j} (-1)^{\alpha_i + \alpha_j} - \cdots \right\} \\ &= 2^{-m+1} \left\{ \sum w(\alpha_i) - \sum_{i<j} w(\alpha_i + \alpha_j) + \cdots \right\}. \end{aligned}$$

If $*$ denotes the componentwise product of words, then

$$(-2)^{m-1} w(a_1 * \cdots * a_m) = \sum_J (-1)^{m-|J|} w\left(\sum_{i \in J} a_i\right),$$

J running over the subsets of $\{1, \dots, m\}$. Thus if $m = e + 1$,

$$w(a_1 * \cdots * a_m) \equiv \sum_J (-1)^{m-|J|} \overline{w} \left(\sum_{i \in J} a_i\right) \pmod 2,$$

and this is the sort of polarization meant. The left side (mod 2) is a G-invariant multilinear form on C. The conclusion is that the degree of $\overline{w}$ is at most $e + 1$, and if it is r, then C has a nonzero r-linear form (the polarization of $\overline{w}$) that is G-invariant. For the general prime, the degree of $\overline{w}$ is $r(p - 1)$ for some nonzero r with $r \leq e + 1$, and C has a nontrivial G-invariant $r(p - 1)$-linear form.

For example, the subcode of the binary Golay code consisting of the words of even weight has for the powers of ς that are not roots the exponents $5 \cdot 2^i$. These correspond to characters in the representation, and a nonzero invariant multilinear form of degree r will require that a sum of r of them be 0 mod 23. The shortest such sum needs three terms: $5(1 + 4 + 64)$. Thus $e \geq 2$ for this code, essentially what was involved in the discussion of minimum weights.

When C is a two-sided ideal that is also a direct summand of FG (so a sum of block ideals), the bound above is exact: $(e+1)(p-1)$ is the least degree of a nonzero invariant multilinear form whose degree is a multiple of $p-1$. Ax's theorem, as well as generalizations by Delsarte and McEliece [**12**], are consequences. Damgård and Landrock [**9**] have investigated further connections between code and module properties; for example, they show that the principal p-block ideal is a divisible code with divisor $|O_{p'}(G)|$.

REFERENCES

1. E. F. Assmus, Jr., H. F. Mattson, Jr., and R. Turyn, *Cyclic codes*, Summary Scientific Report No. 4, Air Force Cambridge Research Laboratories, AFCRL-65-332, 1965.

2. E. F. Assmus, Jr. and H. F. Mattson, Jr., *New 5-designs*, J. Combin. Theory **6** (1969), 122–151.

3. J. Ax, *Zeroes of polynomials over finite fields*, Amer. J. Math. **86** (1964), 255–261.

4. S. D. Berman, *On the theory of group codes*, Cybernetics **3** (1967), 25–31.

5. R. Calderbank and D. B. Wales, *A global code invariant under the Higman-Sims group*, J. Algebra **75** (1982), 233–260.

6. P. J. Cameron and J. H. van Lint, *Graphs, codes, and designs*, London Math. Soc. Lecture Note Ser., vol. 43, Cambridge Univ. Press, Cambridge-New York, 1980.

7. P. Camion, *Abelian codes*, Math. Res. Center, Univ. of Wisconsin, Rept. 1059 (1970).

8. P. Charpin, *The extended Reed-Solomon codes considered as ideals of a modular algebra*, Ann. Discrete Math. **17** (1983), 171–176.

9. I. Damgård and P. Landrock, *Ideals and codes in group algebras*, Preprint.

10. P. Delsarte, *Majority-logic decodable codes derived from finite inversive planes*, Inform. and Control **18** (1971), 319–325.

11. ——, *An algebraic approach to the association schemes of coding theory*, Philips Res. Rep. Suppl. No. 10 (1973).

12. P. Delsarte and R. J. McEliece, *Zeros of functions in finite Abelian group algebras*, Amer. J. Math. **98** (1976), 197–224.

13. P. Delsarte and P. Piret, *Algebraic codes achieving the capacity of the binary symmetric channel*, Ann. Discrete Math. **17** (1983), 265–269.

14. W. Feit, *Some consequences of the classification of finite simple groups*, The Santa Cruz Conference on Finite Groups, Proc. Sympos. Pure Math., vol. 37, Amer. Math. Soc., Providence, R.I., 1980, pp. 175–181.

15. J. W. P. Hirschfeld, *Linear codes and algebraic curves*, Geometrical Combinatorics (F. C. Holroyd and R. J. Wilson, eds.), Res. Notes in Math., vol. 114, Pitman, Boston, 1984, pp. 35–53.

16. J. Justesen, *A class of constructive asymptotically good algebraic codes*, IEEE Trans. Inform. Theory **IT-18** (1972), 652–656.

17. T. Kasami, S. Lin, and W. W. Peterson, *Some results on cyclic codes which are invariant under the affine group and their applications*, Inform. and Control **11** (1968), 475–496.

18. M. Klemm, *Kennzeichnung der erweiterten Quadrate-Codes durch ihre PSL $(2,q)$-Zulässigkeit*, Comm. Algebra **11** (1983), 2051–2068.

19. W. Knapp and P. Schmid, *Codes with prescribed permutation group*, J. Algebra **67** (1980), 415–435.

20. J. H. van Lint, *A survey of perfect codes*, Rocky Mountain J. Math. **5** (1975), 199–224.

21. ——, *Introduction to coding theory*, Graduate Texts in Math., vol. 86, Springer-Verlag, New York-Berlin, 1982.

22. J. H. van Lint and F. J. MacWilliams, *Generalized quadratic residue codes*, IEEE Trans. Inform. Theory **IT-24** (1978), 730–737.

23. J. H. van Lint and R. M. Wilson, *On the minimum distance of cyclic codes*, IEEE Trans. Inform. Theory **IT-32** (1986), 23–40.

24. F. J. MacWilliams and N. J. A. Sloane, *The theory of error-correcting codes*, North-Holland, Amsterdam, 1977.

25. R. J. McEliece, *The theory of information and coding*, Addison-Wesley, Reading, Mass., 1977.

26. D. Newhart, *On minimum weight codewords in QR codes*, Preprint.

27. C. Roos, *A new lower bound for the minimum distance of a cyclic code*, IEEE Trans. Inform. Theory **IT-29** (1983), 330–332.

28. N. J. A. Sloane, *Self-dual codes and lattices*, Relations Between Combinatorics and Other Parts of Mathematics, Proc. Sympos. Pure Math., vol. 34, Amer. Math. Soc., Providence, R.I., 1979, pp. 273–308.

29. T. M. Thompson, *From error-correcting codes through sphere packings to simple groups*, Carus Math. Monograph, vol. 21, Math. Assn. of America, Washington, D.C., 1983.

30. H. N. Ward, *Quadratic residue codes and symplectic groups*, J. Algebra **29** (1974), 150–171.

31. ——, *Divisible codes*, Arch. Math. (Basel) **36** (1981), 485–494.

32. ——, *Multilinear forms and divisors of codeword weights*, Quart. J. Math. Oxford Ser. (2) **34** (1983), 115–128.

UNIVERSITY OF VIRGINIA

Proceedings of Symposia in Pure Mathematics
Volume **47** (1987)

Subgroup Complexes

P. J. WEBB

1. Introduction. We survey the general properties of the simplicial complex of p-subgroups of a finite group G. The partially ordered set of p-subgroups of G together with the conjugation action on this poset carries the information inherent in the p-local structure of G. By forming a simplicial complex out of this poset and considering the group action on it we have a systematic way of dealing with global features of the p-local structure.

The p-subgroups complex is really a geometry for G. Whatever one means by a geometry, there is usually a simplicial complex involved, it is associated to a prime p, and the stabilizers of simplices are treated as analogues of parabolic subgroups. One can take the view that the most canonically defined nontrivial simplicial complex on which G acts, associated to the prime p, is the p-subgroups complex. In all situations where buildings are defined, the p-subgroups complex coincides, in a certain sense, with the building. It is one of the themes of this survey that the p-subgroups complex provides an analogue defined for every finite group and every prime of a building.

Group actions on suitable spaces have long been used to elucidate the group structure, especially with certain classes of infinite groups such as arithmetic groups. The philosophy is to obtain the group structure in terms of the stabilizer subgroups. The action on the p -subgroups complex is eminently suitable for this application and we show how to use the structure of this complex to determine the cohomology of a finite group. This situation with a group action is one where the equivariant cohomology spectral sequence is also used. The approach described here yields stronger results for finite groups and it can be fed back in such a way that the structure of the spectral sequence is made transparent, to the extent that this spectral sequence becomes redundant. It remains to be seen whether a similar approach free of equivariant cohomology can be made to work for infinite groups.

1980 *Mathematics Subject Classification* (1985 *Revision*). Primary 20J06; Secondary 20D15, 20D25,20D30.

We see that the subject of this article enjoys the following close relationships with other areas of group theory:

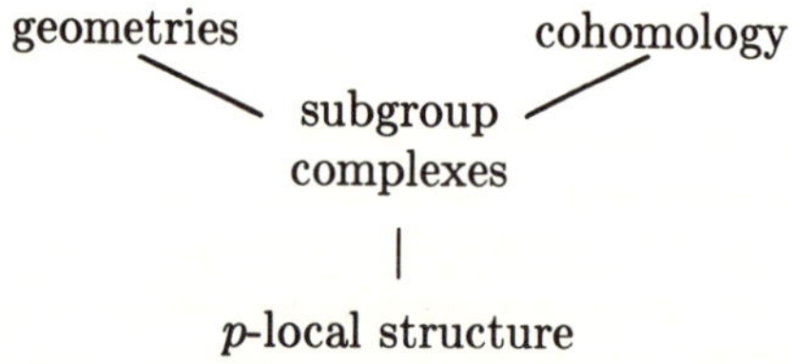

In §2 we describe the p-subgroups complex and also the complexes of elementary abelian p-subgroups, subgroups $H = O_p(N_G(H))$, and normal chains of p-subgroups, which are homotopy equivalent to it. In §3 we discuss the homological structure of the p-subgroups complex and the application to group cohomology. §4 is about Euler characteristics, and in §§5 and 6 we consider a general definition of a Steinberg module. According to this definition, every finite group has a Steinberg module at every prime. §7 is an appendix containing lemmas about chain complexes that are sometimes useful. In the early part of this survey we omit most proofs because the results can almost all be found written up in one or other of the cited references. In §§6 and 7 this is not the case, and we supply proofs.

2. The complexes

2.1. *Construction.* Given a poset $\mathcal{P}$ we form a simplicial complex Δ where the set of n-simplices of Δ is

$$\{x_0 < x_1 < \cdots < x_n \mid x_i \in \mathcal{P}\}.$$

These are chains of elements of $\mathcal{P}$ of length $n+1$, where we require strict inequalities in the chain. The faces of such a simplex are the shorter subchains. If G acts on $\mathcal{P}$ as a group of poset automorphisms then G will act simplicially on Δ.

Example:

Poset Simplicial Complex

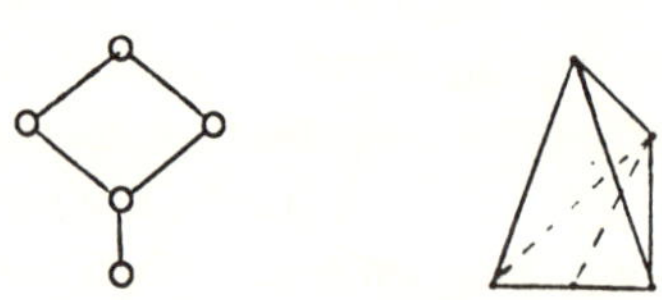

The way the chains in the poset overlap is the way the simplices are glued together.

2.2. *The posets.* Fix a prime number p; we define

$$\mathcal{S}_p(G) = \{\text{all nonidentity } p\text{-subgroups of } G\},$$

$$\mathcal{A}_p(G) = \{\text{nonidentity elementary abelian } p\text{-subgroups of } G\},$$

$$\mathcal{B}_p(G) = \{H \le G \mid H = O_pN_G(H), H \ne 1\}.$$

In this generality the poset $\mathcal{B}_p(G)$ seems to have been first considered by S. Bouc. If H is a subgroup satisfying the condition $H = O_pN_G(H)$ of the last definition, Alperin calls $N_G(H)$ a *parabolic subgroup*. When G is a finite Chevalley group in defining characteristic p, this agrees with the usual definition and the subgroups in $\mathcal{B}_p(G)$ are precisely the unipotent radicals of the parabolic subgroups.

EXAMPLES. 1. $G = S_4, p = 2$.

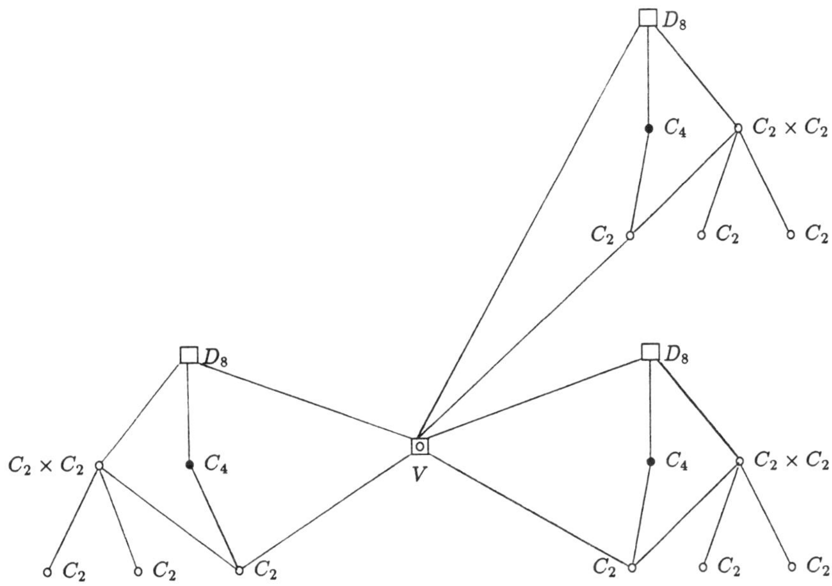

This is the whole of $\mathcal{S}_2(S_4)$. The three wings should be the same, although one of them is drawn above the other two. The subgroups in $\mathcal{A}_2(S_4)$ are shown as open circles ∘, and those in $\mathcal{B}_2(S_4)$ are squares □. Although these are evidently different posets, they are all contractible; indeed $\mathcal{A}_2(S_4)$ and $\mathcal{B}_2(S_4)$ are trees.

2. $G = \mathrm{GL}(3,2)$, $p = 2$. $\mathcal{A}_2(G)$ and $\mathcal{B}_2(G)$ are both isomorphic to the barycentric subdivision of the incidence graph of the projective plane of order 2, but with different subgroups of G as the vertices.

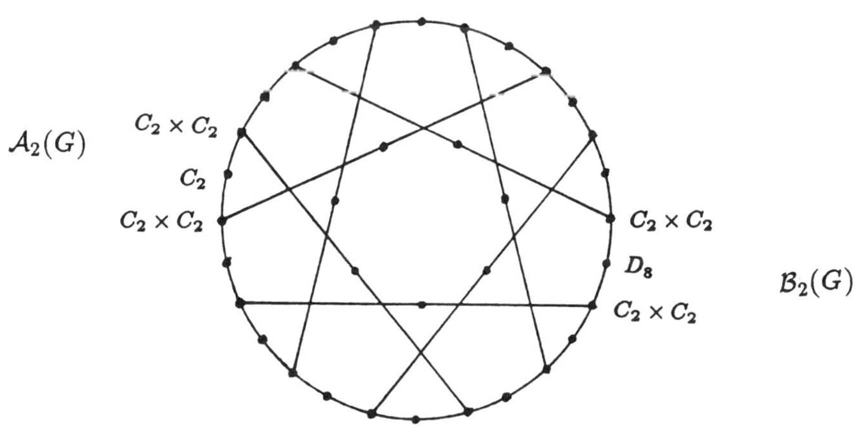

We describe another simplicial complex, which was first considered by G. R. Robinson in his reformulation of Alperin's conjecture. This is denoted $\mathcal{R}_p(G)$ and has as its n-simplices the chains $P_0 < P_1 < \cdots < P_n$ of nonidentity p-subgroups of G such that $P_i \triangleleft P_n$ for all i. The vertices are thus the same as the vertices in the simplicial complex associated to $\mathcal{S}_p(G)$, but there are fewer higher-dimensional simplices. Call a chain of subgroups in which all subgroups are normal in the largest a *normal chain*. We may also consider the complex of normal chains of subgroups in $\mathcal{B}_p(G)$, which we will not bother to name. This complex may also be included in the list in the following theorem.

2.3. THEOREM (QUILLEN [**7**], BOUC [**2**], THÉVENAZ, WEBB). *$\mathcal{S}_p(G)$, $\mathcal{A}_p(G)$, $\mathcal{B}_p(G)$, $\mathcal{R}_p(G)$ are all G-homotopy equivalent.*

The effect of this theorem is that for many purposes it does not matter which of the simplicial complexes we use, but before discussing this let's look at the definitions.

G-Homotopy equivalence. Spaces X_1 and X_2 with G-actions are said to be G-homotopy equivalent if there exist G-equivariant maps $f : X_1 \to X_2$ and $g : X_2 \to X_1$ with $fg \simeq 1$ and $gf \simeq 1$ given by G-equivariant homotopies. Passing to chain complexes, this would imply that the chain complexes of X_1 and X_2 are chain homotopy equivalent in such a way that all of the maps used to establish the chain homotopy equivalence (including the two maps of degree $+1$) are G-homomorphisms. This evidently implies that the homology groups of the spaces are isomorphic as G-modules. It also implies that X_1/G and X_2/G are homotopy equivalent. The following theorem is fundamental:

2.4. THEOREM (BREDON [**3**, §II]). *Let X_1 and X_2 be G-CW-complexes. A G-equivariant map $f : X_1 \to X_2$ is a G-homotopy equivalence if and only if $f|_{X_1^H} : X_1^H \to X_2^H$ is a homotopy equivalence for all subgroups $H \leq G$.*

The equivalent condition here is what is in fact used to prove the G-homotopy equivalence of the four posets. Putting the information together we obtain:

COROLLARY. *In each dimension the homology groups of the four posets $\mathcal{S}_p(G), \mathcal{A}_p(G)$, $\mathcal{B}_p(G)$, $\mathcal{R}_p(G)$ are isomorphic as G-modules. For any subgroup H, the fixed point sets under the action of H are homotopy equivalent.*

2.5. *First properties of the complexes.* We state them just for $\mathcal{S}_p(G)$, and by the above they evidently hold for all the complexes.

1. $\mathcal{S}_p(G_1 \times G_2) \simeq \mathcal{S}_p(G_1) * \mathcal{S}_p(G_2)$ (the topological join) (Quillen [**7**]).

2. For a finite Chevalley group in defining characteristic p, $\mathcal{B}_p(G)$ is the building of G. This is because as a poset it consists of the unipotent radicals of the (nontrivial) parabolic subgroups of G and the inclusion relationship between parabolics is exactly reversed on taking unipotent radicals, so we are looking at the opposite of the poset which normally defines the building. But a poset and its opposite give the same simplicial complex. The force of this remark is that $\mathcal{B}_p(G)$ is a generalization to all finite groups of the building of a Chevalley group.

3. If $O_p(G) \neq 1$ then $\mathcal{S}_p(G)$ is contractible. This is well illustrated in the case of S_4 at $p = 2$. It is appropriate to mention

QUILLEN'S CONJECTURE. *If $\mathcal{S}_p(G)$ is contractible then $O_p(G) \neq 1$.*

Quillen showed in [**7**] that this is true for soluble groups and groups of p-rank 2. It has recently been shown by Thévenaz and Lyons using the classification of finite simple groups that Quillen's conjecture is also true for these groups. This means that if G is a finite simple group, then $\mathcal{S}_p(G)$ is not contractible.

3. The structure theorem and cohomology. Let $\mathbf{Z}_p$ denote the p-adic integers. If

$$D. = \cdots D_{n+1} \xrightarrow{d_{n+1}} D_n \xrightarrow{d_n} D_{n-1} \cdots$$

is a chain complex of $\mathbf{Z}_pG$-modules, we say it is *acyclic split* if it has zero homology and for every n the sequence $0 \to \ker d_n \to D_n \to \operatorname{Im} d_n \to 0$ is split. If $D.$ is augmented by an epimorphism $D_0 \to \mathbf{Z}_p$, by saying $D.$ is *acyclic split augmented* we mean the same condition applied to the augmented complex (i.e., zero reduced homology, all the sequences split, and the augmentation splits).

3.1. STRUCTURE THEOREM (WEBB [**10**]). *Let Δ be the simplicial complex of one of $\mathcal{S}_p(G)$, $\mathcal{A}_p(G)$, $\mathcal{B}_p(G)$, or $\mathcal{R}_p(G)$. Let $C.(\Delta) = C_d(\Delta) \to C_{d-1}(\Delta) \to \cdots \to C_0(\Delta)$ be the chain complex of Δ over $\mathbf{Z}_p$. Then $C.(\Delta)$ has an acyclic split augmented subcomplex $D.$ so that $C.(\Delta)/D.$ is a complex of projective modules.*

This means $C_r(\Delta) = D_r \oplus P_r$ for all r where P_r is projective, and it also follows from the long exact sequence in homology that $C.(\Delta)$ and $C.(\Delta)/D.$ have the same reduced homology. I propose the following:

3.2. CONJECTURE. *The reduced homology groups of $\mathcal{S}_p(G)$, when completed at p, are all projective $\mathbf{Z}_pG$-modules.*

The main evidence for this is the above theorem, which gives the homology as the homology of a complex of projective modules. It is an easy deduction from this theorem that the conjecture holds if $\mathcal{S}_p(G)$ has only one nonzero reduced homology group (as happens, in particular, in the case of a building). It is also the case that for all finite groups G, $\tilde{H}_0(\mathcal{S}_p(G))$ is projective at the prime p, because when this reduced homology is nonzero, G has a strongly p-embedded subgroup H so that $\mathbf{Z}_p\uparrow_H^G \cong \mathbf{Z}_p \oplus \tilde{H}_0(\mathcal{S}_p(G))_p$ (see [**7**]). The nontrivial summand is always projective, and the situation is rather well controlled. Presumably if the conjecture were true, the projectivity in higher dimensions might correspond to other structural statements about G of a similar nature.

Again let Δ be the simplicial complex of one of the above four posets and let Δ_r denote the set of simplices in dimension r. The structure of the chain complex of Δ has the following consequence.

3.3. THEOREM (WEBB [**10**]). *For all integers n and for all $\mathbf{Z}_pG$-modules M there are split exact sequences*

$$0 \to \hat{H}^n(G,M)_p \to \bigoplus_{\sigma\in\Delta_0/G} \hat{H}^n(G_\sigma,M)_p \xrightarrow{(\phi_{\sigma\tau})} \bigoplus_{\tau\in\Delta_1/G} \hat{H}^n(G_\tau,M)_p \to \cdots \to \bigoplus_{\sigma\in\Delta_d/G} \hat{H}^n(G_\sigma,M)_p \to 0$$

and

$$0 \leftarrow \hat{H}^n(G,M)_p \leftarrow \bigoplus_{\sigma\in\Delta_0/G} \hat{H}^n(G_\sigma,M)_p \xleftarrow{(\psi_{\sigma\tau})} \bigoplus_{\tau\in\Delta_1/G} \hat{H}^n(G_\tau,M)_p \leftarrow \cdots \leftarrow \bigoplus_{\sigma\in\Delta_d/G} \hat{H}^n(G_\sigma,M)_p \leftarrow 0,$$

where if σ is a face of τg then

$$\phi_{\sigma\tau} = (-1)^{(\sigma|\tau g)} C_{g^{-1}} \circ \operatorname{res}^{G_\sigma}_{G_{\tau g}}, \qquad \psi_{\sigma\tau} = (-1)^{(\sigma|\tau g)} \operatorname{cores}^{G_\sigma}_{G_{\tau g}} \circ c_g,$$

and otherwise $\phi_{\sigma\tau}$ are zero. Here $(\sigma \mid \tau g)$ denotes the orientation of the embedding of σ in τg.

We will illustrate the theorem in the case $G = \mathrm{GL}(3,2)$, $p = 2$. Let Δ be the building of G. Representatives of the simplex stabilizers are given in the following picture of Δ/G:

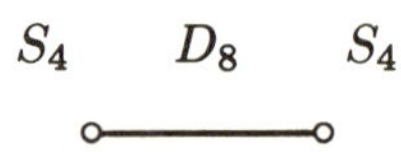

Preciously in §2 we saw that $\mathcal{B}_2(G)$ gives the barycentric subdivision of this complex Δ, but evidently the two are G-homotopy equivalent and the conclusions of Theorem 3.1 are independent of G-homotopy equivalence. The first sequence in Theorem 3.3 is

$$0 \to \hat{H}^n(G,M)_2 \xrightarrow{(-\operatorname{res},\operatorname{res})} \hat{H}^n(S_4,M)_2 \oplus \hat{H}^n(S_4,M)_2 \xrightarrow{\binom{\operatorname{res}}{\operatorname{res}}} \hat{H}^n(D_8,M)_2 \to 0,$$

and by the theorem this is split exact. For example, in the case of $H_1(G,\mathbf{Z}) = \hat{H}^{-2}(G,\mathbf{Z}) \simeq G/G'$ it is

$$1 \to 1 \to C_2 \times C_2 \to C_2 \times C_2 \to 1.$$

Here the restriction homomorphisms are the classical transfer. For the Schur multiplier, $H_2(G,\mathbf{Z}) = \hat{H}^{-3}(G,\mathbf{Z})$, the sequence is

$$1 \to C_2 \to C_2 \times C_2 \to C_2 \to 1.$$

These are the right answers!

Computationally, we would obtain the cohomology of G if we knew all the right-hand terms in these sequences. If we are only interested in isomorphism types of individual groups we do not need to know exactly what the maps are, because of the splitting. These maps become important when we are interested in cohomology ring structure, and then the left-hand morphism from $\hat{H}^n(G,M)_p$ is a ring homomorphism.

You may have noticed in Theorem 3.3 that the restriction sequences with the $\hat{H}^n(G,M)_p$ terms removed constitute the E_1 page of the equivariant cohomology spectral sequence for the action of G on Δ ([**4**, VII.8.1]). The exactness of the sequences implies that the E_2 page only has nonzero terms on the fibre, and the spectral sequence stops there.

PROOF OF 3.3. $\operatorname{Ext}^n_{\mathbf{Z}_pG}(\tilde{C}.(\Delta), M_p)$ is the first sequence (for positive values of n), where $\tilde{C}.(\Delta)$ is the augmented chain complex of Δ, and the second sequence is $\operatorname{Ext}^n_{\mathbf{Z}_pG}(\tilde{C}.(\Delta)^*, M_p)$, where * denotes the dual. This identification is almost the same as [**4**, p. 173]. We apply the structure Theorem 3.3. The Ext groups vanish on the projective modules, and what is left is Ext of a split acyclic complex. This is again split acyclic.

4. Euler characteristics. We take Δ to be the simplicial complex of one of $\mathcal{S}_p(G)$, $\mathcal{A}_p(G)$, $\mathcal{B}_p(G)$, or $\mathcal{R}_p(G)$.

4.1. THEOREM (WEBB [**10**]). *Δ/G is* mod p *acyclic.*

This means the chain complex of Δ/G reduced mod p has the homology of a point. In fact, when G has p-rank 2 this result forces Δ/G to be contractible. The condition p-rank 2 means that the largest elementary abelian p-subgroup is $C_p \times C_p$ and in this case $\mathcal{A}_p(G)$ is a graph. Thus Δ/G is homotopy equivalent to a graph (no matter which simplicial complex Δ is) and it is also connected (because of conjugacy of Sylow p-subgroups). One readily sees that a connected graph which is mod p acyclic is a tree, because the homology of a graph is torsion free of rank depending on the number of cycles in the graph. For example, $\mathcal{A}_2(S_5)/S_5$ is the tree

$$\begin{array}{cccc} C_2 & C_2\times C_2 & C_2 & C_2\times C_2 \\ \circ & \circ & \circ & \circ \end{array}$$

where the vertices are the two conjugacy classes of C_2 subgroups and the two conjugacy classes of $C_2 \times C_2$ subgroups. This quotient complex is some kind of global picture of the p-local structure of the group, and it seems very strange that in general (when G has p-rank 2) it should be a tree.

4.2. CONJECTURE. *If Δ is the simplicial complex of $\mathcal{S}_p(G)$ then Δ/G is contractible.*

If this were true it would be a generalization of what happens for buildings, where Δ/G is a single simplex.

The next result has been proved in many ways since the original proof by K. S. Brown. There is now a more general result of this kind due to Brown and Thévenaz [**11**] which subsumes a lot of earlier work. Special cases of [**11**] have been obtained independently by Hawkes, Isaacs, and Ozaydin.

4.3. COROLLARY (K. S. BROWN). $\chi(\mathcal{S}_p(G)) \equiv 1 (\operatorname{mod} |G|_p)$.

PROOF. From the structure theorem, $C_r(\Delta) = D_r \oplus P_r$ for all r where P_r is a projective $\mathbf{Z}_pG$-module. Thus $\chi(\mathcal{S}_p(G)) = \sum_{r=0}^{d}(-1)^r \operatorname{rank} C_r(\Delta) = \sum_{r=0}^{d}(-1)^r \operatorname{rank} D_r + \sum_{r=0}^{d}(-1)^r \operatorname{rank} P_r$. The first number is 1 because $D.$ has the homology of a point, and the second is congruent to 0 $(\bmod |G|_p)$ because all the modules are projective.

5. The Steinberg module. Let $a(G) = K_0(\mathrm{Latt}_{\mathbf{Z}_pG}, 0)$ be the Green ring of $\mathbf{Z}_pG$-lattices. This is the free abelian group with the set of isomorphism classes of indecomposable $\mathbf{Z}_pG$-lattices as a basis. Because the Krull-Schmidt theorem holds, knowing a lattice as an element of $a(G)$ is equivalent to knowing its isomorphism type. As before, let Δ be the simplicial complex of one of $\mathcal{S}_p(G)$, $\mathcal{A}_p(G)$, $\mathcal{B}_p(G)$, or $\mathcal{R}_p(G)$ and let $C.(\Delta)$ be its chain complex over $\mathbf{Z}_p$. We will see in §6 that it does not matter which simplicial complex we take.

DEFINITION. The Steinberg module of G at the prime p is

$$\mathrm{St}_p(G) = -\mathbf{Z}_p + \sum_{r=0}^{d}(-1)^r C_r(\Delta),$$

the sum taken in $a(G)$.

A similar notation is used by Bouc [**1**] for the corresponding element of the ring of Brauer characters. It may be misleading to call it a module, because by this definition it is only a virtual module. It is intriguing that in all known cases, $\mathrm{St}_p(G)$ is $\pm$ an actual module.

5.1. THEOREM. *$\mathrm{St}_p(G)$ is a virtual projective; that is, there exist projective $\mathbf{Z}_pG$-modules P and Q so that $\mathrm{St}_p(G) = P - Q$ in $a(G)$.*

PROOF. By the structure theorem $C_r(\Delta) = D_r \oplus P_r$ where $D.$ is an acyclic split augmented subcomplex and each P_r is projective. All of the indecomposable summands of the D_r cancel out on taking alternating sums, because of the splitting. Hence $\mathrm{St}_p(G) = \sum_{r=0}^{d}(-1)^r P_r$ is a virtual projective.

5.2. *Properties of* $\mathrm{St}_p(G)$. 1. If G is a finite Chevalley group then $\mathrm{St}_p(G) = (-1)^{\mathrm{rank}(G)-1}\,\mathrm{St}$, where St is the usual Steinberg module and $\mathrm{rank}(G)$ is the rank of the root system. This follows from the Solomon-Tits theorem, which tells us that the top homology of $\mathcal{B}_p(G)$ is St and all other reduced homology is zero. Since we know in advance that St is projective over $\mathbf{Z}_pG$, it follows that it splits off from the chain complex, and in fact using the structure theorem the whole chain complex is the splice of split short exact sequences of $\mathbf{Z}_pG$-lattices. This was also observed by Kuhn and Mitchell [**6**]. The discrepancy with the minus sign between $\mathrm{St}_p(G)$ and St arises because $\mathcal{B}_p(G)$ has dimension $\mathrm{rank}(G) - 1$.

2. $\mathrm{St}_p(G_1 \times G_2) = -\mathrm{St}_p(G_1) \otimes \mathrm{St}_p(G_2)$. This follows from $\mathcal{S}_p(G_1 \times G_2) \simeq \mathcal{S}_p(G_1) * \mathcal{S}_p(G_2)$.

3. Unlike the usual Steinberg character, $\mathrm{St}_p(G)$ is in general not simple. It need not even be indecomposable.

4. If $O_p(G) \neq 1$ then $\mathrm{St}_p(G) = 0$. In particular, p-groups have trivial Steinberg module. If $|G|$ is prime to p then $\mathrm{St}_p(G) = -\mathbf{Z}_p$.

5. $\mathrm{St}_p(G)$ is self-dual.

6. The simple modules whose projective covers appear in the support of $\mathrm{St}_p(G)$ all have dimension at least p. In particular, P_1 never appears in the support of $\mathrm{St}_p(G)$. This follows from the following proposition, which was also used to compute the ensuing table.

5.3. PROPOSITION. *Assume $p\|G|$. For each simple $\mathbf{Z}_pG$-module S, the multiplicity of the projective cover P_S in* $\mathrm{St}_p(G)$ *is*

$$\sum_{\sigma\in\Delta/G}(-1)^{\dim\sigma}(\text{multiplicity of } P_1 \text{ for } G_\sigma \text{ in } S|_{G_\sigma}).$$

We may take Δ to be any of the four simplicial complexes described in §2. In the parenthetic term above we compute how many times the projective cover over $\mathbf{Z}_pG_\sigma$ of the trivial module for $\mathbf{Z}_pZG_\sigma$ is a summand of $S|_{G_\sigma}$. This term is zero if $\dim S < |G_\sigma|_p$, since $\dim P_1 \geq |G_\sigma|_p$. When Δ is the simplicial complex of any of the four posets $\mathcal{S}_p(G)$, $\mathcal{A}_p(G)$, $\mathcal{B}_p(G)$, $\mathcal{R}_p(G)$ we always have $|G_\sigma|_p \geq p$, hence the assertion in 6 that P_S does not occur at all if $\dim S < p$.

Steinberg modules at $p = 2$:

S_3	$P_{(2,1)}$
S_4	0
S_5	$-P_{(3,2)}$
S_6	$-P_{(3,2,1)}$
S_7	$P_{(4,3)}+P_{(4,2,1)}$
A_7	$2P_{20}+P_{14}$
M_{11}	$-3P_{44}-2P_{16}-2P_{\overline{16}}$
M_{12}	$10P_{144}$
M_{24}	P_{1792}

For the symmetric groups the notation P_λ denotes the projective cover of the modular irreducible corresponding to the p-regular partition λ. Otherwise, irreducibles are referred to by their dimensions. I am indebted to L. G. Griffiths, S. E. Rees, M. A. Ronan, and S. Smith for the more difficult calculations.

6. Steinberg module inversion. This is a process like Möbius inversion, which is familiar to workers in several areas. In the context of representations of Chevalley groups it takes the form of a duality between the Steinberg module and the trivial module. We take an approach which was first used in a less general context by Bouc [**1**], and go on to deduce an application to group cohomology, and also Robinson's reformulation of Alperin's conjecture. First the statement of the inversion formula in the language of the previous sections (cf. Proposition 4 of [**1**], but note that Bouc works in the ring of Brauer characters).

6.1. THEOREM. *In $a(G)$,*

$$\mathbf{Z}_p = -\sum_{\{1\leq P\leq G|P\text{a } p\text{-subgroup}\}/G}\mathrm{St}_p(N_G(P)/P)\uparrow_{N_G(P)}^G.$$

The vertical arrow means induction. In this formula we regard $\mathrm{St}_p(N_G(P)/P)$ as a virtual $N_G(P)$-module by inflation. Comparing this with the definition of $\mathrm{St}_p(G)$ we see that each of the trivial module and the Steinberg module can be expressed as a linear combination of terms involving only the other. Notice that the only p-subgroups P which make any contribution to the sum in 6.1 are those with $P = O_p(N_G(P))$ because otherwise the Steinberg module of $N_G(P)/P$ is zero, by property 4 in §5. At the end of this section we give an application to cohomology, but first we explain how the formula may be derived.

We work with the precursor of the Steinberg module which lies in the Burnside ring. We define the *Burnside ring*, $b(G) = K_0(G\text{-sets}, 0)$, to be the Grothendieck group of the category of finite G-sets with respect to relations given by disjoint union decompositions. This is the free abelian group with the set of equivalence classes of transitive G-sets as a basis, two transitive G sets $H\backslash G$ and $K\backslash G$ being equivalent if and only if the stabilizer subgroups H and K are conjugate in G. The basis elements thus biject with the conjugacy classes of subgroups of G. If Δ is a simplicial complex of dimension d arising from a poset on which G acts, we let Δ_r denote the G-set of simplices in dimension r, and let $\Delta_{-1} = 1$ be the trivial G-set. We set

$$\Lambda(\Delta) = \sum_{r=0}^{d} (-1)^r \Delta_r,$$

the *Lefschetz invariant* of Δ, as an element of $b(G)$. We also define the *reduced Lefschetz invariant* to be

$$\tilde{\Lambda}(\Delta) = \sum_{r=-1}^{d} (-1)^r \Delta_r.$$

There is a homomorphism $r : b(G) \to a(G)$ sending each G-set to the corresponding permutation module, and evidently if Δ is the simplicial complex of $\mathcal{S}_p(G)$ then $r(\tilde{\Lambda}(\Delta)) = \mathrm{St}_p(G)$. Burnside showed [**5**, §181, p. 238] that a G-set Ω is completely determined as an element of $b(G)$ by knowing the values of $|\Omega^H|$ for all subgroups $H \leq G$. In our situation this means that $\tilde{\Lambda}(\Delta)$ is completely determined by the values of the (reduced) Euler characteristics

$$\tilde{\chi}(\Delta^H) = \sum_{r=-1}^{d} (-1)^r |\Delta_r^H|.$$

Since these numbers are unchanged under G-homotopy equivalence it follows that if Δ_1 and Δ_2 are G-homotopy equivalent then $\tilde{\Lambda}(\Delta_1) = \tilde{\Lambda}(\Delta_2)$. This approach is taken from [**12**]. Thus we have:

6.2. PROPOSITION. $\tilde{\Lambda}(\mathcal{S}_p(G)) = \tilde{\Lambda}(\mathcal{A}_p(G)) = \tilde{\Lambda}(\mathcal{R}_p(G))$. *It makes no difference which of these four simplicial complexes we use to define* $\mathrm{St}_p(G)$.

We now state the inversion formula in the context of the Burnside ring.

6.3. THEOREM (THÉVENAZ [**12**, 3.3]). *Let G act on a poset $\mathcal{P}$. Then*

$$\Lambda(\mathcal{P}) = -\sum_{x\in\mathcal{P}/G} \tilde{\Lambda}(]x,\ [) \uparrow_{G_x}^{G} = -\sum_{x\in\mathcal{P}/G} \tilde{\Lambda}(]\ ,x[) \uparrow_{G_x}^{G}$$

in the Burnside ring.

The notation $]x,\ [$ means the interval $\{y\in\mathcal{P}\mid x<y\}$, and $]\ ,x[$ means $\{y\in\mathcal{P}\mid y<x\}$. In the statement we are not distinguishing between a poset and its associated simplicial complex, because it looks better. Unfortunately, we will need to make this distinction in the proof, and we denote the simplicial complex associated to the poset $\mathcal{P}$ by $\Delta(\mathcal{P})$. If σ is the simplex $x_0<\cdots<x_n$ we let σ_0 denote the starting vertex x_0.

PROOF.

$$\begin{aligned}
\Lambda(\mathcal{P}) &= \sum_{\sigma\in\Delta(\mathcal{P})/G} (-1)^{\dim\sigma} G_\sigma\setminus G \\
&= \sum_{x\in\mathcal{P}/G}\left(\sum_{\{\sigma\in\Delta(\mathcal{P})\mid\sigma_0=x\}/G_x} (-1)^{\dim\sigma} G_\sigma\setminus G_x\right)\uparrow_{G_x}^{G} \\
&= \sum_{x\in\mathcal{P}/G}\left(G_x\setminus G_x - \sum_{\tau\in\Delta(]x,\ [)/Gx} (-1)^{\dim\tau} G_\tau\setminus G_x\right)\uparrow_{G_x}^{G} \\
&= -\sum_{x\in\mathcal{P}/G} \tilde{\Lambda}(]x,\ [)\uparrow_{G_x}^{G}.
\end{aligned}$$

For the second equality, apply the first to $\Lambda(\mathcal{P}^{op}) = \Lambda(\mathcal{P})$.

The next lemma is 6.1 of [**7**], except that here we have the extra decoration that the homotopy equivalence is an $N_G(P)$-homotopy equivalence.

6.4. LEMMA. *Whenever P is a p-subgroup of G, $N_G(P)$ acts on the subposet $]P,\ [$ of $\mathcal{S}_p(G)$.*

(a) *There is an $N_G(P)$-homotopy equivalence $]P,\ [\simeq \mathcal{S}_p(N_G(P)/P)$, where we regard the second term as an $N_G(P)$-poset by inflation from $N_G(P)/P$.*

(b) $\tilde{\Lambda}(]P,\ [) = \tilde{\Lambda}(\mathcal{S}_p(N_G(P)/P))$ *in $b(N_G(P))$.*

PROOF. (a) There is an inclusion map

$$i : \mathcal{S}_p(N_G(P)/P) \hookrightarrow]P,\ [$$

given by the bijection between p-subgroups of $N_G(P)/P$ and p-subgroups of $N_G(P)$ containing P. By Bredon's Theorem 2.4 it suffices to show that for each subgroup $H\le N_G(P)$ the mapping

$$i^H : \mathcal{S}_p(N_G(P)/P)^H \hookrightarrow]P,\ [^H$$

is a homotopy equivalence. We use Quillen's argument in 6.1 of [**7**]. There is a reverse mapping

$$r^H :]P,\ [^H \to \mathcal{S}_p(N_G(P)/P)^H$$

defined as follows: if Q is an H-stable p-subgroup in $]P, [^H$ then $r^H(Q) = (Q \cap N_G(P))/P = N_Q(P)/P$, an H-stable nontrivial p-subgroup. Then $r^H i^H = \mathrm{id}$ and $i^H r^H(Q) \leq Q$, so it follows from 1.3 of **[7]** that r^H and i^H are homotopy equivalences.

(b) is immediate from (a).

6.5. COROLLARY. *In* $b(G)$,

$$G \backslash G = - \sum_{\{1 \leq P \leq G | P \text{a } p\text{-subgroup}\}/G} \tilde{\Lambda}(\mathcal{S}_p(N_G(P)/P)) \uparrow_{N_G(P)}^G .$$

In the above statement, $G \backslash G$ is, of course, the trivial G-set.

PROOF. We deduce 6.5 from 6.4 and 6.3 by taking $\mathcal{P}$ to be the poset of *all* p-subgroups of G, including $\{1\}$. Then $\mathcal{P}$ has a unique minimal element and is contractible, so $\Lambda(\mathcal{P}) = G \backslash G$.

PROOF OF 6.1. Apply the homomorphism $r : b(G) \to a(G)$ to the formula in 6.5.

ALPERIN'S CONJECTURE. We now show how G. Robinson's reformulation of Alperin's conjecture is an instance of the above identities. Let

$l(G)$ = the number of absolutely irreducible p-modular characters of G.

$f_0(G)$ = the number of blocks of defect zero of G.

Alperin has conjectured the statement

(A) $$l(G) = \sum_{\{1 \leq P \leq G | P \text{a } p\text{-subgroup}\}/G} f_0(N_G(P)/P) \quad \text{for all finite groups } G.$$

Robinson has shown that this is equivalent to the statement

(R) $$f_0(G) = l(G) - \sum_{\sigma \in \Delta/G} (-1)^{\dim \sigma} l(G_\sigma) \quad \text{for all finite groups } G,$$

where Δ is the simplicial complex $\mathcal{R}_p(G)$. Actually, we will see in just a moment that it does not matter if we take Δ to be $\mathcal{S}_p(G)$, $\mathcal{A}_p(G)$, $\mathcal{B}_p(G)$, or $\mathcal{R}_p(G)$. We will deduce the equivalence of (A) and (R) from Corollary 6.5 by regarding l as a function on the Burnside ring. We define

$$l : b(G) \to \mathbf{Z}$$

on the basis elements of $b(G)$ by

$l(H \backslash G)$ = the number of absolutely irreducible p-modular characters of H

and extend by linearity to the whole of $b(G)$. It is immediate that Robinson's reformulation (R) can be stated in the form

(R′) $$f_0(G) = -l(\tilde{\Lambda}(\Delta)) \quad \text{for all finite groups } G,$$

where Δ in the first instance is $\mathcal{R}_p(G)$, but might just as well be $\mathcal{S}_p(G)$, $\mathcal{A}_p(G)$, or $\mathcal{B}_p(G)$ because these all have the same $\tilde{\Lambda}$ by 6.2. We note that as a function on the Burnside ring, l behaves well with respect to induction and inflation when we are extending by a normal p-subgroup:

6.6. LEMMA. *Let H be a subgroup of G and let P be a normal p-subgroup of G. If $x \in b(H)$ and $y \in b(G/P)$ then $l(x) = l(\mathrm{ind}_H^G x)$ and $l(y) = l(\inf_{G/P}^G y)$.*

PROOF. For induction it is that $\mathrm{ind}_H^G(K \setminus H) = K \setminus G$, and the statement for inflation holds because if $P \leq L \leq G$ then L and L/P have the same number of p-modular irreducibles.

Applying the function l to both sides of the equation in Corollary 6.5, and ignoring inductions and inflations by 6.6, we obtain

6.7. COROLLARY.

$$l(G) = - \sum_{\{1 \leq P \leq G | P \text{a } p\text{-subgroup}\}/G} l(\tilde{\Lambda}(\mathcal{S}_p(N_G(P)/P))).$$

PROOF OF THE EQUIVALENCE OF (A) FROM (R′). To deduce (A) from (R′) we substitute (R′) into the formula of Corollary 6.7.

Now assuming (A) we prove (R′) by induction on $|G|$. So assume that (A) holds, and that (R′) holds for smaller values of $|G|$. Then Corollary 6.7 gives

$$\begin{aligned} l(G) &= -l(\tilde{\Lambda}(\mathcal{S}_p(G))) - \sum_{\{1 < P \leq G | P \text{a } p\text{-subgroup}\}/G} l(\tilde{\Lambda}(\mathcal{S}_p(N_G(P)/P))) \\ &= -l(\tilde{\Lambda}(\mathcal{S}_p(G))) + \sum_{\{1 < P \leq G | P \text{a } p\text{-subgroup}\}/G} f_0(N_G(P)/P) \quad \text{by induction} \\ &= \sum_{\{1 \leq P \leq G | P \text{a } p\text{-subgroup}\}/G} f_0(N_G(P)/P) \quad \text{assuming (A).} \end{aligned}$$

Since all of the terms on the two sides of the last equality are the same, except for two of them, we deduce

$$f_0(G) = -l(\tilde{\Lambda}(\mathcal{S}_p(G)))$$

as required. This argument also starts the induction when $p \nmid |G|$.

We conclude with an application of Theorem 6.1 to group cohomology.

6.8. THEOREM. *For all $n \in \mathbf{Z}$, for all finitely generated $\mathbf{Z}G$-modules M,*

$$\hat{H}^n(G, M)_p = - \sum_{\{P \leq G | P \text{a } p\text{-subgroup}\}/G} \mathrm{Hom}_{\mathbf{Z}_p N_G(P)}(\mathrm{St}_p(N_G(P)/P), \hat{H}^n(P, M)_p).$$

This equation holds in the Grothendieck group $K_0(\mathrm{Ab}, 0)$ of finitely generated abelian groups with relations given by direct sum decompositions (in the same way as the analogous theorems in [**8**] and [**9**]). Knowledge of all the terms on the right-hand side determines the isomorphism type of the left-hand side. The suffix p is always used to mean completion at p, but in the case of a finite abelian group this is the same as taking the p-torsion subgroup. In the term in parentheses on the right we regard $\hat{H}^n(P, M)$ as an $N_G(P)$-module by the conjugation action. The whole Hom term then counts the multiplicities of the simple modules at the top of $\mathrm{St}_p(G)$ as composition factors in $\hat{H}^n(P, M)_p$, because $\mathrm{St}_p(G)$ is projective, and also for this reason the calculation here may be done with Brauer characters, which suffice to distinguish composition factors. This formula reduces the Tate cohomology of G to the Tate cohomology of the p-subgroups of G (and only the O_p's of parabolic subgroups contribute) together with knowledge of the action of the normalizers of the p-subgroups.

In the statement of 6.8 the functor $\mathrm{Hom}_{\mathbf{Z}_p N_G(P)}(\ ,\hat{H}^n(P,M)_p)$ is applied to a virtual module. In order to make sense of this we have to regard $\mathrm{Hom}(\ ,X)$ as a homomorphism $a(G) \to K_0(\mathrm{Ab},0)$. Thus if U and V are modules then $\mathrm{Hom}(U-V,X)$ is by definition $\mathrm{Hom}(U,X) - \mathrm{Hom}(V,X)$. We follow this convention with other functors throughout the proof of 6.8.

PROOF OF 6.8. For positive values of n we apply the functor $\mathrm{Ext}^n_{\mathbf{Z}_pG}(\ ,M_p)$ to both sides of the equation in Theorem 6.1. For $n \leq 0$ the result follows from the result for positive n by a dimension shifting argument (cf. [**8**, §2]). Suppose $n > 0$. The term on the left of the equation is

$$\mathrm{Ext}^n_{\mathbf{Z}_pG}(\mathbf{Z}_p, M_p) \cong H^n(G,M)_p$$

(see [**8**] for this isomorphism). The typical term on the right is

$$\begin{aligned} &\mathrm{Ext}^n_{\mathbf{Z}_pG}(\mathrm{St}_p(N_G(P)/P)\uparrow^G_{N_G(P)}, M_p) \cong \mathrm{Ext}^n_{\mathbf{Z}_pN_G(P)}(\mathrm{St}_p(N_G(P)/P), M_p) \\ (*)\qquad &\cong \mathrm{Ext}^n_{\mathbf{Z}_pN_G(P)}(\mathbf{Z}_p, M_p \otimes_{\mathbf{Z}_p} \mathrm{St}_p(N_G(P)/P)^*) \\ &\cong H^n(N_G(P), M_p \otimes_{\mathbf{Z}_p} \mathrm{St}_p(N_G(P)/P)^*). \end{aligned}$$

We apply the Lyndon-Hochschild-Serre spectral sequence for the extension $1 \to P \to N_G(P) \to N_G(P)/P \to 1$. The E_2 page is

$$E_2^{r,s} = H^r(N_G(P)/P, H^s(P, M_p \otimes_{\mathbf{Z}_p} \mathrm{St}_p(N_G(P)/P)^*)).$$

I claim that as $N_G(P)$-modules with the conjugation (or diagonal) action,

$$H^s(P, M_p \otimes_{\mathbf{Z}_p} \mathrm{St}_p(N_G(P)/P)^*) \cong H^s(P, M_p) \otimes_{\mathbf{Z}_p} \mathrm{St}_p(N_G(P)/P)^*.$$

This is because if F is one of the modules in a projective resolution of $\mathbf{Z}_p$ over $\mathbf{Z}_pN_G(P)$ then the conjugation action arises from the diagonal action on $\mathrm{Hom}_{\mathbf{Z}_pP}(F, M_p \otimes_{\mathbf{Z}_p} \mathrm{St}_p(N_G(P)/P)^*)$, which is isomorphic over $\mathbf{Z}_pN_G(P)$ to $\mathrm{Hom}_{\mathbf{Z}_pP}(F, M_p) \otimes_{\mathbf{Z}_p} \mathrm{St}_p(N_G(P)/P)^*$ with the diagonal action because $\mathrm{St}_p(N_G(P)/P)$ is free as a $\mathbf{Z}_p$-module and has the trivial P-action. This isomorphism commutes with taking homology in the definition of $H^*(P, M_p)$ because $\mathrm{St}_p(N_G(P)/P)$ is flat over $\mathbf{Z}_p$. We deduce that $H^s(P, M_p \otimes_{\mathbf{Z}_p} \mathrm{St}_p(N_G(P)/P)^*)$ is a (virtual) projective $\mathbf{Z}_pN_G(P)/P$-module. Hence $E_2^{r,s} = 0$ unless $r = 0$ and then

$$\begin{aligned} E_2^{0,s} &= H^0(N_G(P)/P, H^s(P, M_p \otimes_{\mathbf{Z}_p} \mathrm{St}_p(N_G(P)/P)^*)) \\ &= \mathrm{Hom}_{\mathbf{Z}_pN_G(P)}(\mathbf{Z}_p, H^s(P, M_p) \otimes_{\mathbf{Z}_p} \mathrm{St}_p(N_G(P)/P)^*) \\ &\cong \mathrm{Hom}_{\mathbf{Z}_pN_G(P)}(\mathrm{St}_p(N_G(P)/P), H^s(P, M_p)). \end{aligned}$$

The spectral sequence stops at the E_2 page, where it is concentrated on the fiber. Hence $E_2^{0,n} \cong H^n(N_G(P), M_p \otimes_{\mathbf{Z}_p} \mathrm{St}_p(N_G(P)/P)^*)$. Putting this together with equation $(*)$ yields the statement of the theorem.

7. Appendix: homotopy equivalent complexes. We present some elementary arguments which show that splitting of a complex and the Lefschetz invariant are independent of G-chain homotopy equivalence. The use of this kind of result is that it helps us to transfer between the complexes $\mathcal{S}_p(G)$, $\mathcal{A}_p(G)$, $\mathcal{B}_p(G)$, and $\mathcal{R}_p(G)$ (or some other such set of complexes that we may happen to have in mind). The independence of the Lefschetz invariant was also shown just before Proposition 6.2, but here we give an alternative algebraic proof. First we clarify the notion of splitting. It is convenient to define it in a general situation where the complex may have nonzero homology. This extends the usual definition for short exact sequences. We work with a chain complex

$$(D., d) = \cdots D_{n+1} \xrightarrow{d_{n+1}} D_n \xrightarrow{d_n} D_{n-1} \cdots$$

of R-modules, where R is some ring. In what follows all homomorphisms are supposed to commute with the action of R.

DEFINITION. The chain complex $(D., d)$ is *split* if there is a map $\alpha : D. \to D.$ of degree $+1$ with $d\alpha\, d = d$.

7.1. LEMMA. *Let $(D., d)$ be a chain complex of R-modules. The following are equivalent:*

(i) *$(D., d)$ is split.*

(ii) *For every n, in the factorization of d_n as $D_n \xrightarrow{d_n} \operatorname{Im} d_n \overset{i_n}{\hookrightarrow} D_{n-1}$ where i_n is inclusion, both $d_n : D_n \to \operatorname{Im} d_n$ is a split epimorphism and i_n is a split monomorphism.*

(iii) *For every n we can write $D_n = \operatorname{Im} d_{n+1} \oplus \ker d_n / \operatorname{Im} d_{n+1} \oplus \operatorname{Im} d_n$ so that the mapping*

$$\begin{aligned} D_n &= \operatorname{Im} d_{n+1} \oplus \ker d_n / \operatorname{Im} d_{n+1} \oplus \operatorname{Im} d_n \\ &\xrightarrow{d_n} \operatorname{Im} d_n \oplus \ker d_{n-1} / \operatorname{Im} d_n \oplus \operatorname{Im} d_{n-1} = D_{n-1} \end{aligned}$$

is an isomorphism on the summands $\operatorname{Im} d_n$ and is zero on other summands. Thus $D.$ is the splice of split short exact sequences direct sum its homology.

PROOF. (iii)$\Rightarrow$(i). We define α to be the map which is the inverse of d_n on the summands $\operatorname{Im} d_n$ and zero elsewhere.

(ii)$\Rightarrow$(iii). Both $\ker d_n$ and $\operatorname{Im} d_{n+1}$ are summands of D_n. Since $\operatorname{Im} d_{n+1} \subseteq \ker d_n$, this inclusion also splits by the modular law, giving the required decomposition.

(i)$\Rightarrow$(ii). Suppose there exists a map α of degree $+1$ with $d\alpha\, d = d$. We claim that $\ker(d) = \ker(\alpha d)$ and $\operatorname{Im}(d\alpha) = \operatorname{Im}(d)$ (writing maps on the left). This is because $\ker(d) \subseteq \ker(\alpha d) \subseteq \ker(d\alpha\, d) = \ker(d)$, and $d(D.) = d\alpha\, d(D.) \subseteq d\alpha(D.) \subseteq d(D.)$. Now αd and $d\alpha$ are idempotent mappings from $D.$ to itself, since $\alpha(d\alpha d) = \alpha d$ and $(d\alpha d)\alpha = d\alpha$. Therefore the inclusions $\ker(\alpha d) \subseteq D.$ and $d\alpha(D.) \subseteq D.$ are split, and hence so are the inclusions $\ker d \subseteq D.$ and $\operatorname{Im} d \subseteq D.$ The fact that $\ker d \subseteq D.$ splits means also that $D. \to \operatorname{Im} d$ splits, which is what we need.

For any finite-dimensional complex $(D., d)$ of R-modules we define

$$L(D.) = \sum_{i=0}^{n} (-1)^i D_i \in K_0(\operatorname{mod} R, 0)$$

where $n = \dim D.$.

7.2. PROPOSITION. *Let $f : C. \to D.$ be a chain homotopy equivalence of chain complexes of R-modules. Then*

(i) *$C.$ is split if and only if $D.$ is split.*

(ii) $L(C.) = L(D.)$.

PROOF. (i) Suppose that $C.$ is split by a map α of degree $+1$, so $c\alpha c = c$. Let $g : D. \to C.$ be the chain homotopy inverse, and let $t : D. \to D.$ be the R-homomorphism of degree $+1$ which gives $1_{D.} \simeq fg$. The domains and targets of these mappings are as follows:

$$\begin{array}{ccc} C_n & \underset{\alpha_n}{\overset{c_n}{\rightleftarrows}} & C_{n-1} \\ g_n \uparrow\downarrow f_n & & \\ D_n & \underset{t_n}{\overset{d_n}{\rightleftarrows}} & D_{n-1} \end{array}$$

Using the equations

$$c\alpha c = c, \qquad 1 - fg = t\,d + dt, \qquad df = fc, \qquad g\,d = cg, \qquad d^2 = 0,$$

we calculate

$$df\alpha g\,d = fc\alpha cg = fcg = fg\,d = (1 - (t\,d + dt))\,d = d - dt\,d.$$

Thus $\beta = f\alpha g + t$ splits $D.$, since $d\beta\,d = d$.

(ii) (Proof supplied by P. Eccles). We use similar notation to (i), so that $g : D. \to C.$ is the homotopy inverse, $1 - gf = sc + cs$, and $1 - fg = t\,d + dt$. We show that

$$C_0 \oplus D_1 \oplus C_2 \oplus D_3 \oplus \cdots \cong D_0 \oplus C_1 \oplus D_2 \oplus C_3 \oplus \cdots .$$

Let U denote the left-hand side above, and V the right-hand side. We define maps $\theta : U \to V$ and $\phi : V \to U$ by defining them on each summand of U and V. We define

$$\theta(x) = -c_n(x) + f_n(x) + s_{n+1}(x) \in C_{n-1} \oplus D_n \oplus C_{n+1}, \qquad x \in C_n,$$
$$\theta(y) = d_n(y) - g_n(y) - t_{n+1}(y) \in D_{n-1} \oplus C_n \oplus D_{n+1}, \qquad y \in D_n.$$

We define ϕ be the same formulae. Then

$$\begin{aligned}\phi\theta(x_0, y_1, x_2, y_3, \dots) = (&-x_0, -y_1 + \eta_0(x_0), -x_2 + \eta_1'(y_1) + \varsigma_0(x_0), \\ &- y_3 + \eta_2(x_2) + \varsigma_1'(y_1), -x_4 + \eta_3'(y_3) + \varsigma_2(x_2), \dots),\end{aligned}$$

where

$$\eta_n = f_{n+1}s_{n+1} - t_{n+1}f_n, \qquad \varsigma_n = s_{n+2}s_{n+1},$$

and η_n', ς_n' are defined in the same way with the roles of $C.$ and $D.$ reversed. This uses the chain homotopy equivalence, the fact that $c^2 = 0$, and the fact that f and g are chain maps. This means that $\phi\theta$ is an isomorphism. For

$\phi\theta(D_1 \oplus C_2 \oplus \cdots) \subseteq D_1 \oplus C_2 \oplus \cdots$ so that we have a map of short exact sequences

$$\begin{array}{ccccccccc} 0 & \longrightarrow & D_1 \oplus C_2 \oplus \cdots & \longrightarrow & C_0 \oplus D_1 \oplus C_2 \oplus \cdots & \longrightarrow & C_0 & \longrightarrow & 0 \\ & & \downarrow{\scriptstyle \phi\theta|} & & \downarrow{\scriptstyle \phi\theta} & & \downarrow{\scriptstyle -1} & & \\ 0 & \longrightarrow & D_1 \oplus C_2 \oplus \cdots & \longrightarrow & C_0 \oplus D_1 \oplus C_2 \oplus \cdots & \longrightarrow & C_0 & \longrightarrow & 0. \end{array}$$

Hence $\phi\theta$ will be an isomorphism provided $\phi\theta|$ is, and an inductive argument gives the result.

Similarly $\theta\phi$ is an isomorphism, and so θ and ϕ are isomorphisms.

REFERENCES

1. S. Bouc, *Modules de Möbius*, C.R. Acad. Sci. Paris Sér. I **299** (1984), 9–12.

2. ——, *Homologie de certains ensembles ordonnés*, C.R. Acad. Sci. Paris Sér. I **299** (1984), 49–52.

3. G. E. Bredon, *Equivariant cohomology theories*, Lecture Notes in Math., vol. 34, Springer, Berlin, 1967.

4. K. S. Brown, *Cohomology of groups*, Graduate Texts in Math., vol. 87, Springer, New York, 1982.

5. W. Burnside, *Theory of groups of finite order*, 2nd ed., Cambridge Univ. Press, Cambridge, 1911.

6. N. J. Kuhn and S. A. Mitchell, *The multiplicity of the Steinberg representation of* $\mathrm{GL}_n\mathrm{F}_q$ *in the symmetric algebra*, Preprint.

7. D. Quillen, *Homotopy properties of the poset of nontrivial p-subgroups of a group*, Advances in Math. **28** (1978), 101–128.

8. P. J. Webb, *A local method in group cohomology*, Comment. Math. Helv. (to appear).

9. ——, *Complexes, group cohomology and an induction theorem for the Green ring*, J. Algebra (to appear).

10. ——, *A split exact sequence of cohomology groups* (in preparation).

11. K. S. Brown and J. Thévenaz, *A generalization of Sylow's third theorem* (in preparation).

12. J. Thévenaz, *Permutation representations arising from simplicial complexes*, J. Combin. Theory Ser. A (to appear).

THE UNIVERSITY, MANCHESTER, ENGLAND

II. BLOCK THEORY AND MODULAR REPRESENTATIONS

Proceedings of Symposia in Pure Mathematics
Volume **47** (1987)

Weights for Finite Groups

J. L. ALPERIN

1. Introduction. The theory of finite groups of Lie type has been greatly enriched by its connections and analogies with Lie groups. In particular, there is a well-developed theory of weights [**Cu**] for the modular representations of Lie type groups in the natural characteristic. We shall extend the idea of a weight to representations of arbitrary finite groups in arbitrary characteristic, formulate a sweeping conjecture about these weights, show that it implies a number of known results, hitherto unrelated, and important open conjectures, and describe how it holds in many diverse classes of groups. We fix a finite group G, a prime p, and an algebraically closed field k of characteristic p. Our notation is standard [**A**].

We are now ready to give the key definition. A *weight* for G is a pair (Q, S) where Q is a p-subgroup and S is a simple module for $N(Q)$ which is projective when regarded as a module for $N(Q)/Q$. If we replace S by an isomorphic module we consider this the same weight and we make the same identification when we replace Q by a conjugate subgroup. Since S is simple and Q is a normal p-subgroup of $N(Q)$ it follows that Q acts trivially on S so the definition makes sense. It is also closely related, as we shall see, to the definition of weights in the Lie situation. The main point of this paper is the following:

CONJECTURE. *The number of weights for G equals the number of simple kG-modules.*

Of course, we mean the number of isomorphism classes of simple modules. We also are not suggesting that a natural one-to-one correspondence holds; the evidence, in fact, suggests that this is not the case, in contrast to the situation for Lie type groups. Before proceeding, we pause to record one useful fact which sheds some light on this new concept. Let P be a Sylow p-subgroup of G and let $k[G/P]$ denote the permutation module of G on the cosets of P.

LEMMA 1. *If (Q, S) is a weight then S has vertex Q and the Green correspondent of S is a summand of $k[G/P]$.*

1980 *Mathematics Subject Classification* (1985 *Revision*). Primary 20C20.
Supported in part by National Science Foundation Grant DMS 8421367.

PROOF. Since S is projective as a $k[N(Q)/Q]$-module, it is a summand of the free module $k[N(Q)/Q]$, so is a summand of the $kN(Q)$-module induced from the trivial kQ-module and therefore has vertex contained in Q. On the other hand, the restriction of S to Q is trivial and each indecomposable summand of this restriction is relatively projective for a conjugate of the vertex of S, so the first assertion is valid.

Since S has vertex Q, it is a consequence of the Burry-Carlson-Puig theorem [**BC**] that in order to show the Green correspondent of S is a summand of $k[G/P]$ it suffices to prove that S is a summand of the restriction of this permutation module to $N(Q)$. However, without loss of generality we may assume that P contains Q and that $P \cap N(Q) = R$ is a Sylow p-subgroup of $N(Q) = L$. Since the module $k[G/P]$ is just the kG-module induced from the trivial kP-module, it follows from Mackey's theorem that the restriction of $k[G/P]$ to L has the permutation module $k[L/R]$ of L on the cosets of R as a summand. However, this is just the permutation module M of L/Q on the cosets of its Sylow p-subgroup R/Q regarded as a kL-module since Q is contained in R and normal in L. Since S is a projective module for L/Q, by hypothesis, it therefore is enough to show that S is a homomorphic image of M in order to show that it is a summand of M. But S is simple so we need only prove that $\mathrm{Hom}_{k[L/Q]}(M, S) \neq 0$. However, $\mathrm{Hom}_{k[L/Q]}(M, S) \cong \mathrm{Hom}_{k[R/Q]}(k, S)$, which is certainly not zero as R/Q is a p-group.

Let us now assume that G is a finite group of Lie type and characteristic p; we shall show how to establish quickly that the conjecture holds in this case and determine the exact relation between weights and Lie weights. This was first proved by the author for certain classical groups but the general result is due to Cabanes [**Ca**]. The lemma shows that the number of weights is at most the number of indecomposable summands (up to isomorphism) of the permutation module $k[G/P]$. If V is a simple kG-module then since G is of Lie type we have [**Cu**] that the space of fixed points of the restriction of V to P is of dimension one; hence, $\mathrm{Hom}_{kP}(k, V) \cong k$. Therefore, $\mathrm{Hom}_{kG}(k[G/P], V) \cong k$ and so the quotient of $k[G/P]$ by its radical is the direct sum of all the simple kG-modules each occurring exactly once. In particular, the number of indecomposable summands of $k[G/P]$ is at most the number of simple kG-modules. However, to each simple module V is associated a weight, which is a pair, one component of which is a one-dimensional module for the Borel subgroup $N(P)$. This is associated [**Cu**, p. B38]to a one-dimensional module of a parabolic subgroup containing $N(P)$ which in turn determines the module V. Hence, we now have that the number of weights is at most the total number of one-dimensional modules for parabolic subgroups containing $N(P)$. However, suppose that H is such a subgroup so that $H/O_p(H)$ is also a group of Lie type and has a Steinberg module St_H, which is a simple module for H and projective as a module for $H/O_p(H)$. Hence, the tensor product of St_H with any one-dimensional module for H has the same properties and so is a weight for G. Moreover, since the character of the Steinberg module does not vanish on elements of order coprime to p, distinct one-dimensional modules give distinct weights using this tensor product construction. In particular, the number of simple modules is now at most the number of weights, the reverse of what we have already proved and so the claim is demonstrated. Moreover, the equality we have just shown implies that all the inequalities we used were also equalities and so we have proved, in this case, that the weights, the indecomposable summands of $k[G/P]$, and the one-dimensional modules for parabolic subgroups containing $N(P)$ are all in natural one-to-one correspondences.

The conjecture has also been established for a number of types of groups. It holds for solvable groups by results of Okuyama [**O**]. In a sequel to this paper we shall prove that it holds for the symmetric groups; it is interesting that the proof is an elaborate determination and count of the weights and that the result coincides, without any apparent direct connection, with the known results for the number of simple modules. This, in itself, is quite convincing evidence for the truth of the conjecture. Fong has also proved, using the results for the symmetric groups, that the conjecture holds for the groups $\mathrm{GL}(n,q)$ for primes p that do not divide q, a result which complements the above argument on groups of Lie type in this case.

Anytime one studies modular representations it is reasonable to expect that blocks play a central role. This is certainly the case here as we shall show how to give a stronger, block-theoretic, form of the conjecture. If (Q,S) is a weight then S belongs to a block of $N(Q)$ and this block corresponds with a block of G via the Brauer correspondence; hence, we can say that the weight belongs to that block of G so the weights are partitioned into blocks.

CONJECTURE (BLOCK FORM). *The number of weights in a block of G equals the number of simple modules in the block.*

This is also true for group of Lie type, as is implicit in Cabanes's work [**Ca**], for solvable groups [**O**], for symmetric groups, and in Fong's case. It is also clear that this stronger form of the conjecture implies the original one, as it can be obtained by summing the equalities from the stronger conjecture over the blocks.

2. Synthesis. This section is devoted to showing how the conjectures on weights imply new results, old theorems, and outstanding conjectures; a great deal of material can thus be unified. Hence, we shall assume henceforth in this section that these new conjectures do hold and we shall proceed with a number of consequences. We keep the previous notation and let B be a block of G and D a defect group of B. We let b be the Brauer correspondent of B, the block of $N(D)$ which corresponds to B under the First Main Theorem of block theory. We let $l(B)$ be the number of simple modules in B and define $l(b)$ similarly.

CONSEQUENCE 1. $l(B) \geq l(b)$.

To establish this result, let S be a simple module for $N(D)$ lying in b; it suffices to show that (D,S) is a weight. But S has a vertex contained in D

as b has defect group D so S is relatively D-projective. But the restriction to D is trivial so S is a summand of the $N(D)$-module induced from the trivial kD-module and is projective as a module for $N(D)/D$.

The next result is known to be true for the case of cyclic defect groups as a result of the deep theory of such blocks. However, the statement is much more general and has been suspected for a long time.

CONSEQUENCE 2. *If D is abelian then $l(B) = l(b)$.*

Let (Q, S) be a weight in B; by the above argument it suffices to prove that Q is a defect group of B as we have just proved that the number of weights in B with p-subgroup a defect group is $l(b)$. Certainly, since (Q, S) is in B we have that Q is contained in a defect group of B and, in particular, Q is abelian. Let T be an indecomposable summand of the restriction $S_{C(Q)}$; hence, T is simple and projective as a module for $C(Q)/Q$. Since Q is central in $C(Q)$ and T is the simple module in a block of $C(Q)/Q$ of defect zero we have that the block b_Q of $C(Q)$ containing T has defect group Q. Hence, it follows that $(b_Q)^{N(Q)}$ is the block of $N(Q)$ containing S and since this block corresponds to B we have that $(b_Q)^G = B$. Thus, (Q, b_Q) is a B-subgroup [**A**, **AB**] of the type studied by Brauer [**B**] and so, by Olsson's theorem [**B**] Q must contain the center of a defect group of B, which, in our case, implies that Q is a defect group, as claimed.

Part of the argument just used generalizes immediately to our arbitrary block B and it is important to record this as we shall employ it subsequently. If (Q, b_Q) is a B-subgroup and (Q, S) is a weight in B, with S in $b_Q^{N(Q)}$, then Q is the defect group of b_Q; that is, (Q, b_Q) is a Brauer B-subgroup. Moreover, if (Q, S) is a weight then the B-subgroup (Q, b_Q) used here is determined up to conjugacy in $N(Q)$ since b_Q is a block of $QC(Q)$ covered by the block of $N(Q)$ to which S belongs.

Another case where we can reach the same conclusion is the next result. We say that the block B is a *TI block* provided that every B-subgroup (Q, b_Q) is contained in a unique Sylow B-subgroup.

CONSEQUENCE 3. *If B is a TI block then $l(B) = l(b)$.*

Indeed, suppose that Q is a p-subgroup properly contained in a defect group of B and that there is a weight in B of the form (Q, S). Let b_Q be a block of $QC(Q)$ such that the block of $N(Q)$ containing S is $(b_Q)^{N(Q)}$. Hence (Q, b_Q) is a B-subgroup properly contained in the Sylow B-subgroup (D, b_D) for some choice of defect group D. Hence, if $R = N_D(Q, b_Q)$, then Q is a proper subgroup of R. We now have $(Q, b_Q) \leq (D, b_D)$ so if $x \in N(Q, b_Q)$ then we have that $(Q, b_Q) \leq (D^x, b_D^x)$ and therefore $D = D^x$ and $b_D = b_D^x$. Hence, $R = R^x$, by the definition of R, and $O_p(N(Q, b_Q))$ properly contains Q. Thus, $N(Q, b_Q)$ does not have a simple module which is projective modulo Q as it would be trivial and projective on the quotient $O_p(N(Q, b_Q)/Q$. We assert that this implies that $N(Q)$ does not have a simple module in $(b_Q)^{N(Q)}$ which is projective modulo Q, contradicting the existence of S, and so proving the result.

In order to prove the assertion we shall give a more general result, which holds without any special assumptions, and which provides the necessary connection between $N(Q)$ and $N(Q, b_Q)$.

LEMMA 2. *If (Q, b_Q) is a B-subgroup of G, $H = N(Q, b_Q)$ and $L = N(Q)$ then the categories of kH-modules in $(b_Q)^H$ and kL-modules in $(b_Q)^L$ are equivalent and an equivalence is given by induction.*

This is a typical result of Clifford theory; we know of a very general result of Broué that has this as a special case. However, in this case the argument of Lemma 19.6 of [**A**] readily applies here.

An important concept is that of a nilpotent block [**BP**]; here we can derive one of the key properties of such blocks.

CONSEQUENCE 4. *If B is a nilpotent block then $l(B) = l(b)$.*

Suppose that (Q, S) is a weight in B. Let b_Q be a block of $QC(Q)$ such that S lies in $(b_Q)^{N(Q)}$. Hence, by Lemma 2, there is a simple module T for $H = N(Q, b_Q)$ lying in $(b_Q)^H$ which is projective as a module for H/Q and such that $S \cong T^{N(Q)}$. Let U be the unique simple module in b_Q. We can now apply Clifford theory to the group H and normal subgroup $QC(Q)$. Since U is stable under conjugation by elements of H, as $H = N(Q, b_Q)$, and $H/QC(Q)$ is a p-group, by the nilpotence assumption, there is a unique extension V of U to a kH-module and this is the only simple kH-module whose restriction to $QC(Q)$ is a direct sum of modules isomorphic with U. However, T certainly has this property, as T is simple and T lies in $(b_Q)^H$, so T is this extension. Hence, we have U^H, the induced module, isomorphic with $(T_{QC(Q)})^H \cong T \otimes k^H$. But, since $H/QC(Q)$ is a p-group, every composition factor of k^H is trivial. Moreover, T is projective as a module for H/Q so we must have that U^H is isomorphic with the direct sum of $|H\colon QC(Q)|$ copies of T. However, we have

$$\begin{aligned}\mathrm{Hom}_{QC(Q)}(U, U) &\cong \mathrm{Hom}_{QC(Q)}(U, T_{QC(Q)})\\ &\cong \mathrm{Hom}_{kH}(U^H, T)\\ &\cong \mathrm{Hom}_{kH}(T \oplus \cdots \oplus T, T)\end{aligned}$$

so, by counting dimensions, we deduce that $|H\colon QC(Q)| = 1$. We claim that this implies that Q must be a defect group of B and so we also have that $S \cong U^{N(Q)}$ and B will have a single weight, proving the result.

However, suppose (Q, b_Q) is a proper B-subgroup of the Sylow B-subgroup (D, b_D). Therefore, Q is a proper subgroup of a defect group of b_Q^H. Hence, this defect group intersects $QC(Q)$ in the defect group Q of b_Q (e.g., Theorem 15.1 of [**A**]) so H properly contains $QC(Q)$ and we must have that Q is a defect group as claimed.

The arguments above, in conjunction with Lemma 2, give a practical method for determining the number of weights in B. First, the number of conjugacy classes of B-subgroups (Q, b_Q) must be determined. Second, for each of these B-subgroups, the number of simple modules in b_Q^H, $H = N(Q, b_Q)$, which are projective as modules for $H/QC(Q)$ is to be calculated. Finally, the sum of these numbers equals the number of weights in B. Moreover, if $H/QC(Q)$ is a p-group then the contribution to the total is zero, while if $(Q, b_Q) = (D, b_D)$ then the contribution is the number of simple modules in b_D^N, $N = N(D, b_D)$.

Let us apply these rules in one easy case: let $p = 2$ and assume that D is dihedral of order eight. Let E and F be the two four subgroups of D; let (E, b_E) and (F, b_F) be B-subgroups contained in the Sylow B-subgroup (D, b_D). Since D is a defect group of B, the quotient $|N(D, b_D)\colon DC(D)|$ is of odd order and so it must be one as D has no nonidentity automorphisms of odd order. Moreover, E and F are the only proper subgroups of D which have nonidentity automorphisms of odd order while (E, b_E) and (F, b_F) are not conjugate, since the theorem on local fusion applies to B-subgroups [**AB**] and we may analyze fusion of B-subgroups exactly as we do for a group with a dihedral Sylow 2-subgroup. Hence, the number of weights in B is one plus the contributions from (E, b_E) and (F, b_F). However, if $H = N(E, b_E)$ then $|H\colon EC(E)|$ is divisible by two, since the intersection of D and H must properly contain E, as (E, b_E) is not a Sylow B-subgroup. Hence, this inertial index is either two or six, and in the latter case the quotient $H/EC(E)$ is isomorphic with the symmetric group of degree three, in which case, by an easy argument from Clifford theory, it is easy to see that there is exactly one simple module to be counted, the tensor product of a two-dimensional module for $H/EC(E)$ with an extension of the unique simple module in b_E to H. The same argument applies equally to F so we deduce that the number of weights is one, two, or three according as three does not divide the inertial indices, three divides exactly one of them, and three divides both. This, then, gives the number of simple modules in B in agreement with the known results.

Using these ideas, we can now put several of our previous results together. We say that B is a *controlled block* provided that whenever (Q, b_Q) is a B-subgroup contained in the Sylow B-subgroup (D, b_D) and $g \in G$ with (Q^g, b_Q^g) also contained in (D, b_D) then $g = cn$ where $c \in C(Q)$ and $n \in N(D, b_D)$.

CONSEQUENCE 5. *If B is a controlled block then $l(B) = l(b)$.*

Let (Q, b_Q) be a B-subgroup properly contained in the Sylow B-subgroup (D, b_D). Let $L = N(Q, b_Q)$, $N = N(D, b_D)$, and let (R, b_R) be the normalizer B-subgroup [**AB**] of (Q, b_Q) in (D, b_D); in particular, $R = D \cap L$ and

$$(Q, b_Q) \leq (R, b_R) \leq (D, b_D).$$

Since Q is properly contained in D we have $RC(Q)$ properly containing $QC(Q)$: this is the argument in the second paragraph of the proof of Consequence 4. Moreover, in our case we also have that $RC(Q)$ is *normal* in L: the hypothesis of control yields that $L \leq C(Q)N$ so $L = C(Q)(N \cap L)$ and R is normal in $N \cap L$ since this intersection normalizes (Q, b_Q) and D. Let U be the unique simple module in b_Q and suppose that T is a simple module in b_Q^L which is projective as a module for L/Q; we need only derive a contradiction. However, let W be a simple summand of $T_{RC(Q)}$ so W is projective as a module for $RC(Q)/Q$ and lies in $b_Q^{RC(Q)}$ since $W_{QC(Q)}$ is also the direct sum of modules isomorphic with U. But $|RC(Q) \colon QC(Q)|$ is not one, so the argument proving Consequence 4 applies to give the desired contradiction.

The last result has a partial converse.

PROPOSITION 1. *If B is the principal block $l(B) = l(b)$ and every section of G which has a strongly p-embedded subgroup has a block of defect zero, then B is a controlled block.*

In this case, D is a Sylow p-subgroup of G and, by the identification of B-subgroups, for B the principal block, with p-subgroups [**AB**], B is a controlled block if, and only if, $N(D)$ strongly controls fusion in D. Hence, suppose that $N(D)$ does not have this property; we need only show the existence of a weight in B that comes from a proper subgroup of D, using the existence of blocks of defect zero that we are given. However, by Puig's thesis [**P**], there is an essential subgroup; that is, there is a proper nonidentity subgroup Q of P such that Q is a Sylow p-subgroup of $QC(Q)$ and $N(Q)/QC(Q)$ has a strongly p-embedded subgroup. Let S be a simple module for this quotient which is in a block of defect zero, that is, is a projective module. We claim that S lies in the principal block of $N(Q)$ so that (Q, S) is a weight in the principal block B. Indeed, the restriction of S to $QC(Q)$ is trivial so is in the principal block of $QC(Q)$ and so S is as stated.

Such a converse does not hold for nonprincipal blocks; we shall sketch the structure of an example which shows this. Let G be the semidirect product of an elementary abelian p-group E of order p^3, $p > 2$, by $\mathrm{SL}(2,p)$ with irreducible action. Thus, kG has simple modules of dimension $1, 2, \ldots, p$. Every defect group must contain E while the centralizer of E is the direct product of E and the center of $\mathrm{SL}(2,p)$, a subgroup of order two, so the First Main Theorem implies that E is not a defect group. Similarly, we find that there are exactly two blocks of full defect, the principal block B_0 and one other B_1. The odd-dimensional simple modules, with the exception of the p-dimensional one, lie in the principal block of $G/E \cong \mathrm{SL}(2,p)$, so are in B_0. Now B_0 is not a controlled block so, by the proposition, there must be more simple modules in B_0. However, the even-dimensional simple modules are the simples in a block of $\mathrm{SL}(2,p)$, their restriction to $C(E)$ are not in the principal block of $C(E)$, so they must lie in B_1. Hence, B_0 contains the p-dimensional simple module and so we must have $l(B_1) = (p-1)/2$. However, the block B_1 is also not a controlled block and now $l(B_1)$ coincides with $l(b_1)$ the number of simple modules in its Brauer correspondent b_1.

We shall now turn to characters and results about them that follow from our conjectures. Let $\kappa(B)$ and $\kappa(b)$ be the numbers of irreducible characters in B and b, respectively.

CONSEQUENCE 6. *If D is abelian then $\kappa(B) = \kappa(b)$.*

Let (D, b_D) be a Sylow B-subgroup of G and X a set of representatives for the conjugacy classes in $N = N(D, b_D)$ of elements of D. From the results of §4 of [**AB**] we know that for each x in X there is a block b_x of $C(x)$ with defect group D such that $(x, b_x) \in (D, b_D)$ and that all these Brauer elements are representatives of the conjugacy classes of Brauer B-elements of G. Similarly, there is a block β_x of $N \cap C(x)$ with defect group D such that $(x, \beta_x) \in (D, b_D)$ (where here (D, b_D) is a b-subgroup) and all these Brauer elements are representatives of the conjugacy classes of Brauer b-elements in N. Since $(x, b_x) \in (D, b_D)$ we have $b_D^{C(x)} = b_x$ while $(x, \beta_x) \in (D, b_D)$ yields $b_D^{N\cap C(x)} = \beta_x$. Therefore, by transitivity of the Brauer map, $\beta_x^{C(x)} = b_D^{C(x)} = b_x$. But b_x and β_x have D as a defect group so β_x is the Brauer correspondent of b_x. By Consequence 2 we have that $l(B) = l(b)$. Two applications of the Second Main Theorem of block theory now give us, summing over X,

$$\kappa(B) = \sum l(b_x) = \sum l(\beta_x) = \kappa(b),$$

as required.

Notice that if it is true, as is conjectured, that the number of characters of height zero in B equals the number of characters of height zero in b (which is conjectured without assuming that D is abelian) then this last result implies that each character in B is of height zero since it is known that each character in b is of height zero. Hence, half of the famous conjecture of Brauer on characters of height zero would then follow.

CONSEQUENCE 7. *If B is the principal block and B is controlled then $\kappa(B) = \kappa(b)$.*

The identification of B-subgroups, for B the principal block, with p-subgroups means that this result is equivalent with the following statement: if P is a Sylow p-subgroup of G and if $N(P)$ strongly controls fusion in P then $\kappa(B) = \kappa(b)$, where B is the principal block. We first assert that the hypothesis implies that in each p-local subgroup of G or in each centralizer of a p-element of G the normalizer of a Sylow p-subgroup strongly controls fusion [**P**]; that is, the control property is hereditary in this sense. Indeed, let Q be a nonidentity p-subgroup of G and set $L = N(Q)$. Without any loss of generality we may assume that $Q \leq P$ and that $Np(Q) = S$ is a Sylow p-subgroup of L. Let T be a nonidentity subgroup of S; we have to show that $N_L(T) = C_L(T)N_L(S)$. First, suppose that $Q \leq T$. If $x \in N_L(T)$ then $x = cn$, where $c \in C(T)$ and $n \in N(P)$, by hypothesis, so

$$n = c^{-1}x \in C(T)L \leq C(Q)L \leq L$$

so n normalizes Q as well as P. Therefore, n normalizes S, so $x = cn \in C(T)N_L(S)$ as desired. Next, if T is an arbitrary subgroup of S then we have, as TQ contains Q,

$$N_L(T) \leq N_L(TQ) \leq C_L(TQ)N_L(S) \leq C_L(T)N_L(S),$$

and the assertion is proved for local subgroups. The proof for centralizers is virtually the same.

For any subgroup H of G let $l_0(H)$ be the number of simple modules in the principal block of H. Hence, by the hypothesis and the Second Main Theorem of block theory, we need only prove that $l_0(C(x) = l_0(C(x) \cap N(P))$ for every element x of P. We set $P_1 = C_p(x)$ and define, recursively, $P_{i+1} = N_P(P_i)$ for $i \geq 1$. Since x has a conjugate lying in P which is well-placed [**AG**] in P while $N(P)$ controls fusion, it follows that P_1 is a Sylow p-subgroup of $C = C(x)$ and P_{i+1} is a Sylow p-subgroup of $N_i = N(P_i)$, $i \geq 1$. Since P_1 is a Sylow p-subgroup of C, $N_C(P_1) = N_1 \cap C$ strongly controls fusion in P_1 so, by Consequence 5, we have $l_0(C) = l_0(C \cap N_1)$. We assert that if $i \geq 1$ then $l_0(C \cap N_i) = l_0(C \cap N_{i+1})$. Indeed, $N_i \cap N_{i+1} = N(P_{i+1}) \cap N_i$ is the normalizer in N_i of a Sylow p-subgroup of N_i so $l_0(C \cap N_i) = l_0(C \cap N_i \cap N_{i+1})$. But $C \cap N_{i+1} = C \cap N_i \cap N_{i+1}$ as $C \cap N_{i+1}$ normalizes $\{x\}$ and P_{i+1} and so will normalize all the P_j, $1 \leq j \leq i$. However, for i very large we have $P_i = P$ and $N_i = N(P)$ so it follows that $l_0(C) = l_0(C \cap N)$, as required.

3. Observations. Parabolic subgroups are a prominent feature of the groups of Lie type. In the local analysis of simple groups, local subgroups that appear very much similar to parabolic subgroups commonly arise in a variety of situations. Here, in representation theory, certain local subgroups have been picked out, the subgroups $N(Q)$ where (Q, S) is a weight and $Q \neq 1$. These various subgroups do, in fact, have a number of properties in common.

To see what our subgroups look like, let us start with B now equal to the principal block, since the B-subgroups and p-subgroups can be identified. Suppose that (Q, S) is a weight in the principal block B and $Q \neq 1$. Therefore, S is in the principal block of $N(Q)$, by the Third Main Theorem of block theory. We assert that Q is a Sylow p-subgroup of $O_{p',p}(L)$. First, as the principal block of L covers only the principal block of the normal subgroup $QC(Q)$, we have that Q is a defect group of the principal block of $QC(Q)$; that is, Q is a Sylow p-subgroup of $QC(Q)$ so $QC(Q) = QO_{p'}(L)$. Moreover, as in the argument for Consequence 5, we have $O_p(L/QC(Q)) = 1$, that is, $O_p(L/QO_{p'}(L)) = 1$ and our claim is valid. Thus, L has the properties that characterize the parabolic subgroups of groups of Lie type and that occur in subgroups studied in simple groups. It makes very good sense, therefore, to define a *p-parabolic subgroup* of G to be a subgroup L such that $L = N(Q)$, where Q is a nonidentity p-subgroup such that Q is a Sylow p-subgroup of $O_{p',p}(L)$ and $C_L(Q) \leq O_{p',p}(L)$.

However, letting B be an arbitrary block once more, we can extend this definition to blocks in a way similar to the way B-subgroups generalize p-subgroups. If (Q, b_Q) is Brauer B-subgroup of G such that $O_p(L/QC(Q)) = 1$ then we call L a *B-parabolic subgroup* of G. In particular, if (Q, S) is a weight in B, b_Q is a block of $QC(Q)$ such that $b_Q^{N(Q)}$ is the block of $N(Q)$ containing S then $N(Q, b_Q)$ is a B-parabolic subgroup, by arguments of the last section. In the case of the principal block, this definition coincides with the previous one, as is easy to see from our discussion.

Turning in another direction, let us consider our conjectures from the point of view of the problem to which much of Brauer's work on block theory was devoted: describe the values of the characters of G on the elements of order divisible by p in terms of the p-local subgroups of G. The first step is the result that characters in block of defect zero vanish on such elements. Hence, we are reduced to looking at the submatrix of the character table corresponding to the characters in blocks of positive defect and the columns corresponding to elements of order divisible by p. The columns can be described in terms of local subgroups: this is the author's theorem on local fusion of some twenty years ago. The conjectures we have made here tell us how to determine the number of characters in a block locally since the Second Main Theorem reduces this to determining the number of simple modules in blocks of G and centralizers in G of elements of order divisible by p. Hence, these conjectures would answer Brauer's problem as far as determining the size of the matrix under examination. This leaves a great deal more to do.

Finally, we turn to the thorny question of proving the conjectures. At this point it seems the right question to ask is what sort of structures and objects are even pertinent. We saw at the very beginning that the Green correspondents of the weights were indecomposable summands of the permutation module $k[G/P]$ of G on the cosets of a Sylow p-subgroup P of G. The isomorphism classes of indecomposable summands of $k[G/P]$ correspond one-to-one with the isomorphism classes of projective indecomposable modules for the Hecke algebra of G with respect to P [**S**], the endomorphism algebra of the module $K[G/P]$. This is an algebra which has arisen before and it may be that it should be further studied. Another direction that seems relevant is the simplicial complex of Brown, studied by Quillen and others, which has been used by G. Robinson [**R**] to give a quite different looking reformulation of our conjectures. This complex arises from the study of the action of G on chains of p-subgroups, so again we are dealing with permutation modules in another guise. Finally, the conjecture on characters of height zero in blocks and the conjectures we have made today may be special cases of even more sweeping conjectures; if we had much more general ideas then we might have further clues as to the relevant material to be employed in the desired proofs.

REFERENCES

[**A**] J. L. Alperin, *Local representation theory*, Cambridge University Press, 1987.

[**AB**] J. L. Alperin and M. Broué, *Local methods in block theory*, Ann. Math. (2) **110** (1979), 143–157.

[**AG**] J. L. Alperin and D. Gorenstein, *Transfer and fusion in finite groups*, J. Algebra **6** (1967), 242–255.

[**B**] R. Brauer, *On the structure of blocks of characters of finite groups*, Lecture Notes in Math., vol. 372, Springer-Verlag, 1970, pp. 103–130.

[**BC**] D. Burry and J. F. Carlson, *Restrictions of modules to local subgroups*, Proc. Amer. Math. Soc. **84** (1982), 181–184.

[**BP**] M. Broué and L. Puig, *A Frobenius theorem for blocks*, Invent. Math. **56** (1980), 117–128.

[**Ca**] M. Cabanes, *Brauer morphisms between Hecke algebras*, Preprint.

[**Cu**] C. W. Curtis, *Modular representations of finite groups with split* (B,N)*-pairs*, Lecture Notes in Math., vol. 131, Springer, Verlag, 1970, pp. B1–B39.

[**O**] T. Okuyama, *Module correspondences in finite groups*, Hokkaido Math. J. **10** (1981), 299–318.

[**P**] L. Puig, *Structure locale dans les groupes finites*, Bull. Soc. Math. France Mem. **4** (1976), 1–132.

[**R**] G. Robinson, *A remark on a conjecture of Alperin*, Preprint.

[**S**] L. L. Scott, *Modular permutation representations*, Trans. Amer. Math. Soc. **175** (1973), 101–121.

UNIVERSITY OF CHICAGO

Proceedings of Symposia in Pure Mathematics
Volume **47** (1987)

Some Remarks on the Decomposition Numbers for the Symmetric Groups

DAVE BENSON

1. Introduction. One of the main techniques for analyzing the irreducible representations of finite Chevalley groups in their own characteristic is the sum formula of Jantzen [**14**–**16**]. Recently, Schaper, a student of Jantzen, wrote a thesis [**20**] in which he develops a similar formula for the symmetric groups in arbitrary characteristic. The main purpose of this paper is to present some calculations which I have been performing using this formula in characteristic two and to formulate some conjectures and questions based on the results.

Two particular outcomes of these calculations are worth pointing out in this introduction. One is that it seems to be worth pursuing the 2-modular reductions of the spin representations of the symmetric groups to obtain information about questions not easily answered by other methods. Some explicit conjectures appear in §4. The other outcome worth mentioning concerns an old conjecture of Brauer. The conjecture states that the power of p dividing the degree of a p-modular irreducible character should be at most the power of p dividing the group order [**3**, p. 166]. The first known counterexample to this conjecture appeared in J. Thackray's thesis [**22**] in 1981; namely he used a digital computer to construct an irreducible 2-modular character of MacLaughlin's simple group of degree $7 . 2^9$, whereas the order of the group is only divisible by 2^7. We present here the second known counterexample, a 2-modular irreducible character of $\mathfrak{S}_{15}$ of degree $4096 = 2^{12}$ (corresponding to the partition $(7, 6, 2)$), whereas the order of $\mathfrak{S}_{15}$ is only divisible by 2^{11}. One advantage of our counterexample is that the calculations can be checked (and were made) entirely by hand (some think this is a disadvantage!). It seems plausible that the symmetric groups should provide further counterexamples to the conjecture. In particular, the characters coming from 2-modular reductions of spin representations are all divisible by high powers of two, and the character of degree 4096 may be written in terms of spin

1980 *Mathematics Subject Classification* (1985 *Revision*). Primary: 20C30.
This paper is in final form, and no version of it will be submitted for publication elsewhere.

characters as $\langle\overline{13,2}\rangle - \langle\overline{14,1}\rangle$ (i.e., $\langle\overline{14,1}\rangle$ is equal to the 2-modular irreducible $D^{(8,6,1)}$ of degree 832, while $\langle\overline{13,2}\rangle = D^{(8,6,1)} + D^{(7,6,2)}$).

2. Schaper's thesis. Since Schaper's version of the Jantzen sum formula is not readily available in the literature (I have not seen his thesis myself; I simply guessed the formula and was informed by several sources that it was to be found there!), I think it is worth saying a few words to indicate the nature of this formula.

Let $\lambda = (\lambda_1, \dots, \lambda_s)$ be a partition of n (written $\lambda \vdash n$), i.e., $\lambda_1 \geq \lambda_2 \geq \cdots \geq \lambda_s > 0$ and $\sum_{i=1}^{s} \lambda_i = n$. Denote by S^λ the Specht module [**7**, p. 13], defined over the integers, corresponding to λ. This is given as a submodule of the permutation module M^λ on the cosets of the Young subgroup $\mathfrak{S}_{\lambda_1} \times \cdots \times \mathfrak{S}_{\lambda_s}$ corresponding to λ. The natural symmetric bilinear form on M^λ given by making the permutation basis orthonormal (i.e., length one and inner product zero) gives rise to a nonsingular bilinear form $\langle\ ,\ \rangle$ on S^λ.

Let p be a prime number, and denote by $\overline{S^\lambda}$ the reduction modulo p of S^λ. Following Jantzen [**14–16**], we define a filtration on $\overline{S^\lambda}$ as follows. Denote by $S^\lambda(r)$ the submodule of S^λ consisting of those elements $x \in S^\lambda$ with the property that for all $y \in S^\lambda$, $\langle x, y\rangle$ is divisible by p^r. Denote by $\overline{S^\lambda(r)}$ the image of $S^\lambda(r)$ in $\overline{S^\lambda}$ under reduction modulo p. This gives a filtration

$$\overline{S^\lambda} = \overline{S^\lambda(0)} \geq \overline{S^\lambda(1)} \geq \overline{S^\lambda(2)} \geq \cdots$$

with $\overline{S^\lambda(r)} = 0$ for sufficiently large r. Note that by [**7**, Theorems 4.9 and 11.1], the module $D^\lambda = \overline{S^\lambda}/\overline{S^\lambda(1)}$ is always zero or irreducible, and is zero if and only if λ is p-singular (i.e., has a part repeated p or more times). Every irreducible module for the symmetric group over $\mathbf{F}_p$ arises exactly once this way and is absolutely irreducible. Schaper's formula gives the value of the sum of the Brauer characters $\chi(\overline{S^\lambda(r)})$ of the modules in the above sequence in terms of things we can compute.

Now recall from James and Kerber [**12**, p. 77 ff.] the definition and elementary properties of β-numbers, and their relationship to hooks and skew-hooks (= rim-hooks). In particular, recall that if $\lambda = (\lambda_1, \dots, \lambda_s)$ is a partition, then any sequence of the form $(\lambda_1 - 1 + t, \lambda_2 - 2 + t, \dots, \lambda_s - s + t, -s - 1 + t, \dots, -t + t)$ $(t \geq s)$ is a sequence of β-numbers corresponding to λ. If $(\beta_1, \dots, \beta_t)$ is a sequence of integers, we define $\chi(\beta_1, \dots, \beta_t)$ to be zero if two of the β_i are equal or if any of the β_i are negative, and to equal plus or minus the ordinary character of the representation of the symmetric group corresponding to the partition whose β-numbers are $\{\beta_1, \dots, \beta_t\}$ otherwise. The sign is equal to the signature of the permutation π of $\{1, 2, \dots, t\}$ for which $\beta_{\pi(1)} > \beta_{\pi(2)} > \cdots > \beta_{\pi(t)}$.

If (a, b) is a node in the Young diagram corresponding to λ, we denote by h^λ_{ab} its hook length [**12**, p. 56]. We denote by ν_p the p-adic valuation on integers, namely $\nu_p(p^\alpha g) = \alpha$ if $p \nmid q$.

We may now state Schaper's formula.

THEOREM (SCHAPER [20]). *The Brauer character* $\sum_{r>0}\chi(\overline{S^\lambda(r)})$ *agrees on p-regular conjugacy classes with the ordinary character*

$$\sum_{1\le a\le b\le s}\sum_{c=1}^{\lambda_b}(\nu_p(h^\lambda_{ac})-\nu_p(h^\lambda_{bc}))$$
$$\cdot\chi(h^\lambda_{11},h^\lambda_{21},\dots,h^\lambda_{a-1,1},h^\lambda_{a1}+h^\lambda_{bc},h^\lambda_{a+1,1},\dots,h^\lambda_{b1}-h^\lambda_{bc},\dots,h^\lambda_{s1}).$$

Note that the formula obtained by taking dimensions on both sides of Schaper's formula may be found in James and Murphy [13], where some examples and consequences are also given.

The way the formula is used is as follows. If we know the decomposition numbers for partitions dominated by a given partition (in the sense of James [7, p. 8]), then the formula tells us whether or not the decomposition numbers for the given partition are zero and also gives an upper bound. This may be applied to p-singular as well as p-regular partitions, and indeed often more delicate information is obtained from the p-singular ones. This corresponds to using Jantzen's formula outside the restricted fundamental region.

REMARK. One can show that $\overline{S^\lambda(r)}/\overline{S^\lambda(r+1)}$ is a self-dual module (in Jantzen's case the corresponding subquotient supports a "contravariant form"). However, these subquotients are not always semisimple. For example if $p=2$, $\lambda=(4,1^2)$, $n=6$ then $\overline{S^\lambda(2)}/\overline{S^\lambda(3)}$ is a uniserial module with composition factors $D^{(6)}$, $D^{(4,2)}$, $D^{(6)}$.

3. Decomposition numbers for $\mathfrak{S}_{14}$ and $\mathfrak{S}_{15}$ mod 2**.** Using Schaper's formula together with the following techniques, we established the decomposition matrices given in the appendix. An explanation of these tables may be found in §5.

(i) *Restriction and induction.* As well as the obvious information given by the fact that restriction and induction coefficients must be nonnegative, we also use the information given by the Frobenius reciprocity statement

$$\dim_k \operatorname{Hom}_{kG}(M,N\uparrow^G)=\dim_k\operatorname{Hom}_{kH}(M\downarrow_H,N).$$

This, together with the self-duality of the irreducible modules, can often be used to eliminate possibilities. The idea is that the restriction of a large irreducible $\mathfrak{S}_n$-module to $\mathfrak{S}_{n-1}$ cannot have small modules in the socle, since this would imply that the induced module from the small module would map onto the large module. The same technique may be used with the roles of restriction and induction reversed.

(ii) *Spin representations.* Schur [21] and Morris [18, 19] have investigated the ordinary characters of the proper double covers of $\mathfrak{S}_n$ (there are two isoclinic proper double covers). Of course, these must be expressible as positive linear combinations of the 2-modular irreducible characters of $\mathfrak{S}_n$. This often gives strong information about partitions not easily accessible by other means. See also the conjectures in §4.

(iii) *First row and first column removal.* James [**9**] has shown that the coefficients of $D^{(\lambda_1,\dots,\lambda_s)}$ in $\overline{S}^{(\lambda_1,\mu_2,\dots,\mu_t)}$ is the same as the coefficient of $D^{(\lambda_2,\dots,\lambda_s)}$ in $\overline{S}^{(\mu_2,\dots,\mu_t)}$. This may be seen as saying that the first rows of the partitions may be removed if they are equal. He also shows that the same is true for first column removal. In fact, Donkin [**2**] has recently found a pleasing generalization of this to the removal of arbitrary rectangles in the upper left-hand corner of the partition.

(iv) We have also made use of James's results [**6**] on partitions of the form (λ_1,λ_2) and $(\lambda_1,\lambda_2,1)$.

(v) Occasionally in desperation, we use inner tensor products and the associated reciprocity laws. Thus, for example, the last question to resolve for the decomposition matrix of $\mathfrak{S}_{14}$ mod 2 was: "How many copies of $D^{(13,1)}$ are there in $\overline{S}^{(7,4,2,1)}$?" It is easy to show that the answer is $5+\gamma$ with $\gamma = 0, 1$, or 2. But then we see that $D^{(13,1)} \otimes D^{(8,4,2)}$ has composition factors $D^{(7,4,3)} + D^{(7,4,2,1)} + \gamma \,.\, D^{(13,1)}$. If $\gamma \neq 0$, then by self-duality there is a nonzero homomorphism $D^{(13,1)} \to D^{(13,1)} \otimes D^{(8,4,2)}$ and hence there is a nonzero homomorphism $D^{(13,1)} \otimes D^{(13,1)} \to D^{(8,4,2)}$. This is absurd since $\dim D^{(13,1)} = 12$ and $\dim D^{(8,4,2)} = 2510$.

4. Some conjectures and questions. In some sense, the most important reason for performing extensive calculations is in order to see general patterns and make sensible conjectures. I do not claim that all of the following conjectures are sensible. Some are much more likely to be right than others.

For the purpose of this section, we only work in characteristic two. Most of our conjectures are either meaningless or false in odd characteristic, although some may have analogues.

First, we treat the subject of restriction and induction. The following conjecture seems quite likely to be true.

CONJECTURE 1. Let $\lambda = (\lambda_1,\dots,\lambda_s) \vdash n$. Then $D^\lambda \downarrow_{\mathfrak{S}_{n-1}}$ is irreducible if and only if all the λ_i are congruent modulo two. In this case the restriction is $D^{(\lambda_1-1,\lambda_2,\dots,\lambda_s)}$.

Thus, for example, the module $D^{(7,6,2)}$ for $\mathfrak{S}_{15}$, of dimension 4096, is the restriction of the module $D^{(8,6,2)}$ for $\mathfrak{S}_{16}$, also of dimension 4096.

Reducing to a minimal counterexample by the method of first column removal (note that first column removal holds for restriction coefficients, while first row removal doesn't), and by using the induction-restriction techniques described in §3, one can see that Conjecture 1 follows from the following conjecture.

CONJECTURE 2. Let $n = \lambda_1 + \cdots + \lambda_s + 1$. Suppose $\lambda_1,\dots,\lambda_s$ are all congruent modulo two. Then $D^{(\lambda_1,\dots,\lambda_s)}$ appears in the socle of $D^{(\lambda_1,\dots,\lambda_s,1)} \downarrow_{\mathfrak{S}_{n-1}}$ if and only if the λ_i are even.

There are more complicated conjectures akin to Conjecture 1. For example, we may conjecture the following.

CONJECTURE 3. Let $\lambda = (\lambda_1, \dots, \lambda_s) \vdash n$. Suppose $\lambda_1 \equiv \lambda_2 \equiv \cdots \equiv \lambda_{2r} \not\equiv \lambda_{2r+1} \equiv \cdots \equiv \lambda_s \pmod 2$ for some value of r. Then

$$D^\lambda \downarrow_{\mathfrak{S}_{n-1}} \cong D^{(\lambda_1 - 1, \lambda_2, \dots, \lambda_s)} \oplus D^{(\lambda_1, \dots, \lambda_{2r+1} - 1, \dots, \lambda_s)}.$$

Or more boldly

CONJECTURE 4. Let $\lambda = (\lambda_1, \dots, \lambda_s) \vdash n$. Then $D^\lambda \downarrow_{\mathfrak{S}_{n-1}}$ is completely reducible if and only if there are values $0 < r_1 < \cdots < r_k$ (possibly $k = 0$) such that $\lambda_j \equiv \lambda_{j+1} \pmod 2$ unless and only unless $j \in \{2r_1, \dots, 2r_k\}$. In this case, we have

$$\begin{aligned} D^\lambda \downarrow_{\mathfrak{S}_{n-1}} &\cong D^{(\lambda_1 - 1, \dots, \lambda_s)} \oplus D^{(\lambda_1, \dots, \lambda_{2r_1+1} - 1, \dots, \lambda_s)} \\ &\quad \oplus \cdots \oplus D^{(\lambda_1, \dots, \lambda_{2r_k+1} - 1, \dots, \lambda_s)}. \end{aligned}$$

Thus, for example, according to this conjecture

$$D^{(7,5,4,2,1)} \downarrow_{\mathfrak{S}_{18}} \cong D^{(6,5,4,2,1)} \oplus D^{(7,5,3,2,1)} \oplus D^{(7,5,4,2)}.$$

On the other hand, for induction we make the following conjecture.

CONJECTURE 5. Let $\lambda = (\lambda_1, \dots, \lambda_s) \vdash n$. Then $D^\lambda \uparrow^{\mathfrak{S}_{n+1}}$ is completely reducible if and only if all the λ_i are even. In this case,

$$D^\lambda \uparrow^{\mathfrak{S}_{n+1}} \cong D^{(\lambda_1, \dots, \lambda_s + 1)} \oplus D^{(\lambda_1, \dots, \lambda_s, 1)}.$$

We may form analogous conjectures for the extended decomposition matrices of James [**8**]. It is shown in James [**10**] that the extended character table of $\mathfrak{S}_n$ is in duality with the table of indecomposable direct summands of permutation modules on Young subgroups. It follows that if we restrict or induce an extended modular character of $\mathfrak{S}_n$, we get a positive integral combination of extended modular characters of $\mathfrak{S}_{n-1}$ and $\mathfrak{S}_{n+1}$ respectively. The analogue of Conjecture 1, for example, comes out as follows.

CONJECTURE 6. Let $\lambda = (\lambda_1, \dots, \lambda_s) \vdash n$. If we restrict the irreducible extended modular character of $\mathfrak{S}_n$ corresponding to λ to $\mathfrak{S}_{n-1}$, then the result is

(a) zero if and only if every part of λ occurs with even multiplicity,

(b) an irreducible extended modular character of $\mathfrak{S}_{n-1}$ if and only if the parts of λ occurring with odd multiplicity are congruent modulo two.

In this case, the restriction is the character corresponding to the partition obtained by decreasing by one the last occurrence of the first part to appear with odd multiplicity.

Thus, for example, according to this conjecture, the restriction of the character corresponding to $(8^2 7^2 6^3 4\, 3^2 2)$ is the character corresponding to $(8^2 7^2 6^2 5\, 4\, 3^2 2)$.

Let us turn now to the spin representations. The basic references here are Schur [**21**] and Morris [**18, 19**]. Let Γ_n be one of the two isoclinic proper double covers of $\mathfrak{S}_n$. Then corresponding to a partition $\lambda = (\lambda_1, \dots, \lambda_s)$ of n into s unequal parts, we have either one faithful irreducible representation of Γ_n if $(n - s)$ is even, or two algebraically conjugate irreducible representations of Γ_n if $(n - s)$ is odd. In the latter case, the characters only differ on elements of even order, and so the reductions modulo two will have the same composition factors. Thus we shall not distinguish them for our present purposes, and in each case we shall write $\langle\lambda\rangle = \langle\lambda_1, \ldots, \lambda_s\rangle$ for the character of Γ_n corresponding to $(\lambda_1, \ldots, \lambda_s)$.

We now describe a certain doubling process on partitions, in terms of which our conjectures are formulated. This is a process by which each part λ_i is replaced by two parts to make a new partition. If $\lambda_i = 2k+1$ is odd, then we replace λ_i by parts of lengths $k+1$ and k, while if $\lambda_i = 2k$ is even, then we replace λ_i by parts of lengths $k+1$ and $k-1$. We call a partition "spin regular" if the resulting parts form a partition of n into distinct parts in decreasing order. Thus, for example, $\mathrm{dbl}\langle 11, 6, 1\rangle = (6, 5, 4, 2, 1)$, while $\langle 11, 6, 2\rangle$ is not spin regular because $(6, 5, 4, 2, 2)$ are not distinct, and $\langle 11, 6, 4\rangle$ is not spin regular because $(6, 5, 4, 2, 3, 1)$ are not in decreasing order.

Now if $\langle\lambda_1, \ldots, \lambda_s\rangle$ is a representation of Γ_n, then each composition factor of $\overline{\langle\lambda_1, \ldots, \lambda_s\rangle}$, its reduction modulo two, has the center of Γ_n in its kernel and is hence a representation of $\mathfrak{S}_n$. Thus the composition factors are of the form D^λ. For example, for the basic spin character $\langle n\rangle$, we have $\overline{\langle n\rangle} = D^{\mathrm{dbl}\langle n\rangle}$. The following two theorems were stated as conjectures in the first draft of this paper. Proofs will appear elsewhere.

THEOREM 7. *Let $\langle\lambda\rangle = \langle\lambda_1, \ldots, \lambda_s\rangle$ be a spin regular partition of n. Then the composition factors of $\overline{\langle\lambda_1, \ldots, \lambda_s\rangle}$ consist of $D^{\mathrm{dbl}\langle\lambda\rangle}$ with multiplicity a power of two, and possibly some other composition factors, each of which is of the form D^μ with* $\mathrm{dbl}\langle\lambda\rangle \rhd \mu$ *(for notation see James* [**7**, *p.* 8]*).*

THEOREM 8. *The module D^λ for $\mathfrak{S}_n$ splits on restriction to $\mathfrak{A}_n$ if and only if λ is of the form* $\mathrm{dbl}\langle\mu\rangle$, *with μ a spin regular partition with no part congruent to* 2 (mod 4). *Otherwise it remains irreducible.*

Thus for example, for $n = 17$, the modules which split are $D^{(9,8)}$, $D^{(9,7,1)}$, $D^{(7,6,3,1)}$, $D^{(7,5,3,2)}$, and $D^{(6,5,3,2,1)}$, corresponding to the doubles of $\langle 17\rangle$, $\langle 16, 1\rangle$, $\langle 13, 4\rangle$, $\langle 12, 5\rangle$, and $\langle 11, 5, 1\rangle$.

It is probably slightly more tricky to see exactly what happens, say, for two-part partitions. In order to make our conjecture, we first define some functions.

We define quarter infinite matrices of types V, VI, and VII, analogous to those of types I–IV in James [**6**], as follows.

$$
\begin{array}{ccccccccccc|}
1 &&&&&&&&&& \\
1 & 1 &&&&&&&&& \\
&& 1 &&&&&&&& \\
&& 1 & 1 &&&&&&& \\
1 & 1 &&& 1 &&&&&& \\
1 & 1 &&& 1 & 1 &&&&& \\
&& 1 & 1 &&& 1 &&&& \\
&& 1 & 1 &&& 1 & 1 &&& \\
1 & 1 &&& 1 & 1 &&& 1 && \\
1 & 1 &&& 1 & 1 &&& 1 & 1 & \\
&& 1 & 1 &&& 1 & 1 &&& 1 \\
\hline
\end{array}
$$

type V

$$
\begin{array}{ccccccccccc|}
1 & & & & & & & & & & \\
 & 1 & & & & & & & & & \\
 & & 1 & & & & & & & & \\
 & 1 & & 1 & & & & & & & \\
1 & & & & 1 & & & & & & \\
 & 1 & & 1 & & 1 & & & & & \\
 & & 1 & & & & 1 & & & & \\
 & 1 & & 1 & & 1 & & 1 & & & \\
1 & & & & 1 & & & & 1 & & \\
 & 1 & & 1 & & 1 & & 1 & & 1 & \\
 & & 1 & & & & 1 & & & & 1 \\
\hline
\end{array}
$$

type VI

$$
\begin{array}{ccccccccccc|}
1 & & & & & & & & & & \\
2 & 1 & & & & & & & & & \\
1 & 1 & 1 & & & & & & & & \\
 & & 2 & 1 & & & & & & & \\
1 & 1 & 1 & 1 & 1 & & & & & & \\
2 & 2 & & & 2 & 1 & & & & & \\
1 & 1 & 1 & 1 & 1 & 1 & 1 & & & & \\
 & & 2 & 2 & & & 2 & 1 & & & \\
1 & 1 & 1 & 1 & 1 & 1 & 1 & 1 & 1 & & \\
2 & 2 & & & 2 & 2 & & & 2 & 1 & \\
1 & 1 & 1 & 1 & 1 & 1 & 1 & 1 & 1 & 1 & 1 \\
\hline
\end{array}
$$

type VII

The matrix of type V is formed by tiling $\mathbf{N}^2$ with alternate blocks of four zeros and four ones, and then cutting off everything above the diagonal. The matrix of type VI is formed by filling every fourth leading and trailing diagonal of $\mathbf{N}^2$ with ones, and then cutting off everything above the diagonal. The matrix of type VII is formed from the matrix of type V by adding to each entry the entry immediately above it. A $k \times k$ type V, VI, or VII matrix is given by the intersection of the bottom k rows and the right-hand k columns of the corresponding infinite matrix.

We then define functions ε, ς, and η analogous to James' α, β, γ, and δ as follows. $\varepsilon(x, m, j)$ is the jth entry in the mth row of an x by x type V matrix. $\varsigma(x, m, j)$ and $\eta(x, m, j)$ are defined in the same way, replacing "type V" by "type VI" or "type VII", respectively.

Thus we have

$$\varepsilon(x,m,j) = \begin{cases} 1 & \text{if } 0 < j \leq m \leq x \text{ and either } 4 \mid (m-j) \\ & \qquad\qquad\qquad \text{or } 4 \mid (2x+1-m-j), \\ 0 & \text{otherwise;} \end{cases}$$

$$\varsigma(x,m,j) = \begin{cases} 1 & \text{if } 0 < j \leq m \leq x \text{ and either } 4 \mid (m-j) \\ & \qquad\qquad\qquad \text{or } 4 \mid (2x-m-j), \\ 0 & \text{otherwise;} \end{cases}$$

$$\eta(x,m,j) = \begin{cases} \varepsilon(x,m,j) + \varepsilon(x,m-1,j) & \text{if } 0 < j \leq m \leq x, \\ 0 & \text{otherwise.} \end{cases}$$

CONJECTURE 9. Let $\lambda = \langle \lambda_1, \lambda_2 \rangle$ be a spin regular partition of n into two parts. Then the composition factors of $\overline{\langle \lambda_1, \lambda_2 \rangle}$ are all of the form $D^{\mathrm{dbl}\langle \mu \rangle}$ with $\lambda \trianglerighteq \mu = \langle \mu_1, \mu_2 \rangle$ (note that this is not true for partitions into more than two parts). The multiplicity of $D^{\mathrm{dbl}\langle \mu \rangle}$ in $\overline{\langle \lambda_1, \lambda_2 \rangle}$ is given as follows.

(a) If $n \equiv 1 \pmod 2$ then the multiplicity is $\varepsilon(\frac{n-3}{2}, \lambda_2 + 1, \mu_2 + 1)$.

(b) If $n \equiv 2 \pmod 4$ then the multiplicity is

$$\begin{cases} \varsigma\left(\dfrac{n-2}{2}, \lambda_2 + 1, \mu_2 + 1\right) & \text{if } \lambda_2 = 0 \text{ or } \lambda_2 \text{ is odd,} \\ 2\varsigma\left(\dfrac{n-2}{2}, \lambda_2 + 1, \mu_2 + 1\right) & \text{otherwise.} \end{cases}$$

(c) If $n \equiv 0 \pmod 4$ then the multiplicity is

$$\begin{cases} \eta\left(\dfrac{n-4}{2}, \lambda_2 + 1, \mu_2 + 1\right) & \text{if } \mu_2 = 0 \text{ or } \mu_2 \text{ is odd,} \\ 2\eta\left(\dfrac{n-4}{2}, \lambda_2 + 1, \mu_2 + 1\right) & \text{otherwise.} \end{cases}$$

We now turn to the possibility of finding further formulas like those of Jantzen and Schaper. From the work of Fong and Srinivasan [**4**], one knows that the block structure of the classical Chevalley groups away from the natural characteristic may be described in a way strongly analogous to the Nakayama conjecture for the symmetric groups. James [**11**] has carried this analogy further for the unipotent representations of the general linear groups. Thus we may consider the problem of writing down a formula analogous to the Jantzen-Schaper formulas, for representations of general linear groups away from the natural characteristic. One should probably restrict one's attention to the unipotent representations, which are the ones that behave the most like Specht modules.

CONJECTURE 10. There is an analogue of the Jantzen-Schaper formulae for representations of $\mathrm{GL}_n(q)$ in characteristic $p \nmid q$.

In [**17**], Lusztig writes down a conjecture for the decomposition numbers of Weyl modules in terms of irreducible modules for semisimple algebraic groups in characteristic p. The coefficients are given in terms of the Kazhdan-Lusztig polynomials $P_{y,w}$, evaluated at 1. One should be able to write down a corresponding conjecture for the decomposition numbers of the symmetric groups,

			(14) 1	(13,1) 12	(12,2) 64	(11,3) 208	(10,4) 364	(9,5) 560	(8,6) 64	(11,2,1) 560	(10,3,1) 1300	(9,4,1) 3808	(8,5,1) 4576	(7,6,1) 768	(9,3,2) 3418	(8,4,2) 2510	(7,5,2) 2016	(7,4,3) 10880	(6,5,3) 4704	(8,3,2,1) 23296	(7,4,2,1) 19240	(6,5,2,1) 13312	(6,4,3,1) 11648	(5,4,3,2) 35840
1	(14)	(1^{14})	1																					
13	(131)	(21^{12})	1	1																				
77	(122)	$(2^2 1^{10})$	1	1	1																			
273	(113)	$(2^3 1^8)$	1		1	1																		
637	(104)	$(2^4 1^6)$	1		1	1	1																	
1001	(95)	$(2^5 1^4)$	1	1	1		1	1																
1001	(86)	$(2^6 1^2)$	1	1			1	1	1															
560	(1121)	(321^9)								1														
2002	(1031)	$(32^2 1^7)$	2		2	1	1				1													
4368	(941)	$(32^3 1^5)$								1		1												
6006	(851)	$(32^4 1^3)$	2	1	1		2	1	1				1											
4576	(761)	$(32^5 1)$										1		1										
5733	(932)	$(3^2 21^6)$	3	2	1		1	1			1				1									
12012	(842)	$(3^2 2^2 1^4)$	4	2	2		2	1	1				1		1	1								
14014	(752)	$(3^2 2^3 1^2)$	2	1	1		2	1	2				1		1	1	1							
16016	(743)	$(3^3 21^3)$								1		1		1				1						
15015	(653)	$(3^3 2^2 1)$	3	1	1		1	1	1		1				1	1	1		1					
23296	(8321)	(4321^5)																		1				
42042	(7421)	$(432^2 1^3)$	6	5	3	1	4	4	3				1		2	2	1				1			
36608	(6521)	$(432^3 1)$																		1		1		
50050	(6431)	$(43^2 21^2)$	8	4	4	2	2	2	2		1				1	2	1		1		1		1	
48048	(5432)	$(4^2 321)$								1				1				1						1

The decomposition matrix of $\mathfrak{S}_{14}$ for the prime 2.

and maybe even, following the above ideas, for the decomposition numbers for the classical Chevalley groups away from the natural characteristic. This would give us a good hold on all decomposition numbers for all simple groups modulo all primes. Of course, the sporadic groups would have to be dealt with separately, either on an ad hoc basis using a lot of computer work, or using some sort of geometric theory like that of Smith and Ronan.

5. Explanation of the tables. Our format for decomposition matrices of $\mathfrak{S}_n$ is the same as in James [**7**], except that since the rows corresponding to p-regular partitions determine the rest by a simple algorithm, I only provide these rows of the decomposition matrices. Complete decomposition matrices would take up too much space.

In the decomposition matrix for $\mathfrak{S}_{15}$ mod 2, I have not been able to eliminate one last ambiguity, and so the answers are given in terms of a parameter α, which takes on the value zero or one. It may well be that an extensive check of inner tensor products and symmetrizations will resolve this ambiguity.

			(15) 1	(14,1) 14	(13,2) 90	(12,3) 336	(11,4) 910	(10,5) 1288	(9,6) 1912	(8,7) 128	(1221) 624	(1131) 1300	(1041) 4172	(951) 4576	(861) 832	(1032) 7448	(942) 9346	(852) 32032	(762) 4096	(843) 13390	(753) 4704	(654) 21152	(9321) 31276	(8421) 24260	(7521) 17344	(7431) 76828+2α	(6531) 23296	(6432) 59136	(54321) 292864
1	(15)	(1^{15})	1																										
14	(141)	(21^{13})		1																									
90	(132)	(2^21^{11})			1																								
350	(123)	(2^31^9)		1		1																							
910	(114)	(2^41^7)					1																						
1638	(105)	(2^51^5)		1		1		1																					
2002	(96)	(2^61^3)			1				1																				
1430	(87)	(2^71)		1				1		1																			
715	(1221)	(321^{10})	1		1						1																		
2835	(1131)	(32^21^8)	1				1				1	1																	
7007	(1041)	(32^31^6)	1				1				1	1	1																
11375	(951)	(32^41^4)	1		1				1		1		1	1															
11583	(861)	(32^51^2)	1		1				1				1	1	1														
9100	(1032)	(3^221^7)		2		1		1								1													
22113	(942)	$(3^22^21^5)$	3		2				1		1	1	1	1			1												
32032	(852)	$(3^22^31^3)$																1											
25025	(762)	(3^22^41)	1		1				1				1	1	1		1		1										
35035	(843)	(3^321^4)	3		2				1		1		1	1	1		1			1									
45045	(753)	$(3^32^21^2)$	3		1				1		1	1	1	1	1		1		1	1	1								
30030	(654)	(3^421)		1				1		1						1						1							
42042	(9321)	(4321^6)		5		2		2								1							1						
91000	(8421)	(432^21^4)		4		2		2		1								1					1	1					
108108	(7521)	(432^31^2)		2		1		2		2								1					1	1	1				
135135	(7431)	(43^221^3)	7+2α		3		2		2		1	1	1	1	1		2		1	1	1					1			
128700	(6531)	(43^22^21)		6		3		2		2						1						1	1	1	1		1		
175175	(6432)	(4^2321)	7+2α		3		3		1		1	1			1		1		1	1	1					1		1	
292864	*	(54321)																											1
		Block number:	1	2	1	2	1	2	1	2	1	1	1	1	1	2	1	2	1	1	1	2	2	2	2	1	2	1	3

α = 0 or 1

The decomposition matrix of $\mathfrak{S}_{15}$ for the prime 2.

Γ_4		(4)	(31)
		1	2
2	<4>		1
4	<31>	2	1

Γ_5		(5)	(41)	(32)
		1	4	4
4	<5>			1
6	<41>	2		1
4	<32>		1	

Γ_6		(6)	(51)	(42)	(321)
		1	4	4	16
4	<6>			1	
16	<51>				1
20	<42>	4	2	2	
4	<321>		1		

Γ_7		(7)	(61)	(52)	(43)	(421)
		1	6	14	8	20
8	<7>				1	
20	<61>					1
36	<52>	2		1		1
20	<43>		2		1	
28	<421>			1		

Γ_8		(8)	(71)	(62)	(53)	(521)	(431)
		1	6	14	8	64	40
8	<8>				1		
48	<71>				1		1
112	<62>	4		2			2
112	<53>	4	4	2	2		1
64	<521>					1	
48	<431>		2	2	1		

Γ_9		(9)	(81)	(72)	(63)	(54)	(621)	(531)	(432)
		1	8	26	48	16	78	40	160
16	<9>					1			
56	<81>					1		1	
160	<72>								1
224	<63>		2		1				1
112	<54>	4		2		1		1	
240	<621>	4					2	2	
336	<531>	4		4		2	2	1	
96	<432>				2				

Γ_{10}		(10)	(91)	(82)	(73)	(64)	(721)	(631)	(541)	(532)	(4321)
		1	8	26	48	16	160	198	128	200	768
16	<10>					1					
128	<91>								1		
432	<82>					2				2	
768	<73>										1
672	<64>	8	4	4	2	2				2	
400	<721>	2						1		1	
800	<631>	4	2	4	1	2		2		1	
448	<541>						2		1		
432	<532>	2		4	2	2		1			
96	<4321>				2						

Γ_{11}		(11)	(101)	(92)	(83)	(74)	(65)	(821)	(731)	(641)	(632)	(542)	(5321)
		1	10	44	100	164	32	186	198	144	848	416	1168
32	<11>						1						
144	<101>									1			
560	<92>									1		1	
1200	<83>						1						1
1440	<74>		4		2		1						1
672	<65>	4		2		1						1	
1232	<821>	4							2			2	
3168	<731>		4		2		2				2		1
3168	<641>	8		4		2		4	4	2		2	
2464	<632>		4		6		4				2		
1760	<542>	4				2		4	2	2			
528	<5321>	2				2			1				

Reductions of spin modules modulo two.

Γ_{12}		1 (12)	10 (111)	44 (102)	100 (93)	164 (84)	32 (75)	320 (921)	570 (831)	1408 (741)	288 (651)	1046 (732)	416 (642)	1792 (543)	5632 (6321)	2368 (5421)
32	<12>						1									
320	<111>						1				1					
1408	<102>										2		2			
3520	<93>						1				1		2			1
5280	<84>		8		4		2									2
4224	<75>	8	8	4	4	2	1						2			1
1792	<921>													1		
5600	<831>	4	4		2		2					2	2			1
7776	<741>	8	8	4	4	2	2		2		1	2	2			1
3840	<651>							2		1				1		
5632	<732>														1	
7392	<642>	12	4	4	6	4	4		4		2	2	2			
1760	<543>	4				2			2		1					
3520	<6321>	4	4		6	4	4					2				
2816	<5421>									2						

Γ_{13}		1 (13)	12 (121)	64 (112)	208 (103)	364 (94)	560 (85)	64 (76)	364 (1021)	570 (931)	1572 (841)	288 (751)	2848 (832)	2510 (742)	1728 (652)	2208 (643)	8008 (7321)	3200 (6421)	8448 (5431)
64	<13>							1											
352	<121>							1				1							
1728	<112>														1				
4928	<103>														1			1	
8800	<94>							1				1							1
9504	<85>	8		4		2		1											1
4224	<76>		4		2		1											1	
4992	<1021>											2				2			
18304	<931>	4						2				1		2		2			1
32032	<841>	16		8		4		4		4		2		4					2
27456	<751>	20		8		4		2	4	4	2	2		2		2			1
22464	<832>		4														2	2	
41600	<742>		12		4		6						4		2		2	2	
22464	<652>	20		4		6		4	4	8	2	4		2		2			
18304	<643>		8				6						4		2				
9152	<7321>		2				2										1		
13728	<6421>	12		4		6		4		4	2	2		2					
4576	<5431>	4								2	2	1							

Reductions of spin modules modulo two (cont.)

Γ_{14}		(14) 1	(131) 12	(122) 64	(113) 208	(104) 364	(95) 560	(86) 64	(1121) 560	(1031) 1300	(941) 3808	(851) 4576	(761) 768	(932) 3418	(842) 2510	(752) 2016	(743) 10880	(653) 4704	(8321) 23296	(7421) 19240	(6521) 13312	(6431) 11648	(5432) 35840
64	<14>							1															
768	<131>												1										
4160	<122>							2								2							
13312	<113>																				1		
27456	<104>							2								2						2	
36608	<95>												1										1
27456	<86>	16	8	8	4	4	2	2														2	
6720	<1121>															1		1					
28224	<1031>	4						2							2	1		2				1	
59136	<941>												2				2						1
68992	<851>								4		2		2				2						1
31680	<761>	12	4	8	2	4	1	2		2		1			2			1				1	
40768	<932>	4	2					2							2			1		1		1	
96096	<842>	16	12	8	4	4	6	4						4	4	2				2		2	
91520	<752>	24	10	12	4	10	6	6		4		2		4	4	2		2		1		1	
59904	<743>																		2		1		
40768	<653>	12	8	4		6	6	4		2		1		4	2	2		1					
40768	<8321>		4				4													2			
87360	<7421>	16	12	8	4	12	10	8				2		4	4	2				2			
49918	<6521>								4		6		4				2						
54912	<6431>	16	8	8		12	6	8				4		4	4	2							
9152	<5432>											2											

Γ_{15}		(15) 1	(141) 14	(132) 90	(123) 336	(114) 910	(105) 1288	(96) 1912	(87) 128	(1221) 624	(1131) 1300	(1041) 4172	(951) 4576	(861) 832	(1032) 7448	(942) 9346	(852) 32032	(762) 4096	(843) 13390	(753) 4704	(654) 21152	(9321) 31276	(8421) 24260	(7521) 17344	(7431) 76828 + 2α	(6531) 23296	(6432) 59136	(54321) 292864
128	<15>								1																			
832	<141>													1														
4928	<132>													1				1										
17472	<123>								1															1				
40768	<114>								1															1		1		
64064	<105>													1				1									1	
64064	<96>	4		4		2		1						1													1	
27456	<87>		8		4		2		1																	1		
17600	<1221>																	2		2								
83200	<1131>								2												2			1		1		
202176	<1041>	8												4				2	4	4							2	
292864	<951>																											1
228800	<861>	16		8		4		2		8	4	4	2	4					4	2							2	
137984	<1032>		4						4												2		2			2		
392000	<942>	16-4α		8		4		2						4		4		2	4	2					2		2	
256608	<852>	32 4α		16		8		8		8	8	4	4	4		8		4	4	4					2		2	
246400	<762>		20		12		8		4						4		2				2		2			2		
312000	<843>		16		8		4		2													4	4	2		2		
384384	<753>		20		8		10		6						4		2				2	4	2	2		1		
81536	<654>	8		8				6			4		2			4		2		2								
81536	<9321>	4-2α																		1					1			
224224	<8421>	8-4α		8		4		6					2			4		2							2			
228800	<7521>	16-2α		8		4		6		4	4	6	4	4		4		2	2	2					1			
202176	<7431>		8		4		6		4								2					2	2	1				
145598	<6531>	12		8				6		4	2	6	5	4		4		2	2	1								
64064	<6432>																2											
9152	<54321>												2															

Reductions of spin modules modulo two (cont.)

REFERENCES

1. R. W. Carter and G. Lusztig, *On the modular representations of the general linear and symmetric groups*, Math. Z. **136** (1974), 193–242.

2. S. Donkin, *A note on decomposition numbers for general linear groups and symmetric groups*, Math. Proc. Cambridge Philos. Soc. **97** (1985), 57–62.

3. W. Feit, *The representation theory of finite groups*, North Holland, Amsterdam, 1982.

4. P. Fong and B. Srinivasan, *The blocks of finite general linear and unitary groups*, Invent. Math. **69** (1982), 109–153.

5. J. A. Green, *Polynomial representations of* GL_n, Lecture Notes in Math., No. 830, Springer-Verlag, Berlin, 1980.

6. G. D. James, *Representations of the symmetric group over the field of order* 2, J. Algebra **38** (1976), 280–308.

7. ——, *The representation theory of the symmetric groups*, Lecture Notes in Math., No. 682, Springer-Verlag, Berlin, 1978.

8. ——, *The decomposition of tensors over fields of prime characteristic*, Math. Z. **172** (1980), 161–178.

9. ——, *On the decomposition matrices of the symmetric groups.* III, J. Algebra **71** (1981), 115–122.

10. ——, *Trivial source modules for symmetric groups*, Arch. Math. (Basel) **41** (4) (1983), 294–300.

11. ——, *Representations of general linear groups*, London Math. Soc. Lecture Note Ser., No. 94, Cambridge University Press, Cambridge, 1984.

12. G. D. James and A. Kerber, *The representation theory of the symmetric group*, Encyclopaedia of Mathematics, Vol. 16, Addison-Wesley, Reading, Mass., 1981.

13. G. D. James and G. E. Murphy, *The determinant of the Gram matrix for a Specht module*, J. Algebra **59** (1979), 222–235.

14. J. C. Jantzen, *Darstellungen halbeinfacher Gruppen und Kontravariante Formen*, J. Reine Angew. Math. **290** (1977), 117–141.

15. ——, *Kontravariante Formen auf induzierten Darstellungen halbeinfacher Lie-Algebren*, Math. Ann. **226** (1977), 53–65.

16. ——, *Weyl modules for groups of Lie type*, Finite Simple Groups II, M. Collins, ed., Academic Press, New York, 1980.

17. G. Lusztig, *Some problems in the representation theory of finite Chevalley groups*, Proc. Sympos. Pure Math. **37** (1980).

18. A. O. Morris, *The spin representation of the symmetric group*, Proc. London Math. Soc. (3) **12**, 55–76 (1962).

19. ——, *The spin representation of the symmetric group*, Canad. J. Math. **17** (1965), 543–549.

20. K. D. Schaper, *Characterformeln für Weyl-Moduln und Specht-Moduln in Primcharacteristic*, Diplomarbeit, Universität Bonn, 1981.

21. I. Schur, *Über die Darstellung der symmetrischen und der alternierenden Gruppe durch gebrochene lineare Substitutionen*, J. Reine Angew. Math. **139** (1910), 155–250.

22. J. Thackray, *Modular representations of some finite groups*, Ph.D. Thesis, Cambridge, 1981.

MATHEMATICAL INSTITUTE, ST. GILES, OXFORD

Proceedings of Symposia in Pure Mathematics
Volume 47 (1987)

On Block Induction

HARVEY I. BLAU

We study the situation where a block of a subgroup of a finite group induces to a block of the group (in the classical sense of Brauer), under no particular assumption that the subgroup contains the centralizer of a p-group. Our methods are based in part on work of Okuyama [**3**]

NOTATION. H is a subgroup of G, p is a prime. The G-conjugacy class of $g \in G$ is denoted by g^G. $\hat{S}$ is the sum (in an appropriate group ring) of the elements in a subset S of G.

K is a p-adic number field containing integers R, with residue class field $\overline{R} = R/J(R)$. Assume that K and $\overline{R}$ are splitting fields for all subgroups of G. If M is an RG-module, $\overline{M} := M/J(R)M$. Let ν be the p-adic valuation on K, scaled so that $\nu(p) = 1$. Let $\nu(|G|) = a$, $\nu(|H|) = m$, i.e., $|G|_p = p^a$, $|H|_p = p^m$.

b is a p-block of H with central character $\lambda\colon Z(\overline{R}H) \to \overline{R}$ and defect group $\delta(b)$. $\mathrm{Irr}(b) := \mathrm{Irr}(H) \cap b$. B denotes a p-block of G. The $\overline{R}$-linear function $\lambda^G\colon Z(\overline{R}G) \to \overline{R}$ is defined by $\lambda^G(\widehat{g^G}) = \lambda(\widehat{g^G \cap H})$.

If λ^G is a central character of $\overline{R}G$, Brauer defined the induced block b^G as the block of $\overline{R}G$ corresponding to λ^G ("b^G is defined" in this case).

QUESTION. When is b^G defined, and if so, what block of G is it?

Among Brauer's well-known results are the following:

If $H \geq C_G(\delta(b))$ then b^G is defined.

(THIRD MAIN THEOREM). *Assume b is a block of H with $H \geq C_G(\delta(b))$. Then b^G = principal block of $G \Leftrightarrow b$ = principal block of H.*

In order to try to understand Brauer's definition a little better, we make no assumption on H relative to $C_G(\delta(b))$ or to anything else. One well-known fact is the following:

If $\varphi \in \mathrm{Irr}(b)$ and φ^G is irreducible (or if $(1/n)\varphi^G$ is irreducible for some $n \in \mathbf{Z}^+$) then b^G is defined and $(1/n)\varphi^G \in b^G$.

1980 *Mathematics Subject Classification* (1985 *Revision*). Primary 20C20, 20C15.
The detailed version of this paper has been submitted to the *Journal of Algebra*

Some results on the general situation are found in Broué [**1**], Watanabe [**4**], and Okuyama [**2**, **3**]. The second paper of Okuyama, which provided some of our methods, contains the following result:

(0) Assume b is a block of H such that b^G is defined. Then b^G = principal block of $G \Leftrightarrow b$ = principal block of H.

If ϑ is a rational integral combination of Brauer characters of G, let $\tilde{\vartheta}$ denote the following generalized character of G:

$$\tilde{\vartheta}(g) = \begin{cases} p^a\vartheta(g) & \text{if } g \text{ is a } p'\text{-element,} \\ 0 & \text{otherwise} \end{cases}$$

Note $\tilde{\vartheta}_H = p^{a-m}(\widetilde{\vartheta_H})$.

Let $\mathcal{M}^0(B) = \{\vartheta | \vartheta$ is a rational integral combination of Brauer characters of B and $\nu(\vartheta(1)) = a - d$, where $p^d = |\delta(B)|\}$. So $\mathcal{M}^0(B)$ contains all irreducible ordinary and Brauer characters of height 0 in B. Such ϑ "control" B in the sense that

$$\chi \in \mathrm{Irr}(B) \Rightarrow \nu((\chi, \tilde{\vartheta})_G) = \nu(\chi(1)) \quad \text{for } \vartheta \in \mathcal{M}^0(B),$$
$$\chi \in \mathrm{Irr}(G) - \mathrm{Irr}(B) \Rightarrow (\chi, \tilde{\vartheta})_G = 0.$$

LEMMA A. *Suppose that b^G is defined. The following are equivalent:*
(i) $b^G = B$;
(ii) $\nu((\varphi, \widetilde{\vartheta_H})_H) = \nu(\varphi(1))$ *for some* $\varphi \in \mathrm{Irr}(b)$ *and some* $\vartheta \in \mathcal{M}^0(B)$;
(iii) $\nu((\varphi, \widetilde{\vartheta_H})_H) = \nu(\varphi(1))$ *for all* $\varphi \in \mathrm{Irr}(b)$ *and all* $\vartheta \in \mathcal{M}^0(B)$.

THEOREM B. *Suppose that b^G is defined and equals B, where $|\delta(B)| = p^d$.*
(i) *If M is any RG-module in B such that $\nu(\dim_{\overline{R}} \overline{M}) = a - d$, then some indecomposable summand of M_H belongs to b.*
(ii) *If χ is an ordinary irreducible character of height* 0 *in B, then χ_H contains an irreducible constituent ς of height* 0 *in b, with* $(\chi_H, \varsigma)_H \not\equiv 0 \pmod p$.

The following consequence of Theorem B is a slight generalization of a theorem of Okuyama [**3**]. (0) will follow from it, of course, by setting $\chi = 1_G$.

COROLLARY 1. *Let χ be an ordinary irreducible character in a block B of G, with χ of height* 0, *such that χ_H is an irreducible character in b. Let b' be a block of H such that b'^G is defined.*
(i) *If $b'^G = B$, then $b' = b$.*
(ii) *If b^G is defined, then $b^G = B \Leftrightarrow \chi_H$ is of height* 0 *in b.*

The next result describes the relation between block induction and character induction from a block of defect zero.

THEOREM C. *Suppose that b is a block of H of defect zero. Let φ be the unique character in* $\mathrm{Irr}(b)$. *Then b^G is defined $\Leftrightarrow \varphi^G = n\chi + p\vartheta$, where n is a positive integer prime to p, $\chi \in \mathrm{Irr}(B)$ where B has defect* 0, *and ϑ is either zero or some ordinary character of G.*

EXAMPLE. The following is an instance where a nonzero ϑ occurs as in Theorem C. Let $V = \mathbf{F}_3^2$, $g = \begin{pmatrix} 0 & 1 \\ -1 & 0 \end{pmatrix}$, $G = V\langle g \rangle$. Let H be a subgroup of V of order 3, $\varphi \neq 1_H \in \mathrm{Irr}(H)$, b the 2-block of φ ($p = 2$). Then b^G is defined, but φ^G is not a multiple of a single irreducible character.

The remaining results concern block induction from a normal subgroup.

THEOREM D. *Suppose that $H \triangle G$ and b is a block of H. Then b^G is defined and has defect group in H (i.e., b, b^G have the same defect) $\Leftrightarrow$ b is covered by a unique block B of G and $\delta(B) \leq H$.*

COROLLARY 2. *Suppose that $H \triangle G$, $p \nmid |G : H|$, and b is a block of H. Then b^G is defined $\Leftrightarrow$ b is covered by a unique block of G.*

COROLLARY 3. *Suppose that $H \triangle G$ and b is a block of H of defect 0. Let φ be the unique character in* $\mathrm{Irr}(b)$. *Then b^G is defined $\Leftrightarrow$ $\varphi^G = n\chi$ for some positive integer n, $\chi \in \mathrm{Irr}(G)$.*

REFERENCES

1. M. Broué, *Remarks on blocks and subgroups*, J. Algebra **51** (1978), 228–232.

2. T. Okuyama, *A note on the Brauer correspondence*, Proc. Japan Acad. Ser. A Math. Sci. **54** (1978), 27–28.

3. ——, *On blocks and subgroups*, Hokkaido Math. J. **10** (1981), 555–563.

4. A. Watanabe, *Relations between blocks of a finite group and its subgroup*, J. Algebra **78** (1982), 282–291.

NORTHERN ILLINOIS UNIVERSITY

Proceedings of Symposia in Pure Mathematics
Volume **47** (1987)

Products and Projective Resolutions

JON F. CARLSON

1. Introduction. Let G be a finite group and let K be a field of characteristic $p > 0$. In [**1**], David Benson and the author proved that any finitely generated KG-module M has a projective resolution that is the product of a finite number of periodic complexes (see also [**3**] and Theorem 2.2). The number may be taken to be the complexity of M. Such a resolution presents numerous filtrations and accompanying spectral sequences on cohomology. It is natural to ask whether such spectral sequences commute with cup product.

For example, suppose that $M = K$, the trivial KG-module, and K has a projective resolution, Z, which is the total complex of a tensor product $Z \cong X \otimes Y$ where X and Y are complexes with homology isomorphic to K in degree zero. Then the cohomology $H^*(G, K) \cong H^*(\operatorname{Hom}_{KG}(X \otimes Y, K))$ has a filtration F where

$$F^n = F^n(H^*(G,K)) = H^*\left(\operatorname{Hom}_{KG}\left(X \otimes \left(\sum_{t \geq n} Y_t\right), K\right)\right) \subseteq H^*(G,K).$$

Hence the question is whether the filtration satisfies the condition that $F^m \cdot F^n \subseteq F^{m+n}$. Recall that the cup product can be defined using a chain map $Z \to Z \otimes Z$ which lifts the identity on K. So it is easy to see that the condition would hold if there were chain maps $X \to X \otimes X$ and $Y \to Y \otimes Y$ lifting the identity.

In this paper it is shown that in some circumstances the periodic complexes constructed in [**1**] can be modified to satisfy this last-mentioned condition. Hence any filtration on cohomology determined by such complexes will respect the cup product operation. The necessary hypothesis is that the element $\varsigma \in H^n(G, K)$, which is used to define the periodic complex, must annihilate the cohomology ring of the module L_ς. In the last section we show that this hypothesis is satisfied provided p is an odd prime and the degree of ς is even. (If p is odd and the degree of ς is odd, then $\varsigma^2 = 0$ and hence it is of no interest in this context anyway.)

1980 *Mathematics Subject Classification*. Primary 20C20; Secondary 20J06.
This work was partially supported by the National Science Foundation.

Unfortunately, the proof does not work when p is 2. In fact, this situation is far more complicated as there are counterexamples to Theorem 4.1.

Throughout the paper all KG-modules will be assumed to be finitely generated. If M and N are KG-modules, then $M \otimes N$ denotes the KG-module $M \otimes_K N$ with diagonal G-action. We write $M \cong N \oplus (\text{proj})$ if M is isomorphic to the direct sum of N and a projective module. Also $[n]M$ denotes the direct sum of n copies of M.

2. Notation and preliminaries. Suppose that M is a KG-module and that P is its projective cover. Let $\Omega(M)$ denote the kernel of the surjection $P \to M$. Inductively we define $\Omega^n(M) = \Omega(\Omega^{n-1}(M))$ for $n > 1$. Similarly, if Q is an injective hull for M, then $\Omega^{-1}(M)$ is the cokernel of the injection $M \to Q$ and $\Omega^n(M) = \Omega^{-1}(\Omega^{n+1}(M))$ for $n < -1$.

Let $\cdots \to P_1 \to P_0 \to K \to 0$ be a minimal projective resolution of the trivial KG-module K. Then $\partial(P_n) \cong \Omega^n(M)$ for all n. Suppose that $\varsigma \in \operatorname{Ext}^n_{KG}(K, K) \cong H^n(G, K)$. Then ς is represented by a unique cocycle $\varsigma\colon \Omega^n(K) \to K$ whose kernel we denote by L_ς. We have the following commutative diagram

$$
\begin{array}{ccccccccccccc}
 & & 0 & & 0 & & & & & & & & \\
 & & \downarrow & & \downarrow & & & & & & & & \\
 & & L_\varsigma & & L_\varsigma & & & & & & & & \\
 & & \downarrow & & \downarrow & & & & & & & & \\
0 & \longrightarrow & \Omega^n(K) & \longrightarrow & P_{n-1} & \longrightarrow & P_{n-2} & \longrightarrow \cdots \longrightarrow & P_0 & \longrightarrow & K & \longrightarrow & 0 \\
 & & {\scriptstyle\varsigma}\downarrow & & \downarrow & & \| & & & & \| & & \| \\
D\colon 0 & \longrightarrow & K & \longrightarrow & \Omega^{-1}(L_\varsigma) & \longrightarrow & P_{n-2} & \longrightarrow \cdots \longrightarrow & P_0 & \longrightarrow & K & \longrightarrow & 0 \\
 & & \downarrow & & \downarrow & & & & & & & & \\
 & & 0 & & 0 & & & & & & & &
\end{array}
$$

where the bottom row is the pushout of the middle row. Let $C = C^1$ be the K-dual of the sequence D. That is, C is the sequence

$$0 \longrightarrow K \xrightarrow{\iota} P_0^* \xrightarrow{\partial} \cdots \xrightarrow{\partial} P_{n-2}^* \xrightarrow{\partial} U \xrightarrow{\epsilon} K \longrightarrow 0$$

where $U = (\Omega^{-1}(L_\varsigma))^*$. Let $C^m = C^1 \circ C^{m-1}$ be the splice of m copies of C. That is, the terms of C^m are given by the rules $C^m_{mn} \cong K \cong C^m_{-1}$, $C^m_{km+j} \cong P^*_{n-j-1}$ if $1 \leq j < n$, and $C^m_{km} \cong U$ for $0 \leq k < m$. The boundary homomorphism is $\iota\epsilon$ on $C^m_{km} \cong U$ and coincides with the boundary of C otherwise. Let (C^∞, ∂) denote the complex

$$(C^m, \partial)\colon \ \cdots \to 0 \to 0 \to C^m_{mn} \to \cdots \to C^m_0 \to 0 \cdots$$

and let (C^∞, ∂) be the infinite complex with $C^\infty_r = C^{r+1}_r$ for all $r \geq 0$. In **[1]** we proved the following (see also **[3]**).

THEOREM 2.1. *Suppose that $\varsigma_1, \ldots, \varsigma_r$ are homogeneous elements in $H^*(G, K)$ with the property that $V_G(\varsigma_1) \cap \cdots \cap V_G(\varsigma_r) = \{0\}$. For each $i = 1, \ldots, r$*

let $Y^{(i)}$ be the infinite periodic complex (C^∞, ∂) constructed as above from ς_i. Then the total complex of the tensor product $Y^{(1)} \otimes \cdots \otimes Y^{(r)}$ is a projective resolution of K.

To prove the theorem it is only necessary to show that the terms of the tensored complex are projective. This is a consequence of the condition on the varieties.

Let $(X, \tilde{\partial})$ be the total complex of the tensor product $(C^\infty, \partial) \otimes (C^\infty, \partial)$ where the boundary homomorphism $\tilde{\partial}$ is given by the Künneth tensor formula. Let $\epsilon\colon (C^\infty, K) \to K$ be the map which is ϵ on $C_0^\infty \cong U$ and is zero on C_j^∞ for $j > 0$. Clearly there exist surjective chain maps $1 \otimes \epsilon$ and $\epsilon \otimes 1$ from $(X, \tilde{\partial})$ to (C^∞, ∂). The purpose of this and the next section is to prove the following result.

THEOREM 2.2. *Suppose that ς annihilates $\mathrm{Ext}^*_{KG}(L_\varsigma, L_\varsigma)$, the cohomology ring. Then there exists a chain map $\mu\colon (C^\infty, \partial) \to (X, \tilde{\partial})$ consisting of KG-homomorphisms, such that the diagram*

$$\begin{array}{ccc} (C^\infty, \partial) & \xrightarrow{\ \epsilon\ } & K \\ \big\downarrow{\scriptstyle \mu} & & \big\| \\ (X, \tilde{\partial}) & \xrightarrow{\ \epsilon\otimes\epsilon\ } & K \end{array}$$

commutes and $(1 \otimes \epsilon)\mu = (\epsilon \otimes 1)\mu$ is the identity on (C^∞, ∂).

The proof is mostly given in the next section. Because $C_0 \cong U$ and $X_0 \cong U \otimes U$, the first step is the following result.

PROPOSITION 2.3. *Suppose that ς annihilates the cohomology ring $\mathrm{Ext}^*_{KG}(L_\varsigma, L_\varsigma)$. Then*

$$U \otimes U \cong U \oplus \Omega^{1-n}(U) \oplus (\mathrm{proj}).$$

Moreover, there exists a splitting homomorphism $\phi\colon U \to U \otimes U$ such that $(1 \otimes \epsilon)\phi = (\epsilon \otimes 1)\phi$ is the identity on U.

PROOF. The sequence

$$0 \longrightarrow K \xrightarrow{\ \epsilon^*\ } \Omega^{-1}(L_\varsigma) \longrightarrow \Omega^{n-1}(K) \longrightarrow 0 \tag{2.4}$$

represents the element ς in $\mathrm{Ext}^1_{KG}(\Omega^{n-1}(K), K) \cong \mathrm{Ext}^n_{KG}(K, K)$. Tensoring with L_ς we get the sequence

$$0 \to K \otimes L_\varsigma \longrightarrow \Omega^{-1}(L_\varsigma) \otimes L_\varsigma \longrightarrow \Omega^{n-1}(K) \otimes L_\varsigma \longrightarrow 0,$$

which represents $\varsigma \cdot \mathrm{cls}(\mathrm{Id}) \in \mathrm{Ext}^1_{KG}(\Omega^{n-1}(K) \otimes L_\varsigma, L_\varsigma) \cong \mathrm{Ext}^n_{KG}(L_\varsigma, L_\varsigma)$. Since, by hypothesis, this class is zero we have that $\Omega^{-1}(L_\varsigma)\otimes L_\varsigma \cong (\Omega^{n-1}(K)\otimes L_\varsigma)\oplus L_\varsigma \cong \Omega^{n-1}(L_\varsigma)\oplus L_\varsigma \oplus (\text{proj})$. It follows that

$$\begin{aligned}
U\otimes U &= \Omega^{-1}(L_\varsigma)^*\otimes \Omega^{-1}(L_\varsigma)^* \\
&\cong [\Omega^{-1}(L_\varsigma)\otimes \Omega^{-1}(L_\varsigma)]^* \\
&\cong [\Omega^{-1}(\Omega^{-1}(L_\varsigma)\otimes L_\varsigma)]^*\oplus(\text{proj}) \\
&\cong [\Omega^{-1}(\Omega^{n-1}(L_\varsigma)\oplus L_\varsigma)]^*\oplus(\text{proj}) \\
&\cong [\Omega^{n-1}(\Omega^{-1}(L_\varsigma))]^*\oplus \Omega^{-1}(L_\varsigma)^*\oplus(\text{proj}) \\
&\cong \Omega^{1-n}(\Omega^{-1}(L_\varsigma)^*)\oplus U\oplus(\text{proj}) \\
&\cong \Omega^{1-n}(U)\oplus U\oplus(\text{proj}).
\end{aligned}$$

Now consider the commutative diagram

$$\begin{array}{ccccccccc}
 & & 0 & & 0 & & 0 & & \\
 & & \downarrow & & \downarrow & & \downarrow & & \\
0 & \longrightarrow & \Omega^{1-n}(K)\otimes\Omega^{1-n}(K) & \longrightarrow & \Omega^{1-n}(K)\otimes U & \longrightarrow & \Omega^{1-n}(K) & \longrightarrow & 0 \\
 & & \downarrow & & \downarrow & & \downarrow & & \\
0 & \longrightarrow & U\otimes\Omega^{1-n}(K) & \xrightarrow{1\otimes\sigma} & U\otimes U & \xrightarrow{1\otimes\epsilon} & U & \longrightarrow & 0 \\
 & & {\scriptstyle\tau}\downarrow & & \downarrow{\scriptstyle\epsilon\otimes 1} & & \downarrow{\scriptstyle\epsilon} & & \\
0 & \longrightarrow & \Omega^{1-n}(K) & \xrightarrow{\sigma} & U & \xrightarrow{\epsilon} & K & \longrightarrow & 0 \\
 & & \downarrow & & \downarrow & & \downarrow & & \\
 & & 0 & & 0 & & 0 & &
\end{array}$$

The bottom and right sequences are the duals of sequence (2.4). The two middle sequences are the tensors of the dual of (2.4) with U, and by the above argument they are split. Let ϕ' be a splitting for $1\otimes\epsilon$. Notice that $\epsilon = \epsilon(\epsilon\otimes 1)\phi' = (\epsilon\otimes\epsilon)\phi' = \epsilon(1\otimes\epsilon)\phi'$. Let $\theta = 1-(\epsilon\otimes 1)\phi'$. Clearly $\epsilon\theta = 0$. So $\theta(U)\subseteq \sigma(\Omega^{1-n}(k))$. But $\sigma^{-1}\theta\sigma\colon \Omega^{1-n}(K)\to\Omega^{1-n}(K)$ cannot be an isomorphism since ϵ is not split. But any homomorphism from $\Omega^{1-n}(K)$ to itself that is not an isomorphism must factor through a projective module. So there exists a projective KG-module Q and homomorphism $\gamma\colon \Omega^{1-n}(K)\to Q$ and $\eta\colon Q\to\Omega^{1-n}(K)$ such that $\eta\gamma = \sigma^{-1}\theta\sigma$. Since σ is an injection and Q is injective there exists a homomorphism $\psi\colon U\to Q$ such that $\psi\sigma = \gamma$. Hence $\theta\sigma = \sigma(\sigma^{-1}\theta\sigma) = \sigma\eta\gamma = \sigma\eta\psi\sigma$, and $(\theta-\sigma\eta\psi)\sigma = 0$. It follows that $\theta - \sigma\eta\psi = h\epsilon$ for some $h\colon K\to U$. Also, since Q is projective there exists a map $\mu\colon Q\to U\otimes\Omega^{1-n}(K)$ such that $\tau\mu = \eta$. Then $\sigma\eta\psi = \sigma\tau\mu\psi = (\epsilon\otimes 1)(1\otimes\sigma)\mu\psi$. Since $(\epsilon\otimes 1)\phi' + \theta = \mathrm{Id}_U$, the identity on U,

$$\mathrm{Id}_U = (\epsilon\otimes 1)\phi' + (\epsilon\otimes 1)(1\otimes\sigma)\mu\psi + h\epsilon.$$

Now define $\nu\colon U\to U\otimes U$ by $\nu(u) = u\otimes h(1)$, for all $u\in U$. It can be checked that this is a homomorphism and that $(\epsilon\otimes 1)\nu(u) = (\epsilon\otimes 1)(u\otimes h(1)) = \epsilon(u)\cdot h(1) = h(\epsilon(u)) = h\epsilon(u)$. Also $(1\otimes\epsilon)\nu(u) = u\cdot\epsilon(h(1)) = 0$. So $h(1)\in\sigma(\Omega^{1-n}(K))$.

Consequently

$$\mathrm{Id}_U = (\epsilon\otimes 1)\phi' + (\epsilon\otimes 1)(1\otimes\sigma)\mu\psi + (\epsilon\otimes 1)\nu = (\epsilon\otimes 1)\phi,$$

where $\phi = \phi' + (\epsilon\otimes 1)(1\otimes\sigma)\mu\psi + \nu$. But also $(1\otimes\epsilon)\phi = (1\otimes\epsilon)\phi' = \mathrm{Id}_U$. So we are done.

LEMMA 2.5. *Let X and Y be finitely generated KG-modules. Any exact sequence of the form*

$$0 \longrightarrow X \longrightarrow X\oplus Y \xrightarrow{\beta} Y \longrightarrow 0$$

must split.

PROOF. By a dimension argument the map β^* in the sequence

$$0 \longrightarrow \mathrm{Hom}_{KG}(Y,X) \longrightarrow \mathrm{Hom}_{KG}(Y,X\oplus Y) \xrightarrow{\beta^*} \mathrm{Hom}_{KG}(Y,Y)$$

must be surjective. So there is a splitting homomorphism $\theta\colon Y\to X\oplus Y$ such that $\beta_*(\theta) = \beta\circ\theta = 1$.

LEMMA 2.6. Let U, V, W, X, and Y be KG-modules. Suppose that the diagram

$$\begin{array}{ccccccccc}
0 & \longrightarrow & X\oplus U & \xrightarrow{\theta} & X\oplus Y\oplus V & \xrightarrow{\psi} & Y\oplus W & \longrightarrow & 0\\
 & & {\scriptstyle\alpha}\downarrow & & {\scriptstyle\beta}\downarrow & & {\scriptstyle\gamma}\downarrow & & \\
0 & \longrightarrow & U & \xrightarrow{\sigma} & V & \xrightarrow{\tau} & W & \longrightarrow & 0\\
 & & \downarrow & & \downarrow & & \downarrow & & \\
 & & 0 & & 0 & & 0 & &
\end{array}$$

is commutative with exact rows and columns. Suppose that there exist splitting homomorphisms $\mu_1\colon U\to X\oplus U$, $\mu_3\colon W\to Y\oplus W$ such that $\alpha\mu_1 = \mathrm{Id}_U$ and $\gamma\mu_3 = \mathrm{Id}_W$. Then β is also split, the kernel of β is isomorphic to $X\oplus Y$, and the sequence

$$0 \longrightarrow \mathrm{Ker}\,\alpha \longrightarrow \mathrm{Ker}\,\beta \longrightarrow \mathrm{Ker}\,\gamma \longrightarrow 0$$

is split. Moreover, there exist splitting homomorphisms $\mu_1'\colon U\to X\oplus U$, $\mu_2'\colon V\to X\oplus Y\oplus V$ such that $\alpha\mu_1 = \mathrm{Id}$, $\beta\mu_2 = \mathrm{Id}$, and the diagram

$$\begin{array}{ccccccccc}
0 & \longrightarrow & X\oplus U & \xrightarrow{\theta} & X\oplus Y\oplus V & \xrightarrow{\psi} & Y\oplus W & \longrightarrow & 0\\
 & & {\scriptstyle\mu_1'}\uparrow & & {\scriptstyle\mu_2'}\uparrow & & {\scriptstyle\mu_3}\uparrow & & \\
0 & \longrightarrow & U & \xrightarrow{\sigma} & V & \xrightarrow{\tau} & W & \longrightarrow & 0
\end{array}$$

commutes. (In particular μ_3 may remain the same!)

PROOF. Suppose that $Z = \ker\beta$. Then the sequence of kernels has the form

$$0 \longrightarrow X \longrightarrow Z \xrightarrow{\phi} Y \longrightarrow 0.$$

So we have the commutative diagram:

$$\begin{array}{ccccccc}
 & & & & & & \cdots \\
 & & & & & & \downarrow \\
 & & 0 & & 0 & & \mathrm{Hom}_{KG}(Y,U) \\
 & & \downarrow & & \downarrow & & {\scriptstyle 0}\downarrow \\
\cdots \longrightarrow & & \mathrm{Hom}_{KG}(Y,Z) & \xrightarrow{\phi_*} & \mathrm{Hom}_{KG}(Y,Y) & \longrightarrow & \mathrm{Ext}^1_{KG}(Y,X) \longrightarrow \cdots \\
 & & \downarrow & & \downarrow & & \downarrow \\
\cdots \longrightarrow & & \mathrm{Hom}_{KG}(Y,X\oplus Y\oplus V) & \xrightarrow{\psi_*} & \mathrm{Hom}_{KG}(Y,Y\oplus W) & \longrightarrow & \cdots \\
 & & \downarrow & & \downarrow & & \\
\cdots \longrightarrow & & \mathrm{Hom}_{KG}(Y,V) & \xrightarrow{\tau_*} & \mathrm{Hom}_{KG}(Y,W) & \longrightarrow & \cdots \\
 & & \downarrow & & {\scriptstyle 0}\downarrow & &
\end{array}$$

Because α is split the connecting homomorphism in the upper right corner is the zero map. This implies that the sequence

$$0 \longrightarrow \mathrm{Im}(\phi_*) \longrightarrow \mathrm{Im}(\psi_*) \longrightarrow \mathrm{Im}(\tau_*)$$

is exact. Therefore

$$\begin{aligned}\mathrm{Dim}\,\mathrm{Im}(\psi_*) &= \mathrm{Dim}\,\mathrm{Hom}_{KG}(Y, Y\oplus V) - \mathrm{Dim}\,\mathrm{Hom}_{KG}(Y,U) \\ &= \mathrm{Dim}\,\mathrm{Hom}_{KG}(Y,Y) + \mathrm{Dim}\,\mathrm{Im}(\tau_*),\end{aligned}$$

implying that ϕ_* is onto. So there exists $\mu \in \mathrm{Hom}_{KG}(Y,Z)$ such that $\phi_*(\mu) = \phi\circ\mu$ is the identity. So ϕ is split and $Z \cong X\oplus Y$. By Lemma 2.5, β is also split.

Now let μ_2 be a splitting for β. Now $\gamma(\psi\mu_2 - \mu_3\tau) = \tau\beta\mu_2 - \tau = 0$. Therefore $(\psi\mu_2 - \mu_3\tau)(V) \subseteq \ker\gamma$. Consequently, there exists a homomorphism $f\colon V \to \ker\beta$ such that $\psi\mu_2 - \mu_3\tau = \psi f$ or

$$\psi(\mu_2 - f) = \mu_3\tau.$$

Hence we may let $\mu_2' = \mu_2 - f$. Similarly we can find μ_1'.

3. Proof of Theorem 2.2. Using the notation of the last section, we let $(X^m, \tilde{\partial})$ denote the total complex of the tensor product $(C^m,\partial)\otimes(C^m,\partial)$. Theorem 2.2 is an easy consequence of the following proposition. The four numbered conditions in the proposition are necessary for the induction step in the proof.

THEOREM 3.1. *Suppose ς annihilates the cohomology ring $\mathrm{Ext}^*_{KG}(L_\varsigma, L_\varsigma)$. For every $m > 0$ there exist degree-zero chain maps μ^m and ν^m such that the diagram*

$$\begin{array}{ccc}
(X^m,\tilde{\partial}) & \xrightarrow{\epsilon\otimes\epsilon} & K \\
{\scriptstyle \mu^m}\uparrow\downarrow{\scriptstyle \nu^m} & & \| \\
C^{2m} & \xrightarrow{\epsilon} & K
\end{array}$$

commutes and μ^m_j, ν^m_j are KG-homomorphisms. In addition we have the following.

(1) $\tilde{\partial}^m(X^m_{mn+1}) \cong \Omega^{1-n}(K) \oplus [m-1]\Omega^{1-n}(U) \oplus (\text{proj})$.

(2) $\tilde{\partial}^m(X^m_{mn}) \cong K \oplus [m]U \oplus (\text{proj})$.

(3) *If $j < (m-1)r$ and if C^{2m}_j is identified with $C^{2(m-1)}_j$ and X^m_j is identified with X^{m-1}_j then μ^m_j coincides with μ^{m-1}_j.*

(4) $(\epsilon \otimes 1)\mu^m_j = \text{Id} = (1 \otimes \epsilon)\mu^m_j$ *for all $j < mn$.*

PROOF. If $j = mn$ let $\nu^m_j = \epsilon \otimes 1$. Recall that $C^m_{mn} = K$, so there is a homomorphism $\phi_j \colon C^m_{j-mn} \otimes C^m_{mn} \to C^{2m}_j$ given by $\phi_j(c \otimes a) = a \cdot c$ for $a \in K$ and $c \in C^m_{j-mn} \cong C^{2m}_j$. So for $j \geq mn$, let $\nu^m_j = \phi_j \circ \rho_j$ where $\rho_j \colon X^m_j \longrightarrow C^m_{j-mn} \otimes C^m_{mn}$ is the obvious projection. This defines ν^m. The reader may check that it is a chain map.

Consider first the case $m = 1$. Note that $C^2_0 \cong U$ and $X^1_0 \cong U \otimes U \cong U \oplus \Omega^{1-n}(U) \oplus (\text{proj})$. Let $\mu^1_0 \colon C^2_0 \to X^1_0$ be the homomorphism ϕ given in Proposition 2.3. Then $\mu^1_1, \dots, \mu^1_{n-1}$ can be constructed because $C^2_1, \dots, C^2_{n-1}$ are projective modules. On the other end, $C^2_{2n} \cong K \cong X^1_{2n}$ and we may define $\phi_{2n} \colon C^2_{2n} \to X^1_{2n}$ to be the identity. Then since X^1_{2n-i} is injective for $1 \leq i < n$, there exist maps $\phi_{2n-1} \colon C^2_{2n-i} \to X^1_{2n-i}$ such that $\tilde{\partial}\phi_{2n-i} = \phi_{2n-i-1}\partial$ for $0 \leq i \leq n-2$. By exactness we have that

$$\tilde{\partial}(X^1_{n+1}) \cong \Omega^{1-n}(K) \oplus (\text{proj}) \cong \partial(C^2_{n+1} \oplus (\text{proj}),$$
$$\tilde{\partial}(X^1_n) \cong K \oplus U \oplus (\text{proj}) \cong \partial^n(C^2_n) \oplus U \oplus (\text{proj}),$$

and $C^2_n \cong U$, $X^2_n \cong [2]U \oplus (\text{proj})$. So at the nth term we have the commutative diagram

$$\begin{array}{ccccccccc}
0 & \longrightarrow & \Omega^{1-n}(K) \oplus (\text{proj}) & \longrightarrow & [2]U \oplus (\text{proj}) & \longrightarrow & U \oplus K \oplus (\text{proj}) & \longrightarrow & 0 \\
 & & {\scriptstyle \nu''}\downarrow\uparrow{\scriptstyle \phi'} & & \downarrow{\scriptstyle \nu_n} & & {\scriptstyle \nu'}\downarrow\uparrow{\scriptstyle \mu'} & & \\
0 & \longrightarrow & \Omega^{1-n}(K) & \longrightarrow & U & \longrightarrow & K & \longrightarrow & 0
\end{array}$$

where ν', ν'' are induced by the chain map ν, ϕ' is induced from ϕ_{n+1}, and μ' is induced from μ_{n-1}. Also we have that $\nu''\phi' = \text{Id}$ and $\nu'\mu' = \text{Id}$. So by Lemma 2.6, ν'_n is split by a KG-homomorphism $\mu_n \colon C^2_n \to X^1_n$ such that $\tilde{\partial}\mu_n = \mu_{n-1}\partial$. The last statement in Lemma 2.6 says that we may adjust $\phi_{n+1}, \dots, \phi_{2n}$ to get the desired chain map $\mu \colon C^2 \to X^1$. The reader may verify that $\mu_1, \dots, \mu_{n-1}$ can be chosen to satisfy condition (4).

Now suppose that $m > 1$. We may identify X^m_j with X^{m-1}_j and C^{2m}_j with C^{2m-2}_j whenever $j < (m-1)r$. So let $\mu^m_j = \mu^{m-1}_j$ for $j < (m-1)r$. Let $\sigma = (\sigma_1, \sigma_2, \sigma_3) \colon X^{m-1}_{(m-1)n} \to X^m_{(m-1)n}$ be given by

$$\sigma_1 = \phi \colon C^{m-1}_0 \otimes C^{m-1}_{(m-1)n} \; (\cong U \otimes K) \to C^m_0 \otimes C^m_{(m-1)n} \; (\cong U \otimes U),$$

$$\sigma_2 = \text{Id} \colon \sum_{j=1}^{(m-1)n-1} C^{m-1}_j \otimes C^{m-1}_{(m-1)n-j} \to \sum C^m_j \otimes C^m_{(m-1)n-j},$$

and

$$\sigma_3 = \phi \colon C^{m-1}_{(m-1)n} \otimes C^{m-1}_0 \to C^m_{(m-1)n} \otimes C^m_0.$$

Here C_j^m is identified with C_j^{m-1} for $j < (m-1)n$, and ϕ is given in Proposition 2.3. Now define $\mu^m_{(m-1)n}$ to be the composition of the identification of $C^{2m}_{(m-1)n}$ with $C^{2(m-1)}_{(m-1)n}$, with $\mu^{m-1}_{(m-1)n}$, and with σ. It can be checked that $\nu_{(m-1)n}\ \mu_{(m-1)n} = \mathrm{Id}$ and that $\tilde{\partial}\mu_{(m-1)n} = \mu_{(m-1)n-1}\partial$. Then the maps μ_{mn-j} for $j = n-1, \ldots, 1$ can be defined. Statement (2) is proved inductively using the fact that σ is split and has cokernel $[2]\Omega^{1-n}(U)$.

Now consider the other ends of the sequences. We can make identifications $C^{2(m-1)}_{j-2n} = C^{2m}_j$ and $X^{m-1}_{j-2n} = X^m_j$ for $(m+1)n \leq j \leq 2mn$. So define homomorphisms

$$\phi_j = \mu^{m-1}_{j-2n} \colon C^{2m}_j \to X^m_j$$

in this case. Using techniques analogous to those above we define $\phi_{(m+1)n}$ and $\phi_{(m-1)n-j}$ for $j = 1, \ldots, n-1$. Likewise use statements (1) and (2) and Lemma 2.6 to define μ^m_{mn}. The technique is the same as that in the case in which $m = 1$. Again use Lemma 2.6 to adjust the ϕ_j's and complete the chain map μ^m. Statement (3) is obvious from the construction. Statement (4) can be proved inductively using the fact that it holds when j is a multiple of n. This completes the proof of Proposition (3.1).

The proof of Theorem 2.2 is now immediate, for by making the appropriate identifications we may let $\mu_j = \mu_j^{j+1}$ for all $j \geq 0$.

4. The annihilator of $\mathrm{Ext}^*_{KG}(L_\varsigma, L_\varsigma)$**.** It should be noted that some power of ς always annihilates the cohomology ring of L_ς (see Lemma 2.3 of [**2**]). The purpose of this section is to prove the following.

THEOREM 4.1. *Suppose that p is an odd prime and that $\varsigma \in \mathrm{Ext}^n_{KG}(K, K)$ is not zero. Then ς annihilates $\mathrm{Ext}^*_{KG}(L_\varsigma, L_\varsigma)$ if and only if n is even.*

PROOF. Consider the sequence

$$0 \longrightarrow L_\varsigma \xrightarrow{\ j\ } \Omega^n(K) \xrightarrow{\ \varsigma\ } K \longrightarrow 0$$

and consider the complex $0 \to L_\varsigma \xrightarrow{\ j\ } \Omega^n(K) \to 0$ whose homology is K in degree 0. Tensoring this complex with itself we obtain the exact sequence

$$(4.2)\quad 0 \to L \otimes L \xrightarrow{(j\otimes 1, -1\otimes j)} (\Omega^n(K) \otimes L) \oplus (L \otimes \Omega^n(K)) \xrightarrow{(1\otimes j, j\otimes 1)} \Omega^n(K) \otimes \Omega^n(K) \xrightarrow{\varsigma\otimes\varsigma} K \to 0.$$

The last term is the product of the homologies. Let $H = \langle \sigma | \sigma^2 = 1 \rangle$. We claim that sequence (4.2) can be made into a sequence of $K(G \times H)$-modules. Define the action of H on the terms of the sequence as follows:

$$\sigma(l \otimes l') = -l' \otimes l,$$
$$\sigma(\alpha \otimes l, l' \otimes \alpha') = (\alpha' \otimes l', l \otimes \alpha)$$
$$\sigma(\alpha \otimes \alpha') = \alpha' \otimes \alpha$$

for $l, l' \in L_\varsigma$; $\alpha, \alpha' \in \Omega^n(K)$. Let σ act trivially on K. It is easily checked that this action commutes with the homomorphisms.

If M is a KG-module, let $S(M)$ and $A(M)$ denote, respectively, the symmetric and skew-symmetric tensors in $M \otimes M$. So $M \otimes M \cong S(M) \oplus A(M)$. Now $\Omega^n(K) \otimes \Omega^n(K) \cong S(\Omega^n(K)) \oplus A(\Omega^n(K)) \cong \Omega^{2n}(K) \oplus (\text{proj})$. By a dimension argument it can be seen that $S(\Omega^n(K))$ is projective if n is odd and $A(\Omega^n(K))$ is projective if n is even. This is because $\operatorname{Dim}(S(M)) = \frac{1}{2}(\operatorname{Dim} M)^2 + \frac{1}{2}\operatorname{Dim} M$. So if n is even, then $\operatorname{Dim}\Omega^n(K) \equiv 1 \pmod p$ and p does not divide the dimension of $S(\Omega^n(K))$.

Now consider the subspace of the terms in sequence (4.2) on which σ acts by multiplication by -1. These form an exact subsequence that looks like

$$0 \longrightarrow S(L_\varsigma) \longrightarrow \Omega^n(K) \otimes L_\varsigma \longrightarrow A(\Omega^n(K)) \longrightarrow 0.$$

If n is even then $A(\Omega^n(K))$ is projective, the sequence splits, and $S(L_\varsigma) \cong \Omega^n(L_\varsigma) \oplus (\text{proj})$ since $\Omega^n(K) \otimes L_\varsigma \cong \Omega^n(L_\varsigma) \oplus (\text{proj})$.

Now consider the complex $0 \to \Omega^n(K) \xrightarrow{\varsigma} K \to 0$ whose homology is L_ς in degree 1. Again by the Künneth formula the homology of the tensor of this complex with itself is $L_\varsigma \otimes L_\varsigma$ in degree 2. So we have the exact sequence

$$(4.3)\quad 0 \to L \otimes L \xrightarrow{j \otimes j} \Omega^n(K) \otimes \Omega^n(K) \xrightarrow{(1\otimes\varsigma, \varsigma\otimes 1)} (\Omega^n(K) \otimes K) \oplus (K \otimes \Omega^n(K)) \xrightarrow{(\varsigma,\varsigma)} K \otimes K \to 0.$$

Again $H = \langle \sigma | \sigma^2 = 1\rangle$ acts on the terms by $\sigma(l \otimes l') = -l' \otimes l$, $\sigma(a \otimes a') = -a' \otimes a$, $\sigma(a \otimes b, b' \otimes a') = (a' \otimes b', b \otimes a)$, $\sigma(b \otimes b') = b' \otimes b$ for $l, l' \in L_\varsigma$, $a, a' \in \Omega^n(K)$, $b, b' \in K$. This time consider the $(+1)$-eigenspace of the action of σ. We obtain the sequence

$$0 \longrightarrow A(L_\varsigma) \longrightarrow A(\Omega^n(K)) \longrightarrow \Omega^n(K) \xrightarrow{\varsigma} K \longrightarrow 0.$$

Of course, the kernel of ς is L_ς. So if n is even, $A(\Omega^n(K))$ is projective and we have that $A(L_\varsigma) \cong \Omega(L_\varsigma) \oplus (\text{proj})$. Therefore, if n is even $L_\varsigma \otimes L_\varsigma \cong S(L_\varsigma) \oplus A(L_\varsigma) \cong \Omega^n(L_\varsigma) \oplus \Omega(L_\varsigma) \oplus (\text{proj})$.

If n is odd, then look at the (-1)-eigenspaces of (4.3) under the action of σ. We get the sequence

$$0 \longrightarrow S(L_\varsigma) \longrightarrow S(\Omega^n(K)) \longrightarrow \Omega^n(K) \longrightarrow 0.$$

Since $S(\Omega^n(K))$ is projective in this case $S(L_\varsigma) \cong \Omega^{n+1}(K) \oplus (\text{proj})$. The proof of the theorem is a consequence of the following.

LEMMA 4.4. *Let ς be an element of $H^n(G, K)$. Then ς annihilates* $\operatorname{Ext}^*_{KG}(L_\varsigma, L_\varsigma)$ *if and only if* $L_\varsigma \otimes L_\varsigma \cong \Omega^n(L_\varsigma) \oplus \Omega(L_\varsigma) \oplus (\text{proj})$.

PROOF. Observe that ς annihilates $\operatorname{Ext}^*_{KG}(L_\varsigma, L_\varsigma)$ if and only if it annihilates the identity element $I \in \operatorname{Ext}^0_{KG}(L_\varsigma, L_\varsigma)$. As noted in §2, ς is represented by the extension class

$$0 \longrightarrow K \longrightarrow \Omega^{-1}(L_\varsigma) \longrightarrow \Omega^{n-1}(K) \longrightarrow 0.$$

Then $\varsigma \cdot I$ is represented by the tensor product

$$0 \longrightarrow K \otimes L_\varsigma \longrightarrow \Omega^{-1}(L_\varsigma) \otimes L_\varsigma \longrightarrow \Omega^{n-1}(K) \otimes L_\varsigma \longrightarrow 0$$

in $\mathrm{Ext}^1_{KG}(\Omega^{n-1}(K) \otimes L_\varsigma, L_\varsigma) \cong \mathrm{Ext}^n_{KG}(L_\varsigma, L_\varsigma)$. By Lemma 2.5, $\varsigma \cdot I = 0$ if and only if

$$\Omega^{-1}(L_\varsigma) \otimes L_\varsigma \cong L_\varsigma \oplus \Omega^{n-1}(L_\varsigma) \oplus (\text{proj}).$$

Since $\Omega^{-1}(L_\varsigma) \otimes L_\varsigma \cong \Omega^{-1}(L_\varsigma \otimes L_\varsigma) \oplus (\text{proj})$ we are done.

REFERENCES

1. D. J. Benson and J. F. Carlson, *Complexity and multiple complexes*, Math. Z. (to appear).

2. J. F. Carlson, *The variety of an indecomposable module is connected*, Invent. Math. **77** (1984), 291–299.

3. ——, *Varieties for modules*, this volume.

UNIVERSITY OF GEORGIA

Proceedings of Symposia in Pure Mathematics
Volume 47 (1987)

Extending Self-Dual Modules

CLAUDE CIBILS

Introduction. Let k be a field, G be a finite group, and H be a normal subgroup of G. Let M be a finitely generated left kH-module. If the action of H on M can be extended to G, then M is G-stable, i.e. there exist kH-isomorphisms between M and ${}^sM$, for all $s \in G$ (${}^sM$ is the vector space underlying M, equipped with a new action of H given by $h.m = s^{-1}hsm$ for $h \in H$, $m \in M$).

E. C. Dade [**D**] and J. Thévenaz [**T**] have constructed a short exact sequence associated with a given G-stable kH-module, which is the obstruction to extending the action of H to G (i.e. the exact sequence splits iff the action can be extended).

In this note we give an outline of the proof of the following (see [**C**] for a complete version):

Suppose k is an ordered field and $[G : H]$ is odd. Then all the G-stable absolutely simple kH-modules can be extended to G.

The assumption that k is ordered ensures that the simple modules are self-dual; this is why the result is a particular case of the

THEOREM. *Let k be any field and M be a G-stable and self-dual kH-module such that $\overline{\mathrm{End}}_{kH}M = k$.*

Suppose that $H^2(G/H, k^{\cdot})$ has no elements of order two. If $\mathrm{rEnd}_{kH} M \neq 0$, *suppose in addition that the characteristic of k does not divide $[G{:}H]$. Then the action of H on M can be extended to G.*

REMARKS. $\mathrm{rEnd}_{kH} M = r$ is the Jacobson radical of $\mathrm{End}_{kH} M$, and $\overline{\mathrm{End}}_{kH}M$ is the quotient ring. If $\overline{\mathrm{End}}_{kH}M = k$, then $\mathrm{End}_{kH} M$ is a local ring, so M is an indecomposable module. Of course, if the characteristic of k is 0 or prime to $|H|$ (or if M is simple) then $\mathrm{rEnd}_{kH} M = 0$. But if $\mathrm{char}\, k$ divides $|H|$, there still exist in general nonsimple indecomposable kH-modules such that $\mathrm{rEnd}_{kH} M = 0$.

To prove this theorem we use a description of the cited obstruction, which is a particular case of the results of Dade [**D**]. However, this presentation is useful,

---

1980 *Mathematics Subject Classification* (1985 *Revision*). Primary 20C05; Secondary 20J06.
The detailed version of this paper has been submitted for publication elsewhere.
Supported by a FNSRS (Swiss National Foundation) grant.

in particular to answer the following question: when does there exist a self-dual extension to G of a G-stable self-dual kH-module?

Obstructions. Let M be a G-stable kH-module and α be a set section of the projection $G \twoheadrightarrow G/H = Q$ such that $\alpha(1) = 1$. Consider the set $E_M = \coprod_{x\in Q} \mathrm{Iso}_{kH}({}^{\alpha(x)}M, M)$, where $\mathrm{Iso}_{kH}({}^{\alpha(x)}M, M)$ is the set of the kH-isomorphisms from ${}^{\alpha(x)}M$ to M. Let $\pi : E_M \to Q$ be the natural map which is surjective iff M is G-stable. The set E_M is a group: if $\pi(f) = x$ and $\pi(g) = y$, define $f.g = f \circ g \circ d_{x,y}$, where $d_{x,y}$ is the canonical kH-isomorphism from ${}^{\alpha(xy)}M$ to ${}^{\alpha(x)\alpha(y)}M$.

The exact sequence

$$(o_M) \quad 1 \to \mathrm{Aut}_{kH} M \to E_m \xrightarrow{\pi} Q \to 1$$

is the obstruction to extending the action of H on M to G.

Suppose now that the field is of characteristic p and that $p \nmid [G:H]$. Then (o_M) splits iff

$$(\omega_M) \quad 1 \to \mathrm{Aut}_{kH} M/(1+r) \to E_M/(1+r) \to Q \to 1$$

splits.

If $\overline{\mathrm{End}}_{kH} M = k$, we obtain $\mathrm{Aut}_{kH} M/(1+r) = k^{\cdot}$ and (ω_M) gives an element c_M of $H^2(Q, k^{\cdot})$. Let M^* denote the dual kH-module. Once the equality $c_{M^*} = -c_M$ is verified, the theorem is proved.

DEFINITION. A *kH-form* is a kH-module equipped with a kH-isomorphism from the module to its dual. We denote a kH-form and the underlying kH-module with the same letter.

A kH-form M is isometrically G-stable if there exist kH-isometries from M to ${}^sM$ for all $s \in G$ (the kH-form ${}^sM$ is the kH-module ${}^sM$ equipped with the isomorphism of M with its dual). Clearly, if the action of H on the kH-form M can be extended to G in such a way that the kH-isomorphism $M \to M^*$ is a kG-isomorphism, then M is G-stable isometrically.

Notice that it is possible to deduce from [**QSS**] the following: if k is algebraically closed of characteristic not 2, every kH-form G-stable is isometrically G-stable.

If M is an isometrically G-stable kH-form, we construct a short exact sequence which is the obstruction to extending the action in an isometrical way. Let $O_{kH}({}^{\alpha(x)}M, M)$ be the set of the kH-isometries from ${}^{\alpha(x)}M$ to M. The set $U_M = \coprod_{x\in G} O_{kH}({}^{\alpha(x)}M, M)$ is a subgroup of E_M and

$$(u_M) \quad 1 \to O_{kH}(M, M) \to U_M \to Q \to 1$$

is the required obstruction.

If $\mathrm{rEnd}_{kH} M \neq 0$, suppose $p \nmid [G:H]$. Then (u_M) splits iff

$$(\mu_M) \quad 1 \to O_{kH}(M, M)/(1+r)_\circ \to U_M/(1+r)_\circ \to Q \to 1$$

splits, where $(1+r)_\circ = (1+r) \cap O_{kH}(M, M)$. If $\overline{\mathrm{End}}_{kH} M = k$, the group $O_{kH}(M, M)/(1+r)_\circ$ consists of scalars which are isometries. So this group is $\{-1, +1\} = \mathbf{Z}/2\mathbf{Z}$ when $p \neq 2$ and is $\{1\}$ if $p = 2$.

We obtain the following result:

THEOREM. *Let M be an isometrically G-stable kH-form such that* $\overline{\mathrm{End}}_{kH} M = k$. *If* $\mathrm{rEnd}_{kH} M \neq 0$, *suppose that p does not divide* $[G:H]$.

There is an element of $H^2(G/H, T)$ *which is the obstruction to extending the action in an isometrical way, where* $T = \mathbf{Z}/2\mathbf{Z}$ *if* $p \neq 2$ *and* $T = \{1\}$ *if* $p = 2$. (*Then if* $p = 2$ *or if* $2 \nmid [G:H]$ *the obstruction is trivial.*)

REMARK. Let H be a p-elementary abelian group of rank 2 and G be the semidirect product of H with a cyclic group of order 2 acting on H by taking the inverse of each element. Let k be an algebraically closed field of characteristic p and suppose $p \neq 2$. There exists an infinite family of nonisomorphic kH-forms M such that the obstruction (ω_M) is trivial, so the action can be extended, but never in such a way that the resulting kG-module is self-dual (i.e. μ_M is not trivial).

REFERENCES

[**C**] C. Cibils, *Etendre les représentations autoduales*, Manuscripta Math. **57** (1987), 477–496.

[**D**] E. C. Dade, *Extending irreducible modules*, J. Algebra **72** (1981), 374–403.

[**QSS**] H. G. Quebbemann, W. Scharlau, and M. Schulte, *Quadratic and hermitian forms in additive and abelian categories*, J. Algebra **59** (1979), 264–289.

[**T**] J. Thévenaz, *Extensions of group representations from a normal subgroup*, Comm. Algebra **11** (1983), 391–425.

SECTION DE MATHÉMATIQUES, C.P. 240, 1211 GENÈVE 24, SUISSE

Proceedings of Symposia in Pure Mathematics
Volume **47** (1987)

Order-Theoretic Properties of the Center of a Block

G. H. CLIFF, W. PLESKEN, AND A. WEISS

1. Introduction. The center Λ of a block B of the group ring RG of a finite group over a p-adic ring R can be studied as an R-order. We assume that R is sufficiently large and study the embedding of Λ in the maximal order $\Gamma = \bigoplus_{\chi\in\mathrm{Irr}(B)} \Lambda e_\chi \cong \bigoplus_{\chi\in\mathrm{Irr}(B)} R$ in the semisimple commutative K-algebra $K\Lambda = \bigoplus_{\chi\in\mathrm{Irr}(B)} Ke_\chi$, where K is the field of fractions of R and e_χ is the central primitive idempotent of KG corresponding to the ordinary irreducible character χ in the block B. We compute the conductor of Λ in the maximal order Γ (which is defined as the largest Γ-ideal contained in Λ). As a consequence we deduce that a block with positive defect has at least two ordinary characters of height 0. To study further properties of the embedding of Λ in Γ we use the following chain of intermediate orders:

$$(*)\qquad \Lambda = \Lambda_0 \subsetneqq \Lambda_1 \subsetneqq \cdots \subsetneqq \Lambda_{s-1} \subsetneqq \Lambda_s = \Lambda_{s+1} = \Gamma,$$

where $\Lambda_{i+1} = \{x \in K\Lambda \mid x\,\mathrm{rad}\,\Lambda_i \subseteq \mathrm{rad}\,\Lambda_i\}$ is the order of the radical of Λ_i. We prove that the length s of this chain is equal to the defect of the block B multiplied by the ramification index of K over $\mathbf{Q}_p$. Moreover, the number of characters of height 0 in the block B is a ring-theoretic invariant; namely, for $\chi \in \mathrm{Irr}(B)$, then $e_\chi \in \Lambda_{s-1}$ if and only if χ has positive height. Finally, we compute the index of Λ in its maximal order Γ using Brauer's Second Main Theorem. In doing so we use a basis of Λ for which the multiplication constants turn out to be rational. Thus the center of a block can be defined over the localization of $\mathbf{Z}$ at $p\mathbf{Z}$.

2. Orders in split commutative algebras. Let R be the ring of integers in a local field K and let Λ be an R-order contained in

$$\Gamma = \underbrace{R\oplus\cdots\oplus R}_{k} \subset A = \underbrace{K\oplus\cdots\oplus K}_{k}.$$

1980 *Mathematics Subject Classification* (1985 *Revision*). Primary 20C20, 20C11; Secondary 16A18.

The authors acknowledge financial support from NSERC of Canada and SERC of the U.K.

We shall assume that $k \geq 2$ throughout. Let π be a generator of the maximal ideal of R. Then the conductor $\mathrm{cond}_\Gamma(\Lambda)$ of Λ in Γ, being a Γ-ideal, has the form $\pi^{a_1}R \oplus \cdots \oplus \pi^{a_k}R$, for some nonnegative integers $a_1, \ldots, a_k$. We may assume that $a_1 \geq \cdots \geq a_k$ by rearranging the components of Γ if necessary.

(2.1) LEMMA. $a_1 = a_2$. (A related result appears in [**Ple**, IV.7(iii)].)

PROOF. Since $\pi^{a_2}(1, \ldots, 1) \in \Lambda$ and $0 \oplus \pi^{a_2}R \oplus \cdots \oplus \pi^{a_2}R \subseteq \mathrm{cond}_\Gamma(\Lambda) \subset \Lambda$ we have $(\pi^{a_2}, 0, \ldots, 0) \in \Lambda$ and hence $(\pi^{a_2}, 0, \ldots, 0) \in \mathrm{cond}_\Gamma(\Lambda)$. Therefore $a_2 \geq a_1$, so $a_1 = a_2$ as desired.

We remark that this result cannot be improved in general, since for every k-tuple of nonnegative integers $(a_1, \ldots a_k)$ with $a_1 = a_2 \geq a_3 \geq \cdots \geq a_k$, it can be shown that the conductor of the order $\Lambda = R(1, \ldots, 1) + (\pi^{a_1}R \oplus \cdots \oplus \pi^{a_k}R)$ is exactly $\pi^{a_1}R \oplus \cdots \oplus \pi^{a_k}R$.

As in the introduction, let $\Lambda_0 = \Lambda$ and let

$$\Lambda_{i+1} = \{x \in A \mid x\,\mathrm{rad}(\Lambda_i) \subseteq \mathrm{rad}(\Lambda_i)\}$$

be the order of $\mathrm{rad}\,\Lambda_i$ for $i = 0, 1, 2, \ldots$. Let s be the smallest integer such that $\Lambda_s = \Lambda_{s+1}$. Since $\Lambda_s = \Lambda_{s+1}$, then by [**Rei**, 39.11] Λ_s is extremal, which means that there is no bigger order Λ' in A whose radical contains $\mathrm{rad}\,\Lambda_s$. Since A is commutative, any bigger order Λ' in A satisfies $\mathrm{rad}\,\Lambda_s \subseteq \mathrm{rad}\,\Lambda'$, so Λ_s is a maximal order. Again since A is commutative, Γ is the unique maximal order in A containing Λ, since it is the integral closure of Λ in A (cf. [**Rei**, 8.6]), so $\Lambda_s = \Gamma$. We compute the length s of our chain of orders using a method similar to that in [**ClW**].

(2.2) THEOREM. $\mathrm{cond}_\Gamma(\Lambda_i) = \pi^{-i}\,\mathrm{cond}_\Gamma(\Lambda) \cap \Gamma$ *for* $1 \leq i \leq s$.

PROOF. By splitting off idempotents and induction, it suffices to prove that $\mathrm{cond}_\Gamma(\Lambda_1) = \pi^{-1}\,\mathrm{cond}_\Gamma(\Lambda)$ in the case that $a_k > 0$. Since Γ is commutative, $\mathrm{rad}\,\Lambda = \mathrm{rad}\,\Gamma \cap \Lambda$, and because K is a splitting field for A, then $\mathrm{rad}\,\Gamma = \pi\Gamma$, so $\mathrm{rad}\,\Lambda \subseteq \pi\Gamma$. Then, multiplying by $\pi^{-1}\,\mathrm{cond}_\Gamma(\Lambda)$ we get

$$\pi^{-1}\,\mathrm{cond}_\Gamma(\Lambda)\,\mathrm{rad}\,\Lambda \subseteq \pi^{-1}\,\mathrm{cond}_\Gamma(\Lambda)\pi\Gamma = \mathrm{cond}_\Gamma(\Lambda)\Gamma.$$

Since $\mathrm{cond}_\Gamma(\Lambda)$ is a Γ-ideal, then $\mathrm{cond}_\Gamma(\Lambda)\Gamma = \mathrm{cond}_\Gamma(\Lambda)$. Furthermore, our assumption that $a_k > 0$ implies that $\mathrm{cond}_\Gamma(\Lambda) \subseteq \pi\Gamma$, so we have

$$\pi^{-1}\,\mathrm{cond}_\Gamma(\Lambda)\,\mathrm{rad}\,\Lambda \subseteq \mathrm{cond}_\Gamma(\Lambda) \subseteq \Lambda \cap \pi\Gamma = \mathrm{rad}\,\Lambda.$$

Thus $\pi^{-1}\,\mathrm{cond}_\Gamma(\Lambda) \cap \Gamma$ is a Γ-ideal in Λ_1 and is therefore contained in $\mathrm{cond}_\Gamma(\Lambda_1)$. To prove the reverse inclusion let $L \subseteq \Lambda_1$ be a Γ-ideal. Then $L\,\mathrm{rad}(\Lambda) \subseteq \mathrm{rad}(\Lambda)$ by definition of Λ_1. But $L\,\mathrm{rad}(\Lambda)$ is a Γ-ideal, hence $L\,\mathrm{rad}(\Lambda) \subseteq \mathrm{cond}_\Gamma(\Lambda)$ by definition of $\mathrm{cond}_\Gamma(\Lambda)$. Therefore $L\pi\Lambda \subseteq \mathrm{cond}_\Gamma(\Lambda)$, so $L = L\Lambda \subseteq \pi^{-1}\,\mathrm{cond}_\Gamma(\Lambda)$. Choosing $L = \mathrm{cond}_\Gamma(\Lambda_1)$ gives the result.

(2.3) COROLLARY. *The length s of the chain of orders is equal to a_1.*

3. Applications to centers of blocks. We assume the notation of the introduction; in particular, Λ is the center of a block B. Let ν_π be the exponential valuation of R with $\nu_\pi(\pi) = 1$.

(3.1) PROPOSITION. *The conductor of* Λ *in the maximal order* $\Gamma = \bigoplus_{\chi\in\mathrm{Irr}(B)} Re_\chi$ *is given by* $\bigoplus_{\chi\in\mathrm{Irr}(B)} \pi^{a_\chi} Re_\chi$ *where*

$$a_\chi = \nu_\pi(|G|/(\chi(1)\cdot\gcd\{\chi(g)\mid g\in G\})).$$

PROOF. We have

$$e_\chi = \frac{\chi(1)}{|G|}\sum_{g\in G}\chi(g^{-1})g.$$

Therefore $(|G|/(\chi(1)\cdot\gcd\{\chi(g)\mid g\in G\}))$ is the smallest multiple of the idempotent e_χ which lies in Λ. This completes the proof.

The following result is probably well known.

(3.2) LEMMA. *If* $\chi\in\mathrm{Irr}(B)$ *has height* 0, *then*

$$\nu_\pi(\gcd\{\chi(g)\mid g\in G\}) = 0.$$

PROOF. By the minimax theorem [**Fei**, III,6.10] there exists a class $\mathcal{C}$ of G having the same defect as B, with

$$\frac{\chi(g)|\mathcal{C}|}{\chi(1)} \not\equiv 0 \ (\mathrm{mod}\,\pi), \qquad g\in\mathcal{C}.$$

Therefore $\chi(g)\not\equiv 0\ (\mathrm{mod}\,\pi)$, and the lemma is proved.

Now the first results stated in the introduction follow easily.

(3.3) PROPOSITION. *A block of positive defect has at least two ordinary irreducible characters of height* 0.

PROOF. From (3.1) and (3.2) it follows that the biggest a_χ in the conductor comes from characters of height 0. The result now follows from (2.1).

S. Donkin has pointed out that (3.3) can also be obtained from a formula of Brauer; cf. [**Fei**, V.8.8].

Since we know the essential part of the conductor we also get the length s of our chain $(*)$ of orders from (2.3).

(3.4) THEOREM. *If* Λ *is the center of a block of defect* d, *then the length* s *of the chain of orders* $(*)$ *is equal to* d *multiplied by the ramification index of* K *over* $\mathbf{Q}_p$, *i.e.*

$$s = \nu_\pi(p^d).$$

We also get

(3.5) THEOREM. *For a character* $\chi\in\mathrm{Irr}(B)$ *we have*

$$e_\chi \in \Lambda_{s-1} \quad \textit{if and only if } \chi \textit{ has height } 0.$$

From (3.4) the order chains of Λ and of the corresponding block of the normalizer of the defect group of B coming from Brauer's First Main Theorem have the same length. We have not been able to match these two order chains in any specific way. It is tempting to think that (3.5) could be used to attack the Alperin-McKay conjecture.

The results of §2 can also be applied to the character ring of G (over R). The conductor can be obtained from the orthogonality relations for characters; one finds that $a_i = \nu_\pi(|C_G(g_i)|)$ where the g_i are representatives of the conjugacy classes of G. Then (2.1) implies that G has at least two classes of size prime to p, and (2.3) implies that the length of the order chain is $\nu_\pi(|G|)$.

4. The index of the center of a block in its maximal order. Suppose that L is a submodule of a free R-module M, and that L and M have the same finite rank over R. Then M/L is isomorphic to $\bigoplus R/\pi^{a_i}R$ for some nonnegative integers a_i, and we define the *index* of L in M to be π^a where $a = \sum a_i$.

In order to compute the index of Λ in its maximal order Γ, we define the following bilinear form $\langle\ ,\ \rangle$ on the center $Z(KG)$ of KG. Let $\mathrm{tr} : KG \to K$ be defined by $\mathrm{tr}\left(\sum_{g\in G} a_g g\right) = |G|\alpha_1$; then set

$$\langle x, y\rangle = \mathrm{tr}(xy) \quad \text{for } x, y \in Z(KG).$$

One may then compute indices of R-lattices in $Z(KG)$ by comparing discriminants with respect to this form. The next result follows easily (and is probably well known).

(4.1) PROPOSITION. *The index of the center of RG in its maximal order is π^a where*

$$a = \frac{1}{2}\nu_\pi\left(|G|^{k(G)} \prod_{\text{classes } C} |C| \Big/ \prod_{\chi\in \mathrm{Irr}(G)} \chi(1)^2\right)$$

where $k(G)$ denotes the number of conjugacy classes in G.

PROOF. The set $\{e_\chi\}$ of primitive central idempotents forms an orthogonal basis of the maximal order of $Z(RG)$, and $\langle e_\chi, e_\chi\rangle = \mathrm{tr}(e_\chi) = \chi(1)^2$. On the other hand, the set $\{\hat{C}\}$ of class sums is a basis of $Z(RG)$, and $\langle \hat{C}, \hat{C}'\rangle = 0$ unless C is the class of inverses of C', in which case $\langle \hat{C}, \hat{C}'\rangle = |G||C|$. The result now follows by taking the square root of the quotient of the discriminants.

To get a blockwise version of (4.1) we need a basis of the center Λ of a block B. For a p-element x of G and a Brauer character ϕ of $C_G(x)$, define

$$f_{x,\phi} = \sum_y \phi(y^{-1})\hat{C}_{xy}$$

where y varies over representatives of the p-regular classes of $C_G(x)$ and C_{xy} denotes the class of G containing xy. (Cf. [**Gol**], Chapter 10.) Let S be a set of representatives of conjugacy classes of p-elements of G; then the set

$$\{f_{x,\phi} : x \in S, \phi \in \mathrm{IBr}(C_G(x))\}$$

is an R-basis of $Z(RG)$, since the Brauer character table is invertible over R. Moreover, it follows from Brauer's Second Main Theorem that if e is the block idempotent in RG of B, then

$$ef_{x,\phi} = \begin{cases} f_{x,\phi} & \text{if } \phi \in b \text{ with } b^G = B, \\ 0 & \text{otherwise.} \end{cases}$$

(Again, cf. [**Gol**], Chapter 10.) Therefore the set $\{f_{x,\phi}\}$ with $x \in S$, and $\phi \in \mathrm{IBr}(b)$ where b is a block of $C_G(x)$ with $b^G = B$, is an R-basis of Λ. We can now compute the index of Λ in Γ. Let $k(B)$ denote the number of irreducible ordinary characters in B. For $x \in S$, let $\mathcal{B}(B,x)$ denote the set of blocks b of $C_G(x)$ for which $b^G = B$.

(4.2) PROPOSITION. *The index of* Λ *in its maximal order* Γ *is* $\pi^{a(B)}$ *where*

$$a(B) = \nu_\pi(|G|^{k(B)}) - \frac{1}{2}\nu_\pi\left(\prod_{\chi\in\mathrm{Irr}(B)} \chi(1)^2 \prod_{x\in S}\prod_{b\in\mathcal{B}(B,x)} \det \mathrm{Cart}(b)\right)$$

where $\mathrm{Cart}(b)$ *denotes the Cartan matrix of the block* b.

PROOF. The discriminant of Γ is clearly $\prod_{\chi\in\mathrm{Irr}(B)}\chi(1)^2$, so we need only compute the discriminant of Λ. First of all, $\langle f_{x,\phi}, f_{y,\psi}\rangle = 0$ unless x is conjugate to y^{-1} , since otherwise none of the classes occurring in $f_{x,\phi}$ are the inverses of classes in $f_{y,\psi}$. We have

$$\langle f_{x,\phi}, f_{x^{-1},\psi}\rangle = \mathrm{tr}\, f_{x,\phi} f_{x^{-1},\psi} = \mathrm{tr}\sum_{z,w}\phi(z^{-1})\hat{C}_{xz}\psi(w)\hat{C}_{x^{-1}w^{-1}}$$

where z and w vary over representatives of the p-regular classes of $C_G(x)$. Now unless $w = z$, $\mathrm{tr}\,\hat{C}_{xz}\hat{C}_{x^{-1}w^{-1}} = 0$. So we get

$$\langle f_{x,\phi}, f_{x^{-1},\psi}\rangle = |G|\sum_z \phi(z^{-1})\psi(z)|C_{xz}| = \frac{|G|^2}{|C_G(x)|}\sum_{h\in C_G(x)}{}^{'}\ \phi(h^{-1})\psi(h)$$

where $\sum'$ denotes summation over p-regular elements. Since

$$\frac{1}{|C_G(x)|}\sum_{h\in C_G(x)}{}^{'}\ \phi(h^{-1})\psi(h) = \gamma_{\phi,\psi}$$

where $(\gamma_{\phi,\psi})_{\phi,\psi}$ is the inverse of the Cartan matrix of $C_G(x)$ [**Fei**, IV.3.7.], it follows that $\langle f_{x,\phi}, f_{x^{-1},\psi}\rangle = |G|^2\gamma_{\phi,\psi}$. Let M_x denote the matrix whose rows and columns are indexed by Brauer characters ϕ of $C_G(x)$ in blocks b for which $b^G = B$, and whose ϕ,ψ-entry is $\langle f_{x,\phi}, f_{x^{-1},\psi}\rangle$. Then the discriminant of Λ is equal to $\pm\prod_{x\in S}\det M_x$. Since $\gamma_{\phi,\psi} = 0$ if ϕ,ψ are in different blocks of $C_G(x)$, then $\det M_x = |G|^{2l^x(B)}/\prod_{b\in\mathcal{B}(B,x)}\det\mathrm{Cart}(b)$, where $l^x(B) = \sum_{b\in\mathcal{B}(B,x)} l(b)$ and $l(b)$ is the number of Brauer characters in b. Therefore the discriminant of Λ is equal to

$$|G|^{2k(B)}\Big/\prod_{x\in S}\prod_{b\in\mathcal{B}(B,x)}\det\mathrm{Cart}(b).$$

This completes the proof.

If we multiply the indices of (4.2) over the blocks of G, it is not immediately obvious that we get (4.1). To repair this we shall reinterpret (4.2) using lower defect groups. For a p-subgroup P of G let $m_B(P)$ denote the multiplicity of P as a lower defect group of B (cf. [**Fei**, §V.10]). Let $\mathcal{P}$ be a set of representatives of the conjugacy classes of p-subgroups of G.

(4.3) THEOREM. *The index of* Λ *in its maximal order* Γ *is* $\pi^{a(B)}$ *where*

$$a(B) = \nu_\pi(|G|^{k(B)}) - \frac{1}{2}\nu_\pi\left(\prod_{\chi\in\mathrm{Irr}(B)} \chi(1)^2 \prod_{P\in\mathcal{P}} |P|^{m_B(P)}\right).$$

PROOF. We shall show that

$$\prod_{x\in S}\prod_{b\in\mathcal{B}(B,x)} \det\mathrm{Cart}(b) = \prod_{P\in\mathcal{P}} |P|^{m_B(P)}.$$

The result will then follow from (4.2). For $x \in S$, let $m_B^x(P)$ be the contribution to $m_B(P)$ coming from the x-section of G (cf. [**Fei**, V.10.7]). Then from [**Fei**, V.10.12], the multiplicity of p^m as an elementary divisor of the Cartan matrix of B is equal to $\sum_P m_B^1(P)$ where P varies over all groups in $\mathcal{P}$ of order p^m. For $x \in S$ let $\mathcal{P}^{(x)}$ be a set of representatives of conjugacy classes of p-subgroups of $C_G(x)$. Then

$$\prod_{x\in S}\prod_{b\in\mathcal{B}(B,x)} \det\mathrm{Cart}(b) = \prod_{x\in S}\prod_{b\in\mathcal{B}(B,x)}\prod_{Q\in\mathcal{P}^{(x)}} |Q|^{m_b^1(Q)}.$$

From [**Fei**, V10.10], we have

$$m_B^x(P) = \sum_Q \sum_{b\in\mathcal{B}(B,x)} m_b^1(Q)$$

where Q varies over subgroups in $\mathcal{P}^{(x)}$ which are conjugate to P in G. Then

$$\prod_{x\in S}\prod_{b\in\mathcal{B}(B,x)} \det\mathrm{Cart}(b) = \prod_{x\in S}\prod_{P\in\mathcal{P}} |P|^{m_B^x(P)}.$$

Finally, from [**Fei**, V.10.8(i)] (or by definition), $\sum_{x\in S} m_B^x(P) = m_B(P)$. So we conclude that

$$\prod_{x\in S}\prod_{b\in\mathcal{B}(B,x)} \det\mathrm{Cart}(b) = \prod_{P\in\mathcal{P}} |P|^{m_B(P)}$$

and the proof is complete.

If we sum $m_B(P)$ over all blocks B of G we get the number of conjugacy classes of G whose defect group is conjugate to P. Let $\mathcal{B}$ denote the set of all blocks of G; then we have

$$\nu_\pi\left(\prod_{B\in\mathcal{B}}\prod_{P\in\mathcal{P}} |P|^{m_B(P)}\right) = \nu_\pi\left(|G|^{k(G)} \Big/ \prod_{\text{classes } C} |C|\right).$$

Thus, if we multiply the indices $\pi^{a(B)}$ of (4.2) over all blocks B, we get (4.1). Another way of looking at this is as follows. According to [**Fei**, V.10.1] the classes of G can be assigned to blocks, so that a basis of the center of a block B is given by $\{e\hat{C} : C \in B\}$, where e is the block idempotent of B. Moreover, $m_B(P)$ is the number of classes assigned to B whose defect group is conjugate to P. Then it follows from (4.3) that the index exponent $a(B)$ is equal to

$$\frac{1}{2}\nu_\pi\left(|G|^{k(B)} \prod_{\substack{\text{classes } C\\ \text{in } B}} |C| \Big/ \prod_{\chi\in\mathrm{Irr}(B)} \chi(1)^2\right),$$

which is easily compared to (4.1).

5. Complements. The basis $\{f_{x,\phi}\}$ of Λ which we have used has the following property.

(5.1) THEOREM. *The multiplication constants of the center Λ of a block B with respect to the R-basis $\{f_{x,\phi}\}$ are rational numbers (with denominators prime to p).*

PROOF. Let $|G| = p^a m$ where m is prime to p. Write each

$$f_{x,\phi} = \sum_{\chi \in \mathrm{Irr}(B)} a_{x,\phi;\chi} e_\chi$$

in terms of the primitive central idempotents e_χ, where $a_{x,\phi;\chi} \in R$, $x \in S$, and $\phi \in b$, $b^G = B$. For $\chi \in \mathrm{Irr}(B)$ let ω_χ denote the corresponding central character on KG. Then

$$\begin{aligned} a_{x,\phi;\chi} = \omega_\chi(f_{x,\phi}) &= \frac{1}{\chi(1)} \sum_y \phi(y^{-1})\chi(xy)|C_{xy}| \\ &= \frac{|G|}{|C_G(x)|\chi(1)} \sum_{y,\psi} \phi(y^{-1})\, d^x_{\chi\psi}\psi(y)|C_y^{(x)}| \end{aligned}$$

where $d^x_{\chi\psi}$ is a generalized decomposition number (belonging to $\mathbf{Q}[\exp 2\pi i/p^a]$ by [**Gol**, 10.2]), ψ runs over $\mathrm{IBr}(C_G(x)), y$ runs over representatives of p-regular classes of $C_G(x)$, and $C_y^{(x)}$ is the class of $C_G(x)$ containing y. So we find that

$$a_{x,\phi;\chi} = \frac{|G|}{\chi(1)} \sum_\psi \gamma_{\phi,\psi}\, d^x_{\chi\psi}.$$

Therefore the $a_{x,\phi;\chi}$, as well as the multiplication constants, lie in $\mathbf{Q}[\exp 2\pi i/p^a]$. Now apply the same argument using the original definition of $f_{x,\phi}$ in terms of the class sums; since the values of the Brauer characters lie in $\mathbf{Q}[\exp 2\pi i/m]$, so do the multiplication constants. Hence the multiplication constants lie in $\mathbf{Q}$, as claimed.

The chain of orders for the center of a block $\Lambda = Z(RGe)$ defines a sequence of numbers, namely $m_i = \nu_\pi[\Lambda_i : \Lambda_{i-1}]$, which depends on the block and on the field K. If K is replaced by an extension field K', with ring of integers R', we get a new sequence m'_i for $\Lambda' = Z(R'Ge)$. We have calculated these sequences for several groups; for example, for the principal 2-block of the Mathieu group M_{12} we get

$$\begin{gathered} 7,9,4,7,9,4, \qquad K = \mathbf{Q}_2, \\ 7,9,7,9,4,7,9,4,7,9,4,4, \qquad K = \mathbf{Q}_2[\sqrt{2}], \\ 7,9,7,9,7,9,4,7,9,4,7,9,4,7,9,4,4,4, \qquad K = \mathbf{Q}_2[\sqrt[3]{2}], \\ 7,9,7,9,7,9,7,9,4,7,9,4,7,9,4,7,9,4,7,9,4,4,4,4, \qquad K = \mathbf{Q}_2[\sqrt[4]{2}]. \end{gathered}$$

We can show, in general, that the sequence $\{m'_i\}$ depends only on the ramification index r of K'/K, and not on K'. We conjecture that the sequence $\{m_i\}$

for Λ is the concatenation of certain subsequences, so that the sequence $\{m'_i\}$ for Λ' is the concatenation of these same subsequences, but with each subsequence repeated r times. In the example above, with $K = \mathbf{Q}_2$, the sequence $7, 9, 4, 7, 9, 4$ is the concatenation of $\{7, 9\}$, $\{4, 7, 9\}$, and $\{4\}$. We have verified our conjecture for a block of defect 1; if this block has a nonexceptional and b exceptional ordinary irreducible characters, respectively, then the sequence $\{m_i\}$ is a concatenation of identical subsequences $a, a+1, a+2, \ldots, a+b-1$.

REFERENCES

[ClW] G. Cliff and A. Weiss, *Crossed product and hereditary orders*, Pacific J. Math. **122** (1986), 333–345.

[Fei] W. Feit, *The representation theory of finite groups*, North-Holland, Amsterdam, 1982.

[Gol] D. Goldschmidt, *Lectures on character theory*, Publish or Perish, Berkeley, Calif., 1980.

[Ple] W. Plesken, *Group rings of finite groups over p-adic integers*, Lecture Notes in Math., vol. 1026, Springer-Verlag, 1983.

[Rei] I. Reiner, *Maximal orders*, Academic Press, London 1975.

UNIVERSITY OF ALBERTA, CANADA (CLIFF & WEISS)

QUEEN MARY COLLEGE, ENGLAND (PLESKEN)

Proceedings of Symposia in Pure Mathematics
Volume 47 (1987)

Real 2-Blocks of S_n, A_n and Their Double Covers

R. GOW

Let G be a finite group and let ε be a primitive $|G|$th root of unity in the complex numbers. Let R be the ring of algebraic integers in the field $Q(\varepsilon)$ and let P be a maximal ideal of R that contains the rational prime 2. Let A be the ring of P-local integers in $Q(\varepsilon)$ and let πA be the unique maximal ideal of A, where π is some uniformizing element. The finite field $F = A/\pi A$ is a splitting field of characteristic 2 for G and all its subgroups.

Now let $Z = Z(FG)$ denote the center of the group algebra FG. Let $B \leftrightarrow e \leftrightarrow \lambda$ be a 2-block of G, where e is a primitive central idempotent in FG and λ is the associated central character $Z \to F$, with $\lambda(e) = 1$. The anti-involution $x \to x^{-1}$ of G induces an involutory automorphism * of Z and we are interested here in those central idempotents e that are fixed by *. We say that $B \leftrightarrow e \leftrightarrow \lambda$ is a real block if $e = e^*$. Properties of real blocks were explored in a paper of Brauer [**1**]. A paper of the author [**5**] attempts to develop a theory of extended defect groups for real 2-blocks. We will briefly review some facts proved in the latter paper and then try to see how these results relate to the 2-blocks of the symmetric and alternating groups and their double covers. We will study in more detail real 2-blocks of defect 0 or 1 for S_n and A_n. These blocks are contained in blocks of defect 1 or 2 for the double covers and we will investigate rationality properties of the faithful (spin) characters in these blocks using results derived from real block theory.

1. Some results in real 2-block theory. If $B \leftrightarrow e \leftrightarrow \lambda$ is a 2-block, write

$$e = \sum a_i \hat{K}_i,$$

where the $a_i \in F$ and the $\hat{K}_i$ are class sums corresponding to the conjugacy classes K_i of G. A defect class for B is a class K_i such that $a_i\lambda(\hat{K}_i) \neq 0$.

LEMMA 1. *Let B be a real nonprincipal 2-block of G. Then B has a real nonidentity defect class.*

This is proved in [**5**] and is the starting point for the idea of extended defect groups.

1980 *Mathematics Subject Classification* (1985 *Revision*). Primary 20C15.

Let K be a real class of G. The set of all Sylow 2-subgroups of the extended centralizers $C^*(g)$, as g runs over the elements of K, is denoted by $\delta^*(K)$. The following result is the main theorem of [**5**], and it has some interesting group-theoretic consequences, as indicated in [**5**].

THEOREM 1. *Let $B \leftrightarrow e \leftrightarrow \lambda$ be a real nonprincipal 2-block. Let L be a real defect class for B and let $U \in \delta^*(L)$. Let K be any real class of G for which $\lambda(\hat{K}) \neq 0$. Then U is contained in $C^*(g)$ for some $g \in K$ and if $g^2 \neq 1$, U does not centralize g.*

COROLLARY 1. *Let K, L be real defect classes for a real nonprincipal 2-block. Then $\delta^*(K) = \delta^*(L)$.*

The 2-groups in $\delta^*(L)$ above are called the extended defect groups of B.

Recall that if $\chi \in \mathrm{Irr}(G)$, its Frobenius-Schur indicator $\nu(\chi)$ is defined by

$$\nu(\chi) = |G|^{-1} \sum \chi(g^2).$$

The next result of [**5**] shows the influence of extended defect groups on the problem of computing $\nu(\chi)$ for characters χ of height zero in a real 2-block.

THEOREM 2. *Let B be a real nonprincipal 2-block of G. Suppose that B has an abelian defect group D and let E be an extended defect group of G that contains D. Then B contains a real-valued irreducible character χ of height zero with $\nu(\chi) = -1$ if and only if E is a nonsplit extension of D.*

2. Real 2-blocks and extended defect groups for S_n and A_n. When investigating the 2-blocks of the symmetric group, the following result is crucial (see, for example, the remark after Proposition 3.3 of [**6**]).

LEMMA 2. *The symmetric group S_n has a 2-block of defect 0 if and only if $n = m(m+1)/2$ for some integer $m > 0$ (n is a triangular number). Moreover, if n is such a triangular number, there is a unique 2-block B of defect 0 for S_n.*

A result of Frobenius [**4**] enables us to show that the 2-block B described above has a unique defect class and that the corresponding central idempotent e has a very simple form.

THEOREM 3. *Let $n = m(m+1)/2$ and let $B \leftrightarrow e \leftrightarrow \lambda$ be the unique 2-block of defect 0 of S_n, defined by the irreducible character χ of 2-defect 0. Then we have*

$$e = \hat{K},$$

where K is the conjugacy class of elements whose cycle structure is of the form

$$(2m-1, 2m-5, 2m-9, \dots).$$

In particular, B has a unique defect class. An extended defect group for B is generated by an involution that is the product of r^2 disjoint transpositions if $m = 2r$ and the product of $r(r+1)$ disjoint transpositions if $m = 2r+1$.

PROOF. In the parametrization of irreducible characters of S_n by partitions of n, it is known that the character χ of 2-defect 0 corresponds to the partition of n described by

$$n = m(m+1)/2 = m + m - 1 + \cdots + 1,$$

as this is the unique partition of n having no hooks of length 2. Clearly, χ is reducible on A_n, since the 2-part of $\chi(1)$ exceeds the 2-part of $|A_n|$.

Write

$$\chi = \vartheta_1 + \vartheta_2$$

on A_n. It follows from the work of Frobenius [4] on the characters of A_n that there is a unique conjugacy class K of S_n with the property that if $g \in A_n$ but $g \notin K$, then $\vartheta_1(g) = \vartheta_2(g)$. Thus $\chi(g)$ is an even integer for all $g \notin K$, which implies that K is the unique defect class for B and $e = \hat{K}$ is the corresponding central idempotent of FS_n. The class K is specified by Frobenius in terms of the partition above and consists of elements which have the stated cycle type. The nature of the involution that generates the extended defect group for B is easily deduced from this information. This completes the proof.

In the proof above, the characters ϑ_1, ϑ_2 of A_n determine two 2-blocks of defect zero for A_n and the theorem of Frobenius already used enables us to tell whether or not these blocks are real, since the values $\vartheta_1(g)$ and $\vartheta_2(g)$ for g in the exceptional class K are given (K splits into two classes of A_n). However, in this situation, the reality or otherwise of ϑ_1, ϑ_2 is determined by a simple property of the extended defect group, as the following general result shows.

THEOREM 4. *Let H have index 2 in G. Let $B \leftrightarrow e \leftrightarrow \lambda$ be a real 2-block of defect 0 determined by $\chi \in \mathrm{Irr}(G)$. Put*

$$\chi_H = \vartheta_1 + \vartheta_2$$

where ϑ_1, $\vartheta_2 \in \mathrm{Irr}(H)$. Then ϑ_1, ϑ_2 are real-valued if and only if the extended defect groups of B lie in H.

PROOF. Since the extended defect groups of B are all conjugate, either no extended defect group lies in H or all extended defect groups lie in H. Let the involution t generate an extended defect group for B. Suppose first that t is contained in H, but that ϑ_1, ϑ_2 are not real-valued. In this case, we must have $\vartheta_2 = \overline{\vartheta}_1$ where the bar denotes complex conjugate. If K is any real defect class for B and $g \in K$, g is actually a real element of H, by our assumption that $t \in H$. But then $\vartheta_2(g) = \vartheta_1(g)$ and thus we have

$$\chi(g) = \vartheta_1(g) + \vartheta_2(g) = 2\vartheta_1(g).$$

This shows that $\chi(g)$ is twice an algebraic integer and in particular $\chi(g) \equiv 0 \pmod{\pi A}$. This contradicts the basic fact that $\chi(g) \not\equiv 0 \pmod{\pi A}$ for any element in a defect class for B. Consequently, ϑ_1 and ϑ_2 are real-valued.

Conversely, suppose that ϑ_1, ϑ_2 are real-valued but that t is not contained in H. The characters ϑ_i determine real 2-blocks $B_i \leftrightarrow e_i \leftrightarrow \lambda_i$ of defect 0 for

H, $1 \leq i \leq 2$. Let L be a conjugacy class of H that is a real defect class for B_1 and let $h \in L$. We claim that h does not belong to a real defect class for B. For otherwise h is a real element of 2-defect 0 that must be inverted by a conjugate s, say, of t, by Corollary 1. But as h is real in H, this is easily seen to force $C^*(h)$, and hence s, to lie in H, contrary to the supposition that t and its conjugates do not lie in H. It follows that $\chi(h) \equiv 0 \pmod{\pi A}$. We now have

$$\vartheta_1(h) + \vartheta_2(h) \equiv 0 \pmod{\pi A}$$

and hence

$$\vartheta_1(h) \equiv \vartheta_2(h) \pmod{\pi A}$$

as $2 \in \pi A$. As the central characters λ_i are determined by the ϑ_i, we have

$$\lambda_1(\hat{L}) = \lambda_2(\hat{L}).$$

However, as B_1 is a 2-block of defect 0, the idempotent e_1 is given by

$$e_1 = \sum \vartheta(h_i)\hat{L}_i$$

where h_i is a representative of class L_i and $\vartheta(h_i)$ is the reduction of $\vartheta_1(h_i)$ modulo πA. Thus, we can write

$$e_1 = \sum_i \vartheta(h_i)\hat{L}_i + \sum_j \vartheta(h_j)(\hat{L}_j + \hat{L}_j^*),$$

where the classes L_i appearing in the first sum are real and the classes L_j in the second sum are nonreal ($\hat{L}_j^*$ is the class sum corresponding to the inverse class of L_j). Now we have

$$\lambda_1(e_1) = 1 = \sum \vartheta(h_i)\lambda_1(\hat{L}_i),$$

since $\lambda_1(\hat{L}_j) = \lambda_1(\hat{L}_j^*)$. But we also have $\lambda_2(e_1) = 0 = \sum \vartheta(h_i)\lambda_2(\hat{L}_i)$ by the same argument. This is a contradiction, as we have already obtained $\lambda_1(\hat{L}_i) = \lambda_2(\hat{L}_i)$ for all real defect classes of H. It follows therefore that if ϑ_1, ϑ_2 are real-valued, the extended defect groups of B lie in H. This completes the proof.

COROLLARY 2. *Let $n = m(m+1)/2$ and let χ be the unique irreducible character of S_n of 2-defect* 0. *Write*

$$\chi_{A_n} = \vartheta_1 + \vartheta_2,$$

where ϑ_1, $\vartheta_2 \in \operatorname{Irr}(A_n)$. Then ϑ_1, ϑ_2 are real-valued if and only if $m \not\equiv 2$ (mod 4).

PROOF. This follows from Theorems 3 and 4.

It can be seen directly from the theorem of Frobenius that ϑ_1 and ϑ_2 are real-valued if and only if

$$(2m-1)(2m-5) \times \cdots \equiv 1 \pmod 4,$$

which of course is equivalent to our criterion above.

We show now how Theorem 3 enables us to prove that any 2-block of S_n has a unique defect class. This makes the determination of the structure of the extended defect groups of (real) 2-blocks of S_n particularly simple.

THEOREM 5. *Let n, m be integers with $n > 0$, $m \geq 0$ such that*

$$n - m(m+1)/2 = 2r,$$

for some integer $r \geq 0$. Then S_n has a unique 2-block $B(2r)$ with defect group isomorphic to a Sylow 2-subgroup of S_{2r}. There is a unique defect class for $B(2r)$, which consists of elements having the cycle structure

$$(2m-1, 2m-5, \ldots, 1, \ldots, 1),$$

where there are $2r$ cycles of length 1 if m is even and $2r+1$ if m is odd. (If n is even and $m = 0$, the corresponding block is the principal block, and the defect class is the identity class. The same is true if n is odd and $m = 1$.)

PROOF. We fix n and assume that m has been chosen as above. The subgroup of S_n that fixes elementwise the first $m(m+1)/2$ symbols is clearly isomorphic to S_{2r}, and, by abuse of notation, we will denote this subgroup by S_{2r}. Let D denote a Sylow 2-subgroup of S_{2r}, with N its normalizer in S_n. If the 2-adic decomposition of $2r$ is

$$2r = 2^{a_1} + \cdots + 2^{a_t},$$

where $a_1 > \cdots > a_t > 0$, the orbits of D on the last $2r$ symbols are known to be of size $2^{a_1}, \ldots, 2^{a_t}$, and, in particular, D has no fixed points on these $2r$ symbols. Thus D has $n - 2r$ fixed points, together with orbits of 2-power size. N clearly permutes the fixed points of D among themselves and fixes (setwise) the orbits of D of any given 2-power size (greater than one). Since a Sylow 2-subgroup of a symmetric group is known to be self-normalizing, we see that

$$N \cong S_{n-2r} \times D.$$

By assumption $n - 2r$ is a triangular number and thus S_{n-2r} has a unique 2-block of defect 0. It follows that N has a unique 2-block with defect group D and therefore, by Brauer's First Main Theorem, S_n has a unique 2-block $B(2r)$ with defect group D.

We claim now that we have accounted for all possible 2-blocks of S_n. For, by Nakayama's conjecture [**6**, §3], if λ, μ are partitions of n and χ_λ, χ_μ the corresponding irreducible characters of S_n, χ_λ and χ_μ belong to the same 2-block if and only if λ and μ have the same 2-core. Now the 2-core $\lambda_{(2)}$ of λ is a partition of length at most n having no hooks of length 2, with $|\lambda| - |\lambda_{(2)}|$ an even integer. Since a partition having no hooks of length 2 is a partition of a triangular number $s(s+1)/2$ described by

$$s(s+1)/2 = s + s - 1 + \cdots + 1,$$

we see that there cannot be any more 2-blocks of S_n than those already shown to exist.

Finally, we consider the defect classes for the block $B = B(2r)$ having the defect group D defined above. Let b be the unique block of N having defect group D and let λ be its corresponding central character. Then the central character of B is λ^G defined by

$$\lambda^G(\hat{K}) = \lambda\left(\widehat{K \cap N}\right).$$

Now if K is a defect class for B, $K \cap N$ can be seen to correspond to a single 2-regular class L, say, of defect 0 in S_{n-2r} with $\lambda(\hat{L}) \neq 0$. By Theorem 3, L can only be the class containing elements of cycle structure

$$(2m-1, 2m-5, \dots).$$

Therefore since K contains an element g that commutes with D and hence fixes each of the last $2r$ symbols, K must be the class having the cycle structure stated in the theorem. This completes the proof.

NOTE. The structure of the defect groups of 2-blocks of S_n is well known. It seems that it is only for the prime 2 that such an elementary description of the blocks of S_n can be given.

COROLLARY 3. *An extended defect group of a 2-block of S_n having defect group D is a direct product of D with a cyclic group of order* 2.

We can now determine the defect groups for the 2-blocks of A_n.

COROLLARY 4. *Let $B(2r) \leftrightarrow e \leftrightarrow \lambda$ be a 2-block of S_n having defect greater than* 0. *Then e is a primitive central idempotent of FA_n and hence determines a block $\tilde{B}(2r)$ covered by $B(2r)$. The defect group of $\tilde{B}(2r)$ is isomorphic to a Sylow 2-subgroup of A_{2r}. All 2-blocks of A_n are real unless $n = m(m+1)/2$ with $m \equiv 2 \pmod 4$. In this exceptional case, A_n has exactly two nonreal blocks which are both of defect* 0.

PROOF. Let B denote the block $B(2r)$ described above. As B does not have defect 0, we have seen that a defect group for B is not contained in A_n. Thus the defect class K for B remains a single class in A_n. If e is not primitive in FA_n, it must be a sum of two central idempotents, say

$$e = e_1 + e_2,$$

where e_2 is conjugate to e_1 under the action of S_n. Thus if

$$e_1 = \sum a_i \hat{L}_i,$$

where the L_i are classes of A_n, we have

$$e_2 = \sum a_i \hat{L}_i^t$$

where L_i^t is the class of A_n containing the conjugates of the elements of L_i under the action of a transposition t of S_n. Since $K = K^t$, $\hat{K}$ occurs with the same coefficients in e_1 and e_2. This is impossible as it implies that $\hat{K}$ has coefficient 0 in e. Thus e is primitive in FA_n and defines a block $\tilde{B}$. Different blocks B of S_n give rise to different blocks $\tilde{B}$ of A_n, since there is a unique block of S_n that covers any block of A_n [**2**, Lemma 3.5, p. 199].

On the other hand, if e is the primitive central idempotent that defines a 2-block of defect 0 in S_n, e cannot be primitive, but instead splits into two conjugate idempotents that define blocks of defect 0 in A_n. The remaining statements are clear from earlier work.

COROLLARY 5. *A_n has 2-blocks of defect 0 if and only if $n = m(m+1)/2$ or $n-2 = m(m+1)/2$ for some positive integer m. In the former case, A_n has exactly two blocks of defect 0; in the latter case, it has exactly one.*

3. Some real 2-blocks of the double covers of S_n and A_n. Now we wish to consider real 2-blocks in the double covering groups of S_n and A_n. We first recall that for $n \geq 4$, $n \neq 6$, S_n has two nonisomorphic double covers, which we denote by 2^+S_n and 2^-S_n. The double covers 2^+S_6 and 2^-S_6 are isomorphic. In 2^+S_n, transpositions of S_n lift to involutions, whereas in 2^-S_n, transpositions lift to elements whose squares equal the central involution. For notational convenience, we will generally let $2S_n$ denote either of these double covers. The double cover $2A_n$ of A_n is the unique subgroup of index 2 in $2S_n$. We will call the faithful characters of these double covers spin characters. Given a 2-block of S_n or A_n, having defect d, there is a 2-block of defect $d+1$ of the corresponding double cover that contains the given 2-block and this block always contains irreducible spin characters. Thus each 2-block of $2S_n$ is real and the same holds for $2A_n$, with the exception of two nonreal blocks of defect 1 that occur when $n = m(m+1)/2$ and $m \equiv 2 \pmod 4$, as in Corollary 4. However, work of Schur [**7**] shows that of the groups $2S_n$, only 2^+S_7 and 2^-S_n, $n = 4, 5$, or 9, have the property that all their irreducible characters are real-valued and 2^-S_9 is the single group among these that has a real splitting field.

We want to consider certain rationality questions concerning spin characters that lie in blocks of defect 1 or 2 in the double covers. We note that Schur determined the Frobenius-Schur indicator $\nu(\chi)$ of each faithful character χ of $2S_n$ or $2A_n$ in a remarkable paper [**7**]. We will recover a particular case of Schur's results through the medium of extended defect groups. Our findings seem to indicate that the problem of determining Schur indices of spin characters over p-adic fields may involve arithmetic problems of some complexity.

We begin by assuming that $n = m(m+1)/2$ is a triangular number. Then there is a unique 2-block B of S_n of defect 0, which determines a block $\hat{B}$ of $2S_n$ of defect 1. As is well known, $\hat{B}$ contains exactly two irreducible characters, which define the same irreducible 2-modular character. Both irreducible characters in B are rational-valued, and one character, which we will denote by χ, is a spin character. In the following theorem, V denotes the group $\mathbf{Z}_2 \times \mathbf{Z}_2$, C a cyclic group of order 4.

THEOREM 6. *Let $n = m(m+1)/2$, $m \geq 3$, and let $\hat{B}$ be the unique 2-block of defect 1 of $2S_n$. The extended defect group of $\hat{B}$ is either V or C and is* *determined by the residue m' of m modulo 8, as indicated in the following table.*

m'	0	1	2	3	4	5	6	7
2^+S_n	V	V	V	C	V	C	V	V
2^-S_n	V	V	C	C	V	C	C	V

We have $\nu(\chi) = 1$ if the extended defect group is V, $\nu(\chi) = -1$ otherwise.

PROOF. Let the central involution in $2S_n$ be denoted by z. Let u, v be elements in $2S_n$ that project to disjoint transpositions in S_n. Then we have the known results

$$uv = zvu, \qquad u^2 = v^2 = z^a,$$

where $a = 0$ in 2^+S_n and $a = 1$ in 2^-S_n. It follows that if $u_1, \dots, u_k$ project to pairwise disjoint transpositions, we have

$$(u_1 \cdots u_k)^2 = z^{f(k)},$$

where $f(k) = k(k-1)/2$ in 2^+S_n and $k(k+1)/2$ in 2^-S_n. We have shown in Theorem 3 that an extended defect group of the block B of S_n is generated by an involution that is a product of $m^2/4$ disjoint transpositions if m is even and $(m^2-1)/4$ transpositions if m is odd. Clearly the extended defect group of B is cyclic of order 4 if and only if an involution having this cycle type lifts to an element of order 4. Use of the formula above gives rise to the table. The statement about $\nu(\chi)$ follows from Theorem 2. This completes the proof.

We note in general that the value of $\nu(\chi)$ for any spin character χ is determined by a certain residue modulo 8 [**7**]. In the parametrization of spin characters by partitions of n with distinct parts the character χ of Theorem 6 corresponds to the partition

$$n = m(m+1)/2 = 2m - 1 + 2m - 5 + \cdots.$$

THEOREM 7. *Let $n = m(m+1)/2$ and let $\hat{B}$ be the unique 2-block of defect 1 for $2S_n$. Suppose that the extended defect group of B is cyclic of order 4 (as described in Theorem 6). Then the unique irreducible spin character χ of $2S_n$ in $\hat{B}$ has Schur index 2 over the rationals. Moreover the (finite) primes p for which χ has Schur index 2 over the p-adic numbers are divisors of*

$$(2m-1)(2m-5) \times \cdots$$

and there is an odd number of such primes.

PROOF. If the extended defect group is cyclic, χ has Schur index 2 over the reals. It follows from the Brauer-Speiser theorem that χ also has Schur index 2 over the rationals. Let h be an element of odd order in $2S_n$ that projects onto an element g having cycle structure $(2m-1, 2m-5, \dots)$. If ϑ is the nonfaithful irreducible character in $\hat{B}$, χ and ϑ agree on elements of odd order. Thus

$$\chi(h) = \vartheta(g),$$

where we consider ϑ as a character of S_n. But $\vartheta(g) = \pm 1$ by the theorem of Frobenius used to prove Theorem 3. However, as h has odd order, and g is a rational element in S_n, h is rational in $2S_n$. It follows from a theorem of Feit [**3**, Corollary 3.3] that if χ has Schur index 2 over the p-adic numbers, p divides the order of h. Thus p divides $(2m-1)(2m-5)\cdots$, as required. The fact that there is an odd number of such primes p is a consequence of results of Hasse on division algebras [**3**, Theorem 2.15]. This completes the proof.

NOTE. When $\nu(\chi) = -1$ for both 2^+S_n and 2^-S_n, the primes occurring in Theorem 7 need not be the same. Presumably determination of the appropriate primes requires an analysis of the characters of the normalizer of the cyclic subgroup generated by h in $2S_n$.

As a final application of the method of real 2-blocks, we consider certain spin characters that arise when $n-2$ is a triangular number. We know from Corollary 5 that when $n-2$ is a triangular number, A_n has a unique 2-block b of defect 0. Let ψ be the unique irreducible character of A_n in b. Then ψ extends to two irreducible characters of S_n which define a block B of S_n of defect 1. Similarly b determines a block $\hat{b}$ of $2A_n$ of defect 1, which contains a unique irreducible rational-valued spin character ψ. With this notation, we have an exact analogue of Theorem 7.

THEOREM 8. *Let $n-2 = m(m+1)/2$ and let $\hat{b}$ be the unique 2-block of defect 1 of $2A_n$. Let ψ be the unique irreducible spin character in $\hat{b}$. Then we have $\nu(\psi) = -1$ if and only if $m \equiv 2, 3, 5,$ or $6 \pmod 8$. If m satisfies any of these congruences, ψ certainly has Schur index 2 over the rationals. Moreover, in these cases, the finite primes p for which ψ has Schur index 2 over the p-adic numbers are divisors of $(2m-1)(2m-5)\cdots$, and there is an odd number of these primes.*

We omit the proof, as it is essentially the same as the previous one. We mention that the block B of S_n of defect 1 just described determines a block $\hat{B}$ of $2S_n$ of defect 2. B contains four irreducible characters, all of the same degree, two of which are spin characters. For one of the groups $2S_n$, the spin characters are not real-valued. For the other double cover, the spin characters are real-valued and their Frobenius-Schur indicators are determined by the congruences given in Theorem 8.

REFERENCES

1. R. Brauer, *Some applications of the theory of blocks of group characters*. IV, J. Algebra **17** (1971), 489–521.
2. W. Feit, *The representation theory of finite groups*, North Holland, Amsterdam, 1982.
3. ——, *The computations of some Schur indices*, Israel J. Math. **46** (1983), 274–300.
4. F. G. Frobenius, *Über die Charaktere der Alternirenden Gruppe*, (1901), Gesammelte Abhandlungen, Band III, Springer-Verlag, Berlin, 1968, pp. 167–179.
5. R. Gow, *Real 2-blocks of characters of finite groups*, (submitted for publication).
6. J. Olsson, *McKay numbers and heights of characters*, Math. Scand. **38** (1976), 25–42.
7. I. Schur, *Über die reelen Kollineationsgruppen, die der symmetrischen oder der alternierenden Gruppe isomorph sind*, (1927), Gesammelte Abhandlungen, Band III, Springer-Verlag, Berlin, 1973, pp. 51–67.

UNIVERSITY COLLEGE, DUBLIN, IRELAND

Proceedings of Symposia in Pure Mathematics
Volume 47 (1987)

Hypercohomology Varieties for Complexes of Modules, the Realizability Criterion, and Equivalent Formulations of a Conjecture of Carlsson

NATHAN HABEGGER

Abstract. The notion of variety of a module has been applied to the Steenrod problem of realizing equivariant Moore spaces. For more general G-spaces, the variety of a complex of modules, using hypercohomology, is the appropriate generalization. These behave formally like the variety of a module. In particular, the realizability criterion holds.

G. Carlsson has conjectured the following: any finite free G-CW complex over an elementary abelian p-group G of rank r has total homology of dimension at least 2^r. The hypothesis implies that the complex has zero variety. No known counterexamples to the stronger conjecture—variety zero implies dimension $(H_*) \geq 2^r$—exist. An equivalent formulation can be given in terms of composition series of free modules.

0. Introduction. The notion of cohomology varieties for modules has been studied in [**1**, **3**, **7**, **4**] and applied to the Steenrod problem [**5**, **8**]. In [**2**] these notions have been extended to the category of G-spaces. There, examples of homotopy actions not realizable by topological actions were given on spaces other than Moore spaces.

This suggests that an appropriate setting for defining varieties, including the category of modules, is the category of complexes of modules, or rather its quotient, the so-called "derived category." One simply replaces the notion of group cohomology by the extended notion of group hypercohomology (or see [**6**], "group cohomology with coefficients in a complex of modules"). The definitions are given in §1. It is seen that the variety depends only on the quasi-isomorphism class of the complex, and in fact only on the weaker equivalence relation generated by extension by finite free complexes. Since any (totally finite) complex is equivalent (see 1.10) under this weaker relation to a module, the theory of hypercohomology varieties reduces to that of cohomology varieties for modules. Hence many of the

1980 *Mathematics Subject Classification* (1985 *Revision*). Primary 20J06, 55P91, 57S17; Secondary 18G40, 20C20.

basic properties of the varieties carry over to this larger setting. In particular, the Quillen stratification theorem holds.

A main ingredient in [**5**] is a description of the structure of the variety of a realizable module. In §2, this is shown to hold verbatim with complexes in place of modules. The "projectivity criterion" of [**2**] is a corollary of this.

The hypercohomology spectral sequence relates the (global) hypercohomology of a complex to its (local) homology. If this spectral sequence collapses at E_2, the variety of the complex reduces to the variety of its "total homology." In general, this is not the case, however. For example, the variety of a finite free complex is zero. But every finitely generated module occurs in the homology of some finite free complex (e.g., take a partial resolution of it.) On the other hand, if the (projectivized) varieties of the homology groups are pairwise disjoint, the reduction to "total homology" holds. This is the central ingredient of [**2**] for obtaining nonrealizable complexes (see §2).

It is tempting to speculate on the relevance of the notion of hypercohomology variety to the conjecture of G. Carlsson: any finite free G-CW complex over an elementary abelian p-group G of rank r has total homology of dimension at least 2^r. Since finite free implies zero variety, one might interchange these two notions in the above. (Note that realizability does not seem to play a role here, since the zero variety satisfies the realizability criterion.) We will see in §3 that in fact the above stronger conjecture is equivalent to Carlsson's and that both can be "reduced" to an equivalent statement about composition series for free modules. Some low-dimensional cases of the latter are checked.

1. Varieties for the derived category. G will denote a finite group, K an algebraically closed field of characteristic p. We will consider only cochain complexes $X^{\cdot}$ (degree $d = +1$) of KG modules which are bounded below and are totally finite (i.e., $\sum \operatorname{rank}_K(H^i X^{\cdot})) < \infty$.)

To each such complex $X^{\cdot}$, we define its hypercohomology (with respect to the functor $F = \operatorname{Hom}_{KG}(K, -)$) as follows: Let $X^{\cdot} \to Y^{\cdot}$ be a resolution (by this we mean $H^*(X^{\cdot}) \xrightarrow{\sim} H^*(Y^{\cdot})$) by a complex of F cohomologically trivial modules (i.e., $\operatorname{Ext}^i_{KG}(K, -) = 0$ for $i > 0$). Then

DEFINITION 1.1. $H^*(G; X^{\cdot}) = H^*(\operatorname{Hom}_{KG}(K, Y^{\cdot}))$.

The standard general considerations [**10**] show that this is independent of the choice of resolution. For definiteness we could take

$$Y = EF^{\cdot} \otimes_K X^{\cdot} = \operatorname{Hom}_K(EG., X^{\cdot}) \tag{1.2}$$

where $EG.$ is the standard bar resolution of the trivial module K. Filtering the above double complex by the skeleta of $EG.$ yields a spectral sequence, called the hypercohomology spectral sequence, whose E^2 term is easily computed to be $H^*(G; H^*(X))$ and converges to $H^*(G; X^{\cdot})$

1.3. Consequently a map $X^{\cdot} \to X'^{\cdot}$ inducing an isomorphism on homology (quasi-isomorphism) also induces an isomorphism on hypercohomology. Thus, hypercohomology is defined on the "derived category" (in which quasi-isomorphisms are declared to be isomorphisms.)

1.4. To make the connection with topology, note that if $X^\cdot$ is the cochain complex of a relative G-CW pair (Y, A), then

$$\begin{aligned} H^*(G; X^\cdot) &= H^*(\mathrm{Hom}_{KG}(EG, \mathrm{Hom}_K(C_*(Y,A), K))) \\ &= H^*(EG \overset{*}{\times}_G Y, EG \times_G A; K) = H^*_G(Y, A; K). \end{aligned}$$

Thus hypercohomology is none other than equivariant cohomology.

1.5. Let S denote the direct sum over all simple KG-modules. Set $I_G(X^\cdot) =$ radical of the annihilator of $H^*(G; X^\cdot \otimes_K S)$ (an ideal in $H^*(G; K)$). Let $V_G(X^\cdot) =$ set of maximal ideals containing $I_G(X)$. For equivalent definitions, see [**4**].

1.6. Note that if $G = P$, a p-group, then $S = K$, since there is only one simple module. For $G \neq P$, the above definition is the correct one since we have the following:

1.7. Restriction to a subgroup $H \subset G$ induces a map $t_{H,G}\colon V_H(X^\cdot) \to V_G(X^\cdot)$, natural with respect to inclusions.

LEMMA 1.8. *Let* $0 \to X_1^\cdot \to X_2^\cdot \to X_3^\cdot \to 0$ *be a short exact sequence of complexes. Then*

$$V_G(X_i) \subset V_G(X_j) \cup V_G(X_k), \qquad \{i, j, k\} = \{1, 2, 3\}.$$

PROOF. This follows as for modules, from the long exact sequence of hypercohomology.

COROLLARY 1.9. (1) *If* $V_G(X_i) = 0$, *then* $V_G(X_j) = V_G(X_k)$ $\{i, j, k\} = \{1, 2, 3\}$.

(2) *If* $X^\cdot$ *is a finite free complex, then* $V_G(X^\cdot) = 0$.

(3) $V_G(X^\cdot)$ *is an invariant of the equivalence relation generated by quasi-isomorphisms and extension by finite free complexes.*

PROOF. (1) follows immediately from 1.8,

(2) follows from (1) by filtering $X^\cdot$ by its skeleta (since the variety of a complex concentrated in one degree is the variety of the corresponding module and the variety of a free module is trivial).

(3) follows from 1.3 and (2).

Denote the equivalence relation in 1.9.3 by "free equivalence."

LEMMA 1.10. $X^\cdot$ *is freely equivalent to a module* $M(X)$ *whose composition factors are* $\Omega^i(H^i(X))$ *where* Ω *denotes the looping functor (kernel of the projective cover).*

PROOF. Let r be the first index $i > 0$ with $H^i(X) \neq 0$. Suppose $r < \infty$ and let $P.$ be the minimal projective resolution of $H^r(X^\cdot)$. Let

$$Q^s = \begin{cases} P_{r-s}, & 1 \le s \le r, \\ 0 & \text{otherwise.} \end{cases}$$

Using the acyclicity of $X^\cdot$ in dimensions between 0 and r and the projectivity of $Q^\cdot$, one constructs a map $Q^\cdot \to X^\cdot$ inducing an isomorphism in dimension r. Let $C^\cdot$ denote the mapping cone. One obtains the long exact sequence

$$H^*(Q^\cdot) \to H^*(X^\cdot) \to H^*(C^\cdot) \to H^{*+1}(Q^\cdot)$$

from which one computes that $H^*(C^\cdot) = H^*(X^\cdot)$ if $* \neq r, 0$, $H^r(C) = 0$, and $0 \to H^0(X) \to H^0(C) \to \Omega^r(H^r(X)) \to 0$ is exact.

Since $Q^\cdot$ is finite projective, $C^\cdot$ is equivalent to $X^\cdot$. Continuing in this way we see that $X^\cdot$ is equivalent to a complex whose homology is concentrated in degree zero and has composition series

$$\Omega^n(H^n) - \cdots - \Omega^1(H^1) - H^0.$$

But a complex with one homology group is quasi-isomorphic to its homology.

COROLLARY 1.11 (QUILLEN STRATIFICATION FOR COMPLEXES).

$$V_G(X^\cdot) = \bigcup_E t_{E,G} V_E(X^\cdot).$$

(*Here E runs over the collection of elementary abelian p-subgroups of G.*)

COROLLARY 1.12. $V_G(X^\cdot) \subset V_G(\bigoplus H^i(X))$.

COROLLARY 1.13. *Suppose $V_G(H^i) \cap V_G(H^j) = 0$, $i \neq j$. Then $V_G(X^\cdot) = V_G(\bigoplus H^i(X))$.*

PROOFS OF 1.11, 1.12, 1.13. See [4] for modules and apply 1.10.

2. The structure theorem for varieties of realizable complexes. The structure theorem of [5] for realizable modules extends to the more general setting of complexes.

DEFINITION 2.1. We say the complex $X^\cdot$ is realizable if it (or some suspension, possibly negative) is (freely) equivalent to the relative cochain complex of some G-CW pair.

THEOREM 2.2. *Suppose $X^\cdot$ is realizable. Then there is a subset $\mathcal{E}(X^\cdot)$ of the set of elementary abelian p-subgroups of G so that $V_G(X^\cdot) = \bigcup_{E \subset \mathcal{E}} t_{E,G} V_E$. (Here V_E denotes $V_E(K)$.)*

PROOF. Using the isomorphism of 1.4, $H^*(G, X^\cdot) \simeq H^*_G(Y, A; K)$, which is compatible with the isomorphism $H^*(G; K) = H^*_G(pt)$, one proceeds as in [5] by reducing to a p-Sylow subgroup or alternatively using 1.1 and reducing to the elementary abelian case.

COROLLARY 2.3. *Suppose $V_G(H^i(X^\cdot)) \cap V_G(H^j(X^\cdot)) = 0$, $i \neq j$, and $X^\cdot$ is realizable. Then $V_G(H^i(X)) = \bigcup_{E \subset \mathcal{E}_i} t_{E,G} V_E$ for some collection $\mathcal{E}_i$.*

PROOF. By 1.13 and 2.2, the $V_G(H^i(X))$ are the (projectivized) components of a variety of the stated form. But any component of such a variety has the same form.

The projectivity criterion of [**2**] is a consequence of 2.2 above:

COROLLARY 2.4. *Suppose* $V_C(X^\cdot) = 0$ *for each p-cyclic subgroup* C *of* G *and assume* $X^\cdot$ *is realizable. Then* $V_G(X^\cdot) = 0$.

REMARK 2.5. If the hypercohomology spectral sequence for $H^*(G; X^\cdot \otimes S)$ collapses, then it is easy to see $V_G(X^\cdot) = V_G(\bigoplus H^i(X^\cdot))$. When this holds 2.4 is a criterion for the projectivity of $\bigoplus H^i(X^\cdot)$.

3. Reformulation of Carlsson's conjecture. In this section G denotes an elementary abelian p-group of rank r. The Carlsson conjecture is the following:

CONJECTURE 1. *Let* $X^\cdot$ *be a finite free* KG *complex. Then* $\bigoplus H^i(X^\cdot)$ *has dimension* $\geq 2^r$ *(or zero).*

The following is apparently stronger:

CONJECTURE 2. *Let* $X^\cdot$ *be a* KG*-complex with* $V_G(X^\cdot) = 0$. *Then* $\bigoplus H^i(X^\cdot)$ *has dimension* $\geq 2^r$ *(or zero.)*

We will show the following:

PROPOSITION 3.1. *Conjectures* 1 *and* 2 *are equivalent to the following:*

CONJECTURE 3. *Let* $A_0 - \cdots - A_n$ *be a composition series for a free module. Then dimension* $\bigoplus \Omega^i A_i \geq 2^r$ *(or zero).*

PROOF. Since by 1.9.2 a finite free complex satisfies $V_G(X^\cdot) = 0$, (2) implies (1). Now suppose $A_0 - \cdots - A_n$ is a composition series for a free module F. Construct a finite free KG complex with homology $\Omega^i(A_i)$ as follows: let P^i_* be a partial resolution for A_i (so $H_{i-1}(P^i_*) = \Omega^i(A_i)$). Using these, one constructs on $\bigoplus P^i_* \oplus F$ a differential whose total homology is $\bigoplus \Omega^i(A_i)$. This is a finite free complex, so (1) implies (3).

Given $X^\cdot$ as in 2, construct $M(X^\cdot)$ as in 1.10. $M(X^\cdot)$ has composition series $\Omega^n H^n - \cdots - H^0$. Since $V_G(X^\cdot) = 0$ we have $V_G(M(X^\cdot)) = 0$, hence $M(X^\cdot)$ is free. But then $H^n - \cdots - \Omega^n H^0$ is also a composition series for a free module, so (3) implies (2).

Proposition 3.1 makes it apparent that the Carlsson conjecture can be broken into an infinite sequence of conjectures, one for each length of a composition series. For length 1, the assumption is just that our module A is free, whence $\dim \Omega^i(A) \geq 2^k$ (or zero.)

In what follows we will concentrate on *length* 2. We have modules A_j and A_{j+l} and an exact sequence $0 \to A_{j+l} \to F \to A_j \to 0$ with F free. Hence $A_{j+l} = \Omega^1(A_j)+$ free, whence $\Omega^{j+l}(A_{j+l}) = \Omega^{j+l+1}(A_j) = \Omega^{l+1}(\Omega^j A_j)$. Setting $B = \Omega^j A_j$ we desire to show $\dim B \oplus \Omega^{l+1} B \geq 2^r$ for all $l \geq 1$.

Now an Euler characteristic argument shows

$$\dim B \equiv (-1)^s \dim(\Omega^s B) \bmod |G|.$$

Using this, the conjecture for length 2 is seen to be equivalent to the following:

CONJECTURE 4. *Let B be a module of dimension $< 2^{r-1}$ over an elementary abelian group of rank r. Then*

$$\dim B < \dim \Omega^{2s} B \quad \textit{for all } s > 0.$$

We check some cases of Conjecture 4:

For $r = 1$ conjecture 4 is vacuous.

For $r = 2$ we have dimension $B = 1$ so B is the trivial module K. But $\dim \Omega^{2s} K > 1$.

For $r = 3$ we verify the conjecture if $p = 2$:

Let X_1, X_2, X_3 be generators of G. Then KG is isomorphic to an exterior algebra $E(\theta_1, \theta_2, \theta_3)$ where $\theta_i = 1 + X_i$. Set $J = (\theta_1, \theta_2, \theta_3) \subset E(\theta_1, \theta_2, \theta_3)$. Then $J^4 = 0$.

Consider modules B of dimension ≤ 3. Then $J(B) \neq B$ since $J^4 = 0$. If $JB = 0$, then B is a trivial module so the conjecture holds. If $JB \neq 0$ then we must have $J^2 B \neq JB$. We claim $J^2 B = 0$. For if not, then let $0 \neq \alpha\beta y$ with $\alpha, \beta \in J$. Then it is easy to see that $y, \alpha y, \beta y, \alpha\beta y$ are linearly independent, which is impossible since $\dim B \leq 3$.

Now $\theta_1, \theta_2, \theta_3$ form a basis for J/J^2. Since rank $JB \leq 2$, we must have $\theta B = 0$ for some $\theta \in J/J^2$. Up to automorphism of KG, we may assume $\theta = \theta_1$. This means B is induced from the group $G/\langle X_1 \rangle$. Up to automorphism of $KG/\langle X_1 \rangle$, the only possible irreducible modules of rank ≤ 3 are either $\Omega^{-1}(K)$, $\Omega^1(K)$, or have X_2 acting trivially. In all cases, it can be checked that $\dim \Omega^{2r}(B) > \dim B$. For example, over $KG/\langle X_1 \rangle$ one has $\Omega^2(\Omega^{-1}(K)) = \Omega^1(K)$ (and the rank remains 3) but over KG the rank increases from 3 to 11. (See also the remarks below.)

REMARK. Conjecture 1 has been verified by G. Carlsson [**9**] for $r \leq 3$ and $p = 2$.

Rank > 3. One might try to prove conjecture 4 by induction on the rank r of G. The inequality $\dim \Omega^i(B \downarrow H) \leq \dim \Omega^i(B)(*)$ shows that one need only look at modules with $2^{r-2} \leq \dim B < 2^{r-1}$. For, by induction, those of smaller dimension have been checked on subgroups of rank $r - 1$ and by $*$, the desired inequality holds over G. Furthermore, one can assume B is *indecomposable* on restriction to subgroups of rank $r - 1$. For otherwise at least one component has dimension $< 2^{r-2}$ (and the other has dimension $< 2^{r-1}$ so its dimension cannot decrease on looping).

REFERENCES

1. J. Alperin and L. Evans, *Varieties and elementary abelian groups*, J. Pure Appl. Algebra **26** (1982), 221–227.

2. A. Assadi, *Homotopy actions and cohomology of finite groups* (Proc. on Transformation Groups, Poznan), Lecture Notes in Math., vol 1208, Springer-Verlag.

3. G. Avrunin and L. Scott, *Quillen stratification for modules*, Invent. Math. **66** (1982), 277–286.

4. D. Benson, *Recent trends in modular representation theory*, Lecture Notes in Math., vol. 1081, Springer-Verlag, 1984.

5. D. Benson and N. Habegger, *Varieties for modules and a problem of Steenrod*, Proc. Conf. on Cohomology of Groups, J. Pure. Appl. Algebra **44** (1987), 13–34.

6. K. Brown, *Cohomology of groups*, Graduate Texts in Math., Springer-Verlag, 1982.

7. J. Carlson, *Varieties and the cohomology ring of a module*, J. Algebra **85** (1983), 104–143.

8. G. Carlsson, *A counterexample to a conjecture of Steenrod*, Invent. Math. **64** (1981), 171–174.

9. ——, *Free* $(\mathbf{Z}/2)^3$*-actions on finite complexes*, Preprint.

10. H. Cartan and S. Eilenberg, *Homological algebra*, Princeton University Press, Princeton, N.J., 1956.

UNIVERSITY OF CALIFORNIA, SAN DIEGO

Proceedings of Symposia in Pure Mathematics
Volume 47 (1987)

Blocks of Projective Representations of the Symmetric Groups

JOHN F. HUMPHREYS

For any positive integer n, the group generated by z, $t_1, t_2, \ldots, t_{n-1}$, where z is a central involution and

$$\begin{aligned} t_i^2 &= z, && 1 \le i \le n-1, \\ (t_i t_{i+1})^3 &= z, && 1 \le i \le n-2, \\ t_i t_j &= z t_j t_i, && 1 \le i, j \le n-1 \text{ and } |i-j| \ge 2, \end{aligned}$$

is a group (denoted $\tilde{S}_n$) in which the quotient by $\langle z \rangle$ is the symmetric group S_n. For $n \ge 4$, $\tilde{S}_n$ is a representation group for S_n since z is a commutator. Any irreducible complex representation of $\tilde{S}_n$ restricts to $\langle z \rangle$ to be a multiple of a fixed irreducible representation of $\langle z \rangle$. The negative representations of $\tilde{S}_n$ are those which restrict to be a multiple of the representation taking z to -1. The negative representations of $\tilde{S}_n$ can be regarded as the projective representations of S_n corresponding to a 2-cocycle which is not a coboundary.

In [**2**], Schur showed that the irreducible negative representations of $\tilde{S}_n$ can be indexed by the 2-regular partitions of n (that is, partitions with no repeated parts). However, this indexing is not bijective. If λ is a 2-regular partition of n with k nonzero parts, then when $n-k$ is odd there are two irreducible negative representations corresponding to λ, either of which may be obtained from the other by tensoring with the sign representation of S_n regarded as a representation of $\tilde{S}_n$. In this case the irreducibles are said to be associated. If $n-k$ is even there is a unique irreducible negative representation corresponding to λ.

A useful combinatorial object associated with a 2-regular partition is the shift-symmetric Young diagram defined as follows. First, associate to any partition $\lambda = (\lambda_1, \ldots, \lambda_k)$ a Young diagram with k rows of nodes (denoted by crosses) in which the ith row has λ_i nodes and the first node is in the first column. Next associate to a 2-regular partition λ a shifted Young diagram by indenting the

1980 *Mathematics Subject Classification* (1985 *Revision*). Primary 20C20; Secondary 20C25.
The detailed version of this paper has been submitted for publication elsewhere.

Young diagram so that the first node in the ith row occurs in the ith column. Now form the shift-symmetric Young diagram by adding a further n nodes (denoted by circles), making a new column with λ_1 nodes to the left of the others and thereafter placing λ_i nodes in the ith column underneath the last cross node in this column. As an example, we give the shift-symmetric Young diagram for the partition $(7,6,3,2)$ of 18.

$$\begin{array}{cccccccc}
\circ & \times & \times & \times & \times & \times & \times & \times \\
\circ & \circ & \times & \times & \times & \times & \times & \times \\
\circ & \circ & \circ & \times & \times & \times & & \\
\circ & \circ & \circ & \circ & \times & \times & & \\
\circ & \circ & \circ & \circ & & & & \\
\circ & \circ & & & & & & \\
\circ & \circ & & & & & &
\end{array}$$

The shift-symmetric Young diagram of λ may be used to determine the degree of the irreducible negative representation corresponding to λ. To do this, calculate the length l_{ij} of the hook at each cross node by counting the total number of nodes to the right of the node (including the node itself) and adding the total number of nodes below the node (not including the node). Multiply the numbers l_{ij} together for all cross nodes and divide this number into $n!$. Multiply the resulting integer by 2^r, where r is the integer part of $(n-k)/2$, to obtain the degree of the irreducible associated with λ.

For a Young diagram corresponding to λ, the equivalent rim of the hook at the node (i,j) consists of those nodes on the boundary of the diagram joining the last node in row i to the bottom node in column j. The p-core of λ is obtained by successively deleting the equivalent rims of hooks of length p and is independent of the order in which the rims are deleted. The Nakayama conjecture asserts that for any prime p, two irreducible complex representations D^λ and D^μ of S_n are in the same p-block if and only if λ and μ have the same p-core. In 1965, Morris conjectured that for odd primes p, the p-blocks of negative representations of $\tilde{S}_n$ should be determined by the so-called $\bar{p}$-core of λ. Rather than define this $\bar{p}$-core, we observe that (as was shown by I. G. Macdonald) the $\bar{p}$-core of λ is essentially the p-core of the shift-symmetric Young diagram associated with λ. Thus, in the above example when p is 5, the $\bar{5}$-core is

$$\begin{array}{ccc}
\circ & \times & \times \\
\circ & \circ & \times
\end{array}$$

The following result will appear in the *Journal of the London Mathematical Society.*

THEOREM (J.F. HUMPHREYS). *Let p be an odd prime and λ, μ be 2-regular partitions of n with P^λ and P^μ corresponding irreducible negative representations. If λ and μ have the same $\bar{p}$-core then P^λ and P^μ are in the same p-block unless* (i) *λ and μ are equal,* (ii) *P^λ has zero defect, and* (iii) *P^λ, P^μ are associate representations. If P^λ and P^μ are in the same p-block then λ and μ have the same $\bar{p}$-core.*

The proof of this result is based on the proof of the Nakayama conjecture given by Meier and Tappe in [**1**]. More recently, Olsson has given a proof of the theorem based on the Brauer-Robinson proof of the Nakayama conjecture. Knörr and Olsson have a conjecture on the distribution of negative representations into 2-blocks.

REFERENCES

1. N. Meier and J. Tappe *Ein neuer Beweis der Nakayama-Vermutung über die Blockstruktur symmetrischer Gruppen*, Bull. London Math. Soc. **8** (1976), 34–37.

2. I. Schur, *Uber die darstellung der symmetrischen und der alternierenden Gruppen durch gebrochene lineare Substitutionen*, J. Reine Angew. Math. **139** (1911), 155–250.

UNIVERSITY OF LIVERPOOL, ENGLAND

Proceedings of Symposia in Pure Mathematics
Volume 47 (1987)

A Condition for the Existence of p-Blocks of Defect 0

ZHANG JI-PING

Abstract. In this paper, we prove the following

THEOREM. *If a finite group G possesses a cyclic Sylow p-subgroup, then G has a p-block of defect 0 if and only if $O_p(G) = 1$.*

An important question in modular representation theory is: When is a p-subgroup D of the finite group G a defect group for some p-block of G? By the first main theorem of Brauer, we need only consider the case when D is normal in G and reduce the problem further to questions about blocks of defect 0 of $DC_G(D)/D$ and their inertial subgroups in G. Actually, little is known about the question in general. Brauer and Fowler [**1**] gave a sufficient condition for existence of a p-block of defect 0. Tsushima [**2**] and Wada [**3**] gave further conditions of a similar nature. In this paper, under the assumption that the Sylow p-subgroup of G is cyclic, we prove that G has a p-block of defect 0 if and only if $O_p(G) = 1$. The proof of the result relies heavily on the classification of finite simple groups.

Throughout the paper, all the groups are of finite order and p is a prime number. The notation is standard and follows that of [**5**].

LEMMA 1. *Let G be a sporadic simple group. Then the Sylow r-subgroups of G are either noncyclic or of prime order r for any prime factor r of $|G|$.*

The lemma is proved by considering the 26 sporadic simple groups [**20**].

LEMMA 2. *Let G be a classical simple group of order $q^t \prod_{i=1}^{m}(q^{n(i)} \pm 1)$, where $n(i) \geq 1$, q a prime power. If the Sylow p-subgroups of G are cyclic, then there is at most one i such that $p | q^{n(i)} \pm 1$.*

This is a corollary of the main result of [**8**].

LEMMA 3. *Suppose α is an automorphism of order p of a finite group G. If α fixes every conjugacy class C of G, that is, $C^\alpha = C$, then $p \mid |G|$.*

1980 *Mathematics Subject Classification* (1985 *Revision*). Primary 20C15.
This paper is in final form and no version of it will be submitted for publication elsewhere.

PROOF. Let $C = \{a_1, a_2, \ldots, a_n\}$ be an arbitrary conjugacy class of G; then $a_i^\alpha \in C$. α induces a permutation on set C. If there exists a conjugacy class C of G such that there exist nonfixed points of α on C, write α as a product of cycles. We known that $p|n$. Hence $p|\,|G|$ as $n|\,|G|$. If no such conjugacy class C occurs in G, then $C_G(\alpha) \cap C' \neq \varnothing$ for any conjugacy class C' of G. Therefore $G = \bigcup_{x\in G}(C_G(\alpha))^x$. This forces $G = C_G(\alpha)$, $\alpha = 1$. Contrary to that α is of order p.

THEOREM 4. *Suppose G is an arbitrary finite group with a cyclic Sylow p-subgroup P; then G possesses a p-block of defect* 0 *if and only if $O_p(G) = 1$.*

The proof of Theorem 4 comprises several steps.

I. If G is p-solvable, then the theorem is true. Suppose $O_p(G) = 1$. Set $M = O_{p'}(G)$. Since G is p-solvable, $M \neq 1$. Now P is cyclic; we have $G = MP\langle\rho\rangle$, where $MP \lhd G$, $o(\rho)|p-1$. By a theorem of [**5**, p. 429], we may assume that $G = MP$. Let α be an element of order p of P. If every p-regular conjugacy class C of G is of defect group $D(C) \neq 1$, we may assume $\alpha \in D(C)$. Clearly, $C \subset M$. Suppose $C = C_1 + C_2 + \cdots + C_m$, where C_i is a conjugacy class of M and there exist $x_i \in P$ such that $C_i = C_1^{x_i}$. We may assume that there exists $x \in C_1$ such that $D(C)$ is the Sylow p-subgroup of $C_G(x)$. Then $x^\alpha = x$, $C_1^\alpha = C_1$, $C_i^\alpha = C_1^{x_i\alpha} = C_1^{\alpha x_i} = C_i$. Hence α fixes every conjugacy class of M. Since $O_p(G) = 1$, α is an automorphism of order p of M. By Lemma 3, $p||M|$, contrary to $M = O_{p'}(G)$. Hence G has a p-regular conjugacy class C of defect 0 and G has a p-block of defect 0.

If G has a p-block of defect 0, it is necessary that $O_p(G) = 1$. So we shall only prove that "if" part.

II. If Theorem 4 is true for all simple groups, it is true for all finite groups.

We may assume that G is not p-solvable. Let M be a maximal normal subgroup of G. Then $p \nmid |M|$ or $P \leq M$ [**9**]. If $P \leq M$, then M is not p-solvable. Since $O_p(G) = 1$, then $O_p(M) = 1$. By induction on G, we may assume that M has a p-block of defect 0. It follows that G has a p-block of defect 0. If $p \nmid |M|$, then $M = O_{p'}(G)$. G/M is nonabelian simple. If G/M has a p-block of defect 0, so is G.

III. If G is the alternating group A_n $(n \geq 5)$, then Theorem 4 is true.

If the Sylow p-subgroup P of A_n is cyclic, then $p \leq n \leq 2p-1$. If the symmetric group S_n has an r-block of defect 0 for a prime $r > 2$, then A_n also has an r-block of defect 0. So we need only prove that S_n has a p-block of defect 0. Suppose $\alpha = (\alpha_1, \alpha_2, \ldots, \alpha_t)$ is a partition of n, and $\alpha' = (\alpha_1', \alpha_2', \ldots, \alpha_s')$ the dual partition. There is a one-to-one correspondence between the partitions of n and the irreducible characters of S_n. Let f_α be the character corresponding to the partition α. Then $f_\alpha(1) = n!/\prod_{i,j} h_{ij}^\alpha$, where $h_{ij}^\alpha = \alpha_i - j + \alpha_j' - i + 1$.

If $n = p$, then $p \geq 5$. Set $\alpha = (p-2, 2)$; then $\alpha' = (z^2, 1^{p-4})$, $h_{12}^\alpha \leq h_{11}^\alpha = p-1$, $h_{1j}^\alpha = p-j-1$, $j \geq 3$, $h_{21}^\alpha = 2$, $h_{22}^\alpha = 1$. Hence $p|n!/\prod_{i,j} h_{ij}^\alpha$, that is, $p|f_\alpha(1)$.

If $p < n \leq 2p-1$, set $\alpha = (p, 1^{n-p})$, $\alpha' = (n-p+1, 1^{p-1})$, $h_{11}^\alpha = n$, $h_{1j}^\alpha = p+1-j$, $j \geq 2$, $h_{i1}^\alpha = n-p+2-i$, $i \geq 2$, $h_{1j}^\alpha \leq p-1$, $h_{i1}^\alpha \leq p-1$, $j \geq 2$,

$i \geq 2$. So $p|n!/\prod_{ij} h_{ij}^{\alpha}$, $p|f_\alpha(1)$. This shows that f_α lies in a p-block of defect 0. S_n has a p-block of defect 0.

IV. If G is a simple group other than an alternating group, then Theorem 4 is true for G.

By the Atlas of Finite Groups [**20**], it is easy to verify that the theorem is true for the sporadic groups.

Let G be a simple group of Lie type of characteristic r. If $p = r$, then the Steinberg character of G lies in a p-block of defect 0. In the following, we assume that $p \neq r$.

Let $\overline{G}$ be the group of Lie type such that G is the adjoint group of $\overline{G}$. Then we claim that there exists a unipotent or semisimple character χ of $\overline{G}$ such that the center $Z(\overline{G})$ of $\overline{G}$ lies in the kernel of χ and $p^a|\chi(1)$, where p^a is the order of a Sylow p-subgroup of G. Hence G has a p-block of defect 0.

Here we give only an example to show the method. All facts about the unipotent characters and semisimple characters can be found in [**10**].

For example, let $\overline{G}$ be of type A_{n-1}. Then $G \simeq \mathrm{PSL}(n,q)$, where q is a power of r. Let $(\alpha_1, \alpha_2, \ldots, \alpha_m)$ be a partition of n, $0 \leq \alpha_1 \leq \alpha_2 \leq \cdots \leq \alpha_m$. Set $\lambda_i = \alpha_i + i - 1$, $i = 1, 2, \ldots, m$, $\beta = [\lambda_1, \lambda_2, \ldots, \lambda_m]$. Then $\overline{G}$ has a unipotent character f_β of degree

$$f_\beta(1) = \frac{\prod_{1\leq j<i\leq m}(q^{\lambda_i} - q^{\lambda_j})|\overline{G}|_{r'}(q-1)}{\prod_{i=1}^{m}(\prod_{h=1}^{\lambda_i}(q^h-1))q^{\binom{m-1}{2}+\binom{m-2}{2}+\cdots}}$$

such that $Z(\overline{G}) \leq \ker f_\beta$.

Since P is cyclic, $p \neq 2$. Let e be the smallest integer i such that $p|q^i - 1$. By [**11**] $\mathrm{PSL}(2,q)$ and $\mathrm{PSL}(3,q)$ have p-blocks of defect 0. If $n \geq 4$, by Lemma 2, $e \leq n \leq 2e - 1$. Set $d = n - e$. Then $d < e$ and $n = e + d$. If $d < e - 1$, set $\beta = [\lambda_1, \lambda_2, \lambda_3] = [1, d+1, e+1]$. Then the corresponding unipotent character f_β of $\overline{G}$ is of degree

$$\begin{aligned} f_\beta(1) &= \frac{(q^{e+1}-q^{d+1})(q^{e+1}-q)(q^{d+1}-q)(q-1)(q^2-1)\cdots(q^n-1)}{(q-1)^2(q^2-1)^2(q^3-1)^2\cdots(q^{d+1}-1)^2(q^{d+2}-1)\cdots(q^{e+1}-1)q} \\ &= p^a *, \end{aligned}$$

where $*$ is an integer prime to p. If $d = e - 1$, then $e = d + 1$, $n = 2d + 1$. If d is even, let $d = 2t$. Then $2t + 1$. Set $\beta = [2, 2t-1, 2t+3]$ for $t > 1$ and $\beta = [1, 2, 5]$ for $t = 1$. If d is odd, let $d = 2t + 1$; set $\beta = [2, 2t, 2t+4]$ for $t > 1$ and $\beta = [0, 3, 7]$ for $t = 1$. Then $p^a|f_\beta(1)$. Therefore G has a p-block of defect 0.

Ito [**5**] gave an example showing that if the "cyclic condition" in Theorem 4 is discarded, the theorem is not true.

COROLLARY 5. *Let G be a finite group and P a Sylow p-subgroup of G. If $P/O_p(G)$ is cyclic, then G has a p-block with defect group $O_p(G)$.*

COROLLARY 6 (MICHLER [**19**]). *If a finite simple group G has a cyclic Sylow p-subgroup, then G has nonprincipal p-blocks.*

ACKNOWLEDGMENT. The author would like to thank his supervisors Professor Hsia-Fu Tuan and Associate Professor Shi Sheng-Ming for their helpful comments.

REFERENCES

1. R. Brauer and K. A. Fowler, *On groups of even order*, Ann of Math. (2) **62** (1955), 565–583.

2. Y. Tsushima, *On the weakly regular p-blocks with respect to* $O_{p'}(G)$, Osaka J. Math. **14** (1977), 465–470.

3. T. Wada, *On the existence of p-blocks with given defect groups*, Hokkaido Math. J. **6** (1977), 243–248.

4. Geoffrey R. Robinson, *The number of blocks with a given defect group*, J. Algebra **84** (1983), 493–502.

5. Walter Feit, *The representation theory of finite groups*, North-Holland, Amsterdam, 1982.

6. Robert H. Gilman and Robert L. Griess, Jr., *Finite groups with standard components of Lie type over fields of characteristic two*, J. Algebra **80** (1983), 383–516.

7. Robert L. Griess, Jr., *The friendly giant*, Invent. Math. **69** (1982), 1–102.

8. A. Wier, *Sylow p-subgroups of the classical groups over fields with characteristic prime to p*, Proc. Amer. Math. Soc. **6** (1955), 529–533.

9. Richard Brauer, *On finite groups with cyclic Sylow subgroups*. I, J. Algebra **40** (1976), 556–584.

10. George Lusztig, *Characters of reductive groups over a finite field*, Princeton Univ. Press, Princeton, N. J., 1984.

11. J. S. Frame and W. A. Simpson, *Character tables for* $\mathrm{SL}(3,q)$, $\mathrm{Su}(3,q^2)$, $\mathrm{PSL}(3,q)$ *and* $\mathrm{Psu}(3,q^2)$, Canad. J. Math. **25** (1973), 486–494.

12. R. Burkharat, *Über die Zerlegunszahlen der suziki-gruppen* $\mathrm{Sz}(q)$, J. Algebra **59** (1979), 421–433.

13. Harold N. Ward, *On Ree's series of simple groups*, Trans. Amer. Math. Soc. **121** (1966), 62–89.

14. Morton E. Harris, *On the p-deficiency class of a finite group*, J. Algebra **94** (1985), 411–424.

15. P. Brockhaus and G. O. Michler, *Finite simple groups of Lie type have non-principal p-blocks,* $p \neq 2$, J. Algebra **94** (1985), 113–125.

16. Z. Janko, *A new finite simple group of order* $2^{21} \cdot 3^3 \cdot 5 \cdot 7 \cdot 11^3 \cdot 23 \cdot 29 \cdot 31 \cdot 37 \cdot 43$, J. Algebra **42** (1976), 564–596.

17. Larry Finkelstein, *Finite groups in which the centralizer of an involution is* O, J. Algebra **32** (1974), 173–177.

18. W. Scott and F. Gross, *Proceedings of the conference on finite groups*, Academic Press, New York, 1976.

19. G. O. Michler, *Non-solvable finite groups with cyclic Sylow p-subgroups have non-principal p-blocks*, J. Algebra **83** (1983), 179–188.

20. J. H. Conway et al., *Atlas of finite groups*, Clarendon, Oxford, 1985.

PEKING UNIVERSITY, BEIJING, CHINA

Proceedings of Symposia in Pure Mathematics
Volume **47** (1987)

The Loewy Structure of Projective Indecomposable Modules for Some Modular Group Algebras

SHIGEO KOSHITANI

This is a research announcement of our recent result [**5**].

Let G be a finite group, let F be a field of characteristic $p > 2$ such that F is a splitting field for all subgroups of G, and let FG be the group algebra of G over F. For a finitely generated right FG-module M, let $j(M)$ be the Loewy length of M; that is, $j(M)$ is the least positive integer j such that $MJ^j = 0$ where $J = J(FG)$ is the Jacobson radical of FG. In our previous papers [**2**, **3**], we investigated the Loewy series of the projective indecomposable modules (p.i.m.'s) in p-blocks B of G under the assumption that G is p-solvable and the defect group D of B is extra-special of order p^3 of exponent p, namely

$$\begin{aligned} D &= M(p) = p_+^{1+2} \\ &= \langle a, b, c \mid a^p = b^p = c^p = 1,\ a^{-1}ba = bc,\ a^{-1}ca = c,\ b^{-1}cb = c\rangle; \end{aligned}$$

in particular, we got $j(B) = 4p-3$. Hence, as the next step, we want to know the Loewy series of the p.i.m.'s of the principal 3-blocks of simple groups G whose Sylow 3-subgroups are $M(3)$. So from now on we restrict ourselves to the case $p = 3$. For the past two years, we have been interested in the p.i.m.'s for the Tits simple group

$$T = {}^2F_4(2)' \quad \text{of order } 17971200 (= 2^{11} \cdot 3^3 \cdot 5^2 \cdot 13)$$

in characteristic 3 because $M(3)$ is the Sylow 3-subgroup of T. For this purpose we had to compute the Loewy series of the p.i.m.'s for $\mathrm{SL}(3,3)$ and its automorphism group $M = \mathrm{Aut}(\mathrm{SL}(3,3))$, which is the semidirect product of $\mathrm{SL}(3,3)$ by the cyclic group of order 2, in the same characteristic 3 because M is a unique maximal subgroup of T whose index is prime to 3 (see [**6**]). Though M is maximal in T, its index is 1600, which seems quite large. That means the distance between the p.i.m.'s of FM and those of FT is big. Recently, however, we could

1980 *Mathematics Subject Classification* (1985 *Revision*). Primary 20C05, 16A26.

The final detailed version of this note will be submitted for publication elsewhere.

completely determine the Loewy series of the p.i.m.'s for the Tits simple group $T = {}^2F_4(2)'$ and its automorphism group $\mathrm{Aut}(T) = {}^2F_4(2)$ in characteristic 3.

We write $I = I_G$ for the trivial FG-module of dimension one, and we denote other simple FG-modules by their dimension, together with a subscript if there is more than one simple FG-module of the same dimension which is not the dual of the first one.

By the result of G. Hiss [**1**], all simple FT-modules in the principal 3-block are

$$I,\ 26,\ 26^*,\ 77,\ 124_1,\ 124_2,\ 572,\ 1099_1,\ \text{and } 1099_2,$$

where 26^* is the dual of 26, and the other 3-blocks 27, 27^*, 351_0, 351, 351^*, 675, 702_1, 702_2, and 1728 are of defect 0. Our result is

THEOREM. *The p.i.m.'s in the principal 3-block of the Tits simple group ${}^2F_4(2)'$ have Loewy series*

$$\begin{array}{c} I \\ 77 \\ I\ 572 \\ 77 \\ I \end{array} \qquad \begin{array}{c} 26 \\ 124_1\ 572 \\ 26\ 26^*\ 26^*\ 77\ 1099_2 \\ 124_1\ 124_2\ 124_2\ 572 \\ 26\ 26\ 26^*\ 77\ 1099_1 \\ 124_1\ 572 \\ 26\ 26^*\ 26^* \\ 124_2\ 572 \\ 26 \end{array} \qquad \begin{array}{c} 26^* \\ 124_2\ 572 \\ 26\ 26\ 26^*\ 77\ 1099_1 \\ 124_1\ 124_1\ 124_2\ 572 \\ 26\ 26^*\ 26^*\ 77\ 1099_2 \\ 124_2\ 572 \\ 26\ 26\ 26^* \\ 124_1\ 572 \\ 26^* \end{array}$$

$$\begin{array}{c} 77 \\ I\ 124_1\ 124_2\ 572 \\ 26\ 26^*\ 77\ 77\ 77\ 1099_1\ 1099_2 \\ I\ 124_1\ 124_1\ 124_2\ 124_2\ 572 \\ 26\ 26^*\ 77\ 77\ 1099_1\ 1099_2 \\ 124_1\ 124_2\ 572 \\ 77 \end{array}$$

$$\begin{array}{c} 124_1 \\ 26^*\ 77\ 1099_2 \\ 124_1\ 124_2\ 124_2\ 572 \\ 26\ 26\ 26^*\ 77\ 77\ 1099_1 \\ 124_1\ 124_1\ 124_2\ 572 \\ 26^*\ 77\ 1099_2 \\ 124_2 \\ 26 \\ 124_1 \end{array} \qquad \begin{array}{c} 124_2 \\ 26\ 77\ 1099_1 \\ 124_1\ 124_1\ 124_2\ 572 \\ 26\ 26^*\ 26^*\ 77\ 77\ 1099_2 \\ 124_1\ 124_2\ 124_2\ 572 \\ 26\ 77\ 1099_1 \\ 124_1 \\ 26^* \\ 124_2 \end{array}$$

$$\begin{array}{ccc}
572 & 1099_1 & 1099_2 \\
26\ 26^*\ 77 & 124_2 & 124_1 \\
I\ 124_1\ 124_2\ 572 & 26\ 77 & 26^*\ 77 \\
26\ 26^*\ 77\ 1099_1\ 1099_2 & 124_1\ 572 & 124_2\ 572 \\
124_1\ 124_2\ 572 & 26^*\ 77 & 26\ 77 \\
26\ 26^*\ 77 & 124_2 & 124_1 \\
572 & 1099_1 & 1099_2, \\
26\ 26^* & & \\
572 & &
\end{array}$$

in particular, $j(FT) = 9$.

Let $A = \mathrm{Aut}({}^2F_4(2)') = {}^2F_4(2)$. Then all simple FA-modules in the principal 3-block are

$$I_A,\ 1,\ 52,\ 77_1,\ 77_2,\ 248,\ 572_1,\ 572_2,\ \text{and}\ 2198,$$

where 1 is a simple FA-module of dimension one which is not the trivial FA-module I_A, and other blocks are of defect 0. Since $|A : T| = 2$ and it is prime to 3, from the theorem we automatically get

COROLLARY 1. *We get the Loewy series of all p.i.m.'s for* $A = {}^2F_4(2)$ *in characteristic* 3.

By making use of the theorem, Corollary 1, and our previous results [**2**, **3**, **4**], we obtain

COROLLARY 2. *If* F *is an arbitrary field of characteristic* 3 *and if* G *is one of the types*

(i) 3-*solvable groups with Sylow* 3-*subgroups* $M(3)$,

(ii) $\mathrm{SL}(3,3)$ *or* $\mathrm{Aut}(\mathrm{SL}(3,3))$,

(iii) ${}^2F_4(2)'$ *or* ${}^2F_4(2)$,

then $j(FG) = 9$.

REFERENCES

1. G. Hiss, *The modular characters of the Tits simple group and its automorphism group*, Comm. Algebra **14** (1986), 125–154.

2. S. Koshitani, *On the Loewy series of the group algebra of a finite p-solvable group with p-length* > 1, Comm. Algebra **13** (1985), 2175–2198.

3. ——, *On group algebras of finite groups*, Representation Theory. II: Groups and Orders (Proc. Ottawa 1984) (V. Dlab, P. Gabriel, and G. Michler, eds.), Lecture Notes in Math., vol. 1178, Springer-Verlag, 1986, pp. 109–128.

4. ——, *The Loewy structure of the projective indecomposable modules for* SL(3,3) *and its automorphism group in characteristic* 3, Comm. Algebra (to appear).

5. ——, *The Loewy structure of the projective indecomposable modules for the Tits simple group and its automorphism group in characteristic* 3, Preprint, 1986.

6. R. A. Wilson, *The geometry and maximal subgroups of the simple groups of A. Rudvalis and J. Tits*, Proc. London Math. Soc. (3) **48** (1984), 533–563.

CHIBA UNIVERSITY, JAPAN

Proceedings of Symposia in Pure Mathematics
Volume **47** (1987)

The λ-Structure of the Green Rings of Cyclic p-Groups

FRANK M. KOUWENHOVEN

In this paper we present new results about the exterior powers of certain indecomposable representations of cyclic p-groups in characteristic p, p being a prime number. The proofs, which were omitted in the conference talk, are included here.

The organization of this paper is as follows. In the first section we explain the terminology and notation used, but also the special place cyclic p-groups take among all finite groups is indicated. Moreover a result, not noticed in the literature, is proved which, together with its corollary, constitutes good reason for investigating exterior powers in general, especially in positive characteristics. In the second section are collected results about the group of order p; some of them are new. In the last section we present some new results for the cyclic p-groups of order bigger than p. We finish by formulating a challenging conjecture.

The results below are part of the author's thesis [**6**], dealing with Schur functors (exterior powers are examples) in a more general way as well.

1. The ring- and λ-structure of representation rings, and cyclic p-groups. Let K be a field and G a group (no restrictions for the moment). Let $\mathrm{Rep}_K(G)$ be the abelian category of the left $K[G]$-modules which are finite-dimensional over K. To study this category there are two types of representation rings available:

$R_K(G)$, the Grothendieck ring,

$RS_K(G)$, the Green ring. Recall that for $M \in \mathrm{Rep}_K(G)$ the class $[M]$ of M in $R_K(G)$, resp. $RS_K(G)$, determines M up to composition factors, resp. up to indecomposable summands.

The multiplicative structure on the representation rings comes from the tensor product over K (with diagonal group action) and turns both into commutative rings with unit element.

1980 *Mathematics Subject Classification* (1985 *Revision*). Primary 20C30, 15A69; Secondary 20C20, 15A75.

This paper is in final form and no version of it will be submitted for publication elsewhere.

The λ-structure on the representation rings comes from the exterior powers as follows. Let $M \in \mathrm{Rep}_K(G)$ and $n \in \mathbf{N}_0$, let N_n be the submodule of $M^{\otimes n}$ generated by $(m_1 \otimes m_2 \otimes \cdots \otimes m_n \mid m_i \in M$ all i, $m_i = m_{i+1}$ for some $i)$. Then $\Lambda^n(M) = (M^{\otimes n}/N_n) \in \mathrm{Rep}_K(G)$ is called the nth exterior power of M. Some properties of exterior powers are: $\Lambda^n(M \oplus N) \cong \bigoplus_{i=0}^n \Lambda^i(M) \otimes_K \Lambda^{n-i}(N)$, when $M, N \in \mathrm{Rep}_K(G)$ and $n \in \mathbf{N}_0$, and if $0 \to M \to L \to N \to 0$ is an exact sequence in $\mathrm{Rep}_K(G)$ then $\Lambda^n(L)$ admits a filtration $\Lambda^n(L) = L_0 \supset L_1 \supset \cdots \supset L_{n+1} = 0$ such that $L_i/L_{i+1} \cong \Lambda^i(M) \otimes_K \Lambda^{n-i}(N)$, for $0 \le i \le n$. One can now define operations (i.e., endomaps) λ_n, for $n \in \mathbf{N}_0$, on the representation rings via $\lambda_n[M] = [\Lambda^n M]$, for $M \in \mathrm{Rep}_K(G)$.

Some facts for finite groups G (we will describe λ-rings below):

	$R_K(G)$	$RS_k(G)$
additive structure	finite rank	finite rank $\Leftrightarrow$ the p-Sylow subgroups of G are cyclic
multiplicative structure	no nilpotent elements	up to exceptions in characteristic two: no nilpotent elements $\Leftrightarrow$ the p-Sylow subgroups of G are cyclic
λ-structure	λ-ring	λ-ring $\Leftrightarrow$ $\mathrm{char}(K) \nmid \lvert G \rvert$

From this table it is clear that finite groups with cyclic p-Sylow subgroups take a special place and that, in order to understand the λ-structure of Green rings, we have to understand the case of cyclic p-groups first.

We shall now describe what λ-rings are. Let A be a commutative ring with unit element; then $W(A)\colon = \{x \in A[[t]] \mid x = 1 + \sum_{n=1}^{\infty} a_n t^n$ for $a_n \in A$, all $n \in \mathbf{N}\}$ is called the ring of big Witt vectors over A. The addition on $W(A)$ is the power series multiplication. For a group G, let $\mathrm{Cl}(G)$ be a transversal for the conjugacy classes of G; then there is an additive map: $\mathrm{ch}\,\lambda_t(G)\colon R_K(G) \to \prod_{g \in \mathrm{Cl}(G)} W(K)$, defined by $[M] \mapsto (\sum_{n=0}^{\infty} \mathrm{trace}(\Lambda^n g_M) t^n)$. The multiplication and the λ-operations λ_n ($n \in \mathbf{N}_0$) on $W(A)$ are universally defined, and determined, such that for all G and K the map $\mathrm{ch}\,\lambda_t(G)$ is a ring homomorphism which commutes with the λ-operations. An explicit description of λ-rings appears in [**4**]. Let A be equipped with operations λ_n for $n \in \mathbf{N}_0$. Then $(A, (\lambda_n)_n)$ is called a λ-ring if the map $\lambda_t\colon A \to W(A)$, defined by $a \mapsto \sum_{n=0}^{\infty} \lambda_n(a) t^n$, is a ring homomorphism which commutes with the λ-operations.

It is easy to see that $W(A)$ is a λ-ring. Moreover $W(K)$ is $\mathbf{Z}$-torsion free and has no nilpotent elements. For more about λ-rings see the remarks following Corollary 1.1.

THEOREM 1.1 (NOT REALLY NEW). *$\mathrm{ch}\,\lambda_t(G)\colon R_K(G) \to \prod_{g \in \mathrm{Cl}(G)} W(K)$ is injective.*

PROOF. Any element of $R_K(G)$ is of the form $[M] - [N]$ with $M, N \in \mathrm{Rep}_K(G)$. We may assume that M and N are isomorphic to the direct sum of their composition factors, i.e., M and N are semisimple. In order to prove the injectivity of $\operatorname{ch}\lambda_t(G)$ we may assume that the composition factors of M and of N are absolutely irreducible. To see this, observe that a finite field extension $K \subset L$ induces injections $R_K(G) \to R_L(G)$ and $W(K) \to W(L)$.

Let $I_1, I_2, \ldots, I_r$ be all the (mutually nonisomorphic) irreducibles in $M \oplus N$; then, by the Jacobson density theorem, we can choose, for each $i \in \{1, 2, \ldots, r\}$, an $x_i \in K[G]$ which has trace 1 on I_i and trace 0 on I_j for $j \neq i$.

Say $M \cong \bigoplus_i m_i I_i$ and $N \cong \bigoplus_i n_i I_i$ $(m_i, n_i \in \mathbf{N}_0)$ and suppose $\operatorname{ch}\lambda_t([M] - [N]) = 0$; then $\prod_i (\operatorname{ch}\lambda_t(G)[I_i])^{m_i} = \prod_i (\operatorname{ch}\lambda_t(G)[I_i])^{n_i}$, which means:

$$\prod_i \left(\sum_{n=0}^{\infty} \operatorname{trace}\Lambda^n(g_{I_i})t^n\right)^{m_i} \quad \prod_i \left(\sum_{n=0}^{\infty} \operatorname{trace}\Lambda^n(g_{I_i})t^n\right)^{n_i} \qquad \text{for } g \in G.$$

But $W(K)$ has no $\mathbf{Z}$-torsion so we may assume that

$$\text{g.c.d}(m_1, m_2, \ldots, m_r, n_1, n_2, \ldots, n_r) = 1,$$

and then we can specialize to traces for the x_i's to see $m_i = n_i$ for all i. □

REMARKS. From this theorem we see that the characteristic polynomials take over the role of the characters in positive characteristics (even for infinite groups) because $\sum_n \operatorname{trace}\Lambda^n(g_M)t^n = \det(t \cdot g_M + \mathrm{Id})$. Moreover they generalize Brauer characters then, because we may forget about group elements which act unipotently. For more, see the remarks following Corollary 1.1.

COROLLARY 1.1 (KNOWN). *$R_K(G)$ is a λ-ring which has no nilpotent elements.* □

REMARKS. If you happen to know the standard (combinatorial) definition of a λ-ring, this corollary may seem impressive. However, besides Theorem 1.1, it does not really tell anything extra.

Both the above theorem and its corollary require explanation and therefore constitute good reason for investigating exterior powers. Moreover the exterior power is an example of a Schur functor, the latter being important in the representation theory of the general linear groups.

Let A be equipped with operations λ_n, for $n \in \mathbf{N}_0$. Then operations ψ_n, for $n \in \mathbf{N}$, on A are defined by:

$$\sum_{n=0}^{\infty} (-1)^n \psi_{n+1} t^n = \frac{d}{dt} \log(\lambda_t)$$

(formal logarithm and derivative) or, equivalently,

$$\sum_{i=0}^{n-1} (-1)^i \psi_{n-i} \cdot \lambda_i + (-1)^n n \lambda_n = 0 \quad \text{for all } n \in \mathbf{N}.$$

The operation ψ_n is called the nth Adams operation and one has

$$(A,(\lambda_n)_n) \text{ is a } \lambda\text{-ring} \Rightarrow \begin{cases} \text{for all } a,b\in A \text{ and all } n,m\in\mathbf{N}: \\ \psi_n(a+b)=\psi_n(a)+\psi_n(b), \\ \psi_n(1)=1, \\ \psi_n(a\cdot b)=\psi_n(a)\cdot\psi_n(b), \\ \psi_n(\psi_m(a))=\psi_{nm}(a). \end{cases}$$

In case A is $\mathbf{Z}$-torsion free, the reverse implication holds as well. It is easy to check that $\operatorname{ch}\lambda_t(G)(\psi_m[M]) = \sum_n(\operatorname{trace}\Lambda^n(g_M^m)t^n)$, and by this we see that one can restrict the product $\prod_{g\in\mathrm{Cl}(G)}$ to the product over the generators for a transversal for the conjugacy classes of maximal cyclic subgroups of G!

When $(A,(\lambda_n^A)_n)$ and $(B,(\lambda_n^B)_n)$ are λ-rings then $A\otimes_{\mathbf{Z}} B$ is a λ-ring via the composition

$$A\otimes B \xrightarrow{\lambda_t^A\otimes\lambda_t^B} W(A)\otimes W(B) \xrightarrow{j_A\otimes j_B} W(A\otimes B)\otimes W(A\otimes B) \xrightarrow{\cdot_W} W(A\otimes B),$$

where $j_A\colon A\to A\otimes B$, $j_B\colon B\to A\otimes B$ are the natural homomorphisms and $\cdot_W$ the multiplication in the ring of big Witt vectors.

For more about λ-rings see, for example, [**4**].

So far we have not said anything about the symmetric powers. The symmetric powers, S^n, induce operations s_n for $n\in\mathbf{N}_0$, on both representation rings and they are related with the λ-operations via $\sum_{i=0}^n(-1)^i\lambda_i s_{n-i}=0$, for all $n\in\mathbf{N}$ on $R_K(G)$, for n with $\operatorname{char}(K)\nmid n$ on $RS_K(G)$.

From now on, let q be a power of p, C_q the cyclic group of order q, σ_q a generator of C_q, and K a field of characteristic p.

PROPOSITON 1.2 (FOLKLORE). (a) *The ring homomorphism* $K[X]\to K[C_q]$ *defined by* $X\mapsto\sigma_q$ *induces an isomorphism* $K[X]/(X-1)^q\xrightarrow{\sim}K[C_q]$.

(b) $(K[X]/(X-1)^n|1\le n\le q)$ *is a complete system of mutually inequivalent indecomposable finite-dimensional* $K[X]/(X-1)^q$*-modules.*

PROOF. (a) $(X-1)^q=X^q-1$ in $K[X]$.

(b) Follows from the structure theorem for finitely generated modules over the principal ideal domain $K[X]$. □

REMARK. The element X acts on $K[X]/(X-1)^n$ via a unipotent Jordan block of dimension n; therefore the study of all $\operatorname{Rep}_K(C_q)$ is equivalent to the study of all unipotent (finite-dimensional)matrices over K.

COROLLARY 1.2 (FOLKLORE). (a) $RS_K(C_q)$ *is a free* $\mathbf{Z}$*-module with basis* $([K[X]/(X-1)^n]|1\le n\le q)$, *and the field extension* $\mathbf{F}_p\subset K$ *induces an isomorphism* $RS_{\mathbf{F}_p}(C_q)\xrightarrow{\sim}RS_K(C_q)$.

(b) *The epimorphism* $C_{p\cdot q}\to C_q$, *defined by* $\sigma_{p\cdot q}\mapsto\sigma_q$, *induces an injective homomorphism* $RS_{\mathbf{F}_p}(C_q)\to RS_{\mathbf{F}_p}(C_{p\cdot q})$ *mapping* $[\mathbf{F}_p[X]/(X-1)^n]$ *to* $[\mathbf{F}_p[X]/(X-1)^n]$ *for* $n\le q$. □

Let us denote $RS_{\mathbf{F}_p}(C_q)$ by $RS(q)$. By Corollary 1.2 we may consider the direct limit $RS(\infty) = \varinjlim_q RS(q)$ to deal with all the cyclic p-groups, the λ-structure included, over fields of characteristic p. The class of $\mathbf{F}_p[X]/(X-1)^n$ in $RS(q)$, for $n \leq q$, shall be denoted by V_n. The limit ring $RS(\infty)$ is in fact the Green ring of the continuous representations of the p-adics, as additive group, over $\mathbf{F}_p$.

2. The group of order p. For $n \in \mathbf{N}_0$, let $W_n(X) \in \mathbf{Z}[X]$ be defined by

$$\frac{1}{1-Xt+t^2} = \sum_{n=0}^{\infty} W_n(X)t^n.$$

The polynomial $W_n(2X)$ is the classical Chebyshev polynomial of the second kind: $W_n(2\cos\phi) = \sin((n+1)\phi)/\sin\phi$, ($\phi \in \mathbf{R}$).

PROPOSITION 2.1 (FOLKLORE). *Let $n, m \in \mathbf{N}$, $m \leq n \leq p$.*

(a) $V_n = s_{n-1}(V_2) = W_{n-1}(V_2)$.

(b) $V_2 \cdot V_n = \begin{cases} 2 \cdot V_p & \text{if } n = p, \\ V_{n+1} + V_{n-1} & \text{otherwise.} \end{cases}$

(c) $V_n \cdot V_m = \begin{cases} \sum_{i=1}^{m} V_{n+m+1-2i} & \text{if } n+m \leq p+1, \\ V_{p-n} \cdot V_{p-m} + (n+m-p)V_p & \text{otherwise.} \end{cases}$

(d) *The ring homomorphism $\mathbf{Z}[X] \to RS(p)$, defined by $X \mapsto V_2$ induces an isomorphism $\mathbf{Z}[X]/((X-2) \cdot W_{p-1}(X)) \xrightarrow{\sim} RS(p)$.*

PROOF. (a) $S^{n-1}(\mathbf{F}_p[X]/(X-1)^2) \cong \mathbf{F}_p[X]/(X-1)^n$ follows by direct computation; see [**3**]. Also

$$\frac{1}{1-V_2t+t^2} = \frac{1}{1-\lambda_1(V_2)t+\lambda_2(V_2)t^2} \equiv \sum_{n=0}^{p-1} s_n(V_2) \pmod{t^p}$$

because $\sum_{i=0}^{n}(-1)^i\lambda_i \cdot s_{n-i} = 0$ for $n < p$ on $RS(p)$; therefore $s_{n-1}(V_2) = W_{n-1}(V_2)$.

(b) $V_2 \cdot V_p = 2V_p$: follows by Frobenius reciprocity. Moreover $X \cdot W_n(X) = W_{n+1}(X) + W_{n-1}(X)$, so by (a) we are done.

(c) Follows from (b) by induction.

(d) Follows from (a) and (b) and Corollary 1.2(a). □

REMARK. Part (c) also follows from the Clebsch-Gordan rule for $\mathrm{SL}(2, \mathbf{F}_p)$ because of (a) ($V_n = s_{n-1}(V_2)$).

COROLLARY 2.1. *Let $n \in \mathbf{N}$ and $n \leq p-1$; then V_n is not a zero divisor in $RS(p)$.*

PROOF. The ring homomorphism $i : RS(p) \to \mathbf{Z}[X]/(X-2) \times \mathbf{Z}[X]/W_{p-1}(X)$ defined by $V_n \mapsto (n, W_{n-1}(X))$ is injective by Proposition 2.1(a). Suppose $V_n \cdot y = 0$ for some $y \in RS(p)$; then $i(y) = (0, g(X))$ for a certain $g(X) \in \mathbf{Z}[X]$ of degree at most $p-2$. But then $W_{n-1}(X) \cdot g(X) = W_{p-1}(X) \cdot h(X)$ for some $h(X) \in \mathbf{Z}[X]$. Because $W_{n-1}(2\cos\phi) = \sin(n\phi)/\sin\phi$, the polynomial $W_{n-1}(X)$, for $n \leq p-1$, is relatively prime to $W_{p-1}(X)$ in $\mathbf{Q}[X]$, so $W_{p-1}(X)$ divides $g(X)$. But the latter implies, because of degrees, $g = 0$, i.e., $y = 0$. □

THEOREM 2.2.

(a) [3]. *If* $m, n \in \mathbf{N}$, $m \leq n \leq p$, $m \neq p$ *then:*

$$\lambda_m(V_n) = \frac{V_n \cdot V_{n-1} \cdot \dots \cdot V_{n-m+1}}{V_m \cdot V_{m-1} \cdot \dots \cdot V_1}.$$

(b) [2]. *The span of* $(V_n \mid 1 \leq n \leq p,\ n\,\mathrm{odd})$ *in* $RS(p)$ *makes up a subring,* $RS(p)^{\mathrm{odd}}$, *which is invariant for the* λ*-operations.*

(c) $RS(p)/(V_p - V_{p-1} - V_1)$ *is the biggest* $\mathbf{Z}$*-torsion free quotient of* $RS(p)$ *which is a* λ*-ring via the exterior powers. Moreover* $RS(p)^{\mathrm{odd}}$ *is isomorphic to this ring, as a* λ*-ring.*

PROOF. (c) We shall investigate the Adams operations on $RS(p)$. First of all, because of the natural isomorphisms

$$\Lambda^n(M \oplus N) \cong \bigoplus_{i=0}^{n} \Lambda^i(M) \otimes_{\mathbf{F}_p} \Lambda^{n-i}(N)$$

(for M and N $\mathbf{F}_p[C_p]$-modules and $n \in \mathbf{N}_0$) one has $\lambda_n(a+b) = \sum_{i=0}^n \lambda_i(a) \cdot \lambda_{n-1}(b)$ for all $a, b \in RS(p)$, and all $n \in \mathbf{N}_0$. Therefore

$$\begin{aligned}\sum_{n=0}^{\infty}(-1)^n \psi_{n+1}(a+b)t^n &= \frac{d}{dt}\log(\lambda_t(a+b)) = \frac{d}{dt}\log(\lambda_t(a)\cdot\lambda_t(b)) \\ &= \sum_{n=0}^{\infty}(-1)^n(\psi_{n+1}(a) + \psi_{n+1}(b))t^n,\end{aligned}$$

i.e, ψ_n is additive for $n \in \mathbf{N}$. For $n \in \mathbf{Z}$ the element $X^n + X^{-n} \in \mathbf{Z}[X, X^{-1}]$ is a polynomial over $\mathbf{Z}$ in $X + X^{-1}$, say $X^n + X^{-n} = f_n(X + X^{-1})$. Clearly $(X^n + X^{-n})(X^m + X^{-m}) = X^{n+m} + X^{-n-m} + X^{n-m} + X^{m-n}$, for $n, m \in \mathbf{Z}$. From Proposition 2.1(b) one easily derives, for $n \in \mathbf{Z}$:

$$f_n(V_2) = \begin{cases} 2V_1 & \text{if } n \equiv 0 \bmod 2p, \\ V_{j+1} - V_{j-1} & \text{if } n \equiv j \bmod 2p \text{ with } 1 \leq j \leq p-1, \\ 2(V_p - V_{p-1}) & \text{if } n \equiv p \bmod 2p, \\ V_{p-j+1} - V_{p-j-1} & \text{if } n \equiv (p+j) \bmod 2p \text{ with } 1 \leq j \leq p-1. \end{cases}$$

Put $\mu(n) = f_n(V_2)$, for $n \in \mathbf{Z}$; then $\mu(n) \cdot \mu(m) = \mu(n+m) + \mu(n-m)$ and $\mu(n)$ corresponds to $\mu^n + \mu^{-n}$ in [1]. In [1] a formula is given for $\psi_m(V_n)$, for $1 \leq n \leq p$, which yields after a straightforward calculation:

$$\psi_m(\mu(n)) = \begin{cases} \mu(nm) \text{ if } n \not\equiv p \bmod 2p \text{ or } m \not\equiv p \bmod 2p, \\ \mu(nm) + 4(V_2 - 2V_1) \text{ otherwise, in case } p = 2, \\ \mu(nm) + 2p(V_1 + V_{p-1} - V_p) \text{ otherwise,} \end{cases}$$

for $m \in \mathbf{N}$, $n \in \mathbf{Z}$.

Calculating further we find for $n_1, n_2, m \in \mathbf{N}$ with $n_1, n_2 \le p-1$:

$$\psi_m(\mu(n_1)\cdot\mu(n_2)) - \psi_m(\mu(n_1))\cdot\psi_m(\mu(n_2))$$
$$= \begin{cases} 0 & \text{if } m \not\equiv p \bmod 2p \text{ or } n_1+n_2, n_1-n_2 \not\equiv p \bmod 2p, \\ 4(V_2 - 2V_1) & \text{otherwise, in case } p=2, \\ 2p(V_1 + V_{p-1} - V_p) & \text{otherwise.} \end{cases}$$

Moreover for $m_1, m_2, n \in \mathbf{N}$ with $n \le p-1$:

$$\psi_{m_1}\circ\psi_{m_2}(\mu(n)) - \psi_{m_1\cdot m_2}(\mu(n)) = \begin{cases} 0 & \text{if } m_1 \not\equiv p \bmod 2p \text{ or } m_2, n \not\equiv p \bmod 2p, \\ 4(V_2 - 2V_1) & \text{otherwise, in case } p=2, \\ 2p(V_1 + V_{p-1} - V_p) & \text{otherwise.} \end{cases}$$

In particular,

$$\psi_m(V_p - V_{p-1} - V_1) = \begin{cases} V_p - V_{p-1} - V_1 & \text{if } n \not\equiv 0 \bmod p \text{ and } n \text{ is odd,} \\ 0 & \text{if } (n \not\equiv 0 \bmod p \text{ and } n \text{ is even}) \text{or } n \equiv 0 \bmod 2p, \\ (p-1)(V_1 + V_{p-1} - V_p) & \text{if } n \equiv p \bmod 2p \text{ and } p \text{ is odd,} \\ 2(V_2 - 2V_1) & \text{if } n \equiv 2 \bmod 4 \text{ and } p = 2, \end{cases}$$

for $m \in \mathbf{N}$.

Therefore $RS(p)/(V_p - V_{p-1} - V_1)$ is the biggest $\mathbf{Z}$-torsion free quotient on which the induced Adams operations satisfy the properties needed. However, we have to prove that the λ-operations induce operations on this quotient as well.

CLAIM. $\lambda_t(V_p - V_{p-1}) = 1 + (V_p - V_{p-1})t + \sum_{i=1}^{\infty}(V_1 + V_{p-1} - V_p)(t^{ip} - t^{ip+1})$.

PROOF OF THE CLAIM. $\lambda_t(V_{p-1})\cdot(1 + (V_p - V_{p-1})t + \sum_{i=1}^{\infty}(V_1 + V_{p-1} - V_p)(t^{ip} - t^{ip+1}))$ equals $1 + t^p$ modulo (V_p), by Proposition 2.1(c). But the dimension of the coefficients of this product is all right, so we are done. From the claim it follows that $\lambda_t(V_p - V_{p-1} - V_1) \equiv 1 \bmod (V_p - V_{p-1} - V_1)$. From the formula for $\psi_m(V_p - V_{p-1} - V_1)$ and the relation $\lambda_m = 1/m\sum_{i=1}^{m}(-1)^{i-1}\psi_i\cdot\lambda_{n-i}$ it follows that $\lambda_t(V_n(V_p - V_{p-1} - V_1)) \equiv 1 \bmod (V_p - V_{p-1} - V_1)$, so λ_t factorizes through $RS(p)/(V_p - V_{p-1} - V_1)$, which is therefore a λ-ring. The restriction of the quotient map induces a ring isomorphsim: $RS(p)^{\text{odd}} \xrightarrow{\sim} RS(p)/(V_p - V_{p-1} - V_1)$, because $V_n(V_p - V_{p-1} - V_1) = V_p - V_{p-n} - V_n$ by Proposition 2.1(c), and this isomorphism clearly commutes with the λ-operations. □

REMARKS. In part (a) we need to know Corollary 2.1. From the proof of (c) we learn that the Adams operations on $RS(p)$

$$\text{satisfy} \begin{cases} \psi_m = \psi_{m+2p}, & \\ \psi_m = \psi_{2p-m} & \text{if } m \le p, \\ \psi_m & \text{is a ring homomorphism if } m \not\equiv p \bmod 2p, \\ \psi_m\psi_n = \psi_{mn} & \text{if } m \not\equiv p \bmod 2p, \end{cases}$$

for $m, n \in \mathbf{N}$.

The symmetric powers can be described in a "binomial" way as well (see [**3**]) and for $n \le p$, (see [**2**]):

$$\left(\sum_{i=0}^{\infty}(-1)^i\lambda_i(V_n)t^i\right)\left(\sum_{i=0}^{\infty}s_i(V_n)t^i\right)\begin{bmatrix} 1 & \text{if } n \text{ is odd,} \\ \dfrac{1-(V_p - V_{p-1})t^p}{1-t^p} & \text{if } n \text{ is even.} \end{bmatrix}$$

3. The cyclic p-groups of order bigger than p.

PROPOSITION 3.1. (a) [**7**], *reformulated. Let* $n, m \in \mathbf{N}$, $m \leq n \leq p \cdot q$. *Write* $n = n_0 q + n_1$ *and* $m = m_0 q + m_1$ *with* $n_0, n_1, m_0, m_1 \in \mathbf{N}_0$ *and* $m_1, n_1 \leq q-1$. *Suppose* $V_{n_1} \cdot V_{m_1} = \sum_s a_s(n_1, m_1) V_s$.

If $m + n \leq pq$ *then:*

$$\begin{aligned} V_n \cdot V_m = & \sum_{i=1}^{m_0} \sum_{s=1}^{q-1} a_s(n_1, m_1)(V_{(n_0+m+2-2i)q+s} + V_{(n_0+m_0+2-2i)q-s}) \\ & + \sum_{s=1}^{q-1} a_s(n_1, m_1) V_{(n_0-m_0)q+s} + |n_1 + m_1 - q| \sum_{i=1}^{m_0} V_{(n_0+m_0+1-2i)q} \\ & + |m_1 - n_1| \sum_{i=1}^{m_0} V_{(n_0+m_0+2-2i)q} + \max(0, m_1 + n_1 - q) V_{(n_0+m_0+1)q} \\ & + \max(0, m_1 - n_1) V_{(n_0-m_0)q} \qquad (|\ |\ \textit{denotes absolute value}). \end{aligned}$$

If $m + n > pq$ *then* $V_n \cdot V_m = V_{pq-n} \cdot V_{pq-m} + (n + m - q) V_{pq}$.

(b) [**5**]. *Say* $q = q^k$; *then there is a ring isomorphism* $RS(p)^{\otimes k} \tilde{\to} RS(q)$ *defined by*

$$1 \otimes \cdots \otimes 1 \otimes \underset{\substack{\uparrow \\ i\text{th factor}}}{V_2} \otimes 1 \otimes \cdots \otimes 1 \mapsto V_{2p^{i-1}} - V_{2p^{i-1}-1} + V_1 \qquad (1 \leq i \leq k). \quad \square$$

REMARK. The presentation for $RS(q)$ in [**5**], from which part (b) above follows, can be proved directly.

COROLLARY 3.1. *Say* $q = p^k$; *then the isomorphism in Proposition* 3.1(b) *induces a ring isomorphism* $(RS(p)/(V_p - V_{p-1} - V_1))^{\otimes k} \cong RS(q)/I_q$, *where* I_q *is the ideal generated by the* $V_{p^i} - V_{p^i-1} - V_1$, *for* $1 \leq i \leq k$.

PROOF. If $p = 2$ then $1 \otimes \cdots \otimes 1 \otimes (V_2 - V_1 - V_1) \otimes 1 \otimes \cdots \otimes 1$ is mapped, by the isomorphism in Proposition 3.1(b), to $V_{2^i} - V_{2^i-1} - V_1$, which proves the result for $p = 2$. Let p be odd now; let $Y_i = V_{2p^i} - V_{2p^i-1} + V_1$, $X_i = V_{p^i+1} - V_{p^i-1}$, and $U_i = V_{p^i} - V_{p^i-1}$, for $0 \leq i \leq k-1$. Then $1 \otimes \cdots \otimes 1 \otimes (V_p - V_{p-1} - V_1) \otimes 1 \otimes \cdots \otimes 1$ is mapped to $W_{p-1}(Y_{i-1}) - W_{p-2}(Y_{i-1}) - V_1$ (see Proposition 2.1(a)). By Proposition 3.1(a) we see that $U_i^2 = 1$, $Y_i = U_i \cdot X_i$ and therefore:

$$W_{p-1}(Y_{i-1}) - W_{p-2}(Y_{i-1}) - V_1 = W_{p-1}(X_{i-1}) - U_{i-1} W_{p-2}(X_{i-1}) - V_1.$$

From a straightforward calculation it follows from Proposition 3.1(a) that

$$W_{p-1}(X_{i-1}) - U_{i-1} W_{p-2}(X_{i-1}) - V_1 = V_{(p-1)p^{i-1}+1} - V_{(p-1)p^{i-1}} - V_1.$$

By Proposition 3.1(a) again,

$$V_{(p-1)p^{i-1}+1} - V_{(p-1)p^i} - V_1 = U_i \cdot U_{i-1} - V_1 = (U_i - 1) U_{i-1} + (U_{i-1} - V_1)$$

and

$$U_i - V_1 = \sum_{j=0}^{i-1} [(V_{(p-1)p^j+1} - V_{(p-1)p^j} - V_1) \cdot U_j].$$

By induction on k the result for p is odd now follows. □

Let Res: $RS(\infty) \to RS(\infty)$ be the ring homomorphism induced by the ordinary restriction maps $RS(p \cdot q) \to RS(q)$ (for all q).

Let Ind: $RS(\infty) \to RS(\infty)$ be the additive homomorphism induced by the ordinary induction maps $RS(q) \to RS(p \cdot q)$ (for all q).

PROPOSITION 3.2. (a) [5], *reformulated. Let* $n \in \mathbf{N}$, *write* $n = n_0 \cdot p + n_1$ *with* $n_0, n_1 \in \mathbf{N}_0$, $n_1 \leq p-1$. *Then* $\operatorname{Res} V_n = n_1 \cdot V_{n_0+1} + (p-n_1)V_{n_0}$ *and* $\operatorname{Ind} V_n = V_{p \cdot n}$.

(b) [3]. *Let* $m \in \mathbf{N}$, *write* $m = m_0 \cdot q + m_1$ *with* $m_0, m_1 \in \mathbf{N}_0$, $m_1 \leq q-1$. *Then* $V_m \cdot V_q = m_1 V_{(m_0+1)q} + (q-m_1)V_{m_0 q}$.

(c) *Let* $n \in \mathbf{N}$ *and suppose* $n \leq q$; *then there exists an exact sequence in*

$$\operatorname{Rep}_{\mathbf{F}_p}(C_q)\colon 0 \to \mathbf{F}_p[X]/(X-1)^n \xrightarrow{f_n} \mathbf{F}_p[X]/(X-1)^q \xrightarrow{g_n} \mathbf{F}_p[X]/(X-1)^{q-n} \to 0$$

defined by $f_n(1) = (X-1)^{q-n}$ *and* $g_n(1) = 1$. *In particular,* f_n *is the injective hull of* $\mathbf{F}_p[X]/(X-1)^n$ *in* $\operatorname{Rep}_{\mathbf{F}_p}(C_q)$.

PROOF. (c) The sequence is clearly an exact sequence in $\operatorname{Rep}_{\mathbf{F}_p}(C_q)$. The module $\mathbf{F}_p[X]/(X-1)^q$ is indecomposable and, because it is isomorphic to the group algebra, it is injective as well. □

Let $\Omega_q : RS(q) \to RS(q)$ be the additive homomorphism defined by $V_n \mapsto V_{q-n}$ $(1 \leq n \leq q)$. The connection of Ω_q with injective hulls is described in Proposition 3.2(c). In fact Ω_q is the Heller operation; moreover $\Omega_q(x) \cdot y \equiv x \cdot \Omega_q(y) \equiv \Omega_q(xy) (\operatorname{mod} \mathbf{Z} \cdot V_q)$ and, when $q \geq p$, $\Omega_{qp^{-1}}(\operatorname{Res}(x)) = \operatorname{Res}(\Omega_q(x))$, for all $x, y \in RS(q)$.

PROPOSITION 3.3. *Let* $m, n \in \mathbf{N}_0$, $M \in \operatorname{Rep}_{\mathbf{F}_p}(C_q)$.

(a) *If* $m + nq < pq$ *then* $\Lambda^{nq+m}(M)$ *is a summand of* $\Lambda^{nq}(M) \otimes_{\mathbf{F}_p} \Lambda^m(M)$ *and, if* p *is odd,* $\Lambda^{nq}(M)$ *is a summand of* $\Lambda^n(\Lambda^q M)$.

(b) *If* $d = \dim(M)$ *and* $m \leq d$ *then* $\Lambda^m(M) \cong \Lambda^{d-m}(M)$.

(c) *If* $n \leq q-1$ *then there is an exact sequence in* $\operatorname{Rep}_{\mathbf{F}_p}(C_q)$:

$$0 \to \Lambda^{m-1}(\mathbf{F}_p[X]/(X-1)^n) \xrightarrow{\alpha} \Lambda^m(\mathbf{F}_p[X]/(X-1)^{n+1}) \xrightarrow{\beta} \Lambda^m(\mathbf{F}_p[X]/(X-1)^n) \to 0,$$

where α *is the multiplication by* $(X-1)^n$ *and* β *is induced by the ordinary residue map.*

(d) *Write* $m = m_0 \cdot p^j$ *with* $j \in \mathbf{N}_0$ *such that* $p \nmid m_0$ *if* $p^j < q$ *and* $p^j = q$ *if* $q \mid m$. *Then*

$$\lambda_m(V_q) = \sum_{i=1}^{j} \left[\frac{\binom{q \cdot p^{i-j}}{m_0 \cdot p^i} - \binom{q \cdot p^{i-j-1}}{m_0 \cdot p^{i-1}}}{q \cdot p^{i-j}} V_{q \cdot p^{i-j}} \right] + \frac{\binom{q \cdot p^{-j}}{m_0}}{q \cdot p^{-j}} \cdot V_{q \cdot p^{-j}}.$$

PROOF. (a) $\binom{nq+m}{m} \in \mathbf{F}_p^*$ so the natural map $\Lambda^{nq+m}(M) \to \Lambda^{nq}(M) \otimes_{\mathbf{F}_p} \Lambda^m(M)$ (comultiplication) is split by the multiplication map. Because

$$(nq)!/(q!)^n \cdot n! \in \mathbf{F}_p^*$$

the natural map $\Lambda^{nq}(M) \to \Lambda^n(\Lambda^q M)$ is split by the natural map $\Lambda^n(\Lambda^q M) \to \Lambda^{nq}(M)$.

(b) There is a natural isomorphism $\Lambda^m(M) \tilde{\to} \operatorname{Hom}_{\mathbf{F}_p}(\Lambda^{d-m}(M), \Lambda^d(M))$ (induced by the multiplication). The latter module is isomorphic to

$$\operatorname{Hom}_{\mathbf{F}_p}(\Lambda^{d-m}(M), \mathbf{F}_p) \otimes_{\mathbf{F}_p} \Lambda^d(M).$$

But $\operatorname{Hom}_{\mathbf{F}_p}(\mathbf{F}_p[X]/(X-1)^n, \mathbf{F}_p) \cong \mathbf{F}_p[X]/(X-1)^n$, for all $n \leq q$, and so $\operatorname{Hom}_{\mathbf{F}_p}(\Lambda^{d-m}(M), \mathbf{F}_p) \cong \Lambda^{d-m}(M)$ (by Proposition 1.2(b)). Because σ_q has determinant 1 on M, one has $\Lambda^d(M) \cong \mathbf{F}_p$, and we are done.

(c) Clear.

(d) C_q acts on the set $\{X^{i_1} \wedge X^{i_2} \wedge \cdots \wedge X^{i_m} | i_j \in \mathbf{N}_0,\ 0 \leq i_1 < \cdots < i_m \leq q-1\}$, which constitutes a basis of $\Lambda^m(\mathbf{F}_p[X]/(X-1)^q)$. Now it is straightforward to sort out and count orbits (orbits of length p^n correspond to V_{p^n}'s). □

THEOREM 3.4. *Let $n, m \in \mathbf{N}_0$, $m \leq p-1$, and suppose $q \neq 1$; then:*

$$\lambda_{np+m}(V_{q-1}) = (\Omega_q)^a(\lambda_n V_{qp^{-1}-1}) \operatorname{mod}(\mathbf{Z} \cdot V_q), \text{ where } a = \begin{bmatrix} m & \text{if } p \text{ is odd}, \\ n+m & \text{if } p = 2. \end{bmatrix}$$

PROOF. We shall prove simultaneously the following property for $M = \lambda_{np+m}(V_{q-1})$:

Let $M = \sum_{s=1}^{q} a_s V_s$ in $RS(q), r \in \mathbf{N}$ and

$$k = \begin{cases} p & \text{if } p \text{ is odd}, \\ 4 & \text{if } p = 2. \end{cases}$$

(*)

If $a_r \neq 0$ then $r \equiv 0 \bmod k$ or $r \equiv \pm 1 \bmod k$.
If $a_r \neq 0$ and $r \not\equiv 0 \bmod k$, say $r = r_0 \cdot k + r_1$ ($r_0, r_1 \in \mathbf{N}_0$, $r_1 < k$), then $a_r = 1$ and $a_{r_0 k + s} = 0$ for all $s \in \{1, 2, \ldots, k-1\} \backslash \{r_1\}$.

We may assume that $m = 0$. To see this consider the exact sequence from Proposition 3.3(c)

$$0 \to \Lambda^{np+m}(\mathbf{F}_p[X]/(X-1)^{q-1}) \to \Lambda^{np+m+1}(\mathbf{F}_p[X]/(X-1)^q) \\ \to \Lambda^{np+m+1}(\mathbf{F}_p[X]/(X-1)^{q-1}) \to 0$$

for $0 \leq m \leq p-2$. By Proposition 3.3(d) the middle term is a free $\mathbf{F}_p[C_q]$-module so the first injection factorizes through the injective hull of the first term. By Proposition 3.2(c) we can conclude $\lambda_{np+m+1}(V_{q-1}) \equiv \Omega_q(\lambda_{np+m} V_{q-1}) \operatorname{mod}(\mathbf{Z} \cdot V_q)$, so $\lambda_{np+m}(V_{q-1}) \equiv (\Omega_q)^m(\lambda_{np} V_{q-1}) \operatorname{mod}(\mathbf{Z} \cdot V_q)$. Observe that Ω_q preserves (*). We shall now prove the result for $\lambda_{np}(V_{q-1})$ by induction on n and q in the following way. The cases $q = p$ and $n = 0$ are clear so we may assume $q \geq p^2$ and $n \geq 1$ and that the result and (*) are proved for $\lambda_{bp}(V_{c-1})$ for all divisors c of q and $b \in \mathbf{N}$ such that $c < q$ or ($c = q$ and $b < n$).

By Proposition 3.2(a)

$$\operatorname{Res}(\lambda_{np} V_{q-1}) = \sum_{i=0}^{np} [\lambda_i(V_{qp^{-1}-1}) \cdot \lambda_{np-i}((p-1)V_{qp^{-1}})].$$

LEMMA 1. *Let n, m be as above, $d \in \mathbf{N}_0$; then:*

$$\lambda_{np+m}(d \cdot V_q) \bmod (\mathbf{Z} \cdot V_q) \equiv \begin{cases} \lambda_n(dV_{qp^{-1}}) & \textit{if } m = 0; \\ 0 & \textit{otherwise.} \end{cases}$$

PROOF. If $m \neq 0$ then $\Lambda^{np+m}(N)$ is a summand of $\Lambda^{np+m-1}(N) \otimes_{\mathbf{F}_p} N$ for all $N \in \operatorname{Rep}_{\mathbf{F}_p}(C_q)$ (proof similar to that of Proposition 3.3(a)), so by Proposition 3.2(b) we can derive the result for $m \neq 0$. If $m = 0$ we may assume that $d = 1$ by Proposition 3.2(b) and the result for $m \neq 0$. The case $m = 0$, $d = 1$ follows from Proposition 3.3(d).

By Lemma 1 we see:

$$\begin{aligned}
\lambda_{np}(V_{q-1}) &= \lambda_{np}(V_{qp^{-1}-1}) + \sum_{j=0}^{n-1} \left[\lambda_{jp}(V_{qp^{-1}-1}) \cdot \lambda_{(n-j)p}((p-1)V_{qp^{-1}})\right] \\
&\quad + \sum_{j=0}^{n-1} \sum_{i=1}^{p-1} \left[\lambda_{jp+i}(V_{qp^{-1}-1}) \cdot \lambda_{(n-j)p-i}((p-1)V_{qp^{-1}})\right] \\
&= \lambda_{np}(V_{qp^{-1}-1}) \\
&\quad + \sum_{j=0}^{n-1} \left[\lambda_{jp}(V_{qp^{-1}-1})\right. \\
&\quad \times \left.\left[\lambda_{n-j}(V_{qp^{-2}}) + \frac{\binom{(p-1)qp^{-1}}{(n-j)p} - \binom{(p-1)qp^{-2}}{n-j}}{qp^{-1}} \cdot V_{qp^{-1}}\right]\right] \\
&\quad + \sum_{j=0}^{n-1} \sum_{i=1}^{p-1} \left[\lambda_{jp+i}(V_{qp^{-1}-1}) \cdot \frac{\binom{(p-1)qp^{-1}}{(n-j)p-i}}{qp^{-1}} V_{qp^{-1}}\right].
\end{aligned}$$

By Proposition 3.2(b) this equals:

$$\begin{aligned}
&\lambda_{np}(V_{qp^{-1}-1}) + \sum_j \left[\lambda_{jp}(V_{qp^{-1}-1}) \cdot \lambda_{n-j}(V_{qp^{-2}})\right] \\
&+ \sum_{j=o}^{n-1} \left[\binom{qp^{-1}-1}{jp}) \cdot \frac{\binom{(p-1)qp^{-1}}{(n-j)p} - \binom{(p-1)qp^{-2}}{n-j}}{qp^{-1}} \cdot V_{qp^{-1}}\right] \\
&+ \sum_{j=0}^{n-1} \sum_{i=1}^{p-1} \left[\binom{qp^{-1}-1}{jp+i} \cdot \frac{\binom{(p-1)qp^{-1}}{(n-j)p-i}}{qp^{-1}} \cdot V_{qp^{-1}}\right].
\end{aligned}$$

Now we use the induction hypothesis (for smaller q). If p is odd then we can replace $\lambda_{jp}(V_{qp^{-1}-1})$ by $\lambda_j(V_{qp^{-2}-1}) \bmod (\mathbf{Z}\cdot V_{qp^{-1}})$, so by Proposition 3.2(a):

$$\operatorname{Res}(\lambda_{n\cdot p}V_{q-1}) = \operatorname{Res}\left[\lambda_n(V_{qp^{-1}-1}) + \frac{\binom{q-1}{np} - \binom{qp^{-1}}{n}}{q}\cdot V_q\right].$$

If $p=2$, then by Proposition 3.2(b),

$$\begin{aligned}&\sum_{j=0}^{n-1}[\lambda_{jp}(V_{qp^{-1}-1})\cdot\lambda_{n-j}((p-1)V_{qp^{-2}})]\\ &\qquad\equiv\sum_{j=0}^{n-1}[(\Omega_{qp^{-1}})^j\lambda_j(V_{qp^{-2}-1})\cdot\lambda_{n-j}(V_{qp^{-2}})]\bmod(\mathbf{Z}\cdot V_{qp^{-1}})\\ &\qquad\equiv\sum_{j=0}^{n-1}[\lambda_j(V_{qp^{-2}-1})(\Omega_{qp^{-1}})^j(\lambda_{n-j}V_{qp^{-2}})]\bmod(\mathbf{Z}\cdot V_{qp^{-1}}).\end{aligned}$$

When n is even we now have, by Proposition 3.3(d):

$$\operatorname{Res}(\lambda_{np}V_{q-1}) = \lambda_n(V_{qp^{-2}-1}) + \sum_{j=0}^{n-1}[\lambda_j(V_{qp^{-2}-1})\cdot\lambda_{n-j}(V_{qp^{-2}})]\bmod(\mathbf{Z}\cdot V_{qp^{-1}}).$$

If n is odd then, by Proposition 3.3(d):

$$\begin{aligned}\operatorname{Res}(\lambda_{np}V_{q-1}) &= \Omega_{qp^{-1}}(\lambda_n V_{qp^{-2}-1})\\ &\quad+\sum_{j=0}^{n-1}[\lambda_j(V_{qp^{-2}-1})\cdot\Omega_{qp^{-1}}(\lambda_{n-j}V_{qp^{-2}})]\bmod(\mathbf{Z}\cdot V_{qp^{-1}})\\ &\equiv\Omega_{qp^{-1}}(\lambda_n V_{qp^{-2}-1})\\ &\quad+\sum_{j=0}^{n-1}[\lambda_j(V_{qp^{-2}-1})\cdot\lambda_{n-j}(V_{qp^{-2}})]\bmod(\mathbf{Z}V_{qp^{-1}}).\end{aligned}$$

Therefore in both cases, by Proposition 3.2(a):

$$\operatorname{Res}(\lambda_{np}V_{q-1}) = \operatorname{Res}\left[(\Omega_q)^n(\lambda_n V_{qp^{-1}-1}) + \frac{\binom{q-1}{np} - \binom{qp^{-1}-1}{n}}{q}\cdot V_q\right].$$

So, for all p, the restrictions of the classes of two modules equal and we have to prove that the classes themselves are equal. By Proposition 3.2(a) it is sufficient to prove $(*)$ for $\lambda_{np}(V_{q-1})$.

LEMMA 2. *Let b be a divisor of q, $b\neq 1$; then:*

$$\lambda_b(V_{q-1}) \equiv V_{qb^{-1}+c}\bmod(V_{pqb^{-1}})\quad\textit{where } c=\begin{cases}-1 & \textit{if } p \textit{ is odd}\\ +1 & \textit{if } p=2.\end{cases}$$

REMARK. "$\bmod(V_{qpb^{-1}})$" means, by Proposition 3.2(b), "up to V_n's with $pqb^{-1}|n$."

PROOF. We shall use induction on q for every b except for the cases $q = bp$ and $q = bp^2$, when we use induction on b first. The case $q = b$ is clear, so let $q > b$.

By Proposition 3.2(a),

$$\operatorname{Res}(\lambda_b V_{q-1}) = \lambda_b(V_{qp^{-1}-1}) + \sum_{i=0}^{b-1}[\lambda_i(V_{qp^{-1}-1}) \cdot \lambda_{b-i}((p-1)V_{qp^{-1}})].$$

So, by Proposition 3.3(d) and Proposition 3.2(b),

$$\operatorname{Res}(\lambda_b V_{q-1}) \equiv \lambda_b(V_{qp^{-1}-1}) + V_{qb^{-1}p^{-1}} + \frac{\binom{qb^{-1}}{p} - \binom{qb^{-1}-1}{1}}{qb^{-1}} \cdot V_{qb^{-1}}$$
$$+ \lambda_{bp^{-1}}(V_{qp^{-1}-1} \cdot V_{qb^{-1}} \operatorname{mod}(V_{qb^{-1}p^{-1}}).$$

If p is odd then $\operatorname{Res}(\lambda_b V_{pb-1}) \equiv (p-1)V_1 \operatorname{mod}(V_p)$ and $\operatorname{Res}(\lambda_b V_{p^2b-1}) \equiv \lambda_b(V_{pb-1}) + (p-1)V_p \operatorname{mod}(V_{p^2})$. By Proposition 3.2(a) we conclude $\lambda_b(V_{pb-1}) \equiv V_{p-1} \operatorname{mod}(V_{p^2})$ and therefore (by induction on q), by Proposition 3.2(a) again, the result for p is odd.

If $p = 2$ then $\operatorname{Res}(\lambda_b V_{pb-1}) \equiv V_1 + \lambda_{bp^{-1}}(V_{b-1}) \cdot V_2 \operatorname{mod}(V_4)$. Because clearly $\lambda_2(V_3) = \lambda_1(V_3) = V_3$ we have $\operatorname{Res}(\lambda_4 V_7) \equiv V_1 + V_2 \operatorname{mod}(V_4)$ by Proposition 3.1(a). By Proposition 3.2(a) we now see $\lambda_4(V_7) \equiv V_3 \operatorname{mod}(V_8)$, and by induction on b, and by Proposition 3.2(a), we derive $\lambda_b(V_{pb-1}) \equiv V_3 \operatorname{mod}(V_8)$. In a similar way one can prove $\lambda_b(V_{p^2b-1}) \equiv V_5 \operatorname{mod}(V_8)$. When $q \geq p^3 \cdot b$, $\operatorname{Res}(\lambda_b V_{q-1}) \equiv \lambda_b(V_{qp^{-1}-1}) + V_{qb^{-1}p^{-1}} \operatorname{mod}(V_{qb^{-1}})$, so, by induction on q and by Proposition 3.2(b), we can finish (as for p odd) the proof of the case $p = 2$.

We now show how Lemma 2 can be used to prove (*) for $\lambda_{np}(V_{q-1})$. The case $n \mid q$ is done in Lemma 2. If $n \nmid q$ let b be a nontrivial divisor of q with $b < np < pb$, and let $k = np - b$, $N = \mathbf{F}_p[X]/(X-1)^{q-1}$. Then $\Lambda^{np}(N)$ is a summand of $\Lambda^k(N) \otimes_{\mathbf{F}_p} \Lambda^b(N)$ by Proposition 3.3(a).

By Lemma 2 and Proposition 3.2(b) the class of the latter module is, modulo $(V_{qb^{-1}p})$, equal to $\lambda_k(V_{q-1}) \cdot V_{qb^{-1}+c}$. So, by Proposition 3.1(a), $\lambda_{np}(V_{q-1})$ satisfies (*) because $\lambda_k(V_{q-1})$ does. Since the property (*) for $\lambda_{np}(V_{q-1})$ was the only fact we needed to finish the proof, we are through. □

REMARK. See also Lemma 2 in the proof of Theorem 3.5, ($n \leq q/2$ if $p = 2$!).

THEOREM 3.5. *$\lambda_t(V_{q+1} - V_{q-1}) = 1 + (V_{q+1} - V_{q-1})t + t^2$, especially for $n \in \mathbf{N}$ with $2 \leq n \leq q+1$: $\lambda_n(V_{q+1}) = \lambda_{n-2}(V_{q-1}) + (V_{q+1} - V_{q-1})\lambda_{n-1}(V_{q-1}) + \lambda_n(V_{q-1})$.*

PROOF. We can, and shall, assume that $q \neq 1$ and, in case $p = 2$, because of Proposition 3.3(b), $n \leq q/2$. We shall prove simultaneously the following property for $M = \lambda_n(V_{q+1})$ depending on n and q:

Let $M = \sum_{s=1}^{pq} a_s V_s$ in $RS(pq)$, $r \in \mathbf{N}$, and

$$k = \begin{cases} p & \text{if } p \text{ is odd,} \\ 4 & \text{if } p = 2. \end{cases}$$

If $a_r \neq 0$ then ($r \equiv 0 \bmod k$ or $r \equiv \pm 1 \bmod k$) and $r \leq 2q$.
If $a_r \neq 0$ and $r \not\equiv 0 \bmod k$, say $r = r_0 \cdot k + r_1$ ($r_0, r_1 \in \mathbf{N}_0$, $r_1 < k$), then

$(*)n, q$
$$a_r \leq \max\left(2, \frac{p+1}{2}\right) \quad \text{and} \quad r_1 = \begin{cases} 1 \text{ if } (q < r < 2q; n \text{ is odd}) \text{ or} \\ (r < q \text{ and } n \text{ is even}), \\ k-1 \textit{ otherwise}. \end{cases}$$

LEMMA 1. *Let $n \in \mathbf{N}$, $n \geq 2$ and $\lambda_{n-1}(V_{q-1}) = \sum_{s=1}^{q} b_s V_s$ in $RS(q)$; then:*
$$\lambda_n(V_{q-1}) + (V_{q+1} - V_{q-1})\lambda_{n-1}(V_{q-1}) + \lambda_{n-2}(V_{q-1})$$
$$\equiv \lambda_m(V_{q-1}) + \sum_{s=1}^{q} b_s V_{s+q} \bmod (V_q),$$
where
$$m = \begin{cases} n-2 & \textit{if } p \nmid n \\ n & \textit{if } p \mid n. \end{cases}$$

PROOF. The result follows directly from Theorem 3.4 and Proposition 3.1(a).

We shall prove the theorem, and $(*)_{n,q}$, for $\lambda_n(V_{q+1})$ by induction on n and q in the following way. The cases $q = 1$, $q = 2$, or $n \leq 1$ are clear so we may assume $n \geq 2$, $q \geq p$, $q \neq 2$ and that the theorem and $(*)_{m,b}$ are proved for $\lambda_m(V_{b+1})$ for all divisors b of q and $m \in \mathbf{N}_0$ such that $b < q$ or ($b = q$ and $m < n$).

By Proposition 3.2(a)
$$\operatorname{Res}(\lambda_n V_{q+1}) = \lambda_n(V_{qp^{-1}+1}) + \sum_{i=0}^{n-1} [\lambda_i(V_{qp^{-1}+1}) \cdot \lambda_{n-i}((p-1) \cdot V_{qp^{-1}})]$$
and
$$\begin{aligned}\operatorname{Res}[\lambda_n(V_{q-1}) &+ (V_{q+1} - V_{q-1}) \cdot \lambda_{n-1}(V_{q-1}) + \lambda_{n-2}(V_{q-1})] \\ &= \lambda_n(V_{qp^{-1}-1}) + (V_{qp^{-1}+1} - V_{qp^{-1}-1}) \cdot \lambda_{n-1}(V_{qp^{-1}-1}) + \lambda_{n-2}(V_{qp^{-1}-1}) \\ &+ \sum_{i=0}^{n-1} [(\lambda_i(V_{qp^{-1}-1}) + (V_{qp^{-1}+1} - V_{qp^{-1}-1}) \cdot \lambda_{i-1}(V_{qp^{-1}-1}) \\ &\qquad + \lambda_{i-2}(V_{qp^{-1}-1})) \cdot \lambda_{n-i}((p-1)V_{qp^{-1}})].\end{aligned}$$
Here we have put $\lambda_m = 0$ if $m < 0$ for convenience. By the induction hypothesis the sum $\sum_{i=0}^{n-1}[\ ]$ is equal to $\sum_{i=0}^{n-1}[\lambda_i(V_{qp^{-1}+1}) \cdot \lambda_{n-i}((p-1)V_{qp^{-1}})]$. By Proposition 3.2(a) it is sufficient to prove $(*)_{n,q}$ for both $\lambda_n(V_{q+1})$ and $\lambda_n(V_{q-1}) + (V_{q+1} - V_{q-1}) \cdot \lambda_{n-1}(V_{q-1}) + \lambda_{n-2}(V_{q-1})$ to establish the theorem.

LEMMA 2. *Let $n \in \mathbf{N}_0$, $n \leq q-1$; then if $q \neq 1$ there is a unique $m \in \mathbf{N}$ such that (V_m has multiplicity one in $\lambda_n(V_{q-1})$ and $m \not\equiv 0 \bmod k$). Moreover, if b is the divisor of q such that $b \leq n < bp$ then*
$$m \pmod{qb^{-1}} \equiv \begin{cases} -1 & \textit{if } n \textit{ is odd} \\ 1 & \textit{if } n \textit{ is even.} \end{cases}$$

PROOF. This follows directly from Theorem 3.4 by induction on q and n.

By Lemmas 1 and 2, $\lambda_n(V_{q-1}) + (V_{q+1} - V_{q-1})\lambda_{n-1}(V_{q-1}) + \lambda_{n-2}(V_{q-1})$ satisfies $(*)_{n,q}$.

We shall now prove that $\lambda_n(V_{q+1})$ satisfies $(*)_{n,q}$. Because of Proposition 3.2(a) we already know the implication $a_r \neq 0 \Rightarrow r \leq 2q$. If $p \nmid n$ then $\Lambda^n(\mathbf{F}_p[X]/(X-1)^{q+1})$ is a summand of

$$\Lambda^{n-1}(\mathbf{F}_p[X]/(X-1)^{q+1}) \otimes_{\mathbf{F}_p} \mathbf{F}_p[X]/(X-1)^{q+1}$$

(proof similar to that of Proposition 3.3(a)). By the induction hypothesis we know $\Lambda^n(\mathbf{F}_p[X]/(X-1)^{q+1})$ well enough (Lemmas 1 and 2!) to conclude, by Proposition 3.1(a), that $\lambda_n(V_{q+1})$ satisfies $(*)_{n,q}$.

LEMMA 3. *Let b be a divisor of $q, b \neq 1$, then:*

$$\lambda_b(V_{q+1}) \operatorname{mod}(V_{qb^{-1}p}) \equiv \begin{bmatrix} V_{q+1} + V_{qb^{-1}-1} & \text{if } p \text{ is odd,} \\ V_{q+qb^{-1}-1} + V_{qb^{-1}+1} & \text{if } p = 2. \end{bmatrix}$$

PROOF. We use induction on q for every b except for the case $q = bp$ with $p = 2$, when we use induction on b first. The case $q = b$ is clear, so let $q > b$. By Proposition 3.2(a),

$$\operatorname{Res}(\lambda_b V_{q+1}) = \lambda_b(V_{qp^{-1}+1}) + \sum_{i=0}^{b-1} [\lambda_i(V_{qp^{-1}+1}) \cdot \lambda_{b-i}((p-1)V_{qp^{-1}})],$$

so by Proposition 3.3(d) and Proposition 3.2(b):

$$\operatorname{Res}(\lambda_b V_{q+1}) \equiv \lambda_b(V_{qp^{-1}+1}) + (p-1)V_{qb^{-1}p^{-1}} + \frac{\binom{qb^{-1}}{p} - \binom{qb^{-1}p^{-1}}{1}}{qb^{-1}} \cdot V_{qb^{-1}}$$
$$+ \lambda_{bp^{-1}}(V_{qp^{-1}+1}) \cdot V_{qb^{-1}} \operatorname{mod}(V_{qb^{-1}p}).$$

If p is odd $\operatorname{Res}(\lambda_b V_{pb+1}) \equiv V_{b+1} + (p-1)V_1 \operatorname{mod}(V_p)$, by Proposition 3.3(b), and $\operatorname{Res}(\lambda_b V_{p^2 b+1}) \equiv \lambda_b(V_{pb+1}) + V_{p-1} \operatorname{mod}(V_{p^2})$.

So, by Proposition 3.2(a), $\lambda_b(V_{pb+1}) \equiv V_{pb+1} + V_{p-1} \operatorname{mod}(V_{p^2})$. And now, by induction on q, the result for p odd follows.

If $p = 2$ then $\operatorname{Res}(\lambda_b V_{pb+1}) \equiv V_{b+1} + V_1 + \lambda_{bp^{-1}}(V_{b+1}) \cdot V_2 \operatorname{mod}(V_4)$, by Propositions 3.3(b) and 3.1(a), especially $\operatorname{Res}(\lambda_2 V_5) \equiv V_3 + V_1 + V_2 + V_4 (\operatorname{mod} V_4)$. By Proposition 3.2(a) we conclude $\lambda_2(V_5) = V_3 + V_7$ and by induction on b, and by Propositions 3.2(a) and 3.1(a), we see $\lambda_b(V_{pb+1}) \equiv V_{pb+3} + V_3 (\operatorname{mod} V_4)$. If $q > pb$ then, by Proposition 3.2(d), $\operatorname{Res}(\lambda_b V_{q+1}) \equiv \lambda_b(V_{qb^{-1}+1}) + V_{qb^{-1}p^{-1}} \operatorname{mod}(V_{qb^{-1}})$, so by induction on q and by Proposition 3.2(a), we can prove the result for $p = 2$.

We can now prove $(*)_{n,q}$ for $\lambda_n(V_{q+1})$ in the case $p \mid n$. The case $n \mid q$ follows from Lemma 3, so let b be a divisor of q such that $b < n < pb$, say $n = n_0 b + n_1$ with $n_0, n_1 \in \mathbf{N}_0$, $n_1 \leq b-1$. Let N be $\mathbf{F}_p[X]/(X-1)^{q+1}$; then $\Lambda^n(N)$ is a summand of $\Lambda^{n_0 \cdot b}(N) \otimes_{\mathbf{F}_p} \Lambda^{n_1}(N)$ by Proposition 3.3(a).

If $p = 2$ then $\lambda_{n_0}(V_{q+1})$ is described by Lemma 3 and $\lambda_{n_1}(V_{q+1})$ by Lemmas 1 and 2, so $(*)_{n,q}$ follows from Proposition 3.1(a).

If p is odd then $\Lambda^{n_0 \cdot b}(N)$ is a summand of $\Lambda^{n_0}(\Lambda^b N)$ by Proposition 3.3(a) and, by Lemma 3 and Propositions 3.3(a) and 3.2(b),

$$\lambda_{n_0}(\lambda_b V_{q+1}) \equiv \sum_{i=0}^{n_0} [\lambda_i(V_{q+1}) \cdot \lambda_{n_0-i}(V_{qb^{-1}-1})] \bmod (V_{qb^{-1}}).$$

Now $\lambda_i(V_{q+1})$ is described in Lemmas 1 and 2 and $\lambda_{n_0-i}(V_{qb^{-1}-1})$ is described in Lemma 2, so we can describe the sum $\sum_{i=0}^{n_0}$ modulo $(V_{qb^{-1}})$, by Proposition 3.1(a). We then see that $\lambda_{n_0 b}(V_{q+1})$ satisfies $(*)_{n_0 b,q}$. By this $\lambda_{n_0 b}(V_{q+1})$ is as in the theorem (so is described in Lemmas 1 and 2) and we can now establish $(*)_{n,q}$ for $\lambda_{n_0 b+n_1}(V_{q+1})$ by computing $\lambda_{n_0 \cdot b}(V_{q+1}) \cdot \lambda_{n_1}(V_{q+1})$, which can be done modulo (V_p), by Proposition 3.1(a), $(\lambda_{n_1}(V_{q+1})$ is also described by the Lemmas 1 and 2). □

REMARK. Theorem 3.5 generalizes the (trivial) result $\lambda_t(V_2) = 1 + V_2 \cdot t + t^2$ ($q = 1$!) which implied Theorem 2.2(a) (because of Proposition 2.1(a) and Corollary 2.1). In fact, Theorem 2.2(a) generalizes as well. Let $V_n(q)$ denote $W_{n-1}(V_{q+1} - V_{q-1})$ for $n \in \mathbf{N}$ with $n \leq p$; then for $n \leq p$, $m \leq \min(n, p-1)$:

$$\lambda_m(V_n(q)) = \frac{V_n(q) \cdot V_{n-1}(q) \cdot \dots \cdot V_{n-m+1}(q)}{V_m(q) \cdot V_{m-1}(q) \cdot \dots \cdot V_1(q)}.$$

Its proof can be given in a similar way to that for Proposition 2.1(a) in [**3**]. And Corollary 2.1 can be generalized! (Use Res and Proposition 3.2(a).)

CONJECTURE. The exterior powers induce operations on $RS(q)/I_q$ (see Cor. 3.1) and the ring isomorphism $RS(p)/(V_p - V_{p-1} - V_1))^{\otimes k} \tilde{\to} RS(q)$, which is defined by

$$\underset{(i\text{th}\uparrow\text{factor})}{1 \otimes \dots \otimes 1 \otimes V_2 \otimes 1 \otimes \dots \otimes 1} \mapsto V_{p^{i-1}+1} - V_{p^{i-1}-1}$$

(because $V_{p^{i-1}+1} - V_{p^{i-1}-1} \equiv V_{2p^{i-1}} - V_{2p^{i-1}-1} - V_1 \bmod I_q$), commutes with the λ-operations. In particular, $RS(q)/I_q$ is a λ-ring by Proposition 2.2(c).

REFERENCES

1. G. Almkvist, *Representation of* **Z**/p**Z** *in characteristic p and reciprocity theorems*, J. Algebra **68** (1981), 1–27.

2. ——, *Some formulas in invariant theory*, J. Algebra **77** (1982), 338–359.

3. G. Almkvist and R. M. Fossum, *Decomposition of exterior and symmetric powers of indecomposable* **Z**/p**Z**-*modules in characteristic p and relations to invariants*, Sém. d'Algèbre, P. Dubreil 1976/1977, Springer Lecture Notes in Math., No. 641, Springer-Verlag, pp. 1–111.

4. M. F. Atiyah and D. Tall, *Group representations, λ-rings and the J-homomorphism*, Topology **8** (1969) 253–297.

5. R. M. Fossum, *Decompositions revisited*, Sém. d'Algèbre P. Dubreil/ M. P. Malliavin 1981, Springer Lecture Notes in Math., No. 924, Springer-Verlag, pp. 261–295.

6. F. M. Kouwenhoven, *Universal operations in the representation theory of groups*, Ph.D. Thesis, Utrecht, 1986.

7. J. C. Renaud, *The decomposition of products in the modular representation ring of a cyclic group of prime power order*, J. Algebra **58** (1979), 1–11.

UNIVERSITEIT TWENTE, ENSCHEDE, THE NETHERLANDS

Proceedings of Symposia in Pure Mathematics
Volume 47 (1987)

On the Cohomology of p-Constrained Groups

P. A. LINNELL AND U. STAMMBACH

1. Introduction. Let G be a finite group, let p be a prime dividing the group order, and let k be a field of characteristic p. As usual we denote by kG the group algebra of G with coefficients in k. Let V, W be simple kG-modules. It is a well-known and easy-to-prove fact that $\mathrm{Ext}^j_{kG}(V, W) = 0$ for $j =, 0, 1, 2, \ldots$ if V, W are in different blocks of kG. The question of when $\mathrm{Ext}^j_{kG}(V, W) \neq 0$ for some value of j, if V, W are in the *same* block, has been asked by many people. Some partial results have been known for some time. They mostly concern the case $V = k$ (and W in the principal block). Since $\mathrm{Ext}^j_{kG}(V, W) = H^j(G, W)$ the question in this case is whether the cohomology of G with coefficients in W is nontrivial. This has been answered affirmatively in the following cases; in each of them it has been shown that the cohomology is nontrivial in infinitely many dimensions.

(i) For arbitrary G and $W = k$ by Swan [**8**].

(ii) By Ogawa [**6**] for G p-solvable with abelian p-Sylow subgroup and arbitrary W.

(iii) By Diethelm [**1**] for G in a class of groups containing the groups of p-length 1 and arbitrary W.

(iv) By Linnell [**3**] for G p-solvable, $p \neq 2, 3$, and W arbitrary.

(v) By Sasaki [**7**] for arbitrary G and W with abelian vertex and trivial source.

In the negative direction we mention the following well-known example. If G is the simple group of order 60 and if k is a large enough field of characteristic 2, then the two 2-dimensional simple modules M_1, M_2 satisfy $\mathrm{Ext}^j_{kG}(M_1, M_2) = 0$ for $j = 0, 1, 2, \ldots$.

2. Main result. Our main theorem is an extension of Linnell's result.

THEOREM. (a) *Let G be p-constrained and let V, W be simple kG-modules in the principal block of kG. Then* $\mathrm{Ext}^j_{kG}(V, W) \neq 0$ *for infinitely many values of* j.

1980 *Mathematics Subject Classification* (1985 *Revision*). Primary 20C05, 20C20; Secondary 18G15, 18G40.

Key words and phrases. Block, solvable group, spectral sequence, Ext.

(b) *Let G be p-solvable and let V, W be simple kG-modules in the same block of kG. Then* $\mathrm{Ext}^j_{kG}(V,W) \neq 0$ *for infinitely many values of j unless V (or W) is projective, in which case $V \cong W$ and* $\mathrm{Ext}^0_{kG}(V,W) \neq 0$.

We shall give a short summary of the proof of the theorem in §3; for details the reader may consult [**4,5**]. We note that examples show that the word "simple" in the statement of the theorem cannot be replaced by "indecomposable nonprojective." Also, the conclusion of part (a) of the theorem does not hold in general for nonprincipal blocks.

As an application of the theorem we mention the following Proposition 1. We note that Jackowski and Marciniak have obtained results in the same direction by completely different methods (see [**2**]).

PROPOSITION 1. *Let H be a p-solvable group with $O_{p'}H = e$. If ϕ is an automorphism of H, with $(|\phi|, |H|) = 1$, then*

$$1 \neq \phi^* \colon H^*(H, \mathbf{Z}/p\mathbf{Z}) \to H^*(H, \mathbf{Z}/p\mathbf{Z}).$$

PROOF. Set $G = H \lhd C_q$, $C_q = \langle \phi \rangle$. Obviously we may assume that q is prime. Choose a field k of characteristic p admitting a nontrivial C_q-action. Regard k with this action as a kG-module $\tilde{k}$. Since $O_{p',p}G$ acts trivially on $\tilde{k}$ and since G is p-solvable, $\tilde{k}$ is in the principal block of kG. By our theorem we have

$$0 \neq H^*(G, \tilde{k}) = H^0(C_q, H^*(H, \tilde{k})) = \mathrm{Hom}_{C_q}(H_*(H, k), \tilde{k}).$$

Thus C_q cannot act trivially on $H_*(H, k)$. From this it easily follows that C_q does not act trivially on $H^*(H, \mathbf{Z}/p\mathbf{Z})$. However, this action is given by ϕ^*.

3. The proof. The Fong-reductions allow us to reduce part (b) of the theorem to part (a) (see [**4**]). A further reduction shows that it suffices to consider p-constrained groups G with $O_{p'}G = e$. Such a group satisfies $O_pG \neq e$; this property plays a crucial role in the rest of the proof. The following Proposition 2 (see [**5**, Proposition 2.3]), which is of independent interest, implies part (a) of our theorem.

PROPOSITION 2. *Let P be a normal p-subgroup of G and let H be the centralizer in G of $P/\Phi P$ where ΦP denotes the Frattini subgroup of P. Then if M is any kG/H-module, $M \neq 0$, then $H^j(G, M) \neq 0$ for infinitely many values of j.*

The proof of Proposition 2 is by induction on the length of the fastest descending central series of P with elementary abelian factors. If P is elementary abelian and if Q is any projective kG/H-module then one shows that Q is a direct summand in $H^*(H, k)$. This step uses the fact that Q is a direct summand in $H^*(P, k)$ and an argument involving Even's norm map. By an elementary homological argument it is then shown that this implies $H^*(G, M) \neq 0$ if M is any kG/H-module, $M \neq 0$. The induction step consists of a delicate analysis of the Lyndon-Hochschild-Serre spectral sequence associated with the extension corresponding to the last nontrivial term of the fastest descending central series of P with elementary abelian factors. This step makes use of the detailed knowledge available on the product structure and module stucture of that spectral sequence.

We finally show how Proposition 2 implies statement (a) of our theorem. We set $P = O_pG$. Since G is p-constrained we have $H = P$. Consider $M = V^* \otimes W$. As V and W are simple, we can assume that M is a kG/H-module. Thus Proposition 2 yields $0 \neq H^*(G, M) = \mathrm{Ext}^*_{kG}(V, W)$.

REFERENCES

1. T. Diethelm, *On the cohomology of groups of p-length* 1, Comment. Math. Helv. **58** (1983), 379–387.
2. S. Jackowski and Z. Marciniak, *Group automorphisms inducing the identity map on cohomology*, J. Pure Appl. Algebra **44** (1987), 241–250.
3. P. A. Linnell, *Cohomology of finite soluble groups*, J. Algebra **107** (1987), 53–62.
4. P. A. Linnell and U. Stammbach, *The block structure and* Ext *of p-soluble groups*, J. Algebra (to appear).
5. ——, *The cohomology of p-constrained groups*, J. Pure. Appl. Algebra (to appear).
6. Y. Ogawa, *On the cohomology of some finite groups of p-length* 1, Japan. J. Math. **8** (1982), 211–216.
7. H. Sasaki, *On the nontriviality of cohomology of finite groups*, Hokkaido Math. J. **11** (1982), 234–237.
8. R. G. Swan, *The nontriviality of the restriction map in the cohomology of groups*, Proc. Amer. Math. Soc. **11** (1960), 885–887.

VIRGINIA POLYTECHNIC INSTITUTE AND STATE UNIVERSITY

EIDGENÖSSISCHE TECHNISCHE HOCHSCHULE, SWITZERLAND

Proceedings of Symposia in Pure Mathematics
Volume **47** (1987)

Block Invariants in A_n, $\hat{A}_n$, and $\hat{S}_n$

JØRN B. OLSSON

Abstract. This note describes some of the results in three forthcoming papers [**4**, **7**, **8**], where the proofs will be presented.

We consider the following groups: S_n, the symmetric group of order $n!$, A_n, the alternating group of order $n!/2$, $\hat{S}_n$, a covering group of S_n of order $2n!$, and $\hat{A}_n$, the covering group of A_n of order $n!$

PROBLEM 1. Describe the distribution of the ordinary irreducible characters of these groups into p-blocks, p a prime.

2. Describe the invariants of a p-block B in these groups. We are interested in $\Delta(B)$, a defect group of B, $k(B)$, the number of characters in B, and $k_i(B)$, the number of characters of height i in B.

3. Verify the following conjectures for the above blocks:

(I) $k(B) \leq |\Delta(B)|$,

(II) $k_0(B) \leq |\Delta(B)'|$,

(III) $k(B) = k_0(B) \Leftrightarrow \Delta(B)$ is abelian.

1. The blocks of S_n. The irreducible characters $[\lambda]$ of S_n are indexed canonically by the partitions λ of n. Given a prime p, a partition λ determines and is uniquely determined by its *p-core* $\lambda_{(p)}$ (obtained by removing inductively all p-hooks from λ) and its *p-quotient*

$$\lambda^{(p)} = (\lambda_0, \lambda_1, \ldots, \lambda_{p-1}),$$

a p-tuple of partitions (see, e.g., [**3**, §2.7]). We have $|\lambda| = |\lambda_{(p)}| + pw_p(\lambda)$, where $w_p(\lambda) = |\lambda_0| + \cdots + |\lambda_{p-1}|$ is the *p-weight* of λ. The importance of the p-quotient lies in the following statement.

LEMMA 1. *There is a canonical correspondence between the set of tp-hooks in λ and the union of the set of t-hooks in $\lambda_0, \lambda_1, \ldots, \lambda_{p-1}$, which is compatible with the removal of hooks.*

The importance of the p-core lies in the following:

THEOREM 2 (THE NAKAYAMA CONJECTURE). *Let λ, μ be partitions of n. Then $[\lambda]$ and $[\mu]$ are in the same p-block, if and only if $\lambda_{(p)} = \mu_{(p)}$.*

Several proofs of this are known, the latest due to Macdonald.

1980 *Mathematics Subject Classification* (1985 *Revision*). Primary 20C30; Secondary 20C20.

Let B be a p-block of S_n. If $[\lambda] \in B$, we define $w(B) = w_p(\lambda)$, the *p-weight* of B, which is an invariant of B. Also we call $\lambda_{(p)}$ the *core* of B.

If $w = w(B)$, then $\Delta(B)$ is a p-Sylow subgroup of S_{pw}. Moreover, $k(B) = k(p,w)$ is the number of p-quotients of weight w. Using Lemma 1, $k_i(B) = k_i(p,w)$ may be computed [**5**] and it is dependent on the p-weight only and not on the p-core. Conjectures I and II were verified in [**6**] and Conjecture III in [**5**].

2. The blocks of A_n. Let λ be a partition of n. Then $[\lambda]|_{A_n}$ is irreducible if and only if $\lambda \neq \lambda^0$ (i.e., λ is not symmetric). We call a p-quotient $\lambda^{(p)} = (\lambda_0, \lambda_1, \dots, \lambda_{p-1})$ *symmetric* if $\lambda^{(p)} = (\lambda_{p-1}^0, \lambda_{p-2}^0, \dots, \lambda_0^0)$. Then λ is symmetric if and only if $\lambda_{(p)}$ and $\lambda^{(p)}$ are symmetric. It is quite easy to show:

PROPOSITION 3. *If B is a p-block of S_n of weight $w > 0$ then B covers a unique block $\tilde{B}$ of A_n.*

Whereas $k_i(B)$, $k(B)$ are independent of the core, this is not the case of $k_i(\tilde{B})$, $k(\tilde{B})$, when p is odd.

Suppose p *is odd.* Then $\Delta(B) = \Delta(\tilde{B})$ and we have

PROPOSITION 4. (1) *If the core is nonsymmetric then*

$$k(B) = k(\tilde{B}) \quad \text{and} \quad k_i(B) = k_i(\tilde{B}).$$

(2) *If the core is symmetric then*

$$k(\tilde{B}) = 2k^s(p,w) + \tfrac{1}{2}k^a(p,w)$$

where $k^s(p,w)$ ($k^a(p,w)$) is the number of symmetric (nonsymmetric) p-quotients of weight w. A similar statement holds for the $k_i(\tilde{B})$'s.

Suppose $p = 2$. Here the core is insignificant for the block invariants, but the parity of w play a role. We omit the details. Again there are two possibilities.

Using some combinatorial arguments, conjectures I, II, and III can be deduced from the corresponding results for S_n mentioned above.

3. The blocks of $\hat{S}_n$. The group $\hat{S}_n$ has a central subgroup Z of order 2, such that $\hat{S}_n/Z \cong S_n$. An irreducible character of $\hat{S}_n$, which does not have Z in its kernel, is called a *spin character.* A *bar partition* (2-regular) of n is a partition

$$\lambda = (a_1, a_2, \dots, a_m), \qquad a_1 > a_2 > \dots > a_m > 0,$$

of n with all parts different. We define $\varepsilon(\lambda) = (-1)^{n-m}$, the *sign* of λ. The spin characters are indexed canonically by the bar partitions, such that λ indexes one character if $\varepsilon(\lambda) = 1$ and two characters if $\varepsilon(\lambda) = -1$. Let $\langle\lambda\rangle$ be a spin character indexed by λ. If λ is a bar partition as above, it is convenient to visualize λ in a shifted young diagram $S(\lambda)$ containing a_i nodes in the ith row, $i = 1, 2, \dots, m$. To each node of $S(\lambda)$ one associates a bar length, such that the bar lengths of the nodes in the ith row are the integers in the set

$$(\{1, 2, \dots, a_i\} \cup \{a_i + a_j | j > i\}) \backslash \{a_i - a_j | j > i\}$$

written in decreasing order.

EXAMPLE. $= (6, 4, 2)$

	• • • • • •	Bar lengths	10 8 6 5 3 1	
$S(\gamma)$	• • • •		6 4 3 1	
	• •		2 1	

To each node in $S(\lambda)$ one also associates a subdiagram, a *bar*, consisting of a number of nodes indicated by the bar length. Then it is possible to define the *removal* of a bar in formal analogy with the removal of a hook in the case of partitions. (For details see, e.g., §2 of [**7**].) If p is a prime the p-bar core $\lambda_{(\bar{p})}$ of λ is obtained by removing inductively all bars of length p from λ. It is also possible to define a *p-bar quotient* for *p odd*

$$\lambda^{(\bar{p})} = (\lambda_0, \lambda_1, \dots, \lambda_t), \qquad t = (p-1)/2,$$

where λ_0 is a bar partition and $\lambda_1, \dots, \lambda_t$ are partitions, such that we have [**7**]:

LEMMA 5 (p odd). *There is a canonical correspondence between the set of tp-bars in λ and the union of the set of t-bars in λ_0 with the sets of t-hooks in $\lambda_1, \lambda_2, \dots, \lambda_t$, which is compatible with the removal of bars (hooks).*

The *p-bar weight* $w_{\bar{p}}(\lambda)$ is defined as

$$w_{\bar{p}}(\lambda) = |\lambda_0| + |\lambda_1| + \cdots + |\lambda_t|$$

and Lemma 5 shows that

$$|\lambda| = |\lambda_{(\bar{p})}| + p w_{\bar{p}}(\lambda).$$

We have

THEOREM 6 (THE MORRIS CONJECTURE). *Let p be odd and let λ, μ be bar partitions of n, which are not p-bar cores. Then $\langle\lambda\rangle$ and $\langle\mu\rangle$ are in the same p-block if and only if $\lambda_{(\bar{p})} = \mu_{(\bar{p})}$. If λ is a p-bar core, then $\langle\lambda\rangle$ forms a p-block of defect* 0.

A first proof of this was given by Humphreys [**2**]. Another proof, due to the author [**8**], uses central characters and has been inspired by the original proof of the Nakayama conjecture. It emphasizes the importance of the "canonical character" for these questions. A third proof of Theorem 6, using the theory of subpairs, may be found in [**1**].

When p is odd, a p-block of $\hat{S}_n$ contains only spin characters or no spin characters at all. For $p = 2$ there are "mixed" blocks. For a conjecture about what happens in this case as well as some examples, see [**7**].

Assume again that p is odd. Lemma 5 makes it possible to compute the exact power of p dividing the degree $\langle\lambda\rangle(1)$ and to compute the height of $\langle\lambda\rangle$ in the p-block containing it. (It turns out that the defect group of a p-block of p-bar weight w is a p-Sylow subgroup of $\hat{S}_{pw}$ [**8**].) Then $k(\hat{B})$, $k_i(\hat{B})$ may be computed when $\hat{B}$ is a block of spin characters. These numbers depend not only on the weight of $\hat{B}$ but also on the *sign* of the corresponding p-bar core, defined above. We write $k(\hat{B}) = \hat{k}^{\varepsilon}(p, w)$ where ε is a sign and w the weight and proceed to describe $\hat{k}^{\varepsilon}(p,w)$. If $(\lambda_0, \lambda_1, \ldots, \lambda_t)$ is a p-bar quotient of weight w, its sign is defined as $(-1)^{n-m_0}$, when m_0 is the number of parts in λ_0. If $q^{\varepsilon}(p,w)$ is the number of p-bar quotients of weight w and sign ε, then

$$\hat{k}^{\varepsilon}(p,w) = q^{\varepsilon}(p,w) + 2q^{-\varepsilon}(p,w).$$

We get a similar decomposition for the $k_i(\hat{B})$.

Again using combinatorial arguments, the proofs of Conjectures I, II, and III may be reduced to the case of S_n. (For instance, $\hat{k}^{\varepsilon}(p,w) \leq k(p,w)$).

4. The blocks of $\hat{A}_n$. If $\hat{B}$ is a block of spin characters in $\hat{S}_n$, of weight $w > 0$, then $\hat{B}$ covers a unique block $\hat{B}'$ of $\hat{A}_n$. If ε is the sign of the p-bar core of $\hat{B}$, then

$$k(\hat{B}') = \hat{k}^{-\varepsilon}(p,w).$$

A corresponding result holds for the $k_i(\hat{B}')$'s. Then the proof of the conjectures in this case is identical to the case of $\hat{S}_n$.

ACKNOWLEDGMENT. This work has been supported by Deutsche Forschungsgemeinschaft. The author thanks the University of Chicago and the University of Illinois at Chicago for their hospitality.

REFERENCES

1. M. Cabanes, *Local structure of the p-blocks of $\tilde{S}_n$*, Preprint.

2. J. F. Humphreys, *Blocks of projective representations of the Symmetric groups*, J. London Math. Soc. (2) **33** (1986), 441–452.

3. G. James and A. Kerber, *The representation theory of the symmetric group*, Addison-Wesley, Reading, Mass., 1981.

4. A. O. Morris and J. B. Olsson, *p-quotients for spin characters*, Preprint series, no. 6, Københavns Universitet, Matematisk Institut, 1987.

5. J. B. Olsson, *McKay numbers and heights of characters*, Math. Scand. **38** (1976), 25–42.

6. ——, *On the number of characters in blocks of finite general linear, unitary and symmetric groups*, Math. Z. **186** (1984), 41–47.

7. ——, *Frobenius symbols for partition and degrees of spin characters*, Preprint Series, no. 18, Københavns Universitet, Matematisk Institut, 1986.

8. ——, *On the p-blocks of symmetric and alternating groups and their covering groups*, (in preparation).

KØBENHAVNS UNIVERSITET, DENMARK

Proceedings of Symposia in Pure Mathematics
Volume **47** (1987)

On the Loewy Structure of the Modular Group-Algebra over a Group of Small p-Length

REINER STASZEWSKI

Let G be a finite group and let K be a field of characteristic p such that $p \mid |G|$. Then the group algebra KG is not semisimple by Maschke's theorem and thus we consider the Loewy factors $J(KG)^n/J(KG)^{n+1}$, where $J(KG)^n$ denotes the nth power of the Jacobson radical of KG. Our aim is to decompose the semisimple Loewy factors into the irreducible KG-modules.

At first, we want to consider the easiest case, namely that G is a p-group. Then KG has only one irreducible module, namely the trivial module denoted by K. Therefore, the only thing to determine remains the number

$$c_n := \dim_K(J(KG)^n/J(KG)^{n+1}).$$

The answer was given by S. Jennings in 1941 in terms of a certain central series.

THEOREM ([**2**], CF. ALSO [**1**, VIII, §1–2]). *Let G be a finite p-group and K an arbitrary field of characteristic p. Define the central series*

$$G = \kappa_1(G) \geq \kappa_2(G) \geq \cdots \geq \kappa_1(G) > \kappa_{l+1}(G) = 1$$

by

$$\kappa_i(G) := [\kappa_{i-1}(G), G] \cdot \kappa_m(G)^p,$$

where m denotes the least integer such that $pm \geq i$ $(i > 1)$. If $|\kappa_i(G){:}\,\kappa_{i+1}(G)| =: p^{d_i}$ $(1 \leq i \leq l)$ and s denotes the natural number such that $J(KG)^s > J(KG)^{s+1} = 0$, we have

$$\prod_{i=1}^{l}(1 + t^i + \cdots + t^{i(p-1)})^{d_i} = \sum_{n=0}^{s} c_n t^n.$$

In particular, the Loewy series is symmetric, namely $c_n = c_{s-n}$.

We have $c_0 = 1$, and as $\kappa_2(G) = \Phi(G)$, we have $c_1 = d_1 =$ minimal number of generators. Note that it is possible that $\kappa_{i-1}(G) = \kappa_i(G) > \kappa_{i+1}(G)$, because,

1980 *Mathematics Subject Classification* (1985 *Revision*). Primary 20C05; Secondary 16A26.

for

EXAMPLE. Let G be an abelian p-group. Then $\kappa_i(G) = G^{p^k}$ for $p^{k-1} < i \le p^k$. Thus, we have for a cyclic p-group G that $d_1 = d_p = d_{p^2} = \cdots = 1$ and $c_0 = c_1 = \cdots = c_{|G|-1} = 1$. If G is an elementary abelian p-group, then $|G| = p^{d_1}$ and all other d_i's are 0.

Now, for some years, the question circulated whether the Loewy series of a p-group is monotonic, i.e., $c_{n-1} \le c_n$ for $1 \le n \le s/2$. By applying Jennings's theorem, we could prove the monotony of the Loewy series for some classes of p-groups.

PROPOSITION [5]. *The Loewy series is monotonic if one of the following holds:*

(a) *G is an abelian p-group.*

(b) *G is a regular p-group* (*in the sense of P. Hall*) *and $G' \le G^p$.*

(c) *G is an extra-special p-group.*

But the Loewy series of a p-group is not monotonic in general.

EXAMPLES.

(a) [5]. Let
$$G = \left\{ \begin{pmatrix} 1 & x & y \\ & 1 & x^3 \\ & & 1 \end{pmatrix} \middle| \, x, y \in GF(3^3) \right\}.$$
Then G is a 3-group of order 3^6 and of exponent 3; moreover,
$$Z(G) = G' = \left\{ \begin{pmatrix} 1 & 0 & y \\ & 1 & 0 \\ & & 1 \end{pmatrix} \middle| \, y \in GF(3^3) \right\}.$$
With the notation above, we have $d_1 = d_2 = 3$ and
$$(1+t+t^2)^3 \cdot (1+t^2+t^4)^3 = \cdots + 87t^8 + 85t^9 + 87t^{10} + \cdots.$$

(b) [6, 8]. Let G be a 3-group of maximal class and of order 3^5. Then $d_1 = 2$ and $d_2 = d_3 = d_6 = 1$. Then
$$(1+t+t^2)^2(1+t^2+t^4)(1+t^3+t^6)(1+t^6+t^{12})$$
$$= \cdots + 14t^{12} + 13t^{13} + 14t^{14} + \cdots.$$

(c) [8]. Let G be a 5-group of maximal class and of order 5^7. Then $d_1 = 2$ and $d_2 = d_3 = d_4 = d_5 = d_{10} = 1$ and the Loewy series is not monotonic.

(d) Seeing the last two examples, one might guess that a p-group of maximal class of order p^{p+2} could be a counterexample for each prime p, but this is not the case; using a computer to calculate Jennings's polynomial, we considered p-groups G of maximal class with $7^9 \mid |G| \mid 7^{16}$, $11^{13} \mid |G| \mid 11^{19}$, $13^{15} \mid |G| \mid 13^{19}$, and $|G| = 17^{19}$. But none of them was a counterexample.

(e) Calculating 3-groups of maximal class with $3^5 \mid |G| \mid 3^{13}$, it turned out that the groups of order 3^5, 3^7, 3^9, 3^{11}, and 3^{13} are counterexamples; if the order of the group increases, the number of dents in the Loewy series of KG increases.

(f) Calculating 5-groups of maximal class with $5^7 \mid |G| \mid 5^{15}$, it turned out that the groups of order 5^7, 5^{11}, and 5^{15} are counterexamples; if the order of the group increases, the number of dents in the Loewy series is always 1, but the depth of the dents increases.

(g) Let $p = 2$. As the p-groups of maximal class are dihedral, semidihedral, or quaternion, their Loewy factors have the dimensions $1, 2, 2, \ldots, 2, 1$. J. Neubüser considered certain p-groups of coclass 2 and 3, respectively: let $\mathbf{Z}_4(D_4)$ act on $\mathbf{Z} \times \mathbf{Z}$ and factor out a normal subgroup of the basic group. The behavior of those groups turned out to be in a way similar to that of the 3-groups of maximal class.

(h) (J. Neubüser). Calculating 3-groups, 5-groups, and in particular 7-groups of coclass 2, which are constructed similarly to the 2-groups of coclass 2, it turned out that their Loewy series is monotonic.

No other counterexamples were found up to now (as far as I know). Analyzing counterexamples (e) and (f), we could verify the observation stated there.

THEOREM [**6**]. *Let G be a 3-group of maximal class and of order 3^n, where $n \geq 5$ is odd. Then $s = 3^{(n+1)/2} - 1$ and for $(s+4)/3 \leq i \leq s/2$ we have*

$$c_i = \begin{cases} s/2 + 1 & \text{for } i \text{ even,} \\ s/2 & \text{for } i \text{ odd.} \end{cases}$$

In particular, the number of dents is not bounded.

THEOREM [**6**]. *Let G be a 5-group of maximal class and of order 5^{8n+3} $(n \in \mathbf{N})$. Then $c_{s/2-2} - c_{s/2-1} = (5^{2n} - 1)/4$. In particular, the depth of the dents is not bounded.*

So far the negative results on this topic. But even if the Loewy series is not monotonic, there is, let's say, a tendency for the c_n's to increase from the edge to the middle.

THEOREM [**6**]. *Let $a := [(p-1)(d-1)/2]$ and $b := [(p-1)d/2]$, where d denotes the minimal number of generators of G.*

(a) *If $d > 1$, then $c_0 < c_1 < \cdots < c_b$.*

(b) *If $d > 1$, then $c_a < c_i$ for $a < i < s - a$. In particular, $d \leq c_i$ for $1 \leq i \leq s-1$.*

We now want to finish the part dealing with p-groups. The next case is to consider a p-nilpotent group G.

THEOREM. *Let G be p-nilpotent and K an arbitrary field of characteristic p.*

(a) (*R.Brauer*). *Any block B of KG contains exactly one irreducible KG-module V.*

(b) (*W. Schwarz*). *If $P(V)$ denotes the projective cover of V, then $m_n(V, P(V)) = c_n(\delta(B))$, where $m_n(V, P(V))$ denotes the multiplicity of V in $P(V)J(KG)^n/P(V)J(KG)^{n+1}$ and $\delta(B)$ the defect group of B.*
In particular, $m_n(V, B) = \dim V \cdot c_n(\delta(B))$ if K is algebraically closed.

The next groups one should consider are the groups with a normal Sylow p-subgroup. But in this situation, no analog for statement (b) of the last theorem exists for the projective indecomposables, as one can see, for example, by considering the group S_3 with $p = 3$. Nevertheless, the conclusion "in particular" remains true.

THEOREM [**4**]. *Let G contain a normal Sylow p-subgroup Q. Moreover, let B be a block of KG and V any irreducible KG-module in B. Then Q is the defect group of B and $m_n(V,B) = \dim V \cdot c_n(Q)$.*

As a consequence of this theorem, we obtain by applying Fong's reduction theorems

THEOREM [**4**]. *Let K be an algebraically closed field of characteristic p and let G be a p-solvable group. Let B be a block of KG with defect group $\delta(B)$ and let V be an irreducible KG-module in B. Moreover, suppose that one of the following conditions holds:*

(a) *G has p-length* 1.

(b) *$\delta(B)$ is abelian.*

(c) *$\delta(B)$ is a T. I. subgroup of G.*

Then $m_n(V,B) = \dim V \cdot c_n(\delta(B))$.

Considering $G = A_5$ and $p = 5$, we see that the conclusion of the last theorem doesn't hold if the group is not p-solvable, even if (b) and (c) are fulfilled. At last, we want to have a look at some examples of groups of p-length 2 and see what might be proved about those groups concerning the Loewy factors.

EXAMPLES.

(a) Let $G = S_4$ and let K be an algebraically closed field of characteristic $p = 2$. Then KG has one block and two irreducible modules K and V of dimensions 1 and 2, respectively, and the Loewy structure of the projective indecomposables is

$$\begin{array}{ccc} & K & \\ K & \oplus & V \\ K & \oplus & V \\ & K & \end{array} \qquad \begin{array}{ccc} & V & \\ K & \oplus & V \\ & K & \\ & V & \end{array}.$$

So, we get

$$\begin{aligned} &m_0(K,KG) = 1, \quad m_1(K,KG) = 3, \quad m_2(K,KG) = 3, \quad m_3(K,KG) = 1, \\ &m_0(V,KG) = 2, \quad m_1(V,KG) = 3, \quad m_2(V,KG) = 1, \quad m_3(V,KG) = 2. \end{aligned}$$

On the other hand, we mention the Loewy structure of the group algebra of the defect group $\delta(KG) \cong D_4$:

$$\begin{array}{ccc} & K & \\ K & \oplus & K \\ K & \oplus & K \\ K & \oplus & K \\ & K & \end{array}$$

We see that the smallest example of p-length 2 doesn't fulfill the statement of the last theorem; even the Loewy length of a block can be smaller than the Loewy length of its defect group. But at least the estimation

$$m_n(K, KG) + m_n(V, KG) \leq (\dim K + \dim V) \cdot c_n(D_4).$$

(b) Another example of a group of p-length 2 is $G := (Z_3 \times Z_3) \cdot \mathrm{SL}(2,3)$ the Loewy series of which was calculated by S. Koshitani in [**3**]. In this example, we also have $\sum_{V \in \mathrm{Irr}(KG)} m_n(V, KG) \leq \sum_{V \in \mathrm{Irr}(KG)} \dim V \cdot c_n(D)$, where D is the defect group of the two-sided indecomposable group algebra KG.

(c) Let $G := \Gamma(p^p)$ be the group of affine semilinear mappings over $GF(p^p)$ and let K be an algebraically closed field of characteristic p. Then G has $p-1$ simple KG-modules of dimension 1, say $K_0, \ldots, K_{p-1}$, where $K_0 = K$ denotes the trivial KG-module, and $p^{p-1}-1$ simple KG-modules of dimension p denoted by $V_{i,j}$ $(1 \leq i \leq (p^{p-1}-1)/(p-1),\ 0 \leq j \leq p-2)$, where $V_{i,j} = K_j \otimes V_{i,0}$. In [**4**], an algorithm is given for computing the Loewy factors of the projective indecomposables and executed for $p = 3$. For a suitable chosen enumeration we have that $V_{1,1} = V \otimes_{GF(p)} K$, where $V = O_p(G)$ is considered as a $GF(p)G$-module. The Loewy structures for $P(K_0)$ and $P(V_{1,1})$ are, for example,

$$\begin{array}{c}
K_0 \\
K_0 \quad V_{1,1} \\
K_0 \quad V_{1,1} \quad V_{2,0} \quad V_{3,0} \\
K_1 \quad V_{1,1} \quad V_{2,0} \quad V_{2,1} \quad V_{3,0} \quad V_{4,1} \\
K_1 \quad V_{1,0} \quad V_{2,0} \quad V_{2,1} \quad V_{3,0} \quad V_{4,0} \quad V_{4,1} \\
K_1 \quad V_{1,0} \quad V_{2,1} \quad V_{3,1} \quad V_{4,0} \quad V_{4,1} \\
K_0 \quad V_{1,0} \quad V_{3,1} \quad V_{4,0} \\
K_0 \quad V_{3,1} \\
K_0
\end{array}$$

$$\begin{array}{c}
V_{1,1} \\
V_{2,0} \quad V_{3,0} \quad V_{3,0} \\
K_1 \quad V_{1,1} \quad V_{2,1} \quad V_{2,1} \quad V_{4,1} \quad V_{4,1} \\
K_1 \quad V_{1,0} \quad V_{2,0} \quad V_{3,0} \quad V_{4,0} \quad V_{4,0} \\
K_1 \quad V_{1,0} \quad V_{1,1} \quad V_{3,1} \quad V_{4,1} \\
K_0 \quad V_{1,0} \quad V_{3,1} \\
K_0 \quad V_{2,1} \quad V_{3,1} \\
K_0 \quad V_{2,0} \quad V_{4,0} \\
V_{1,1}
\end{array}$$

Since all $P(V_{i,j})$ have p p-dimensional simple modules in their second Loewy layer and all $P(K_j)$ have K_j and $K_j \otimes V_{1,1}$ in their second Loewy layer, we get $\sum_{V \in \mathrm{Irr}(KG)} m_1(V, KG) = p\, p(p^{p-1}-1) + (p-1) \cdot 2 = p^{p+1} - p^2 + 2p - 2$. As all blocks of KG have maximal defect, $D := Z_p \sim Z_p$ is a defect group each block, so that

$$\sum_{V \in \mathrm{Irr}(KG)} \dim V \cdot c_1(D) = (p \cdot (p^{p-1} - 1) + (p - 1)) \cdot 2 = 2p^p - 2.$$

ACKNOWLEDGMENT. I would like to express my thanks to the "Freunde der Universität Mainz," whose generous support made it possible for me to join the conference in Arcata.

REFERENCES

1. B. Huppert and N. Blackburn, *Finite groups*. II, Berlin, 1982.

2. S. Jennings, *The structure of the group ring of a p-group over a modular field*, Trans. Amer. Math. Soc. **50** (1941), 175–185.

3. S. Koshitani, *On the Loewy series of a group algebra of a finite p-solvable group with p-length* > 1, Comm. Algebra **13** (1985), 2175–2198.

4. O. Manz, U. Stammbach, and R. Staszewski, *On the Loewy-series of the group algebra of groups of small p-length*, Comm. Algebra (submitted).

5. O. Manz and R. Staszewski, *On the number of generators and the modular group ring of a finite p-group*, Proc. Amer. Math. Soc. **98** (1986), 189–195.

6. ——, *On the Loewy-series of the modular group algebra of a finite p-group*, J. Reine Angew. Math. **368** (1986), 108–118.

7. W. Schwarz, *Die Struktur modularer Gruppenringe endlicher Gruppen der p-Länge* 1, J. Algebra **60** (1979), 51–75.

8. U. Stammbach and M. Stricker, *A remark on the Loewy-layers of the modular group ring of a finite p-group*, Topology Appl. **25** (1987), 121–124.

UNIVERSITÄT MAINZ, MAINZ, WEST GERMANY

Proceedings of Symposia in Pure Mathematics
Volume 47 (1987)

Blocks of Defect Zero and Degree Problems

W. WILLEMS

1. Introduction. A series of recent papers concerning degree problems of classical or modular type ends up by checking the considered problem through the list of all finite simple groups. One of the most significant results herein is Michler's proof [**7**] that a Sylow p-subgroup of G is normal and abelian (normal) if and only if p doesn't divide the degree of any irreducible ordinary (modular) character. For p-solvable groups this is known as Ito's theorem. The crucial part of Michler's paper is attached to the existence of p-blocks of defect zero for all finite groups of Lie type in odd characteristics p. (Here there is no need to consider the prime 2, since in this case a beautiful argument applying a result of R. Baer leads to the extension of Ito's theorem for arbitrary groups; see for instance [**9**].) Nevertheless, an answer for the prime 2 is doubtless of some interest, first from the point of view of applications in degree problems, and second since almost no condition of "practical use" is known which guarantees the existence of p-blocks of defect zero in general.

2. Blocks of defect zero. Using the ideas developed by Michler in [**7**] we are able to extend [**7**, §3 and §4] to the prime 2. Thus altogether we have the following general result.

THEOREM. *Let L be a finite simple group of Lie type and let p be an arbitrary prime. Then L has a p-block of defect zero.*

Applying the CAS character table system [**8**] or the group atlas [**2**] and well-known results on alternating groups, the answer for finite simple groups concerning the existence of p-blocks of defect zero is as follows.

COROLLARY. *Let G be any finite simple nonabelian group and let p be an arbitrary prime.*

(a) *If $p = 2$, then G has a 2-block of defect zero unless G is isomorphic to*

(i) $M_{12}, M_{22}, M_{24}, J_2, HS, Suz, Ru, C1, C3, BM$, *or*

(ii) A_n, *where n is not of the form $n = \binom{k}{2} + s$ with $s \in \{0, 2\}$ and $k \in \mathbf{N}$.*

1980 *Mathematics Subject Classification* (1985 *Revision*). Primary 20C20.

A final version of the theorem of this paper will be submitted for publication elsewhere.

(b) *If $p = 3$, then G has a 3-block of defect zero unless G is isomorphic to*
(i) *$Suz, C3$, or*
(ii) *A_n, where n is not of the form $n = t^2 + s^2 - st + s$ with $s, t \in \mathbf{N} \cup \{0\}$.*

(c) *For $p \geq 5$, G has a p-block of defect zero, if G is not isomorphic to an alternating group. Here we do not know the exceptions in the alternating series.*

During the conference the author was informed that recently A. L. O. Atkin and J. B. Olsson proved the existence of 5-, 7-, and 11-blocks of defect zero for the alternating groups. At this stage it might be conjectured that for $p \geq 5$ a finite simple group always has a p-block of defect zero.

Before starting with applications in degree problems let us indicate the main line of the proof of the above theorem and illuminate the situation by some remarks. Unfortunately, a deeper insight is still missing, so that a general proof without using the list of simple groups of Lie type fails. In addition, we are not yet able to count the number of p-blocks of defect zero. The problem which arises here lies in the fact that characters occur with mixed Jordan decomposition in the sense of Lusztig (see [**1**, pp. 391–392] and the remark at the end of this section).

SKETCH OF THE PROOF OF THE THEOREM. Let L be a finite simple group of Lie type defined over a finite field $GF(q)$ with $q = r^m$ elements (r a prime) and let G be a simply connected simple linear algebraic group with endomorphism σ such that $L \simeq (G_\sigma/Z(G_\sigma))'$, where G_σ denotes the set of fixed points of σ on G and $Z(G_\sigma) = Z$, the center of G_σ.

According to the result of Michler in [**7**] and the existence of the Steinberg character, we may assume that the prime of the underlying field F is 2 and different from r. For each type of groups the proof runs as follows:

Find a maximal σ-stable torus T of G with the following three properties.

(i) $|T_\sigma|_2 = |Z|_2$.
(ii) There exists a character $\theta \in \hat{T}_\sigma = \mathrm{Hom}(T_\sigma, \mathbf{C}^x)$ which is in general position, i.e., $|\{\theta^w \mid w \in W(T)_\sigma\}| = |W(T)_\sigma|$, where $W(T)$ denotes the Weyl group of T.
(iii) $Z \subseteq \ker\theta$ for θ in (ii).

If this is done, consider the Deligne-Lusztig character $R_{T,\theta}$. According to (ii), $R_{T,\theta}$ defines up to a sign an irreducible character of G_σ, say χ (see [**1**, 7.3.5]). By [**1**, 7.2.8], we see that $\chi \in \mathrm{Irr}_{\mathbf{C}}(G_\sigma/Z)$. Since $\chi(1) = |G_\sigma/Z : T_\sigma/Z|_{r'}$ [**1**, 7.5.1] and $2 \in r'$, condition (i) implies that $\chi(1)_2 = |G_\sigma/Z|_2$. Thus χ belongs to a 2-block of defect zero for the group G_σ/Z. A constituent of the restriction $\chi|_{(G_\sigma/Z)'}$ to the commutator group then gives the desired 2-block of defect zero for L.

REMARKS. In both cases, p odd and p even, the characters of defect zero which have been constructed explicitly are either of semisimple or unipotent type. Unfortunately, there are irreducible characters belonging to blocks of defect zero which came from characters of mixed Jordan decomposition in groups of adjoint type. For instance, such a character can be constructed for the exceptional group $E_8(q)$.

3. Applications to degree problems.

A. For every prime p define $\omega_p(G) = \max\{\nu_p(\chi(1)) \mid \chi \in \mathrm{Irr}_{\mathbf{C}}(G)\}$, where $\nu_p(n)$ denotes the usual valuation with respect to p for the integer n. In **[4]** Huppert and Manz proved that in the case $\omega_p(G) = 1$ for all primes p (this means that all the degrees are square-free) the group G is either solvable or $G = G_0 \times A_7$ with a solvable group G_0. The fact that A_7 is the only nonsolvable composition factor does not depend on the strong assumption on all primes p; indeed it is a consequence on the assumption on the prime 2.

PROPOSITION A. *Assume $\omega_2(G) = 1$. Then either G is solvable or G has a solvable normal subgroup with factor group isomorphic to A_7.*

B. A group is called a PPC-group if all ordinary irreducible character degrees are powers of primes. For solvable groups this class was studied by Manz in his thesis. In the case of nonsolvable groups G, he could show that under the hypothesis of one part of the height zero conjecture G is up to an abelian factor isomorphic to PSL(2,4) or PSL(2,8) (see **[6]**). Using the theorem of §2, we can drop the hypothesis he needs in his proof, i.e.,

PROPOSITION B. *If G is a nonsolvable PPC-group, then $G = G_0 \times A$, where A is abelian and $G_0 \in \{\mathrm{PSL}(2,4), \mathrm{PSL}(2,8)\}$.*

C. It is well known that for each of the groups $\mathrm{PSL}(2,2^n)$, $\mathrm{PSp}(4,2^n)$, and $\mathrm{Sz}(2^{2n+1})$ the degrees of the irreducible Brauer characters in characteristic 2 are powers of 2. R. Gow **[3]** pointed out that these are the only finite groups of Lie type defined over $GF(p^r)$ for which the degrees of all irreducible Brauer characters in p are powers of p. Again using the theorem of §2 to handle the case that the characteristic of the underlying field differs from the characteristic of the Lie group, we have

PROPOSITION C. *Let G be a finite simple group of Lie type defined over $GF(q^r)$, q a prime. Assume that $\beta(1)$ is a p-power for all $\beta \in \mathrm{Irr}_{\mathbf{C}}(G)$. Then $p = q = 2$ and $G \in \{\mathrm{PSL}(2,2^n), \mathrm{PSp}(4,2^n), \mathrm{Sz}(2^{2n+1})\}$.*

D. The character degree graph $\Gamma(G)$ of a finite group G is defined in the following way. The points are represented by the different $\chi(1) \neq 1$, where $\chi \in \mathrm{Irr}_{\mathbf{C}}(G)$. We connect two points $\chi(1)$ and $\psi(1)$ if there is a prime p dividing $(\chi(1), \psi(1))$. For solvable groups this graph has at most 2 components (see **[5]**). In a forthcoming paper we shall study $\Gamma(G)$ for arbitrary groups. This can be done since we have good information on suitable character degrees for groups of Lie type according to our theorem.

REFERENCES

1. R. W. Carter, *Finite groups of Lie type, conjugacy classes and complex characters*, J. Wiley, Chichester, 1985.

2. J. H. Conway, R. T. Curtis, S. P. Norton, F. A. Parker, and R. A. Wilson, *Atlas of finite groups*, Clarendon Press, Oxford, 1985.

3. R. Gow, private communication (1986).

4. B. Huppert and O. Manz, *Degree problems.* I, *Square-free character degrees*, Arch. Math. **45** (1985), 125–132.

5. O. Manz, *Degree problems.* II: *π-separable character degrees*, Comm. Algebra **13** (1985), no. 11, 2421–2431.

6. ——, *Endliche nicht-auflösbare Gruppen, deren sämtliche Charaktergrade Primzahlpotenzen sind*, J. Algebra **96** (1985), 114–119.

7. G. Michler, *A finite simple group of Lie type has p-blocks with different defects, $p \neq 2$*, J. Algebra **104** (1986), 220–230.

8. J. Neubüser, H. Pahlings, and W. Plesken, *CAS, design and use for the handling of characters of finite groups*, Proc. Conf. on Computational Group Theory, Academic Press, London, 1984, pp. 195–247.

9. T. Okuyama, *On a problem of Wallace*, (to appear).

UNIVERSITÄT ESSEN, WEST GERMANY

Proceedings of Symposia in Pure Mathematics
Volume **47** (1987)

Degrees of Brauer Characters in p-solvable Groups

THOMAS R. WOLF

This paper is a report on a joint work [**4**] of Olaf Manz and myself. We let $\mathrm{Irr}(G)$ and $\mathrm{IBr}_p(G)$ denote the ordinary irreducible characters and the irreducible Brauer characters (in the prime characteristic p) of a finite group G. A well-known theorem states that $p \nmid \chi(1)$ for all $\chi \in \mathrm{Irr}(G)$ if and only if G has an abelian normal Sylow p-subgroup. This was proved for p-solvable G by Ito and extended to the general case by Michler [**5**] via the classification of finite simple groups. Also, G has a normal Sylow p-subgroup if and only if $p \nmid \phi(1)$ for all $\phi \in \mathrm{IBr}_p(G)$. This is due to Okuyama [**7**] for $p = 2$ and Michler [**6**] for $p \neq 2$.

We now assume that q is a prime different from p and proceed to investigate the structure of G when $q \nmid \phi(1)$ for all $\phi \in \mathrm{IBr}_p(G)$. This study was initiated by Manz [**3**] for groups of odd order and we extend this to p-solvable groups. Our main results are to show that $O^{q'}(G)$ is solvable, to bound the q-length of G, and to bound the p-length of $O^{q'}(G)$. Also, the derived length of a Sylow q-subgroup of G is bounded.

Suppose $q \nmid \phi(1)$ for all $\phi \in \mathrm{IBr}_p(G)$ and that G is p-solvable. Let $L \trianglelefteq K \trianglelefteq G$ with K/L a chief factor of G. Either K/L is a p-group or a p'-group. In the latter case, if $\mu \in \mathrm{Irr}(K/L)$ then μ is also an irreducible Brauer character of K and each $\phi \in \mathrm{IBr}_p(G|\mu)$ is of q'-degree. It follows from Clifford's theorem that $q \nmid |G\colon I_G(\mu)|$. As a consequence, we study some unusual permutation actions and module actions of groups. The first such is a corollary of a theorem of D. Gluck [**2**].

LEMMA 1. *Let (G, Ω) be a solvable and primitive permutation group. Assume that $q \,|\, |G|$ and that for each $\Delta \subseteq \Omega$, the stabilizer of Δ contains a Sylow q subgroup. Then*

(a) $G \simeq D_6$, $q = 2$, $|\Omega| = 3$;

(b) $G \simeq D_{10}$, $q = 2$, $|\Omega| = 5$; *or*

(c) $G \simeq J$, $q = 3$, $|\Omega| = 8$; *(Here J is the group of all semilinear mappings on $GF(8)$ and is a solvable, 2-transitive group on* 8 *letters with order* 168.)

1980 *Mathematics Subject Classification.* Primary 20Cxx.

0082-0717/87 \$1.00 + \$.25 per page

Using the above lemma and the "classification," we get the following

THEOREM 2. *Assume G is p-solvable and $q \nmid \phi(1)$ for all $\phi \in \mathrm{IBr}_p(G)$. Then $O^{q'}(G)$ is solvable.*

SKETCH. WLOG, $G = O^{q'}(G)$.

Choose a nonsolvable chief factor M/N of G with $|G : M|$ minimal. By p-solvability, $p \nmid |M/N|$. Thus $q \nmid \beta(1)$ for all $\beta \in \mathrm{Irr}(M/N)$. By the Ito-Michler theorem $q \nmid |M/N|$. We consider two cases:

(1) M/N is simple.

Since $O^{q'}(G) = G$ and $q \nmid |M/N|$, it follows from the "classification" that $G/M \neq 1$ is a q-group which must leave every irreducible character of M/N invariant. Thus a Sylow q-subgroup of G/N acts trivially on M/N, contradicting the choice of M. This case is impossible.

(ii)$M/N = S_1 \times \cdots \times S_n$ for isomorphic nonsolvable S_i that are primitively and faithfully permuted by G/C, where $M \leq C \triangleleft G$ and $n > 2$. The condition on characters implies each $\Delta \subseteq \{S_1, \ldots, S_n\}$ is left invariant by a Sylow q-subgroup of G, whence Lemma 1 applies to the action of G/C on $\{S_1, \ldots, S_n\}$. But then one can argue further that c.d. $(S_1) \leq 2$, contradicting the nonsolvablity of S_1. □

Consequently we study solvable groups G and the action of G on $\mathrm{Irr}(M/N)$ for abelian chief sections of G.

LEMMA 3. *Assume V is a faithful irreducible G-module and finite, that $q||G|$ and $q \nmid |G : C_G(x)|$ for all $x \in V$. Assume G is solvable and $C \trianglelefteq G$ is maximal w.r.t V_C is not homogeneous. Let $V_1, \ldots, V_n$ be the homogeneous components of V_C. Then*

(a) *G/C primitively permutes the V_C;*

(b) *$n = 3,\ 5,$ or 8 and $q = 2,\ 3,\ 3$ (resp.);*

(c) *$G/C \simeq D_6,\ D_{10},$ or J (resp.);*

(d) *C acts transitively on each $V_r^{\#}$; and*

(e) *C is metabelian or $|V_r| = 3^2,\ 3^4,\ 5^2,\ 7^2,\ 11^2,$ or 23^2.*

PROPOSITION 4. *If G acts on a finite vector space V such that $2||G|$ but $2 \nmid |G : C_G(x)|\ \forall x \in V$, then $\mathrm{char}(V) = 2$.*

In particular, the exceptional cases in Lemma 3 occur only when $q = 3$.

THEOREM 5. *Assume that G is solvable, $O^{q'}(G) = G$, and that V is a faithful and finite G-module satisfying $q \nmid |G : C_G(x)|\ \forall x \in V$. Then, if V_N is homogeneous for all N char G, V is an irreducible G-module and*

(i) *$O^q(G)$ is cyclic;*

(ii) *$G \simeq \mathrm{SL}(2,3)$ and $|V| = 9$; or*

(iii) *$q = 2 = |G : O^q(G)|$, $O^q(G)$ is extra-special of order 3^3 and exponent 3, and $|V| = 4^3$.*

The proof of the above uses an inductive argument as well as the known structure of primitive solvable groups. Using an inductive argument, it may be concluded that $O^q(G)$ is a q'-group and the unique maximal characteristic subgroup. Employing Lemma 3 and the inductive hypotheses, it can be further reduced to where $q \leq 3$. Further use of Lemma 3 and a similar theorem of Gluck and Wolf [**2**] is also employed (this theorem of [**2**] concerns only the case where $q^2 \nmid |G|$ and has different homogeneity hypotheses). Finally, estimates are needed for $|\operatorname{Syl}_q(G)|$ and $|C_V(Q)|$ for $Q \in \operatorname{Syl}_q(G)$.

We now state our results concerning the situation where $q \nmid \phi(1)\ \forall \phi \in \operatorname{IBr}_p(G)$. First, we note that if $q = 2$, then $O^{q'}(G)$ is a $\{2,p\}$-group. This, which follows from Proposition 4, was first observed by Gow.

THEOREM 6. *Assume G is p-solvable, $q \neq p$, and $q \nmid \phi(1)\ \forall \phi \in \operatorname{IBr}_p(G)$. Then $O^{q'}(G)$ is solvable and*

(i) $l_q(G) \leq 2$;

(ii) *If $q \nmid p-1$ and $(p,q) \neq (2,3)$, then $l_q(G) \leq 1$;*

(iii) *If $H = O^{q'}(G)$, then $l_p(H/O_p(H)) \leq 1$ except possibly when $(q,p) = (3,7)$ and $l_p(H/O_p(H)) = 2$.*

SKETCH. WLOG $G = O^{q'}(G)$ is solvable.

In a minimal counterexample $F(G) = O_r(G)$, for a prime $r \neq p$. Choose $1 = K_0 \leq K_1 \leq \cdots \leq K_n = F(G)$ a chief series of G. Set $W_i = \operatorname{Irr}(K_i/K_{i-1})$ and $C_i = C_G(W_i) = C_G(K_i/K_{i-1})$. Then each $w \in W_i$ is invariant under some Sylow q-subgroup of G. By Lemma 3 and Theorem 5, we may bound the q-length and p-length of each G/C_i. Since $\bigcap C_i = F(G) = O_r(G)$, $l_q(G)$ and $l_p(G)$ are bounded. □

COROLLARY 7. *Under the same hypotheses, a Sylow q-subgroup is metabelian.*

COROLLARY 8. *Assume that G is p-solvable and each $\phi \in \operatorname{IBr}_p(G)$ has p-power degree. Then G is solvable, and if $K = O^p(G)$, then $K/O_p(K)$ has p-length at most one.*

REFERENCES

1. D. Gluck, *Trivial set stabilizers in finite permutation groups*, Canad. J. Math. **35** (1983), 59–67.

2. D. Gluck and T. R. Wolf, *Defect groups and character heights in blocks of solvable groups. II*, J. Algebra **87** (1984), 222–246.

3. O. Manz, *On the modular version of Ito's theorem on character degrees for groups of odd order*, Nagoya Math. J. **105** (1987), 121–128.

4. O. Manz and T. R. Wolf, *Brauer characters of q'-degree in p-solvable groups*, J. Algebra (accepted).

5. G. Michler, *Brauer's conjectures and the classification of finite simple groups*, Representation Theory. II, Springer-Verlag, 1985, pp. 129–142.

6. ——, *A finite simple group of Lie type has p-blocks with different defects, $p \neq 2$*, J. Algebra **104** (1986), 220–230.

7. T. Okuyama, *On a problem of Wallace*, (to appear).

OHIO UNIVERSITY

ABCDEFGHIJ-8987